10 Group	11 Group IB	12 Group IIB	13 Group IIIA	14 Group IVA	15 Group VA	16 Group VIA	17 Group VIIA	VIIIA 2 He 4.002602
			5 B 10.811	6 C 12.011	7 N 14.0067	8 O 15.9994	9 F 18.998403	10 Ne 20.179
			13 Al 26.98154	14 Si 28.0855	15 P 30.97376	16 S 32.066	17 Cl 35.453	18 Ar 39.948
28 Ni 58.6934	29 Cu 63.546	30 Zn 65.38	31 Ga 69.723	32 Ge 72.59	33 As 74.9216	34 Se 78.96	35 Br 79.904	36 Kr 83.80
46 Pd 106.42	47 Ag 107.8682	48 Cd 112.41	49 In 114.818	50 Sn 118.710	51 Sb 121.757	52 Te 127.60	53 I 126.9045	54 Xe 131.29
78 Pt 195.08	79 Au 196.9665	80 Hg 200.59	81 Tl 204.383	82 Pb 207.2	83 Bi 208.9804	84 Po (209)	85 At (210)	86 Rn (222)

Metals ← → Nonmetals

63 Eu 151.96	64 Gd 157.25	65 Tb 158.9254	66 Dy 162.50	67 Ho 164.9304	68 Er 167.26	69 Tm 168.9342	70 Yb 173.04	71 Lu 174.967
95 Am (243)	96 Cm (247)	97 Bk (247)	98 Cf (251)	99 Es (252)	100 Fm (257)	101 Md (258)	102 No (259)	103 Lr (260)

PREPARATORY CHEMISTRY

FOURTH EDITION

H. Stephen Stoker
WEBER STATE UNIVERSITY

MACMILLAN PUBLISHING COMPANY
NEW YORK
Maxwell Macmillan Canada
TORONTO
Maxwell Macmillan International
NEW YORK OXFORD SINGAPORE SYDNEY

Copyright © 1993 by Macmillan Publishing Company, a division of Macmillan, Inc.

PRINTED IN THE UNITED STATES OF AMERICA

All rights reserved. No part of this book may be reproduced or transmitted in any form or by any means, electronic or mechanical, including photocopying, recording, or any information storage and retrieval system, without permission in writing from the publisher.

Earlier editions copyright © 1990, 1986, and 1983 by Macmillan Publishing Company.

Material reprinted from *Introduction to Chemical Principles*, Fourth Edition, by H. Stephen Stoker, copyright © 1993 by Macmillan Publishing Company.

Macmillan Publishing Company
866 Third Avenue, New York, New York 10022

Macmillan Publishing Company is part of the Maxwell Communication Group of Companies.

Maxwell Macmillan Canada, Inc.
1200 Eglinton Avenue East
Suite 200
Don Mills, Ontario M3C 3N1

LIBRARY OF CONGRESS CATALOGING-IN-PUBLICATION DATA

Stoker, H. Stephen (Howard Stephen),
  Preparatory chemistry/H. Stephen Stoker.—4th ed.
    Includes index.
      p. cm.
    ISBN 0-02-417722-9
    1. Chemistry.   I. Title.
QD31.2.S763 1993   92-13832
540—dc20   CIP

Printing: 1 2 3 4 5 6 7 8   Year: 3 4 5 6 7 8 9 0 1 2

Preface

Preparatory Chemistry is a text for students who have had little or no previous instruction in chemistry or who had such instruction long enough ago that a thorough review is needed. The text's purpose is to give students the background (and confidence) needed for a subsequent successful encounter with a main sequence college level general chemistry course.

Many texts written for preparatory chemistry courses are simply "watered down" versions of general chemistry texts; they treat almost all topics found in the general chemistry course, but at a superficial level. *Preparatory Chemistry* does not fit this mold. The author's philosophy is that it is better to treat fewer topics extensively and have the student understand those topics in greater depth. The very real temptation to include "lots and lots" of additional concepts in this new edition of the text was resisted. Instead the focus for the revision was on rewriting selected portions to improve the clarity of presentation.

Important Features of This Textbook

1. Because of the varied degrees of understanding of chemical principles possessed by students taking a preparatory chemistry course, development of each topic in this text starts at "ground level" and continues step by step until the level of sophistication required for a further chemistry course is attained.
2. Problem solving receives major emphasis. Over twenty years of teaching experience indicate to the author that student "troubles" in general chemistry courses are almost always centered in the inability to set up and solve problems. Whenever possible, dimensional analysis is used in problem solving. This method, which requires no mathematics beyond arithmetic and elementary algebra, is a powerful and widely applicable problem-solving tool. Most important, it is a method that an average student can master with an average amount of diligence. Mastering dimensional analysis also helps build the confidence that is so valuable for future chemistry courses.
3. Significant figure concepts are emphasized in all problem-solving situations.

Routinely, electronic calculators display answers that contain more digits than are needed or acceptable. In all worked-out examples, students are reminded about these "unneeded digits" by the appearance of two answers to the example: the calculator answer (which does not take into account significant figures) and, in color, the correct answer (which is the calculator answer adjusted to the correct number of significant figures).

4. Numerous worked-out example problems are found within the textual material with detailed commentary accompanying each such example. In addition an unworked practice exercise is coupled to each example. It is intended that students will work this exercise immediately after "working through" the example. For immediate feedback, the answer to each practice exercise follows the exercise.

5. All end-of-chapter exercises occur in "matched pairs". In essence, each chapter has two independent, but similar, problem sets. Counting subparts to problems, there are over 5000 questions and problems available for a student to use in his or her "struggle" to become proficient at problem solving. Answers to all of the odd-number problems are found at the end of the text. Thus, two problem sets exist, one with answers and one without answers.

6. Each end-of-chapter problem set, except for Chapters 1 and 2, is divided into four sections: (1) Practice Problems, (2) Additional Problems, (3) Cumulative Problems, and (4) Grid Problems. The practice problems are categorized by topic and are arranged in the same sequence as the chapter's textual material. These problems, which are always single-concept, are "drill" problems which most students will find "routine". The additional problem section contains problems that involve more than one concept from the chapter and are usually more difficult than the practice problems. The cumulative-skills section draws not only on materials from the current chapter but also on concepts discussed in previous chapters. The working of problems in this third group allows students to continue to use, rather than forget, problem-solving techniques presented earlier. Finally, each chapter ends with four grid problems. Here students apply the principles from the chapter in a setting where multiple concepts are examined at the same time. Answers are selected from a three-by-three grid of choices; most often multiple answers are correct.

Supplements

A Student Solutions Manual has been prepared to accompany this text, and an Instructor's Resource Manual with Test Bank is available to adopters.

Acknowledgments

As always, the valuable contributions of reviewers are gratefully acknowledged: Paul E. Jacobson, Tacoma Community College; L. A. Kuprenas, California State University, Long Beach; and George Schenk, Wayne State University.

H. S. S.

Contents

CHAPTER ONE

The Science of Chemistry 1
1.1 Chemistry—A Scientific Discipline 1
1.2 Scientific Disciplines and Technology 2
1.3 The Scope of Chemistry and Chemical Technology 3
1.4 How Chemists Discover Things—The Scientific Method 4
Key Terms 9 · Practice Problems 9

CHAPTER TWO

Numbers from Measurements 11
2.1 The Importance of Measurement 11
2.2 Accuracy, Precision, and Error 12
2.3 Significant Figures—A Method for Handling Uncertainty in Measurements 13
2.4 Significant Figures and Calculated Quantities 19
2.5 Scientific Notation 27
2.6 Scientific Notation and Mathematical Operations 33
Key Terms 40 · Practice Problems 40 · Additional Problems 45 · Grid Problems 47

CHAPTER THREE

Unit Systems and Dimensional Analysis 48
3.1 The Metric System of Units 48
3.2 Metric Units of Length 50
3.3 Metric Units of Mass 52

- 3.4 Metric Units of Volume 54
- 3.5 Units in Mathematical Operations 57
- 3.6 Conversion Factors 59
- 3.7 Dimensional Analysis 62
- 3.8 Density 74
- 3.9 Percent as a Conversion Factor 78
- 3.10 Temperature Scales 81
- 3.11 Types and Forms of Energy 85
- 3.12 Heat Energy and Specific Heat 87

  Key Terms 93 · Practice Problems 94 · Additional Problems 98 · Cumulative Problems 99 · Grid Problems 100

CHAPTER FOUR

Basic Concepts About Matter 102

- 4.1 Chemistry—The Study of Matter 102
- 4.2 Physical States of Matter 103
- 4.3 Properties of Matter 104
- 4.4 Changes in Matter 106
- 4.5 Mixtures and Pure Substances 109
- 4.6 Types of Mixtures: Heterogeneous and Homogeneous 111
- 4.7 Types of Pure Substances: Elements and Compounds 113
- 4.8 Discovery and Abundance of the Elements 116
- 4.9 Names and Symbols of the Elements 117
- 4.10 Provisional Names for New Elements 119

  Key Terms 121 · Practice Problems 121 · Additional Problems 125 · Cumulative Problems 125 · Grid Problems 126

CHAPTER FIVE

Atoms, Molecules, Formulas, and Subatomic Particles 128

- 5.1 The Atom 128
- 5.2 The Molecule 130
- 5.3 Natural and Synthetic Compounds 133
- 5.4 Chemical Formulas 133
- 5.5 Subatomic Particles: Protons, Neutrons, and Electrons 136
- 5.6 Evidence Supporting the Existence and Arrangement of Subatomic Particles 139
- 5.7 Atomic Number and Mass Number 143
- 5.8 Isotopes 145
- 5.9 Atomic Masses 148

  Key Terms 156 · Practice Problems 157 · Additional Problems 161 · Cumulative Problems 162 · Grid Problems 162

CHAPTER SIX

Electronic Structure and Chemical Periodicity — 164

- 6.1 The Periodic Law 164
- 6.2 The Periodic Table 165
- 6.3 The Energy of an Electron 169
- 6.4 Electron Shells 170
- 6.5 Electron Subshells 172
- 6.6 Electron Orbitals 173
- 6.7 Writing Electron Configurations 176
- 6.8 Orbital Diagrams 180
- 6.9 Electron Configurations and the Periodic Law 183
- 6.10 Electron Configurations and the Periodic Table 184
- 6.11 Classification Systems for the Elements 191
- 6.12 Chemical Periodicity 193

Key Terms 196 · Practice Problems 197 · Additional Problems 200 · Cumulative Problems 202 · Grid Problems 202

CHAPTER SEVEN

Chemical Bonding — 204

- 7.1 Chemical Bonds 204
- 7.2 Valence Electrons and Electron-Dot Structures 205
- 7.3 The Octet Rule 207
- 7.4 Ions and Ionic Bonds 208
- 7.5 Ionic Compound Formation 211
- 7.6 Formulas for Ionic Compounds 215
- 7.7 Structure of Ionic Compounds 217
- 7.8 Polyatomic Ions 218
- 7.9 The Nature of a Covalent Bond 220
- 7.10 Electronegativities and Bond Polarities 225
- 7.11 Multiple Covalent Bonds 231
- 7.12 Coordinate Covalent Bonds 234
- 7.13 Resonance Structures 235
- 7.14 A Systematic Method for Determining Electron-Dot Structures 236
- 7.15 Molecular Polarity 242
- 7.16 Predicting Molecular Geometries (VSEPR Theory) 246

Key Terms 254 · Practice Problems 255 · Additional Problems 261 · Cumulative Problems 262 · Grid Problems 263

CHAPTER EIGHT

Chemical Nomenclature 265

- 8.1 Nomenclature Classifications for Compounds 265
- 8.2 Types of Binary Ionic Compounds 266
- 8.3 Nomenclature for Binary Ionic Compounds 268
- 8.4 Nomenclature for Ionic Compounds Containing Polyatomic Ions 273
- 8.5 Nomenclature for Binary Molecular Compounds 276
- 8.6 Nomenclature for Acids 279
- 8.7 Nomenclature Rules—A Summary 284

Key Terms 287 · Practice Problems 288 · Additional Problems 291 · Cumulative Problems 292 · Grid Problems 293

CHAPTER NINE

Chemical Calculations: The Mole Concept and Chemical Formulas 295

- 9.1 The Law of Definite Proportions 295
- 9.2 Calculation of Formula Masses 298
- 9.3 Percent Composition 301
- 9.4 The Mole: The Chemist's Counting Unit 303
- 9.5 The Mass of a Mole 307
- 9.6 Counting Particles by Weighing 313
- 9.7 The Mole and Chemical Formulas 316
- 9.8 The Mole and Chemical Calculations 318
- 9.9 Determination of Empirical and Molecular Formulas 324

Key Terms 332 · Practice Problems 332 · Additional Problems 338 · Cumulative Problems 339 · Grid Problems 340

CHAPTER TEN

Chemical Calculations Involving Chemical Equations 342

- 10.1 The Law of Conservation of Mass 342
- 10.2 Writing Chemical Equations 344
- 10.3 Balancing Chemical Equations 345
- 10.4 Special Symbols Used in Equations 350
- 10.5 Patterns in Chemical Reactivity 351
- 10.6 Classes of Chemical Reactions 353
- 10.7 Chemical Equations and the Mole Concept 356
- 10.8 Calculations Based on Chemical Equations—Stoichiometry 359
- 10.9 The Limiting Reactant Concept 365
- 10.10 Yields: Theoretical, Actual, and Percent 370
- 10.11 Simultaneous and Consecutive Reactions 372

Key Terms 375 · Practice Problems 375 · Additional Problems 382 · Cumulative Problems 383 · Grid Problems 385

CHAPTER ELEVEN

States of Matter 387

- **11.1** Physical States of Matter 387
- **11.2** Property Differences of Physical States 389
- **11.3** The Kinetic Molecular Theory 390
- **11.4** The Solid State 392
- **11.5** The Liquid State 392
- **11.6** The Gaseous State 393
- **11.7** A Comparison of Solids, Liquids, and Gases 394
- **11.8** Endothermic and Exothermic Changes of State 395
- **11.9** Temperature Change as a Substance Is Heated 395
- **11.10** Energy and Changes of State 397
- **11.11** Heat Energy Calculations 400
- **11.12** Evaporation of Liquids 404
- **11.13** Vapor Pressure of Liquids 407
- **11.14** Boiling and Boiling Point 408
- **11.15** Intermolecular Forces in Liquids 410
- **11.16** Water—A Most Unusual (Unique) Substance 413
- **11.17** Types of Solids 419

    Key Terms 422 · Practice Problems 423 · Additional Problems 426 · Cumulative Problems 427 · Grid Problems 427

CHAPTER TWELVE

Gas Laws 429

- **12.1** Properties of Some Common Gases 429
- **12.2** Gas Law Variables 430
- **12.3** Boyle's Law: A Pressure–Volume Relationship 435
- **12.4** Charles's Law: A Temperature–Volume Relationship 440
- **12.5** Gay-Lussac's Law: A Temperature–Pressure Relationship 444
- **12.6** The Combined Gas Law 445
- **12.7** Standard Conditions for Temperature and Pressure 449
- **12.8** Gay-Lussac's Law of Combining Volumes 450
- **12.9** Avogadro's Law: A Volume–Quantity Relationship 453
- **12.10** Molar Volume of a Gas 455
- **12.11** The Ideal Gas Law 460
- **12.12** Equations Derived from the Ideal Gas Law 463
- **12.13** Gas Laws and Chemical Equations 466
- **12.14** Dalton's Law of Partial Pressures 469

    Key Terms 475 · Practice Problems 475 · Additional Problems 482 · Cumulative Problems 483 · Grid Problems 484

CHAPTER THIRTEEN

Solutions 486

- 13.1 Types of Solutions 486
- 13.2 Terminology Used in Describing Solutions 488
- 13.3 Solution Formation 489
- 13.4 Solubility Rules 491
- 13.5 Solution Concentrations 493
- 13.6 Concentration: Percentage of Solute 494
- 13.7 Concentration: Parts per Million and Parts per Billion 498
- 13.8 Concentration: Molarity 500
- 13.9 Concentration: Molality 506
- 13.10 Dilution 509
- 13.11 Molarity and Chemical Equations 512

Key Terms 516 · Practice Problems 517 · Additional Problems 522 · Cumulative Problems 524 · Grid Problems 525

CHAPTER FOURTEEN

Acids, Bases, and Salts 527

- 14.1 Arrhenius Acid–Base Theory 527
- 14.2 Brønsted–Lowry Acid–Base Theory 528
- 14.3 Conjugate Acids and Bases 530
- 14.4 Polyprotic Acids 533
- 14.5 Strengths of Acids and Bases 535
- 14.6 Salts 538
- 14.7 Ionic and Net Ionic Equations 539
- 14.8 Reactions of Acids 544
- 14.9 Reactions of Bases 546
- 14.10 Reactions of Salts 546
- 14.11 Dissociation of Water 551
- 14.12 The pH Scale 553
- 14.13 Hydrolysis of Salts 561
- 14.14 Buffer Solutions 564
- 14.15 Buffers in the Human Body 565
- 14.16 Acid–Base Titrations 567
- 14.17 Acid–Base Titration Calculations Using Normality 569
- 14.18 Acid and Base Stock Solutions 574

Key Terms 575 · Practice Problems 576 · Additional Problems 583 · Cumulative Problems 584 · Grid Problems 584

CHAPTER FIFTEEN

Oxidation and Reduction 586

- 15.1 Oxidation-Reduction Terminology 586
- 15.2 Oxidation Numbers 588

15.3	Types of Chemical Reactions 594	
15.4	Balancing Oxidation–Reduction Equations 596	
15.5	Using the Oxidation Number Method to Balance Redox Equations 597	
15.6	Using the Ion-Electron Method to Balance Redox Equations 603	
15.7	Disproportionation Reactions 611	
15.8	Some Important Oxidation–Reduction Reactions 614	

Key Terms 622 · Practice Problems 623 · Additional Problems 627 · Cumulative Problems 628 · Grid Problems 629

APPENDIX

Mathematical Review   A.1

A.1 Basic Mathematical Operations   A.1
A.2 Fractions   A.2
A.3 Positive and Negative Numbers   A.6
A.4 Solving Algebraic Equations   A.8

ANSWERS TO ODD-NUMBERED PROBLEMS   A.13

INDEX   I.1

CHAPTER ONE

The Science of Chemistry

1.1 Chemistry—A Scientific Discipline

During the entire time of their existence on Earth, human beings have been concerned with and fascinated by their surroundings. This desire to understand their surroundings—an attribute that distinguishes them from all other living organisms—has led them to accumulate vast amounts of information concerning themselves, their world, and the universe. **Science** is the study in which humans attempt to organize and explain in a systematic and logical manner knowledge about themselves and their surroundings.

The enormous range of types of information covered by science, the sheer amount of accumulated knowledge, and the limitations of human mental capacity relative to mastering such a large and diverse body of knowledge have led to the division of the whole of science into smaller subdivisions called scientific disciplines. A **scientific discipline** is a branch of scientific knowledge limited in its size and scope to make it more manageable. Chemistry is one of the scientific disciplines. Astronomy, botany, geology, physics, and zoology are some of the others.

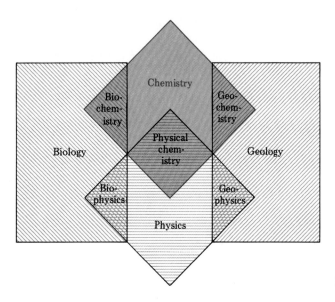

FIGURE 1.1 Interrelationship of scientific disciplines.

In a sense, the boundaries between scientific disciplines are artificial, because all scientific disciplines borrow information and methods from each other. No scientific discipline is totally independent. This overlap requires that scientists, in addition to having in-depth knowledge of a selected discipline, also have limited knowledge of other disciplines. Problems scientists have encountered in the last decade have particularly pointed out the interdependence of disciplines. For example, chemists attempting to solve the problem of chemical contamination of the environment find that they need some knowledge of geology, zoology, and botany. Because of this overlap, it is now common to talk not only of chemists, but also of geochemists, biochemists, chemical physicists, and so on. Figure 1.1 shows how chemistry merges with some of the other scientific disciplines.

The overlap of the various scientific disciplines affects not only practicing scientists but also today's college students. It means that students must necessarily study in disciplines other than those of primary interest to them. The knowledge they gain from such studies is often useful and even essential if they are to be competent in their chosen field. Many students in courses for which this textbook is written are studying chemistry because of its applicability to other disciplines in which they have a more specific interest.

1.2 Scientific Disciplines and Technology

Scientific disciplines represent abstract bodies of knowledge. The abstractness of such knowledge is modified by technology. **Technology** is the physical application of scientific knowledge to the production of new products to improve human survival, comfort, and quality of life. Technology manipulates nature for advantage. Technological advances began affecting our society about 200 years ago, and new advances still continue, at an accelerating pace, to have a major impact on human society. Section 1.3 considers numerous contributions of chemical technology to human well-being.

Technology, like science, involves human activities. Whether or not a given piece of scientific knowledge is technologically used for good or evil purposes depends on the motives of those men and women, whether in industry or government, who have the decision-making authority. In democratic societies, citizens (the voters) can influence many technological decisions. Therefore, it is important for everyone to be informed about scientific and technological issues.

1.3 The Scope of Chemistry and Chemical Technology

Although chemistry is concerned with only a part of the scientific knowledge that has been accumulated, it is in itself an enormous and broad field. Chemistry touches all parts of our lives.

Many of the clothes we wear are made from synthetic fibers produced by chemical processes. Even natural fibers, such as cotton or wool, are the products of naturally occurring chemical reactions within living systems. Our transportation usually involves vehicles powered with energy obtained by burning chemical mixtures, such as gasoline. The drugs used to cure many of our illnesses are the result of much chemical research. The paper on which this textbook is printed was produced through a chemical process, and the ink used in printing the words and illustrations is a mixture of many chemicals. The movies we watch are possible because of synthetic materials called film. The images on film are produced through the interaction of selected chemicals. Almost all of our recreational pursuits involve objects made of materials produced by chemical industries. Skis, boats, basketballs, bowling balls, musical instruments, and television sets all contain materials that do not occur naturally, but are products of human technological expertise.

Our bodies are a complex mixture of chemicals. Principles of chemistry are fundamental to an understanding of all processes of the living state. Chemical secretions (hormones) produced within our bodies help determine our outward physical characteristics such as height, weight, and appearance. Digestion of food involves a complex series of chemical reactions. Food itself is an extremely complicated array of chemical substances. Chemical reactions govern our thought processes and how knowledge is stored in and retrieved from our brains. In short, chemistry runs our lives.

A formal course in chemistry can be a fascinating experience because it helps us understand ourselves and our surroundings. We cannot truly understand or even know very much about the world we live in or about our own bodies without being conversant with the fundamental ideas of chemistry.

1.4 How Chemists Discover Things— The Scientific Method

The word *chemistry* conjures up images of people in white lab coats peering at instruments and shaking test tubes or other similar apparatus. Why is this generally valid image associated with chemists? The reason is simple. Chemists, as well as all other scientists, discover the general principles that govern the physical world (both its seen and unseen parts) through experimentation and observation. (see Fig. 1.2).

A majority of the scientific and technological advances of the twentieth century are the result of systematic experimentation using a method of problem solving known as the scientific method. The **scientific method** is a set of procedures used to acquire knowledge and explain phenomena. The procedural steps in the scientific method are

1. Identify the problem, break it into small parts, and carefully plan procedures to obtain information about all aspects of this problem.
2. Collect data concerning the problem through observation and experimentation.
3. Analyze and organize the data in terms of general statements (generalizations) that summarize the experimental observations.
4. Suggest probable explanations for the generalizations.
5. Experiment further to prove or disprove the proposed explanations.

Occasionally a great discovery is made by accident, but the majority of scientific discoveries are the result of the application of these five steps over long periods of time. There are no instantaneous steps in the scientific method; applying them requires considerable amounts of time. Even in those situations where luck is involved,

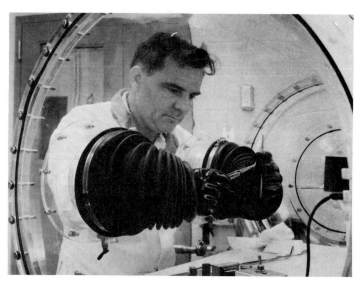

FIGURE 1.2 Experimentation and observation— the basis for discovering the general principles that govern the physical world. This research chemist is working with mildly radioactive materials. Such materials can be handled directly and safely in a glove box. [Brookhaven National Laboratory]

it must be remembered that "chance favors the prepared mind." To take full advantage of an accidental discovery, a person must be well trained in the procedures of the scientific method.

The imagination, creativity, and mental attitude of a scientist using the scientific method are always major factors in scientific success. The procedures of the scientific method must always be enhanced with the abilities of a thinking scientist.

There are special vocabulary terms associated with the scientific method and its use. This vocabulary includes the terms *experiment, fact, law, hypothesis,* and *theory.* An understanding of the relationships among these terms is the key to a real understanding of how to obtain chemical knowledge.

The beginning step in the search for chemical knowledge is the identification of a problem concerning some chemical system that needs study. After determining what other chemists have already learned about the selected problem, a chemist sets up experiments for obtaining more information. An **experiment** is a well-defined, controlled procedure for obtaining information about a system under study. The exact conditions under which an experiment is carried out must always be noted because conditions such as temperature and pressure affect results.

New facts about the system under study are obtained by actually carrying out the experimental procedures. A **fact** is a valid observation about some natural phenomenon. Facts are reproducible pieces of information. If a given experiment is repeated under exactly the same conditions, the same facts should be obtained. To be acceptable, all facts must be verifiable by anyone who has the time, means, and knowledge needed to repeat the experiments that led to their discovery. It is important that scientific data be published so that other scientists have the opportunity to critique and double-check both the data and experimental design.

It is interesting to contrast the differing ways in which scientific facts and the results of technology (Sec. 1.2) are shared. Scientists publish their observations (facts) as widely, openly, and quickly as possible. Technological breakthroughs, on the other hand, are usually kept secret by an individual or company until patent rights for the new process or product can be obtained.

As a next step, the chemist makes an effort to determine ways in which the facts about a given chemical system relate both to each other and to facts known about similar chemical systems. Repeating patterns often emerge among the collected facts. These patterns lead to generalizations that are called laws about how chemical systems behave under specific conditions. A **law** is a generalization that summarizes facts about natural phenomena.

Do not assume that laws are easy to discover. Often, many years of work and thousands of facts are needed before the true relationships among variables in the area under study emerge.

A law is a description of what happens in a given type of experiment. No new understanding of nature results from simply stating a law. A law merely summarizes already known observations (facts).

A law can be expressed either as a verbal statement or as a mathematical equation. An example of a verbally stated law is "If hot and cold pieces of metal are placed in contact with each other, the temperature of the hot piece always decreases and the temperature of the cold piece always increases."

It is important to distinguish between the use of the word *law* in science and its use in a societal context. Scientific laws are *discovered* by research (see Fig. 1.2), and researchers have *no control* over what the laws turn out to be. Societal laws, which are designed to control aspects of human behavior, are *arbitrary conventions* agreed upon (in a democracy) by the majority of those to whom the laws apply. These laws *can be* and *are changed* when necessary. For example, the speed limit for a particular highway (a societal law) can be decreased or increased for various safety or political reasons.

There is no mention in a law as to why the occurrence described happens. The law simply summarizes experimental observations, without attempting to clarify the reasons for the occurrence. Chemists, and other scientists, are not content with such a situation. They want to know *why* a certain type of observation is always made. Thus, after a law is discovered, scientists work out *plausible, tentative* explanations of the behavior encompassed by the law. These explanations are called *hypotheses*. A **hypothesis** is a tentative model or statement that offers an explanation for a law.

Once a hypothesis has been proposed, experimentation begins again. Scientists run more experiments, under varied conditions, to test the reliability of the proposed explanation. The hypothesis must be able to predict the outcome of as-yet-untried experiments. The validity of the hypothesis depends upon its predictions being true.

It is much easier to disprove a false hypothesis than to prove a true one. A negative result from an experiment indicates that the hypothesis is not valid as formulated and must be modified. Obtaining positive results supports the hypothesis, but it does not definitely prove it. There is always the chance that someone will carry out a new type of experiment, one that was not previously thought of, that disproves the hypothesis.

In practice, scientists usually start with a number of alternative hypotheses for a given law. Evaluation proceeds by demonstrating that certain proposals are *not* valid. A successful experiment is one in which one or more of the alternative hypotheses are demonstrated to be inconsistent with experimental observation and are thus rejected. Scientific progress is made in the same way a marble statue is; unwanted bits of marble are chipped away. Example 1.1 contains a simple illustration of this "chipping away" principle.

EXAMPLE 1.1

Suppose you are dealing with a situation involving two closed doors and four alternative hypotheses, which are (1) there is a tiger behind the door on the left, (2) there is a tiger behind the door on the right, (3) there is a tiger behind each door, and (4) there is no tiger behind either door. What evaluative information about these hypotheses can be obtained by opening the door on the right and having a tiger jump out at you?

Solution

This experiment (opening the right door) disproves hypothesis 4; it does not prove that *only one* of the hypotheses is true, but rather demonstrates that one of them is not true. The fact that a tiger is behind the door on the right does not rule out the possibility that there is also a tiger behind the door on the left.

PRACTICE EXERCISE 1.1

Based on the same "two-door, four-hypotheses" situation stated in Example 1.1, what evaluative information about the hypotheses is obtained from the single observation that there is no tiger behind the door on the left?

Ans: Hypotheses 1 and 3 are eliminated

Similar exercises: Problems 1.9 and 1.10

As further experimentation continues to validate a particular hypothesis, its acceptance in scientific circles increases. If, after extensive testing, the reliability of a hypothesis is still very high, confidence in it increases to the extent that it is accepted by the scientific community at large. After more time has elapsed and more positive support has accumulated, the hypothesis assumes the status of a theory. A **theory** is a hypothesis that has been tested and validated over a long period of time. The dividing line between a hypothesis and a theory is arbitrary and cannot be precisely defined. There is no set number of supporting experiments that must be performed in order to give theory status to a hypothesis. Figure 1.3 shows the interplay that must occur between hypotheses and experimentation before an acceptable theory is obtained.

Theories serve two important purposes: (1) they allow scientists to predict what will happen in experiments that have not yet been run, and (2) they simplify the very

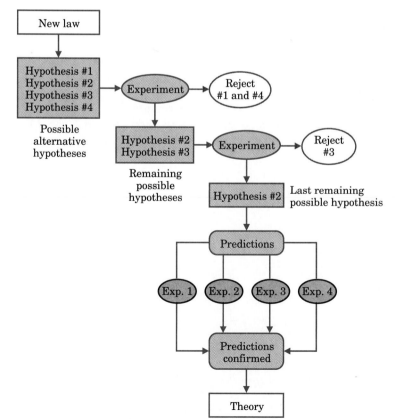

FIGURE 1.3 A number of alternative hypotheses are proposed to explain a new law; experiments are carried out to eliminate invalid hypotheses; predictions are made based on the surviving hypothesis and further experiments are carried out to test these predictions.

real problem of being able to remember all the scientific facts that have already been discovered.

Theories must often undergo modification. As scientific tools, particularly instrumentation, become more sophisticated, there is an increasing probability that some experimental observations will not be consistent with all aspects of a given theory. A theory affected in this manner must either be modified to accommodate the new results or be restated in such a way that scientists know where it is useful and where it is not. Most theories in use have known limitations. These "imperfect" theories are simply the best ideas anyone has found so *far* to describe, explain, and predict what happens in the world in which we live. Theories with limitations are generally not abandoned until a better theory is developed.

Scientists do not view scientific theories as "absolute truth." All theories in science are considered provisional—subject to change in the light of new experimental observations. A science is like a living organism; it continues to grow and change. Science develops through a constant interplay between theory and experimentation.

Facts that have been verified by repeated experiments will never be changed, but the theories that were invented to explain these facts are subject to change. In this sense, facts are more important than the theories devised to explain them. It is a mistake to believe that if you know all the laws and theories that are derived from experimental observations, you need not know the experimental facts. New theories can only be developed by people who have a wide knowledge of the facts relating to a particular field, especially those facts that have not been satisfactorily accounted for by existing theories.

The term *theory* is often misused by nonscientists in everyday contexts. "I have a theory that such and such is the case" is a frequently heard comment. In this case, "theory" means a "speculative guess," which is not what a theory is. A more appropriate term to describe this speculation is *hypothesis*.

Figure 1.4 summarizes the sequence of steps scientists generally use when applying the scientific method to a given research problem. It also shows, as did Figure 1.3, the central role that experimentation plays in the scientific method.

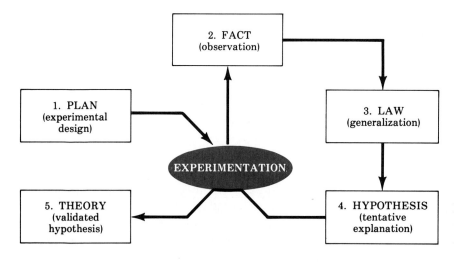

FIGURE 1.4 The central role of experimentation in the scientific method.

Key Terms

The new terms or concepts defined in this chapter are

experiment (Sec. 1.4) A well-defined, controlled procedure for obtaining information about a system under study.
fact (Sec. 1.4) A valid observation about some natural phenomenon.
hypothesis (Sec. 1.4) A tentative model or statement that offers an explanation for a law.
law (Sec. 1.4) A generalization that summarizes in a concise way facts about natural phenomena.
science (Sec. 1.1) The study in which humans attempt to organize and explain in a systematic and logical manner knowledge about themselves and their surroundings.
scientific discipline (Sec. 1.1) A branch of scientific knowledge limited in size and scope to make it more manageable.
scientific method (Sec. 1.4) A set of procedures for acquiring knowledge and explaining phenomena.
technology (Sec. 1.2) The physical application of scientific knowledge to the production of new products to improve human survival, comfort, and quality of life.
theory (Sec. 1.4) A hypothesis that has been tested and validated over a long period of time.

Practice Problems

The Scientific Method (Sec. 1.4)

1.1 Arrange the following steps in the scientific method in the sequence in which they normally occur.
  (a) Suggest probable explanations for generalizations obtained from data.
  (b) Collect data concerning a problem through observation and experimentation.
  (c) Identify a problem and carefully plan procedures to obtain information about all aspects of this problem.
  (d) Experiment further to prove or disprove proposed explanations.
  (e) Analyze and organize data in terms of general statements that summarize experimental observations.

1.2 Arrange the following terms associated with the scientific method in the order in which they are normally encountered as the scientific method is applied to a problem.
  (a) law   (b) fact
  (c) theory   (d) experiment
  (e) hypothesis

1.3 Indicate whether each of the following pairings of terms is correct or incorrect.
  (a) theory—validated hypothesis
  (b) hypothesis—unvalidated law
  (c) law—tentative theory
  (d) fact—reproducible observation

1.4 Indicate whether each of the following pairings of terms is correct or incorrect.
  (a) hypothesis—tentative explanation
  (b) law—validated theory
  (c) fact—well-defined experiment
  (d) theory—generalization that summarizes facts

1.5 Indicate whether each of the following statements is true or false.
  (a) A theory is a summary of experimental observations.
  (b) A hypothesis is a summary of experimental facts
  (c) A theory is subject to modification in light of new experimental observations
  (d) An experiment is a well-defined, controlled procedure for obtaining facts

1.6 Indicate whether each of the following statements is true or false.
  (a) A theory is a hypothesis that has not yet been subjected to experimental testing.
  (b) It is much easier to disprove a false hypothesis than it is to prove a valid one.
  (c) Established theories eventually become laws.
  (d) A law is an explanation of why a particular natural phenomenon occurs.

1.7 Classify each of the following statements as a fact, a law, or a hypothesis.
  (a) Metal objects, when thrown out a window, always fall to the ground.
  (b) The boiling point of water decreases with increasing altitude because air is colder at higher altitudes.
  (c) A yellow flame is always produced when table salt is placed in the flame of a gas burner.
  (d) Since the car will not start, its battery is probably dead.

1.8 Classify each of the following statements as a fact, a law, or a hypothesis.
  (a) The explosion was probably caused by a natural gas leak.
  (b) Analysis of the water sample showed the presence of 6.2 parts per million of arsenic.
  (c) The boiling point of water always decreases as atmospheric pressure decreases.
  (d) At an elevation of 4390 feet, pure water boils at a temperature of 95.6 °C.

1.9 Assume that you have four pennies with unknown mint dates and four hypotheses concerning these dates: (1) all dates are the same, (2) two different dates are present, (3) three different dates are present, and (4) all dates are different. Which of the listed hypotheses could be eliminated by determining that
  (a) two pennies have the same date?
  (b) two pennies have different dates?
  (c) two of three pennies have the same date?
  (d) three pennies have different dates?

1.10 Assume that you have four red balls of equal size and four hypotheses concerning the masses of the balls: (1) each ball has a different mass, (2) there are balls of two masses, (3) balls of three different masses are present, and (4) all balls have the same mass. Which of the listed hypotheses could be eliminated by determining that
  (a) two balls have the same mass?
  (b) three balls have the same mass?
  (c) there are two masses among three balls?
  (d) there are two masses among four balls?

1.11 Why is it important that scientific data be published?

1.12 Why is it useless to conduct an experiment under uncontrolled conditions?

1.13 What are the differences between a scientific law and a societal law?

1.14 What is the reason for repeating experiments several times before developing a law based on the experiments?

1.15 The phrase "it has been proved scientifically" is rarely used by scientists. Explain why.

1.16 Most theories in current use have known limitations. Explain why.

1.17 What is the difference between a hypothesis and a theory?

1.18 What are two important purposes that theories serve?

CHAPTER TWO

Numbers from Measurements

2.1 The Importance of Measurement

It would be extremely difficult for a carpenter to build cabinets without being able to use tools such as hammers, saws, and drills. They are a carpenter's "tools of the trade." Chemists also have "tools of the trade." Their most used tool is the one called *measurement*. Understanding measurement is indispensable in the study of chemistry. Questions such as "how much . . . ?," "how long . . . ?," and "how many . . . ?" simply cannot be answered without resorting to measurements.

Most of the concepts now considered to be the basic principles of chemistry had their origin in extensive tabulations of experimental data obtained by making measurements. The concepts were "discovered" as these data tabulations (measurements) were subjected to the procedures of the scientific method (Sec. 1.4).

It is the purpose of this chapter and the next to help students acquire the necessary background to deal properly with measurement. Almost all the material of these two chapters is mathematical. An understanding of this mathematics is a necessity for students of chemistry who want their encounters with the subject to be successful. The following analogy is appropriate for the situation. Physical exertion in sports can be fun, relaxing, and challenging if you are in good physical shape. If you are not in good physical condition, such exertion is not satisfying and may even be downright painful (especially the day after). Being "in good shape" mathematically has the same effect on the study of chemistry. It can cause that study to be a very satisfying and enjoyable experience. On the other hand, a lack of the necessary mathematical skills can cause "chemical exercise" to be somewhat painful. The message should be clear. The contents of this chapter (and Chapter 3) must be taken very seriously. Skimming over this material is a sure invitation to frustration and struggle with the chemical topics that follow.

2.2 Accuracy, Precision, and Error

It is important that measurements made by scientists be precise and accurate. Although the terms *precise* and *accurate* are used somewhat interchangeably in nonscientific discussion, they have distinctly different meanings in science. **Precision** refers to how close multiple measurements of the same quantity come to each other. **Accuracy** refers to how close a measurement (or the average of multiple measurements) comes to the true or accepted value. A simple analogy not directly involving measurement—shooting at a target—illustrates nicely the difference between these two terms (see Fig. 2.1). Accuracy depends on how close the shots are to the center (bull's-eye) of the target. Precision depends on how close the shots are to each other.

The preciseness of a measurement is directly related to the actual physical measuring device used; that is, precision is an inherent part of any measuring device. You would expect, and it is the case, that the reproducibility (preciseness) of temperature readings obtained from a thermometer with a scale marked in tenths of a degree would be greater than readings obtained from a thermometer whose scale has only degree marks. A stopwatch whose dial shows tenths of a second is preferred over one that shows only seconds if a precise time measurement is to be made.

In contrast to precision, accuracy depends not only on the measuring device used but also on the technical skill of the person making the measurement. How well can that person read the numerical scale of the instrument? How well can that person calibrate the instrument before its use?

Good accuracy
Good precision

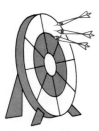
Poor accuracy
Good precision

Poor accuracy
Poor precision

FIGURE 2.1 The difference between precision and accuracy.

Normally, high precision results in high accuracy. However, high precision and low accuracy are also possible. Results obtained using a high-precision, poorly calibrated instrument would give precision but not accuracy. All measurements would be off by a constant amount as a result of the improper calibration.

Even the most accurate and precise measurements involve some amount of error. It is impossible to have a 100% accurate measurement. Flaws in measuring-device construction, improper calibration of an instrument, and the skills (or lack of skills) possessed by a person using a measuring device all contribute to error.

Errors in measurement can be classified as either random errors or systematic errors. **Random errors** are errors originating from uncontrolled variables in an experiment. Such errors result in experimental values that fluctuate about the true value. Variance in the angle from which a measurement scale is viewed will cause random error. Momentary changes in air currents, atmospheric pressure, or temperature near a sensitive balance for weighing would cause random errors. The net result of random errors, which can never be completely eliminated, is a decrease in the precision of measurements.

Systematic errors are errors originating from controlled variables in an experiment. They are "constant" errors that occur again and again. A flaw in a piece of equipment, such as a chipped weight in a balance, would cause systematic error. All readings would be off by a specific amount because of the improper calibration. Systematic errors affect the accuracy of measurements. Results are consistently either too high or too low compared to the true value.

2.3 Significant Figures—A Method for Handling Uncertainty in Measurement

Two kinds of numbers exist—those that are *counted* or *defined* and those that are *measured*. The difference between them is that we may know the exact values of counted or defined numbers but can never know the exact values of measured numbers.

You can count the number of peaches in a bushel of peaches or the number of toes on your left foot with absolute certainty. Counting does not involve reading the scale of a measuring device, and thus counted numbers are not subject to the uncertainties inherent in a measurement.

An example of a defined number is the number of objects in a dozen—twelve. By definition, 12 and exactly 12 (not 12.01 or 12.02) objects make a dozen. There are exactly 24 hours in a day, never 24.07 hours. A square has 4 sides, never 3.75 or 3.83 sides. Thus, a defined number always has one exact value.

In contrast to counted or defined numbers, every measured number carries with it a degree of uncertainty or error (as previously noted in Sec. 2.2). Even when very elaborate and expensive measuring devices are used some degree of uncertainty will always be present. Let us look at the origin of this uncertainty in more detail.

Consider how two different thermometer scales, illustrated in Figure 2.2, are used to measure a given temperature. Determining the temperature involves determining the height of the mercury column in the thermometer. The scale on the left in Figure

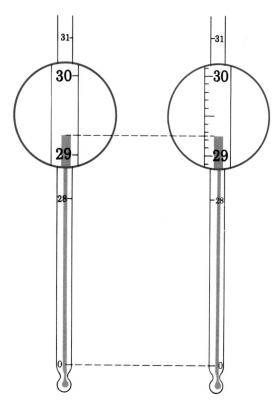

FIGURE 2.2 Measuring a temperature. A portion of the degree scale on each of the two differently scaled thermometers has been magnified.

2.2 is marked off in one-degree intervals. Using this scale we can say with certainty that the temperature is between 29 and 30 degrees. We can further say that the actual temperature is closer to 29 degrees than to 30 and estimate it to be 29.2 degrees. The scale on the right has more subdivisions, being marked off in tenths of a degree rather than in degrees. Using this scale we can definitely say that the temperature is between 29.2 and 29.3 degrees and can estimate it to be 29.25 degrees. Note how both temperature readings contain some digits (all those except the last one) that are exactly known and one digit (the last one) that is estimated. Note also that the uncertainty in the second temperature reading is less than that in the first reading—an uncertainty in the hundredths place compared to an uncertainty in the tenths place. We say that the scale on the right is *more precise* than the one on the left.

Because measurements are never exact, anytime a scientist writes down a numerical value for a measurement, two kinds of information must be conveyed: (1) the magnitude of the measurement and (2) the precision or reliability of the measurement. The digit values give the magnitude. Precision is indicated by the number of significant figures recorded.

Significant figures are the digits in any measurement that are known with certainty plus one digit that is uncertain. Only one estimated digit is ever recorded as part of a measurement. It would be incorrect for a scientist to report that the height of the mercury column in Figure 2.2, as read on the scale on the right, corresponds to a temperature of 29.247 degrees. The value 29.247 contains two estimated digits (the 4 and the 7) and would indicate a measurement of greater precision than is actually obtainable with that particular measuring device.

Section 2.3 · Significant Figures—A Method For Handling Uncertainty in Measurement **15**

The magnitude of the uncertainty in the last significant digit in a measurement (the estimated digit) may be indicated using a "plus-minus" notation. The following three time measurements illustrate this notation.

$$15 \pm 1 \text{ seconds}$$
$$15.3 \pm 0.1 \text{ seconds}$$
$$15.34 \pm 0.03 \text{ seconds}$$

Most often the uncertainty in the last significant digit is one unit (as in the first two time measurements), but it may be larger (as in the third time measurement). In this text we will follow the almost universal practice of dropping the "plus-minus" notation if the magnitude of the uncertainty is one unit. Thus, in the absence of "plus-minus" notation you will be expected to assume that there is an uncertainty of one unit in the last significant digit. A measurement reported simply as 27.3 inches means 27.3 ± 0.1 inches. Only in the situation where the uncertainty is greater than one unit in the last significant digit will the amount of the uncertainty be explicitly shown.

The precision of a measurement is determined by the number of significant figures in the measurement. A measured length of 2.453 cm (centimeters) for an object is more precise than a measured length of 2.45 cm for the same object. Thus, the term *precision* refers not only to the degree of reproducibility of repeated measurements (Sec. 2.2) but also to the number of significant figures in a measurement. Example 2.1 relates measurement preciseness to actual measuring device scales.

EXAMPLE 2.1

How many significant figures should be reported in each of the measurements?

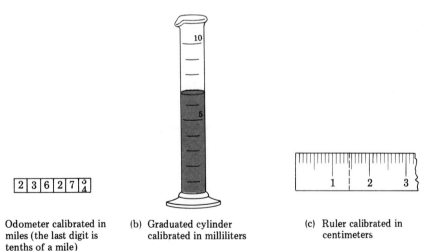

(a) Odometer calibrated in miles (the last digit is tenths of a mile)

(b) Graduated cylinder calibrated in milliliters

(c) Ruler calibrated in centimeters

Solution

(a) We know definitely that the mileage shown on the odometer is between 23,627.3 and 23,627.4 miles. We estimate the final digit (hundredths of a mile) to be a 5, giving a reading of 23,627.35 miles. Thus, *seven* siginificant figures are reportable.

16 Chapter 2 · Numbers from Measurements

(b) The level of the liquid is between the 6 and 7 milliliter marks. We estimate the level to be at 6.8 milliliters. *Two* significant figures are to be recorded.

(c) The ruler is calibrated in tenths of a centimeter. The broken line is definitely between 1.4 and 1.5 cm, being closer to 1.4. Estimating the next digit to be 2 we get a measurement of 1.42 cm. Thus, a measurement involving *three* significant figures is obtained.

PRACTICE EXERCISE 2.1

How many significant figures should be reported in each of the following volume measurements? *Ans.* (a) 2; (b) 3; (c) 4.

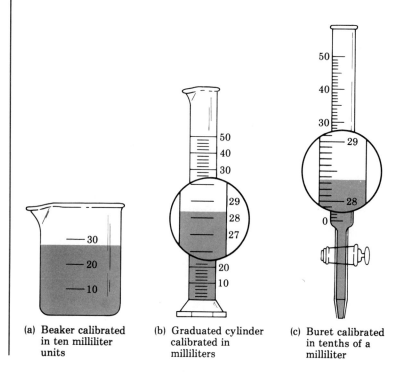

(a) Beaker calibrated in ten milliliter units

(b) Graduated cylinder calibrated in milliliters

(c) Buret calibrated in tenths of a milliliter

Determining the number of significant figures in a measurement is not always as straightforward as Example 2.1 would lead one to believe. In this example you knew the type of instrument used for each measurement and its limitations because you made the measurement. Quite often when someone else makes a measurement, such information is not available. All that is known is the reported final result—the numerical value of the measured quantity. In this situation questions do arise about the "significance" of various digits in the measurement. For example, consider the published value of the distance from the Earth to the sun, which is 93,000,000 miles. Intuition tells you that it is highly improbable that this distance is known to the closest mile. You suspect that this is an estimated distance. To what digit has this number been estimated? Is it to the nearest million miles, the closest hundred thousand miles, the nearest ten thousand miles, or what?

A set of guidelines has been developed to aid scientists in interpreting the significance of reported measurements or values calculated from measurements. Four rules constitute the guidelines—one rule for the digits 1 through 9 and three rules for the digit 0. A zero in a measurement may or may not be significant depending upon its location in the sequence of digits forming the numerical value for the measurement. There is a rule for each of three classes of zeros—leading zeros, confined zeros, and trailing zeros.

RULE 1 The digits 1 through 9 inclusive (all of the nonzero digits) always count as significant figures.

$$14.232 \quad \text{five significant figures}$$
$$3.11 \quad \text{three significant figures}$$
$$244.6 \quad \text{four significant figures}$$

RULE 2 *Leading zeros* are zeros that occur at the start of a number, that is, zeros that precede all nonzero digits. Such zeros do not count as significant figures. Their function is simply to indicate the position of the decimal point.

$$0.000\underline{45} \quad \text{two significant figures}$$
$$0.0\underline{113} \quad \text{three significant figures}$$
$$0.0000000\underline{72} \quad \text{two significant figures}$$

Leading zeros are always to the left of the first nonzero digit.

RULE 3 *Confined zeros* are zeros between nonzero digits. Such zeros always count as significant figures.

$$2.075 \quad \text{four significant figures}$$
$$6007 \quad \text{four significant figures}$$
$$0.0\underline{3007} \quad \text{four significant figures}$$

RULE 4 *Trailing zeros* are zeros at the end of a number. They are significant if (a) there is a decimal point present in the number or (b) they carry overbars. Otherwise trailing zeros are not significant.

The following numbers, all containing decimal points, illustrate condition (a) of rule 4.

$$62.00 \quad \text{four significant figures}$$
$$24.70 \quad \text{four significant figures}$$
$$0.0\underline{2000} \quad \text{four significant figures}$$
$$4300.00 \quad \text{six significant figures}$$

By condition (b) trailing zeros in numbers lacking an explicitly shown decimal point become significant when marked with a small bar above the zero(s).

$$36,\overline{000} \quad \text{five significant figures}$$
$$36,\overline{00}0 \quad \text{four significant figures}$$
$$36,\overline{0}00 \quad \text{three significant figures}$$
$$10,02\overline{0} \quad \text{five significant figures}$$

In cases involving trailing zeros where neither a decimal point nor overbar(s) are present the trailing zeros are not significant.

93,000,000	two significant figures
360,000	two significant figures
330,300	four significant figures
6310	three significant figures

Another method, more convenient than rule 4, for dealing with the significance of trailing zeros is to express the number in scientific notation. In this notation, to be presented in Section 2.5, only significant digits are shown.

EXAMPLE 2.2

Determine the number of significant figures in the following measured values.

(a) 2763   (b) 0.000367   (c) 3001.27
(d) 10.0   (e) 3300   (f) 0.01070

Solution

(a) There are *four* significant figures. All nonzero digits are significant (rule 1).
(b) There are *three* significant figures. Leading zeros (rule 2) are not significant.
(c) There are *six* significant figures. Confined zeros (rule 3) are significant.
(d) There are *three* significant figures. Trailing zeros in numbers containing a decimal point are significant (rule 4).
(e) There are *two* significant figures. Trailing zeros in numbers without an explicitly shown decimal point or overbar are not significant (rule 4).
(f) There are *four* significant figures. This number contains all three types of zeros. The leading zeros are not significant (rule 2), the confined zero is significant (rule 3), and the trailing zero is significant (rule 4).

PRACTICE EXERCISE 2.2

Determine the number of significant figures in the following measured values.

(a) 10,423   (b) 1230.0   (c) 0.0032
(d) 2.527   (e) 150   (f) 0.00300300

Ans. (a) 5; (b) 5; (c) 2; (d) 4; (e) 2; (f) 6

Similar exercises: Problems 2.13–2.18

EXAMPLE 2.3

What meaning, in terms of significant figures and magnitude of uncertainty, is conveyed by each of the following notations for a measurement involving the number two hundred sixty thousand?

(a) 260,000   (b) 26$\overline{0}$,000   (c) 260,$\overline{0}$00
(d) 260,0$\overline{0}$0   (e) 260,00$\overline{0}$ (f) 260,000.

Solution

(a) *Two* significant figures are present in this number: the two nonzero digits. Since the last significant digit, the 6, is located in the fifth place to the left of the understood decimal point (the ten-thousands place), the uncertainty is $\pm 10{,}000$.

(b) This number has *three* significant figures since only one of the four zeros is significant, as indicated by the bar over it. The last significant digit, the zero with the overbar, occupies the thousands place in the number. Thus, the uncertainty is ± 1000.

(c) Here we have *four* significant figures since two of the four zeros are significant. The uncertainty now lies with the hundreds digit (± 100).

(d) With *five* significant figures present (the 2, 6, and three zeros) the uncertainty is reduced to the tens digit (± 10).

(e) All digits are now significant. The uncertainty lies in the last digit, which occupies the ones place (± 1).

(f) Explicitly placing a decimal point at the end of the number makes all the trailing zeros significant. The meaning is thus the same as in part (**e**).

PRACTICE EXERCISE 2.3

What meaning, in terms of significant figures and magnitude of uncertainty, is conveyed by each of the following notations for measurements?

(a) 15$\bar{0}$ (b) 2$\overline{00}$0 (c) 56$\overline{0{,}0}$00 (d) 3,000,1$\bar{0}$0,000

Ans. (a) 3, ± 1; (b) 3, ± 10; (c) 4, ± 100; (d) 6, $\pm 10{,}000$.

Similar exercises: Problems 2.19 and 2.20

2.4 Significant Figures and Calculated Quantities

Most experimental measurements are not end results in themselves. Instead they function as intermediates in the calculation of other quantities. For example, we might measure in a laboratory the height, width, depth, and mass of a rectangular solid object and from this information calculate its volume and density.

In doing a calculation using experimental data, a scientist must give major consideration to the number of significant figures in the computed result. Correct calculations never improve (or decrease) the precision of experimental measurements.

Concern about the number of significant figures in a calculated number is particularly critical when an electronic calculator is used to do the arithmetic of the calculation. Hand calculators now in common use are not programmed to take significant figures into account. Consequently, the digital readouts on them more often than not display more digits than are needed. It is a mistake to record these extra digits, since they are meaningless; that is, they are not significant figures (see Fig. 2.3).

In order to record correctly the numbers obtained through calculations, students must be able to (1) adjust (usually decrease) the number of digits in a number to give it the correct number of significant figures and (2) determine the allowable

FIGURE 2.3 The digital readout on the average electronic calculator usually shows more digits than are needed or justified. Electronic calculators are not programmed to take significant figures into account. [Courtesy Texas Instruments]

number of significant figures in the result of any mathematical operation. We will consider these skills in the order listed.

Rounding off is the process of deleting unwanted (nonsignificant) digits from a calculated number. Three simple rules govern the process.

RULE 1   If the first digit to be dropped is less than 5, that digit and all digits that follow it are simply dropped.

Thus, 62.312 rounded off to three significant figures becomes 62.3.

RULE 2   If the first digit to be dropped is a digit greater than 5, or a 5 followed by digits other than all zeros, the excess digits are all dropped and the last retained digit is increased in value by one unit.

Thus, 62.782 and 62.558 rounded off to three significant figures become, respectively, 62.8 and 62.6.

RULE 3   If the first digit to be dropped is a 5 not followed by any other digit or a 5 followed only by zeros, an odd-even rule applies. Drop the 5 and any zeros that follow it and then
(a) increase the last retained digit by one unit if it is *odd*, or
(b) leave the last retained digit the same if it is *even*.

Thus, 62.650 and 62.350 rounded to three significant figures become, respectively, 62.6 (even rule) and 62.4 (odd rule). The number zero as a last retained digit is always considered an even number; thus, 62.050 rounded to three significant figures becomes 62.0.

These rounding rules must be modified slightly when digits to the left of the decimal point are to be dropped. In order to maintain the position of the decimal point in such situations, zeros must replace all of the dropped digits that are to the left of the decimal point. Parts (c) and (f) of Example 2.4 illustrate this point.

EXAMPLE 2.4

Round off each of the following numbers to three significant figures.

(a) 423.87   (b) 423.17   (c) 655,234
(d) 0.03315   (e) 352.50   (f) 162,500

Solution

(a) Rule 2 applies. The last retained digit (the 3) is increased in value by one unit.

$$423.87 \quad \text{becomes} \quad 424$$

(b) Rule 1 applies. The last retained digit remains the same (the 3), and all digits that follow it are simply dropped.

$$423.17 \quad \text{becomes} \quad 423$$

(c) Since the first digit to be dropped is a 2, rule 1 applies.

$$655{,}234 \quad \text{becomes} \quad 655{,}000$$

Note that to maintain the position of the decimal, zero must replace all of the "dropped" digits. This will always be the case when digits to the left of the decimal place are "dropped."

(d) Rule 3 applies. This first and only digit to be dropped is a 5. The last retained digit (the 1) is an odd number, so, using the odd-even rule, its value is increased by one unit.

$$0.03315 \quad \text{becomes} \quad 0.0332$$

(e) Rule 3 applies again. This time the last digit retained is even (the 2), so its value is not changed.

$$352.50 \quad \text{becomes} \quad 352$$

(f) This is again a rule 3 situation. Since an even digit (the 2) occupies the third significant figure place, its value is not changed.

$$162{,}500 \quad \text{becomes} \quad 162{,}000$$

Note again that zeros must take the place of all digits to the left of the decimal place that are "dropped."

PRACTICE EXERCISE 2.4

Round off each of the following numbers to two significant figures.

(a) 257   (b) 253   (c) 0.43276
(d) 655,000   (e) 625,000   (f) 12.53

Ans. (a) 260; (b) 250; (c) 0.43; (d) 660,000; (e) 620,000; (f) 13.

Similar exercises: Problems 2.27 and 2.28

The number of significant figures in the result of any mathematical calculation is limited by the least precise measurement used in the calculation. Calculations cannot improve the precision of measurements. Two operational rules exist to help ensure that calculations do not increase measurement precision. One rule covers the operations of multiplication and division and the other the operations of addition and subtraction.

22   Chapter 2 · Numbers from Measurements

RULE 1 *Multiplication and Division.* In multiplication and division the number of significant figures in the product or quotient is the same as in the number in the calculation that contains the fewest significant figures.

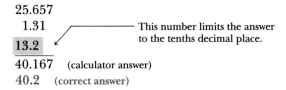

$$6.038 \times \boxed{2.57} = 15.51766 \quad \text{(calculator answer)}$$
$$= 15.5 \quad \text{(correct answer)}$$

This number limits the answer to three significant figures.

RULE 2 *Addition and Subtraction.* In addition and subtraction the last digit retained in the sum or difference should correspond to the first doubtful (estimated) decimal position in the series of numbers.

```
  25.657
   1.31          ← This number limits the answer
  13.2              to the tenths decimal place.
  ─────
  40.167   (calculator answer)
  40.2     (correct answer)
```

Note that for multiplication and division (rule 1) significant figures are counted and that for addition and subtraction (rule 2) decimal places are counted. The answers from either addition or subtraction can have more or fewer significant figures than any of the numbers that have been added or subtracted, as is shown in Example 2.7.

EXAMPLE 2.5

Indicate the number of significant figures that should be present in the answer to each of the following multiplications. Assume that all numbers are measured quantities.

(a) $2.0 \times 2.0 \times 3.0$
(b) $2.00 \times 2.00 \times 3.0$
(c) $2.000 \times 2.00 \times 3.00$
(d) $2.0000 \times 2.00 \times 3.0000$

Solution
The number of significant figures in a product is the same as in the number in the multiplication that contains the fewest significant figures. The desired answers can be obtained without actually doing the multiplications.

(a) Each number to be multiplied contains two significant figures. This limits the answer to **two** significant figures.
(b) Two input numbers contain three significant figures and the third number only two significant figures. Only **two** significant figures should be present in the answer.
(c) The input numbers contain, respectively, four, three, and three significant figures. Thus, **three** significant figures should be present in the answer.
(d) Even though two input numbers contain five significant figures, the third input number has only three and this number limits the answer to **three** significant figures.

PRACTICE EXERCISE 2.5

Indicate the number of significant figures that should be present in the answer to each of the following multiplications. Assume that all numbers are measured quantities.

(a) 6.00 × 6.00 × 6.0
(b) 0.6 × 0.600 × 0.60
(c) 5.000 × 0.5000 × 0.5000
(d) 3.0 × 2.0 × 4.0

Ans: (a) 2; (b) 1; (c) 4; (d) 2.

Similar exercises: Problems 2.33 and 2.34

EXAMPLE 2.6

Perform the following computations, all of which involve multiplication and/or division. Express your answers to the proper number of significant figures. Assume that all numbers are measured quantities.

(a) 6.7321×0.0021

(b) $\dfrac{16{,}240}{23.42}$

(c) $\dfrac{120.0}{4.000}$

(d) $\dfrac{5.444 \times 8.670}{2.321 \times 3.27}$

Solution

(a) The calculator answer to this problem is

$$6.7321 \times 0.0021 = 0.01413741$$

The input number with the least number of significant figures is 0.0021.

$$\underbrace{6.7321}_{\text{five significant figures}} \times \underbrace{0.0021}_{\text{two significant figures}}$$

Thus the calculator answer must be rounded off to two significant figures

$$\underbrace{0.01413741}_{\text{calculator answer}} \;\text{becomes}\; \underbrace{0.014}_{\text{correct answer}}$$

(b) The calculator answer to this problem is

$$\dfrac{16{,}240}{23.42} = 693.42442$$

Both input numbers contain four significant figures. Thus the correct answer will also contain four significant figures and is obtained by rounding the calculator answer to four significant figures.

$$\underbrace{693.42442}_{\text{calculator answer}} \;\text{becomes}\; \underbrace{693.4}_{\text{correct answer}}$$

(c) The calculator answer to this problem is

$$\dfrac{120.0}{4.000} = 30$$

Both input numbers contain four significant figures. Thus the correct answer must also contain four significant figures.

$$30 \text{ becomes } 30.00$$

calculator answer → ↗ ↖ — correct answer

Note here how the calculator gave too few significant figures. Most calculators cut off zeros after the decimal point even if they are significant. Using too few significant figures is just as wrong as using too many.

(d) This problem involves both multiplication and division. The calculator answer is

$$\frac{5.444 \times 8.670}{2.321 \times 3.27} = 6.2189107$$

The input with the least number of significant figures is 3.27, which contains three significant figures.

four significant figures ⟶ ↘   ↙ ⟵ four significant figures

$$\frac{5.444 \times 8.670}{2.321 \times \boxed{3.27}}$$

four significant figures ⟶ ↗   ↖ ⟵ three significant figures

Thus the calculator answer must be rounded off to three significant figures.

$$6.2189107 \text{ becomes } 6.22$$

calculator answer ⟶ ↗   ↖ correct answer

PRACTICE EXERCISE 2.6

Perform the following computations, all of which involve multiplication and/or division. Express your answers to the proper number of significant figures. Assume that all numbers are measured quantities.

(a) 3.751×0.42 (b) $\dfrac{1{,}810{,}000}{3.1453}$

(c) $\dfrac{1800.0}{6.000}$ (d) $\dfrac{3.130 \times 3.140}{3.15}$

Ans. (a) 1.6; (b) 575,000; (c) 300.0; (d) 3.12.   Similar exercises: Problems 2.35–2.40

EXAMPLE 2.7

Perform the following computations, all of which involve addition or subtraction. Express your answers to the proper number of significant figures. Assume that all numbers are measured quantities.

(a) $28.7 + 7.01 + 22$ (b) $8.3 + 1.2 + 1.7$
(c) $0.4378 - 0.4367$ (d) $4200 + 14.7$

Solution

(a) The calculator answer to this problem is

$$28.7 + 7.01 + 22 = 57.71$$

The input number 28.7 has an uncertainty in the tenths position, the input number 7.01 an uncertainty in the hundredths position, and the input number 22 an uncertainty in the ones position. Thus the number 22 has the greatest uncertainty, and the last retained digit in the answer should reflect this uncertainty (the ones digit).

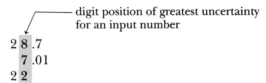

Hence, the calculator answer 57.71 must be rounded off to the ones place.

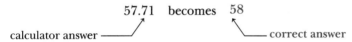

(b) The calculator answer to this problem is

$$8.3 + 1.2 + 1.7 = 11.2$$

All three input numbers have uncertainty in the tenths place. Thus, the last retained digit in the correct answer will be that of tenths. In this particular problem the calculator answer and the correct answer are the same, a situation which does not occur very often. The correct answer is 11.2.

Note how each of the numbers in the sum in this problem contains two significant figures and yet the correct answer has *three* significant figures. Why? Again, the number of significant figures is not the determining factor in addition and subtraction (rule 2) as it is in multiplication and division (rule 1).

(c) Both numbers are known to the ten-thousandths place. Thus the answer is reportable to the ten-thousandths place.

```
                    ┌─── digit position of greatest uncertainty
                    │      for an input number
                    ↓
          0.437 8
         −0.436 7
         ─────────
          0.001 1    (calculator and correct answer)
```

Again, the calculator and correct answer are one and the same. Note that two significant figures were "lost" in the subtraction. The answer has two significant figures. The two input numbers each have four significant figures.

(d) The calculator answer for this problem is

$$4200 + 14.7 = 4214.7$$

The uncertainty in the input number 4200 is the hundreds place (it has only two significant figures), and the uncertainty in the input number 14.7 is in the tenths place. The uncertainty in the number 4200 will be the "limiting factor" for the answer

```
          ┌──── digit position of greatest uncertainty
          │     for an input number
       4 2 00
         14.7
       ───────
       4 2 14.7   (calculator answer)
```

Hence, the calculator answer of 4214.7 must be rounded to the hundreds position.

$$4214.7 \quad \text{becomes} \quad 4200$$

calculator answer ─── ↑   ↑ ─── correct answer

The correct answer 4200 is the same as one of the input numbers. From the standpoint of significant figures, the number 14.7 is negligible when added to the number 4200.

PRACTICE EXERCISE 2.7

Perform the following computations, all of which involve addition or subtraction. Express your answers to the proper number of significant figures. Assume that all numbers are measured quantities.

(a) $13.01 + 13.001 + 13.001$ (b) $10.2 + 3.4 - 6.01$
(c) $0.6700 - 0.6644$ (d) $34.7 + 0.0007$

Ans. (a) 39.01; (b) 7.6; (c) 0.0056; (d) 34.7.

Similar exercises: Problems 2.41–2.44

Not all numbers in a calculation have to originate from a measurement Sometimes *exact numbers* (counted or defined numbers) are part of a calculation. Since there is no uncertainty in an exact number, it is considered to possess an infinite number of significant figures. Consequently, exact numbers, when they appear in a calculation, will never limit the number of significant figures allowable in the answer. We do not consider them when determining the number of allowable significant figures. The allowable number is determined in the usual way, taking into account only those numbers that were experimentally determined by measurement.

EXAMPLE 2.8

A nickel is found to weigh 5.0715 g (grams). What would be the mass in grams of 13 such coins?

Solution

Since the coin has a mass of 5.0715 g, the mass of 13 such coins will be

$$13 \times 5.0715 \text{ g} = 65.9295 \text{ g} \quad \text{(calculator answer)}$$

The number 13 is an exact number (a counted number) and can be considered to have an infinite number of significant figures. Thus, the correct answer will have five significant figures, the number of significant figures in the measurement 5.0715.

$$65.9295 \text{ g} \quad \text{becomes} \quad 65.930 \text{ g}$$

calculator answer ⟶   ⟵ correct answer

PRACTICE EXERCISE 2.8

What would be the total combined length, in centimeters, of 7 new pencils, each of which has a length of 19.13 cm?

Ans. 133.9 cm.

Similar exercises: Problems 2.47 and 2.48

2.5 Scientific Notation

In scientific work, very large and very small numbers are frequently encountered. For example, in one cup of water—an "ordinary" amount of water—there are approximately

7,910,000,000,000,000,000,000,000 molecules

(A molecule of water is the smallest possible unit of water; see Sec. 5.2). A single water molecule has a mass of

0.0000000000000000000000299 gram

Such large and small numbers as these are difficult to use. Recording them is not only time consuming but also very prone to errors—too many or too few zeros recorded. Also such numbers are awkward to work with in calculations. Consider the problem of multiplying the above two numbers together. Handling all of those zeros is "mind boggling."

A method exists for expressing cumbersome many-digit numbers, such as the two just shown, in compact form. Called scientific notation, this method eliminates the need to write all the zeros. **Scientific notation** is a system in which an ordinary decimal number is expressed as a product of a number between 1 and 10 multiplied by 10 raised to a power. The two previously cited numbers, dealing with molecules of water, expressed in scientific notation are, respectively,

$$7.91 \times 10^{24} \text{ molecules} \quad \text{and} \quad 2.99 \times 10^{-23} \text{ gram}$$

Note that in scientific notation each number requires only seven digits, compared to 25 in normal arithmetic notation.

Exponents

An understanding of exponents is the key to handling scientific notation. A brief review of exponents and their use is in order before we consider the rules for converting numbers from ordinary decimal notation to scientific notation and vice versa.

In mathematics a convenient method exists for showing that a number has been multiplied by itself two or more times. It involves the use of an exponent. An **exponent** is a number written as a superscript following another number and indicates how many times the first number is to be multiplied by itself. The following examples illustrate the use of exponents.

$$6^2 = 6 \times 6 = 36$$
$$3^5 = 3 \times 3 \times 3 \times 3 \times 3 = 243$$
$$10^3 = 10 \times 10 \times 10 = 1000$$

Exponents are also frequently referred to as *powers* of numbers. Thus, 6^2 may be verbally read as "six to the second power," and 3^5 as "three to the fifth power." Raising a number to the second power is often called "squaring," and raising it to the third power "cubing."

Scientific notation exclusively uses powers of ten. *When ten is raised to a positive power, its decimal equivalent is the number 1 followed by as many zeros as the power.* This one-to-one correlation between power magnitude and number of zeros is shown in color in the following examples.

$$10^2 = 100 \qquad \text{(two zeros and a power of 2)}$$
$$10^4 = 10,000 \qquad \text{(four zeros and a power of 4)}$$
$$10^6 = 1,000,000 \qquad \text{(six zeros and a power of 6)}$$

The notation 10^0 is a defined quantity.

$$10^0 = 1$$

The above generalization easily explains why. Ten to the zero power is the number 1 followed by no zeros, which is simply 1.

All of the examples of exponential notation presented so far have had positive exponents. This is because each example represented a number of magnitude greater than one. Negative exponents are also possible. They are associated with numbers of magnitude less than one.

A negative sign in front of an exponent is interpreted to mean that the number and the power to which it is raised are in the denominator of a fraction in which 1 is the numerator. The following examples illustrate this interpretation.

$$10^{-1} = \frac{1}{10^1} = \frac{1}{10} = 0.1$$

$$10^{-2} = \frac{1}{10^2} = \frac{1}{10 \times 10} = \frac{1}{100} = 0.01$$

$$10^{-3} = \frac{1}{10^3} = \frac{1}{10 \times 10 \times 10} = \frac{1}{1000} = 0.001$$

When the number 10 is raised to a negative power, the absolute value of the power (the value ignoring the minus sign) is always one more than the number of zeros between the decimal point and the one. This correlation between power magnitude and number of zeros is shown in color in the following examples.

$$10^{-2} = 0.01 \qquad \text{(one zero and a power of } -2\text{)}$$
$$10^{-4} = 0.0001 \qquad \text{(three zeros and a power of } -4\text{)}$$
$$10^{-6} = 0.000001 \qquad \text{(five zeros and a power of } -6\text{)}$$

Differences in magnitude between numbers that are powers of 10 are often described with the phrase "orders of magnitude." An **order of magnitude** is a single exponential value of the number 10. Thus, 10^6 is four orders of magnitude larger than 10^2, and 10^7 is three orders of magnitude larger than 10^4.

Writing Numbers in Scientific Notation

A number written in scientific notation has two parts: (1) a *coefficient*, written first, which is a number between 1 and 10, and (2) an *exponential term*, which is 10 raised to a power. The coefficient part is always multiplied by the exponential term. Using the scientific notation form of the number 703 as an example, we have

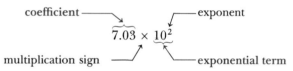

Rules for converting numbers from decimal to scientific notation are very simple.

RULE 1   The value of the exponent is determined by counting the number of places the original decimal point must be moved to give the coefficient.

RULE 2   If the decimal point is moved to the *left* to get the coefficient, the exponent is a *positive number*. If the decimal point is moved to the *right* to get the coefficient, the exponent is a *negative number*.

Numerous applications of these two rules are found in Examples 2.9 and 2.10.

When a number is expressed in scientific notation, *only significant digits become part of the coefficient.* Because of this there is never any confusion (ambiguity) in determining the number of significant figures in a number expressed in scientific notation. There are five possible precision interpretations for the number 10,000 (1, 2, 3, 4, or 5 significant figures). In scientific notation each of these interpretations assumes a different form.

$$1 \times 10^4 \qquad \text{(10,000 with one significant figure)}$$
$$1.0 \times 10^4 \qquad \text{(10,000 with two significant figures)}$$
$$1.00 \times 10^4 \qquad \text{(10,000 with three significant figures)}$$
$$1.000 \times 10^4 \qquad \text{(10,000 with four significant figures)}$$
$$1.0000 \times 10^4 \qquad \text{(10,000 with five significant figures)}$$

Most scientists use the preceding scientific notation method for designating significant figures in an "ambiguous" number instead of the overbar notation discussed in Section 2.3. The overbar notation is used only in those situations where there is reason for not expressing the number in exponential notation.

EXAMPLE 2.9

Express the following numbers in scientific notation.

(a) 2305.7   (b) 112.00   (c) 0.000656   (d) 220,000

Solution

(a) The coefficient, which must be a number between 1 and 10, is obtained by moving the decimal point in 2305.7 three places to the left.

$$2\,305.7$$
decimal point shift

This gives us 2.3057 as the coefficient. The value of the exponent in the exponential term is $+3$, since the decimal point was shifted three places to the left. A shift of the decimal point to the left will always result in a positive exponent. Multiplying the coefficient by the exponential term 10^3 gives us the scientific notation form of the number.

$$2.3057 \times 10^3$$

(b) The coefficient, obtained by moving the decimal point two places to the left, is 1.1200.

$$1\,12.00$$
decimal point shift

Note that the two zeros after the decimal point in the original number must be shown in the coefficient. Dropping the zeros gives 1.12, which contains only three significant figures. The original number had five significant figures. You can never lose or gain significant figures in converting a number from decimal notation to scientific notation.

In obtaining the coefficient, the decimal point was moved two places to the left. Such a decimal point movement sets the power of ten at $+2$. Thus, the scientific notation form of the number is

$$1.1200 \times 10^2$$

(c) Since this number has a value less than one, the decimal must be moved to the right in order to obtain a coefficient with a value between one and ten.

$$0.0006\,56$$
decimal point shift

Movement of the decimal point to the right means the exponent will have a negative value, in this case -4. Thus, in scientific notation we have

$$6.56 \times 10^{-4}$$

(d) The coefficient, obtained by moving the decimal point five places to the left, is 2.2

$$2\underset{\uparrow\!\!\text{—————}\!\!\rfloor}{\,}20,000$$
$$\text{decimal point shift}$$

The coefficient is 2.2 rather than 2.20000 because the trailing zeros (Sec. 2.3) are not significant digits. The original number contains only two significant figures.

The value of the exponent in the exponential term is +5, since the decimal point was moved five places to the left. Thus, the scientific notation form of the number 220,000 is

$$2.2 \times 10^5$$

PRACTICE EXERCISE 2.9

Express the following numbers in scientific notation.

(a) 11.013 (b) 3,001,000 (c) 0.000100 (d) 0.100055

Ans. (a) 1.1013×10^1; (b) 3.001×10^6; (c) 1.00×10^{-4}; (d) 1.00055×10^{-1}

Similar exercises: Problems 2.55–2.58

Scientific notation is not often used to express numbers that are more easily written in decimal form. For example, the numbers 2.4, 0.911, and 57 are simpler in their original form, and we would probably not use scientific notation for them. Although there are no fixed rules as to when scientific notation should be used, it is preferable to use it for numbers greater than 10^3 or less than 10^{-1}.

Converting from Scientific Notation to Decimal Notation

To convert a number in scientific notation, such as 6.02×10^{23}, into a regular decimal number, we start by examining the exponent. The value of the exponent tells how many places the decimal point must be moved. If the exponent is positive, movement is to the right to give a number greater than one; if it is negative, movement is to the left to give a number less than one. Zeros may have to be added to the number as the decimal point is moved.

EXAMPLE 2.10

Write the following exponential numbers in ordinary decimal notation.

(a) 3.78×10^{-6} (b) 7.23×10^6 (c) 3.0000×10^3
(d) 7.500×10^{-2}

Solution

(a) The exponent -6 tells us the decimal is to be located six places to the left of where it is in 3.78. Leading zeros will have to be added to accommodate the decimal point change. Remember, when the exponent is negative, the decimal number will always be less than one.

$$0\,000003.78 \times 10^{-6} = 0.00000378$$

with added zeros and decimal point shift six places to the left.

(b) The exponent $+6$ tells us the decimal is to be located six places to the right of where it is in 7.23. Again, zeros must be added to mark the new decimal place correctly. These added "trailing" zeros are not significant zeros. Only three significant figures are present in the number.

$$7.230000 \times 10^6 = 7{,}230{,}000$$

with added zeros and decimal point shift.

(c) The decimal point is moved three places to the right of where it is in 3.0000. This gives the number 3000.0. The zero to the right of the decimal point in 3000.0 must be retained. Otherwise the number of significant digits would drop from five to four as a result of the notation change. The number of significant figures must always remain the same unless specific instructions relative to rounding off are given.

$$3.000\,0 \times 10^3 = 3000.0$$

with decimal point shift.

(d) The decimal point is to be moved two places to the left. Leading zeros must be added to mark the new decimal place correctly.

$$0\,07.500 \times 10^{-2} = 0.07500$$

with added zeros and decimal point shift.

PRACTICE EXERCISE 2.10

Write the following exponential numbers in ordinary decimal notation.

(a) 6.24×10^{-4} (b) 3.4×10^5 (c) 3.000×10^2 (d) 7.5×10^{-3}

Ans. (a) 0.000624; (b) 340,000; (c) 300.0; (d) 0.0075.

Similar exercises: Problems 2.59–2.62

2.6 Scientific Notation and Mathematical Operations

A major advantage of writing numbers in scientific notation is that it greatly simplifies the mathematical operations of multiplication and division.

Multiplication in Scientific Notation

Multiplication of two or more numbers expressed in scientific notation involves two separate operations or steps.

STEP 1 Multiply the coefficients (the decimal numbers between 1 and 10) together in the usual manner.

STEP 2 *Add* algebraically the exponents of the powers of ten to obtain a new exponent.

In general terms, we can represent the multiplication of two scientific notation numbers as follows.

$$(a \times 10^x) \times (b \times 10^y) = ab \times 10^{x+y}$$

EXAMPLE 2.11

Carry out the following multiplications in scientific notation. Be sure to take into account significant figures in obtaining your final answer.

(a) $(2.543 \times 10^3) \times (2.003 \times 10^6)$ (b) $(1.15 \times 10^{-2}) \times (4.52 \times 10^{-3})$
(c) $(4.210 \times 10^{-5}) \times (2.0 \times 10^2)$ (d) $(5.329 \times 10^{-1}) \times (3.11 \times 10^9)$

Solution

(a) Multiplying the two coefficients together gives

$$2.543 \times 2.003 = 5.093629 \quad \text{(calculator answer)}$$

Since each of the input numbers for the multiplication has four significant figures, the answer should also have four significant figures, not seven as given by the calculator.

5.093629 becomes 5.094

calculator answer ⎯⎯⎯⎯⎯⎯⎯⎯⎯⎯⎯⎯ correct answer

Multiplication of the two powers of ten to give the exponential part of the answer requires adding the exponents to give a new exponent.

$$10^3 \times 10^6 = 10^{3+6} = 10^9$$

Combining the new coefficient with the new exponential term gives the answer.

$$5.094 \times 10^9$$

(b) Multiplying the two coefficients together gives

$$1.15 \times 4.52 = 5.198 \quad \text{(calculator answer)}$$

Since both input numbers for the multiplication contain three significant figures, the calculator answer must be rounded to three significant figures.

$$5.198 \quad \text{becomes} \quad 5.20$$

calculator answer ─────╱   ╲───── correct answer

Next, the exponents of the powers of ten are added to generate the new power of ten.

$$10^{-2} \times 10^{-3} = 10^{(-2)+(-3)} = 10^{-5}$$

(To add two numbers of the same sign—positive or negative—just add the numbers and place the common sign in front of the sum.) Combining the coefficient and the exponential term gives the answer of

$$5.20 \times 10^{-5}$$

(c) Multiplying the two coefficients together gives

$$4.210 \times 2.0 = 8.42 \quad \text{(calculator answer)}$$

Because the input number 2.0 contains only two significant figures, the answer is also limited to two significant figures. Thus,

$$8.42 \quad \text{becomes} \quad 8.4$$

calculator answer ─────╱   ╲───── correct answer

In combining the exponential terms we will have to add exponents with different signs, -5 and $+2$. To do this, we first determine the larger number (5 is larger than 2) and then subtract the smaller number from it ($5 - 2 = 3$). The sign is always the sign of the larger number (minus in this case). Thus

$$(-5) + (+2) = (-3)$$

and

$$10^{-5} \times 10^{+2} = 10^{(-5)+(+2)} = 10^{-3}$$

Combining the coefficient and the exponential parts gives

$$8.4 \times 10^{-3}$$

(d) Multiplying coefficients gives

$$5.329 \times 3.11 = 16.57319 \quad \text{(calculator answer)}$$

The answer must be rounded to three significant figures because the input number 3.11 has only three significant figures.

$$16.57319 \quad \text{becomes} \quad 16.6$$

calculator answer ─────╱   ╲───── correct answer

Again, in combining the exponential terms we have exponents of different signs. The smaller exponent (1) is subtracted from the larger exponent (9), and the sign of the larger exponent (+) is used.

$$(+9) + (-1) = (+8)$$
$$10^9 \times 10^{-1} = 10^8$$

Combining the coefficient and the exponential term gives

$$16.6 \times 10^8$$

This answer has something wrong with it; it is not in correct scientific notation form. The coefficient is not a number between one and ten. This problem is corrected by recognizing that 16.6 is equal to 1.66×10^1, making this substitution for 16.6, and then combining exponential terms.

$$16.6 \times 10^8 = (1.66 \times 10^{①}) \times 10^{⑧} = 1.66 \times 10^{⑨}$$

PRACTICE EXERCISE 2.11

Carry out the following multiplications in scientific notation. Be sure to take significant figures into account in obtaining your final answer.

(a) $(1.113 \times 10^3) \times (7.200 \times 10^5)$ (b) $(2.05 \times 10^{-3}) \times (1.19 \times 10^{-7})$
(c) $(4.21 \times 10^{-9}) \times (2.107 \times 10^6)$ (d) $(7.92 \times 10^{10}) \times (2.3 \times 10^{-4})$

Ans. (a) 8.014×10^8; (b) 2.44×10^{-10}; (c) 8.87×10^{-3}; (d) 1.8×10^7.

Similar exercises: Problems 2.65–2.68

Division in Scientific Notation

Division of two numbers expressed in scientific notation involves two separate operations or steps.

STEP 1 Divide the coefficients (the decimal numbers between 1 and 10) in the usual manner.

STEP 2 *Subtract* algebraically the exponent in the denominator (bottom) from the exponent in the numerator (top) to give the exponent of the new power of ten.

Note in multiplication we add exponents, and in division we subtract exponents.

EXAMPLE 2.12

Carry out the following divisions in scientific notation. Be sure to take significant figures into account in obtaining your final answer.

(a) $\dfrac{3.76 \times 10^9}{1.23 \times 10^6}$ (b) $\dfrac{9.98 \times 10^{-3}}{2.341 \times 10^{-7}}$

(c) $\dfrac{5.1 \times 10^{-10}}{8.76 \times 10^7}$ (d) $\dfrac{3.43 \times 10^5}{3.93 \times 10^{-4}}$

Solution

(a) Performing the indicated division involving the coefficients gives

$$\frac{3.76}{1.23} = 3.0569105 \quad \text{(calculator answer)}$$

Since both input numbers for the division have three significant figures, the calculator answer must be rounded off to three significant figures.

$$3.0569105 \quad \text{becomes} \quad 3.06$$

The division of exponential terms involves the algebraic subtraction of exponents.

$$\frac{10^9}{10^6} = 10^{(+9)-(+6)} = 10^3$$

Algebraic subtraction involves changing the sign of the number to be subtracted and then following the rules for addition (as outlined in Example 2.11). In this problem the number to be subtracted (+6) becomes, upon changing the sign, −6. Then we add +9 and −6. The answer is +3, as shown in the preceding equation. Combining the coefficient and the exponential term gives

$$3.06 \times 10^3$$

(b) The new coefficient is obtained by dividing 2.341 into 9.98.

$$\frac{9.98}{2.341} = 4.2631354 \quad \text{(calculator answer)}$$

The correct answer will contain only three significant figures, the same number as in the input number 9.98.

$$4.2631354 \quad \text{becomes} \quad 4.26$$

calculator answer ⎯⎯⎯⎯⎯⎯⎯⎯⎯⎯⎯⎯⎯⎯⎯⎯⎯⎯⎯ correct answer

The exponential part of the answer is obtained by subtracting −7 from −3. Changing the sign of the number to be subtracted gives +7. Adding +7 and −3 gives +4. Therefore;

$$\frac{10^{-3}}{10^{-7}} = 10^{(-3)-(-7)} = 10^4$$

Combining the coefficient and the exponential term gives

$$4.26 \times 10^4$$

(c) One input number for the coefficient division has two significant figures, and the other has three significant figures. Therefore, the answer should contain two significant figures.

$$\frac{5.1}{8.76} = 0.58219178 \quad \text{(calculator answer)}$$

$$= 0.58 \quad \text{(correct answer)}$$

The exponential-term division involves subtracting $+7$ from -10. Changing the sign of the number to be subtracted gives a -7. Adding -7 and -10 gives -17.

$$\frac{10^{-10}}{10^7} = 10^{(-10)-(+7)} = 10^{-17}$$

Combining the two parts of the problem yields the number

$$0.58 \times 10^{-17}$$

which is not in correct scientific notation form because the coefficient is a number less than one. This problem is remedied by recognizing that 0.58 is equal to 5.8×10^{-1}, making this substitution for 0.58, and then combining the two exponential terms.

$$0.58 \times 10^{-17} = (5.8 \times 10^{-1}) \times 10^{-17} = 5.8 \times 10^{-18}$$

(d) The new coefficient, obtained by dividing 3.43 by 3.93, should contain three significant figures, the same number as in both of the input numbers.

$$\frac{3.43}{3.93} = 0.87277353 \quad \text{(calculator answer)}$$

$$= 0.873 \quad \text{(correct answer)}$$

Performing the exponential-term division by subtracting the powers of the exponential terms gives

$$\frac{10^5}{10^{-4}} = 10^{(+5)-(-4)} = 10^9$$

Combining the coefficient and the exponential term gives

$$0.873 \times 10^9$$

As in part (c), this answer is not in correct scientific notation form because the coefficient is a number less than one. The change to correct form is effected by rewriting 0.873 as 8.73×10^{-1}, making this substitution for 0.873, and then combining the two exponential terms.

$$0.873 \times 10^9 = (8.73 \times 10^{-1}) \times 10^9 = 8.73 \times 10^8$$

PRACTICE EXERCISE 2.12

Carry out the following divisions in scientific notation. Be sure to take significant figures into account in obtaining your final answer.

(a) $\dfrac{2.05 \times 10^5}{1.19 \times 10^3}$ (b) $\dfrac{7.200 \times 10^{-3}}{1.113 \times 10^{-7}}$

(c) $\dfrac{4.21 \times 10^{-9}}{2.107 \times 10^6}$ (d) $\dfrac{3.92 \times 10^{10}}{9.1 \times 10^{-4}}$

Ans. (a) 1.72×10^2; (b) 6.469×10^4; (c) 2.00×10^{15}; (d) 4.3×10^{13}.   Similar exercises: Problems 2.71–2.74

Addition and Subtraction in Scientific Notation

To add or subtract numbers written in scientific notation, *the power of ten for all numbers must be the same.* More often than not, one or more exponents must be adjusted. Adjusting the exponent requires rewriting the number in a form where the coefficient is a number greater than 10 or less than 1. With exponents all the same, the *coefficients are then added or subtracted and the exponent is maintained at its now common value.* Although any of the exponents may be changed, changing the smaller numbers to larger numbers will usually result in an accompanying coefficient that is a number between 1 and 10.

EXAMPLE 2.13

Perform the following additions or subtractions with all numbers expressed in scientific notation. Answers will need to be checked for the correct number of significant figures and for correct scientific notation form.

(a) $(2.723 \times 10^4) + (6.045 \times 10^4)$ (b) $(3.73 \times 10^3) - (3.73 \times 10^2)$
(c) $(9.97 \times 10^{-2}) + (5.89 \times 10^{-4})$

Solution

(a) The exponents are the same to begin with. Therefore, we can proceed with the addition immediately.

$$\begin{array}{r} 2.723 \times 10^4 \\ + 6.045 \times 10^4 \\ \hline 8.768 \times 10^4 \end{array}$$ (calculator and correct answer)

Note that the exponent values are not added.

(b) The exponents are different, so before subtracting we must change one of the exponents. Let us change 10^2 to 10^3.

$$10^2 \quad \text{can be written as} \quad 10^{-1} \times 10^3$$

Then, by substitution, we have

$$3.73 \times 10^2 = 3.73 \times 10^{-1} \times 10^3$$

The coefficient and the first exponent are then combined to give a new coefficient.

$$(3.73 \times 10^{-1}) \times 10^3 = 0.373 \times 10^3$$

We are now ready to make the subtraction called for in the original statement of the problem.

```
                            ┌──── common exponent
    3.73 × 10³ =    3.73  × 10³
  − 3.73 × 10² = − 0.373 × 10³
                  ─────────────
                    3.357 × 10³    (calculator answer)
```

The calculator answer must be adjusted for significant figures. On the common exponent basis of 10^3, the uncertainty in 3.73 lies in the hundredths place and that in 0.373 in the thousandths place. The correct answer, therefore, is limited to an uncertainty of hundredths. (Recall, from Example 2.6, the rules on addition and significant figures.) Thus,

$$3.357 \times 10^3 \quad \text{becomes} \quad 3.36 \times 10^3$$

(c) The exponents are 10^{-2} and 10^{-4}. Since -4 is smaller than -2, let us have 10^{-2} as the common exponent.

$$10^{-4} \quad \text{can be rewritten as} \quad 10^{-2} \times 10^{-2}$$

Then by substitution we have

$$5.89 \times 10^{-4} = 5.89 \times 10^{-2} \times 10^{-2}$$

The coefficient and the first exponent are then combined to give a new coefficient.

$$5.89 \times 10^{-2} \times 10^{-2} = 0.0589 \times 10^{-2}$$

We are now ready to make the addition called for in the original statement of the problem.

```
                              ┌──── common exponent
    9.97 × 10⁻² =  9.97   × 10⁻²
    5.89 × 10⁻⁴ =  0.0589 × 10⁻²
                  ─────────────
                   10.0289 × 10⁻²   (calculator answer)
```

The calculator answer must be adjusted for significant figures and also changed into correct scientific notation since the coefficient has a value greater than 10. The significant figure adjustment rounds the answer to hundredths, giving

$$10.03 \times 10^{-2}$$

The correct scientific notation adjustment involves rewriting 10.03 as a power of ten and then simplifying the resulting expression.

$$10.03 \times 10^{-2} = (1.003 \times 10^1) \times 10^{-2}$$
$$= 1.003 \times 10^{-1}$$

PRACTICE EXERCISE 2.13

Perform the following additions or subtractions with all numbers expressed in scientific notation. Check answers for the correct number of significant figures and for correct scientific notation form.

(a) $(2.661 \times 10^3) + (3.011 \times 10^3)$ (b) $(2.66 \times 10^4) - (1.03 \times 10^3)$
(c) $(9.98 \times 10^{-3}) + (8.04 \times 10^{-5})$

Ans. (a) 5.672×10^3; (b) 2.56×10^4; (c) 1.006×10^{-2}.

Similar exercises: Problems 2.79–2.82

Key Terms

The new terms or concepts defined in this chapter are

accuracy (Sec. 2.2)   How close a measurement (or the average of multiple measurements) comes to the true or accepted value.

exponent (Sec. 2.5)   A number written as a superscript following another number that indicates how many times the first number is to be multiplied by itself.

order of magnitude (Sec. 2.5)   A single exponential value of the number ten.

precision (Sec. 2.2)   How close multiple measurements of the same quantity come to each other.

random error (Sec. 2.2)   An error originating from uncontrolled variables in an experiment.

rounding off (Sec. 2.4)   The process of deleting unwanted (nonsignificant) digits from a calculated number.

scientific notation (Sec. 2.4)   A system in which an ordinary decimal number is expressed as a number between 1 and 10 multiplied by 10 raised to a power.

significant figures (Sec. 2.3)   Digits in any measurement that are known with certainty plus one digit that is uncertain.

systematic error (Sec. 2.2)   An error originating from controlled variables in an experiment.

Practice Problems

Accuracy and Precision (Sec. 2.2)

2.1 With a very precise volumetric measuring device, the volume of a liquid sample is determined to be 6.321 L (liters). Three students are asked to determine the volume of the same liquid sample using a less precise measuring device. How do you evaluate the following work of the three students with regard to precision and accuracy?

		Students	
Trials	A	B	C
1	6.35 L	6.31 L	6.36 L
2	6.31 L	6.32 L	6.36 L
3	6.38 L	6.32 L	6.35 L
4	6.32 L	6.32 L	6.36 L

Practice Problems

2.2 With a very precise measuring device, the length of an object is determined to be 13.452 mm (millimeters). Three students are asked to determine the length of the same object using a less precise measuring device. How do you evaluate the following work of the three students with regard to precision and accuracy?

	Students		
Trials	A	B	C
1	13.6 mm	13.4 mm	13.8 mm
2	13.7 mm	13.5 mm	13.8 mm
3	13.3 mm	13.5 mm	13.3 mm
4	13.4 mm	13.4 mm	14.1 mm

2.3 Can a series of measurements of the same quantity be precise and not accurate? Explain.

2.4 Can a series of measurements of the same quantity be accurate and not precise? Explain.

Uncertainty in Measurement (Sec. 2.3)

2.5 Indicate whether each of the following statements contains an exact number or a measured (estimated) quantity within it.
(a) The beehive contains 14,000 bees.
(b) There were 97 people in attendance at the meeting.
(c) A 20 lb bag of potatoes was purchased at the grocery store.
(d) There are 4 quarts in a gallon.

2.6 Indicate whether each of the following statements contains an exact number or a measured (estimated) quantity within it.
(a) Americans consume 75 acres of pizza per year.
(b) A sheet of paper is 0.0042 inch thick.
(c) There are 12 eggs in a dozen.
(d) The maximum speed of a three-toed sloth is 0.15 mile per hour.

2.7 What is the difference in meaning between the times 3.3 seconds and 3.30 seconds?

2.8 What is the difference in meaning between the lengths 0.54 inch and 0.540 inch.

2.9 Express the uncertainty for each of the following numbers assuming the uncertainty in the last digit is ± 1.
(a) 2563 (b) 134,001
(c) 0.00357 (d) 12.00

2.10 Express the uncertainty for each of the following numbers assuming the uncertainty in the last digit is ± 1.
(a) 362,009 (b) 0.1234
(c) 10.023 (d) 0.000007

Significant Figures (Sec. 2.3)

2.11 When is a zero considered to be a significant digit, and when is it not?

2.12 Indicate whether each of the following "types" of zeros is always significant, never significant, or sometimes significant.
(a) leading zero (b) confined zero
(c) trailing zero

2.13 Determine the number of significant figures in each of the following measured values.
(a) 45.67 (b) 3.3338
(c) 6.003 (d) 300,081

2.14 Determine the number of significant figures in each of the following measured values.
(a) 453.1 (b) 1.0303
(c) 65.03 (d) 23,609

2.15 Determine the number of significant figures in each of the following measured values.
(a) 0.00043 (b) 0.0220022
(c) 0.30303030 (d) 0.03030303

2.16 Determine the number of significant figures in each of the following measured values.
(a) 0.111101 (b) 0.0000007
(c) 0.013013013 (d) 0.130130130

2.17 Determine the number of significant figures in each of the following measured values.
(a) 4700 (b) 37,540
(c) 67.010 (d) 3000.01

2.18 Determine the number of significant figures in each of the following measured values.
(a) 4000 (b) 4.000
(c) 67,000,100 (d) 43,200

2.19 Determine the number of significant figures and the magnitude of uncertainty in each of the following measured values.
(a) 1,0$\overline{0}$0,000 (b) 1000.
(c) 67$\overline{0}$0 (d) 1$\overline{0}$0

2.20 Determine the number of significant figures and the magnitude of uncertainty in each of the following measured values.
(a) 63$\overline{0,000}$ (b) 650.
(c) 3330 (d) 57$\overline{0}$0

2.21 A balance has a precision of ± 0.0001 gram. A sample that weighs about 20 grams is weighed on this balance. How many significant figures should be reported for this measurement?

2.22 A balance has a precision of ± 0.01 gram. A sample that weighs about 2 grams is weighed on this balance.

How many significant figures should be reported for this measurement?

2.23 Determine the number of significant figures in each of the following measured values.
(a) 2600 ± 10 (b) 1.375 ± 0.001
(c) 42 ± 1 (d) $73,000 \pm 1$

2.24 Determine the number of significant figures in each of the following measured values.
(a) 700 ± 100 (b) 700 ± 10
(c) 43.57 ± 0.01 (d) $64,000 \pm 1$

2.25 What is the magnitude of the uncertainty conveyed by each of the following notations for a measurement involving the number fifty-seven hundred?
(a) 5700.00 (b) 5700
(c) 57$\overline{0}$0 (d) 5700.

2.26 What is the magnitude of the uncertainty conveyed by each of the following notations for a measurement involving the number six hundred fifty thousand?
(a) 650,000.0 (b) 65$\overline{0}$,000
(c) 650,000 (d) 650,000.

Rounding Off (Sec. 2.4)

2.27 Round off each of the following numbers to the number of significant figures indicated in parentheses.
(a) 0.350763 (3) (b) 653,899 (4)
(c) 22.5550 (5) (d) 1000. (2)

2.28 Round off each of the following numbers to the number of significant figures indicated in parentheses.
(a) 3883 (2) (b) 4.4050 (3)
(c) 4,000,300 (4) (d) 0.277654 (5)

2.29 Using proper rounding techniques, decrease by two the number of significant figures in each of the following numbers.
(a) 0.03455 (b) 2.5003
(c) 1,456,000 (d) 100.0

2.30 Using proper rounding techniques, decrease by two the number of significant figures in each of the following numbers.
(a) 0.50505 (b) 2,000,567
(c) 2.335 (d) 1234.5

2.31 Rewrite each of the following numbers so that it contains two significant figures.
(a) 0.123 (b) 123,000
(c) 12.3 (d) 0.000123

2.32 Rewrite each of the following numbers so that it contains two significant figures.
(a) 21.000 (b) 21$\overline{0}$,000
(c) 0.0210 (d) 2.100

Significant Figures in Multiplication and Division (Sec. 2.4)

2.33 Without actually solving the problems, indicate the number of significant figures that should be present in the answers to the following multiplications.
(a) $0.3 \times 0.3 \times 0.3$
(b) $0.300 \times 0.300 \times 0.3000$
(c) $1.2 \times 1.22 \times 1.222$
(d) $3.000 \times 0.300 \times 3.003$

2.34 Without actually solving the problems, indicate the number of significant figures that should be present in the answers to the following divisions.
(a) $\dfrac{6.0}{33}$ (b) $\dfrac{6.000}{33}$ (c) $\dfrac{6.00}{33.0}$ (d) $\dfrac{6.00}{33.000}$

2.35 Carry out the following multiplications, expressing your answers to the correct number of significant figures. Assume that all numbers are measured quantities.
(a) 4.1567×0.00345 (b) 434.6×3.11
(c) 8.0×435.678 (d) 0.0073×3700

2.36 Carry out the following multiplications, expressing your answers to the correct number of significant figures. Assume that all numbers are measured quantities.
(a) $320,000 \times 0.0110$ (b) 435.67×3.203
(c) 3.116×8.700 (d) 3.78888×11

2.37 Carry out the following divisions, expressing your answers to the correct number of significant figures. Assume that all numbers are measured quantities.
(a) $\dfrac{3380}{4.444}$ (b) $\dfrac{1100}{23}$ (c) $\dfrac{5.01}{5.10}$ (d) $\dfrac{40}{16.0}$

2.38 Carry out the following divisions, expressing your answers to the correct number of significant figures. Assume that all numbers are measured quantities.
(a) $\dfrac{4.670}{210}$ (b) $\dfrac{30.0}{0.030}$
(c) $\dfrac{530,000}{465,200}$ (d) $\dfrac{8.0}{2.20}$

2.39 Carry out the following mathematical operations, expressing your answers to the correct number of significant figures. Assume that all numbers are measured quantities.
(a) $\dfrac{4.5 \times 6.3}{7.22}$ (b) $\dfrac{5.567 \times 3.0001}{3.45}$
(c) $\dfrac{37 \times 43}{4.2 \times 6.0}$ (d) $\dfrac{112 \times 20}{30 \times 63}$

2.40 Carry out the following mathematical operations, expressing your answers to the correct number of significant figures. Assume that all numbers are measured quantities.

(a) $\dfrac{2.322 \times 4.00}{3.200 \times 6.73}$ (b) $\dfrac{7.403}{3.220 \times 5.000}$

(c) $\dfrac{11.2 \times 11.2}{3.3 \times 6.5}$ (d) $\dfrac{5600 \times 300}{22 \times 97.1}$

Significant Figures in Addition and Subtraction (Sec. 2.4)

2.41 Perform the following additions or subtractions. Report your results to the proper number of significant figures. Assume that all numbers are measured quantities.
(a) $12 + 23 + 127$ (b) $3.111 + 3.11 + 3.1$
(c) $1237.6 + 23 + 0.12$ (d) $43.65 - 23.7$

2.42 Perform the following additions or subtractions. Report your results to the proper number of significant figures. Assume that all numbers are measured quantities.
(a) $237 + 37 + 7$ (b) $4.000 + 4.002 + 4.20$
(c) $235.45 + 37 + 36.4$ (d) $4.111 - 3.07$

2.43 Perform the following additions or subtractions. Report your results to the proper number of significant figures. Assume that all numbers are measured quantities.
(a) $999.0 + 1.7 - 43.7$ (b) $345 - 6.7 + 4.33$
(c) $1200 + 43 + 7$ (d) $132 - 0.0073$

2.44 Perform the following additions or subtractions. Report your results to the proper number of significant figures. Assume that all numbers are measured quantities.
(a) $1237.6 + 1237.4$ (b) $1237.6 - 1237.4$
(c) $23,000 + 457 + 23$ (d) $3.12 - 0.00007$

2.45 The addition of two numbers, each of which contains three significant figures and is a number between 400 and 600, can give an answer that has three or four significant figures. Explain how both cases can arise, and provide an example of each.

2.46 The subtraction of two numbers, each of which contains two significant figures and is a number between 40 and 60, can give an answer that has one or two significant figures. Explain how both cases can arise, and provide an example of each.

Significant Figures and Exact Numbers (Sec. 2.4)

2.47 A chemistry textbook is found to have a thickness of 1.34 inches. What would be the height, in inches, of a stack of 7 such identical books?

2.48 A round marble is found to weigh 1.672 grams. What would be the total mass, in grams, of 322 such identical marbles?

2.49 Each of the following calculations involves the numbers 4.3, 230, 20, and 13.00. The numbers 4.3 and 13.00 are measurements; 230 and 20 are exact numbers. Express each answer to the proper number of significant figures.
(a) $4.3 + 230 + 20 + 13.00$
(b) $4.3 \times 230 \times 20 \times 13.00$
(c) $4.3 + 230 - 20 - 13.00$
(d) $\dfrac{4.3 \times 230}{20 \times 13.00}$

2.50 Each of the following calculations involves the numbers 200, 17, 24, and 40. The numbers 200 and 17 are exact; 24 and 40 are measurements. Express each answer to the proper number of significant figures.
(a) $200 + 17 + 24 + 40$ (b) $200 \times 17 \times 24 \times 40$
(c) $200 - 17 - 24 - 40$ (d) $\dfrac{200 \times 17}{24 \times 40}$

Exponents and Orders of Magnitude (Sec. 2.5)

2.51 Express each of the following exponential terms as a product or fraction.
(a) 3^4 (b) 2^6 (c) 10^5 (d) 10^{-2}

2.52 Express each of the following exponential terms as a product or fraction.
(a) 4^2 (b) 3^2 (c) 10^{-5} (d) 10^8

2.53 Perform the indicated changes on the following exponential terms.
(a) 10^4; increase by three orders of magnitude
(b) 10^6; decrease by seven orders of magnitude
(c) 10^{-5}; increase by two orders of magnitude
(d) 10^{-3}; decrease by four orders of magnitude

2.54 Perform the indicated changes on the following exponential terms.
(a) 10^2; increase by two orders of magnitude
(b) 10^3; decrease by four orders of magnitude
(c) 10^{-4}; increase by three orders of magnitude
(d) 10^{-1}; decrease by five orders of magnitude

Scientific Notation (Sec. 2.5)

2.55 Express the following numbers in scientific notation.
(a) 23.45 (b) 120.7
(c) 0.0034 (d) 0.1005

2.56 Express the following numbers in scientific notation.
(a) 12.03 (b) 3407
(c) 0.1174 (d) 0.0000073

2.57 Express the following numbers in scientific notation.
(a) 231.00 (b) 231,000,000
(c) 0.0001000 (d) 0.10010

2.58 Express the following numbers in scientific notation.
(a) 230,000 (b) 231,000,100
(c) 0.100003 (d) 0.5000

2.59 Express the following numbers in decimal notation.
(a) 1.70×10^{-6} (b) 5.73×10^2
(c) 5.55×10^{-1} (d) 6.0012×10^4

2.60 Express the following numbers in decimal notation.
(a) 3.41×10^{-2} (b) 3.1445×10^2
(c) 6.19×10^{10} (d) 1.11×10^{-9}

2.61 Express the following numbers in decimal notation.
(a) 2.3×10^4 (b) 2.30×10^4
(c) 2.300×10^4 (d) 2.300000×10^4

2.62 Express the following numbers in decimal notation.
(a) 4.31×10^2 (b) 4.31000×10^2
(c) 4.31×10^{-2} (d) 4.31000×10^{-2}

Multiplication and Division in Scientific Notation
(Sec. 2.6)

2.63 Carry out the following multiplications of exponential terms
(a) $10^5 \times 10^3$ (b) $10^{-5} \times 10^{-3}$
(c) $10^5 \times 10^{-3}$ (d) $10^{-5} \times 10^3$

2.64 Carry out the following multiplications of exponential terms.
(a) $10^7 \times 10^4$ (b) $10^{-7} \times 10^{-4}$
(c) $10^7 \times 10^{-4}$ (d) $10^{-7} \times 10^4$

2.65 Carry out the following multiplications, expressing each answer in scientific notation to the correct number of significant figures.
(a) $(3.0 \times 10^2) \times (2.0 \times 10^{10})$
(b) $(3.71 \times 10^{-4}) \times (1.117 \times 10^{-1})$
(c) $(2.37 \times 10^{-8}) \times (1.79 \times 10^9)$
(d) $(1.171 \times 10^6) \times (2.555 \times 10^2) \times (1.00 \times 10^{-4})$

2.66 Carry out the following multiplications, expressing each answer in scientific notation to the correct number of significant figures.
(a) $(2.45 \times 10^3) \times (1.97 \times 10^3)$
(b) $(4.098 \times 10^{-3}) \times (1.114 \times 10^4)$
(c) $(9 \times 10^{-9}) \times (1.111 \times 10^{-1})$
(d) $(3.076 \times 10^{-7}) \times (2.00 \times 10^{-8}) \times (8.0 \times 10^{13})$

2.67 Carry out the following multiplications, making sure that your answer is expressed in correct scientific notation form and to the correct number of significant figures.
(a) $(3.20 \times 10^9) \times (4.50 \times 10^{-7})$
(b) $(6.55 \times 10^{-1}) \times (5.66 \times 10^{-3})$
(c) $(3.000 \times 10^{-10}) \times (9.00 \times 10^8)$
(d) $(9.8 \times 10^2) \times (8.80 \times 10^2) \times (7.000 \times 10^2)$

2.68 Carry out the following multiplications, making sure that your answer is expressed in correct scientific notation form and to the correct number of significant figures.
(a) $(4.300 \times 10^{-5}) \times (2.700 \times 10^{-6})$
(b) $(2.111 \times 10^2) \times (9.7703 \times 10^9)$
(c) $(3 \times 10^6) \times (7.32118 \times 10^{-1})$
(d) $(8.77 \times 10^5) \times (7.03 \times 10^{-3}) \times (6.44 \times 10^{-2})$

2.69 Carry out the following divisions of exponential terms.
(a) $\dfrac{10^5}{10^3}$ (b) $\dfrac{10^5}{10^{-3}}$ (c) $\dfrac{10^{-5}}{10^3}$ (d) $\dfrac{10^{-5}}{10^{-3}}$

2.70 Carry out the following divisions of exponential terms.
(a) $\dfrac{10^2}{10^3}$ (b) $\dfrac{10^2}{10^{-3}}$ (c) $\dfrac{10^{-2}}{10^3}$ (d) $\dfrac{10^{-2}}{10^{-3}}$

2.71 Carry out the following divisions, expressing each answer in scientific notation to the correct number of significant figures.
(a) $\dfrac{6.0 \times 10^5}{3.0 \times 10^3}$ (b) $\dfrac{3.711 \times 10^5}{2.710 \times 10^{-4}}$
(c) $\dfrac{9.87 \times 10^6}{3.001 \times 10^2}$ (d) $\dfrac{3.777 \times 10^{-5}}{2 \times 10^6}$

2.72 Carry out the following divisions, expressing each answer in scientific notation to the correct number of significant figures.
(a) $\dfrac{8.2 \times 10^{-3}}{4.1 \times 10^{-2}}$ (b) $\dfrac{5.000 \times 10^{-5}}{2.00 \times 10^4}$
(c) $\dfrac{3.5436 \times 10^7}{2.11111 \times 10^{-7}}$ (d) $\dfrac{8 \times 10^{-6}}{6.51123 \times 10^2}$

2.73 Carry out the following divisions, expressing each answer in correct scientific notation form to the correct number of significant figures.
(a) $\dfrac{1.000 \times 10^2}{5.00 \times 10^{-4}}$ (b) $\dfrac{8.70 \times 10^8}{9.33 \times 10^7}$
(c) $\dfrac{5.621 \times 10^{-3}}{5.833 \times 10^{-6}}$ (d) $\dfrac{4.300 \times 10^1}{4.10 \times 10^6}$

2.74 Carry out the following divisions, expressing each answer in correct scientific notation form to the correct number of significant figures.
(a) $\dfrac{2.000 \times 10^3}{3.0000 \times 10^5}$ (b) $\dfrac{3.20 \times 10^9}{9.51167 \times 10^3}$
(c) $\dfrac{9.8876 \times 10^{-3}}{8.0000 \times 10^{-2}}$ (d) $\dfrac{8.0000 \times 10^{-5}}{9.8876 \times 10^{-2}}$

2.75 Perform the following mathematical operations involving exponential terms.
(a) $\dfrac{10^2 \times 10^3}{10^4}$ (b) $\dfrac{10^{-2} \times 10^{-3}}{10^{-4}}$
(c) $\dfrac{10^6}{10^{-5} \times 10^{-9}}$ (d) $\dfrac{10^{-3} \times 10^2 \times 10^5}{10^{-6} \times 10^8}$

2.76 Perform the following mathematical operations involving exponential terms.

(a) $\dfrac{10^4 \times 10^5}{10^6 \times 10^3}$

(b) $\dfrac{10^{-3} \times 10^{-3} \times 10^{-3}}{10^{-6}}$

(c) $\dfrac{10^2 \times 10^3 \times 10^4}{10^{-2} \times 10^{-3} \times 10^{-4}}$

(d) $\dfrac{10^{-6} \times 10^4}{10^3 \times 10^{-5}}$

2.77 Perform the following mathematical operations. Be sure your answer contains the correct number of significant figures and that it is in correct scientific notation form.

(a) $\dfrac{(6.0 \times 10^3) \times (5.0 \times 10^3)}{2.0 \times 10^7}$

(b) $\dfrac{2.0 \times 10^7}{(6.0 \times 10^3) \times (5.0 \times 10^3)}$

(c) $\dfrac{(3.571 \times 10^{-5}) \times (4.5113 \times 10^{-9})}{(5.10 \times 10^{-6}) \times (3.71300 \times 10^{10})}$

(d) $\dfrac{(5 \times 10^{10}) \times (6.0 \times 10^7) \times (3.111 \times 10^{-5})}{(3 \times 10^3) \times (4.00 \times 10^{-6})}$

2.78 Perform the following mathematical operations. Be sure your answer contains the correct number of significant figures and that it is in correct scientific notation form.

(a) $\dfrac{(3.00 \times 10^5) \times (6.00 \times 10^3) \times (5.00 \times 10^6)}{2.00 \times 10^7}$

(b) $\dfrac{4.1111 \times 10^{-3}}{(3.003 \times 10^{-6}) \times (9.8760 \times 10^{-5})}$

(c) $\dfrac{(6 \times 10^5) \times (6 \times 10^{-5})}{(3 \times 10^2) \times (1 \times 10^{-10})}$

(d) $\dfrac{(3.00 \times 10^6) \times (2.7 \times 10^3) \times (8.50 \times 10^3)}{(2.22 \times 10^2) \times (8.504 \times 10^6)}$

Addition and Subtraction in Scientific Notation (Sec. 2.6)

2.79 Carry out the following additions, expressing, each answer in correct scientific notation form to the correct number of significant figures.

(a) $(2.731 \times 10^3) + (3.70 \times 10^3)$
(b) $(3.00 \times 10^3) + (2.00 \times 10^2)$
(c) $(3.7111 \times 10^{-3}) + (2.340 \times 10^{-7})$
(d) $(6.00 \times 10^6) + (5.25 \times 10^5) + (4.32 \times 10^4)$

2.80 Carry out the following additions, expressing each answer in correct scientific notation form to the correct number of significant figures.

(a) $(9.765 \times 10^4) + (8.700 \times 10^4)$
(b) $(2.000 \times 10^{-3}) + (2.000 \times 10^{-2})$
(c) $(6.32 \times 10^2) + (5.0 \times 10^6)$
(d) $(7.00 \times 10^4) + (7.00 \times 10^3) + (7.00 \times 10^2)$

2.81 Carry out the following subtractions, expressing each answer in correct scientific notation form to the correct number of significant figures.

(a) $(3.103 \times 10^{-3}) - (1.17 \times 10^{-3})$
(b) $(3.45 \times 10^4) - (9.67 \times 10^3)$
(c) $(3.71114 \times 10^4) - (6.2 \times 10^3)$
(d) $(5.11 \times 10^1) - (6.32 \times 10^{-6})$

2.82 Carry out the following subtractions, expressing each answer in correct scientific notation form to the correct number of significant figures.

(a) $(9.571 \times 10^7) - (6.00 \times 10^7)$
(b) $(9.571 \times 10^7) - (6.00 \times 10^6)$
(c) $(9.571 \times 10^7) - (6.00 \times 10^2)$
(d) $(3.003 \times 10^{-5}) - (2.0 \times 10^{-7})$

2.83 Carry out the following addition-subtraction problems, expressing each answer in correct scientific notation form to the correct number of significant figures.

(a) $(8.934 \times 10^5) + (5.236 \times 10^5) - (1.23 \times 10^4)$
(b) $(8.934 \times 10^5) - (5.236 \times 10^5) - (1.23 \times 10^4)$
(c) $(8.934 \times 10^5) - (5.24 \times 10^5) + (1.23 \times 10^4)$
(d) $(8.934 \times 10^5) + (7.343 \times 10^6) + (1.213 \times 10^4)$

2.84. Carry out the following addition-subtraction problems, expressing each answer in correct scientific notation form to the correct number of significant figures.

(a) $(7.123 \times 10^4) + (3.266 \times 10^4) - (4.71 \times 10^3)$
(b) $(7.123 \times 10^4) - (3.266 \times 10^4) - (4.71 \times 10^3)$
(c) $(7.123 \times 10^4) - (3.27 \times 10^4) + (4.71 \times 10^3)$
(d) $(7.123 \times 10^4) + (5.971 \times 10^5) + (1.200 \times 10^3)$

Additional Problems

2.85 Round off the number 3.6305023 to the indicated numbers of significant figures.
(a) seven   (b) five   (c) four   (d) three

2.86 Round off the number 363,545,525 to the indicated numbers of significant figures.
(a) eight   (b) six   (c) four   (d) two

2.87 How many significant figures does each of the following numbers have?
(a) 7770
(b) 7.770
(c) 7.770×10^{-3}
(d) 7.770×10^{3}

2.88 How many significant figures does each of the following numbers have?
(a) 0.4500
(b) 4.500
(c) 4.500×10^{-3}
(d) 4.500×10^{3}

2.89 In which of the following pairs of numbers do both members of the pair contain the same number of significant figures?
(a) 11.0 and 11.00
(b) 600.0 and 6.000×10^{3}
(c) 6300 and 6.3×10^{3}
(d) 0.300045 and 0.345000

2.90 In which of the following pairs of numbers do both members of the pair contain the same number of significant figures?
(a) 0.000066 and 660,000
(b) 1.500 and 1.500×10^{2}
(c) 54,000 and 5400.0
(d) 0.05700 and 0.0570

2.91 Write each of the following numbers in scientific notation to the number of significant figures indicated in parentheses.
(a) 632,567 (4)
(b) 0.312546 (3)
(c) 63,000,023 (4)
(d) 0.500000 (4)

2.92 Write each of the following numbers in scientific notation to the number of significant figures indicated in parentheses.
(a) 0.00300300 (3)
(b) 936,000 (2)
(c) 23.5003 (3)
(d) 450,000,001 (6)

2.93 Perform the following mathematical operations, expressing your answers to the correct number of significant figures. Assume all numbers are measured quantities.
(a) $4.0 \times (2.3 + 4.5)$
(b) $3.0 \times (3.7 - 3.4)$
(c) $6.0 \times (34 - 4.23)$
(d) $7.02 \times (0.0001 + 0.01)$

2.94 Perform the following mathematical operations, expressing your answers to the correct number of significant figures. Assume all numbers are measured quantities.
(a) $5.5 \times (12.3 - 3.2)$
(b) $6.0 \times (3.7 + 0.001)$
(c) $2.0 \times (3.43 - 2.415)$
(d) $3.0 \times (37 - 3.7)$

2.95 What is wrong with the statement "The number of objects is 12.00 exactly"?

2.96 What is wrong with the statement "Through counting, it was determined that the basket contained 6.70×10^{1} peaches"?

2.97 Each of the following calculations contains the numbers 4.2, 5.30, 11, and 28. The numbers 4.2 and 28 are measured quantities, and 5.30 and 11 are exact numbers. Express each answer to the proper number of significant figures.
(a) $(4.2 + 5.30) \times (11 + 28)$
(b) $4.2 \times 5.30 \times (11 - 28)$
(c) $\dfrac{28 - 4.2}{5.30 \times 11}$
(d) $\dfrac{28 - 4.2}{11 - 5.30}$

2.98 Each of the following calculations contains the numbers 3.111, 5.03, 100, and 33. The numbers 3.111 and 100 are exact numbers, and 5.03 and 33 are measurements. Express each answer to the proper number of significant figures.
(a) $(3.111 + 5.03) \times (100 + 33)$
(b) $3.111 \times 5.03 \times (100 + 33)$
(c) $\dfrac{3.111 + 5.03}{100 \times 33}$
(d) $\dfrac{5.03 - 3.111}{100 + 33}$

2.99 Arrange the following sets of numbers in ascending order (from smallest to largest).
(a) 2.07×10^{2}, 243, 1.03×10^{3}
(b) 0.0023, 3.04×10^{-2}, 2.11×10^{-3}
(c) 23,000, 2.30×10^{5}, 9.67×10^{4}
(d) 0.00013, 0.000014, 1.5×10^{-4}

2.100 Arrange the following sets of numbers in ascending order (from smallest to largest).
(a) 350, 3.51×10^{2}, 3.522×10^{1}
(b) 0.000234, 2.341×10^{-3}, 2.3401×10^{-4}
(c) 965,000,000, 9.76×10^{8}, 2.03×10^{8}
(d) 0.00010, 0.00023, 3.4×10^{-2}

2.101 Perform the following mathematical operations, expressing your answer in scientific notation.
(a) $0.00000000000023 \times 0.000000000000000067$
(b) $930,000,000,000 \times 0.00000000000000000000062$
(c) $(9.3111 \times 10^{12}) \times 300,100,000$
(d) $(6.352 \times 10^{-3}) \times 0.0000345000$

2.102 Perform the following mathematical operations, expressing your answer in scientific notation.
(a) $333,000 \times 0.000000000000000000937$
(b) $700,000,000,000,000 \times 63,000,000,000,000,000$
(c) $(9.45 \times 10^{-9}) \times 0.00000000945$
(d) $(3.0001 \times 10^{-17}) \times 654,730,000,000,000$

Grid Problems

2.103 Select from the grid *all* correct responses for each situation.

1.	2.	3.
0.00305	0.0120	0.1307
4.	5.	6.
1.670	10.20	107.0
7.	8.	9.
1100	17,001	223,000

(a) numbers that contain three significant figures
(b) numbers that have a precision of at least ±0.01
(c) numbers that contain one or more "confined" zeros
(d) numbers in which all zeros are significant
(e) numbers for which the scientific notation form would have a coefficient containing four digits
(f) numbers for which the scientific notation form would have a coefficient that does not contain zeros

2.104 Select from the grid *all* correct responses for each situation.

1.	2.	3.
6.2×10^{-1}	6.20×10^{-1}	6.200×10^{-1}
4.	5.	6.
6.2×10^{1}	6.20×10^{1}	6.200×10^{1}
7.	8.	9.
6.2×10^{3}	6.20×10^{3}	6.200×10^{3}

(a) numbers that contain three significant figures
(b) numbers that have a magnitude of less than one
(c) pairs of numbers that have the same precision
(d) numbers with an uncertainty greater than ±1
(e) numbers that when multiplied by 1.000×10^3 would give an answer possessing three significant figures
(f) numbers that would be considered negligible when added to 1.000×10^3

2.105 Select from the grid *all* correct responses for each situation.

1.	2.	3.
10^{-9}	10^{-6}	10^{-3}
4.	5.	6.
10^{0}	10^{3}	10^{6}
7.	8.	9.
10^{9}	10^{12}	10^{15}

(a) pairs of numbers that when multiplied together give 10^6
(b) pairs of numbers that when multiplied together give 10^{-3}
(c) numbers that differ in value from 10^2 by four or five orders of magnitude
(d) pairs of numbers that differ from each other by six orders of magnitude
(e) numbers that when multiplied by themselves (squared) given an even power of ten
(f) numbers that will divide into 10^{15} to give an even power of ten

2.106 Select from the grid *all* correct responses for each situation.

1.	2.	3.
4.300	43.00	430.0
4.	5.	6.
4300	43$\overline{0}$0	4$\overline{30}$0
7.	8.	9.
4.3×10^{2}	4.300×10^{2}	4.30×10^{3}

(a) numbers that possess three significant figures
(b) numbers that have the same magnitude as 4.30×10^2
(c) numbers that have the same precision as 430
(d) pairs of numbers that have the same magnitude and same precision
(e) numbers that are correct answers for the problem $(1.00 \times 10^{-2}) \times (4.30 \times 10^5)$
(f) numbers that are correct answers for the problem $(414.0) + (16.0)$

CHAPTER THREE

Unit Systems and Dimensional Analysis

3.1 The Metric System of Units

All measurements consist of three parts: a number that tells the amount of the quantity measured, an error that affects the accuracy of that amount, and a unit that tells the nature of the quantity being measured. Chapter 2 dealt with the interpretation and manipulation of the number and error parts of a measurement. We now turn our attention to units.

A unit is a "label" that describes (or identifies) what is being measured (or counted). It can be almost anything: quarts, dimes, dozen frogs, bushels, inches, or pages, for example. Having units for a measurement is an absolute necessity. If you were to ask a neighbor to lend you six sugar, the immediate response would be "how much sugar?" You would then have to indicate that you wished to borrow six pounds, six ounces, six cups, six teaspoons, or whatever amount of sugar you needed.

Two formal systems of units of measurement are used in the United States today. Common measurements in commerce—in supermarkets, lumberyards, gas stations, and so on—are made in the **English system**. The units of this system include the familiar inch, foot, pound, quart, and gallon. A second system, the **metric system**, is used in scientific work. Units in this system include the gram, meter, and liter. The United States is one of only a very few countries that use different unit systems in commerce and scientific work. On a worldwide basis, almost universally, the metric system is used in both areas.

The metric system is slowly coming into use in the United States for commercial purposes. Metric units now appear on many consumer products (see Fig. 3.1). Soft drinks can now be bought in 2-liter containers, and automobile engine sizes are often given in liters. Road signs in some states display distances in both miles and kilometers. Canned and packaged goods (cereals, mixes, fruits, etc.) on grocery store shelves now have the content masses listed in grams as well as ounces or pounds.

The metric system is superior to the English system. Its superiority lies in the area of interrelationships between units of the same type (volume, length, etc.). Metric unit interrelationships are less complicated than English unit interrelationships because the metric system is a decimal-unit system. In the metric system conversion from one unit size to another can be accomplished simply by moving the decimal point to the right or left an appropriate number of places. The metric system is no more precise than the English system, but it is more convenient.

The most recent modification of the metric system is called the International System of Units and is abbreviated SI (after the French name Système International). Since SI units are still not in universal use, we shall use the more traditional metric system in this text.

Because the SI and metric systems are very similar, switching to the SI system sometime in the future, when its use is more extensive, will present no major difficulties to the student who properly understands the traditional metric system.

(a)

(b)

FIGURE 3.1 Metric units are becoming increasingly evident on highway signs and consumer products. [(a) © *George Gerster/Photo Researchers, Inc.*]

TABLE 3.1 Metric System Prefixes and Their Mathematical Meanings

Prefix	Symbol	Mathematical Meaning	Origin	Original Meaning
Tera-	T	$1{,}000{,}000{,}000{,}000 = 10^{12}$	Greek	monstrous
Giga-	G	$1{,}000{,}000{,}000 = 10^{9}$	Greek	gigantic
Mega-	M	$1{,}000{,}000 = 10^{6}$	Greek	great
Kilo-	k	$1{,}000 = 10^{3}$	Greek	thousand
Hecto-	h	$100 = 10^{2}$	Greek	hundred
Deca-	da	$10 = 10^{1}$	Greek	ten
		$1 = 10^{0}$		
Deci-	d	$0.1 = 10^{-1}$	Latin	tenth
Centi-	c	$0.01 = 10^{-2}$	Latin	hundredth
Milli-	m	$0.001 = 10^{-3}$	Latin	thousandth
Micro-	μ[a]	$0.000\,001 = 10^{-6}$	Greek	small
Nano-	n	$0.000\,000\,001 = 10^{-9}$	Greek	very small
Pico-[b]	p	$0.000\,000\,000\,001 = 10^{-12}$	Spanish	extremely small

[a] This is the Greek letter mu (pronounced "mew").
[b] This prefix is pronounced "peek-o."

Metric System Prefixes

In the metric system there is one base unit for each type of measurement—length, volume, mass, etc. These base units are then multiplied by appropriate powers of ten to form smaller or larger units. The names of the larger and smaller units are constructed from the base unit name by attaching to it a prefix that tells which power of ten is involved. These prefixes are given in Table 3.1, along with their symbols or abbreviations and mathematical meanings. The prefixes in color are those most frequently used.

The use of numerical prefixes should not be new to you. Consider the use of the prefix tri- in the following words: triangle, tricycle, trio, trinity, triple. Every one of these words conveys the idea of three of something. We will use the metric system prefixes in the same way.

The meaning of a prefix always remains constant; it is independent of the base it modifies. For example, a kilosecond is a thousand seconds; a kilowatt, a thousand watts; and a kilocalorie, a thousand calories. The prefix kilo- will always mean a thousand.

3.2 Metric Units of Length

The **meter** is the basic unit of length in the metric system. (*Metre* has been adopted as the preferred international spelling for this unit, but *meter* is the spelling used in the United States and in this book.)

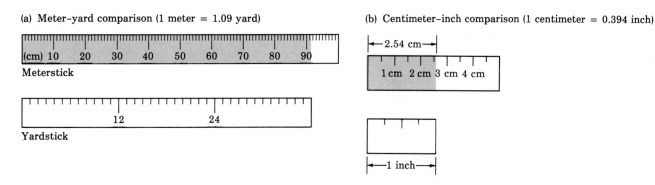

FIGURE 3.2 A comparison of metric and English units of length.

Other units of length in the metric system are derived from the meter by using the prefixes listed in Table 3.1. The kilometer (km) is 1000 meters, whereas the centimeter (cm) and millimeter (mm) are, respectively, $\frac{1}{100}$ and $\frac{1}{1000}$ of a meter. Note that the abbreviation for meter is m.

A comparison of metric lengths with the commonly used English system lengths of mile, yard, and inch gives us some size perspective. A kilometer is approximately three fifths (60%) of a mile, a meter is slightly larger (109%) than a yard (see Fig. 3.2a), and a centimeter is about two fifths (40%) of an inch (see Fig. 3.2b).

A better "feel" for lengths equivalent to a meter, centimeter, and millimeter can perhaps be obtained by relating these metric units to objects and situations we encounter in everyday life. This is done in Figure 3.3 in terms of a coin, a paper clip, some chalk, and a basketball player.

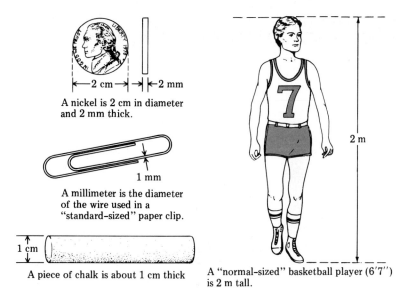

FIGURE 3.3 Metric units of length and the "everyday realm."

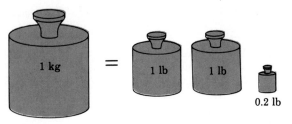

1 kilogram = 2.2 pounds

FIGURE 3.4 A comparison of the metric kilogram with the English pound.

3.3 Metric Units of Mass

Mass is a measurement of the total quantity of matter present in an object. A **gram** is the basic unit of mass in the metric system. The gram is relatively small compared to the commonly used English mass units of ounce and pound. It takes approximately 28 grams to equal an ounce and nearly 454 grams to equal a pound. Because of the small size of the gram, the kilogram (kg) is a very commonly used unit. (The abbreviation for gram is g). A kilogram is equal to approximately 2.2 pounds, as shown in Figure 3.4.

To give you a "feel" for metric mass unit magnitude, Figure 3.5 relates the mass units of milligram (mg), gram, and kilogram to everyday objects. Comparisons involve a single staple, a postage stamp, two thumbtacks, a nickel, a quart of milk in a paper carton, and a "medium-sized" football player.

The terms *mass* and *weight* are frequently used interchangeably in discussions. Although in most cases this practice does no harm, technically it is incorrect to interchange the terms. Mass and weight refer to different properties of matter, and their difference in meaning should be understood.

Mass is a measure of the total quantity of matter in an object. Weight is a measure of the force exerted on an object by the pull of gravity. The mass of a substance is a constant; the weight of an object is a variable dependent upon the geographical location of that object.

A single staple has a mass of about 25 mg.

A postage stamp has a mass of 80 mg.

A nickel has a mass of about 5 g.

Two thumbtacks have a mass of 1 g.

1 quart of milk in a cardboard container has a mass of 1 kg.

A 220-lb football player has a mass of 100 kg.

FIGURE 3.5 Metric units of mass and the "everyday realm."

Figure 3.6 As astronaut on a "space walk" is weightless but not massless. He or she has exactly the same mass as on Earth.

Matter at the equator weighs less than it would at the North Pole because the Earth is not a perfect sphere but bulges at the equator. As a result, an object at the equator is farther from the center of the Earth. It therefore weighs less because the magnitude of gravitational attraction (the measure of weight) is inversely proportional to the distance between the centers of the attracting objects; that is, the gravitational attraction is larger when the objects' centers are closer together and smaller when the objects' centers are farther apart. Gravitational attraction also depends on the masses of the attracting bodies; the greater the masses, the greater the attraction. For this reason, an object would weigh much less on the moon than on Earth because of the smaller size of the moon and the correspondingly lower gravitational attraction. Quantitatively, a 22.0-pound mass weighing 22.0 pounds at the Earth's North Pole would weigh 21.9 pounds at the Earth's equator and only 3.7 pounds on the moon. In outer space an astronaut may be weightless but never massless. In fact, he or she has the same mass in space as on Earth (see Fig. 3.6).

Different types of equipment are used to determine mass and weight. A *balance* is used to determine the mass of an object. Figure 3.7 shows three common types of balances. With balances such as these, the actual mass of the object can be obtained by balancing the object against objects ("weights") of known mass. The object whose mass is to be determined is placed on the left pan of the platform or triple-beam balance (Figs. 3.7a and 3.7b), and objects ("weights") of known mass are placed on the right-hand pan and/or sliding scale to counterbalance the object. The analytical balance shown in Figure 3.7c has only one pan. The counterbalancing masses for this balance are located within the balance case and are put in place by manipulating knobs and levers. Gravitational attraction will have the same effect on the object whose mass is to be determined and the counterbalancing masses, thus making the measurements independent of geographical location.

Scales, rather than balances, are used to determine weight. Weight is found by noting the magnitude of the stretch of a calibrated spring. The extent of stretching of the spring is dependent on the gravitational attraction of the object for the Earth. Bathroom scales and grocery store scales operate on this principle.

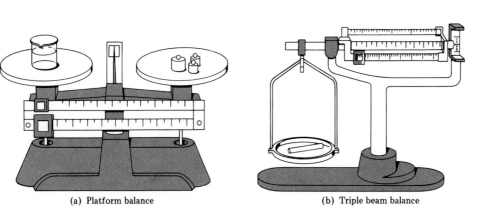

FIGURE 3.7 Common laboratory balances, used for determining mass, are (a) the platform balance (double pan balance), (b) the triple beam balance, and (c) the single pan analytical balance.

The processes of determining mass and weight are both referred to as "weighing" since the word "massing" is not an accepted part of the English language (although it should be). This dual usage of the term weighing leads to the practice (somewhat common) of using mass and weight interchangeably. Although some textbooks follow this practice, this one will not. A mass is not a weight. A mass will be called a mass. Chemists always use balances rather than scales. Hence they always measure the mass of an object.

3.4 Metric Units of Volume

Before specific metric units of volume are considered, a quick review of how the quantities area and volume are calculated is in order.

Area is a measure of the extent of a surface. The units for area are squared units of length. Common area units include square inches (in.^{2}), square feet (ft^2), square meters (m^2), and square centimeters (cm^2). Note that a squared unit is just that unit multiplied by itself. The unit cm^2 means centimeter × centimeter in the same way that 3^2 means 3 × 3. Figure 3.8 lists the formulas needed for determining the areas of commonly encountered geometrical shapes. Notice that in each case the formula involves taking the product of two lengths. Note also that the constant pi (π) is needed in calculating the area of a circle. Its value to three, four, and five significant figures, respectively, is 3.14, 3.142, and 3.1416.

Volume is a measure of the amount of space occupied by an object. It is a three-dimensional measure and thus involves units that have been cubed: in.^{3}, ft^3, m^3, cm^3, etc. Again, a cubed unit is just that unit multiplied by itself three times in the same manner that 3^3 is 3 × 3 × 3. Figure 3.9 lists volume formulas for commonly encountered three-dimensional shapes. Note particularly the manner in which the

Square	Area = side × side $A = s \times s$ $\quad = s^2$		
Rectangle	Area = length × width $A = l \times w$		
Circle	Area = π × (radius)² $A = \pi \times r^2$	$\pi = 3.14$	
Triangle	Area = ½ × base × height $A = \tfrac{1}{2} \times b \times h$		**FIGURE 3.8** Formulas for calculating the areas of various geometrical figures.

volume of a cube is calculated—side × side × side. This is the key to understanding metric units of volume whose definitions are cube-related.

A **liter** is the basic unit of volume in the metric system. (As with meter, we will use the United States spelling rather than the international spelling, which is litre.) A liter is abbreviated L, to avoid the confusing of lowercase l with the number 1.

Cube	$V = s \times s \times s$ $\quad = s^3$		
Rectangular solid	$V = l \times w \times h$	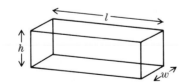	
Cylinder	$V = \pi \times r^2 \times h$	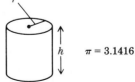 $\pi = 3.1416$	
Sphere	$V = \tfrac{4}{3}\pi \times r^3$	 $\pi = 3.1416$	**FIGURE 3.9** Formulas for calculating the volumes of various three-dimensional shapes.

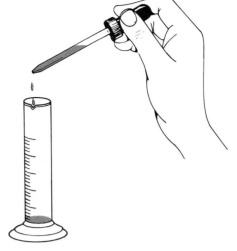

1 teaspoon of liquid is approximately 5 mL.

A ½-ounce-size container holds approximately 15 mL.

20 drops from an eyedropper is about 1 mL.

FIGURE 3.10 The metric volume unit the milliliter and the "everyday realm."

A liter is a volume equal to that occupied by a perfect cube that is 10 centimeters (or 1 decimeter) on each side. Since the volume of a cube is calculated by multiplying length × width × height (which are all the same for a cube) we have

$$1\text{ L} = \text{volume of a 10-cm-edged cube}$$
$$= 10\text{ cm} \times 10\text{ cm} \times 10\text{ cm}$$
$$= 1000\text{ cm}^3$$

The abbreviation cc is sometimes used for the unit cubic centimeter, most frequently in medically oriented situations.

As with the units of length and mass, the basic unit of volume is modified with prefixes to represent smaller or larger units. The most commonly used prefixed volume unit is the milliliter (mL), which is $\frac{1}{1000}$ L. A milliliter is much smaller than a fluid ounce; it takes approximately 30 mL to equal 1 fluid ounce. A quart (32 fl oz) contains 946 mL. Figure 3.10 shows some comparisons that will give you a better "feel" for the size of a milliliter.

A liter is equal to 1000 milliliters. We have also just seen that it is equal to 1000 cubic centimeters. Therefore,

$$1000\text{ mL} = 1000\text{ cm}^3$$

Dividing both sides of this equation by a thousand, we find that

$$1\text{ mL} = 1\text{ cm}^3$$

Consequently, the units mL and cm^3 are interchangeable. In practice, mL is usually used for volumes of liquids and gases and cm^3 for volumes of solids. Figure 3.11 shows the relationship between 1 mL (1 cm^3) and its parent unit, liter, in terms of cubic measurement.

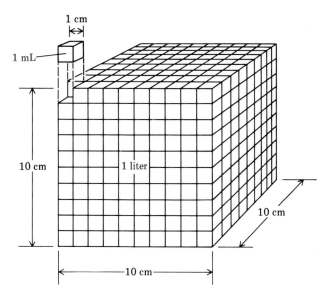

FIGURE 3.11 The relationship between the units liter and milliliter in terms of cubic centimeters.

3.5 Units in Mathematical Operations

Just as numbers can have exponents associated with them (3^2, 6^3, 10^2, 10^4, etc.), so also can units have exponents associated with them. In Section 3.4 in the discussion of volume, the unit cubic centimeter (cm^3) was introduced. As noted, cm^3 denotes the unit centimeter multiplied by itself three times (cm × cm × cm). Other commonly encountered units with exponents include

$in.^2$ (square inches)
ft^3 (cubic feet)
cm^2 (square centimeters)
m^2 (square meters)

In mathematical problems the powers on units are manipulated in the same manner as powers of ten (Sec. 2.6). Exponents are added during multiplication and subtracted during division. Just as

$$10^2 \times 10^4 = 10^6$$

so likewise

$$km^2 \times km^4 = km^6$$

Similarly,

$$\frac{10^2}{10^4} = 10^{-2}$$

and

$$\frac{km^2}{km^4} = km^{-2}$$

EXAMPLE 3.1

Carry out the following multiplications and divisions of measurements and units. Remember that significant figures are a consideration for all measurements.

(a) 6.2 ft × 3.7 ft   (b) 3.01 mm × 6.007 mm^2

(c) $\dfrac{5.000 \text{ yd}^3}{2.00 \text{ yd}}$ (d) $\dfrac{7.0 \text{ cm}^2 \times 6.0 \text{ cm}^3}{3.0 \text{ cm}^4}$

Solution

Each of these problems can be considered in two stages. First, the numerical parts of the measurements are multiplied or divided as indicated. Then the units are combined in the appropriate way.

(a) 6.2 ft × 3.7 ft = (6.2 × 3.7) × (ft × ft)
= 22.94 × ft^2
= 22.94 ft^2 (calculator answer)
= 23 ft^2 (correct answer)

(b) 3.01 mm × 6.007 mm^2 = (3.01 × 6.007) × (mm × mm^2)
= 18.08107 × mm^3
= 18.08107 mm^3 (calculator answer)
= 18.1 mm^3 (correct answer)

Note that the exponents on the units (1 and 2) were added to give mm^3.

(c) $\dfrac{5.000 \text{ yd}^3}{2.00 \text{ yd}} = \dfrac{5.000}{2.00} \times \dfrac{\text{yd}^3}{\text{yd}}$
= 2.5 × yd^2
= 2.5 yd^2 (calculator answer)
= 2.50 yd^2 (correct answer)

In this case, according to the rule for division of exponential terms (Sec. 2.6), the unit exponents were subtracted.

(d) $\dfrac{7.0 \text{ cm}^2 \times 6.0 \text{ cm}^3}{3.0 \text{ cm}^4} = \dfrac{7.0 \times 6.0}{3.0} \times \dfrac{\text{cm}^2 \times \text{cm}^3}{\text{cm}^4}$
= 14 × cm
= 14 cm   (calculator and correct answer)

PRACTICE EXERCISE 3.1

Carry out the following multiplications and divisions of measurements and units.

(a) 2.4 cm × 3.6 cm   (b) 3.6 km^2 × 6.0 km

(c) $\dfrac{6.3 \text{ ft}^3}{1.0 \text{ ft}}$ (d) $\dfrac{6.0 \text{ in.}^2 \times 3.0 \text{ in.}^2}{2.0 \text{ in.}}$

Ans. (a) 8.6 cm^2; (b) 22 km^3; (c) 6.3 ft^2; (d) 9.0 in.^{3}.

Similar exercises: Problems 3.15 and 3.16

3.6 Conversion Factors

Many times a need arises to change the units of a quantity or measurement to different units. In this section we deal with this topic. The new units needed may be in the same measurement system as the old ones, or they may be in a different system. With two unit systems in common use in the United States, the need to change measurements in one system to their equivalent in the other frequently occurs.

The mathematical tool we will use to accomplish the task of changing units is a general method of problem solving called *dimensional analysis.* By this method, unit conversion is accomplished by multiplying a given quantity or measurement by one or more conversion factors to obtain the desired quantity or measurement.

$$\text{Given quantity} \times \text{conversion factor(s)} = \text{desired quantity}$$

Prior to discussing the mechanics involved in using dimensional analysis to solve "unit-conversion problems," some comments concerning conversion factors are in order. A proper understanding of conversion factors is the key to being able to solve problems "comfortably" with dimensional analysis.

Formally defined, a conversion factor is an equality, expressed in fractional form, that is used as a multiplier to convert a quantity in one unit into its equivalent in another unit.

Multiplication by one or more conversion factors does not change the value of a measurement or quantity because *all conversion factors have a numerical value equal to unity* (*one*). Multiplication by unity does not change the value of an expression.

A fixed relationship between two quantities gives sufficient information to construct a conversion factor. Let us construct some conversion factors to see (1) how they originate and (2) why they always have a value of unity.

The quantities "1 minute" and "60 seconds" both describe the same amount of time. We may write an equation describing this fact.

$$1 \text{ min} = 60 \text{ sec}$$

This fixed relationship can be used to construct a pair of conversion factors that relate seconds and minutes. (Conversion factors always occur in pairs, as will become obvious shortly.)

Dividing both sides of our minute–second equation by the quantity "1 minute" gives

$$\frac{1 \text{ min}}{1 \text{ min}} = \frac{60 \text{ sec}}{1 \text{ min}}$$

Since the numerator and denominator of the fraction on the left are identical, this fraction has a value of unity.

$$1 = \frac{60 \text{ sec}}{1 \text{ min}}$$

The fraction on the right side of the equation is our conversion factor. Its value is one. Note that the numerator and denominator of the conversion factor describe the same "amount" of time.

Two conversion factors are always obtainable from any given equality. For the equality we are considering (1 min = 60 sec) the second conversion factor is

$$\frac{1 \text{ min}}{60 \text{ sec}}$$

It is obtained by dividing both sides of the equality by "60 seconds" instead of "1 minute." The relationship between the two conversion factors is that of reciprocals.

$$\frac{1 \text{ min}}{60 \text{ sec}} \quad \text{and} \quad \frac{60 \text{ sec}}{1 \text{ min}}$$

In general, we will always be able to construct a set of two conversion factors, each with a value of unity, from any two terms that describe the same "amount" of whatever we are considering. The two conversion factors will always be reciprocals of each other. (Conversion factors can also be constructed from two quantities that are equivalent rather than equal. We will consider this situation in Sec. 3.8.)

For convenience in discussing additional information about conversion factors we will classify them into three general categories: (1) metric-to-metric, (2) English-to-English, and (3) metric-to-English or English-to-metric.

Metric-to-Metric Conversion Factors

Metric-to-metric conversion factors are used to change a measurement in one metric unit into another metric unit. Both the numerator and the denominator of such conversion factors involve metric-system units.

Conversion factors relating a prefixed metric unit to the base unit for that type of measurement are derived from the meaning of the prefix of concern. For example, the set of conversion factors involving kilogram and gram is derived from the meaning of the prefix kilo-, which is 10^3. The two conversion factors are

$$\frac{10^3 \text{ m}}{1 \text{ km}} \quad \text{and} \quad \frac{1 \text{ km}}{10^3 \text{ m}}$$

Note again the reciprocal relationship between the two conversion factors of the set. The conversion factors relating centimeter and meter involve 10^{-2}, the mathematical equivalent of centi-, and are

$$\frac{1 \text{ cm}}{10^{-2} \text{ m}} \quad \text{and} \quad \frac{10^{-2} \text{ m}}{1 \text{ cm}}$$

For conversion factors of the type now under discussion, the numerical equivalent of the prefix always goes with the base unit.

All metric-to-metric conversion factors are derived from exact definitions. Thus, they all contain an unlimited number of significant figures. This observation about significant figures will be of major importance when these conversion factors become part of an actual calculation.

English-to-English Conversion Factors

English-to-English conversion factors also contain an unlimited number of significant figures, since they also result from defined equalities. Twelve inches equals 1 foot—exactly. Three feet equal 1 yard—exactly. The majority of English-to-English conversion factors should be "second nature" to most students. You have used these relationships all your life, even though you may not consciously have thought of them as conversion factors. The following are representative of these "second-nature" conversion factors.

$$\frac{2 \text{ pt}}{1 \text{ qt}} \quad \frac{36 \text{ in.}}{1 \text{ yd}} \quad \frac{1 \text{ mi}}{5280 \text{ ft}} \quad \frac{16 \text{ oz}}{1 \text{ lb}}$$

Metric-to-English and English-to-Metric Conversion Factors

Conversion factors that relate metric units to English units or vice versa must be established by measurement since they involve two different unit systems. One of the numbers associated with each such conversion factor must be determined experimentally. Since this number will not be *exact*, concern about the number of significant figures present arises.

To avoid confusion about the accuracy of such English-to-metric conversion factors, scientists in English-speaking countries have agreed on common "experiment-based" *definitions* for English units in terms of metric units. They have also agreed that these definitions should be taken to be *exact*. The three *exact* definitions we consider are

$$1 \text{ inch} = 2.54 \text{ centimeters}$$
$$1 \text{ pound} = 453.59237 \text{ grams}$$
$$1 \text{ quart} = 0.94633343 \text{ liter}$$

The presence of only three digits in the inch-to-centimeter definition derives from the experimental value being 2.540005 cm. For simplicity, the mass and volume definitions are most often rounded off to four digits.

$$1 \text{ pound} = 453.6 \text{ grams}$$
$$1 \text{ quart} = 0.9463 \text{ liter}$$

In these rounded forms, the definitions are not exact.

Table 3.2 lists some commonly encountered relationships between metric and English system units. These few factors are sufficient to solve most of the problems that we will encounter. In fact, later in this chapter we will see that only one factor of a given type (volume, mass, length) is sufficient to solve most problems.

TABLE 3.2 Conversion Factors That Relate the English and Metric Systems of Measurement to Each Other

	Factor to Convert from Metric to English Unit
Length	
Inch and centimeter	$\dfrac{1 \text{ in.}}{2.54 \text{ cm}}$
Inch and meter	$\dfrac{1 \text{ in.}}{0.0254 \text{ m}}$
Mile and kilometer	$\dfrac{1 \text{ mi}}{1.609 \text{ km}}$
Mass	
Pound and gram	$\dfrac{1 \text{ lb}}{453.6 \text{ g}}$
Pound and kilogram	$\dfrac{1 \text{ lb}}{0.4536 \text{ kg}}$
Ounce and gram	$\dfrac{1 \text{ oz}}{28.34 \text{ g}}$
Volume	
Quart and liter	$\dfrac{1 \text{ qt}}{0.9463 \text{ L}}$
Pint and liter	$\dfrac{1 \text{ pt}}{0.4732 \text{ L}}$
Fluid ounce and milliliter	$\dfrac{1 \text{ fl oz}}{29.57 \text{ mL}}$

3.7 Dimensional Analysis

Formally defined, **dimensional analysis** is a general problem-solving method that uses the units associated with numbers as a guide in setting up the calculation. This method treats units in the same way as numbers, that is, they can be multiplied, divided, canceled, etc. For example, just as

$$3 \times 3 = 3^2 \quad (3 \text{ squared})$$

we have

$$\text{km} \times \text{km} = \text{km}^2 \quad (\text{km squared})$$

Also, just as the two's cancel in the expression

$$\frac{\cancel{2} \times 3 \times 6}{\cancel{2} \times 5}$$

the inches cancel in the expression

$$\frac{\cancel{\text{inch}} \times \text{cm}}{\cancel{\text{inch}}}$$

Like units found in the numerator and denominator of a fraction will always cancel just as like numbers do.

The steps followed in setting up a problem by dimensional analysis are as follows.

STEP 1 *Identify the known or given quantity (both a numerical value and units) and the units of the new quantity to be determined.*

This information, which serves as the starting point for setting up the problem, willl always be found in the statement of the problem. Write an equation with the given quantity on the left and the units of the desired quantity on the right.

STEP 2 *Multiply the given quantity by one or more conversion factors in a manner such that the unwanted (original) units are canceled out, leaving only the new desired unit.*

The general format for the multiplications is

Information given × conversion factor(s) = information sought

The number of conversion factors used depends on the individual problem. Except in the simplest of problems, it is a good idea to predetermine formally the sequence of unit changes to be used. This sequence will be called the unit "pathway."

STEP 3 *Perform the mathematical operations indicated by the conversion factor setup.*

In performing the calculation you need to double check that all units except the desired set have canceled out. You also need to check the numerical answer to see that it contains the proper number of significant figures.

Now let us work a number of sample problems using dimensional analysis and the steps just outlined. Our first two examples involve only metric system units and thus will involve only metric–metric conversion factors. The first problem (Example 3.2) is very simple. Even if you are able to do this problem in your head, let us formally set it up using the steps of dimensional analysis. It is better to cut our teeth on something soft than on something hard. Much can be learned about dimensional analysis from this simple problem.

EXAMPLE 3.2

One drop of water has a volume of 0.000050 L. What is its volume in milliliters?

Solution

STEP 1 The given quantity is 0.000050 L, the volume of one drop. The unit of the desired quantity is milliliters.

$$0.000050 \text{ L} = ? \text{ mL}$$

STEP 2   Only one conversion factor will be needed to convert from liters to milliliters—one that relates liters to milliliters. Two forms of this factor exist.

$$\frac{1 \text{ mL}}{10^{-3} \text{ L}} \quad \text{and} \quad \frac{10^{-3} \text{ L}}{1 \text{ mL}}$$

The first factor is used because it allows for the cancellation of the liter units, leaving us with milliliters as the new units.

$$0.000050 \text{ L} \times \frac{1 \text{ mL}}{10^{-3} \text{ L}}$$

For cancellation, a unit must appear in both the numerator and the denominator. Because the given quantity (0.000050 L) has liters in the numerator, the conversion factor used must be the one with liters in the denominator.

If the other conversion factor had been used, we would have

$$0.000050 \text{ L} \times \frac{10^{-3} \text{ L}}{1 \text{ mL}}$$

No unit cancellation is possible in this setup. Multiplication gives L^2/mL as the final units, which is certainly not what we want. In all cases, only one of the two conversion factors of a reciprocal pair will correctly fit into a dimensional-analysis setup.

STEP 3   Step 2 takes care of the units. All that is left is to combine numerical terms to get the final answer; we still have to do the arithmetic. Collecting the numerical terms gives

$$\frac{0.000050 \times 1}{10^{-3}} \text{ mL} = 5 \times 10^{-2} \text{ mL} \quad \text{(calculator answer)}$$

$$= 5.0 \times 10^{-2} \text{ mL} \quad \text{(correct answer)}$$

Since the conversion factor used in this problem is derived from a definition, it contains an unlimited number of significant figures and will not limit in any way the allowable number of significant figures in the answer. Therefore the answer should have two significant figures, the same number as in the given quantity.

PRACTICE EXERCISE 3.2

A vitamin C tablet is found to contain 0.00500 g of vitamin C. How many milligrams of vitamin C does this tablet contain?   *Ans.* 5.00 mg.

Similar exercises: Problems 3.27 and 3.28

EXAMPLE 3.3

An industrial air pollution control device collected 1.6×10^9 µg of dust in an hour. How many kilograms of dust were collected?

Solution

STEP 1 The given quantity is 1.6×10^9 µg, and the units of the desired quantity are kilograms.

$$1.6 \times 10^9 \text{ µg} = ? \text{ kg}$$

STEP 2 In dealing with metric–metric unit changes where both the original and desired units carry prefixes (which is the case in this problem), it is recommended that you always channel units through the base unit (unprefixed unit). If you do that, you will not need to deal with any conversion factors other than those resulting from prefix definitions. Following this recommendation, the unit pathway for this problem is

$$\underset{\text{prefixed unit}}{\text{µg}} \longrightarrow \underset{\text{base unit}}{\text{g}} \longrightarrow \underset{\text{prefixed unit}}{\text{kg}}$$

In the setup for this problem we will need two conversion factors, one for the micrograms-to-grams change and one for the grams-to-kilograms change.

$$1.6 \times 10^9 \text{ µg} \times \frac{10^{-6} \text{ g}}{1 \text{ µg}} \times \frac{1 \text{ kg}}{10^3 \text{ g}}$$

This conversion factor converts µg to g.

This conversion factor converts g to kg.

Note how all the units cancel except for kilograms.

STEP 3 Carrying out the indicated numerical calculation gives

$$\frac{1.6 \times 10^9 \times 10^{-6} \times 1}{1 \times 10^3} \text{ kg} = 1.6 \text{ kg} \quad \text{(calculator and correct answer)}$$

numbers from first conversion factor

numbers from second conversion factor

The correct answer is the same as the calculator answer for this problem. The given quantity has two significant figures, and both conversion factors are exact. Thus, the correct answer should contain two significant figures.

PRACTICE EXERCISE 3.3

The average diameter of a coronary artery is 0.32 cm. What is this diameter in nanometers? *Ans.* 3.2×10^6 nm.

Similar exercises: Problems 3.29 and 3.30

Our next two worked-out example problems involve the use of English–English conversion factors. The conversion factors themselves should pose no problems for you. The examples are intended to give you further insights into dimensional analysis, particularly in the area of setting up the pathway for unit change. In addition, Example 3.5 exposes you to the complication of having to deal with "compound" units.

EXAMPLE 3.4

The estimated amount of recoverable oil from the field at Prudhoe Bay in Alaska is 4.0×10^{11} gal. What is this amount of oil in fluid ounces (fl oz)?

Solution

STEP 1 The given quantity is 4.0×10^{11} gal, and the unit of the desired quantity is fluid ounces.

$$4.0 \times 10^{11} \text{ gal} = ? \text{ fl oz}$$

STEP 2 The logical pathway to follow to accomplish the desired unit change is

$$\text{gal} \longrightarrow \text{qt} \longrightarrow \text{fl oz}$$

Always use logical steps in setting up the pathway for a unit change. It does not have to be done in one big jump. Use smaller steps for which you know the conversion factors. Big steps usually get you involved with unfamiliar conversion factors. Most people do not carry around in their head the number of fluid ounces in a gallon, but they do know that there are 32 fl oz in a quart.

The setup for this problem will require two conversion factors—gallons to quarts and quarts to fluid ounces.

$$4.0 \times 10^{11} \text{ \cancel{gal}} \times \frac{4 \text{ \cancel{qt}}}{1 \text{ \cancel{gal}}} \times \frac{32 \text{ fl oz}}{1 \text{ \cancel{qt}}}$$

$$\text{gal} \longrightarrow \text{qt} \longrightarrow \text{fl oz}$$

The units all cancel except for the desired fluid ounces.

STEP 3 Performing the indicated multiplications, we get

$$\frac{4.0 \times 10^{11} \times \boxed{4} \times \boxed{32}}{\boxed{1} \times \boxed{1}} \text{ fl oz} = 5.12 \times 10^{13} \text{ fl oz} \quad \text{(calculator answer)}$$

$$= 5.1 \times 10^{13} \text{ fl oz} \quad \text{(correct answer)}$$

numbers from first conversion factor

numbers from second conversion factor

The calculator answer contains too many significant figures. The correct answer should contain only two significant figures, the number in the given quantity. Again, both conversion factors involve exact definitions and will not enter into significant figure considerations.

PRACTICE EXERCISE 3.4

If a person's stomach produces 87 fl oz of gastric juice in a day, what is the equivalent volume of the gastric juice in gallons?   *Ans.* 0.68 gal.

Similar exercises: Problems 3.33 and 3.34

EXAMPLE 3.5

The speed of sound in air at room temperature is about 767 mi/hr. What is this speed in feet per second?

Solution

STEP 1 The given quantity is 767 mi/hr, and the units of this quantity are to be changed to ft/sec.

$$767 \frac{\text{mi}}{\text{hr}} = ? \frac{\text{ft}}{\text{sec}}$$

STEP 2 This problem is more complex than previous examples because two different types of units are involved: length and time. However, the approach we use is the same, other than that we must change two units instead of one. The miles must be converted to feet and the hours to seconds.

The logical pathway for the length change is the direct one-step path

$$\text{mi} \longrightarrow \text{ft}$$

For time, it is logical to make the change in two steps.

$$\text{hr} \longrightarrow \text{min} \longrightarrow \text{sec}$$

It does not matter whether time or length is handled first in the conversion factor setup. We will arbitrarily choose to handle time first. The setup becomes

$$767 \frac{\text{mi}}{\text{hr}} \times \frac{1 \text{ hr}}{60 \text{ min}} \times \frac{1 \text{ min}}{60 \text{ sec}}$$

The units at this point are mi/min.   The units at this point are mi/sec.

Note that in the first conversion factor, hours had to be in the numerator in order to cancel the hours in the denominator of the given quantity.

We are not done yet. The time conversion from hours to seconds has been accomplished, but nothing has been done with length. To take care of length we do not start a new conversion factor setup. Rather, an additional conversion factor is tacked onto those we already have in place.

$$767 \frac{\text{mi}}{\text{hr}} \times \frac{1 \text{ hr}}{60 \text{ min}} \times \frac{1 \text{ min}}{60 \text{ sec}} \times \frac{5280 \text{ ft}}{1 \text{ mi}}$$

The miles in the denominator of the last factor cancel the miles in the numerator of the given quantity. With this cancellation the units now become ft/sec.

STEP 3 Collecting the numerical factors and performing the indicated math gives

$$\frac{767 \times 1 \times 1 \times 5280}{60 \times 60 \times 1} \frac{\text{ft}}{\text{sec}} = 1124.9333 \text{ ft/sec} \quad \text{(calculator answer)}$$

$$= 1120 \text{ ft/sec} \quad \text{(correct answer)}$$

The original speed, 767 mi/hr, involves three significant figures. Rounding the calculator answer to three significant figures gives a speed of 1120 ft/sec. Again, none of the conversion factors plays a role in significant figure considerations since they all originated from definitions.

PRACTICE EXERCISE 3.5

The average human heart pumps blood at the rate of 6.8 fl oz/sec. What is this pump rate in gallons per hour?

Ans. 190 gal/hr.

Similar exercises: Problems 3.35–3.38

We will now consider some sample problems that involve both the English and metric systems of units. Some of these experimentally determined factors were given in Table 3.2.

Instead of trying to remember all of the conversion factors listed in Table 3.2, memorize only one factor for each type of measurement (mass, volume, length). Knowing only one factor of each type is sufficient information to work metric–English or English–metric conversion problems. The relationships that are the most useful for you to memorize are the following.

Length: 1 in. = 2.54 cm   (exact)
Mass: 1 lb = 453.6 g   (4 significant figures)
Volume: 1 qt = 0.9463 L   (4 significant figures)

These three equalities can be considered "bridge relationships" connecting English and metric system measurement units of various types. These bridge relationships are depicted in Figure 3.12.

These bridge relationships are always applicable in problem solving. For example, no matter what mass units are given or asked for in a problem, we can convert to pounds or grams (bridge units), cross the bridge with our memorized conversion factor, and then convert to the desired final unit. The only advantage that would be gained by memorizing all the factors in Table 3.2 would be that you could work some problems with fewer conversion factors, usually only one factor less. The reduction in the number of conversion factors used is usually not worth the added complication of keeping track of the additional conversion factors.

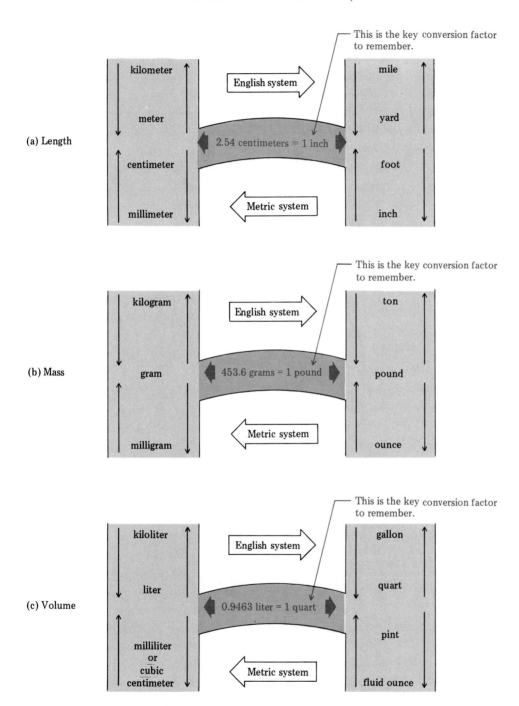

FIGURE 3.12 Measurement bridges connecting the metric and English systems of measurement for (a) length, (b) mass, and (c) volume.

EXAMPLE 3.6

A piece of copper metal has a mass of 17.62 lb. What is the copper mass in kilograms?

Solution

STEP 1 The given quantity is 17.62 lb, and the units of the desired quantity are kilograms.

$$17.62 \text{ lb} = ? \text{ kg}$$

STEP 2 This is an English–metric mass-unit conversion problem. The mass-unit "measurement bridge" (Fig. 3.12b) involves pounds and grams. Since pounds are the given units, we are at the bridge to start with. Pounds are converted to grams (crossing the bridge), and then the grams are converted to kilograms.

$$\text{lb} \longrightarrow \text{g} \longrightarrow \text{kg}$$

The conversion factor setup for this problem is

$$17.62 \text{ lb} \times \frac{453.6 \text{ g}}{1 \text{ lb}} \times \frac{1 \text{ kg}}{10^3 \text{ g}}$$

The units all cancel except for kilograms.

STEP 3 Performing the indicated arithmetic gives

$$\frac{17.62 \times 453.6 \times 1}{1 \times 10^3} \text{ kg} = 7.992432 \text{ kg} \quad \text{(calculator answer)}$$

$$= 7.992 \text{ kg} \quad \text{(correct answer)}$$

The calculator answer must be rounded off to four significant figures, since the given quantity (17.62 lb) has only four significant figures. The first conversion factor also has four significant figures, and the second one is exact.

An alternative pathway for working this problem makes use of the conversion factor in Table 3.2 involving pounds and kilograms. If we use this conversion factor, we can go directly from the given to the desired unit in one step. The setup is

$$17.62 \text{ lb} \times \frac{1 \text{ kg}}{2.205 \text{ lb}} = 7.9909297 \text{ kg} \quad \text{(calculator answer)}$$

Rounding the calculator answer to four significant figures gives 7.991 lb.
Note that the calculator answers obtained from the two setups for this problem are different: 7.992432 and 7.9909297. The correct answers also differ in the last significant digit: 7.992 versus 7.991. These differences arise from the fact that the English-to-metric conversion factors used are not exact definitions. The two conversion factors used have different rounding-off errors in them. Both the answers 7.992 and 7.991

are considered correct. They differ only in the last significant figure, which by definition is an estimated figure (Sec. 2.3).

As is the case here, for most problems there is more than one pathway that can be used to get the answer. When alternative pathways exist, it cannot be said that one way is more correct than another. The important concept is that you select a pathway, choose a correct set of conversion factors consistent with that pathway, and get the answer.

Although both pathways used in this problem are "correct," we prefer the first solution because it uses our mass measurement bridge. Use of the measurement bridge system cuts down on the number of conversion factors you are required to know (or look up in a table).

PRACTICE EXERCISE 3.6

A steel ball bearing has a mass of 0.0004231 kg. What is this mass in pounds?

Ans. 9.328×10^{-4} lb.

Similar exercises: Problems 3.39–3.44

EXAMPLE 3.7

An average-size adult woman has about 4500 mL of blood flowing through her blood vessels. What is the volume, in gallons, of this amount of blood?

Solution

STEP 1   The given quantity is 4500 mL, and the units of the desired quantity are gallons.

$$4500 \text{ mL} = ? \text{ gal}$$

STEP 2   The "bridge relationship" for volume (Fig. 3.12c) involves liters and quarts. Thus, we need to convert the milliliters to liters, cross the bridge to quarts, and then convert quarts to gallons.

$$\text{mL} \longrightarrow \text{L} \longrightarrow \text{qt} \longrightarrow \text{gal}$$

Following the pathway, the setup becomes

$$4500 \text{ mL} \times \frac{10^{-3} \text{ L}}{1 \text{ mL}} \times \frac{1 \text{ qt}}{0.9463 \text{ L}} \times \frac{1 \text{ gal}}{4 \text{ qt}}$$

STEP 3   The numerical calculation involves the following collection of numbers.

$$\frac{4500 \times 10^{-3} \times 1 \times 1}{1 \times 0.9463 \times 4} \text{ gal} = 1.1888407 \text{ gal} \quad \text{(calculator answer)}$$

$$= 1.2 \text{ gal} \quad \text{(correct answer)}$$

The calculator answer must be rounded to two significant figures because the input number 4500 has only two significant figures.

Again, as in Example 3.6, to illustrate that there is more than one way to set up almost any unit conversion problem, let us work this problem two alternative ways using the other two volume conversion factors from Table 3.2.

To use the conversion factor based on 1 L = 2.113 pt requires a pathway of

$$\text{mL} \longrightarrow \text{L} \longrightarrow \text{pt} \longrightarrow \text{qt} \longrightarrow \text{gal}$$

The conversion factor setup is

$$4500 \text{ mL} \times \frac{10^{-3} \text{ L}}{1 \text{ mL}} \times \frac{2.113 \text{ pt}}{1 \text{ L}} \times \frac{1 \text{ qt}}{2 \text{ pt}} \times \frac{1 \text{ gal}}{4 \text{ qt}} = 1.1885625 \text{ gal}$$

(calculator answer)
= 1.2 gal
(correct answer)

The conversion factor based on 29.57 mL = 1 fl oz requires a pathway of

$$\text{mL} \longrightarrow \text{fl oz} \longrightarrow \text{qt} \longrightarrow \text{gal}$$

The conversion factor setup for this pathway is

$$4500 \text{ mL} \times \frac{1 \text{ fl oz}}{29.57 \text{ mL}} \times \frac{1 \text{ qt}}{32 \text{ fl oz}} \times \frac{1 \text{ gal}}{4 \text{ qt}} = 1.1889161 \text{ gal}$$

(calculator answer)
= 1.2 gal (correct answer)

Note that all three methods give the same correct answer. Although all three setups are correct, we still prefer the method of having a bridge relationship for volume, mass, and length to use in crossing the metric-to-English bridge.

PRACTICE EXERCISE 3.7

A typical normal loss of water through sweating per day for a human is 450 mL. What is the volume, in pints, of this amount of water?   *Ans.* 0.95 pt.

Similar exercises: Problems 3.39–3.44

EXAMPLE 3.8

A slab of concrete 1.0 ft thick is laid in an area of dimensions 6.0 ft by 8.0 ft. Find the volume of concrete used in

(a) cubic feet   **(b)** cubic centimeters

Solution

(a) From Figure 3.9 we note that the formula for the volume of a rectangular object is

$$V = \text{length} \times \text{width} \times \text{height}$$

In our case, the height will be the thickness of the concrete.

The volume of the concrete thus is

$$V = 8.0 \text{ ft} \times 6.0 \text{ ft} \times 1.0 \text{ ft}$$
$$= 48 \text{ ft}^3 \quad \text{(calculator and correct answer)}$$

Note that the units must be multiplied together in a manner similar to that for numbers. In the above calculation we have

$$\text{ft} \times \text{ft} \times \text{ft} = \text{ft}^3$$

(b) From part (a) the volume is known to be 48 ft³. A unit change from ft³ to cm³ is now needed. The steps of dimensional analysis are appropriate for making this change.

STEP 1 The given quantity is 48 ft³ and the units for the desired quantity are cm³.

$$48 \text{ ft}^3 = ? \text{ cm}^3$$

STEP 2 If this were a problem involving length rather than (length)³, the pathway would be

$$\text{ft} \longrightarrow \text{in.} \longrightarrow \text{cm}$$

The pathway for (length)³ is just an adaptation of this pathway.

$$\text{ft}^3 \longrightarrow \text{in.}^3 \longrightarrow \text{cm}^3$$

The conversion factors needed for this pathway are simply those for length raised to the third power, that is,

$$48 \text{ ft}^3 \times \left(\frac{12 \text{ in.}}{1 \text{ ft}}\right)^3 \times \left(\frac{2.54 \text{ cm}}{1 \text{ in.}}\right)^3$$

Note that the entire conversion factor, in each case, must be cubed, not just the units on the conversion factor. The first conversion factor is

$$\left(\frac{12 \text{ in.}}{1 \text{ ft}}\right)^3$$

not

$$\frac{12 \text{ in.}^3}{1 \text{ ft}^3}$$

The quantity $\left(\frac{12 \text{ in.}}{1 \text{ ft}}\right)^3$ means that the conversion factor within the parentheses is multiplied by itself three times. Thus,

$$\left(\frac{12 \text{ in.}}{1 \text{ ft}}\right)^3 = \frac{12 \text{ in.}}{1 \text{ ft}} \times \frac{12 \text{ in.}}{1 \text{ ft}} \times \frac{12 \text{ in.}}{1 \text{ ft}}$$

$$= \frac{1728 \text{ in.}^3}{1 \text{ ft}^3}$$

The complete setup for this problem, showing the removal of parentheses and cancellation of units, is

$$48 \text{ ft}^3 \times \left(\frac{12 \text{ in.}}{1 \text{ ft}}\right)^3 \times \left(\frac{2.54 \text{ cm}}{1 \text{ in.}}\right)^3 = ? \text{ cm}^3$$

$$48 \text{ ft}^3 \times \frac{1728 \text{ in.}^3}{1 \text{ ft}^3} \times \frac{16.4 \text{ cm}^3}{1 \text{ in.}^3} = ? \text{ cm}^3$$

Note that in going from $(2.54)^3$ to 16.4, the calculator answer was 16.387064, which, rounded to three significant figures, is 16.4.

STEP 3 The numerical setup is

$$\frac{48 \times 1728 \times 16.4}{1 \times 1} \text{ cm}^3 = 1.3602816 \times 10^6 \text{ cm}^3 \quad \text{(calculator answer)}$$

$$= 1.4 \times 10^6 \text{ cm}^3 \quad \text{(correct answer)}$$

PRACTICE EXERCISE 3.8

A rectangular swimming pool is of dimensions 12.0 ft wide × 16.0 ft long × 8.00 ft deep. Express the water capacity of this pool in

(a) cubic feet (b) cubic meters

Ans. (a) 1540 ft³; (b) 43.6 m³.

Similar exercises: Problems 3.47 and 3.48

Although the emphasis in this section has been on using conversion factors to change units within the English or metric systems or from one to the other, the applications of conversion factors go far beyond this type of activity. We will resort to using conversion factors time and time again throughout this textbook in solving problems. What has been covered in this section is only the "tip of the iceberg" relative to dimensional analysis and conversion factors.

3.8 Density

Density is the ratio of the mass of an object to the volume occupied by that object, that is

$$\text{Density} = \frac{\text{mass}}{\text{volume}} \quad (3.1)$$

The most frequently encountered density units in chemistry are grams per cubic centimeter (g/cm³) for solids, grams per milliliter (g/mL) for liquids, and grams per liter (g/L) for gases. Use of these units avoids the problem of having density values that are extremely small or extremely large numbers. Table 3.3 gives density values for a number of substances.

People often speak of one substance being "heavier" or "lighter" than another. For example, it is said that "lead is a heavier metal than aluminum." What is actu-

Section 3.8 · Density

TABLE 3.3 Densities of Selected Solids, Liquids, and Gases

Solids	Density (g/cm³ at 25°C)[a]	Liquids	Density (g/mL at 25°C)[a]	Gases	Density (g/L at 25°C, 1 atm)[a]
Gold	19.3	Mercury	13.55	Chlorine	3.17
Lead	11.3	Milk	1.028–1.035	Carbon dioxide	1.96
Copper	8.93	Blood plasma	1.027	Oxygen	1.42
Aluminum	2.70	Urine	1.003–1.030	Air (dry)	1.29
Table salt	2.16	Water	0.997	Nitrogen	1.25
Bone	1.7–2.0	Olive oil	0.92	Methane	0.66
Table sugar	1.59	Ethyl alcohol	0.79	Hydrogen	0.08
Wood, pine	0.30–0.50	Gasoline	0.56		

[a] Density changes with temperature. (In most cases it decreases with increasing temperature, since almost all substances expand when heated.) Consequently, the temperature must be recorded along with a density value. In addition, the pressure of gases must be specified.

ally meant by this statement is that lead has a higher density than aluminum; that is, there is more mass in a specific volume of lead than there is in the same volume of aluminum. The density of an object is a measure of how tightly the object's mass is packed into a given volume. Even though the density of lead (11.3 g/cm³) is greater than that of aluminum (2.70 g/cm³), 1 lb of lead weighs exactly the same as 1 lb of aluminum—1 lb is 1 lb. Because the aluminum is less dense than the lead, the mass in the 1 lb of aluminum will occupy a larger volume than the mass in the 1 lb of lead. Said another way, if equal-volume samples of lead and aluminum are weighed, the lead will have the greater mass. When we say lead is "heavier" than aluminum, we actually mean that lead is more dense than aluminum.

The densities of solids and liquids are often compared to the density of water. Anything less dense ("lighter") than water floats on it, and anything more dense ("heavier") sinks. In a similar vein, densities of gases are compared to that of air. Any gas less dense ("lighter") will rise in air, and anything more dense ("heavier") will sink in air.

To calculate on object's density, we must make two measurements; one involves determining the object's mass, and the other its volume.

EXAMPLE 3.9

Osmium is the densest of all metals. What is its density in grams per cubic centimeter if 50.00 g of the metal occupies a volume of 2.22 cm³?

Solution

Using the formula

$$\text{Density} = \frac{\text{mass}}{\text{volume}}$$

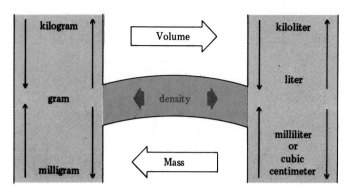

FIGURE 3.13 Density, a measurement bridge connecting mass and volume.

we have

$$\text{Density} = \frac{50.00 \text{ g}}{2.22 \text{ cm}^3} = 22.522522 \text{ g/cm}^3 \quad \text{(calculator answer)}$$

$$= 22.5 \text{ g/cm}^3 \quad \text{(correct answer)}$$

Since the input volume contains only three significant figures, the calculator answer must be rounded to three significant figures to give the correct answer.

PRACTICE EXERCISE 3.9

The mass of a piece of copper rod is 21.09 g. The volume of this same piece of copper rod is found to be 2.361 cm³. What is the density; in grams per cubic centimeter, of copper?

Ans. 8.993 g/cm³.

Similar exercises: Problems 3.51 and 3.52

In a mathematical sense, density can be thought of as a conversion factor that relates the volume and mass of an object. Interpreting density in this manner enables us to calculate a substance's volume from its mass and density or its mass from its volume and density. Density is thus a "bridge" connecting mass and volume (see Fig. 3.13).

EXAMPLE 3.10

The density of gold is 19.3 g/cm³. What is the mass, in grams, of 7.83 cm³ of gold?

Solution

We will use density as a conversion factor in solving this problem by dimensional analysis.

STEP 1 The given quantity is 7.83 cm³ of gold. The unit of the desired quantity (mass) is grams. Thus,

$$7.83 \text{ cm}^3 = ? \text{ g}$$

STEP 2 The pathway going from cubic centimeters to grams involves a single step, since density, used as a conversion factor, directly relates grams and cubic centimeters.

$$7.83 \text{ cm}^3 \times \frac{19.3 \text{ g}}{1 \text{ cm}^3}$$

STEP 3 Doing the indicated math gives the following answer.

$$\frac{7.83 \times 19.3}{1} \text{ g} = 151.119 \text{ g} \quad \text{(calculator answer)}$$

$$= 151 \text{ g} \quad \text{(correct answer)}$$

The correct answer is the calculator answer rounded to three significant figures, the number in each input number.

PRACTICE EXERCISE 3.10

The density of octane, a component of gasoline, is 0.702 g/mL. What is the mass, in grams, of 875 mL of octane?

Ans. 614 g.

Similar exercises: Problems 3.57 and 3.58

EXAMPLE 3.11

The density of ethyl alcohol is 0.789 g/mL. What volume of ethyl alcohol, in milliliters, would have a mass of 25.0 g?

Solution

STEP 1 The given quantity is 25.0 g of ethyl alcohol. The unit of the desired quantity (volume) is milliliters. Thus,

$$25.0 \text{ g} = ? \text{ mL}$$

STEP 2 The pathway going from grams to milliliters involves a single step, since density, used as a conversion factor, directly relates grams and milliliters.

$$25.0 \text{ g} \times \frac{1 \text{ mL}}{0.789 \text{ g}}$$

Note that the conversion factor used is actually the density in its reciprocal form. Use of the inverted form is necessary in order for the gram units to cancel.

STEP 3 Doing the indicated math gives the following answer.

$$\frac{25.0 \times 1}{0.789} \text{ mL} = 31.685678 \text{ mL} \quad \text{(calculator answer)}$$

$$= 31.7 \text{ mL} \quad \text{(correct answer)}$$

The correct answer is the calculator answer rounded to three significant figures, the number in each input number.

PRACTICE EXERCISE 3.11

Common table sugar has a density of 1.587 g/cm^3. What would be the volume, in cubic centimeters, of 1.000 g of table sugar?   *Ans.* 0.6301 cm^3.

Similar exercises: Problems 3.59 and 3.60

Density is a conversion factor of a different type than those previously used in this chapter. It is a conversion factor whose numerator and denominator are *equivalent* rather than *equal*. All previously used conversion factors have been fractions where the numerator and denominator have been the same quantity under two different names. Twelve inches and one foot are different names for the same distance. The mass and volume involved in density are not different names for the same thing, but related equivalent quantities.

A major difference between *equivalence* conversion factors and *equality* conversion factors is that the former have applicability only in the particular problem setting for which they were derived, whereas the latter are applicable in all problem-solving situations. Many different gram-to-cubic centimeter (mass-to-volume) relationships (densities) exist, but only one foot-to-inch relationship exists. Mathematically, equivalence conversion factors can be used the same way equality conversion factors are, and we use the equal sign for both relationships. Equivalence conversion factors will be used often in later chapters.

3.9 Percent as a Conversion Factor

The word percent is actually a way of expressing a conversion factor. **Percent** means parts per hundred; that is, it is the number of items of a specified type in a group of 100 items. The quantity 45% means 45 items per 100 total items.

Let us look at the mechanics of writing a percentage as a conversion factor, paying particular attention to the units involved. The percent value in the statement "A gold alloy is found upon analysis to contain 77% gold by mass" will be our focus point for the discussion. What are the mass units associated with the value 77%? The answer is that many mass units could be appropriate. The following statements, written as fractions (conversion factors) are all consistent with our 77% gold analysis.

$$\frac{77 \text{ oz of gold}}{100 \text{ oz of gold alloy}} \qquad \frac{77 \text{ lb of gold}}{100 \text{ lb of gold alloy}}$$

$$\frac{77 \text{ g of gold}}{100 \text{ g of gold alloy}} \qquad \frac{77 \text{ kg of gold}}{100 \text{ kg of gold alloy}}$$

Thus, in writing a percent as a conversion factor, the choice of unit (for mass in this case) is arbitrary. In practice, the complete context of the problem in which the percentage is found usually makes obvious the appropriate unit choice.

In conversion factors derived from percentages, confusion about cancellation of units frequently arises. Both numerator and denominator contain the same units, for example, ounces. Yet the ounces do not cancel. Not considering the *complete unit* is what causes confusion in the minds of students. In the conversion factor

$$\frac{77 \text{ oz of gold}}{100 \text{ oz of gold alloy}}$$

the numerator and denominator dimensions are not simply "ounces." The dimensions are, respectively, "ounces of gold" and "ounces of gold alloy." Ounces cannot be canceled because they are only a part of the complete dimension. Cancellation is possible only when *complete* units are identical.

To avoid the problem of mistakenly canceling dimensions that are not the same, always write complete dimensions. Percent will always involve the same dimensional units (pounds, grams, meters) of different things. The identities of the different things, gold and gold alloy in our example, must always be included as part of the units.

EXAMPLE 3.12

A sample of a copper alloy is found to be 38.7% copper by mass. How many pounds of copper are present in a 30.2 lb sample of the alloy?

Solution

STEP 1 The given quantity is a 30.2 lb sample of copper alloy, and the desired quantity is pounds of copper.

$$30.2 \text{ lb copper alloy} = ? \text{ lb copper}$$

STEP 2 This is a one-conversion-factor problem with the conversion factor, obtained from the given percentage, being

$$\frac{38.7 \text{ lb copper}}{100 \text{ lb copper alloy}}$$

The setup for the problem is

$$30.2 \text{ lb copper alloy} \times \frac{38.7 \text{ lb copper}}{100 \text{ lb copper alloy}}$$

STEP 3 The numerical calculation involves the following arrangement of numbers.

$$\frac{30.2 \times 38.7}{100} \text{ lb copper} = 11.6874 \text{ lb copper} \quad \text{(calculator answer)}$$

$$= 11.7 \text{ lb copper} \quad \text{(correct answer)}$$

The original percentage (38.7%) and the alloy mass (30.2 lb) both have three significant figures. Therefore, the correct answer should also contain three significant figures. The number 100 in the setup is an exact number, since it originates from the definition of percent.

PRACTICE EXERCISE 3.12

A 75.0 g solution containing table sugar and water is 7.0% sugar by mass. What is the mass, in grams, of sugar in the solution?

Ans. 5.2 g.

Similar exercises: Problems 3.65 and 3.66

EXAMPLE 3.13

A study involving the inhabitants of a small town (population 1821) in Weber County in the state of Utah includes the following facts: (1) 21.3% of the inhabitants are adult males; (2) 32.3% of the adult males are bald-headed; and (3) 98.3% of the bald-headed adult males are handsome. How many handsome bald-headed adult males live in the town?

Solution

STEP 1 The given quantity is 1821 inhabitants, with the units of the desired quantity being handsome bald-headed adult males.

$$1821 \text{ inhabitants} = ? \text{ handsome bald-headed adult males}$$

STEP 2 When conversion factors, particularly unusual ones, are given in word form in the statement of a problem, it is suggested that you first extract them from the problem statement and write them down before starting to solve the problem. You then have them available to use as you solve the problem. The given conversion factors in this problem are

$$\frac{21.3 \text{ adult males}}{100 \text{ inhabitants}}$$

$$\frac{32.3 \text{ bald-headed adult males}}{100 \text{ adult males}}$$

$$\frac{98.3 \text{ handsome bald-headed adult males}}{100 \text{ bald-headed adult males}}$$

The unit-conversion pathway for this problem will be

$$\text{Inhabitants} \longrightarrow \text{adult males} \longrightarrow \text{bald-headed adult males} \longrightarrow \text{handsome bald-headed adult males}$$

The setup for the problem is

$$1821 \text{ inhabitants} \times \frac{21.3 \text{ adult males}}{100 \text{ inhabitants}} \times \frac{32.3 \text{ bald-headed}}{100 \text{ adult males}} \times \frac{98.3 \text{ handsome}}{100 \text{ bald-headed}}$$

$$= ? \text{ handsome bald-headed adult males}$$

All of the units cancel except those in the numerator of the last conversion factor.

STEP 3   Performing the indicated arithmetic gives

$$\frac{1821 \times 21.3 \times 32.3 \times 98.3}{100 \times 100 \times 100} \text{ handsome bald-headed adult males}$$

= 123.15316 handsome bald-headed adult males   (calculator answer)

= 123 handsome bald-headed adult males   (correct answer)

The calculator answer must be rounded off to *three* significant figures since all three of the given percentages contain only three significant figures.

PRACTICE EXERCISE 3.13

A 5-lb box of chocolates contains $12\overline{0}$ chocolates. Dark chocolates are more prevalent than light chocolates—$7\overline{0}$% versus $3\overline{0}$%. Exactly 25% of the dark chocolates are cream-filled. How many cream-filled dark chocolates are there in the 5-lb box of chocolates?   *Ans.* 21 cream-filled dark chocolates.

Similar exercises: Problems 3.67 and 3.68

3.10 Temperature Scales

Temperature, a concept with which we are all familiar, is a measure of the hotness or coldness of an object. The most common instrument for measuring temperature is the mercury-in-glass thermometer, which consists of a glass bulb containing mercury sealed to a slender glass capillary tube. The higher the temperature the farther the mercury will rise in the capillary tube. Graduations on the capillary tube indicate the height of the mercury column in terms of defined units, usually called *degrees*. A tiny superscript circle is used as the symbol for a degree.

Three different temperature scales are in common use—Celsius, Kelvin, and Fahrenheit. Both the Celsius and Kelvin scales are part of the metric measurement system, and the Fahrenheit scale belongs to the English measurement system. Different degree sizes and different reference points are what produce the various temperature scales.

The Celsius scale, named after Anders Celsius (1701–1744), a Swedish astronomer, is the scale most commonly encountered in scientific work. On this scale the boiling and freezing points of water serve as reference points, with the former having a value of 100 °C (degrees Celsius) and the latter 0 °C. Thus, there are 100 degree intervals between the two reference points. The Celsius scale was formerly called the centigrade scale.

The Kelvin scale is a close relative of the Celsius scale. The size of the temperature unit is the same on both scales, as is the interval between the reference points. The two scales differ only in the numerical values assigned to the reference points and the names of the units. On the Kelvin scale the boiling point of water is at 373.15 K (kelvins), and the freezing point of water at 273.15 K. This scale, proposed by the British mathematician and physicist William Kelvin (1824–1907) in 1848, is particularly useful when working with relationships between temperature and pressure–volume behavior of gases (see Sec. 12.3).

A unique feature of the Kelvin scale is that negative temperature readings never occur. The lowest possible temperature thought to be obtainable occurs at 0 on the Kelvin scale. This temperature, known as *absolute zero*, has never been produced experimentally, although scientists have come within a fraction of a degree of reaching it.

The Fahrenheit scale was designed by the German physicist Gabriel Fahrenheit (1686–1736) in the early 1700s. After proposing several scales he finally adopted a system that used a salt–ice mixture and boiling mercury as the reference points. These were the two extremes in temperature available to him at that time. A reading of 0 was assigned to the salt–ice mixture and 600 to the boiling mercury. The distance between these two points was divided into 600 equal parts or degrees. On this scale, water freezes at 32 °F (degrees Fahrenheit) and boils at 212 °F. Thus, there are 180 degrees between the freezing and boiling points of water on this scale as contrasted to 100 degrees on the Celsius scale and 100 kelvins on the Kelvin scale. Figure 3.14 shows a comparison of the three temperature scales.

When changing a temperature reading on one scale to its equivalent on another scale, we must take two factors into consideration: (1) the size of the unit on the two scales may differ, and (2) the zero points on the two scales will not coincide.

Difference in unit size will be a factor any time the Fahrenheit scale is involved in a conversion process. The conversion factors necessary to relate the size of the Fahrenheit degree to the size of the Celsius degree or the kelvin are obtainable from the information in Figure 3.14. From that figure we see that 180 Fahrenheit degrees are equivalent to 100 Celsius degrees or kelvins. Using this relationship and the fact that $180/100 = 9/5$, we obtain the following equalities.

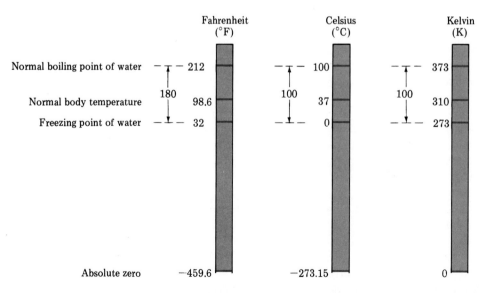

FIGURE 3.14 The relationships among the Fahrenheit, Celsius, and Kelvin temperature scales.

5 Celsius degrees = 9 Fahrenheit degrees

5 kelvins = 9 Fahrenheit degrees

Conversion factors derived from these equalities will contain an infinite number of significant figures; that is, they are exact conversion factors.

Adjustment for differing zero-point locations is carried out by considering how many degrees above or below the freezing point of water (the ice point) the original temperature is. Examples 3.14 and 3.15 show how this zero-point adjustment is carried out in addition to illustrating the use of temperature-scale conversion factors.

EXAMPLE 3.14

Milk is pasteurized at 145 °F to kill germs. What is the equivalent temperature on the Celsius scale?

Solution

First, we determine the number of degrees between the ice point (freezing point of water) and the given temperature on the original scale.

$$145\,°F - 32\,°F = 113\,°F$$

Second, we convert from Fahrenheit units to Celsius units.

$$113 \text{ Fahrenheit degrees} \times \frac{5 \text{ Celsius degrees}}{9 \text{ Fahrenheit degrees}} = 62.777777 \text{ Celsius degrees}$$
(calculator answer)

$$= 62.8 \text{ Celsius degrees}$$
(correct answer)

Third, taking into account the ice point on the new scale, we determine the new temperature. On the Celsius scale, the temperature will be 62.8 degrees above the ice point. Since the ice point is 0 °C, the new temperature reading is 62.8 °C.

PRACTICE EXERCISE 3.14

A comfortable temperature for bath water is 95 °F. What is the equivalent temperature on the Celsius scale?

Ans. 35 °C.

Similar exercises: Problems 3.71 and 3.72

EXAMPLE 3.15

The temperature at the bottom of a blast furnace used in the production of iron (steel making) is measured at 1935 °C. What is the equivalent temperature on the Fahrenheit scale?

Solution

First, we determine how many degrees there are between the original temperature and the ice point. On the Celsius scale this will always be equal numerically to the original temperature, since the ice point is 0 °C.

Second, we change this number of degrees—1935 in this case—from Celsius to Fahrenheit.

$$1935 \text{ \sout{Celsius degrees}} \times \frac{9 \text{ Fahrenheit degrees}}{5 \text{ \sout{Celsius degrees}}} = 3483 \text{ Fahrenheit degrees}$$

(calculator and correct answer)

Third, taking into account the ice point on the new scale, we determine the new temperature. The new temperature will be 3483 °F above the ice point, which is 32 °F.

$$1935 \,°\text{C} = 3483 \,°\text{F} + 32 \,°\text{F} = 3515 \,°\text{F}$$

PRACTICE EXERCISE 3.15

An oven for baking pizza operates at approximately 275 °C. What is the equivalent temperature on the Fahrenheit scale?   *Ans.* 527 °F.

Similar exercises: Problems 3.73 and 3.74

Examples 3.14 and 3.15 point out that temperature-scale conversions are a little more complicated than the unit conversions of the last three sections. Not only is multiplication by a conversion factor required, but also addition and subtraction.

Since the size of the unit is the same, the relationship between the Kelvin and Celsius scales is very simple. No conversion factors are needed. All that is required is an adjustment for the differing zero points. This adjustment involves the number 273, the number of units by which the two scales are offset from each other. The adjustment factor is specifically 273, 273.2, or 273.15 depending on the precision of the temperature measurement. Since temperatures are most often stated in terms of a whole number of units (31 °C, 43 °C, etc.), 273 is the most used adjustment factor. However, the other two factors are needed when dealing with temperatures involving tenths or hundredths of a unit (31.5 °C, 452.72 °C, etc.).

To change a Celsius temperature to the Kelvin scale we add the adjustment factor 273.

$$K = °C + 273$$

To change a Kelvin temperature to Celsius we subtract this same adjustment factor.

$$°C = K - 273$$

Note that the symbol for the kelvin is K, not °K.

The relationship between the Fahrenheit scale and the Celsius scale can also be stated in an equation format.

$$°F = \frac{9}{5}(°C) + 32$$

$$°C = \frac{5}{9}(°F - 32)$$

Some students prefer to use these equations rather than the dimensional-analysis approach used in Examples 3.14 and 3.15. The use of these equations is illustrated in Example 3.16.

EXAMPLE 3.16

A hospital patient with a fever has a temperature of 39.4 °C. What is this temperature on (a) the Fahrenheit scale? (b) the Kelvin scale?

Solution

(a) Substituting into the Celsius-to-Fahrenheit equation, we get

$$°F = \frac{9}{5}(°C) + 32 = \frac{9}{5}(39.4) + 32$$
$$= 70.9 + 32$$
$$= 102.9 °F \quad \text{(calculator and correct answer)}$$

Both the multiplication and addition significant figure rules were needed to obtain the answer 102.9 °F (correct answer). The multiplication $\frac{9}{5}(39.4)$ gives the calculator answer 70.92, which is rounded to 70.9 (three significant figures). Adding 32 (an exact number) to this gives 102.9, which is both a calculator and correct answer.

(b) Substituting into the Celsius-to-Kelvin equation, we get

$$K = °C + 273.2 = 39.4 + 273.2$$
$$= 312.6 \text{ K} \quad \text{(calculator and correct answer)}$$

Note that we used the adjustment factor 273.2 rather than simply 273 because the given temperature involved tenths of a degree.

PRACTICE EXERCISE 3.16

A person suffering from heat stroke is found to have a body temperature of 41.1 °C. What is this temperature on (a) the Fahrenheit scale? (b) the Kelvin scale?

Ans. (a) 106.0 °F; (b) 314.3 K.

Similar exercises: Problems 3.73–3.76

3.11 Types and Forms of Energy

On some days you wake up feeling "very energetic." On these days you usually accomplish a great deal. By the end of the day you are tired, you have "no energy left," and you do not feel like doing any more "work." The scientific definition for energy closely parallels the ideas presented in these three sentences. **Energy** is the capacity to do work.

Energy can exist in any of several forms. Common forms include radiant (light) energy, chemical energy, thermal (heat) energy, electrical energy, and mechanical energy. These forms of energy are interconvertible. The heating of a home is a process that illustrates energy interconversion. As the result of burning natural gas or some other fuel, chemical energy is converted into heat energy. In large conventional power plants that are used to produce electricity, the heat energy obtained from burning coal is used to change water into steam, which can then turn a turbine (mechanical energy) to produce electricity (electrical energy).

During energy interconversions, the law of conservation of energy is always obeyed. The **law of conservation of energy** states that in any chemical or physical change, energy can be converted from one form to another, but it is neither created nor destroyed. Studies of energy changes in numerous systems have shown that no system acquires energy except at the expense of energy possessed by another system.

Almost all of our energy on Earth originates from the sun in the form of radiant or light energy. Green plants convert this radiant energy into chemical energy by means of a process called photosynthesis. The chemical energy is stored within the living plant. **Chemical energy** is energy stored in a substance that can be released during a chemical change. Energy for the human body is obtained, either directly or indirectly, from plants when they are consumed as food.

Chemical change can be used to produce other forms of energy. Chemical changes in an automobile battery produce the electrical energy needed to start a car. Chemical changes that occur in a magnesium flash bulb generate the light energy needed for photographic purposes. The burning of fuels releases both heat and light energy. The energy that "runs" our life processes—for example, breathing, muscle contraction, and blood circulation—is produced by chemical changes occurring within the cells of the body. The energy required for or generated by chemical changes will be an important point of focus in many discussions in later chapters.

In additional to the various *forms* of energy, there are two *types* of energy: potential energy and kinetic energy. The basis for determining energy types depends on whether the energy is available but not being used or is actually in use. **Potential energy** is stored energy that results from an object's position, condition, and/or composition. Water that is backed up behind a dam represents potential energy because of its position. When the water is released, it can be used to produce electrical energy at a hydroelectric plant. A compressed spring can spontaneously expand and do work as the result of potential energy associated with condition. Chemical energy, such as that stored in gasoline, is potential energy arising from composition. This stored energy is released when the gasoline is burned.

Kinetic energy is energy that matter possesses because of its motion. An object that is in motion has the capacity to do work. If it collides with another object, it will do work on that object. A hammer held in the air above an object possesses potential energy of position. As the hammer moves downward toward a nail, this potential energy becomes kinetic energy that can be used to drive the nail into a board. When water behind a dam is released and allowed to flow, its potential energy of position becomes kinetic energy. During the operation of a hydroelectric plant, some of this kinetic energy becomes mechanical and electrical energy.

Figure 3.15 summarizes the forms and types of energy discussed in the preceding paragraphs.

The concepts of potential and kinetic energy will play a part in many discussions in future chapters. For example, in Chapter 11, we will explain the differences be-

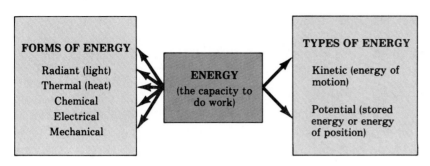

FIGURE 3.15 Energy may be classified by form and by type.

tween the solid, liquid, and gaseous states of matter in terms of relative amounts of potential and kinetic energy. The pressure that a gas exerts (Sec. 12.4) is related to kinetic energy.

3.12 Heat Energy and Specific Heat

The form of energy that is most often required for or released by the chemical and physical changes considered in this text is heat energy. For this reason, we will consider further particulars about this form of energy.

Units of Heat Energy

The most commonly used unit of measurement for heat energy is the joule. The joule (pronounced *jool*, as in pool)—after the English physicist James Prescott Joule (1819–1889) who studied the energy concept—is a derived rather than a fundamental measurement unit. Its derivation involves the mass of an object (kg) and the square of the object's velocity (m^2/sec^2). A **joule** is an energy unit obtained from the base units kilogram, meter, and second. Mathematically, the equation for a joule is

$$1 \text{ joule} = \frac{1 \text{ kg} \cdot m^2}{sec^2}$$

The joule unit, abbreviated as J, is suitable for measuring all types of energy, not just heat energy.

An older unit for heat energy, whose use is rapidly decreasing, is the calorie (cal). For many years it was commonly used by chemists. A **calorie** is the amount of heat energy needed to raise the temperature of one gram of water by one degree Celsius measured between 14.5 and 15.5 °C. The relationship between the calorie and the joule is

$$1 \text{ cal} = 4.184 \text{ J} \quad \text{(exact definition)}$$

Both the joule and the calorie involve relatively small amounts of energy, so the kilojoule (kJ) and the kilocalorie (kcal) are often used instead. It follows from the relationship between joules and calories that

$$1 \text{ kcal} = 4.184 \text{ kJ}$$

In discussions involving nutrition, the energy content of foods, or dietary tables, the term *Calories* (spelled with a capital C) is used. The dietetic Calorie is actually 1 kilocalorie (1000 cal). The statement that an oatmeal raisin cookie contains 60 Calories means that 60 kcal (60,000 cal) of energy is released when the cookie is metabolized (undergoes chemical change) within the body.

EXAMPLE 3.17

Using dimensional analysis, convert the energy measurement 4.32 kJ to (a) joules, (b) kilocalories, and (c) calories.

Solution

(a) The relationship between joules and kilojoules is

$$1000 \text{ J} = 1 \text{ kJ}$$

Therefore, using dimensional analysis, we have

$$4.32 \text{ kJ} \times \frac{10^3 \text{ J}}{1 \text{ kJ}} = 4320 \text{ J} \quad \text{(calculator and correct answer)}$$

(b) The relationship between kilocalories and kilojoules is

$$1 \text{ kcal} = 4.184 \text{ kJ}$$

The one-step dimensional analysis setup for this problem is

$$4.32 \text{ kJ} \times \frac{1 \text{ kcal}}{4.184 \text{ kJ}} = 1.0325048 \text{ kcal} \quad \text{(calculator answer)}$$

$$= 1.03 \text{ kcal} \quad \text{(correct answer)}$$

(c) There will be two conversion factors in this part, derived, respectively, from the relationship 1 kcal = 4.184 kJ and 1000 cal = 1 kcal. The pathway will be

$$\text{kJ} \longrightarrow \text{kcal} \longrightarrow \text{cal}$$

$$4.32 \text{ kJ} \times \frac{1 \text{ kcal}}{4.184 \text{ kJ}} \times \frac{10^3 \text{ cal}}{1 \text{ kcal}} = 1032.5048 \text{ cal} \quad \text{(calculator answer)}$$

$$= 1030 \text{ cal} \quad \text{(correct answer)}$$

There can be only three significant figures in the correct answer.

PRACTICE EXERCISE 3.17

Using dimensional analysis, convert the energy measurement 7.21 kcal to (a) calories, (b) joules, and (c) kilojoules.   *Ans.* (a) 7210 cal; (b) 30,200 J; (c) 30.2 kJ.

Similar exercises: Problems 3.93 and 3.94

Specific Heat

For every pure substance, in a given state (solid, liquid, or gas), we can measure a physical property called the *specific heat* of that substance. **Specific heat** is the amount of heat needed to raise the temperature of 1 g of a substance in a specific physical state by 1 °C. The units most commonly used for specific heat, in scientific work, are joules per gram per degree Celsius [J/(g · °C)]. Specific heats for a number of substances in various states are given in Table 3.4. Note the three different entries for water in this table—one for each of the physical states. As these entries point out, the magnitude of the specific heat for a substance changes when the physical state of the substance changes. (Water is one of the few substances that we routinely encounter in all three physical states; hence the three specific-heat values in the table for this substance.)

The higher the specific heat of a substance, the less its temperature will change when it absorbs a given amount of heat. For liquids, water has a relatively high specific heat; it is thus a very effective coolant. The moderate climates of geographical areas where large amounts of water are present—for example, the Hawaiian Islands—are related to water's ability to absorb large amounts of heat without undergoing drastic temperature changes. Desert areas, areas that lack water, are the areas where the extremes of high temperature are encountered on Earth. The temperature of a living organism remains relatively constant because of the large amounts of water present in it.

The amount of heat energy needed to cause a fixed amount of a substance to undergo a specific temperature change (within a range that causes no change of state) can easily be calculated if the substance's specific heat is known. The specific heat [in J/(g · °C)] is multiplied by the mass (in grams) and by the temperature

TABLE 3.4 Specific Heats of Selected Pure Substances

Substance	Physical State	Specific Heat [J/(g · °C)]
Aluminum	solid	0.908
Copper	solid	0.382
Ethyl alcohol	liquid	2.42
Gold	solid	0.13
Iron	solid	0.444
Nitrogen	gas	1.0
Oxygen	gas	0.92
Silver	solid	0.24
Sodium chloride	solid	0.88
Water (ice)	solid	2.1
Water	liquid	4.18
Water (steam)	gas	2.0

change (in degrees Celsius) to eliminate the units of g and °C and obtain the unit calories.

$$\text{Heat absorbed} = \text{specific heat} \times \text{mass} \times \text{temperature change}$$

$$= \frac{J}{\cancel{g} \cdot \cancel{°C}} \times \cancel{g} \times \cancel{°C}$$

$$= J$$

The temperature change, denoted as ΔT, is always calculated as a positive number; the lower temperature is always subtracted from the higher temperature.

EXAMPLE 3.18

Calculate the number of joules of heat energy needed to increase the temperature of each of the following amounts of substance from 81 to 93 °C.

(a) 3.4 g of ethyl alcohol   (b) 3.4 g of water

Solution

(a) From Table 3.4 we determine that the specific heat of liquid ethyl alcohol is 2.42 J/(g · °C). Using this value, the given mass in grams, and the value of ΔT in degrees Celsius, which is 12 °C, we calculate the heat needed as follows.

$$\text{Heat absorbed} = \frac{2.42 \text{ J}}{\cancel{g} \cdot \cancel{°C}} \times 3.4 \cancel{g} \times 12 \cancel{°C}$$

$$= 98.736 \text{ J} \quad \text{(calculator answer)}$$

$$= 99 \text{ J} \quad \text{(correct answer)}$$

(b) From Table 3.4 we determine that the specific heat of liquid water is 4.18 J/(g · °C). Note that three specific heats for water are given in Table 3.4—one for each of the three physical states. In general, as is the caase for water, a substance's specific heat as a solid is different from that as a liquid, which in turn is different from that as a gas.

$$\text{Heat absorbed} = \frac{4.18 \text{ J}}{\cancel{g} \cdot \cancel{°C}} \times 3.4 \cancel{g} \times 12 \cancel{°C}$$

$$= 170.544 \text{ J} \quad \text{(calculator answer)}$$

$$= 171 \text{ J} \quad \text{(correct answer)}$$

The heat-carrying capacity of water is almost twice that of ethyl alcohol when both are liquids.

PRACTICE EXERCISE 3.18

Calculate the number of joules of heat energy needed to increase the temperature of 50.0 g of copper metal from 21.0 to 80.0 °C.

Ans. 1130 J.

Similar exercises: Problems 3.97 and 3.98

EXAMPLE 3.19

A one-half cup serving of zucchini has a caloric value of approximately 15 Cal (15,000 cal). What would be the temperature change in a cup of water (237 g) at 20 °C if the same amount of energy was added to it?

Solution

The equation

$$\text{Heat absorbed} = \text{specific heat} \times \text{mass} \times \text{temperature change}$$

is rearranged to isolate temperature change on the left side of the equation.

$$\text{Change in temperature (°C)} = \frac{\text{heat absorbed (J)}}{\text{mass (g)} \times \text{specific heat [J/(g} \cdot \text{°C)]}}$$

In this equation the heat absorbed will be the heat energy (caloric value) supplied by the zucchini. Since this is given in calories, we will need to change it to joule units.

$$15{,}000 \text{ cal} \times \frac{4.184 \text{ J}}{\text{cal}} = 62{,}760 \text{ J} \quad \text{(calculator answer)}$$

$$= 63{,}000 \text{ J} \quad \text{(correct answer)}$$

The mass is 237 g and the specific heat of water (from Table 3.4) is 4.18 J/(g · °C). Substituting the known values into the equation gives

$$\text{Change in temperature (°C)} = \frac{63{,}000 \text{ J}}{237 \text{ g} \times 4.18 \text{ J/(g} \cdot \text{°C)}}$$

$$= 63.593968 \text{ °C} \quad \text{(calculator answer)}$$

$$= 64 \text{ °C} \quad \text{(correct answer)}$$

The temperature of the water would increase from 20 to 84 °C.

PRACTICE EXERCISE 3.19

What will be the temperature change in a 43-g bar of silver metal at 43 °C if 275 J of heat energy is removed from it?

Ans. 27 C°.

Similar Exercises: Problems 3.99 and 3.100

A quantity closely related to specific heat is that of heat capacity. **Heat capacity** is the amount of heat needed to raise the temperature of a given quantity of a substance in a specific physical state by 1 °C. The relationship between heat capacity and specific heat is

$$\text{Heat capacity} = \text{grams} \times \text{specific heat}$$

Heat capacity refers to a property of a whole object (its entire mass), while specific heat refers to the heat capacity per unit mass (1 g). If either heat capacity or specific heat is known, the other quantity can always be calculated from the known quantity.

EXAMPLE 3.20

A 80.0-g sample of the metal gold has a heat capacity of 10.5 J/°C. What is the specific heat of gold?

Solution
The relationship between heat capacity and specific heat is

$$\text{Heat capacity} = \text{mass} \times \text{specific heat}$$

Rearranging this equation to isolate specific heat on one side gives

$$\text{Specific heat} = \frac{\text{heat capacity}}{\text{mass}}$$

$$= \frac{10.5 \text{ J/°C}}{80.0 \text{ g}}$$

$$= 0.13125 \text{ J/(g} \cdot \text{°C)} \quad \text{(calculator answer)}$$

$$= 0.131 \text{ J/(g} \cdot \text{°C)} \quad \text{(correct answer)}$$

PRACTICE EXERCISE 3.20

A 35.0-g sample of the metal iron has a heat capacity of 15.5 J/°C. What is the specific heat of iron?

Ans. 0.443 J/(g · °C)

Similar exercises: Problems 3.105 and 3.106

Another common type of heat energy calculation involves the transfer of heat from one substance to another. The following two generalizations always apply to such situations.
1. Heat always flows from the hotter body to the colder body.
2. The heat lost by the hotter body is equal to the heat gained by the colder body.

EXAMPLE 3.21

How many grams of copper can be heated from 20 to 30 °C by the heat energy released when 245 g of aluminum cools from 80 to 50 °C?

Solution
The specific heats for the two metals are given in Table 3.4. The heat lost by the aluminum will equal the heat gained by the copper.

$$\text{Heat lost (aluminum)} = \text{heat gained (copper)}$$

Recalling, from Example 3.19, that the general equation for heat lost or heat gained is

$$\text{Heat lost or gained} = \text{specific heat} \times \text{mass} \times \Delta T$$

we have for our present situation

$$0.908 \frac{J}{g \cdot °C} \times 245 \text{ g} \times 30 \text{ °C} = 0.382 \frac{J}{g \cdot °C} \times \text{g copper} \times 10 \text{ °C}$$

$$\underbrace{\phantom{0.908 \frac{J}{g \cdot °C} \times 245 \text{ g} \times 30 \text{ °C}}}_{\text{heat lost by aluminum}} \qquad \underbrace{\phantom{0.382 \frac{J}{g \cdot °C} \times \text{g copper} \times 10 \text{ °C}}}_{\text{heat gained by copper}}$$

Solving this equation for grams of copper gives

$$\text{g copper} = \frac{0.908 \frac{J}{g \cdot °C} \times 245 \text{ g} \times 30 \text{ °C}}{0.382 \frac{J}{g \cdot °C} \times 10 \text{ °C}} = 1747.068 \text{ g} \quad \text{(calculator answer)}$$

$$= 1700 \text{ g} \quad \text{(correct answer)}$$

PRACTICE EXERCISE 3.21

How many grams of aluminum can be heated from 10 to 50 °C by the heat energy released when 125 g of copper cools from 45 to 25 °C? *Ans.* 26 g.

Similar exercises: Problems 3.109 and 3.110

The topic of heat energy and heat energy calculations will be revisited in Section 11.9 when the energetics for changes of physical state are considered.

Key Terms

The new terms or concepts defined in this chapter are

area (Sec. 3.4)   A measure of the extent of a surface.

calorie (Sec. 3.12)   The amount of heat energy needed to raise the temperature of 1 g of water by 1 °C measured between 14.5 and 15.5 °C.

chemical energy (Sec. 3.11)   Energy stored in a substance that can be released during a chemical change.

conversion factor (Sec. 3.6)   An equality, expressed in fractional form, that is used as a multiplier to convert a quantity in one unit into its equivalent in another unit.

density (Sec. 3.8)   Ratio of the mass of an object to the volume occupied by that object.

dimensional analysis (Sec. 3.7)   A general problem-solving method that uses the units associated with numbers as a guide in setting up the calculation.

energy (Sec. 3.11)   The capacity to do work.

gram (Sec. 3.3)   Basic unit of mass in the metric system.

heat capacity (Sec. 3.12)   The amount of heat needed to raise the temperature of a given quantity of a substance in a specific physical state by 1 °C.

joule (Sec. 3.12)   An energy unit obtained from the base units kilogram, meter, and second.

kinetic energy (Sec. 3.11)   Energy that matter possesses because of its motion.

law of conservation of energy (Sec. 3.11)   In any chemical or physical change, energy can be converted from one form to another, but it is neither created nor destroyed.

liter (Sec. 3.4)   The basic unit of volume in the metric system.

mass (Sec. 3.3)   A measure of the total quantity of matter in an object.

meter (Sec. 3.2)   The basic unit of length in the metric system.

percent (Sec. 3.9)   Parts per hundred, that is, the number of items of a specified type in a group of 100 items.

potential energy (Sec. 3.11) Stored energy that results from an object's position, condition, and/or composition.

specific heat (Sec. 3.12) The amount of heat needed to raise the temperature of 1 g of a substance in a specific physical state by 1 °C.

temperature (Sec. 3.10) A measure of the hotness or coldness of an object.

volume (Sec. 3.4) A measure of the amount of space occupied by an object.

weight (Sec. 3.3) A measure of the force exerted on an object by the pull of gravity.

Practice Problems

Metric System Units (Sec. 3.1–3.4)

3.1 Write the name of the metric system prefix associated with each of the following mathematical terms.
(a) 10^3 (b) 10^{-6}
(c) $1/1000$ (d) 10^{-2}

3.2 Write the name of the metric system prefix associated with each of the following mathematical terms.
(a) 10^{-3} (b) $1/100$
(c) 10^6 (d) 10^{-9}

3.3 What is the metric system abbreviation for each of the following units?
(a) millimeter   (b) microliter
(c) kilogram   (d) megagram

3.4 What is the metric system abbreviation for each of the following units?
(a) centimeter   (b) nanogram
(c) decaliter   (d) deciliter

3.5 Use the appropriate metric prefix abbreviation to replace the power of ten in each of the following values.
(a) 6.8×10^{-9} m   (b) 3.2×10^{-6} L
(c) 7.23×10^3 L   (d) 6.5×10^9 g

3.6 Use the appropriate metric prefix abbreviation to replace the power of ten in each of the following values.
(a) 4.1×10^{-3} L   (b) 9.9×10^{-12} g
(c) 8.721×10^{-2} g   (d) 4.4×10^6 m

3.7 Write out the names of the metric system units having the following abbreviations.
(a) cL   (b) μm   (c) Gm   (d) mg

3.8 Write out the names of the metric system units having the following abbreviations.
(a) mL   (b) pg   (c) dL   (d) km

3.9 What type of quantity (length, mass, area, or volume) do each of the following units represent?
(a) cm^3 (b) mm   (c) ML   (d) km^2

3.10 What type of quantity (length, mass, area, or volume) do each of the following units represent?
(a) L   (b) cm^2 (c) kg   (d) km^3

3.11 Arrange each of the following groups of units in an increasing size sequence, from smallest to largest.
(a) milligram, decigram, centigram
(b) kilometer, millimeter, micrometer
(c) cg, kg, ng
(d) GL, TL, pL

3.12 Arrange each of the following groups of units in an increasing size sequence, from smallest to largest.
(a) microliter, nanoliter, deciliter
(b) picometer, centimeter, decimeter
(c) hm, km, dm
(d) kg, Gg, Mg

3.13 For each of the pairs of units listed, indicate which quantity is larger.
(a) 1 centimeter, 1 inch   (b) 1 meter, 1 yard
(c) 1 gram, 1 pound   (d) 1 liter, 1 gallon

3.14 For each of the pairs of units listed, indicate which quantity is larger.
(a) 1 kilometer, 1 mile
(b) 1 milliliter, 1 fluid ounce
(c) 1 kilogram, 1 pound
(d) 1 liter, 1 quart

Units and Mathematical Operations (Secs. 3.4 and 3.5)

3.15 Carry out the following mathematical operations, being sure to treat the units in the proper mathematical manner.
(a) 8.2 mm × 2.4 mm   (b) 13.2 cm × 4.03 cm^2
(c) $\dfrac{5.0 \text{ km}^3}{3.0 \text{ km}}$ (d) $\dfrac{4.0 \text{ m}^3 \times 4.0 \text{ m}}{3.0 \text{ m}^2 \times 1.5 \text{ m}^2}$

3.16 Carry out the following mathematical operations being sure to treat the units in the proper mathematical manner.
(a) 6.0 cm^2 × 5.0 cm^2 (b) 1.3 mm × 3.4 mm^3
(c) $\dfrac{9.7 \text{ km}^3}{9.3 \text{ km}^2}$ (d) $\dfrac{3.3 \text{ cm}^2 \times 4.1 \text{ cm}}{9.5 \text{ cm}^2 \times 1.2 \text{ cm}}$

3.17 Calculate the area of the following surfaces.
  (a) a square surface whose side is 4.52 cm
  (b) a rectangular surface whose width is 3.5 m and whose length is 9.2 m
  (c) a circle whose radius is 4.579 mm
  (d) a triangle whose height is 3.0 mm and whose base is 5.5 mm

3.18 Calculate the area of the following surfaces.
  (a) a rectangular surface whose dimensions are 24.3 m and 32.1 m
  (b) a circle of radius 2.7213 cm
  (c) a triangular object whose base is 12.0 mm and whose height is 8.00 mm
  (d) a square surface with sides of 6.7 cm

3.19 Calculate the volume of each of the following objects, each of which has a regular geometrical shape.
  (a) a copper bar 5.4 cm long, 0.52 cm high, and 3.4 cm wide
  (b) a cylindrical piece of cheese that has a height of 7.5 mm and a radius of 2.4 mm
  (c) a spherical piece of styrofoam with a radius of 87.32 mm
  (d) a piece of gold in the shape of a cube whose edge is 7.2 cm

3.20 Calculate the volume of each of the following objects, each of which has a regular geometrical shape.
  (a) a cube of cheese whose edge is 3.5175 mm
  (b) a spherical marble with a radius of 1.212 cm
  (c) a bar of iron 6.0 m long, 0.10 m wide, and 0.20 m high
  (d) a cylindrical rod of copper whose length is 62 mm and whose radius is 3.2 mm

Conversion Factors (Sec. 3.6)

3.21 Explain what is meant by the statement "metric–metric conversion factors are exact."

3.22 Explain what is meant by the statement "metric–English conversion factors are not exact."

3.23 Give both forms of the conversion factor that you would use to relate the units of each set to each other.
  (a) 1 milligram and grams
  (b) 1 kiloliter and liters
  (c) 1 pint and liters
  (d) 1 inch and centimeters

3.24 Give both forms of the conversion factor that you would use to relate the units of each set to each other.
  (a) 1 micrometer and meters
  (b) 1 nanogram and grams
  (c) 1 quart and liters
  (d) 1 ounce and grams

3.25 Indicate how each of the following conversion factors should be interpreted in terms of significant figures present.
  (a) $\dfrac{1.609 \text{ km}}{1 \text{ mi}}$
  (b) $\dfrac{10^{-2} \text{ m}}{1 \text{ cm}}$
  (c) $\dfrac{28.34 \text{ g}}{1 \text{ oz}}$
  (d) $\dfrac{12 \text{ in.}}{1 \text{ ft}}$

3.26 Indicate how each of the following conversion factors should be interpreted in terms of significant figures present.
  (a) $\dfrac{2.54 \text{ cm}}{1 \text{ in.}}$
  (b) $\dfrac{453.6 \text{ g}}{1 \text{ lb}}$
  (c) $\dfrac{2.113 \text{ pt}}{1 \text{ L}}$
  (d) $\dfrac{10^{-9} \text{ m}}{1 \text{ nm}}$

Dimensional Analysis—Conversions Within Unit Systems (Sec. 3.7)

3.27 A nickel has a mass of 4.832 g. Express this mass in the following units.
  (a) micrograms   (b) gigagrams
  (c) kilograms   (d) centigrams

3.28 A standard basketball has a volume of 7.47 L. Express this volume in the following units.
  (a) milliliters   (b) nanoliters
  (c) kiloliters   (d) megaliters

3.29 Convert each of the following measurements to kilometers.
  (a) 0.00312 Mm   (b) 33 nm
  (c) 13,000 cm   (d) 3×10^8 mm

3.30 Convert each of the following measurements to milliliters.
  (a) 33,000 μL   (b) 33,000 cL
  (c) 0.0357 kL   (d) 4.37 nL

3.31 For each of the following sets of measurements, determine which quantity is larger.
  (a) 300 mL or 3 cL
  (b) 3 km or 3×10^5 mm
  (c) 3 cg or 3×10^6 μg
  (d) 30 mg or 3×10^{-3} kg

3.32 For each of the following sets of measurements, determine which quantity is larger.
  (a) 300 mg or 0.00003 kg
  (b) 0.03 mL or 3,000 nL
  (c) 3 Mg or 300 kg
  (d) 3,000,000 μm or 30 cm

3.33 The Earth has an estimated mass of 6.6×10^{21} tons. What is the mass of the Earth in ounces?

3.34 The length of a football field, between goal lines, is 100.0 yards. What is this length in inches?

3.35 The current speed limit on rural interstate highways in the United States is 65 mi/hr. What is the speed in inches per second?

3.36 The speed of light is 1.18×10^{10} in./sec. What is this speed in feet per hour?

3.37 The concentration of sugar in a solution is found to be 2.30 μg/L. What is the sugar concentration in grams per milliliter?

3.38 The concentration of salt in a solution is found to be 4.5 mg/mL. What is this salt concentration in micrograms per liter?

Dimensional Analysis—Conversions Between Unit Systems (Sec. 3.7)

3.39 The diameter of a human hair is about 0.040 mm. What is this diameter in inches?

3.40 The distance from Earth to the moon is 3.9×10^5 km. What is this distance in feet?

3.41 An individual has a mass of 87.5 kg. What is the person's mass in pounds?

3.42 A 100-mL sample of blood is found to contain 9.7 mg of calcium. How many ounces of calcium are present in this sample of blood?

3.43 What volume of water, in gallons, would be required to fill a 5435-mL container?

3.44 What volume of gasoline, in liters, would be required to fill a 17.0-gal gasoline tank?

3.45 A regular-issue U.S. postage stamp is 2.1 cm wide and 2.5 cm long. Express the surface area of this postage stamp in (a) square centimeters and (b) square inches.

3.46 A rectangular piece of concrete has dimensions of 3.6 m and 1.2 m. Express its surface area in (a) square meters and (b) square yards.

3.47 The luggage compartment of an automobile has the dimensions 95 cm × 105 cm × 145 cm. What is the volume of this compartment in cubic feet?

3.48 A copper bar is 65 cm long, 3.0 cm high, and 4.0 cm wide. What is the volume of this bar in cubic inches?

3.49 Using dimensional analysis, convert the following measurements to gallons.
(a) 4.67 L   (b) 4.670 L
(c) 4.6700 L   (d) 4.67000 L

3.50 Using dimensional analysis, convert the following measurements to pounds.
(a) 4.67 g   (b) 4.670 g
(c) 4.6700 g   (d) 4.67000 g

Density (Sec. 3.8)

3.51 A sample of mercury is found to have a mass of 524.5 g and a volume of 38.72 cm³. What is its density in grams per cubic centimeter?

3.52 A sample of sand is found to have a mass of 12.0 g and a volume of 2.69 cm³. What is its density in grams per cubic centimeter?

3.53 A small bottle contains 2.171 mL of a red liquid. The total mass of the bottle and liquid is 5.261 g. The empty bottle weighs 3.006 g. What is the density, in grams per milliliter, of the liquid?

3.54 A piece of metal weighing 187.6 g is placed in a graduated cylinder containing 225.2 mL of water. The combined volume of solid and liquid is 250.3 mL. From these data, calculate the density, in grams per milliliter, of the metal.

3.55 Calculate the density of table salt, in grams per milliliter, given that 1.000 lb of salt occupies a volume of 209.7 cm³.

3.56 Calculate the density of table sugar, in grams per cubic centimeter, given that 4.000 mL of sugar weighs 1.400×10^{-2} lb.

3.57 Acetone, the solvent in nail polish remover, has a density of 0.791 g/mL. What is the mass, in grams, of 60.0 mL of acetone?

3.58 The density of homogenized milk is 1.03 g/mL. How much does 1 cup (236 mL) of homogenized milk weigh, in grams?

3.59 A sample of dry air is found to have a density of 1.29 g/L. What is the volume, in liters, of a 35.0 g sample of this air?

3.60 The density of silver is 10.40 g/cm³. What will be the volume, in cubic centimeters, of a pure silver bar whose mass is 100.0 g?

3.61 An automobile gasoline tank holds 13.0 gal when full. How many pounds of gasoline will it hold, if the gasoline has a density of 0.56 g/mL?

3.62 Liquid sodium metal has a density of 0.93 g/cm³. How many pounds of liquid sodium are needed to fill a container whose capacity is 15.0 L?

3.63 What mass of the metal chromium (density = 7.18 g/cm³) occupies the same volume as 100.0 g of aluminum (density = 2.70 g/cm³)?

3.64 What volume of the metal nickel (density = 8.90 g/cm³) has the same mass as 100.0 cm³ of lead (density = 11.4 g/cm³)?

Percentage as a Conversion Factor (Sec. 3.9)

3.65 How many grams of water are contained in 65.3 g of a mixture of alcohol and water that is 34.2% water by mass?

3.66 How many grams of alcohol are contained in 467 g of a mixture of alcohol and water that is 23.0% alcohol by mass?

3.67 Consider the following facts about a candy mixture containing "Gummi bears" and "Gummi worms":
(1) 30.9% of the 661 items present are Gummi bears;
(2) 23.0% of the Gummi bears are orange; and (3) 6.4% of the orange Gummi bears have only one ear. How many one-eared orange Gummi bears are present in the candy mixture?

3.68 An analysis of the makeup of a beginning chemistry class gives the following facts: (1) 47.1% of the 87 students are female; (2) 43.9% of the female students are married; and (3) 33.3% of the married female students are sophomores. How many students in the class are sophomore female students who are married?

3.69 A solution of table salt in water contains 15.3% by mass of table salt. If 437 g of solution is evaporated to dryness, how many grams of table salt will remain?

3.70 A solution of table salt in water contains 15.3% by mass of table salt. In a 542-g sample of this solution, how many grams of water are present?

Temperature Scales (Sec. 3.10)

3.71 On a hot summer day the temperature outside may reach 105 °F. What is this temperature in degrees Celsius?

3.72 A recommended temperature setting for household hot-water heaters is 140 °F. What is this temperature in degrees Celsius?

3.73 Oxygen, the gas necessary to sustain life, freezes to a solid at −218.4 °C. What is this temperature in degrees Fahrenheit?

3.74 The body temperature for a hypothermia victim is found to have dropped to 29.1 °C. What is this temperature in degrees Fahrenheit?

3.75 Convert each of the following temperature readings to the Kelvin scale.
(a) 231 °C (b) 231.7 °C
(c) 231.74 °C (d) 37.3 °F

3.76 Convert each of the following temperature readings to the Kelvin scale.
(a) 137 °C (b) 137.2 °C
(c) 137.23 °C (d) 79.0 °F

3.77 Birds have higher body temperatures than humans. For example, the chickadee maintains a body temperature of 41.0 °C. How many Fahrenheit degrees higher than normal human body temperature (98.6 °F) is the body temperature of the chickadee?

3.78 Mercury freezes at −38.9 °C. What is the coldest temperature, in degrees Fahrenheit, that can be measured using a mercury thermometer?

3.79 Which is the higher temperature, −10 °C or 10 °F?

3.80 Which is the higher temperature, −15 °C or 4 °F?

3.81 Which is the lower temperature, 223 K or −60 °F?

3.82 Which is the lower temperature, 381 K or 98 °C?

Types and Forms of Energy (Sec. 3.11)

3.83 What is the scientific definition for energy?

3.84 State the law of conservation of energy.

3.85 List the predominant *forms* of energy produced when each of the following processes occurs.
(a) An electric light bulb is turned on.
(b) A log is burned in a fireplace.
(c) A green plant grows.
(d) A burner on an electric stove is turned on.

3.86 List the predominant *forms* of energy produced when each of the following processes occurs.
(a) A bicycle is pedaled.
(b) A flashlight is turned on.
(c) A photographer's flash bulb goes off.
(d) A candle is lighted.

3.87 What is the difference between potential energy and kinetic energy?

3.88 What is the difference between potential energy and chemical energy?

3.89 Identify the principal *type* of energy (kinetic or potential) that is exhibited by each of the following.
(a) a car parked on a hill
(b) a car traveling at 65 mi/hr
(c) an elevator stopped at the 35th floor
(d) water behind a dam

3.90 Identify the principal *type* of energy (kinetic or potential) that is exhibited by each of the following.
(a) a piece of coal
(b) a falling rock
(c) a compressed metal spring
(d) a rolling soccer ball

Heat Energy and Specific Heat (Sec. 3.12)

3.91 What is the relationship between a joule and a calorie?

3.92 What is the relationship between a kilojoule and a kilocalorie?

3.93 The energy content of a chicken sandwich is found to be 2290 kJ. Specify this amount of energy in
(a) joules (b) kilocalories
(c) calories (d) Calories

3.94 The energy content of a quarter-pound hamburger is found to be 211,000 J. Specify this amount of energy in
(a) kilojoules (b) kilocalories
(c) calories (d) Calories

3.95 The same quantity of heat is absorbed by equal masses of iron and gold that are at the same temperature. Use Table 3.4 to determine which metal will have the higher temperature after the heat absorption.

3.96 The same quantity of heat is absorbed by equal masses of silver and aluminum that are at the same temperature. Use Table 3.4 to determine which will have the lower temperature after the heat absorption.

3.97 With the help of Table 3.4, calculate the amount of heat added or removed to effect the following changes.
(a) raise the temperature of a 48.0-g aluminum bar from 25.0 to 55.0 °C
(b) raise the temperature of 2500 g of liquid ethyl alcohol from 70 to 85 °C
(c) decrease the temperature of 50.0 g of nitrogen gas from 75 to 43 °C
(d) decrease the temperature of 25.0 g of water (ice) from −5 to −17 °C

3.98 With the help of Table 3.4, calculate the amount of heat added or removed to effect the following changes.
(a) raise the temperature of a 50.0-g copper bar from 23.3 to 45.5 °C
(b) raise the temperature of 1200 g of liquid water from 25 to 37 °C
(c) decrease the temperature of 2.00 g of oxygen gas from −152 to −167 °C
(d) decrease the temperature of 37.5 g of water (steam) from 112 to 101 °C

3.99 What will be the temperature change, in Celsius degrees, if 145 J of heat energy is added to 40.0 g of solid aluminum?

3.100 What will be the temperature change, in Celsius degrees, if 175 J of heat energy is added to 60.0 g of solid silver?

3.101 What is the mass, in grams, of a piece of aluminum if its temperature changes from 20.0 to 315°C when it absorbs 422 J of heat energy?

3.102 What is the mass, in grams, of a piece of copper if its temperature changes from 20.0 to 55.6 °C when it absorbs 22.7 J of heat energy?

3.103 What is the heat capacity of 40.0 g of each of the following substances?
(a) solid gold (b) solid copper
(c) gaseous oxygen (d) liquid water

3.104 What is the heat capacity of 75.2 g of each of the following substances?
(a) solid silver (b) solid iron
(c) gaseous nitrogen (d) gaseous water

3.105 Calculate the specific heat, in joules per gram degree Celsius, for each of the following metals from the given data.
(a) 79.0 g of metal has a heat capacity of 35.1 J/°C
(b) 582 g of metal has a heat capacity of 139 J/°C

3.106 Calculate the specific heat, in joules per gram degree Celsius, for each of the following metals from the given data.
(a) 1.57 g of metal has a heat capacity of 0.20 J/°C
(b) 75.0 g of metal has a heat capacity of 28.7 J/°C

3.107 Calculate the specific heat of a metal given that 46.9 J of heat is required to raise the temperature of 40.0 g of the metal by 3.0 °C.

3.108 Calculate the specific heat of a metal given that 221 J of heat is required to raise the temperature of 55.0 g of the metal by 4.0 °C.

3.109 How many grams of copper can be heated from 30 to 50 °C when 1.0 g of liquid water cools from 100 to 15 °C?

3.110 How many grams of aluminum can be heated from 20 to 45 °C when 20.0 g of oxygen gas cools from 285 to 15 °C?

Additional Problems

3.111 Five blueberries of differing sizes have the following masses: 375 mg, 0.500 g, 0.000200 kg, 25.0 cg, and 1.00 dg. What is the combined mass, in grams, of the five blueberries?

3.112 Which of the following has (a) the smallest mass and (b) the largest mass? (1) a 1400-mg cockroach, (2) a 0.0022-kg bumble bee, (3) a 1,100,000-μg grasshopper, or (4) a 0.0000030-Mg giant ant.

3.113 Are you exceeding the speed limit if you are driving 102 km/hr in a 65-mi/hr zone?

3.114 A motorcycle averages 195 mi/gal of fuel consumed. What is this fuel consumption in kilometers per gallon?

3.115 Light traveling at a speed of 3.0×10^{10} cm/sec takes 8.3 min to travel from the sun to Earth. Calculate the distance, in miles, from the sun to Earth.

3.116 How many seconds would it take to send radio signals from Earth to the moon—a distance of 2.4×10^5 mi— if the signals travel at the speed of light, which is 3.0×10^{10} cm/sec?

3.117 If your blood has a density of 1.05 g/mL at 20 °C, how many grams of blood would you lose if you donated exactly 1 pint of blood?

3.118 If your urine has a density of 1.030 g/mL at 20 °C, how many pounds of urine would you lose if you eliminated exactly $\frac{1}{2}$ pint of urine?

3.119 A certain mix of concrete is found to have a density of 2.9 g/cm^3. What would be the mass, in pounds, of 1.0 yd^3 of this concrete?

3.120 A particular bone fragment has a density of 1.82 g/cm^3. What would be the volume, in cubic inches, of this fragment if it has a mass of 41.3 g?

3.121 A copper rod has a radius of 0.75 cm and a volume of 75.6 cm^3. How long, in inches, is the copper rod?

3.122 A rectangular bar of lead has a cross-sectional area of 4.34 mm^2 and a volume of 37.2 cm^3. How long, in feet, is the iron bar?

3.123 A certain brand of household disinfectant contains 2.9% by mass of active ingredient. How many grams of active ingredient are present in 200.0 mL of disinfectant that has a density of 1.09 g/mL?

3.124 Gold jewelry is usually made of 14-karat gold, a gold alloy that is 58.33% gold by mass. What would be the mass of gold, in grams, present in a 50.0 cm^3 sample of 14-karat gold of density 14.9 g/cm^3?

3.125 Assume that a milliliter of water contains 20 drops. How long, in hours, will it take you to count the number of drops of water in 1.0 gal of water at a counting rate of 10 drops/sec?

3.126 Seedless grapes are on sale in a grocery store for 49¢/lb. It is found that there are, on average, 205 grapes per kilogram. How much money, in dollars, would you have to pay to purchase 1 million (1.00×10^6) grapes?

3.127 Levels of blood glucose higher than 400 mg/dL are life threatening. Is either of the following laboratory-measured glucose levels life threatening?
(a) 5000 µg/mL   (b) 0.5 g/L

3.128 Levels of blood glucose lower than 40 mg/dL are life threatening. Is either of the following laboratory-measured glucose levels life threatening?
(a) 2000 µg/L   (b) 20,000,000 ng/cL

3.129 A sheet of $8\frac{1}{2} \times 11$ in. typewriter paper is found to have a thickness of 0.10 mm and a mass of 4.72 g. What is the density, in grams per cubic centimeter, of the typewriter paper?

3.130 An economy-size package of aluminum foil contains 202 ft^2 of aluminum foil, and the foil has a mass of 32 oz. Assuming a density of 2.70 g/cm^3 for the foil, what is its thickness, in millimeters?

3.131 A sample of metal weighing 183 g is heated to 143.0 °C and then dropped in 342 g of water at 27.0 °C. If the final temperature of the water and metal is 37.6 °C, what is the specific heat of the metal? Assume that no heat is lost to the surroundings.

3.132 A sample of metal weighing 204 g is heated to 167.2 °C and then dropped in 413 g of water at 35.0 °C. If the final temperature of the water and metal is 47.4 °C, what is the specific heat of the metal? Assume that no heat is lost to the surroundings.

3.133 The body contains approximately 5.7 L of blood. Assuming that the density of blood is 1.06 g/mL and that the specific heats of blood and water are the same, how many kilojoules of energy are required to raise the temperature of this amount of blood by 1.0 °C?

3.134 Assuming that the density of blood is 1.06 g/mL and that the specific heats of blood and water are the same, how many liters of blood are present in a human body if 36.0 kJ of energy raises the temperature of the blood present by 1.5 °C?

Cumulative Problems

3.135 A sample of a colorless liquid has a mass of 2 g and a volume of 4 mL. Calculate the liquid's density using the following precision specifications and express your answers in scientific notation.

(a) 2.000 g and 4.00 mL
(b) 2.00 g and 4.0 mL
(c) 2.0000 g and 4.0000 mL
(d) 2.000 g and 4.0000 mL

3.136 A 1-g sample of a powdery white solid is found to have a volume of 2 cm³. Calculate the solid's density using the following precision specifications and express your answers in scientific notation.
(a) 1.0 g and 2.0 cm³
(b) 1.000 g and 2.00 cm³
(c) 1.0000 g and 2.0000 cm³
(d) 1.000 g and 2.0000 cm³

3.137 A small rectangular box measures 10 cm wide, 200 cm long, and 4 cm high. Calculate the volume of the box, in cubic centimeters, given that all of the dimensions are known to
(a) the closest centimeter
(b) the closest tenth of a centimeter
(c) the closest hundredth of a centimeter
(d) two significant figures

3.138 A small rectangular room has a width of 9 m and a length of 21 m. Calculate the area of this room, in square meters, given that both dimensions are known to
(a) the closest meter
(b) the closest tenth of a meter
(c) the closest hundredth of a meter
(d) three significant figures

3.139 A container contains 23 *identical* marbles, each with a mass of *exactly* one g. Calculate the total mass of these marbles, in pounds, to the following number of significant figures. Express all answers in scientific notation.
(a) two   (b) four   (c) six   (d) eight

3.140 Six *identical* ball bearings are found to have a total mass of *exactly* ten g. Calculate the mass of a single ball bearing, in pounds, to the following number of significant figures. Express all answers in scientific notation.
(a) two   (b) four   (c) six   (d) eight

3.141 The measurements 325.0 cm, 3.2500 m, and 0.00325 km are equivalent ways of expressing the same length. Which of the measurements is the most precise?

3.142 The measurements 35.300 mL, 3.530 cL, and 0.0353 L are equivalent ways of expressing the same volume. Which of the measurements is the most precise?

Grid Problems

3.143 Select from the grid *all* correct responses for each situation.

1. d	2. M	3. c
4. k	5. p	6. h
7. m	8. μ	9. n

(a) metric system prefixes that increase the size of the base unit
(b) metric system prefixes whose power of ten value is greater than 10^2
(c) metric system prefixes whose power of ten value is 10^{-6} or less
(d) pairs of metric system prefixes whose magnitude differs by 6 orders of magnitude
(e) pairs of metric system prefixes whose magnitude differs by a factor of one million
(f) pairs of metric system prefixes whose magnitudes are the reciprocal of each other (e.g. 100 and $\frac{1}{100}$)

3.144 Select from the gird *all* correct responses for each situation.

1. $\dfrac{4.184 \text{ J}}{1 \text{ cal}}$	2. $\dfrac{4.82 \text{ g}}{1 \text{ mL}}$	3. $\dfrac{24 \text{ hr}}{1 \text{ day}}$
4. $\dfrac{1 \text{ ft}}{12 \text{ in.}}$	5. $\dfrac{10^{-3} \text{ L}}{1 \text{ mL}}$	6. $\dfrac{453.6 \text{ g}}{1 \text{ lb}}$
7. $\dfrac{1 \text{ in.}}{2.54 \text{ cm}}$	8. $\dfrac{5\,C°}{9\,F°}$	9. $\dfrac{1 \text{ km}}{10^3 \text{ m}}$

(a) conversion factors that are exact
(b) conversion factors that do not involve the English system of units.
(c) conversion factors that have a specific rather than general validity
(d) conversion factors that would limit a calculation to three significant figures
(e) conversion factors that involve an equivalence rather than an equality
(f) conversion factors that when used as written would take you from the metric to English system of units

3.145 Select from the grid *all* correct responses for each situation.

1. g/mL	2. m³	3. J/(g · °C)
4. mL	5. lb/ft³	6. in.²
7. L²	8. kg · m²/sec²	9. J/°C

(a) units (in total) that specify volume
(b) units (in total) that specify density
(c) units (in total) that specify specific heat
(d) units (in total) that specify energy
(e) units (in total) that specify area
(f) units (in total) that specify heat capacity

3.146 Select from the grid *all* corret responses for each situation.

1. $\dfrac{10^{-3} \text{ g}}{1 \text{ mg}}$	2. $\dfrac{10^{-6} \text{ g}}{1 \text{ μg}}$	3. $\dfrac{10^{-9} \text{ g}}{1 \text{ ng}}$
4. $\dfrac{1 \text{ mg}}{10^{-3} \text{ g}}$	5. $\dfrac{1 \text{ μg}}{10^{-6} \text{ g}}$	6. $\dfrac{1 \text{ ng}}{10^{-9} \text{ g}}$
7. $\dfrac{1 \text{ g}}{10^{-12} \text{ pg}}$	8. $\dfrac{1 \text{ g}}{10^{2} \text{ cg}}$	9. $\dfrac{1 \text{ kg}}{10^{3} \text{ g}}$

(a) conversion factors that are incorrect as written
(b) conversion factors that relate the units of mg and g
(c) conversion factors that when used as written would decrease mass unit size
(d) conversion factors that would be needed in making the change g → ng
(e) conversion factors that would be needed in making the change μg → g → ng
(f) conversion factors that would change unit size by a factor of a thousand

CHAPTER FOUR

Basic Concepts About Matter

4.1 Chemistry—The Study of Matter

Chemistry is the branch of science that is concerned with matter and its properties. What is matter? What is it that chemists study?

Intuitively, most people have a general feeling for the meaning of the word matter. They consider matter to be the materials of the physical universe—that is, the "stuff" from which the universe is made. Such an interpretation is a correct one.

Formally defined, **matter** is anything that has mass and occupies space. A substance need not be visible to the naked eye to be labeled as matter as long as it meets the two qualifications of having mass and occupying space. Wood, paper, stone, the food we eat, the air we breathe, the fluids we drink, our bodies themselves, our clothing, and our shelter are all examples of matter.

The question may be asked, "What, then, does matter not include?" Various forms of energy, such as heat, light, and electricity, are not included. Neither are wisdom, friendship, ideas, thoughts, and emotions (such as anger and love) included.

Consider the extraordinary breadth of our definition of matter. This concept encompasses all objects we know about from the largest objects in outer space to the most minute objects seen under a microscope to objects so small they cannot be seen with any known type of instrumentation. Individuals first encountering this breadth automatically suppose that the study of matter will be a most complicated subject because of the literally millions of different types of matter that exist. One purpose of this chapter is to show that all matter can be classified into a surprisingly small number of categories. The naturally occurring materials of the universe and the synthetic materials humans have fashioned from them are, indeed, much simpler in makeup than they outwardly appear.

4.2 Physical States of Matter

All matter can be conveniently classified into three physical states: solid, liquid, and gas. This classification scheme, familiar to everyone, relates to the definiteness of the shape and volume of the sample of matter. Matter in the **solid state** is found always to have a *definite* shape and a *definite* volume. In powdered or granular solids each individual particle has a definite shape and a definite volume; however, a quantity of such a solid can take the shape of its container because of the small particle size. The **liquid state** is recognized by its properties of an *indefinite* shape and a *definite* volume. A liquid always takes the shape of its container to the extent it fills it. The **gaseous state** is characterized by both an *indefinite* shape and an *indefinite* volume. A gas always completely fills its container, adopting both its volume and shape. Figure 4.1 summarizes the shape and volume characteristics of the three physical states of matter.

In nature we find matter in all three physical states. Rocks and minerals are in the solid state, water and petroleum are usually in the liquid state, and air, a mixture of many gases, represents the gaseous state.

The state of matter observed for a particular substance is always dependent on the temperature and pressure under which the observation is made. Because we live on a planet characterized by relatively narrow temperature extremes, we tend to fall into

(a)

(b)   (c)

FIGURE 4.1 Shape and volume characteristics of the three states of matter. (a) A solid has its own shape and volume, which are independent of the shape of the container; (b) a liquid takes the shape of its container, but has a volume independent of the container volume; (c) a gas takes the shape of its container and also always completely fills the container.

Figure 4.2 During the making of steel, temperatures are high enough for the metal iron (the main ingredient in steel) to be in the liquid state. [Inland Steel Co. photo]

the error of believing that the commonly observed states of substances are the only states in which they occur. Under laboratory conditions, states other than the "natural" ones can be attained for almost all substances. Oxygen, which is nearly always thought of as a gas, can be obtained in the liquid and solid states at very low temperatures. People seldom think of the metal iron as being a gas, its state at extremely high temperatures (above 3000 °C). At intermediate temperatures (1535–3000 °C) iron is a liquid (see Fig. 4.2). Water is one of the very few substances familiar to everyone in all three of its physical states: solid ice, liquid water, and gaseous steam.

Not all matter can exist in all three physical states as do oxygen, iron, and water. Some types of matter decompose when an attempt is made to change their physical states. For example, paper does not exist in the liquid state. Attempts to change paper to a liquid through heating always result in charring and decomposition.

Chapter 11 will consider in detail further properties of the different physical states of matter, changes from one state to another, and the question of why some substances decompose. Suffice it to say at present that physical state is one of the qualities by which matter can be classified.

4.3 Properties of Matter

How are the various kinds of matter differentiated from each other? The answer is simple—by their properties. **Properties** are the distinguishing characteristics of a substance used in its identification and description. Just as we recognize a friend by characteristics such as hair color, walk, tone of voice, or shape of nose, we recognize various chemical substances by how they look and behave. Each chemical substance has a unique set of properties that distinguishes it from all other substances. If two samples of pure materials have identical properties, they must necessarily be the same substance.

Knowledge of the properties of substances is useful in:

1. *Identifying an unknown substance.* Identifying a confiscated drug as cocaine involves comparing the properties of the drug to those of known cocaine samples.
2. *Distinguishing between different substances.* A dentist can quickly tell the difference between a real tooth and a false tooth because of property differences.
3. *Characterizing a newly discovered substance.* Any new substance must have a unique set of properties different from those of any previously characterized substance.
4. *Predicting the usefulness of a substance for specific applications.* Water-soluble substances obviously should not be used in the manufacture of bathing suits.

There are two general categories of properties of matter: physical and chemical. **Physical properties** are properties that can be observed without changing a substance into another substance. Color, odor, taste, size, physical state, boiling point, melting point, and density are all examples of physical properties.

The physical appearance of a substance may change while a physical property is being determined, but the substance's identity will not. For example, the melting point of a solid cannot be measured without melting the solid, changing it to a liquid. Although the liquid's appearance is much different from that of the solid, the substance is still the same. Its chemical identity has not changed. Hence, melting point is a physical property.

Chemical properties are properties that matter exhibits as it undergoes changes in chemical composition. Most often such composition changes result from the interaction (reaction) of the matter with other substances, or they may simply occur in the presence of heat or light (a process called decomposition). When copper objects are exposed to moist air for long periods of time, they turn green; this is a chemical property of copper. The green coating formed on the copper is a new substance; it results from the reaction of copper metal with the oxygen, carbon dioxide, and water in air. The chemical properties of this green coating are very different from those of metallic copper. The substance hydrogen peroxide, in the presence of either heat or light decomposes into the substances water and oxygen.

The *failure* of a substance to undergo change in the presence of heat, light, or another substance is also considered a chemical property. Flammability and non-flammability are both chemical properties.

Most often, in describing chemical properties, conditions such as temperature and pressure are specified because they can and do influence interactions between two or more substances. For example, two substances may interact explosively at an elevated temperature and yet not interact at all at room temperature.

Selected physical and chemical properties of water are contrasted in Table 4.1. Note how the chemical properties of water cannot be described without reference to other substances. It does not make sense to say simply that a substance reacts. The substance that it interacts with must be specified because it might interact with many different substances.

TABLE 4.1 Selected Physical and Chemical Properties of Water

Physical Properties	Chemical Properties
1. Colorless	1. Reacts with bromine to form a mixture of two acids
2. Odorless	2. Can be decomposed by means of electricity to form hydrogen and oxygen
3. Boiling point = 100 °C	3. Reacts vigorously with the metal sodium to produce hydrogen
4. Freezing point = 0 °C	4. Does not react with gold even at high temperatures
5. Density = 1.000 g/mL at 4 °C	5. Reacts with carbon monoxide at elevated temperatures to produce carbon dioxide and hydrogen

4.4 Changes in Matter

Changes in matter are common and familiar occurrences. Many occur spontaneously, independent of human influence. Others must be forced to occur. Representative of the wide variety of known change processes are melting of snow, digestion of food, burning of wood, detonation of dynamite, rusting of iron, and sharpening of a pencil.

Like properties (Sec. 4.3), changes in matter can be classified as physical or chemical. A **physical change** is a process that does not alter the basic nature (chemical composition) of the substance under consideration. No new substance is ever formed as a result of a physical change. A **chemical change** is a process that involves a change in the basic nature (chemical composition) of the substance. Such changes always involve conversion of the material or materials under consideration into one or more new substances with distinctly different properties and composition from those of the original materials (see Fig. 4.3).

The comparison of two sources of light—a standard light bulb and a flash cube—illustrates nicely the fundamental difference between physical and chemical changes.

FIGURE 4.3 Chemical changes produce new substances that often have properties very different from the starting material. The strong steel with which this automobile fender was made has been converted by oxidation to powdered rust. [Photo by John Schultz, PAR/NYC]

When an electric light bulb is turned on, light is emitted by a glowing filament inside the bulb. When the electricity is turned off, the filament ceases to glow. The filament almost instantaneously reacquires the characteristics (appearance) it had prior to its use in supplying light. This process can be repeated numerous times with the same results. The composition of the filament is unchanged—the characteristic of physical changes. Contrast this situation with that of a throw-away flash cube for a camera. Prior to use, a small filament is present within the flash cube. After use, a white powdery substance has been formed from the filament. The bulb will not flash again. The filament no longer exists in its original form; a chemical change has occurred.

A change in physical state is the most common type of physical change. The melting of ice, the freezing of liquid water, the conversion of liquid water into steam (evaporation), the condensation of steam to water, the sublimation of ice in cold weather, and the formation of snow crystals in clouds in the winter (deposition) all represent changes of state. The terminology used in describing changes of state, with the exception of the terms sublimation and deposition, should be familiar to almost everyone. Although the processes of sublimation and deposition—going from a solid directly to the gaseous state or vice versa—are not common, they are encountered in everyday life. Dry ice sublimes, as do mothballs placed in a clothing storage area. As mentioned previously, ice or snow forming in clouds is an example of deposition. Figure 4.4 summarizes the terminology used in describing changes of state.

In any change of state, the composition of the substance undergoing change remains the same even though its physical state and outward appearance have changed. The melting of ice does not produce a new substance. The substance is water before and after the change. Similarly, the steam produced from boiling water is still water. Changes such as these illustrate that matter can change in appearance without undergoing a change in chemical composition.

Changes in size, shape, and state of subdivision are examples of physical changes that are not changes of state. Pulverizing a lump of coal into a fine powder and tearing a piece of aluminum foil into small pieces are physical changes that involve only the solid state.

The creation of one or more new substances is always a characteristic of a chemical change. Carbon dioxide and water are two new substances produced when the chemical change associated with the burning of gasoline occurs. Ashes are among the new

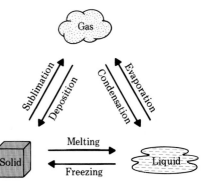

FIGURE 4.4 Terminology associated with changes of state.

TABLE 4.2 Classification of Changes as Physical or Chemical

Change	Classification
Rusting of iron	chemical
Melting of snow	physical
Sharpening a pencil	physical
Digesting food	chemical
Taking a bite of food	physical
Burning gasoline	chemical
Slicing an onion	physical
Detonation of dynamite	chemical
Souring of milk	chemical
Breaking of glass	physical

substances produced when wood is burned. Chemical changes are often called chemical reactions. A **chemical reaction** is a process in which at least one new substance is produced as a result of chemical change.

Table 4.2 classifies a number of changes for matter as being either physical or chemical.

Most changes for matter can easily be classified as either physical or chemical. However, not all changes are "black" or "white." There are some "gray" areas. For example, the formation of certain solutions falls in the "gray" area. Common salt dissolves easily in water to form a solution of saltwater. The salt can easily be recovered by the physical process of evaporating the water. When gaseous hydrogen chloride is dissolved in water, again a solution results; but in this case the starting materials cannot be recovered by evaporation. The formation of saltwater is considered a physical change because the original components can be recovered in an unchanged form using physical methods. The second solution presents classification problems because of the possibility that a chemical reaction took place.

The changes involved in the cooking of an egg also present classification problems. The cooked egg contains the same structural units as the uncooked egg. However, some changes in structural arrangement have taken place, so, is the change physical or chemical? Despite the existence of "gray" areas, we shall continue to use the concepts of physical and chemical change because their usefulness far outweighs the problems created by a few exceptions.

The difference between the word *change* (the topic of this section) and the word *property* (Sec. 4.3) must be kept clear. A change always involves a transformation from one form to another. The new form may or may not be a new substance depending on whether the change was chemical or physical. Properties distinguish one substance from other substances.

Note that the term *physical*, when used to modify another term, as in physical property or physical change, always conveys the idea that the composition of the substances involved does not change. Similarly, the term *chemical* is always associated with the concept of change in composition.

The terms *physical* and *chemical* are commonly used to qualify the meaning of general scientific terms. For example, techniques used to accomplish physical change are called physical methods or physical means. Chemical methods and chemical means are used to bring about chemical change. A physical separation would be a separation process where none of the components experienced composition changes. Composition changes would be part of a chemical separation process. The message of the "modifiers" physical and chemical is constant: *Physical* denotes no change in composition, and *chemical* denotes change in composition.

4.5 Mixtures and Pure Substances

All samples of matter can be divided into two categories: mixtures and pure substances. Most natural samples of matter are mixtures. Quite often, recognition of a substance as a mixture is easy; the distinct properties of various components are clearly visible. Other times, as we will see shortly (Sec. 4.6), the components of a mixture cannot be visually distinguished. Natural pure substances are not common; they are seldom encountered. Gold and diamond are two of the few naturally occurring pure substances. Despite the lack of their natural occurrence, many, many pure substances are known. They are obtained from natural mixtures using numerous types of separation techniques.

Formally defined, a **mixture** is a physical combination of two or more pure substances in which the pure substances retain their identity. The chemical identity of individual components is retained in a mixture because the components are combined physically rather than chemically. Consider, for example, a mixture of salt and pepper (see Fig. 4.5). Close examination of such a mixture will show distinct particles of salt and pepper with no obvious interaction between them. The salt particles in the mixture are identical in properties and composition to the salt particles in the salt container, and the pepper particles in the mixture are no different from those in the pepper bottle.

Once a particular mixture is made up, its composition is constant. However, mixtures of the same components with different compositions can also be made up; thus, mixtures are considered to have variable compositions. Consider the large number of salt and pepper mixtures that could be produced by varying the amounts of the two substances present.

An additional characteristic of any mixture is that its components can often be retrieved intact from the mixture by physical means, that is, without a chemical change. In many cases, the differences in properties of the various components make the separation relatively easy. For example, in our salt–pepper mixture, if the pepper grains were large enough, the two components could be separated manually by picking out all the pepper grains. Alternatively, the separation could be carried out by dissolving the salt particles in water, removing the insoluble pepper particles, and then evaporating the water to recover the salt.

Most mixture separations are not as easy as that for a salt–pepper mixture; expensive instrumentation and numerous separation steps are required. Image the logistics involved in the separation of the components of blood, a water-based mixture of

Figure 4.5 The individual particles of salt and pepper are easily recognizable in a mixture of the two substances because a mixture is a physical rather than chemical combination of substances.

varying amounts of proteins, sugar (glucose), salt (sodium chloride), oxygen, carbon dioxide, and other components.

A pure substance is exactly what its name implies—a single uncontaminated type of matter. All samples of a pure substance contain only that pure substance and nothing else. Pure water is water and nothing else. Pure table salt (sodium chloride) contains only that substance and nothing else.

A pure substance is defined in terms of its properties. A **pure substance** is a form of matter that always has a definite and constant composition. This constancy of composition is reflected in the properties of the pure substance. They never vary, always being the same under a given set of conditions. All samples of a pure substance, no matter what their source, must have the same properties under the same conditions. Collectively, these definite and constant physical and chemical properties of a pure substance form a set not duplicated by any other pure substance. This unique set of properties provides the identification for the pure substance.

It is important to note that there is a significant difference between the terms substance and pure substance. Substance is a general term used to denote any variety of matter. Pure substance is a specific term that applies only to matter with those characteristics we have just noted.

Figure 4.6 summarizes the differences between mixtures and pure substances. Further considerations about mixtures are found in Section 4.6, and Section 4.7 contains further details about pure substances.

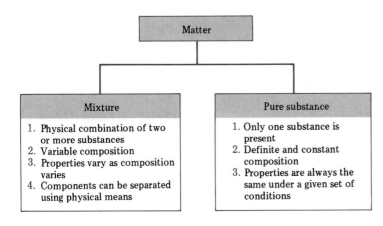

FIGURE 4.6 A comparison of the characteristics of mixtures and pure substances.

4.6 Types of Mixtures: Heterogeneous and Homogeneous

Visual identification of the components of a mixture is not always possible. Sometimes the separate ingredients of a mixture are not visible, even with the aid of a microscope. A mixture prepared by dissolving sugar in water falls in this category. Outwardly, this mixture has the same appearance as pure water.

Mixtures can be classified as being either heterogeneous or homogeneous on the basis of the visual recognizability of the components present. A **heterogeneous mixture** contains visibly different parts, or phases, each of which has different properties. The phases present in a heterogeneous mixture have distinct boundaries and are usually easily observed. A pepperoni and cheese pizza is obviously a heterogeneous mixture that has numerous identifiable components. Even a piece of pepperoni contains a number of phases. Common materials such as rocks and wood are also heterogeneous mixtures; different parts of these materials clearly have different properties, such as hardness and color (see Fig. 4.7).

The phases in a heterogeneous mixture may or may not be in the same physical state. Set concrete contains a number of phases, all of which are in the solid state. A mixture of sand and water contains two phases, and each is in a different state (solid and liquid). It is possible to have heterogeneous mixtures in which all components are liquids. In order for these mixtures to occur, the mixed liquids must have limited or no solubility in each other. When this is the case, the mixed liquids form separate layers with the least dense liquid on top. An oil-and-vinegar salad dressing is an example of such a liquid–liquid mixture (see Fig. 4.8a). Oil-and-vinegar dressing consists of two phases (oil and vinegar) regardless of whether the mixture consists of two separate layers or of oil droplets dispersed throughout the vinegar, a condition caused by shaking the mixture (see Fig. 4.8b). All of the oil droplets together are considered to be a single phase.

A **homogeneous mixture** contains only one visibly distinct phase, which has uniform properties throughout. This type of mixture can have only one set of properties, which are associated with the single phase present. A spoonful of sugar water taken

(a) (b)

FIGURE 4.7 Common materials such as rocks and wood are heterogeneous mixtures. [(a) John Schultz, PAR/NYC; (b) Grant Heilman]

from the surface of a homogeneous sugar–water mixture is just as sweet to the taste as one taken from the bottom of the container. If this were not the case the mixture would not be truly homogeneous.

Homogeneous mixtures are possible only when all components present are in the same physical state. Homogeneous mixtures for all three physical states are common. Air is a homogeneous mixture of gases; motor oil and gasoline are multicomponent homogeneous mixtures of liquids; and metal alloys such as 14-karat gold (a mixture of copper and gold) are examples of solid homogeneous mixtures.

A thorough intermingling of the components in a homogeneous mixture is required in order for a single phase to exist. Sometimes this occurs almost instantaneously during the preparation of the mixture, as in the addition of alcohol to water. At other times, an extended period of mixing or stirring is required. For example, when a hard sugar cube is added to a container of water, it does not

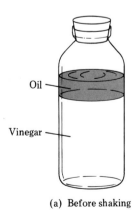

Oil
Vinegar

(a) Before shaking

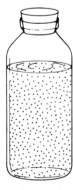

(b) After shaking

FIGURE 4.8 Oil and vinegar salad dressing is a two-phase heterogeneous mixture. Shaking the mixture does not change the number of phases present.

instantaneously dissolve to give a homogeneous solution. Only after much stirring does the sugar completely dissolve. Prior to that point, the mixture was heterogeneous because a solid phase (the undissolved sugar cube) was present.

4.7 Types of Pure Substances: Elements and Compounds

There are two kinds of pure substances: elements and compounds. An **element** is a pure substance that cannot be broken down into simpler substances by ordinary chemical means (a reaction, an electric current, a beam of light, etc.). Elements resist all attempts to fragment them into simpler pure substances. The metals gold, silver, and copper are all elements. A **compound** is a pure substance that can be broken down into two or more simpler substances by chemical means. Water is a compound. By means of electric current water can be broken down into the gases hydrogen and oxygen, both of which are elements. The properties of the simpler substances obtained from compound breakdown are always distinctly different from those of the parent compound.

Ultimately the products from the breakdown of any compound are elements. In practice the breakdown often occurs in steps, with simpler compounds resulting from the intermediate steps, as illustrated in Figure 4.9.

Presently 109 pure substances are classified as elements. These elements, which are the simplest known substances, are considered the building blocks of all other types of matter. Every object, regardless of its complexity, is a collection of substances that are made up of the 109 elements.

Compared to the total number of compounds characterized by chemists, the number of known elements is extremely small. Over 6 million different compounds are known. Each is a definite chemical combination of two or more of the known elements. Figure 4.10 summarizes the characteristics of elements and compounds.

Before a substance can be classified as an element, all possible attempts must be made chemically to subdivide it into simpler substances. If a sample of pure substance, S, is subjected to a chemical process and two new substances, X and Y, are produced, S would be classified as a compound. If, on the other hand, a number of attempts made chemically to subdivide S proved unsuccessful, we might correctly call

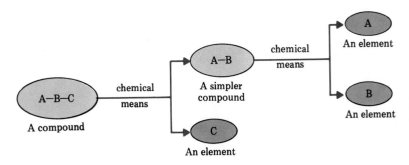

FIGURE 4.9 Stepwise breakdown of a compound containing three elements (A, B, and C) to yield the constituent elements.

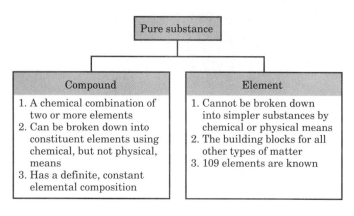

FIGURE 4.10 A comparison of the characteristics of elements and compounds.

it an element, but until all possible reactions have proved unsuccessful, such a classification could be in error.

Samples of elements or compounds are homogeneous, except if two physical states (solid, liquid, gas) are present. Elements are homogeneous, since they consist of just one component throughout. Compounds are also homogeneous, even though two or more elements are present in a chemical combination, because they have a definite composition, which ensures fixed proportions of all components throughout a sample.

A heterogeneous sample of a pure substance is possible. For this situation to occur the pure substance must be present in two or more states. An ice cube floating in a container of water is such a system. The solid phase (ice) obviously has some physical properties different from those of the liquid phase (liquid water). An ice–water system, although heterogeneous, is not a mixture. Only one substance is present—water. Mixtures require the presence of two substances.

Figure 4.11 summarizes the overall classification scheme for matter developed in Sections 4.5–4.7.

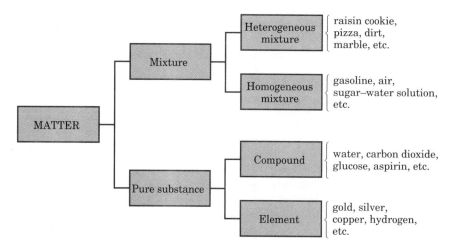

FIGURE 4.11 Categories of classification for matter.

Students frequently have trouble with the concept that compounds are not mixtures even though two or more simple substances can be obtained from compound decomposition. It is very important that the distinction between a mixture and a compound be understood. There are three distinct areas of difference between these two classifications of matter.

1. Compounds always have properties distinctly different from those of the substances (elements or compounds) used to produce the compound. This is because the combining substances are *bound* together into discrete units; the substances are *chemically* combined. The substances in a mixture retain their individual identities because they are *not bound* together; hence, their individual properties are still manifest; these substances are *physically* combined.
2. Compounds have a definite composition, a property of all pure substances. Mixtures have a variable composition.
3. The individual components of a mixture can be physically separated from each other. The elements in a compound cannot be separated from each other using physical means since they are bound together. They can only be separated by chemical means through which the compound is destroyed.

Example 4.1, which involves a comparison between two breakfast cereals, illustrates nicely the difference between compounds and mixtures.

EXAMPLE 4.1

Consider the characteristics of the two breakfast cereals "Crispy Wheat 'N Raisins" and "Crispix." The first cereal contains wheat flakes and raisins. The second cereal contains a fused two-layered flake, one side of which is rice and the other side corn. Characterize the properties of these two cereals that make one an analogy for a mixture and the other an analogy for a compound.

Solution

Crispy Wheat 'N Raisins represents the mixture. Two samples drawn from a box of this cereal would not have the same composition; the ratio of flakes to raisins would vary. It would be fairly easy to separate the flakes from the raisins and eat each separately.

Crispix represents the compound. Two samples drawn from a box of this cereal would have the same composition; each flake is one-half rice and one-half corn. You could not separate the rice and corn without destroying (breaking apart) each individual flake.

PRACTICE EXERCISE 4.1

Consider two boxes with the following contents: the first contains 30 bolts and 30 nuts which fit the bolts; the second contains the same number of bolts and nuts with the difference that each bolt has a nut screwed on it. Which box has contents that would be an analogy for a mixture and which has contents that would be an analogy for a compound?

Ans. first box, mixture; second box, compound.

Similar exercises: Problems 4.41 and 4.42

Table 4.3 Number of Elements Discovered During Various Time Periods

Time Period	Number of Elements Discovered During Time Period	Total Number of Elements Known at End of Time Period
Ancient–1700	13	13
1701–1750	3	16
1751–1800	18	34
1801–1850	25	59
1851–1900	23	82
1901–1950	16	98
1951–present	11	109

4.8 Discovery and Abundance of the Elements

The discovery and isolation of the 109 elements have taken place over several centuries. Discovery, for the most part, has occurred since 1700, and in particular during the 1800s. Table 4.3 shows how the number of known elements has increased dramatically since 1750.

Eighty-eight of the 109 elements occur naturally, and 21 are synthetic, having been produced in the laboratory from naturally occurring elements. It is generally accepted by scientists that no more naturally occurring elements will be found. The last of the naturally occurring elements was discovered in 1925. It is possible however, that additional synthetic elements will be prepared. The synthetic elements are unstable (radioactive) and usually change rapidly into naturally occurring elements as the result of radioactive decay.

The naturally occurring elements are not evenly distributed in our world and universe. What is startling is the degree of inequality in the distribution. A very few elements completely dominate. In answering the question of which elements are most common, one must define the area to be considered. The abundances of the elements in the Earth's crust are considerably different from those for the Earth as a whole; they are even more different from abundances for the universe. When living organisms such as vegetation or the human body are considered, an altogether different perspective emerges. Table 4.4 gives information on element abundances. The abundance of each element is the percentage of the total elemental units (atoms; see Sec. 5.1) present that are units of that element. Only element abundances greater than 1% are listed in any given analysis.

Note from Table 4.4 that the most abundant elements in the universe are not the most abundant ones on Earth. The cosmic figures reflect the composition of stars, which are almost entirely hydrogen and helium. Significant differences also exist between the composition of the Earth as a whole and that of its crust. The figures for the Earth's crust—taken to mean the Earth's waters, atmosphere, and outer

TABLE 4.4 Abundance of the Elements in Various Realms of Matter

Element	Abundance (atom %)	Element	Abundance (atom %)
Universe		*Atmosphere*	
Hydrogen	91	Nitrogen	78.3
Helium	9	Oxygen	21.0
Earth (including core)		*Hydrosphere*	
Oxygen	49.3	Hydrogen	66.4
Iron	16.5	Oxygen	33
Silicon	14.5	*Human body*	
Magnesium	14.2	Hydrogen	63
Earth's crust		Oxygen	25.5
Oxygen	60.1	Carbon	9.5
Silicon	20.1	Nitrogen	1.4
Aluminum	6.1	*Vegetation*	
Hydrogen	2.9	Hydrogen	49.8
Calcium	2.6	Oxygen	24.9
Magnesium	2.4	Carbon	24.9
Iron	2.2		
Sodium	2.1		

covering to a depth of 10 miles—do not take into account the composition of the Earth's core, which is mostly iron. Note also that only two elements occur in large amounts (greater than 1%) in both the atmosphere and hydrosphere and that carbon is more abundant and hydrogen less abundant in vegetation than in the human body.

4.9 Names and Symbols of the Elements

Each element has a unique name, which in most cases was selected by its discoverer. A wide variety of rationales for choosing a name is found when name origins are studied. Some elements bear geographical names. Germanium was named after the native country of its German discoverer. The elements francium and polonium acquired names in a similar manner. The elements mercury, uranium, neptunium, and plutonium are all named for planets. Helium gets its name from the Greek word *helios* for sun, since it was first observed spectroscopically in the sun's corona during an eclipse. Some elements carry names that relate to specific properties of the element or compounds containing it. Chlorine's name is derived from the Greek *chloros* denoting greenish yellow, the color of chlorine gas. Iridium gets its name from the Greek *iris* meaning rainbow because of the various colors of the compounds from which it was isolated.

TABLE 4.5 The Elements and Their Chemical Symbols

Ac	actinium	Cr	chromium	K	potassium	Pb	lead	Tb	terbium
Ag	silver	Cs	cesium	Kr	krypton	Pd	palladium	Tc	technetium
Al	aluminum	Cu	copper	La	lanthanum	Pm	promethium	Te	tellurium
Am	americium	Dy	dysprosium	Li	lithium	Po	polonium	Th	thorium
Ar	argon	Er	erbium	Lu	lutetium	Pr	praseodymium	Ti	titanium
As	arsenic	Es	einsteinium	Lr	lawrencium	Pt	platinum	Tl	thallium
At	astatine	Eu	europium	Md	mendelevium	Pu	plutonium	Tm	thulium
Au	gold	F	fluorine	Mg	magnesium	Ra	radium	U	uranium
B	boron	Fe	iron	Mn	manganese	Rb	rubidium	Une	unnilennium
Ba	barium	Fm	fermium	Mo	molybdenum	Re	rhenium	Unh	unnilhexium
Be	beryllium	Fr	francium	N	nitrogen	Rh	rhodium	Uno	unniloctium
Bi	bismuth	Ga	gallium	Na	sodium	Rn	radon	Unp	unnilpentium
Bk	berkelium	Gd	gadolinium	Nb	niobium	Ru	ruthenium	Unq	unnilquadium
Br	bromine	Ge	germanium	Nd	neodymium	S	sulfur	Uns	unnilseptium
C	carbon	H	hydrogen	Ne	neon	Sb	antimony	V	vanadium
Ca	calcium	He	helium	Ni	nickel	Sc	scandium	W	tungsten
Cd	cadmium	Hf	hafnium	No	nobelium	Se	selenium	Xe	xenon
Ce	cerium	Hg	mercury	Np	neptunium	Si	silicon	Y	yttrium
Cf	californium	Ho	holmium	O	oxygen	Sm	samarium	Yb	ytterbium
Cl	chlorine	I	iodine	Os	osmium	Sn	tin	Zn	zinc
Cm	curium	In	indium	P	phosphorus	Sr	strontium	Zr	zirconium
Co	cobalt	Ir	iridium	Pa	protactinium	Ta	tantalum		

In the early 1800s chemists adopted the practice of assigning chemical symbols to the elements. **Chemical symbols** are abbreviations for the names of the elements. These chemical symbols are used more frequently in referring to the elements than are the names themselves. The system of chemical symbols now in use was first proposed in 1814 by the Swedish chemist Jöns Jakob Berzelius (1779–1848).

A list of elements and their symbols is given in Table 4.5. The symbols and names of the more frequently encountered elements are shown in color in this table. You would do well to learn the symbols of these more common elements. Learning them is a key to having a successful experience in studying chemistry.

Fourteen elements have one-letter symbols, 89 have two-letter symbols, and 6 have three-letter symbols. If a symbol consists of a single letter, it is capitalized. In all two- and three- letter symbols, only the first letter is capitalized. Double-letter symbols usually, but not always, start with the first letter of the element's name. The second letter of the symbol is frequently, but not always, the second letter of the name. Consider the elements terbium, technetium, and tellurium, whose symbols are respectively Tb, Tc, and Te. Obviously, a variety of choices of second letters is necessary because the first two letters are the same in all three elements' names.

The six elements with triple-letter symbols are all synthetic elements and were the last six elements to be produced. These symbols come from a recently introduced provisional system for naming elements that is the subject of Section 4.10.

TABLE 4.6 Elements Whose Symbols Are Derived from a Non-English Name of the Element

English Name of Element	Non-English Name of Element	Symbol
Symbols from Latin		
Antimony	stibium	Sb
Copper	cuprum	Cu
Gold	aurum	Au
Iron	ferrum	Fe
Lead	plumbum	Pb
Mercury	hydrargyrum	Hg
Potassium	kalium	K
Silver	argentum	Ag
Sodium	natrium	Na
Tin	stannum	Sn
Symbol from German		
Tungsten	wolfram	W

Eleven elements have symbols that bear no relationship to the element's English language name. In ten of these cases, the symbol is derived from the Latin name of the element; in the case of tungsten a German name is the symbol source. Most of these elements have been known for hundreds of years and date back to the time when Latin was the language of scientists. Table 4.6 shows the relationship between the symbol and the non-English name of these eleven elements.

The symbols of the elements are also found on the inside front and back covers of this book. The chart of elements on the inside front cover is called a *periodic table*. More will be said about it in later chapters. Both cover listings also give other information about the elements. This additional information will be discussed in Chapter 5.

4.10 Provisional Names for New Elements

As noted in Sections 4.7 and 4.8, the total number of known elements now stands at 109. This number is subject to change. It is entirely possible that additional elements will be produced.

Beginning with elements 104 and 105, produced for the first time in the years 1969 and 1970, respectively, the naming of elements became a problem. Both American and Russian teams of scientists claimed to have discovered the elements, and both teams proposed names. The Russian proposals were khurchatovium (Ku) and nielsbohrium (Nl), respectively; the American suggestions were rutherfordium (Rf) and hahnium (Ha), respectively. (All four names are derived from the names of famous deceased scientists.) Both sets of names began to be used in the chemical literature—not a desirable situation. The resolution of this problem became the responsibility of the International Union of Pure and Applied Chemistry (IUPAC), an international chemical organization that "officially approves" a discoverer's suggestions for the name and the symbol of a new element.

In the cases of elements 104 and 105, resolving the priority and validity of the discoveries is still a "tough problem" for IUPAC. Neither set of names has yet been officially approved by the organization, and it is over 20 years since the elements were discovered.

In 1979, to prevent this type of problem from occurring again and also "temporarily" to solve the 104 and 105 problem, IUPAC decided that *provisional* names and symbols, derived from a systematic set of rules, should be given to all new elements. These provisional names would then be used until the discovery and discoverer of a new element had been established beyond all doubt in the general scientific community. Once "discovery rights" were established, the definitive name suggested by the discoverer would be adopted. Furthermore, it was decided that these provisional name rules should be retroactively applied to elements 104 and 105.

The provisional names are Latin–Greek hybrids for the identifying number of the element. (Each element has an identifying number, called an atomic number, a concept that is discussed in detail in Sec. 5.6.) The following numerical roots are used in assigning these names.

$$
\begin{array}{lll}
0 = \text{nil} & 4 = \text{quad} & 7 = \text{sept} \\
1 = \text{un} & 5 = \text{pent} & 8 = \text{oct} \\
2 = \text{bi} & 6 = \text{hex} & 9 = \text{enn} \\
3 = \text{tri} & &
\end{array}
$$

To form the name of the element, the roots are put together in order of the digits that make up the identifying number of the element and are terminated with the suffix *-ium*. Thus, the provisional name for element 104 is unnilquadium. Provisional names and symbols for elements 104 through 110 are given in Table 4.7. To avoid

TABLE 4.7 Provisional Names Assigned to Elements 104–110

Identifying Number	Name	Symbol
104	unnilquadium	Unq
105	unnilpentium	Unp
106	unnilhexium	Unh
107	unnilseptium	Uns
108	unniloctium	Uno
109	unnilennium	Une
110	ununnilium	Uun

duplication of existing two-letter symbols of elements, all symbols for provisional names have three letters. The symbols are derived from the initial letters of each of the numerical roots that make up the name. At the present time provisional names are in use for elements 104 through 109.

Key Terms

The new terms or concepts defined in this chapter are

chemical change (Sec. 4.4)   A process that involves a change in the basic nature (chemical composition) of a substance.

chemical properties (Sec. 4.3)   Properties that matter exhibits as it undergoes changes in chemical composition.

chemical reaction (Sec. 4.4)   A process in which at least one new substance is produced as a result of chemical change.

chemical symbol (Sec. 4.9)   An abbreviation for the name of an element.

chemistry (Sec. 4.1)   The branch of science that is concerned with matter and its properties.

compound (Sec. 4.7)   A pure substance that can be broken down into two or more simpler substances by chemical means.

element (Sec. 4.7)   A pure substance that cannot be broken down into simpler substances by ordinary chemical means.

gaseous state (Sec. 4.2)   A state characterized by both an indefinite shape and an indefinite volume.

heterogeneous mixture (Sec. 4.6)   A mixture that contains visibly different parts, or phases, each with different properties.

homogeneous mixture (Sec. 4.6)   A mixture that contains only one visibly distinct phase with uniform properties throughout.

liquid state (Sec. 4.2)   A state characterized by its properties of an indefinite shape and a definite volume.

matter (Sec. 4.1)   Anything that has mass and occupies space.

mixture (Sec. 4.5)   A physical combination of two or more pure substances in which the pure substances retain their identity.

physical change (Sec. 4.4)   A process that does not alter the basic nature (chemical composition) of a substance.

physical properties (Sec. 4.3)   Properties that can be observed without changing a substance into another substance.

properties (Sec. 4.3)   Distinguishing characteristics of a substance used in its identification and description.

pure substance (Sec. 4.5)   A form of matter that always has a definite and constant composition.

solid state (Sec. 4.2)   A state characterized by its properties of a definite shape and a definite volume.

Practice Problems

Physical States of Matter (Sec. 4.2)

4.1 Use the word *definite* or *indefinite* to characterize the shape of
  (a) solids   (b) liquids   (c) gases

4.2 Use the word *definite* or *indefinite* to characterize the volume of
  (a) solids   (b) liquids   (c) gases

4.3 Gold has a melting point of 1063 °C and a boiling point of 2966 °C. Specify the physical state of gold at each of the following temperatures.
  (a) 500 °C   (b) 1000 °C
  (c) 2000 °C   (d) 3000 °C

4.4 Oxygen has a melting point of −218 °C and a boiling point of −183 °C. Specify the physical state of oxygen at each of the following temperatures.

(a) $-250\,°C$ (b) $-200\,°C$
(c) $-100\,°C$ (d) $100\,°C$

4.5 Criticize the statement "All solids will melt if heated to a high enough temperature."

4.6 Criticize the statement "Carbon dioxide is a gaseous substance."

Properties of Matter (Sec. 4.3)

4.7 How does a physical property differ from a chemical property?

4.8 What do physical and chemical properties have in common?

4.9 Classify each of the following observations about a substance as either a physical or chemical property.
(a) physical state at room temperature
(b) electrical conductivity
(c) melting point
(d) reactivity with oxygen

4.10 Classify each of the following observations about a substance as either a physical or chemical property.
(a) flammability   (b) thermal conductivity
(c) color   (d) density

4.11 The following are properties of the element magnesium. Classify them as physical or chemical properties.
(a) silvery white in color
(b) finely divided form burns in oxygen with a dazzling white flame
(c) has a density of $1.738\ g/cm^3$ at $20\,°C$
(d) does not react with cold water

4.12 The following are properties of the element lithium. Classify them as physical or chemical properties.
(a) can be cut with a sharp knife
(b) light enough to float on water
(c) in the liquid state reacts spontaneously with its glass container, producing a hole in the container
(d) changes from silvery gray to black when placed in moist air

4.13 Indicate whether each of the following statements describes a physical or chemical property.
(a) An iron nail is attracted to a magnet.
(b) Diamonds are very hard substances.
(c) Chlorine gas has a yellowish green color.
(d) Beryllium metal vapor is extremely toxic to humans.

4.14 Indicate whether each of the following statements describes a physical or chemical property.
(a) Lithium burns in oxygen with a bright red flame.
(b) Magnesium is a solid at room temperature.

(c) Gold will not dissolve in either hydrochloric or nitric acid.
(d) Aspirin tablets can be pulverized with a hammer.

Changes in Matter (Sec. 4.4)

4.15 What is the meaning associated with the word *physical* when it is used in the phrase *physical change*?

4.16 What is the meaning associated with the word *chemical* when it is used in the phrase *chemical change*?

4.17 Classify each of the following changes as physical or chemical.
(a) crushing a dry leaf
(b) hammering a metal into a thin sheet
(c) burning your chemistry book
(d) slicing a ham

4.18 Classify each of the following changes as physical or chemical.
(a) evaporation of water from a lake
(b) "scabbing over" of a skin cut
(c) rusting of an iron nail
(d) melting of some wax

4.19 Indicate whether each of the following methods for obtaining various substances involves physical or chemical change.
(a) Sodium chloride (salt) is obtained from salt water by evaporation of the water.
(b) Nitrogen gas is obtained from air by letting the nitrogen boil off from liquid air.
(c) Oxygen gas is obtained by decomposition of the oxygen-containing compound potassium chlorate.
(d) Water is obtained by the high-temperature reaction of gaseous hydrogen with gaseous oxygen.

4.20 Indicate whether each of the following methods for obtaining various substances involves physical or chemical change.
(a) Mercury is obtained by decomposing a mercury–oxygen compound, liberating the oxygen and leaving the mercury behind.
(b) Sand is obtained from a sand–sugar mixture by adding water to the mixture and pouring off the resulting sugar–water solution.
(c) Ammonia is obtained by the high-temperature high-pressure reaction between hydrogen and nitrogen.
(d) Water is obtained from a sugar–water solution by evaporating off and then collecting the water.

4.21 Give the name of the change of state associated with each of the following processes.
(a) Water is made into ice cubes.
(b) The inside of your car window fogs up.

(c) Mothballs in the clothes closet disappear with time.
(d) Perspiration dries.

4.22 Give the name of the change of state associated with each of the following processes.
(a) Dry ice disappears without melting.
(b) Snowflakes form.
(c) Dew on the lawn disappears when the sun comes out.
(d) Ice cubes in a soft drink disappear with time.

Pure Substances and Mixtures (Secs. 4.5 and 4.6)

4.23 Consider the following classes of matter: heterogeneous mixture, homogeneous mixture, and pure substance.
(a) In which of these classes must two or more substances be present?
(b) In which of these classes must the composition be uniform throughout?

4.24 Consider the following classes of matter: heterogeneous mixture, homogeneous mixture, and pure substance.
(a) Which of these classes could not possibly have a variable composition?
(b) Which of these classes could not possibly be separated into simpler substances using physical means?

4.25 Classify each of the following as a heterogeneous mixture, a homogeneous mixture, or a pure substance.
(a) paint (stirred) (b) cough syrup
(c) table salt (d) concrete

4.26 Classify each of the following as a heterogeneous mixture, a homogeneous mixture, or a pure substance.
(a) paint (unstirred) (b) pork chop
(c) table sugar (d) pulp-free orange juice

4.27 Assign each of the following descriptions of matter to one of the following categories: heterogeneous mixture, homogeneous mixture, pure substance.
(a) two substances present, two phases present
(b) two substances present, one phase present
(c) three substances present, one phase present
(d) three substances present, three phases present

4.28 Assign each of the following descriptions of matter to one of the following categories: heterogeneous mixture, homogeneous mixture, pure substance.
(a) one substance present, one phase present
(b) one substance present, two phases present
(c) one substance present, three phases present
(d) three substances present, two phases present

4.29 Classify each of the following as a heterogeneous mixture, a homogeneous mixture, or a pure substance. Also indicate how many phases are present. (In each case the substances are present in the same container.)
(a) water and dissolved salt
(b) water and dissolved sugar
(c) water and sand
(d) water and oil

4.30 Classify each of the following as a heterogeneous mixture, a homogeneous mixture, or a pure substance. Also indicate how many phases are present. (In each case the substances are present in the same container.)
(a) liquid water and ice
(b) liquid water, oil, and ice
(c) carbonated water (soda water) and ice
(d) oil, ice, saltwater solution, sugar–water solution, and pieces of copper metal

4.31 What characteristics do a heterogeneous mixture and a homogeneous mixture have in common?

4.32 What characteristics do a homogeneous mixture and a pure substance have in common?

Elements and Compounds (Sec. 4.7)

4.33 Explain the difference between an element and a compound.

4.34 Is it possible to have a heterogeneous sample of an element or compound? Explain.

4.35 Based on the information given, classify each of the pure substances A through D as elements or compounds, or indicate that no such classification is possible because of insufficient information.
(a) Analysis with an elaborate instrument indicates that substance A contains two elements.
(b) Substance B decomposes upon heating.
(c) Heating substance C to 1000 °C causes no change in it.
(d) Heating substance D to 500 °C causes it to change from a solid to a liquid.

4.36 Based on the information given, classify each of the pure substances A through D as elements or compounds, or indicate that no such classification is possible because of insufficient information.
(a) Substance A cannot be broken down into simpler substances by chemical means.
(b) Substance B cannot be broken down into simpler substances by physical means.
(c) Substance C readily dissolves in water.
(d) Substance D readily reacts with the element chlorine.

4.37 Indicate whether each of the following statements is true or false.
(a) Both elements and compounds are pure substances.
(b) A compound results from the physical combination of two or more elements.
(c) In order for matter to be heterogeneous, at least two compounds must be present.

(d) Compounds, but not elements, can have a variable composition.

4.38 Indicate whether each of the following statements is true or false.
(a) Compounds can be separated into their constituent elements using chemical means.
(b) Elements can be separated into their constituent compounds using physical means.
(c) A compound must contain at least two elements.
(d) A compound is a physical mixture of different elements.

4.39 Based on the information given in the following equations, classify each of the pure substances A through G as elements or compounds, or indicate that no such classification is possible because of insufficient information.
(a) $A + B \rightarrow C$
(b) $D \rightarrow E + F + G$

4.40 Based on the information given in the following equations, classify each of the pure substances A through G as elements or compounds, or indicate that no such classification is possible because of insufficient information.
(a) $A \rightarrow B + C$
(b) $D + E \rightarrow F + G$

4.41 Consider two boxes with the following contents: the first contains 50 individual paper clips and 50 individual rubber bands; the second contains the same number of paper clips and rubber bands with the difference that each paper clip is interlocked with a rubber band. Which box has contents that would be an analogy for a mixture and which has contents that would be an analogy for a compound?

4.42 Consider two boxes with the following contents: the first contains 25 ball-point pens each with its cap on; the second contains 25 ball-point pens without caps and 25 ball-point pen caps. Which box has contents that would be an analogy for a mixture and which has contents that would be an analogy for a compound?

4.43 Is it possible to have a mixture of two elements and also to have a compound of the same two elements? Explain.

4.44 Compounds are classified as pure substances even though two or more elements are present. Explain why.

Discovery and Abundance of the Elements (Sec. 4.8)

4.45 How many elements were discovered in each of the following time periods?
(a) ancient–1700
(b) 1701–1800
(c) 1801–1900
(d) 1901–present

4.46 What percentage of the 109 known elements were known before
(a) 1700
(b) 1800
(c) 1900
(d) 1950

4.47 What are the two most abundant elements in each of the following realms of matter?
(a) universe
(b) Earth (including core)
(c) Earth's crust
(d) human body

4.48 In which of the seven realms of matter listed in Table 4.4 is each of the following the most abundant element?
(a) oxygen
(b) hydrogen
(c) nitrogen
(d) carbon

Names and Symbols of the Elements (Secs. 4.9 and 4.10)

4.49 Name the element each of the following chemical symbols represents.
(a) Ag
(b) Au
(c) Ca
(d) Na
(e) P
(f) S

4.50 Name the element each of the following chemical symbols represents.
(a) Cl
(b) Fe
(c) He
(d) Hg
(e) Ni
(f) Pb

4.51 What are the chemical symbols of the following elements?
(a) tin
(b) copper
(c) aluminum
(d) boron
(e) barium
(f) argon

4.52 What are the chemical symbols of the following elements?
(a) zinc
(b) silicon
(c) beryllium
(d) fluorine
(e) cobalt
(f) magnesium

4.53 List the elements whose chemical symbols contain only one letter.

4.54 List the elements whose chemical symbols begin with a letter other than the first letter of the English name for the element.

4.55 Each of the following names of elements is spelled incorrectly. Correct the misspellings.
(a) flourine
(b) zink
(c) potasium
(d) sulfer

4.56 Each of the following names of elements is spelled incorrectly. Correct the misspellings.
(a) phosphorous
(b) murcury
(c) clorine
(d) argone

4.57 What would be the provisional name and provisional symbol for the element with each of the following identifying numbers?
(a) 107
(b) 113
(c) 115
(d) 121

4.58 What would be the provisional name and provisional symbol for the element with each of the following identifying numbers?
(a) 108
(b) 110
(c) 119
(d) 132

Additional Problems

4.59 Assign each of the following descriptions of matter to one of the following categories: element, compound, mixture.
  (a) one substance present, one phase present, substance can be decomposed by chemical means
  (b) two substances present, one phase present
  (c) one substance present, two elements present
  (d) two elements present, composition is variable

4.60 Assign each of the following descriptions of matter to one of the following categories: element, compound, mixture.
  (a) one substance present, one phase present, substance cannot be decomposed by chemical means
  (b) one substance present, three elements present
  (c) two substances present, two phases present
  (d) two elements present, composition is definite and constant

4.61 Indicate whether each of the following samples of matter is a heterogeneous mixture, a homogeneous mixture, a compound, or an element.
  (a) A colorless single-phased liquid that when boiled away (evaporated) leaves behind a solid white residue.
  (b) A uniform red liquid with a boiling point of 59 °C that cannot be broken down into simpler substances using chemical means.
  (c) A nonuniform white crystalline substance, part of which dissolves in water and part of which does not.
  (d) A colorless liquid that completely evaporates without decomposition when heated and produces a gas that can be separated into simpler components using physical means.

4.62 Indicate whether each of the following samples of matter is a heterogeneous mixture, a homogeneous mixture, a compound, or an element.
  (a) A colorless gas, only part of which reacts with hot iron.
  (b) A "cloudy" liquid that separates into two layers upon standing for two hours.
  (c) A green solid, all of which melts at the same temperature to produce a liquid that decomposes upon further heating.
  (d) A colorless gas that cannot be separated into simpler substances using physical means and that reacts with copper to produce both a copper–nitrogen compound and a copper–oxygen compound.

4.63 In which of the following sequences of elements do each of the elements have a two-letter symbol.
  (a) magnesium, nitrogen, phosphorus
  (b) bromine, iron, calcium
  (c) aluminum, copper, chlorine
  (d) boron, barium, beryllium

4.64 In which of the following sequences of elements do each of the elements have a symbol that starts with a letter not the first letter of the element's English name?
  (a) silver, gold, mercury
  (b) copper, helium, neon
  (c) cobalt, chromium, sodium
  (d) potassium, iron, lead

Cumulative Problems

4.65 Specify the physical state of a pure substance at each of the following conditions or indicate that state determination is not possible from the information given.
  (a) 10 °C below its freezing point
  (b) 30 °C above its melting point
  (c) after sublimation has taken place
  (d) at its boiling point

4.66 Specify the physical state of a pure substance at each of the following conditions or indicate that state determination is not possible from the information given.
  (a) 10 °C below its melting point
  (b) 30 °C above its freezing point
  (c) after decomposition has taken place
  (d) after deposition has taken place

4.67 The following density determination data was obtained by three students for their unknowns.

Student I:	mass = 4.32 g	volume = 3.78 mL
Student II:	mass = 5.73 g	volume = 5.02 mL
Student III:	mass = 1.52 g	volume = 1.33 mL

  (a) Is it likely that the students were working with different unknowns or that they were working with the same substance?
  (b) Is it possible to tell from the given data whether the unknowns were elements or compounds?

4.68 The following density determination data was obtained by three students for their unknowns.

Student I:	mass = 27.2 g	volume = 23.6 mL
Student II:	mass = 30.3 g	volume = 28.3 mL
Student III:	mass = 55.6 g	volume = 42.5 mL

(a) Is it likely that the students were working with different unknowns or that they were working with the same substance?
(b) Is it possible to tell from the given data whether the unknowns were elements or compounds?

4.69 Three samples of a substance were subjected to analysis with each sample analyzed by a different technique. The results were:

Technique I:	34.1% of A and 65.9% of B
Technique II:	34.12% of A and 65.88% of B
Technique III:	34.12497% of A and 65.87503% of B

Is the substance that was analyzed likely an element, a compound, or a mixture?

4.70 Three samples of a substance were subjected to analysis with each sample analyzed by a different technique. The results were:

Technique I:	34.2 of C and 65.8% of D
Technique II:	36.32% of C and 63.68% of D
Technique III:	37.2111% of C and 62.7889% of D

Is the substance that was analyzed likely an element, a compound, or a mixture?

4.71 The specific heat of substance A is 0.88 J/(g · °C) and that of substance B is 2.1 J/(g · °C). You are given an unknown that could be pure substance A, pure substance B, or a homogeneous mixture of A and B. In the laboratory you determine that it requires 59.5 J of heat energy to raise the temperature of a 35.0 g sample of the unknown by 1.0 °C. What conclusions can you make about the identity of your unknown from this data?

4.72 The specific heat of substance C is 0.93 J/(g · °C) and that of substance D is 1.8 J/(g ·°C). You are given an unknown that could be pure substance C, pure substance D, or a homogeneous mixture of C and D. In the laboratory you determine that it requires 23.3 J of heat energy to raise the temperature of a 25.0-g sample of the unknown by 1.0 °C. What conclusions can you make about the identity of your unknown from this data?

Grid Problems

4.73 Select from the grid *all* correct responses for each situation.

1. change of state	2. physical property	3. chemical property
4. change in state of subdivision	5. physical change	6. chemical change
7. chemical reaction	8. new substances produced	9. change in composition

(a) the process of burning of a newspaper
(b) the fact that metallic copper is insoluble in water
(c) the process of melting of ice
(d) the fact that ice can be chipped into smaller pieces
(e) the process of decomposing hydrogen peroxide
(f) the process of subliming moth balls

4.74 Select from the grid *all* correct responses for each situation.

1. mixture	2. homogeneous mixture	3. heterogeneous mixture
4. pure substance	5. element	6. compound
7. one phase	8. two phases	9. three or more phases

(a) characterizations for chicken-vegetable soup
(b) characterizations for soil (dirt)
(c) characterizations for human blood
(d) characterizations for mercury
(e) characterizations for clean dry air
(f) characterizations for boiling water

4.75 Select from the grid *all* correct responses for each situation.

1. melted butter	2. partially melted butter	3. refrigerated butter
4. ice	5. ice in water	6. carbonated drink
7. fresh strawberry milkshake	8. aluminum foil	9. popped popcorn

(a) substances that are homogeneous mixtures
(b) substances that contain two or more phases
(c) substances that contain a single element
(d) substances that contain two or more states
(e) substances that are heterogeneous
(f) substances that contain a single compound

4.76 Select from the grid *all* correct responses for each situation.

1. argon aluminum	2. calcium copper	3. barium beryllium
4. neon nickel	5. gold silver	6. nitrogen sodium
7. hydrogen iodine	8. sulfur zinc	9. lead phosphorus

(a) both elements have one-letter symbols
(b) both elements have two-letter symbols
(c) both elements have symbols that are the first two letters of the element's name
(d) both elements have symbols that come from a non-English name of the element
(e) both elements have symbols that start with the same letter as the element's name
(f) both elements have symbols in which the first letter is capitalized

CHAPTER FIVE

Atoms, Molecules, Formulas, and Subatomic Particles

5.1 The Atom

If you took a sample of the element gold and started to break it into smaller and smaller and smaller pieces, it seems reasonable that you would eventually reach a "smallest possible piece" of gold that could not be divided further and still be called gold. This smallest possible unit of gold would be a gold atom. An **atom** is the smallest particle of an element that can exist and still have the properties of the element.

The concept of an atom is an old one, dating back to ancient Greece. Records indicate that around 460 B.C. Democritus, a Greek philosopher, suggested that continued subdivision of matter ultimately would yield small indivisible particles which he called atoms (from the Greek word *atomos* meaning "uncut or indivisible"). Democritus's ideas about matter were, however, lost (forgotten) during the Middle Ages, as were the ideas of many other people.

It was not until the beginning of the nineteenth century that the concept of the atom was "rediscovered." In a series of papers published in the period 1803–1807, the English chemist John Dalton (1766–1844) again proposed that the fundamental building block for all kinds of matter was an atom. This time, however, there was a firm basis for the proposal. Dalton's proposal was based on experimental observations. This is in marked contrast to the early Greek concept of atoms, which was based solely on philosophical speculation. Because of its experimental basis, Dalton's idea got wide attention and stimulated new work and thought concerning the ultimate building blocks of matter.

Additional research, carried out by many scientists, has now validated Dalton's basic conclusion that the building blocks for all types of matter are atomic. Some of the details of Dalton's original proposals have had to be modified in the light of later, more sophisticated experiments, but the basic concept of atoms remains.

Today, among scientists, the concept that atoms are the building blocks for matter is a foregone conclusion. The large accumulated amount of supporting evidence for atoms is most impressive. Key concepts about atoms, in terms of current knowledge, are found in what is known as the atomic theory of matter. The **atomic theory of matter** is a set of five statements that summarizes modern-day scientific thought about atoms. These five statements are

1. All matter is made up of small particles called atoms, of which 109 different "types" are known, with each "type" corresponding to a different element.
2. All atoms of a given type are similar to one another and significantly different from all other types.
3. The relative number and arrangement of different types of atoms contained in a pure substance (its composition and structure) determine its identity.
4. Chemical change is a union, separation, or rearrangement of atoms to give new substances.
5. Only whole atoms can participate in or result from any chemical change, since atoms are considered indestructible during such changes.

Just how small is an atom? Atomic dimensions and masses, although not directly measurable, are known quantities obtained by calculation. The data used for the calculations come from measurements made on macroscopic amounts of pure substances.

The diameter of an atom is on the order of 10^{-8} cm. If one were to arrange atoms of diameter 1×10^{-8} cm in a straight line, it would take 10 million of them to extend a length of 1 mm and 254 million of them to reach 1 in. (see Fig. 5.1). Indeed, atoms are very small.

The mass of an atom is also a very small quantity. The mass of a uranium atom, which is one of the heaviest of the known kinds of naturally occurring atoms, is 4×10^{-22} g or 9×10^{-25} lb. To produce a mass of 1 lb would require 1×10^{24} atoms of uranium. The number 1×10^{24} is so large it is difficult to visualize its magnitude.

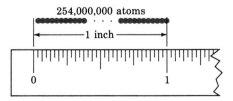

FIGURE 5.1 Comparison of atomic diameters with the common measuring unit of 1 inch.

130   Chapter 5 · Atoms, Molecules, Formulas, and Subatomic Particles

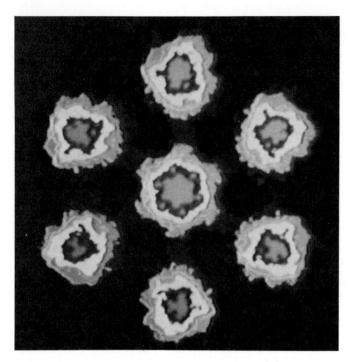

FIGURE 5.2 A uranyl acetate cluster on a very thin carbon substrate. The individual uranium atoms are the roundish spots with darker gray centers. [Courtesy of M. Isaacson, Cornell University, and M. Ohtsuki, The University of Chicago]

The following comparison "hints" at this number's magnitude. If each of the 5 billion people on Earth were made a millionaire (receiving 1 million $1 bills), we would still need 200 million other worlds, each inhabited by the same number of millionaires, to have 1×10^{24} dollar bills in circulation.

Atoms are incredibly small particles. No one has seen or ever will see an atom with the naked eye. The question may thus be asked: "How can you be absolutely sure that something as minute as an atom really exists?" The achievements of twentieth-century scientific instrumentation have gone a long way toward removing any doubt about the existence of atoms. Electron microscopes, capable of producing magnification factors in the millions, have made it possible to photograph "images" of individual atoms. In 1976 physicists at The University of Chicago were successful in obtaining motion pictures of the movement of single atoms. One of these pictures is shown in Figure 5.2.

5.2 The Molecule

Free isolated atoms are rarely encountered in nature. Instead, under normal conditions of temperature and pressure, atoms are almost always found together in aggregates or clusters ranging in size from two atoms to numbers too large to count. When the group or cluster of atoms is relatively small and bound together tightly,

Section 5.2 · The Molecule

the resulting entity is called a molecule. Thus a **molecule** is a group of two or more atoms that functions as a unit because the atoms are tightly bound together. This resultant "package" of atoms behaves in many ways as a single, distinct particle would. The forces that hold the atoms of a molecule together (chemical bonds) will be discussed in Chapter 7.

A *diatomic molecule*, a molecule containing just two atoms, is the simplest type of molecule. Next in complexity are *triatomic* (three atom) and *tetraatomic* (four atom) molecules. Numerous examples of all three of these types of molecules are known. The molecular unit present in water, the most common of all compounds, is triatomic, containing two hydrogen atoms and one oxygen atom. The molecules present in the poisonous gas carbon monoxide are diatomic—one carbon atom and one oxygen atom; tetraatomic molecules are found in the common household cleaning agent called ammonia—one nitrogen atom and three hydrogen atoms. Substances with a much larger molecular unit than two, three, or four atoms can and do exist. Tetrahydrocannabinol (THC), the active ingredient in marijuana, has a molecular unit containing 44 atoms: 14 carbon, 25 hydrogen, and 5 oxygen atoms. Figure 5.3 gives space-filling models for selected types of di-, tri-, and tetraatomic molecules.

The atoms contained in a molecule may all be of the same kind, or two or more kinds may be present. On the basis of this observation molecules are classified into two categories: homoatomic and heteroatomic. **Homoatomic molecules** are molecules in which all atoms present are the same kind. A substance containing homoatomic molecules is, thus, an element. **Heteroatomic molecules** are molecules in which two or more different kinds of atoms are present. Substances containing heteroatomic molecules must be compounds. All of the molecules in Figure 5.3 are heteroatomic, as are the previously mentioned molecules of water, carbon monoxide, ammonia, and THC.

The fact that homoatomic molecules exist indicates that individual atoms are not always the preferred structural unit for an element. Oxygen is an element that exists in molecular form. Almost all of the oxygen present in air is in the form of diatomic molecules. The elements hydrogen, nitrogen, and chlorine are most commonly encountered with their atoms in groups of two, that is, as diatomic molecules. Under most conditions sulfur atoms collect together in groups of eight; phosphorus atoms readily form tetraatomic molecules. Some guidelines for determining which elements have individual atoms as their basic unit and which exist in molecular form will be given in Section 7.2.

Many, but not all, compounds have heteroatomic molecules as their basic structural unit. Those compounds that do are called *molecular compounds*. Some compounds in the liquid and solid state, however, are not molecular; that is, the atoms present are not collected together into discrete heteroatomic molecules. These nonmolecular compounds still contain atoms of at least two kinds (a necessary requirement for a compound), but the form of aggregation is different. It involves an extended three-dimensional assembly of positively and negatively charged particles called *ions* (Sec. 7.4). Compounds that contain ions are called *ionic compounds*. The familiar substances sodium chloride (table salt) and calcium carbonate (limestone) are ionic compounds. The reasons some compounds have ionic rather than molecular structure are considered in Section 7.7.

(a) A diatomic molecule containing one atom of A and one atom of B

(b) A triatomic molecule containing two atoms of A and one atom of B

(c) A tetraatomic molecule containing two atoms of A and two atoms of B

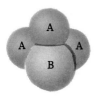

(d) A tetraatomic molecule containing three atoms of A and one atom of B

FIGURE 5.3 Space-filling models of simple molecules. Spheres of different colors are used to represent different kinds of atoms.

For molecular compounds, the molecule is the smallest particle of the compound capable of a stable independent existence. It is the limit of physical subdivision for the compound. Consider the molecular compound sucrose (table sugar). Continued subdivision of a quantity of table sugar to yield smaller and smaller amounts would ultimately lead to the isolation of one single particle of table sugar—a molecule of table sugar. This particle of sugar could not be broken down any further and still maintain the physical and chemical properties of table sugar. The sugar molecule could be broken down further by chemical (not physical) means to give atoms, but if that occurred we would no longer have sugar. The *molecule* is the limit of *physical* subdivision. The *atom* is the limit of *chemical* subdivision.

Every molecular compound has as its smallest characteristic unit a *unique* molecule. If two samples had the same molecule as a basic unit, both would have the same properties; thus, they would be one and the same compound. An alternative way of stating the same conclusion is: There is only one kind of molecule for any given molecular substance.

Since every molecule in a sample of a molecular compound is the same as every other molecule in the sample, it is commonly stated that molecular compounds are made up of a single kind of particle. Such terminology is correct as long as it is remembered that the particle referred to is the molecule. In a sample of a molecular compound there are at least two kinds of atoms present but only one kind of molecule.

The properties of molecules are very different from the properties of the atoms that make up the molecules. Molecules do not maintain the properties of their constituent elements. Table sugar is a white crystalline molecular compound with a sweet taste. None of the three elements present in table sugar (carbon, hydrogen, and oxygen) is a white solid or has a sweet taste. Carbon is a black solid, and hydrogen and oxygen are colorless gases.

Figure 5.4 summarizes the relationships between hetero- and homoatomic molecules and elements, compounds, and pure substances.

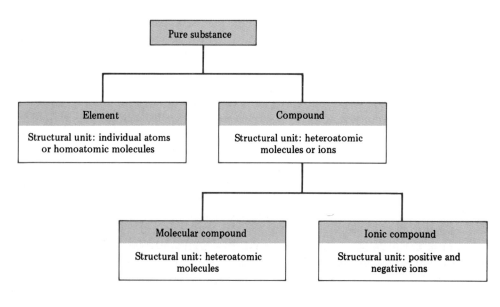

FIGURE 5.4 Structural units for various types of pure substances.

5.3 Natural and Synthetic Compounds

Approximately 6 million chemical compounds are now known, with more being characterized daily. No end appears to be in sight as to the number of compounds that can and will be prepared in the future. At present, approximately 7000 new chemical substances are registered every week with Chemical Abstracts Service, a clearing house for new information concerning chemical substances.

Many compounds, perhaps the majority now known, are not naturally occurring substances. These synthetic (laboratory-produced) compounds are legitimate compounds and should not be considered "second class" or "unimportant" simply because they lack the distinction of being natural. Many of the plastics, synthetic fibers, and prescription drugs now in common use are synthetic materials produced through controlled chemical change carried out on an industrial scale.

We have noted that chemists can produce compounds not found in nature. The reverse is also true. Nature is capable of making many compounds, especially those found in living systems, that chemists are not yet able to prepare in the laboratory.

There is a middle ground also. Many compounds that exist in nature can also be produced in the laboratory. A fallacy exists in the thinking of some people concerning these compounds that have "dual origins." A belief still persists that there is a difference between compounds prepared in the laboratory and samples of the same compounds found in nature. This is not true for pure samples of compound. The message of the law of definite proportions is that all pure samples of a compound, regardless of their origin, have the same composition. Since compositions are the same, properties will also be the same. There is no difference, for example, between a laboratory-prepared vitamin and a "natural" vitamin, if both are pure samples of the same vitamin, despite frequent claims to the contrary.

5.4 Chemical Formulas

A most important piece of information about a compound is its composition. Chemical formulas represent a concise means of specifying compound compositions. A **chemical formula** is a notation made up of the symbols of the elements present in a compound with numerical subscripts (located to the right of each symbol) that indicate the number of atoms of each element present in a formula unit.

The chemical formula for the compound we call aspirin in $C_9H_8O_4$. This formula provides us with the following information about an aspirin molecule: three elements are present—carbon (C), hydrogen (H), and oxygen (O); and 21 atoms are present—9 carbon atoms, 8 hydrogen atoms, and 4 oxygen atoms.

When only one atom of a particular element is present in a molecule of a compound, the element's symbol is written without a numerical subscript in the formula of the compound. In the formula for rubbing alcohol, C_3H_6O, for example, the subscript 1 for the element oxygen is not written.

To write formulas correctly, it is necessary to strictly follow the capitalization rules for elemental symbols (Sec. 4.9). Making the error of capitalizing the second letter of an element's symbol can dramatically alter the meaning of a chemical formula. The

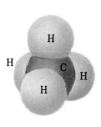
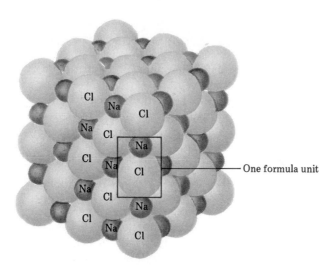

(a) A molecule of the molecular compound methane (CH_4)

(b) A formula unit of the ionic compound sodium chloride (NaCl)

FIGURE 5.5 Comparison of a molecule of a molecular compound and a formula unit of an ionic compound. A molecule can exist as a separate unit, whereas a formula unit is simply two or more ions "plucked" from a much larger array of ions.

formulas $CoCl_2$ and $COCl_2$ illustrate this point; the symbol Co stands for the element cobalt, whereas CO stands for one atom of carbon and one atom of oxygen. The properties of the compounds $CoCl_2$ and $COCl_2$ are dramatically different. The compound $CoCl_2$ is a blue crystalline solid with a melting point of 724 °C and a boiling point of 1029 °C. The compound $COCl_2$ is a highly toxic colorless gas with a melting point of −118 °C and a boiling point of 8 °C.

For molecular compounds, chemical formulas give the composition of the molecules making up the compounds. For ionic compounds, which have no molecules, a chemical formula gives the ion ratio found in the compound. For example, the ionic compound sodium oxide contains sodium ions and oxygen ions in a two-to-one ratio—twice as many sodium ions as oxygen ions. The formula of this compound is Na_2O, which expresses the ratio between the two types of ions present. The term *formula unit* is used to describe this smallest ratio between ions. The distinction between the formula unit of an ionic compound and the molecule of a molecular compound is graphically portrayed in Figure 5.5.

Sometimes chemical formulas contain parentheses, an example being $Al_2(SO_4)_3$. The interpretation of this formula is straightforward; in a formula unit there are present two aluminum (Al) atoms and three SO_4 groups. The subscript following the parentheses always indicates the number of units in the formula of the multi-atom entity inside the parentheses. As another example, consider the compound $Pb(C_2H_5)_4$. Four units of C_2H_5 are present. In terms of atoms present, the formula $Pb(C_2H_5)_4$ represents 29 atoms—1 lead (Pb) atom, $4 \times 2 = 8$ carbon (C) atoms, and $4 \times 5 = 20$ hydrogen (H) atoms. The formula could be (but is not) written as PbC_8H_{20}. Both versions of the formula convey the same information in terms of atoms present. However, $Pb(C_2H_5)_4$ gives the additional information that the C and H are present as C_2H_5 units and is therefore the preferred way of writing the formula. Further information concerning the use of parentheses (when and why) will be presented in Section 7.8. The important concern now is being able to interpret formulas that contain parentheses in terms of total atoms present. Example 5.1 deals with this skill in greater detail.

EXAMPLE 5.1

Interpret each of the following formulas in terms of how many atoms of each element are present in one structural unit of the substance.

(a) $C_8H_{10}N_4O_2$ (caffeine, an addictive central nervous system stimulant)
(b) $(NH_4)_3PO_4$ (ammonium phosphate, an ingredient in some lawn fertilizers)
(c) $Ca(NO_3)_2$ (calcium nitrate, used in fireworks to give a reddish color)

Solution

(a) We simply look at the subscripts following the symbols for the elements. This formula indicates that 8 carbon atoms, 10 hydrogen atoms, 4 nitrogen atoms, and 2 oxygen atoms are present in one molecule of the compound.
(b) The subscript following the parentheses, 3, indicates that 3 NH_4 units are present. Collectively, in these three units, we have 3 nitrogen atoms and $3 \times 4 = 12$ hydrogen atoms. In addition, 1 phosphorus atom and 4 oxygen atoms are present.
(c) The amounts of nitrogen and oxygen are affected by the subscript 2 following the parentheses. In one unit of compound we have 1 calcium atom, $2 \times 1 = 2$ nitrogen atoms, and $2 \times 3 = 6$ oxygen atoms.

PRACTICE EXERCISE 5.1

How many atoms of each element are present in one structural unit of the substance represented by each of the following formulas?

(a) H_2SO_4 (b) $Ca(HCO_3)_2$ (c) $(NH_4)_2CO_3$

Ans. (a) 2 atoms of H, 1 atom of S, and 4 atoms of O; (b) 1 atom of Ca, 2 atoms of H, 2 atoms of C, and 6 atoms of O; (c) 2 atoms of N, 8 atoms of H, 1 atom of C, and 3 atoms of O.

Similar Exercises: Problems 5.15–5.18

The phrases *atoms of an element* and *molecules of an element* are concepts often confused by students. An atom of an element is the smallest particle of that element that can combine to form a compound. A molecule of an element, if the element exists in a molecular form, is the preferred structural unit for the element as it is found in nature. It is not the preferred structural unit, however, for the element in compounds. Thus, the symbols and numbers found in chemical formulas deal with the number of *atoms* of various elements present; they have nothing to do with molecules of the elements. The formula for a molecule of oxygen is O_2. A molecule of carbon dioxide, CO_2, does not contain an oxygen molecule; it contains two oxygen atoms.

In addition to formulas, compounds have names. Naming compounds is not as simple as naming elements. Although the nomenclature of elements (Sec. 4.9) has been largely left up to the imagination of their discoverers, extensive sets of systematic rules exist for naming compounds. Rules must be used because of the large number of compounds that exist. Chapter 8 is devoted to compound nomenclature, and we will not worry about this problem until then. For the time being, our focus will be on the meaning and significance of chemical formulas; knowing how to name the compounds that the formulas represent is not a prerequisite for understanding the meaning of formulas.

5.5 Subatomic Particles: Protons, Neutrons, and Electrons

Until the closing decades of the nineteenth century, scientists believed that atoms were solid indivisible spheres without substructure. Today this concept is known to be incorrect. Evidence from a variety of sources, some of which will be discussed in Section 5.6, indicates that atoms themselves are made up of smaller, more fundamental particles called *subatomic particles*. **Subatomic particles**, particles smaller than atoms, are the building blocks from which all atoms are made.

Three major types of subatomic particles exist: the proton, the neutron, and the electron. The properties of these subatomic particles that are of most concern to us are mass and electrical charge. A **proton** is a subatomic particle that possesses one unit of positive charge and has a mass of 1.673×10^{-24} g. Protons were discovered in 1886 by the German physicist Eugene Goldstein (1850–1930). An **electron** is a subatomic particle that possesses one unit of negative charge and has a mass of 9.109×10^{-28} g. Electrons were discovered in 1897 by the English physicist Joseph John Thomson (1856–1930). A **neutron** is a subatomic particle that is neutral (i.e., has no charge) and has a mass of 1.675×10^{-24} g. The neutron, the last of the three subatomic particle types to be identified, was characterized in 1932 by the English physicist James Chadwick (1891–1974). Table 5.1 summarizes the mass and electrical charge characteristics for subatomic particles. In this table relative masses are given in addition to actual masses.

Using the relative mass values from Table 5.1, we see that the neutron has a mass value only slightly greater than that of the proton. For most purposes their masses can be considered equal. Both neutrons and protons are very massive particles compared to an electron, being nearly 2000 times heavier. The mass of an electron is almost negligible compared to that of the other two types of subatomic particles.

Electrons and protons, the two types of *charged* subatomic particles, possess the same amount of electrical charge; the character of the charge is, however, opposite

TABLE 5.1 Charges and Masses of the Major Subatomic Particles

	Electron	Proton	Neutron
Charge	−1	+1	0
Actual mass (g)	9.109×10^{-28}	1.673×10^{-24}	1.675×10^{-24}
Relative mass (based on the electron being one unit)	1	1837	1839
Relative mass (based on the neutron being one unit)	0 (1/1839)	1	1

(negative versus positive). The fact that these subatomic particles are charged is most important because of the way in which charged particles interact. *Particles of opposite or unlike charge attract each other; particles of like charge repel each other.* This behavior of charged particles will be of major concern in many of the discussions in later portions of the text.

The arrangement of subatomic particles within an atom is not haphazard. As is shown in Figure 5.6, the atom can be considered to be composed of two regions: (1) a nuclear region and (2) an extranuclear region, which is everything except the nuclear region.

At the center of every atom is a nucleus. The **nucleus** is the center region (core) of an atom and contains within it all protons and neutrons present in the atom. Because of their presence in the nucleus, neutrons and protons are often collectively called *nucleons*. Almost all (over 99.9%) of the mass of an atom is concentrated in its nucleus; all of the heavy subatomic particles (protons and neutrons) are there. A nucleus always carries a positive charge because of the presence of the positively charged protons.

The extranuclear region contains all of the electrons. It is an extremely large region compared to the nucleus. It is mostly empty space. It is a region in which the electrons move rapidly about the nucleus. The motion of the electrons in this extranuclear

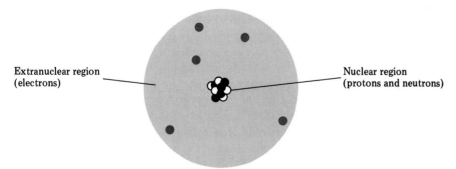

FIGURE 5.6 Arrangement of subatomic particles in an atom (not to scale).

region determines the volume (size) of the atom in the same way as the blade of a fan determines an area by its motion. The volume occupied by the electrons is sometimes referred to as the *electron cloud*. Since electrons are negatively charged, the electron cloud is said to be negatively charged.

An atom as a whole is neutral. How can it be that an entity possessing positive charge (the nuclear region) and negative charge (the extranuclear region or electron cloud) can end up neutral overall? For this to occur the same amount of positive and negative charge must be present in the atom; equal amounts of positive and negative charge cancel each other. Atom neutrality thus requires that there be the same number of electrons and protons present in an atom, which is always the case.

It is important that the size relationships between the parts of an atom be correctly visualized. The nucleus is extremely small (supersmall) compared to the total atomic size. A nonmathematical conceptual model to help visualize this relationship involves clouds. Visualize a large spherical or nearly spherical fluffy cloud in the sky. Let the cloud represent the negatively charged extranuclear region of the atom. Buried deep within the cloud would be the positively charged nucleus, the size of a small pebble. As another example, consider the electron cloud to be the size of a major league baseball park. The nucleus would be no larger than a small fly located somewhere in the region behind second base. As a more quantitative example, consider enlarging (magnifying) the nucleus until it is the size of a baseball (7.4 cm in diameter). If the nucleus were this large, the whole atom would have a diameter of approximately 2.5 mi. The nuclear volume is only 1/100,000 that of the atom's total volume; that is, almost all (over 99.9%) of the volume of an atom is occupied by the electron cloud.

The concentration of almost all of the mass of an atom in the nucleus can best be illustrated by using an example. If a copper penny contained copper nuclei (copper atoms stripped of their electrons) rather than copper atoms (which are mostly empty space), the penny would weigh 190,000,000 tons. Nuclei are indeed very dense matter.

Our just-completed discussion of the makeup of atoms in terms of subatomic particles is based on the existence of three types of subatomic particles: protons, neutrons, and electrons. Actually this model of the atom is an oversimplification. In recent years, as the result of research carried out by nuclear physicists, our picture of the atom has lost its simplicity. Experimental evidence now available indicates that protons and neutrons themselves are made up of even smaller particles. Numerous other particles, with names such as leptons, mesons, and baryons, have been discovered. No theory is yet available that can explain all of these new discoveries relating to the complex nature of the nucleus.

Despite the existence of other nuclear particles we will continue to use the three-subatomic-particle model of the atom. It readily explains almost all chemical observations about atoms. We will have no occasion to deal with any of the recently discovered types of subatomic (or sub-subatomic) particles. Protons, neutrons, and electrons will meet all our needs.

We will also continue to use the concept that atoms are the fundamental building blocks for all types of matter (Sec. 5.1) despite the existence of protons, neutrons, and electrons. This is because under normal conditions subatomic particles do not lead an independent existence for any appreciable length of time. The only way they gain stability is by joining together to form an atom.

5.6 Evidence Supporting the Existence and Arrangement of Subatomic Particles

A significant collection of evidence is consistent with and supports the existence, nature, and arrangement of subatomic particles as given in Section 5.5. Two historically important types of experiments illustrate some of the sources of this evidence. *Discharge-tube experiments* resulted in the original concept that the atom contained negatively and positively charged particles. *Metal-foil experiments* provided evidence for the existence of a nucleus within the atom.

Discharge Tube Experiments

Neon signs, fluorescent lights, and television tubes are all basic ingredients of our modern technological society. The forerunner for all three of these developments was the *gas discharge tube*. Gas discharge tubes also provided some of the first evidence that an atom consisted of still smaller particles (subatomic particles).

The principle behind the operation of a gas discharge tube—that gases at low pressure conduct electricity—was discovered in 1821 by the English chemist Humphry Davy (1778–1829). Subsequently, gas discharge tube studies were carried out by many scientists.

A simplified diagram of a gas discharge tube is shown in Figure 5.7. The apparatus consists of a sealed glass tube containing two metal disks called *electrodes*. The glass tube also has a side arm for attachment to a vacuum pump. During operation, the electrodes are connected to a source of electrical power. (The electrode attached to the positive side of the electrical power source is called the *anode*; the one attached to the negative side is known as the *cathode*.) Use of the vacuum pump allows the amount of gas within the tube to be varied. The smaller the amount of gas present, the lower the pressure within the tube.

Early studies with gas discharge tubes showed that when the tube was almost evacuated (low pressure), electricity flowed from one electrode to the other and the residual gas became luminous (it glowed). Different gases in the tube gave different colors to the glow. After the pressure in the tube was reduced to still lower levels (very

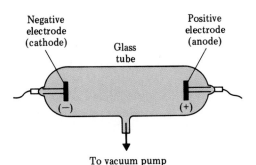

FIGURE 5.7 A simplified version of a gas discharge tube.

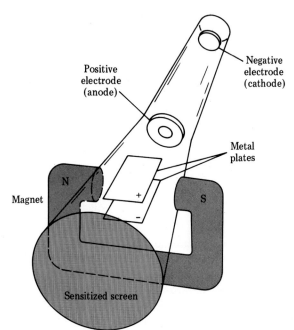

FIGURE 5.8 J. J. Thomson's cathode ray tube involved the use of both electrical and magnetic fields.

little gas remaining), it was found that the luminosity disappeared but the electrical conductance continued, as shown by a greenish glow given off by the tube's glass walls. This glow was the initial discovery of what became known as *cathode rays*. Their discovery marked the beginning of nearly 40 years of discharge tube experimentation that ultimately led to the discovery of both the electron and the proton.

The term *cathode rays* comes from the observation that when an obstacle is placed between the negative electrode (cathode) and the opposite glass wall, a sharp shadow the shape of the obstacle is cast on that wall. This indicates that the rays are coming from the cathode.

Further studies showed that these cathode rays caused certain minerals such as sphalerite (zinc sulfide) to glow. Glass plates were coated with sphalerite and observed under high magnification while being bombarded with cathode rays. The light emitted by the sphalerite coating consisted of many pinpoint flashes. This observation suggested that cathode rays were in reality a stream of extremely small particles.

Joseph John Thomson (1856–1940), an English physicist, provided many facts about the nature of cathode rays. Using a variety of materials as cathodes, he showed that cathode ray production was a general property of matter. By using a specially designed cathode ray tube (see Fig. 5.8), he also found that cathode rays could be deflected by charged plates or a magnetic field. The rays were repelled by the north pole or negative plate and attracted to the south pole or positive plate, thus indicating that they were negatively charged. In 1897, Thomson concluded that cathode rays were streams of negatively charged particles, which today we call *electrons*. Further experiments by others proved that his conclusions were correct.

In 1886, a German physicist, Eugene Goldstein (1850–1930), showed that positive particles were also present in discharge tubes. He used a discharge tube in which the

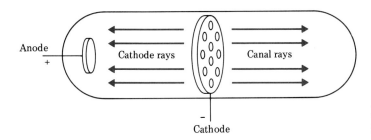

Figure 5.9 A gas discharge tube showing canal rays.

cathode was a metal plate with a large number of holes drilled in it. The usual cathode rays were observed to stream from cathode to anode. In addition, rays of light appeared to stream from each of the holes in the cathode in a direction opposite to that of the cathode rays (see Fig. 5.9). Because these rays were observed streaming through the holes or channels in the cathode, Goldstein called them *canal rays*.

Further research showed that canal rays were of many different types, in contrast to cathode rays, which are only one type, and that the particles making up canal rays were much heavier than those of cathode rays. The type of canal rays produced depended upon the gas in the tube. The simplest canal rays were eventually identified as the particles now called protons.

Canal rays are now known to be gas atoms that have lost one or more electrons. Their origin and behavior in a discharge tube can be understood as follows. Electrons (cathode rays) emitted from the cathode collide with residual gas molecules (air) on the way to the anode. Some of these electrons have enough energy to knock electrons away from the gas molecules, leaving behind a positive particle (the remainder of the gas molecule). These positive particles are attracted to the cathode, and some of them pass through the holes or channels. The fact that atoms, under certain conditions, can lose electrons will be discussed further in Section 7.4.

On the basis of discharge tube experiments, Thomson proposed in 1898 that the atom was composed of a sphere of positive electricity containing most of the mass, and that small negative electrons were attached to the surface of the positive sphere. He postulated that a high voltage could pull off surface electrons to produce cathode rays. Thomson's model of the atom, sometimes referred to as the "raisin muffin" or "plum pudding" model—with the electrons as the raisins—is now known to be incorrect. Its significance is that it set the stage for an experiment, commonly called the gold-foil experiment, that led to the currently accepted arrangement of protons and electrons in the atom.

Metal Foil Experiments

In 1911 Ernest Rutherford (1871–1937) designed an experiment to test the Thomson model of the atom. In this experiment thin sheets of metal foil were bombarded by alpha particles from a radioactive source. Alpha particles, which are positively charged, are ejected at high speeds from some radioactive materials. The phenomenon of radioactivity had been discovered in 1896 and gave further evidence that electrical charges existed within the atom. Gold was chosen as the target metal because it is

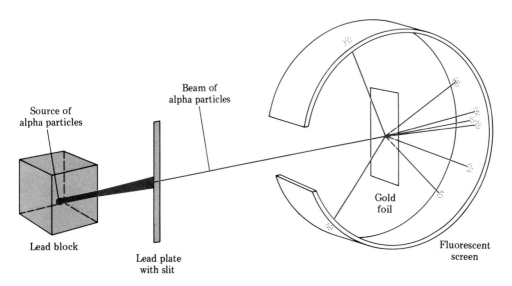

FIGURE 5.10 Rutherford's gold foil-alpha particle experiment. Most of the alpha particles went straight through the foil, but a few were deflected at large angles.

easily hammered into very thin sheets. The experimental setup for Rutherford's experiment is shown in Figure 5.10. Alpha particles do not appreciably penetrate lead, so a lead plate with a slit was used to produce a narrow alpha particle beam. Each time an alpha particle hit the fluorescent screen, a flash of light was produced.

Rutherford expected that all the alpha particles, since they were so energetic, would pass straight through the thin gold foil. His reasoning was based on the Thomson model, in which the mass and positive charge of the gold atoms were distributed uniformly through each atom. As each positive particle neared the foil, Rutherford assumed that it would be confronted by a uniform positive charge. All particles would be affected the same way (no deflection), which would support the Thomson model of the atom.

The results from the experiment were very surprising. Most of the particles—more than 99%—went straight through as expected. A few, however, were appreciably deflected by something that had to be much heavier than the alpha particles themselves. A very few particles were deflected almost directly back toward the alpha particle source. Similar results were obtained when elements other than gold were used as targets.

Extensive study of the results of his experiments led Rutherford to propose the following explanation.

1. A very dense, small nucleus exists in the center of the atom. This nucleus contains most of the mass of the atom and all of the positive charge.
2. Electrons occupy most of the total volume of the atom and are located outside the nucleus.
3. When an alpha particle scores a direct hit on a nucleus, it is deflected back along the incoming path.

4. A near miss of a nucleus by an alpha particle results in repulsion and deflection.
5. Most of the alpha particles pass through without any interference, because most of the atomic volume is empty space.
6. Electrons have so little mass that they do not deflect the much larger alpha particles (an alpha particle is almost 8000 times heavier than an electron).

Many other experiments have since verified Rutherford's conclusion that at the center of an atom there is a nucleus that is very small and very dense.

5.7 Atomic Number and Mass Number

The characteristic of an atom that identifies what kind of atom it is is the atomic number. The **atomic number** is equal to the number of protons in the nucleus of an atom. Atomic numbers are always integers (whole numbers). All atoms of a given element must contain the same number of protons. If two atoms differ in the number of protons present, they must be atoms of two different elements.

Since an atom has the same number of electrons as protons, the atomic number also specifies the number of electrons present.

Atomic number = number of protons = number of electrons

A second necessary quantity in specifying atomic identities is the mass number. The **mass number** is equal to the number of protons plus neutrons in the nucleus of the atom, that is, the total number of nucleons present. The mass of an atom is almost totally accounted for by the protons and neutrons (Sec. 5.5); hence the term mass number. Like the atomic number, the mass number is always an integer.

Mass number = number of protons + number of neutrons

Knowing the atomic number and mass number of an atom uniquely specifies the atom's makeup in terms of subatomic particles. The following equations show the relationship between subatomic particles and the two numbers.

Number of protons = atomic number
Number of electrons = atomic number
Number of neutrons = mass number − atomic number

EXAMPLE 5.2

An atom has an atomic number of 12 and a mass number of 26. Determine

(a) the number of protons present (b) the number of neutrons present
(c) the number of electrons present

Solution

(a) The number of protons is 12. The atomic number is always equal to the number of protons present.

(b) The number of neutrons is 14. The number of neutrons is always obtained by subtracting the atomic number from the mass number.

$$\underset{\text{mass number}}{(\text{Protons + neutrons})} - \underset{\text{atomic number}}{\text{protons}} = \text{neutrons}$$

$$26 - 12 = 14$$

(c) The number of electrons is 12. In a neutral atom the number of protons and the number of electrons are always equal.

PRACTICE EXERCISE 5.2

For an atom having an atomic number of 6 and a mass number of 14, determine

(a) the number of protons present (b) the number of neutrons present
(c) the number of electrons present

Ans. (a) 6 protons; (b) 8 neutrons; (c) 6 electrons.

Similar exercises: Problems 5.49 and 5.50

An alphabetical listing of the 109 currently known elements, along with selected information about each element, is printed on the inside back cover of this book. One of the pieces of information given for each element is its atomic number. If you carefully checked the atomic number data, you would find that an element exists for each atomic number in the numerical sequence 1 through 109; no numbers in the sequence are missing. The existence of an element that corresponds to each of these numbers is an indication of the order existing in nature.

The atomic numbers of the elements are also found inside the front cover of the text. The diagram there is a *periodic table*, a graphical presentation of selected characteristics of the elements. (The periodic table will be considered in detail in Chapter 6.) We note at this time that each box in the periodic table designates an element, with that element's symbol in the center of the box. The number above the symbol is the element's atomic number. The elements are arranged in the periodic table in order of increasing atomic number.

Mass numbers are not tabulated in a manner similar to atomic numbers because, as we learn in the next section, most elements lack a unique mass number.

The chemical properties of an atom, which are the basis for its identification, are determined by the number and arrangement of the electrons about the nucleus. That electrons are the determining factor for chemical properties is a logical conclusion if one considers that when two atoms interact, the outer part (electrons) of the one will interact with the outer part (electrons) of the other. The small nuclear centers will never come in contact with each other.

The number of electrons about a nucleus is determined by the number of protons in the nucleus; there must be charge balance. Hence, the number of protons (which is the atomic number) characterizes an atom. All atoms with the same atomic number will have the same chemical properties and will be atoms of the same element.

Previously, in Section 4.7, an element was defined as a pure substance that cannot be broken down into simpler substances by ordinary chemical means. Although this is a good historical definition for an element, we can now give a more rigorous definition using the concept of atomic number. An element is a pure substance in which all atoms present have the same atomic number; that is, all atoms have the same number of protons.

5.8 Isotopes

Though all atoms of a given element have the same atomic number and therefore the same number of protons (and electrons), they need not all be identical. They can differ in the number of neutrons present. The presence of one or more additional neutrons in the tiny nucleus of an atom has essentially no effect on the way it behaves chemically. For example, all oxygen atoms have eight protons and eight electrons. Most oxygen atoms also contain eight neutrons. Some oxygen atoms exist, however, that contain nine neutrons, and a few exist that contain ten neutrons. Thus, three different kinds of oxygen atoms exist, all with the same chemical properties. Three oxygen isotopes are said to exist. Formally defined, **isotopes** are atoms that have the same atomic number and differing mass numbers. Isotopes will always have the same number of protons (and electrons) and differing numbers of neutrons.

When it is necessary to distinguish between isotopes, the following notation is used.

$$\text{(mass number)} \, ^A_Z \text{Symbol} \, \text{(atomic number)}$$

The atomic number, whose general symbol is Z, is written as a subscript to the left of the elemental symbol for the atom. The mass number, whose general symbol is A, is also written to the left of the elemental symbol, but as a superscript. By this symbolism, the three previously mentioned oxygen isotopes would be designated, respectively, as

$$^{16}_{8}O \quad ^{17}_{8}O \quad \text{and} \quad ^{18}_{8}O$$

The existence of isotopes adds clarification to the wording used in some of the statements of atomic theory (Sec. 5.1). Statement 1 reads: "All matter is made up of small particles called atoms, of which 109 different 'types' are known." It should now be apparent why the word types was put in quotation marks. Because of the existence of isotopes, atoms of each type are similar, but not identical. Atoms of a given element are similar in that they have the same atomic number, but not identical since they may have different mass numbers.

Statement 2 reads: "All atoms of a given type are similar to one another and significantly different from all other types." All atoms of an element are similar in chemical properties and differ significantly from atoms of other elements with different chemical properties.

Most elements occurring naturally are mixtures of isotopes. The various isotopes of a given element are of varying abundance; usually one isotope is predominant. Typical of this situation is the element magnesium, which exists in nature in three isotopic forms: $^{24}_{12}Mg$, $^{25}_{12}Mg$, and $^{26}_{12}Mg$. The percentage abundances for these three

Chapter 5 · Atoms, Molecules, Formulas, and Subatomic Particles

TABLE 5.2 Naturally Occurring Isotopic Abundances for Some Common Elements

Element	Isotope	Percent Natural Abundance	Isotopic Mass (amu)
Hydrogen	$^{1}_{1}H$	99.985	1.0078
	$^{2}_{1}H$	0.015	2.0141
Carbon	$^{12}_{6}C$	98.89	12.0000
	$^{13}_{6}C$	1.11	13.0033
Nitrogen	$^{14}_{7}N$	99.63	14.0031
	$^{15}_{7}N$	0.37	15.0001
Oxygen	$^{16}_{8}O$	99.759	15.9949
	$^{17}_{8}O$	0.037	16.9991
	$^{18}_{8}O$	0.204	17.9992
Sulfur	$^{32}_{16}S$	95.0	31.9721
	$^{33}_{16}S$	0.76	32.9715
	$^{34}_{16}S$	4.22	33.9679
	$^{36}_{16}S$	0.014	35.9671
Chlorine	$^{35}_{17}Cl$	75.53	34.9689
	$^{37}_{17}Cl$	24.47	36.9659
Copper	$^{63}_{29}Cu$	69.09	62.9298
	$^{65}_{29}Cu$	30.91	64.9278
Titanium	$^{46}_{22}Ti$	7.93	45.95263
	$^{47}_{22}Ti$	7.28	46.9518
	$^{48}_{22}Ti$	73.94	47.94795
	$^{49}_{22}Ti$	5.51	48.94787
	$^{50}_{22}Ti$	5.34	49.9448
Uranium	$^{234}_{92}U$	0.0057	234.0409
	$^{235}_{92}U$	0.72	235.0439
	$^{238}_{92}U$	99.27	238.0508

isotopes are, respectively, 78.70%, 10.13%, and 11.17%. Percentage abundances are number percentages (number of atoms) rather than mass percentages. A sample of 10,000 magnesium atoms would contain 7870 $^{24}_{12}Mg$ atoms, 1013 $^{25}_{12}Mg$ atoms, and 1117 $^{26}_{12}Mg$ atoms. Table 5.2 gives natural isotopic abundances and isotopic masses for selected elements. The units used for specifying the mass of the various isotopes (last column of Table 5.2) will be discussed in Section 5.9.

The percentage abundances of the isotopes of an element may vary slightly in samples obtained from different locations, but such variations are ordinarily extremely small. We will assume in this text the isotopic composition of an element is a constant.

Isotopic masses, although not whole numbers, have values that are very close to whole numbers. This fact can be verified by looking at the numbers in the last column of Table 5.2. If an isotopic mass is rounded off to the closest whole number, this value is the same as the mass number of the isotope. This statement can be verified by comparing the second and fourth columns in Table 5.2.

Isotopes do not exist for all elements. Twenty-three elements have only one naturally occurring form. Of the simpler elements (atomic number of 18 or less) those with only one form are

$$^{9}_{4}Be \quad ^{19}_{9}F \quad ^{23}_{11}Na \quad ^{27}_{13}Al \quad ^{31}_{15}P$$

It is possible for isotopes of two different elements to have the same mass number. For example, the element iron (atomic number 26) exists in nature in four isotopic forms, one of which is $^{58}_{26}Fe$. The element nickel, with an atomic number two units greater than that of iron, exists in nature in five isotopic forms, one of which is $^{58}_{28}Ni$. Thus, atoms of both iron and nickel exist with a mass number of 58. Thus, mass numbers are not unique for elements as are atomic numbers. Atoms of different elements that have the same mass number are called isobars; $^{58}_{26}Fe$ and $^{58}_{28}Ni$ are isobars. **Isobars** are atoms that have the same mass number but different atomic numbers. Even though atoms of two *different* elements can have the same mass number (isobars), they cannot have the same atomic number. All atoms of a given atomic number must necessarily be atoms of the same element.

Figure 5.11 gives information about number of isotopes and mass number identity for the simpler elements (atomic numbers 1–20). The patterns in both number of isotopes and mass number are relatively simple for the first 15 elements. For the first seven elements the number of isotopes varies in the pattern

$$2\text{–}2\text{–}2\text{–}1\text{–}2\text{–}2\text{–}2$$

FIRST SEVEN ELEMENTS—NUMBER OF ISOTOPES (2–2–2–1–2–2–2)

H	He	Li	Be	B	C	N
1	3	(no 5)	(no 8)	10	12	14
2	4	6	9	11	13	15
		7				

NEXT EIGHT ELEMENTS—NUMBER OF ISOTOPES (3–1–3–1–3–1–3–1)

O	F	Ne	Na	Mg	Al	Si	P
16	19	20	23	24	27	28	31
17		21		25		29	
18		22		26		30	

NEXT FIVE ELEMENTS—NUMBER OF ISOTOPES (irregular pattern)

S	Cl	Ar	K	Ca
32	35	36	39	40
33	37	38	40	42
34		40	41	43
36				44
				46
				48

FIGURE 5.11 Variance in mass numbers and number of isotopes for elements 1–20. (The mass numbers that occur for an element are listed underneath its symbol with the most abundant isotope in color.)

For the next eight elements the pattern is

$$3-1-3-1-3-1-3-1$$

Mass numbers are sequential for the first 15 elements with the exception that mass numbers 5 and 8 are skipped. (There are no naturally occurring atoms with mass numbers 5 and 8.) Element 16 (S) is the first element with nonsequential mass numbers for its isotopes. The first isobars occur at mass number 36 (S and Ar).

EXAMPLE 5.3

Indicate whether the members of each of the following pairs are isotopes, isobars, or neither.

(a) $^{27}_{13}X$ and $^{28}_{14}Y$ (b) $^{46}_{20}X$ and $^{48}_{20}Y$ (c) $^{36}_{16}X$ and $^{36}_{18}Y$
(d) an atom X with eight protons and nine neutrons and an atom Y with nine protons and ten neutrons

Solution

(a) These atoms are not isotopes or isobars. Isotopes must have the same atomic number and isobars must have the same mass number. Neither is the case here.
(b) These atoms are isotopes. Both atoms have the same atomic number of 20. Isotopes differ from each other only in neutron count, which is the case here. Atom X has 26 neutrons, and atom Y has 28 neutrons.
(c) These atoms are isobars. They have the same mass number (36) and differing atomic numbers (16 and 18).
(d) These atoms are not isotopes or isobars. They are not isotopes because a differing number of protons means differing atomic numbers. They are not isobars because the mass numbers differ; 17 for atom X and 19 for atom Y.

PRACTICE EXERCISE 5.3

Indicate whether the members of each of the following pairs are isotopes, isobars, or neither.

(a) $^{42}_{20}X$ and $^{43}_{20}Y$ (b) $^{40}_{19}X$ and $^{40}_{20}Y$ (c) $^{44}_{20}X$ and $^{45}_{21}Y$
(d) an atom X with 20 protons and 21 neutrons and an atom Y with 19 protons and 22 neutrons.

Ans. (a) isotopes; (b) isobars; (c) neither; (d) neither.

Similar exercises: Problems 5.59 and 5.60

5.9 Atomic Masses

The mass of an atom of a specific element can have one of several values if the element exists in isotopic forms. For example, oxygen atoms can have any one of three masses, since three isotopes of oxygen exist. Because of the existence of isotopes, it might seem necessary to specify isotopic identity every time atomic masses must be dealt with; however, this is not the case. In practice, isotopes are seldom mentioned in discussions involving atomic masses; instead, the atoms of an element are treated as if they all had a single mass. The mass used is an average relative mass that takes into account the existence of isotopes. The use of this average relative mass concept reduces the number of masses needed for calculations from many hundreds to 109—one for each element. These average masses are called *atomic masses*.

An **atomic mass** is the relative mass of an average atom of an element on a scale using the $^{12}_{6}C$ atom as the reference. The key to understanding this definition of atomic mass involves understanding clearly what is meant by the two terms *relative mass* and *average atom*. Let us consider in detail their meanings.

Relative Mass

The usual standards of mass, such as grams or pounds, are not convenient for use with atoms, because very small numbers are always encountered. For example, the mass in grams of a $^{238}_{92}U$ atom, one of the heaviest atoms known, is 3.95×10^{-22}. To avoid repeatedly encountering such small numbers scientists have chosen to work with relative rather than actual mass values.

A relative mass value for an atom is the mass of that atom relative to some standard rather than the actual mass value of the atom in grams. The term relative means "as compared to." The choice of the standard is arbitrary; this gives scientists control over the magnitude of the numbers on the relative scale, thus avoiding very small numbers.

For most purposes in chemistry, relative mass values serve just as well as actual mass values. Knowing how many times heavier one atom is than another, information obtainable from a relative mass scale, is just as useful as knowing the actual mass values of the atoms involved. Example 5.4 illustrates the procedures involved in constructing a relative mass scale and also points out some of the characteristics of such a scale.

EXAMPLE 5.4

Construct a relative mass scale for the hypothetical atoms A, B, and C given the following information about them.

1. Atoms of A are four times heavier than those of B.
2. Atoms of B are three times heavier than those of C.

Solution
Atoms of C are the lightest of the three types of atoms. We will arbitrarily assign atoms of C a mass value of one unit. The unit name can be anything we wish, and we shall choose "snurk." On this basis, one atom of C has a mass value of 1

snurk. Atoms of C will be our scale reference point. Atoms of B will have a mass value of 3 snurks (three times as heavy as C) and A atoms a value of 12 snurks (four times as heavy as B).

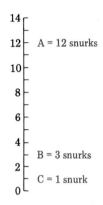

The name chosen for the mass unit was arbitrary. The assignment of the value 1 for the mass of C, the reference point on the scale, was also arbitrary. What if we had chosen to call the unit a "bleep" and had chosen a value of 3 for the mass of an atom of C? If this had been the case, the resulting relative scale would have appeared as

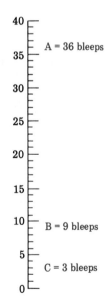

Which of the above relative scales is the "best" scale? The answer is that the scales are equivalent. The relationships between the masses of A, B, and C are the same on the two scales, even though the reference points and unit names

differ. On the "snurk" scale, for example, A is four times heavier than B (12/3); on the "bleep" scale, A is also four times heavier than B (36/9).

Notice that we did not need to know the actual masses of A, B, and C to set up either the "snurk" or "bleep" scale. All that is needed to set up a relative scale is a set of interrelationships among quantities. One value—the reference point—is arbitrarily assigned, and all other values are determined by using the known interrelationships.

The information given at the start of this example is sufficient to set up an infinite number of relative mass scales. Each scale would differ from the others in choice of reference point and unit name. All the scales would, however, be equivalent to each other, and each scale would provide all of the mass relationships obtainable from an actual mass scale except for actual mass values.

PRACTICE EXERCISE 5.4

Construct a relative mass scale for the hypothetical atoms A, B, and C given the following information about them.

(1) Atoms of A are two times as heavy as those of B.
(2) Atoms of B are four times as heavy as those of C.
(3) The reference point for the scale is C = 3.00 sloops.

Ans. C = 3.00 sloops, B = 12.00 sloops, and A = 24.00 sloops.

Similar exercises: Problems 5.67 and 5.68

A relative scale of atomic masses has been set up in a manner similar to that used in Example 5.4. The unit is called the *atomic mass unit*, abbreviated amu. The arbitrary reference point involves a particular isotope of carbon, $^{12}_{6}C$. The mass of this isotope is set at 12.00000 amu. The masses of all other atoms are then determined relative to that of $^{12}_{6}C$. For example, if an atom is twice as heavy as a $^{12}_{6}C$ atom, its mass is 24.00000 amu on the scale, and if an atom weighs half as much as a $^{12}_{6}C$ atom, its scale mass is 6.00000 amu.

The masses of all atoms have been determined relative to each other experimentally. Actual values for the masses of selected isotopes on the $^{12}_{6}C$ scale are given in Figure 5.12.

Reread the formal definition of atomic mass given at the start of this section. Note how $^{12}_{6}C$ is mentioned explicitly in the definition because of the central role it plays in the setting up of the relative atomic mass scale.

On the basis of the values given in Figure 5.12 it is possible to state, for example: $^{238}_{92}U$ is 4.256 times as heavy as $^{56}_{26}Fe$ (238.05 amu/55.93 amu = 4.256), and $^{56}_{26}Fe$ is 2.798 times as heavy as $^{20}_{10}Ne$ (55.93 amu/19.99 amu = 2.798). We do not need to know the actual masses of the atoms involved to make such statements; relative masses are sufficient to calculate the information.

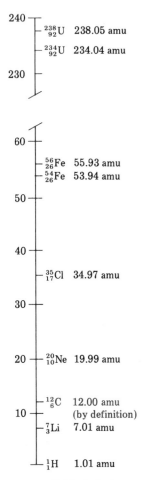

FIGURE 5.12 Relative masses of selected isotopes on the $^{12}_{6}C$ atomic mass scale.

Average Atom

Since isotopes exist, the mass of an atom of a specific element can have one of several values. For example, oxygen atoms can have any one of three masses, since three isotopes exist: $^{16}_{8}O$, $^{17}_{8}O$, and $^{18}_{8}O$. Despite mass variances among isotopes, the atoms of an element are treated as if they all had a single common mass. The common mass value used is a *weighted average mass*, which takes into account the natural abundances and atomic masses of the isotopes of an element.

The validity of the weighted average mass concept rests on two points. First, extensive studies of naturally occurring elements have shown that the percent abundance of the isotopes of a given element is generally constant. No matter where the element sample is obtained on Earth, it generally contains the same percentage of each isotope. Because of these constant isotopic ratios, the mass of an "average atom" does not vary. Second, chemical operations are always carried out with very large numbers of atoms. The tiniest piece of matter visible to the eye contains more atoms than can be counted by a person in a lifetime. The numbers are so great that any collection of atoms a chemist works with will be representative of naturally occurring isotopic ratios.

Weighted Averages

Atomic masses are weighted averages calculated from the following three pieces of information.

1. The *number* of isotopes that exist for the element.
2. The *isotopic mass* for each isotope, that is, the relative mass of each isotope on the $^{12}_{6}C$ scale.
3. The *percent abundance* of each isotope.

Table 5.2 gives these data for selected elements.

Examples 5.5 and 5.6 illustrate the operations needed to calculate weighted averages. Example 5.5 is a general exercise concerning weighted averages, and Example 5.6 illustrates the calculation of an atomic mass by the method of weighted averages.

EXAMPLE 5.5

Five aspirin tablets are examined in the laboratory. Two of the tablets are found to contain $21\overline{0}$ mg of aspirin, two to contain $2\overline{00}$ mg of aspirin, and one to contain $17\overline{0}$ mg of aspirin. What is the average aspirin content, in milligrams, of the tablets examined?

Solution

Let us solve this problem using two different methods. The first method involves procedures familiar to you—the "normal" way of taking an average. By this method, the average is found by dividing the sum of the numbers by the number of values summed.

$(21\overline{0} + 21\overline{0} + 2\overline{00} + 2\overline{00} + 17\overline{0})$ mg $= 990$ mg   (calculator answer)

$\phantom{(21\overline{0} + 21\overline{0} + 2\overline{00} + 2\overline{00} + 17\overline{0}) \text{ mg }}= 99\overline{0}$ mg   (correct answer)

$$\frac{99\overline{0} \text{ mg}}{5} = 198 \text{ mg} \quad \text{(calculator and correct answer)}$$

Now let us solve this same problem again, this time treating it as a "weighted average" problem. To do this, we organize the given information in a different way. In our list of given information we have three kinds of aspirin tablets: those that contain $21\overline{0}$, $2\overline{00}$ and $17\overline{0}$ mg of aspirin.

Two of the five tablets (40.0%) contain $21\overline{0}$ mg of aspirin.
Two of the five tablets (40.0%) contain $2\overline{00}$ mg of aspirin.
One of the five tablets (20.0%) contains $17\overline{0}$ mg of aspirin.

We will use the data in this "percent form" for our weighted average calculation.

To find the weighted average we multiply each number ($21\overline{0}$, $2\overline{00}$, and $17\overline{0}$) by its fractional abundance, that is, its percentage expressed in decimal form, and then we sum the products from the multiplications.

$0.400 \times 21\overline{0}$ mg $= 84.0$ mg
$0.400 \times 2\overline{00}$ mg $= 80.0$ mg
$0.200 \times 17\overline{0}$ mg $= 34.0$ mg
$\phantom{0.200 \times 17\overline{0} \text{ mg } =\ } \overline{198.0 \text{ mg}}$ (same average value as before)

This averaging method, although it appears somewhat more involved than the "normal" method, is the one that must be used in calculating atomic masses because of the form in which data about isotopes are obtained. The percent abundances of isotopes are experimentally determinable quantities. The total number of atoms of various isotopes present in nature, a prerequisite for using the "normal" average method, is not easily determined. Therefore, we use the "percent" method.

PRACTICE EXERCISE 5.5

Twenty aspirin tablets are examined in the laboratory. It is found that 8 of them (40.0%) contain $21\overline{0}$ mg of aspirin, 5 of them (25.0%) contain $2\overline{00}$ mg of aspirin, and 7 of them (35.0%) contain $17\overline{0}$ mg of aspirin. What is the average aspirin content, in milligrams, for these tablets?   *Ans.* 194 mg

Similar exercises: Problems 5.69 and 5.70

EXAMPLE 5.6

Silicon occurs in nature in three isotopic forms: $^{28}_{14}$Si (92.21% abundance), $^{29}_{14}$Si (4.70% abundance), and $^{30}_{14}$Si (3.09% abundance). Their relative masses are 27.977, 28.976, and 29.974 amu, respectively. Calculate the atomic mass of silicon.

Solution

The atomic mass of an element is calculated using the weighted average method illustrated in Example 5.5. Each of the isotopic masses is multiplied by the fractional abundance associated with that mass, and then the products are summed.

$$0.9221 \times 27.977 \text{ amu} = 25.797591 \text{ amu} = 25.80 \text{ amu}$$
$$0.0470 \times 28.976 \text{ amu} = 1.361872 \text{ amu} = 1.36 \text{ amu}$$
$$0.0309 \times 29.974 \text{ amu} = 0.9261966 \text{ amu} = 0.926 \text{ amu}$$

Note that the method for converting percentages to fractional abundances is always the same. The decimal point in the percentage is moved two places to the left. For example,

$$92.21\% \text{ becomes } 0.9221$$

Significant figures are always an important part of an atomic mass calculation. Both isotopic masses and percent abundances are experimentally determined numbers.

Summing, to obtain the atomic mass, gives

$$(25.80 + 1.36 + 0.926) \text{ amu} = 28.086 \text{ amu} \quad \text{(calculator answer)}$$
$$= 28.09 \text{ amu} \quad \text{(correct answer)}$$

Since the numbers 25.80 and 1.36 are known only to the hundredths place, the answer can be expressed only to the hundredths place.

The above calculation involved an element that exists in three isotopic forms. An atomic mass calculation for an element having four isotopic forms would be carried out in an almost identical fashion. The only difference would be four products to calculate (instead of three) and four terms in the resulting sum.

PRACTICE EXERCISE 5.6

Magnesium occurs in nature in three isotopic forms: $^{24}_{12}\text{Mg}$ (78.70% abundance), $^{25}_{12}\text{Mg}$ (10.13% abundance), and $^{26}_{12}\text{Mg}$ (11.17% abundance). Their relative masses are 23.985, 24.986, and 25.983 amu, respectively. Calculate the atomic mass of magnesium.   *Ans.* 24.31 amu.

Similar exercises: Problems 5.71–5.74

In Example 5.6 the atomic mass of silicon was calculated to be 28.09 amu. How many silicon atoms have a mass of 28.09 amu? The answer is none. Silicon atoms have a mass of 27.977 amu, 28.976 amu, or 29.974 amu, depending on which isotope they are. The mass 28.09 amu is the mass of an "average" silicon atom. It is this average value that is used in calculations even though no silicon atoms have masses equal to this average value. Only in the case where there are no isotopes of an element (all atoms are the same kind) will the isotopic mass and the atomic mass be the same.

Atomic masses are subject to change and they do change. Every two years an updated atomic mass listing is published by the International Union of Pure and Applied Chemistry (IUPAC). This update, produced by an international committee of chemists, takes into account all new research on isotopic abundances.

New atomic mass values are almost always more precise than the older values they replace. This is a reflection of the increasingly sophisticated instrumentation available to current researchers, which enables them to make more precise measurements.

Illustrative of atomic mass changes that occur are those found in the 1989 and 1991 IUPAC reports. The atomic masses of five elements were updated.

1989 changes:	Nickel	58.6934 amu instead of 58.69 amu
	Antimony	121.757 amu instead of 121.75 amu
1991 changes:	Indium	114.818 amu instead of 114.82 amu
	Tungsten	183.84 amu instead of 183.85 amu
	Osmium	190.23 amu instead of 190.2 amu

Atomic mass revisions such as these explain why various textbooks (and periodic tables) often differ in a few atomic mass values. The differing values come from different biannual IUPAC reports on atomic masses.

The precision of atomic mass values varies from element to element. For example, we have

B	10.811	amu
F	18.9984032	amu
Si	28.0855	amu
Pb	207.2	amu

What causes such precision variance? The key factor is the constancy of isotopic percentage abundance measurements between various samples of an element. Although all elements have an essentially constant set of isotopic percentage abundances, there is slight variance between samples obtained from different sources. All variances are small; however, for some elements they are greater than for others. When only one form of an element occurs in nature (no isotopes), such as F, maximum precision can be obtained for atomic mass.

An atomic mass cannot be calculated for all elements. Recall, from Section 4.8, that not all elements are naturally occurring substances. Twenty-one of the known elements are "synthetic," having been produced in the laboratory from naturally occurring elements. Obviously a weighted average atomic mass cannot be calculated for these laboratory-produced elements, since the amount of each isotope produced is arbitrary, governed by the needs of the scientists involved in the production work.

Tabulations of atomic masses do contain entries for the synthetic elements. Such entries are the mass number of the most stable isotope of the synthetic element. (All isotopes of all synthetic elements are unstable.) Such mass numbers are always enclosed in parentheses to distinguish them from calculated atomic masses. Note the presence of such entries in both the atomic mass listing inside the back cover and the periodic table inside the front cover.

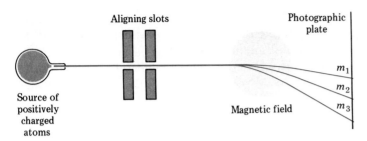

FIGURE 5.13 A simplified diagram of a mass spectrometer, an instrument used to obtain information about isotopic masses and abundances.

Before leaving the subject of atomic masses, we need to consider one additional question. How do scientists determine the abundances and masses of the various isotopes of an element? A device known as a *mass spectrometer* is the key to obtaining such information.

A schematic diagram of a mass spectrometer is given in Figure 5.13. The important components of this instrument are a source of charged particles of the element under investigation, an aligning system to create a narrow beam of the charged particles, a magnetic field to affect the path of the charged particles, and a means of detecting the charged particles (such as a photographic plate).

Suppose oxygen gas containing all three isotopes is admitted to the instrument. The gaseous molecules are bombarded with an energetic electron beam, producing positively charged oxygen species. The aligning system produces a narrow beam of these charged atoms, which then enters the magnetic field. The most massive particles (heaviest isotope) are not deflected by the magnetic field as much as the less massive ones, so the charged atoms are divided into separate beams that strike the photographic plate at different points depending on their masses. The more abundant isotopes will create more intense lines on the plate. The relative intensities of the lines correlates exactly with the relative abundances of the isotopes.

Key Terms

The new terms or concepts defined in this chapter are

atom (Sec. 5.1) The smallest particle of an element that can exist and still have the properties of the element.

atomic number (Sec. 5.7) The number of protons in the nucleus of an atom.

atomic theory of matter (Sec. 5.1) A set of five statements that summarize modern-day scientific thought about atoms.

atomic mass (Sec. 5.9) The relative mass of an average atom of an element on a scale using the $^{12}_{6}C$ atom as the reference.

chemical formula (Sec. 5.4) A notation made up of the symbols of the elements present in a compound with numerical subscripts (located to the right of each symbol) that indicate the number of atoms of each element present in a formula unit.

electron (Sec. 5.5) Subatomic particle, located outside the nucleus of an atom, that has a negative charge and a mass of 9.109×10^{-28} g.

element (Sec. 5.7) A pure substance in which all atoms present have the same atomic number; that is, all atoms have the same number of protons.

heteroatomic molecule (Sec. 5.2) Molecule in which two or more different kinds of atoms are present.

homoatomic molecule (Sec. 5.2) Molecule in which all atoms present are the same kind.
isobars (Sec. 5.8) Atoms that have the same mass number but different atomic numbers.
isotopes (Sec. 5.8) Atoms that have the same atomic number but different mass numbers.
mass number (Sec. 5.7) The number of protons plus neutrons in the nucleus of an atom, that is, the total number of nucleons present.
molecule (Sec. 5.2) A group of two or more atoms that functions as a unit because the atoms are tightly bound together.

neutron (Sec. 5.5) Subatomic particle, located within the nucleus of an atom, that has no charge and has a mass of 1.675×10^{-24} g.
nucleus (Sec. 5.5) The center region of an atom that contains within it all protons and neutrons present in the atom.
proton (Sec. 5.5) Subatomic particle, located within the nucleus of an atom, that has a positive charge and a mass of 1.673×10^{-24} g.
subatomic particle (Sec. 5.5) A particle smaller than an atom that is a building block from which the atom is made.

Practice Problems

Atoms and Molecules (Secs. 5.1 and 5.2)

5.1 Which of the following concepts are not consistent with the statements of modern-day atomic theory?
 (a) Atoms are the basic building blocks for all kinds of matter.
 (b) Different "types" of atoms exist.
 (c) All atoms of a given "type" are identical.

5.2 Which of the following concepts are not consistent with the statements of modern-day atomic theory?
 (a) Only whole atoms can participate in chemical reactions.
 (b) Atoms change identity during chemical change processes.
 (c) 113 different "types" of atoms are known.

5.3 How does an atom differ from a molecule?

5.4 How do molecules of elements differ from molecules of compounds?

5.5 Indicate whether each of the following statements is *true* or *false*. If a statement is false, change it to make it true. (Such a rewriting should involve more than merely converting the statement to the negative of itself.)
 (a) Molecules must contain three or more atoms.
 (b) The atom is the limit of chemical subdivision for an element.
 (c) All compounds have molecules as their basic structural unit.
 (d) A molecule of a compound must be heteroatomic.
 (e) There is only one kind of molecule for any given molecular substance.

5.6 Indicate whether each of the following statements is *true* or *false*. If a statement is false, change it to make it true. (Such a rewriting should involve more than merely converting the statement to the negative of itself.)
 (a) A molecule of an element may be homoatomic or heteroatomic depending on which element is involved.
 (b) The limit of chemical subdivision for a molecular compound is a molecule.
 (c) Heteroatomic molecules do not maintain the properties of their constituent elements.
 (d) Only one kind of atom may be present in a homoatomic molecule.
 (e) The main difference between molecules of elements and molecules of compounds is the number of atoms they contain.

5.7 What is the approximate size, in centimeters, of an atom?

5.8 What is the approximate mass, in grams, of an atom?

Chemical Formulas (Sec. 5.4)

5.9 The notation HF stands for a compound (hydrogen fluoride) rather than an element. How can you tell this from the notation used?

5.10 The notation Hf stands for an element (hafnium) rather than a compound. How can you tell this from the notation used?

5.11 On the basis of its formula, classify each of the following substances as an element or compound.
 (a) $NaClO_2$ (b) CO
 (c) S_8 (d) Al

5.12 On the basis of its formula, classify each of the following substances as an element or compound.
 (a) AlN (b) CO_2
 (c) Co (d) O_3

5.13 Write formulas for the following substances using the information given about a molecule of each substance.
(a) A molecule of vitamin A contains 20 atoms of carbon, 30 atoms of hydrogen, and 1 atom of oxygen.
(b) A molecule of sulfuric acid contains 2 atoms of hydrogen, 1 atom of sulfur, and 4 atoms of oxygen.
(c) A molecule of elemental phosphorus is tetraatomic.
(d) A molecule of hydrogen cyanide, a compound that contains hydrogen, carbon, and nitrogen, is triatomic.

5.14 Write formulas for the following substances using the information given about a molecule of each substance.
(a) A molecule of nicotine contains 10 atoms of carbon, 14 atoms of hydrogen, and 2 atoms of nitrogen.
(b) A molecule of ethyl alcohol contains 2 carbon atoms, 6 hydrogen atoms, and 1 oxygen atom.
(c) A molecule of ozone (a form of oxygen) is triatomic.
(d) A formula unit of sodium hydroxide, a compound that contains sodium, oxygen, and hydrogen, is triatomic.

5.15 How many sulfur atoms are present in one formula unit of each of the following substances?
(a) $Na_2S_2O_3$ (b) $(NH_4)_2SO_4$
(c) Al_2S_3 (d) $Al_2(SO_4)_3$

5.16 How many phosphorus atoms are present in one formula unit of each of the following substances?
(a) P_4O_{10} (b) PH_3
(c) $Na_3P_3O_{10}$ (d) $Ba_3(PO_4)_2$

5.17 In each of the following pairs of formulas select the formula that denotes the greater number of atoms.
(a) NH_4ClO_4 and NH_4NO_3
(b) $Al_2(SO_4)_3$ and $Al(NO_3)_3$
(c) $Ba(ClO)_2$ and $Ba(ClO_2)_2$
(d) K_3PO_4 and $Ca(OH)_2$

5.18 In each of the following pairs of formulas select the formula that denotes the greater number of atoms.
(a) KNO_2 and KCN (b) $CaSO_4$ and $Ca(NO_3)_2$
(c) $Mg(ClO)_2$ and $NaClO_2$
(d) $Be_3(PO_4)_2$ and $Be(C_2H_3O_2)_2$

5.19 The following molecular formulas are incorrectly written. Rewrite each formula in the correct manner.
(a) H3PO4
(b) SICL₄ (a silicon-chlorine compound)
(c) NOO
(d) 2HO (two H atoms and two O atoms)

5.20 The following molecular formulas are incorrectly written. Rewrite each formula in the correct manner.
(a) H2CO3
(b) ALBR₃ (an aluminum-bromine compound)
(c) HSH
(d) 2NO₂ (two N atoms and four O atoms)

Subatomic Particles (Sec. 5.5)

5.21 Compare the relative masses of the three subatomic particles.

5.22 Compare the relative charges of the three subatomic particles.

5.23 Which subatomic particle accounts for the charge on an atom's nucleus?

5.24 Which subatomic particles account for the mass of an atom's nucleus?

5.25 What relationship, if any, exists between the *numbers* of electrons and protons in an atom?

5.26 What relationship, if any, exists between the *numbers* of electrons and neutrons in an atom?

5.27 Why is the mass of the electron listed as zero in tabulations of subatomic particle masses?

5.28 Why are relative masses of subatomic particles frequently used rather than absolute masses?

5.29 Indicate which subatomic particle (proton, neutron, or electron) correctly matches each of the following statements. More than one particle can be used as an answer.
(a) possesses a negative charge
(b) has a mass slightly less than that of a neutron
(c) can be called a nucleon
(d) is the heaviest of the three particles

5.30 Indicate which subatomic particle (proton, neutron, or electron) correctly matches each of the following statements. More than one particle can be used as an answer.
(a) has no charge
(b) has a charge equal to but opposite in sign to that of an electron.
(c) is not found in the nucleus
(d) has a positive charge

5.31 Indicate whether each of the following statements about the nucleus of an atom is *true* or *false*.
(a) The nucleus of an atom is neutral.
(b) The nucleus of an atom contains only neutrons.
(c) The number of nucleons present in the nucleus is equal to the number of electrons present outside the nucleus.

5.32 Indicate whether each of the following statements about the nucleus of an atom is *true* or *false*.
(a) The nucleus accounts for almost all of the mass of an atom.

(b) The nucleus accounts for almost all of the volume of an atom.
(c) The nucleus can be positively or negatively charged, depending on the identity of the atom.

5.33 Explain what is meant by the statement "Atoms as a whole are neutral."

5.34 Explain why atoms are considered to be the fundamental building blocks of matter even though smaller subatomic particles exist.

Evidence Supporting the Existence of Subatomic Particles (Sec. 5.6)

5.35 In gas discharge tube experiments, what experimental evidence indicates that cathode rays originate at the cathode?

5.36 In gas discharge tube experiments, what experimental evidence indicates that cathode rays have a negative charge?

5.37 Describe the experimental setup used to detect canal rays.

5.38 Describe how canal rays are produced in a discharge tube.

5.39 What subatomic particles were identified as the result of gas discharge tube experiments?

5.40 Contrast the properties of cathode rays and canal rays.

5.41 Describe the experimental setup for Rutherford's gold-foil experiment.

5.42 Describe the observations obtained from Rutherford's gold-foil experiment.

5.43 What was Thomson's proposal, based on gas discharge tube observations, for the structure of an atom?

5.44 What was Rutherford's proposal, based on metal-foil experiment observations, for the structure of an atom?

Atomic Number, Mass Number, and Isotopes
(Secs. 5.7 and 5.8)

5.45 What information about the subatomic makeup of an atom is given by the following?
(a) atomic number
(b) mass number − atomic number

5.46 What information about the subatomic makeup of an atom is given by the following?
(a) mass number
(b) mass number + atomic number

5.47 Determine the atomic number and mass number for atoms with the following subatomic makeups.

(a) 2 protons, 2 neutrons, and 2 electrons
(b) 4 protons, 5 neutrons, and 4 electrons
(c) 5 protons, 4 neutrons, and 5 electrons
(d) 28 protons, 30 neutrons, and 28 electrons

5.48 Determine the atomic number and mass number for atoms with the following subatomic makeups.
(a) 1 proton, 1 neutron, and 1 electron
(b) 10 protons, 12 neutrons, and 10 electrons
(c) 12 protons, 10 neutrons, and 12 electrons
(d) 50 protons, 69 neutrons, and 50 electrons

5.49 Determine the number of protons, neutrons, and electrons present in atoms with the following characteristics.
(a) atomic number = 8 and mass number = 16
(b) mass number = 18 and $Z = 8$
(c) atomic number = 20 and $A = 44$
(d) $A = 257$ and $Z = 100$

5.50 Determine the number of protons, neutrons, and electrons present in atoms with the following characteristics.
(a) atomic number = 5 and mass number = 10
(b) $A = 11$ and atomic number = 5
(c) mass number = 110 and $Z = 48$
(d) $Z = 92$ and $A = 238$

5.51 The notations $^{15}_{7}N$ and ^{15}N can be used interchangeably, but the notations $^{15}_{7}N$ and $_{7}N$ cannot. Explain why and why not?

5.52 Does the notation $_{11}Na$ or ^{24}Na give the greater amount of information about a sodium atom? Explain your answer.

5.53 With the help of the information listed inside the front or back cover, write complete symbols for the four isotopes of iron with 28, 30, 31, and 32 neutrons, respectively.

5.54 With the help of the information listed inside the front or back cover, write complete symbols for the four isotopes of chromium with 26, 28, 29, and 30 neutrons, respectively.

5.55 Determine the number of protons, neutrons, and electrons present in each of the following atoms.
(a) $^{53}_{24}Cr$ (b) $^{103}_{44}Ru$ (c) $^{256}_{101}Md$ (d) $^{34}_{16}S$

5.56 Determine the number of protons, neutrons, and electrons present in each of the following atoms.
(a) $^{67}_{30}Zn$ (b) $^{9}_{4}Be$ (c) $^{40}_{20}Ca$ (d) $^{3}_{1}H$

5.57 Write complete symbols, with the help of the information listed inside the front or back cover, for atoms with the following characteristics.
(a) contains 15 electrons and 16 neutrons
(b) oxygen atom with 10 neutrons
(c) chromium atom with a mass number of 54
(d) gold atom that contains 276 subatomic particles

5.58 Write complete symbols, with the help of the information listed inside the front or back cover, for atoms with the following characteristics.
(a) contains 20 electrons and 24 neutrons
(b) radon atom with a mass number of 211
(c) silver atom that contains 157 subatomic particles
(d) beryllium atom that contains 9 nucleons

5.59 Indicate whether the members of each of the following pairs of atoms are isotopes, isobars, or neither.
(a) $^{24}_{12}X$ and $^{26}_{12}Y$ (b) $^{71}_{31}X$ and $^{71}_{32}Y$
(c) $^{56}_{25}X$ and $^{54}_{24}Y$ (d) $^{57}_{27}X$ and $^{60}_{27}Y$

5.60 Indicate whether the members of each of the following pairs of atoms are isotopes, isobars, or neither.
(a) $^{64}_{30}X$ and $^{64}_{29}Y$ (b) $^{20}_{10}X$ and $^{22}_{10}Y$
(c) $^{36}_{16}X$ and $^{35}_{17}Y$ (d) $^{60}_{28}X$ and $^{60}_{26}Y$

5.61 Identify the chemical symbol that X represents in each of the following notations for stable atoms.
(a) ^{6}X (b) ^{11}X (c) ^{19}X (d) ^{27}X

5.62 Identify the chemical symbol that X represents in each of the following notations for stable atoms.
(a) ^{4}X (b) ^{15}X (c) ^{16}X (d) ^{30}X

5.63 Specify the mass number(s) that apply in each of the following situations.
(a) lowest mass number for which no stable atoms exist
(b) mass number sequence for the first element (lowest-atomic-numbered) for which two stable isotopes exist

5.64 Specify the mass number(s) that apply in each of the following situations.
(a) mass number for atoms of the lowest-atomic-numbered element for which all atoms are identical (no isotopes)
(b) mass number sequence for the first element (lowest-atomic-numbered) for which three stable isotopes exist

5.65 What are the relationships among the number of protons present, number of neutrons present, and number of electrons present in each of the following situations?
(a) atoms of different isotopes of the same element
(b) atoms of isobars where the atomic numbers differ by 1

5.66 What are the relationships among the number of protons present, number of neutrons present, and number of electrons present in each of the following situations?
(a) atoms of the same element with different mass numbers
(b) atoms of isobars where the atomic numbers differ by 2

Atomic Masses (Sec. 5.9)

5.67 Construct a relative mass scale for the hypothetical atoms A, B, and C, given that atoms of A are two times as heavy as those of B, atoms of B are two times as heavy as atoms of C, and the reference point for the scale is B = 4.00 bebs.

5.68 Construct a relative mass scale for the hypothetical atoms A, B, and C, given that atoms of A are three times as heavy as those of B, atoms of B are four times as heavy as atoms of C, and the reference point for the scale is C = 2.50 bobs.

5.69 A football team has the following distribution of players: 22.0% are defensive linemen with an average mass of 271 lb. 19.0% are defensive backs with an average mass of 175 lb. 26.0% are offensive linemen with an average mass of 263 lb. 15.0% are offensive backs with an average mass of 182 lb. and 19.0% are specialty team members with an average mass of 191 lb. What is the average mass of a football player on this team?

5.70 The assortment of automobiles on a used car lot is categorized as follows: 18.0% are 1 year old, 10.0% are 2 years old, 33.0% are 3 years old, 4.0% are 4 years old, 31.0% are 5 years old, and 4.0% are 6 years old. What is the average age of a car on this used car lot?

5.71 Naturally occurring chlorine consists of two isotopes with the atomic masses and abundances given in Table 5.2. Using these data, calculate the atomic mass of chlorine.

5.72 Naturally occurring copper consists of two isotopes with the atomic masses and abundances given in Table 5.2. Using these data, calculate the atomic mass of copper.

5.73 Naturally occurring titanium consists of five isotopes with the atomic masses and abundances given in Table 5.2. Using these data, calculate the atomic mass of titanium.

5.74 Naturally occurring sulfur consists of four isotopes with the atomic masses and abundances given in Table 5.2. Using these data, calculate the atomic mass of sulfur.

5.75 A certain isotope of silver is 8.91 times as heavy as $^{12}_{6}C$. What is the mass of this silver isotope on the amu scale?

5.76 A certain isotope of silicon is 2.42 times as heavy as $^{12}_{6}C$. What is the mass of this silicon isotope on the amu scale?

5.77 How many times as heavy, on the average, is an atom of iron than an atom of beryllium?

5.78 How many times as heavy, on the average, is an atom of iron than an atom of fluorine?

5.79 The arbitrary standard for the atomic mass scale is the exact number 12 for the mass of $^{12}_{6}C$. Why, then, is the atomic mass of carbon listed as 12.011?

5.80 The atomic mass of fluorine is 18.9984 amu and that of copper is 63.546 amu. All fluorine atoms have a mass of 18.9984 amu, and not a single copper atom has a mass of 63.546 amu. Explain.

Additional Problems

5.81 Give the name and the atomic number of each kind of atom in the following formulas.
(a) ClF$_5$ (b) Ca$_3$N$_2$ (c) Ni(IO$_3$)$_2$ (d) Na$_2$S

5.82 Give the name and the atomic number of each kind of atom in the following formulas.
(a) ICl$_3$ (b) MgS   (c) Cu$_3$PO$_4$ (d) Be(CN)$_2$

5.83 Three naturally occurring isotopes of potassium exist: $^{39}_{19}$K, $^{40}_{19}$K, and $^{41}_{19}$K. The atomic mass of potassium is 39.102 amu. Which of the three potassium isotopes is most abundant? Explain your answer.

5.84 Naturally occurring boron is a mixture of two isotopes: $^{10}_{5}$B and $^{11}_{5}$B. Given that the atomic mass of boron is 10.811 amu, estimate the abundances of the two isotopes to the nearest 10%.

5.85 Write the complete symbol for the isotope of boron with each of the following characteristics.
(a) contains two fewer neutrons than $^{10}_{5}$B
(b) contains three more subatomic particles than $^{9}_{5}$B
(c) contains the same number of neutrons as ^{14}N
(d) contains the same number of subatomic particles as $^{14}_{7}$N

5.86 Write the complete symbol for the isotope of chromium with each of the following characteristics.
(a) contains two more neutrons than $^{55}_{24}$Cr
(b) contains two fewer subatomic particles than $^{52}_{24}$Cr
(c) contains the same number of neutrons as $^{60}_{29}$Cu
(d) contains the same number of subatomic particles as $^{60}_{29}$Cu

5.87 Arrange the five isotopes $^{42}_{20}$Ca, $^{39}_{19}$K, $^{44}_{21}$Sc, $^{37}_{18}$Ar, and $^{43}_{22}$Ti in order of
(a) increasing number of electrons
(b) decreasing number of neutrons
(c) increasing number of protons
(d) decreasing mass

5.88 Arrange the five isotopes $^{92}_{40}$Zr, $^{89}_{39}$Y, $^{95}_{41}$Nb, $^{87}_{38}$Sr, and $^{93}_{42}$Mo in order of
(a) decreasing number of electrons
(b) increasing number of neutrons
(c) increasing number of nucleons
(d) decreasing number of subatomic particles

5.89 The following are selected properties for the most abundant isotope of a particular element. Which of these properties would be the same for the second most abundant isotope of the element?
(a) Mass number is 70.   (b) Melting point is 29.8 °C.
(c) Isotope reacts with chlorine to give the compound XCl$_3$.

5.90 The following are selected properties for the most abundant isotope of a particular element. Which of these properties would be the same for the second most abundant isotope of the element?
(a) Atomic number is 31.
(b) Boiling point is 2403 °C.
(c) Isotopic mass is 69.92 amu.

5.91 Can the atomic number of an element be larger than its mass number? Explain.

5.92 What is the significance of the mass number of an atom being double the atomic number of the same atom?

5.93 For which numbers in the atomic number sequence 1 through 20 are
(a) all naturally occurring atoms identical (no isotopes)
(b) there three naturally occurring isotopes

5.94 For which numbers in the atomic number sequence 1 through 20 are
(a) there two naturally occurring isotopes
(b) there four naturally occurring isotopes

5.95 For which numbers in the mass number sequence 1 through 40 are there
(a) no stable atoms
(b) two types of stable atoms

5.96 For which numbers in the mass number sequence 1 through 40 are there
(a) one type of stable atom
(b) three types of stable atoms

5.97 Suppose it was decided to redefine the atomic mass scale by choosing as an arbitrary reference point a value of 20.000 amu to represent the naturally occurring mixture of nickel isotopes. What would be the atomic mass of the following elements on the new atomic mass scale?
(a) silver   (b) gold

5.98 Suppose it was decided to redefine the atomic mass scale by choosing as an arbitrary reference point a value of 40.000 amu to represent the mass of fluorine atoms. (All fluorine atoms are identical; that is, fluorine is monoisotopic.) What would be the atomic mass of the following elements on the new atomic mass scale?
(a) sulfur   (b) platinum

5.99 Using the data found in Table 5.2, calculate how many $^{17}_{8}$O and $^{18}_{8}$O atoms you would find in an oxygen sample containing 1 million (1.0 × 10^6) atoms.

5.100 Using the data found in Table 5.2, calculate how many $^{234}_{92}$U and $^{235}_{92}$U atoms you would find in a uranium sample containing 1 million (1.0 × 10^6) atoms.

5.101 How many protons are present in nine molecules of the compound $C_{12}H_{22}O_{11}$ (table sugar)?

5.102 How many electrons are present in twelve molecules of the compound $C_6H_{12}O_6$ (glucose, blood sugar)?

5.103 The isotopic mass for a particular isotope of nickel is 63.9280 amu. What is the total number of subatomic particles present in an atom of this isotope?

5.104 The isotopic mass for a particular isotope of chromium is 53.9389 amu. What is the total number of subatomic particles present in an atom of this isotope?

Cumulative Problems

5.105 The density of gold is 19.3 g/cm³, and the mass of a single gold atom is 3.27×10^{-22} g. How many gold atoms are present in a piece of gold whose volume is 3.22 cm³?

5.106 The density of copper is 8.93 g/cm³, and the mass of a single copper atom is 1.06×10^{-22} g. How many copper atoms are present in a copper bar whose dimensions are 2.00 cm × 3.00 cm × 5.00 cm?

5.107 In 1.00 g of fluorine atoms there are 3.17×10^{22} fluorine atoms. If you lined these atoms up side by side, how many miles long would the line of fluorine atoms be? The diameter of a fluorine atom is 1.44×10^{-8} cm.

5.108 In 1.00 g of fluorine atoms there are 3.17×10^{22} fluorine atoms. If you started counting these atoms at the rate of 10 per second, how many years would it take to count all the atoms in the 1.00 g sample?

5.109 An electron has a mass of 5.5×10^{-4} amu. What percent of the total mass of a Pb atom with a mass of 207 amu is due to its electrons?

5.110 How many electrons, with a mass of 5.5×10^{-4} amu, would it take to equal the mass of one proton, which has a mass of 1.0073 amu?

5.111 The diameter of an atom is approximately 10^5 times as large as the diameter of the nucleus. If the nucleus was enlarged to the size of a Ping-Pong ball (1.5 in. in diameter), what would be the diameter of the atom, in miles?

5.112 The diameter of an atom is approximately 10^{-8} cm, and that of the nucleus is 10^{-13} cm. How many times larger is the volume of the atom than the volume of the nucleus? The formula for the volume of a sphere is $V = 0.524 \times d^3$ (where d is the diameter of the sphere).

Grid Problems

5.113 Select from the grid *all* correct responses for each situation.

1. HF	2. Hf	3. S_8
4. NH_4Cl	5. KNO_3	6. H_2S
7. $Ba(OH)_2$	8. $AlPO_4$	9. $Ca(C_2H_3O_2)_2$

(a) formulas that represent elements
(b) formulas that involve pentaatomic units
(c) formulas in which there are equal numbers of two kinds of atoms (other atoms may also be present)
(d) formulas in which there are four oxygen atoms
(e) pairs of formulas that contain an equal number of hydrogen atoms
(f) pairs of formulas in which there are the same number of atoms and also the same number of elements

5.114 Select from the grid *all* correct responses for each situation.

1. IF_5 IF_3O_2	2. NaCN NaSCN	3. NH_3 HN_3
4. $KClO_2$ BaS_2O_3	5. $SOCl_2$ SO_2Cl_2	6. Li_2SO_4 $Li_2S_2O_3$
7. H_2 H_2S	8. $Fe(OH)_2$ $Fe(OH)_3$	9. NO_2 N_2O

(a) the two paired formulas contain equal numbers of atoms of two or more different elements
(b) the two paired formulas differ in composition in the same way as Cl_2O and $SOCl_2$
(c) the two paired formulas collectively contain the same number of different elements as do NaOH and NaH
(d) the two paired formulas are related to each other in the same way as SO_2 and S_2O
(e) the two paired formulas have subscript sums that are equal
(f) the two paired formulas each have an element (not necessarily the same one) that accounts for 50% or more of the atoms in the formula unit

5.115 Select from the grid *all* correct responses for each situation.

1. $^{4}_{2}He$	2. $^{9}_{4}Be$	3. $^{14}_{7}N$
4. $^{28}_{14}Si$	5. $^{31}_{15}P$	6. $^{35}_{17}Cl$
7. $^{36}_{18}Ar$	8. $^{38}_{18}Ar$	9. $^{39}_{19}K$

(a) atoms that contain more neutrons than protons
(b) atoms that contain equal numbers of protons and electrons
(c) atoms that contain equal numbers of all three kinds of subatomic particles
(d) pairs of atoms that contain an equal number of electrons
(e) pairs of atoms that contain an equal number of neutrons
(f) pairs of atoms in which the total number of subatomic particles is four greater for one member than the other

5.116 Select from the grid *all* correct responses for each situation.

1. beryllium	2. boron	3. nitrogen
4. oxygen	5. sodium	6. aluminum
7. phosphorus	8. sulfur	9. potassium

(a) elements for which all stable atoms are identical
(b) elements whose stable isotopes have sequential mass numbers
(c) pairs of elements for which the number of stable isotopes is the same
(d) elements for which all stable atoms have more neutrons than protons
(e) elements for which some, but not all, stable atoms have three more neutrons than protons
(f) elements for which the stable isotope with the lowest mass number is the most abundant isotope

CHAPTER SIX

Electronic Structure and Chemical Periodicity

6.1 The Periodic Law

During the early part of the nineteenth century, an abundance of chemical facts became available from detailed studies of the then known elements. With the hope of providing a systematic approach to the study of chemistry, scientists began to look for some form of order in the increasing amount of chemical information. They were encouraged in their search by the unexplained but well-known fact that certain groups of elements had very similar properties. Numerous attempts were made to explain these similarities and to use them as a means for arranging or classifying the elements.

In 1869, these efforts culminated in the discovery of what is now called the *periodic law*. The **periodic law** states that when elements are arranged in order of increasing atomic number, elements with similar properties occur at periodic (regularly recurring) intervals. Proposed independently by both the Russian chemist Dmitri Ivanovich Mendeleev (1834–1907) and the German chemist Julius Lothar Meyer (1830–1895), the periodic law is one of the most important of all chemical laws.

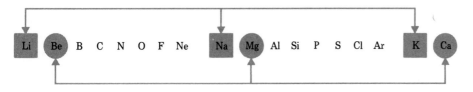

FIGURE 6.1 An illustration of the periodicity in properties that occurs when elements are arranged in order of increasing atomic number. Li, Na, and K have similar chemical properties, as do Be, Mg, and Ca.

The preceding statement of the periodic law is in modern-day language. It differs from the original 1869 statements in that the phrase *atomic number* has replaced "atomic mass." The use of "atomic mass" in the original statements reflected theories prevalent in 1869. According to these theories, the masses of atoms were their most important distinguishing properties—a knowledge of subatomic particles (Sec. 5.5) was still 30 years away. When the details of subatomic structure were finally discovered, it became obvious that the properties of atoms were related not to their masses but to the number and arrangement of their electrons. Thus, the periodic law was modified to reflect this new knowledge.

Figure 6.1 shows parts of the repeating pattern for chemical properties (periodic law) for the sequence of elements with atomic numbers 3 through 20. The elements within similar geometric symbols (circles and squares) have similar chemical properties. For the sake of simplicity, only two of the periodic relationships (repeating patterns) are shown; for the elements listed, similar properties are found in every eighth element.

The concept of a periodic variation or pattern should not be new to you, as numerous everyday examples of this exist. Most, but not all, involve time. Mondays occur at regular intervals of 7 days. Office workers get paid at regular intervals, such as every two weeks. The red-green-yellow light sequence for a traffic semaphore repeats itself in a periodic fashion.

To be useful, the relationships generated by the periodic law must be easily visualized. This is the purpose of periodic tables, the subject of Section 6.2.

6.2 The Periodic Table

A **periodic table** is a graphical representation of the behavior described by the periodic law. In this table, the elements are arranged according to increasing atomic number in such a way that the similarities predicted by the periodic law become readily apparent.

The most commonly used form of the periodic table is the one shown in Figure 6.2 (and also inside the front cover of this book). It is an arrangement of the elements in which vertical columns contain elements with similar chemical properties. Note how Li, Na, and K—elements with similar chemical properties (Fig. 6.1)—are all

Chapter 6 · Electronic Structure and Chemical Periodicity

1 Group IA																	18 Group VIIIA
1 H 1.00794	2 Group IIA											13 Group IIIA	14 Group IVA	15 Group VA	16 Group VIA	17 Group VIIA	2 He 4.002602
3 Li 6.941	4 Be 9.01218											5 B 10.811	6 C 12.011	7 N 14.0067	8 O 15.9994	9 F 18.998403	10 Ne 20.179
11 Na 22.98977	12 Mg 24.305	3 Group IIIB	4 Group IVB	5 Group VB	6 Group VIB	7 Group VIIB	8 Group	9 Group VIIIB	10 Group	11 Group IB	12 Group IIB	13 Al 26.98154	14 Si 28.0855	15 P 30.97376	16 S 32.066	17 Cl 35.453	18 Ar 39.948
19 K 39.0983	20 Ca 40.078	21 Sc 44.9559	22 Ti 47.88	23 V 50.9415	24 Cr 51.9961	25 Mn 54.9380	26 Fe 55.847	27 Co 58.9332	28 Ni 58.6934	29 Cu 63.546	30 Zn 65.38	31 Ga 69.723	32 Ge 72.59	33 As 74.9216	34 Se 78.96	35 Br 79.904	36 Kr 83.80
37 Rb 85.4678	38 Sr 87.62	39 Y 88.9059	40 Zr 91.22	41 Nb 92.9064	42 Mo 95.94	43 Tc (98)	44 Ru 101.07	45 Rh 102.9055	46 Pd 106.42	47 Ag 107.8682	48 Cd 112.41	49 In 114.818	50 Sn 118.710	51 Sb 121.757	52 Te 127.60	53 I 126.9045	54 Xe 131.29
55 Cs 132.9054	56 Ba 137.33	57 La 138.9055	72 Hf 178.49	73 Ta 180.9479	74 W 183.84	75 Re 186.207	76 Os 190.23	77 Ir 192.22	78 Pt 195.08	79 Au 196.9665	80 Hg 200.59	81 Tl 204.383	82 Pb 207.2	83 Bi 208.9804	84 Po (209)	85 At (210)	86 Rn (222)
87 Fr (223)	88 Ra 226.0254	89 Ac 227.0278	104 Unq (261)	105 Unp (262)	106 Unh (263)	107 Uns (262)	108 Uno (265)	109 Une (266)									

Metals ← → Nonmetals

58 Ce 140.12	59 Pr 140.9077	60 Nd 144.24	61 Pm (145)	62 Sm 150.36	63 Eu 151.96	64 Gd 157.25	65 Tb 158.9254	66 Dy 162.50	67 Ho 164.9304	68 Er 167.26	69 Tm 168.9342	70 Yb 173.04	71 Lu 174.967
90 Th 232.0381	91 Pa 231.0359	92 U 238.0289	93 Np 237.0482	94 Pu (244)	95 Am (243)	96 Cm (247)	97 Bk (247)	98 Cf (251)	99 Es (252)	100 Fm (257)	101 Md (258)	102 No (259)	103 Lr (260)

FIGURE 6.2 The most commonly used form of the periodic table.

in the same vertical column in Figure 6.2. Other elements with properties similar to those of these three elements are H, Rb, Cs, and Fr.

Within the periodic table each element is represented by a rectangular box. Within each box are given the symbol, atomic number, and atomic mass of the element, as shown in Figure 6.3. It should be noted that some periodic tables, but not the ones in this text, also give the element's name within the box.

Special chemical terminology exists for specifying the position (location) of an element within the periodic table. This terminology involves the use of the words

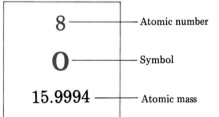

FIGURE 6.3 Arrangement of information about elements within the "boxes" of the periodic table.

group and *period*. A **group** in the periodic table is a vertical column of elements. There are two notations in use for designating individual periodic table groups. In the first notation, which has been in use for many years, groups are designated by Roman numerals and the letters A and B. The second notation, which has been recently recommended for use by an international scientific commission, uses the Arabic numbers 1 through 18. Notice that both group notations are given at the top of each group in the periodic table of Figure 6.2. The elements with atomic numbers 8, 16, 34, 52, and 84 (O, S, Se, Te, and Po) constitute group VIA (old notation) or group 16 (new notation). Because it may be some time before the new group numbering system is widely accepted, we will use the old notation.

A **period** in the periodic table is a horizontal row of elements. For identification purposes, the periods are numbered sequentially, with Arabic numbers, starting at the top of the periodic table. (These period numbers are not explicitly shown on the periodic table.) Period 3 is the third row of elements, period 4 the fourth row of elements, and so forth. The elements Na, Mg, Al, Si, P, S, Cl, and Ar are all members of period 3 (see Fig. 6.2). Period 1 has only two elements—H and He.

The location of any element in the periodic table is specified by giving its group number and its period number. The element gold (Au), with an atomic number of 79, belongs to group IB and is in period 6. Nitrogen (N), with an atomic number of 7, belongs to group VA and is in period 2.

There is one area in the table where the practice of arranging the elements according to increasing atomic number seems to be violated. This is the area of elements 57 and 89, which are both in group IIIB. Shown next to them, in group IVB, are elements 72 and 104, respectively. The missing elements, 58–71 and 90–103, are located in two rows at the bottom of the periodic table. Technically, these elements should be included in the body of the table, as shown in Figure 6.4a. However, to have a more compact table, they are placed in the position shown in Figure 6.4b. This arrangement should present no problems to the user of the periodic table as long as it is recognized for what it is—a space-saving device.

Students often assume that a corollary of the fact that the elements are arranged in the periodic table in order of increasing atomic number is that they are also arranged in order of increasing atomic mass. This latter conclusion is correct most of the time; however, there are exceptions.

The following element sequences are exceptions to the generalization that atomic mass increases with increasing atomic number.

$_{18}$Ar	39.95 amu	$_{27}$Co	58.93 amu	$_{52}$Te	127.60 amu
$_{19}$K	39.10 amu	$_{28}$Ni	58.69 amu	$_{53}$I	126.90 amu
$_{20}$Ca	40.08 amu	$_{29}$Cu	63.55 amu	$_{54}$Xe	131.29 amu

The inversions in atomic mass result from isotopic abundances (Sec. 5.8). Let us consider such data for the Ar–K–Ca triad.

^{36}Ar	0.34%	^{39}K	93.10%	^{40}Ca	96.97%
^{38}Ar	0.07%	^{40}K	0.01%	^{42}Ca	0.64%
^{40}Ar	99.60%	^{41}K	6.88%	^{43}Ca	0.14%
				^{44}Ca	2.06%
				^{46}Ca	0.003%
				^{48}Ca	0.18%

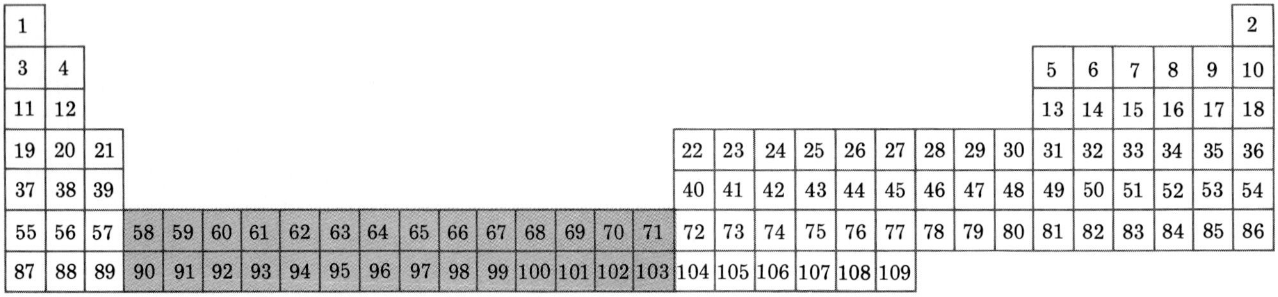

FIGURE 6.4 "Long" and "short" forms of the periodic table: (a) periodic table with elements 58–71 and 90–103 (in color) in their proper positions—the "long" form; (b) periodic table modified to conserve space by placing elements 58–71 and 90–103 (in color) below the rest of the elements—the "short" form.

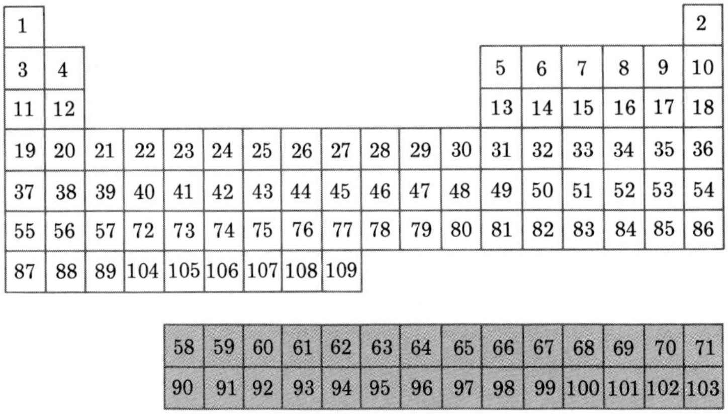

For argon the heaviest isotope ($A = 40$) is most abundant, and for potassium the lightest isotope ($A = 39$) is most abundant; this results in potassium having a lower atomic mass than argon. The atomic masses of argon and calcium are almost equal (39.95 and 40.08 amu, respectively); the dominant isotope for both elements has a mass number of 40.

Later in this chapter we will find that a periodic table conveys much more information about the elements than the symbols, atomic numbers, and atomic masses that are printed on it. Information about the arrangement of electrons in an atom is "coded" into the table, as is information concerning some physical and chemical property trends. Indeed, the periodic table is considered to be the single most useful study aid available for organizing information about the elements.

For many years after the formulation of the periodic law and the periodic table, both were considered to be empirical. The law worked and the table was very useful, but there was no explanation available for the law or for why the periodic table had the shape it had. It is now known that the theoretical basis for both the periodic law and the periodic table lies in electronic theory. The properties of the elements repeat themselves in a periodic manner because the arrangement of electrons about the nucleus of an atom follows a periodic "pattern." Electron arrangements and an explanation of the periodic law and periodic table in terms of electronic theory are the subject matter for most of the remainder of this chapter.

6.3 The Energy of an Electron

In Section 5.5 we learned the following facts about electrons.

1. They are one of the three fundamental subatomic particles.
2. They have very little mass in comparison to protons and neutrons.
3. They are located outside the nucleus of the atom.
4. They move rapidly about the nucleus in a volume that defines the size of the atom.

Much more must be known about electrons to understand the chemical behavior of the various elements, why some elements have similar chemical properties, and the theoretical basis for the periodic law.

More specific information concerning the behavior and arrangement of electrons within the extranuclear region of an atom is derived from a complex mathematical model for electron behavior called *quantum mechanics*. All of the early work on this subject was done by physicists rather than chemists.

During the early part of the twentieth century (1910–1930) a major revolution occurred in the field of physics, a revolution that profoundly affected chemistry. During this time it became clear that the "established" laws of physics, the laws that had been used for many years to predict the behavior of macroscopic objects, could not explain the behavior of extremely small objects such as atoms and electrons. The work of a number of European physicists led to this conclusion and also beyond it. These same physicists were able to formalize new laws that did apply to small objects. It is from these new laws, collectively called quantum mechanics, that information concerning the arrangement and behavior of electrons about a nucleus came.

A major force in the development of quantum mechanics was the Austrian physicist Erwin Schrödinger (1887–1961). In 1926 he showed that the laws of quantum mechanics could be used to characterize the motion of electrons. Most of the concepts of this chapter come from solutions to equations developed by Schrödinger.

A formal discussion of quantum mechanics is beyond the scope of this course (and many other chemistry courses also) because of the rigorous mathematics involved. The answers obtained from quantum mechanics are, however, simple enough to be understood to a surprisingly large degree at the level of an introductory chemistry course. A consideration of these quantum-mechanical answers will enable us to develop a system for specifying electron arrangements around a nucleus and further to develop basic rules governing compound formation. This latter topic will occupy our attention in Chapter 7.

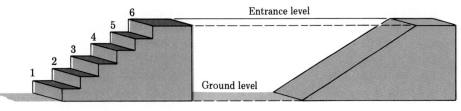

FIGURE 6.5 A stairway with quantized position levels versus a ramp with continuous position levels.

Present-day quantum mechanical theory describes the arrangement of an atom's electrons in terms of their energies. Indeed, the energy of an electron is the property that is most important to any consideration of its behavior about the nucleus.

The energy of an electron is manifested primarily in its velocity. The higher its energy, the higher its average velocity. The faster an electron travels, the farther it tends to move from the nucleus with which it is associated.

A most significant characteristic of an electron's energy is that it is a quantized property. A **quantized property** is a property that can have only certain values; that is, not all values are allowed. Since an electron's energy is quantized, an electron can have only certain specific energies.

Quantization is a phenomenon not commonly encountered in the macroscopic world. Somewhat analogous is the process of a person climbing a flight of stairs. In Figure 6.5a you see six steps between ground level and the level of the entrance door. As a person climbs these stairs there are only six permanent positions he or she can occupy (with both feet together). Thus the person's position (height above ground level) is quantized; only certain positions are allowed. The opposite of quantization is continuousness. A person climbing a ramp up to the entrance (Fig. 6.5b) would be able to assume a continuous set of heights above ground level; all values are allowed.

The energy of an electron determines its behavior about the nucleus. Since electron energies are quantized, only certain behavior patterns are allowed. Descriptions of electron behaviors involve the use of the terms *shell*, *subshell*, and *orbital*. Sections 6.4–6.6 consider the meaning of these terms and the mathematical interrelationships between them.

6.4 Electron Shells

It was mentioned in Section 6.3 that electrons with higher energy will be found farther from the nucleus than those with lower energy. Based on energy–distance-from-the-nucleus considerations, electrons can be grouped into shells or main energy levels. An **electron shell** is an energy level in which the electrons have approximately the same energy and spend most of their time at approximately the same distance from the nucleus.

Two different methods are used for identifying electron shells. The older method uses letters of the alphabet beginning with the letter K and then continuing sequentially. Shell K is the shell of electrons closest to the nucleus (the lowest in energy). The next closest shell is designated L; then come M, N, and so on. The more modern method designates shells by a number n, which may have the values 1, 2, 3, ... The shell with the lowest energy is assigned an n value of 1, the next higher 2, then 3, and so on. Comparing the two systems, we see that the $n = 1$ shell and the K shell are the same. Similarly, $n = 2$ and L are designations for the same shell. Only values of $n = 1$–7 (or letters K–Q) are used at present to designate electron shells. No known atom has electrons farther from the nucleus than the seventh main energy level ($n = 7$ or letter Q).

The maximum number of electrons in an electron shell varies; the higher the shell energy, the more electrons the shell can accommodate. The farther electrons are from the nucleus (a higher energy shell), the greater the volume of space available for them; hence the more electrons there can be in the shell. (Conceptually, electron shells may be considered to be nested one inside another, somewhat like the layers of flavors inside a jawbreaker or similar type of candy.)

The lowest energy shell ($n = 1$) accommodates a maximum of 2 electrons. In the second, third, and fourth shells, 8, 18, and 32 electrons, respectively, are allowed. A very simple mathematical equation can be used to calculate the maximum number of electrons allowed in any given shell.

$$\text{Shell electron capacity} = 2n^2 \quad (\text{where } n = \text{shell number})$$

For example, when $n = 4$ the value $2n^2 = 2(4^2) = 32$, which is the number previously given for the number of electrons allowed in the fourth shell. Although there is a maximum electron occupancy for each shell or main energy level, a shell may hold fewer then the allowable number of electrons in a given situation.

Table 6.1 summarizes concepts presented in this section concerning electron shells or electron main energy levels.

TABLE 6.1 Important Characteristics of Electron Shells

Shell	Number Designation (n)	Letter Designation	Electron Capacity ($2n^2$)
1st	1	K	$2 \times 1^2 = 2$
2nd	2	L	$2 \times 2^2 = 8$
3rd	3	M	$2 \times 3^2 = 18$
4th	4	N	$2 \times 4^2 = 32$
5th	5	O	$2 \times 5^2 = 50$[a]
6th	6	P	$2 \times 6^2 = 72$[a]
7th	7	Q	$2 \times 7^2 = 98$[a]

[a] The maximum number of electrons in this shell has never been attained in any element now known.

6.5 Electron Subshells

All electrons in a shell do not have the same energy. Their energies are all close to each other in magnitude, but they are not identical. The range of energies for electrons in a shell is due to the existence of electron subshells or electron *energy sublevels*. An **electron subshell** is an energy level within an electron shell in which the electrons all have the same energy.

The number of subshells within a shell varies. A shell contains the same number of subshells as its own shell number; that is,

Number of subshells in a shell = n (where n = shell number)

Thus, each successive shell has one more subshell than the previous one. Shell 3 contains three subshells, shell 4 contains four subshells, and shell 5 contains five subshells.

Subshells are identified by a number and a letter. The number indicates the shell to which the subshell belongs. The letters are *s*, *p*, *d*, or *f* (all lower case letters), which, in that order, denote subshells of increasing energy within a shell. The lowest energy subshell within a shell is always the *s* subshell, the next higher the *p* subshell, then the *d*, and finally the *f*. Shell 1 has only one subshell, the 1*s*. Shell 2 has two subshells, the 2*s* and 2*p*. The 3*s*, 3*p*, and 3*d* subshells are found in shell 3. Table 6.2 gives information concerning subshells for the first seven shells. Note that number-and-letter designations are not given for all the subshells in shells 5, 6, and 7; some of these subshells are not needed to describe the electron arrangements for the 109 known elements. Why they are not needed will become evident in Section 6.7.

The unusual sequence of letters (*s*, *p*, *d*, and *f*) used to denote the four subshell types has a historical origin. The early development of quantum mechanical theory (Sec. 6.3) was closely linked to the study of spectral emissions produced by atoms. The letters *s*, *p*, *d*, and *f* are derived, respectively, from old spectroscopic terminology, which describes spectral lines as *s*harp, *p*rincipal, *d*iffuse, and *f*undamental.

The maximum number of electrons that a subshell can hold varies from 2 to 14 depending on the type of subshell—*s*, *p*, *d*, or *f*. An *s* subshell can accommodate only 2 electrons. What shell the *s* subshell is located in does not affect the maximum electron occupancy figure: that is, the 1*s*, 2*s*, 3*s*, 4*s*, 5*s*, 6*s*, and 7*s* subshells all have a

TABLE 6.2 Subshell Arrangements Within Shells

Shell Number (n)	Subshells					
1	1*s*					
2	2*s*	2*p*				
3	3*s*	3*p*	3*d*			
4	4*s*	4*p*	4*d*	4*f*		
5	5*s*	5*p*	5*d*	5*f*	—	
6	6*s*	6*p*	6*d*	—	—	—
7	7*s*	—	—	—	—	—

TABLE 6.3 Distribution of Electrons Within Subshells

Shell	Number of Subshells Within Shell	Maximum Number of Electrons Within Each Subshell				Maximum Number of Electrons Within Shell ($2n^2$)
		s	p	d	f	
$n = 1$ (K)	1	2				2
$n = 2$ (L)	2	2	6			8
$n = 3$ (M)	3	2	6	10		18
$n = 4$ (N)	4	2	6	10	14	32

maximum electron occupancy of 2. Subshells of the p, d, and f types can accommodate maximums of 6, 10, and 14 electrons, respectively. Again the maximum numbers of electrons in these types of subshells depend only on the subshell types and are independent of shell number. Table 6.3 summarizes the information just presented for shells 1 through 4. Notice the consistency between the numbers in columns 3 and 4 in Table 6.3. Within a shell, the sum of the subshell electron occupancies is the same as the shell electron occupancy ($2n^2$). For example, in shell 4, an s subshell containing 2 electrons, a p subshell containing 6 electrons, a d subshell containing 10 electrons, and an f subshell containing 14 electrons add up to a total of 32 electrons, which is the maximum occupancy of shell 4 as calculated by the $2n^2$ formula.

6.6 Electron Orbitals

The last and most basic of the three terms used in describing electron arrangements about nuclei is *orbital*. An **electron orbital** is a region of space around a nucleus where an electron with a specific energy is most likely to be found.

An analogy for the relationship between shells, subshells, and orbitals can be found in the physical layout of a high-rise condominium complex. A shell is the counterpart of a floor of the condominium in our analogy. Just as each floor will contain apartments (of different sizes), a shell contains subshells (of different sizes). Further, just as apartments contain rooms, subshells contain orbitals. An apartment is a collection of rooms; a subshell is a collection of orbitals. A floor of a condominium building is a collection of apartments; a shell is a collection of subshells.

The characteristics of orbitals, the rooms in our "electron apartment house," include

1. The number of orbitals in a subshell varies, being one for an s subshell, three for a p subshell, five for a d subshell, and seven for an f subshell.
2. The maximum number of electrons in an orbital does not vary. It is always 2.
3. The notation used to designate orbitals is the same as that used for subshells. Thus, orbitals in the $4f$ subshell (there are seven of them) are called $4f$ orbitals.

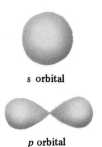

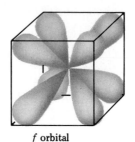

FIGURE 6.6 Shapes of atomic orbitals. To improve the perspective, the *f* orbital is shown within a cube with one lobe pointing to each corner of the cube. Only two electrons may occupy an orbital regardless of the number of lobes present in its shape.

We have already noted (Sec. 6.5) that all electrons in a subshell have the same energy. Thus all electrons in orbitals of the same subshell will have the same energy. This means that *shell and subshell designations are sufficient to specify the energy of an electron.* This statement will be of great importance in the discussions of Section 6.7.

Orbitals have a definite size and shape related to the type of subshell in which they are found. (Remember, an orbital is a region of space. We are not talking about the size and shape of an electron, but rather the size and shape of a region of space where an electron is found.) At any given time an electron can be at only one point in an orbital, but because of its rapid movement throughout the orbital it "occupies" the entire orbital. An analogy would be the "definite" volume occupied by a rotating fan blade. Typical *s*, *p*, *d*, and *f* orbital shapes are given in Figure 6.6. Notice that the shapes increase in complexity in the order *s*, *p*, *d*, and *f*. Some of the more complex *d* and *f* orbitals have shapes related, but not identical, to those shown in Figure 6.6.

Orbitals within the same subshell differ mainly in orientation. For example, the three 2*p* orbitals look the same but are aligned in different directions—along the *x*, *y*, and *z* axes in a Cartesian coordinate system (see Fig. 6.7).

Orbitals of the same type but in different shells (for example, 1*s*, 2*s*, and 3*s*) have the same general shape but differ in size (volume) (see Fig. 6.8).

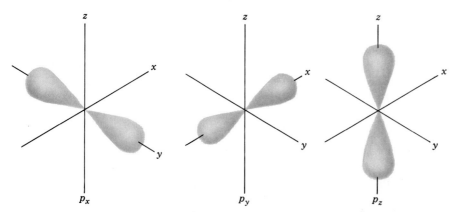

FIGURE 6.7 The orientation of the three orbitals in a *p* subshell.

1s   2s   3s   **FIGURE 6.8** Size relationships among *s* orbitals in different shells.

EXAMPLE 6.1

Determine the following for the fourth electron shell (fourth main electron energy level) of an atom.

(a) the number of subshells it contains
(b) the designation used to describe each subshell
(c) the number of orbitals in each subshell
(d) the maximum number of electrons that could be contained in each subshell
(e) the maximum number of electrons that could be contained in the shell

Solution

(a) This whole problem deals with shell number 4. The number of subshells in a shell is the same as the shell number. Thus, shell 4 ($n = 4$) will contain four subshells.
(b) The lowest of the four subshells in terms of energy is designated 4s, the next higher 4p, the next 4d, and the final subshell (highest energy) is 4f.
(c) The number of orbitals in a given type (s, p, d, or f) of subshell is independent of the shell number. Each s subshell (1s, 2s, 3s, etc.) contains one orbital; each p subshell, three orbitals; each d subshell, five orbitals, and each f subshell, seven orbitals.
(d) The maximum number of electrons in an orbital is always two. Therefore the 4s subshell (one orbital) contains 2 electrons; the 4p subshell (three orbitals), 6 electrons; the 4d subshell (five orbitals), 10 electrons; and the 4f subshell (seven orbitals), 14 electrons.
(e) The maximum number of electrons in a shell is given by the formula $2n^2$, where n is the shell number. Since $n = 4$ in this problem, $2n^2 = 2 \times 4^2 = 32$. Alternatively, from part (c), we note that shell 4 contains 16 orbitals $(1 + 3 + 5 + 7)$. Since each orbital can hold two electrons, the maximum number of electrons will be 32.

PRACTICE EXERCISE 6.1

Determine the following for the third electron shell (third main energy level) of an atom.

(a) the number of subshells it contains
(b) the designation used to describe each subshell
(c) the number of orbitals in each subshell

(d) the maximum number of electrons that could be contained in each subshell
(e) the maximum number of electrons that could be contained in the shell

Ans. (a) 3; (b) 3s, 3p, and 3d; (c) 1 for 3s, 3 for 3p, and 5 for 3d; (d) 2 for 3s, 6 for 3p, and 10 for 3d; (e) 18.

Similar exercises: Problems 6.17 and 6.18

6.7 Writing Electron Configurations

An **electron configuration** is a statement of how many electrons an atom has in each of its subshells. Since subshells group electrons according to energy (Sec. 6.5), electron configurations indicate how many electrons an atom has of various energies.

Electron configurations are not written out in words; a shorthand system with symbols is used. Subshells containing electrons, listed in order of increasing energy, are designated using number–letter combinations ($1s$, $2s$, $2p$, etc.). A superscript following each subshell designation indicates the number of electrons in that subshell. The electron configuration for oxygen using this shorthand notation is

$$1s^2 2s^2 2p^4 \qquad \text{(read "one-s-two, two-s-two, two-p-four")}$$

An oxygen atom thus has an electron arrangement of two electrons in the $1s$ subshell, two electrons in the $2s$ subshell, and four electrons in the $2p$ subshell.

To determine the electron configuration for an atom, a procedure called the *Aufbau principle* (German *aufbauen*, to build) is used. The **Aufbau principle** is that electrons normally occupy the lowest energy subshell available. This guideline brings order to what could be a very disorganized situation. Many orbitals exist about the nucleus of any given atom. Electrons do not occupy these orbitals in a random, haphazard fashion; a very predictable pattern, governed by the Aufbau principle, exists for electron orbital occupancy. Orbitals are filled in order of increasing energy.

Use of the Aufbau principle requires knowledge concerning the electron capacities of orbitals and subshells (which we already have; see Sec. 6.6) and knowledge concerning the relative energies of subshells (which we now consider).

Figure 6.9 gives the order in which electron subshells about a nucleus acquire electrons. Note from studying this figure that the sequence of subshell filling is not as simple a pattern as you probably would have predicted. All subshells within a given shell do not necessarily have lower energies than all subshells of higher numbered shells. Because of energy overlaps, beginning with shell 4, one or more lower energy subshells of a specific shell have energies lower than the upper subshells of a preceding shell and thus acquire electrons first. For example, the $4s$ subshell acquires electrons before the $3d$ subshell does (see Fig. 6.9). As another example, the s subshell of the sixth energy level fills before the d subshell of the fifth energy level or the f subshell of the fourth energy level (again refer to Fig. 6.9).

The sequence in which subshells acquire electrons must be learned before electron configurations can be written. A useful mnemonic (memory) device, called an *Aufbau diagram*, helps considerably with this learning process. As can be seen from Figure 6.10, an **Aufbau diagram** is a listing of subshells in the order in which electrons occupy them. All s subshells are located in column 1, all p subshells in column 2, and so on. Subshells belonging to the same shell are found in the same row. The

Section 6.7 · Writing Electron Configurations   **177**

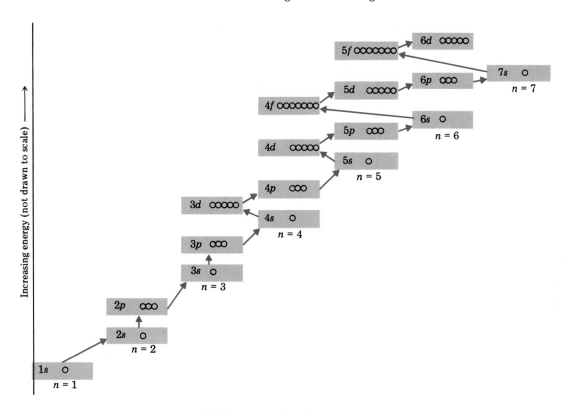

FIGURE 6.9 Relative energies and filling order for the electron subshells.

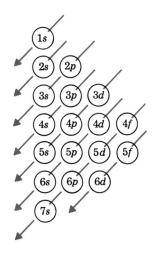

FIGURE 6.10 The Aufbau diagram—an aid to remembering subshell filling order.

order of subshell filling is given by following the diagonal arrows, starting with the top one. The 1s subshell fills first. The second arrow points to (goes through) the 2s subshell. It fills next. The third arrow points to both the 2p and 3s subshells. The 2p fills first, followed by the 3s. Any time a single arrow points to more than one subshell, start at the tail of the arrow and work to its head to determine the proper filling sequence. The 3p subshell fills next, and so on. An Aufbau diagram is an easy way to catalog the information given in Figure 6.9.

We are now ready to write electron configurations. Let us systematically consider electron configurations for the first few elements in the periodic table.

Hydrogen ($Z = 1$) has only one electron, which goes into the 1s subshell, which has the lowest energy of all subshells. Hydrogen's electron configuration is written as

$$1s^1$$

Helium ($Z = 2$) has two electrons, both of which occupy the 1s subshell. (Remember, an s subshell contains one orbital, and an orbital can accommodate two electrons.) Helium's electron configuration is

$$1s^2$$

For lithium ($Z = 3$), with three electrons, the third electron cannot enter the 1s subshell since its maximum capacity is two electrons. (All s subshells are completely filled with two electrons; see Sec. 6.6.) The third electron is placed in the next highest energy subshell, the 2s. The electron configuration for lithium is thus

$$1s^2 2s^1$$

With beryllium ($Z = 4$) the additional electron is placed in the 2s subshell, which is now completely filled, giving beryllium the electron configuration

$$1s^2 2s^2$$

In boron ($Z = 5$), the 2p subshell, the subshell of next highest energy (check Fig. 6.9 or 6.10 to be sure) becomes occupied for the first time. Boron's electron configuration is

$$1s^2 2s^2 2p^1$$

A p subshell can accommodate six electrons since there are three orbitals within it (Sec. 6.6). The 2p subshell can thus accommodate the additional electrons found in C, N, O, F, and Ne. The electron configurations of these elements are

$$\begin{aligned} \text{C } (Z=6): & \quad 1s^2 2s^2 2p^2 \\ \text{N } (Z=7): & \quad 1s^2 2s^2 2p^3 \\ \text{O } (Z=8): & \quad 1s^2 2s^2 2p^4 \\ \text{F } (Z=9): & \quad 1s^2 2s^2 2p^5 \\ \text{Ne } (Z=10): & \quad 1s^2 2s^2 2p^6 \end{aligned}$$

With sodium ($Z = 11$) the 3s subshell acquires an electron for the first time.

$$\text{Na } (Z=11): \quad 1s^2 2s^2 2p^6 3s^1$$

Note the pattern that is developing in the electron configurations we have written so far. Each element has an electron configuration the same as the one just before it with the addition of one electron.

Electron configurations for other elements are obtained by simply extending the principles we have just illustrated. Electrons are added to subshells, always filling one of lower energy before adding electrons to the next highest subshell, until the correct number of electrons has been accommodated.

EXAMPLE 6.2

Write the electron configurations for **(a)** potassium ($Z = 19$) and **(b)** bromine ($Z = 35$).

Solution

(a) The number of electrons in a potassium atom is 19. Remember, the atomic number (Z) gives the number of electrons (Sec. 5.7). We will need to fill subshells, in order of increasing energy, until 19 electrons have been accommodated.

The 1s, 2s, and 2p subshells fill first, accommodating a total of 10 electrons among them

$$1s^2 2s^2 2p^6 \ldots$$

Next, according to Figure 6.9 or 6.10, the $3s$ fills and then the $3p$ subshell.

$$1s^2 2s^2 2p^6 3s^2 3p^6 \ldots$$

We have accommodated 18 electrons at this point. We still need to add one more electron to get our desired number of 19.

The nineteenth electron is added to the $4s$ subshell

$$1s^2 2s^2 2p^6 3s^2 3p^6 4s^1$$

The $4s$ subshell can accommodate 2 electrons, but we do not want it filled to capacity, because that would give us too many electrons.

(b) To write the electron configuration for bromine, we continue along the same lines as in part (a), remembering that the maximum electron subshell populations are $s = 2$, $p = 6$, and $d = 10$.

The first 18 electrons, as with potassium (part a), will fill the $1s$, $2s$, $2p$, $3s$, and $3p$ subshells.

$$1s^2 2s^2 2p^6 3s^2 3p^6 \ldots$$

The $4s$ subshell fills next, accommodating 2 electrons, followed by the $3d$ subshell, which accommodates 10 electrons.

$$1s^2 2s^2 2p^6 3s^2 3p^6 4s^2 3d^{10} \ldots$$

We now have a total of 30 electrons.

Five more electrons are needed, which are added to the next higher subshell in energy, the $4p$.

$$1s^2 2s^2 2p^6 3s^2 3p^6 4s^2 3d^{10} 4p^5$$

The $4p$ subshell can accommodate 6 electrons, but we do not want it filled to capacity because that would give us too many electrons.

To double check that we have the correct number of electrons, 35, we add the superscripts in our final electron configuration.

$$2 + 2 + 6 + 2 + 6 + 2 + 10 + 5 = 35$$

The sum of the superscripts in any electron configuration should add up to the atomic number if the configuration is for a neutral atom.

PRACTICE EXERCISE 6.2

Write the electron configurations for (a) chlorine ($Z = 17$) and (b) antimony ($Z = 51$).

Ans. (a) $1s^2 2s^2 2p^6 3s^2 3p^5$;
(b) $1s^2 2s^2 2p^6 3s^2 3p^6 4s^2 3d^{10} 4p^6 5s^2 4d^{10} 5p^3$.

Similar exercises: Problems 6.33–6.36

It should be noted that for a few elements in the middle of the periodic table the actual distribution of electrons within subshells differs slightly from that obtained using the Aufbau principle and Aufbau diagram. These exceptions are caused by very small energy differences between some subshells and are not important in the uses we shall make of electronic configurations.

Abbreviated electron configurations that give shell electron occupancy rather than subshell electron occupancy are sometimes written. Such configurations are simply a listing of numbers. Example 6.3 gives a comparison between a regular electron configuration (subshell occupancy) and an abbreviated electron configuration (shell occupancy).

EXAMPLE 6.3

The electron configuration for Cl is

$$1s^2 2s^2 2p^6 3s^2 3p^5$$

Write this configuration in an abbreviated form that gives only shell occupancy information.

Solution

Chlorine atoms have three electron shells that contain electrons.

Shell 1 contains 2 electrons: $1s^2$

Shell 2 contains 8 electrons: $2s^2 2p^6$

Shell 3 contains 7 electrons: $3s^2 3p^5$

The abbreviated electron configuration for Cl is

2, 8, 7

In such notations the identity of each shell is not explicitly shown. It is assumed that you know that the listed numbers are ordered in terms of increasing shell number ($n = 1, 2, 3, \ldots$).

Chlorine has 2 electrons in the first shell, 8 in the second shell, and 7 in the third shell.

PRACTICE EXERCISE 6.3

The electron configuration for Ca is

$$1s^2 2s^2 2p^6 3s^2 3p^6 4s^2$$

Write this configuration in an abbreviated form that gives only shell occupancy information.

Ans. 2, 8, 8, 2.   Similar exercises: Problems 6.39 and 6.40

6.8 Orbital Diagrams

The arrangement of electrons about a nucleus can be specified in terms of shell occupancy, subshell occupancy, or orbital occupancy. The notation for specifying electron arrangements in terms of shell occupancy (abbreviated electron configurations) and subshell occupancy (regular electron configurations) has been previously considered (Sec. 6.7). In this section we consider electron arrangements at the orbital level.

Two principles are needed to specify orbital occupancy for electrons.

1. The Aufbau principle.
2. Hund's rule.

The Aufbau principle was used extensively in Section 6.7 in writing electron configurations. Hund's rule has not been encountered previously. The namesake for Hund's rule is the German physicist Frederick Hund (1896–), an early worker in the field of quantum mechanics.

Hund's rule states that when electrons are placed in a set of orbitals of equal energy (the orbitals of a subshell), the order of filling for the orbitals is such that each orbital will be occupied by one electron before any orbital receives a second electron. Such a pattern of orbital filling minimizes repulsions between electrons. Numerous applications of Hund's rule are required in stating electron arrangements in terms of orbital occupancies, that is, in drawing orbital diagrams.

Orbital diagrams are diagrams that show the electron occupancy of each orbital about a nucleus. In drawing orbital diagrams, we use circles to represent orbitals and arrows with a single barb to denote electrons. The orbital diagram for hydrogen, with its one electron, is

$$\text{H:} \quad \underset{1s}{\textcircled{\uparrow}}$$

A helium atom contains two electrons, both of which occupy the 1s orbital. The orbital diagram for helium is

$$\text{He:} \quad \underset{1s}{\textcircled{\uparrow\downarrow}}$$

Note the notation used to denote two electrons in the same orbital—one arrow points up and the other points down. The two electrons are said to be *paired*. When only one electron is present in an orbital it is said to be *unpaired*.

Orbital diagrams for the next two elements, lithium and beryllium, are drawn according to reasoning similar to that followed for H and He. The two electrons in the 2s orbital of Be are paired.

$$\text{Li:} \quad \underset{1s}{\textcircled{\uparrow\downarrow}} \quad \underset{2s}{\textcircled{\uparrow}}$$

$$\text{Be:} \quad \underset{1s}{\textcircled{\uparrow\downarrow}} \quad \underset{2s}{\textcircled{\uparrow\downarrow}}$$

Boron has the electron configuration $1s^2 2s^2 2p^1$. The fifth electron in boron must enter a 2p orbital, since both the 1s and 2s orbitals are full. Boron's orbital diagram is

$$\text{B:} \quad \underset{1s}{\textcircled{\uparrow\downarrow}} \quad \underset{2s}{\textcircled{\uparrow\downarrow}} \quad \underset{2p}{\textcircled{\uparrow}\bigcirc\bigcirc}$$

Note that it does not matter which one of the three $2p$ orbitals of boron is shown as containing an electron. All three orbitals have the same energy. Each of the following notations is correct.

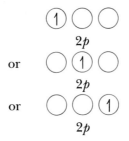

With carbon, element 6, we encounter the use of Hund's rule for the first time. Carbon has two electrons in the $2p$ subshell ($1s^2 2s^2 2p^2$). Do the two electrons go into the same orbital (paired) or do they go into separate equivalent orbitals (unpaired)? Hund's rule indicates that the latter is the case. The orbital diagram for carbon is

C: (1↓)(1↓)(1↓)(1)(1)〇
    1s   2s   2p

Again, it does not matter which two of the three $2p$ orbitals of carbon are shown as containing electrons since all three orbitals have the same energy.

Nitrogen, with the electron configuration $1s^2 2s^2 2p^3$, contains three unpaired electrons.

N: (1↓)(1↓)(1)(1)(1)
    1s   2s   2p

With oxygen ($1s^2 2s^2 2p^4$) two of the four $2p$ electrons must pair up, leaving two unpaired electrons. Fluorine, the next element in the periodic table, has only one unpaired electron. Finally, with neon, all of the electrons are paired up.

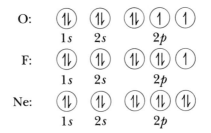

EXAMPLE 6.4

Draw an orbital diagram for the electrons in the element manganese. The electron configuration for manganese is

$$1s^2 2s^2 2p^6 3s^2 3p^6 4s^2 3d^5$$

Solution

All of the occupied subshells of manganese are completely filled except for the $3d$. There are five orbitals in the $3d$ subshell, and there are five electrons present. Following Hund's rule, we will put one electron in each of the five $3d$ orbitals, giving us five unpaired electrons.

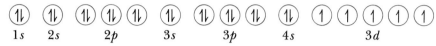

PRACTICE EXERCISE 6.4

Draw an orbital diagram for the electrons in the element sulfur. The electron configuration for sulfur is

$$1s^2 2s^2 2p^6 3s^2 3p^4$$

Ans. ⥮ ⥮ ⥮⥮⥮ ⥮ ⥮↑↑
    1s 2s   2p   3s   3p

Similar exercises: Problems 6.41–6.44

Atoms may be classified as paramagnetic or diamagnetic on the basis of unpaired electrons. A **paramagnetic atom** has an electron arrangement containing one or more unpaired electrons. The presence of unpaired electrons causes paramagnetic materials to be slightly attracted to a magnet. Measurement of paramagnetism provides experimental verification of the presence of unpaired electrons. A **diamagnetic atom** has an electron arrangement in which all electrons are paired.

6.9 Electron Configurations and the Periodic Law

A knowledge of electron configurations for the elements provides an explanation for the periodic law. Recall, from Section 6.1, that the periodic law points out that the properties of the elements repeat themselves in a regular manner when the elements are ordered in sequence of increasing atomic number. Those elements with similar chemical properties are placed one under another in vertical columns (groups) in a periodic table.

Groups of elements have similar chemical properties because of similarities that exist in the electron configurations of the elements of the group. *Chemical properties repeat themselves in a regular manner among the elements because electron configurations repeat themselves in a regular manner among the elements.*

To illustrate this correlation between similar chemical properties and similar electron configurations, let us look at the electron configurations of two groups of elements known to have similar chemical properties.

We begin with the elements lithium, sodium, potassium, and rubidium—all members of group IA of the periodic table. The electron configurations for these similarly propertied elements are

$$_3\text{Li}: \quad 1s^2 2s^1$$
$$_{11}\text{Na}: \quad 1s^2 2s^2 2p^6 3s^1$$
$$_{19}\text{K}: \quad 1s^2 2s^2 2p^6 3s^2 3p^6 4s^1$$
$$_{37}\text{Rb}: \quad 1s^2 2s^2 2p^6 3s^2 3p^6 4s^2 3d^{10} 4p^6 5s^1$$

We see that each of these elements has one outer electron (shown in color), the last one added by the Aufbau principle, in an s subshell. It is this similarity in outer-shell electron arrangements that causes these elements to have similar chemical properties. It is found in general that elements with similar outer shell electron configurations have similar chemical properties.

Let us consider another group of elements known to have similar chemical properties: the elements fluorine, chlorine, bromine, and iodine of group VIIA of the periodic table. The electron configurations for these four elements are

$$_9\text{F}: \quad 1s^2 2s^2 2p^5$$
$$_{17}\text{Cl}: \quad 1s^2 2s^2 2p^6 3s^2 3p^5$$
$$_{35}\text{Br}: \quad 1s^2 2s^2 2p^6 3s^2 3p^6 4s^2 3d^{10} 4p^5$$
$$_{53}\text{I}: \quad 1s^2 2s^2 2p^6 3s^2 3p^6 4s^2 3d^{10} 4p^6 5s^2 4d^{10} 5p^5$$

Once again similarities in electron configurations are readily apparent. This time the repeating pattern is the seven electrons (in color) in the outermost s and p subshells.

Section 7.2 will consider in depth the fact that the electrons most important in controlling chemical properties are those found in the outermost shell of an atom.

6.10 Electron Configurations and the Periodic Table

One of the strongest pieces of supporting evidence for the assignment of electrons to shells, subshells, and orbitals is the periodic table itself. The basic shape and structure of this table, which was determined many years before electrons were even discovered, is consistent with and can be explained by electron configurations. Indeed, the specific location of an element in the periodic table can be used to obtain information about its electron configuration.

The concept of distinguishing electrons is the key to obtaining "electron configuration information" from the periodic table. The **distinguishing electron** for an element is the last electron added to its electron configuration when the configuration is written according to the Aufbau principle. This last electron added is the one that causes an element's electron configuration to differ from that of the element immediately preceding it in the periodic table; hence the term *distinguishing electron*.

As the first step in linking electron configurations to the periodic table, let us analyze the general shape of the periodic table in terms of columns of elements. As shown in Figure 6.11, we have on the extreme left of the table two columns of ele-

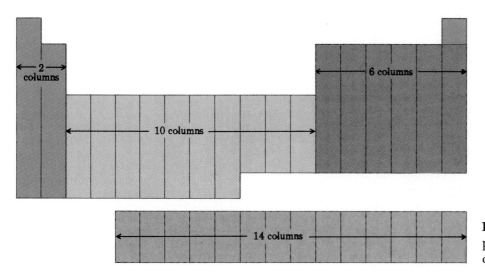

FIGURE 6.11 Structure of the periodic table in terms of columns.

ments, in the center an area containing ten columns of elements, to the right a block of six columns of elements, and at the bottom of the table, in two rows, fourteen columns of elements. These numbers of columns of elements in the various regions of the periodic table—2, 6, 10, and 14—are the same as the maximum numbers of electrons that the various types of subshells can accommodate. We will see shortly that this is a very significant observation; the number matchup is no coincidence. The various columnar regions of the periodic table are called the *s* area (two columns), the *p* area (six columns), the *d* area (ten columns), and the *f* area (fourteen columns), as shown in Figure 6.12.

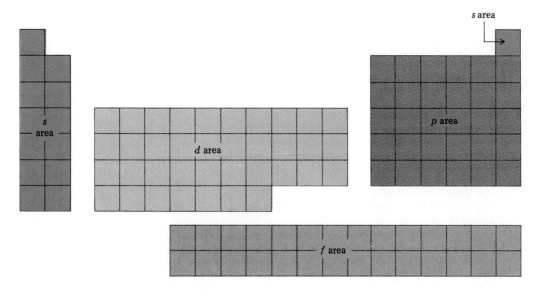

FIGURE 6.12 Areas of the periodic table.

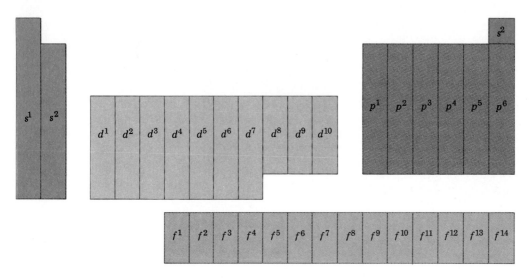

FIGURE 6.13 Extent of subshell filling as a function of periodic table location.

For all elements located in the s area of the periodic table the distinguishing electron is always found in an s subshell. All p area elements have distinguishing electrons in p subshells. Similarly, elements in the d and f areas of the periodic table have, respectively, distinguishing electrons located in d and f subshells. Thus, the area location of an element in the periodic table can be used to determine the type of subshell that contains the distinguishing electron. Note that the element helium is considered to belong to the s rather than the p area of the periodic table even though its table position is on the right-hand side. (The reason for this placement of helium will be explained in Sec. 7.3.)

The extent of filling of the subshell containing an element's distinguishing electron can also be determined from the element's position in the periodic table. All elements in the first column of a specific area contain only one electron in the subshell, all elements in the second column contain two electrons in the subshell, and so on. Thus, all elements in the first column of the p area (group IIIA) have an electron configuration ending in p^1. Elements in the second column of the p area (group IVA) have electron configurations ending in p^2, and so forth. Similar relationships hold in other areas of the table, as shown in Figure 6.13. A few exceptions to these generalizations do exist in the d and f areas (because of irregular electron configurations; see Sec. 6.7), but we will not be concerned with them in this text.

We can also use the periodic table to determine the shell in which the distinguishing electron is located. The relationship used involves the number of the period in which the element is found. In the s and p areas, the period number gives the shell number directly. In the d area, the period number minus one is equal to the shell number. (Remember that the $3d$ subshell is filled during the fourth period.) For similar reasons, in the f area the period number minus two equals the shell number. Thus, the subshell that contains the distinguishing electron for elements of period

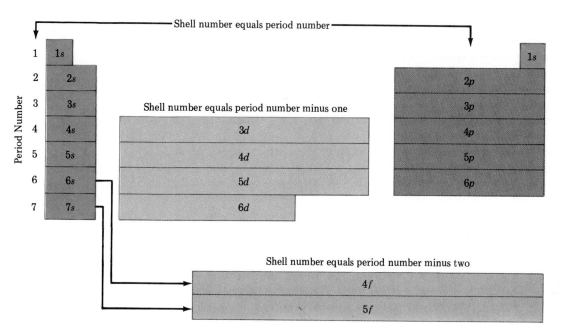

FIGURE 6.14 Relationships of period numbers to shell numbers for distinguishing electrons.

6 may be the 6s, 6p, 5d (period number minus one), or 4f (the period number minus two), depending on the location of the element in period 6. It must be remembered that even though the f area is located at the bottom of the table, it correctly belongs in periods 6 and 7. The complete matchup between period number and shell number for the distinguishing electron is given in the periodic table of Figure 6.14.

EXAMPLE 6.5

Using the periodic table and Figures 6.12–6.14, determine the following for the elements magnesium ($Z = 12$), vanadium ($Z = 23$), and iodine ($Z = 53$).

(a) the type of subshell in which the distinguishing electron is found
(b) the extent of filling of the subshell containing the distinguishing electron
(c) the shell in which the subshell containing the distinguishing electron is found

Solution

(a) Knowing the area of the periodic table in which an element is found is sufficient to determine the type of subshell in which the distinguishing electron is found.

   Mg: Since this element is found in the s area of the periodic table, the distinguishing electron will be in an **s subshell.**

V: Since this element is found in the *d* area of the periodic table, the distinguishing electron will be in a *d* **subshell**.

I: Since this element is found in the *p* area of the periodic table, the distinguishing electron will be in a *p* **subshell**.

(b) The extent of filling of the subshell containing the distinguishing electron is determined by noting the column in the area that the element occupies.

Mg: Since this element is in the second column of the *s* area, the *s* subshell involved contains **two electrons** (s^2).

V: Since this element is in the third column of the *d* area, the *d* subshell involved contains **three electrons** (d^3).

I: Since this element is in the fifth column of the *p* area, the *p* subshell involved contains **five electrons** (p^5).

(c) The shell number of the subshell containing the distinguishing electron is obtained from the period number, sometimes directly and sometimes with modifications.

Mg: Since this element is in period 3, the *s* subshell involved is the 3*s*; therefore, the electron configuration for Mg ends in $3s^2$.

V: Since this element is in period 4, the *d* subshell involved is the 3*d* (period number minus one); therefore, the electron configuration for V ends in $3d^3$.

I: Since this element is in period 5, the *p* subshell involved is the 5*p*; therefore, the electron configuration for I ends in $5p^5$.

PRACTICE EXERCISE 6.5

Using the periodic table and Figures 6.12–6.14, determine the following for the elements cesium ($Z = 55$) and silver ($Z = 47$).

(a) the type of subshell in which the distinguishing electron is found
(b) the extent of filling of the subshell containing the distinguishing electron
(c) the shell in which the subshell containing the distinguishing electron is found

Ans. (a) *s* for Cs, *d* for Ag; (b) s^1 for Cs, d^9 for Ag; (c) $6s^1$ for Cs, $4d^9$ for Ag.

Similar exercises: Problems 6.53–6.56

In order to write complete electron configurations, you must know the order in which the various electron subshells are filled. Up until now, we have obtained this filling order by using an Aufbau diagram (Fig. 6.10). We can also obtain this information directly from the periodic table. To obtain it, we merely follow a path of increasing atomic number through the table, noting the various subshells as we encounter them.

Figure 6.15 illustrates this way of using the periodic table. Note particularly how the *f* area of the periodic table is worked into the scheme. The results of working our way through the periodic table in terms of atomic numbers are summarized in Table 6.4.

Section 6.10 · Electron Configurations and the Periodic Table

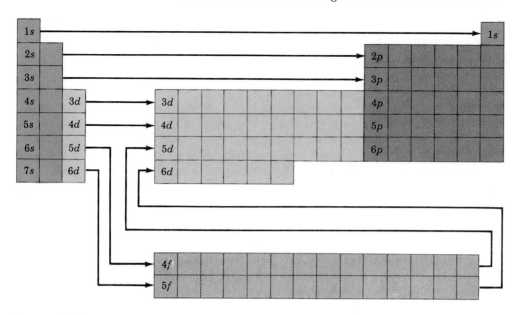

FIGURE 6.15 Using the periodic table as a guide to obtain the order in which subshells are occupied by electrons.

TABLE 6.4 Subshell Filling Order from the Periodic Table

Atomic Numbers	Subshell Involved
1–2	1s
3–4	2s
5–10	2p
11–12	3s
13–18	3p
19–20	4s
21–30	3d
31–36	4p
37–38	5s
39–48	4d
49–54	5p
55–56	6s
57	5d
58–71	4f
72–80	5d
81–86	6p
87–88	7s
89	6d
90–103	5f
104–109	6d

EXAMPLE 6.6

Write the complete electron configuration of $_{80}$Hg using the periodic table as your sole guide.

Solution

The needed information will be obtained by "working our way" through the periodic table. We will start at hydrogen and go from box to box, in order of increasing atomic number, until we arrive at position 80, mercury. Every time we traverse the s area we will fill an s subshell; the p area, a p subshell; and so on. We will remember that the shell number for s and p subshells is the period number; for d subshells, the period number minus one; and for f subshells, the period number minus two.

Let us begin our journey through the periodic table. As we cross period 1 we encounter H and He, both $1s$ elements. We thus add the $1s$ electrons.

$$\text{Hg: } 1s^2\ldots$$

In traversing period 2 we pass through the s area (elements 3 and 4) and the p area (elements 5–10) in that order. We add $2s$ and $2p$ electrons.

$$\text{Hg: } 1s^2 2s^2 2p^6\ldots$$

Our trip through period 3 is very similar to that of period 2. Only s-area elements (11 and 12) and p-area elements (13–18) are encountered. The message is to add s and p electrons—$3s$ and $3p$ because we are in period 3.

$$\text{Hg: } 1s^2 2s^2 2p^6 3s^2 3p^6\ldots$$

In passing through period 4 we go through the s area (elements 19 and 20), the d area (elements 21–30), and the p area (elements 31–36). Electrons to be added are those in the $4s$, $3d$ (period number minus one), and $4p$ subshells, in that order.

$$\text{Hg: } 1s^2 2s^2 2p^6 3s^2 3p^6 4s^2 3d^{10} 4p^6\ldots$$

In period 5 we encounter scenery similar to that in period 4—the s area (elements 37 and 38), the d area (elements 39–48), and the p area (elements 49–54). Hence, the $5s$, $4d$ (period number minus one), and $5p$ subshells are filled in order.

$$\text{Hg: } 1s^2 2s^2 2p^6 3s^2 3p^6 4s^2 3d^{10} 4p^6 5s^2 4d^{10} 5p^6\ldots$$

Our journey ends in period 6. We go through the $6s$ area (elements 55 and 56), the $4f$ area (elements 58–71), and the $5d$ area (elements 57 and 72–80). We will completely fill the $6s$, $4f$ (period number minus two), and $5d$ (period number minus one) subshells.

$$\text{Hg: } 1s^2 2s^2 2p^6 3s^2 3p^6 4s^2 3d^{10} 4p^6 5s^2 4d^{10} 5p^6 6s^2 4f^{14} 5d^{10}$$

PRACTICE EXERCISE 6.6

Write the complete electron configuration of $_{54}$Xe using the periodic table as your sole guide.

Ans. $1s^2 2s^2 2p^6 3s^2 3p^6 4s^2 3d^{10} 4p^6 5s^2 4d^{10} 5p^6$.

Similar exercises: Problems 6.63 and 6.64

Again, let us mention that a few slightly irregular electronic configurations are encountered in the *d* and *f* areas of the periodic table. The generalizations in this section do not address this problem. We will be working mostly with *s* and *p* area elements in future chapters. There are no irregularities in electron configurations for these elements.

6.11 Classification Systems for the Elements

The elements can be classified in several ways. The two most common classification systems are

1. A system based on the electron configurations of the elements, in which elements are described as *noble gas, representative, transition,* or *inner transition* elements.
2. A system based on selected physical properties of the elements, in which elements are described as *metals* or *nonmetals*.

The classification scheme based on electron configurations of the elements is depicted in Figure 6.16. This type of classification system is used in numerous discussions in subsequent chapters.

The **noble gases** are found in the far right column of the periodic table. They are all gases at room temperature, and they have little tendency to form chemical compounds. With one exception, the distinguishing electron for a noble gas completes the *p* subshell. Therefore, they have electron configurations ending in p^6. The exception is helium, in which the distinguishing electron completes the first shell—a shell that has only two electrons. Helium's electron configuration is $1s^2$.

The **representative elements** are all of the elements of the *s* and *p* areas of the periodic table with the exception of the noble gases. The distinguishing electron in these elements partially or completely fills an *s* subshell or partially fills a *p* subshell. The representative elements include most of the more common elements.

The **transition elements** are all of the elements of the *d* area of the periodic table. The common feature in the electronic configurations of the transition elements is the presence of the distinguishing electron in a *d* subshell.

The **inner transition elements** are all of the elements of the *f* area of the periodic table. The characteristic feature of their electronic configurations is the presence of the distinguishing electron in an *f* subshell. There is very little variance in the properties of these elements.

On the basis of selected physical properties of the elements, the second of the two classification schemes divides the elements into the categories of metals and nonmetals. A **metal** is an element that has the characteristic properties of luster, thermal conductivity, electrical conductivity, and malleability. With the exception of mercury,

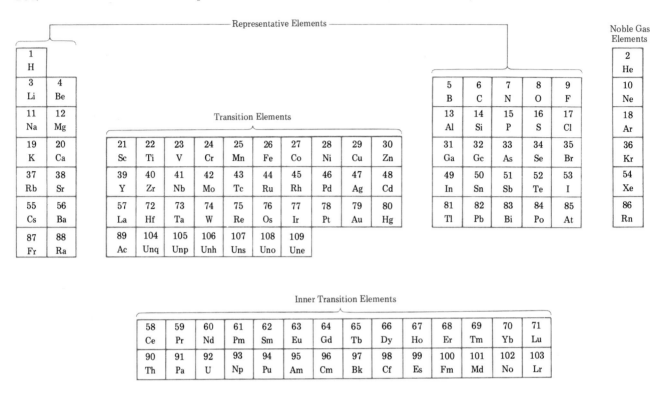

FIGURE 6.16 Elemental classification scheme based on the electron configurations of the elements.

all metals are solids at room temperature (25 °C). Metals are good conductors of heat and electricity. Most metals are ductile (they can be drawn into wires) and malleable (they can be rolled into sheets). Most metals have high luster (shine), high density, and high melting points. Among the more familiar metals are the elements iron, aluminum, copper, silver, and gold.

A **nonmetal** is an element characterized by the absence of the properties of luster, thermal conductivity, electrical conductivity, and malleability. Many of the nonmetals, such as hydrogen, oxygen, nitrogen, and the noble gases are gases at room temperature (25 °C). The only nonmetal that is a liquid at room temperature is bromine. Solid nonmetals include carbon, sulfur, and phosphorus. In general, the nonmetals have lower densities and melting points than metals.

The majority of the elements are metals; only 22 elements are nonmetals; the rest (87) are metals. It is not necessary to memorize which elements are nonmetals and which are metals. As can be seen from Figure 6.17, the location of an element in the periodic table correlates directly with its classification as a metal or nonmetal. The steplike heavy line that runs through the *p* area of the periodic table separates the metals from the nonmetals; metals are on the left and nonmetals on the right. Note that the element hydrogen is a nonmetal even though it is located on the left side of the periodic table.

The fact that the vast majority of elements are metals in no way indicates that metals are more important than nonmetals. Most nonmetals are relatively abundant

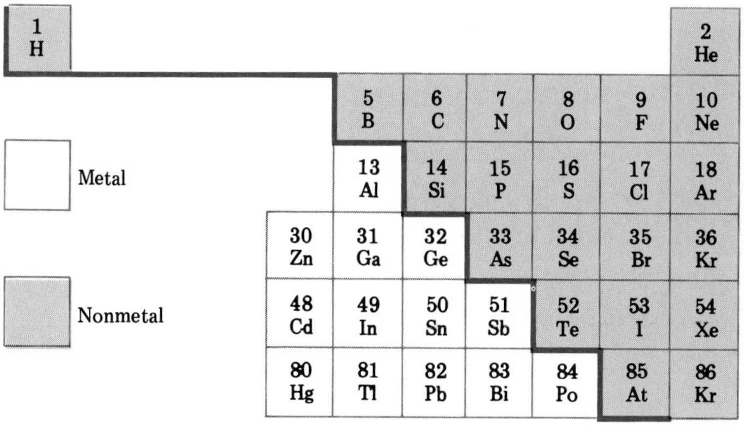

FIGURE 6.17 A portion of the periodic table showing the dividing line between metals and nonmetals. All elements that are not shown are metals.

and are found in many important compounds. For example, water (H_2O) is a compound involving two nonmetals. An analysis of the previously given abundances of the elements in the Earth's crust (Table 4.4) in terms of metals and nonmetals shows that three of the four most abundant elements, which account for 83% of all atoms, are nonmetals—oxygen, silicon, and hydrogen. The four most abundant elements in the human body (Table 4.4), which constitute more than 99% of all atoms in the body, are nonmetals—hydrogen, oxygen, carbon, and nitrogen.

6.12 Chemical Periodicity

Chemical periodicity is the variation in properties of elements with their positions in the periodic table. In this section we consider two properties of elements that exhibit chemical periodicity—metallic–nonmetallic character and atomic size (atomic radius)—and then in Section 7.10 we consider a third property exhibiting chemical periodicity—electronegativity.

The physical properties that distinguish metals and nonmetals were considered in Section 6.11. In general, these properties are opposites for these two classes of substances.

Not all metals possess metallic properties to the same extent; they have them, but to varying degrees. For example, some metals are better conductors of electricity than other metals. Similarly, not all nonmetals possess nonmetallic properties to the same extent.

Metallic and nonmetallic character for the various elements—the extent to which they possess metallic and nonmetallic properties—can be correlated with periodic table position for the elements.

1. Metallic character increases from right to left across a given period in the periodic table.
2. Metallic character increases from top to bottom down a group in the periodic table.

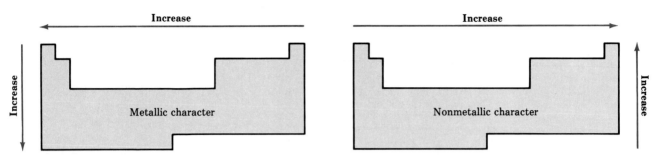

FIGURE 6.18 Chemical periodicity is associated with the intensities of metallic and nonmetallic character.

Thus, among the period 3 elements, Na is more metallic than Mg, which in turn is more metallic than Al. For the group IA elements, K is more metallic than Na, since it is farther down in group IA.

Similar, but opposite, trends exist for nonmetallic character.

1. Nonmetallic character increases from left to right across a given period in the periodic table.
2. Nonmetallic character increases from bottom to top up a group in the periodic table.

Figure 6.18 summarizes the chemical periodicity associated with the properties of metallic and nonmetallic character. The combined effect of these two trends is that the most nonmetallic elements are those in the upper right portion of the periodic table, a generalization consistent with Figure 6.17 in the preceding section of this chapter.

In Section 6.11 a "dividing line" was shown for classifying elements as metals or nonmetals (Fig. 6.17). Elements to the far left of this line have dominant metallic character. Elements to the far right of this line have dominant nonmetallic character. Most elements whose periodic table positions "touch" this metal–nonmetal dividing line actually show some properties that are characteristic of both metals and nonmetals. These elements are frequently given a classification of their own: metalloids (or semimetals). A **metalloid** is an element with properties intermediate between those of metals and nonmetals. Figure 6.19 identifies the metalloid elements. Several of the metalloids, including silicon, germanium, and antimony, are semiconductors. A **semiconductor** is an element that does not conduct electrical current at room temperature but does so at higher temperatures. Semiconductor elements are very important in the electronics industry.

The quantum mechanical model for an atom (Sec. 6.3) suggests that the shape of an atom is approximately spherical and that the size of the atom can be expressed in terms of the radius of a sphere. The most commonly used unit for expressing atomic sizes (atomic radii) is the picometer (Sec. 3.1). In this unit most atomic radii fall in the range of 50–200 pm.

Atomic sizes exhibit chemical periodicity, as is shown in Figure 6.20, which gives atomic radii for the representative elements. General periodic trends for atomic radii

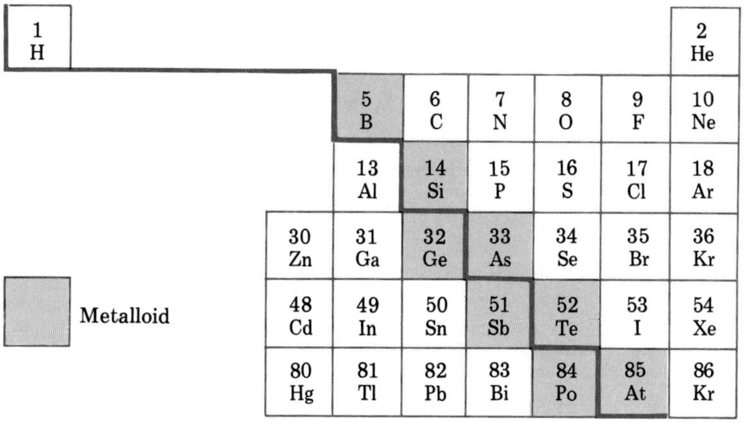

Section 6.12 · Chemical Periodicity

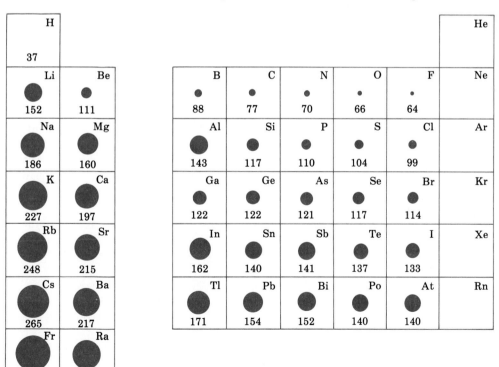

FIGURE 6.19 A portion of the periodic table showing the metalloids (in color), elements that show some properties characteristic of metals and some other properties characteristic of nonmetals.

are

1. Atomic radii tend to decrease from left to right within a period of the periodic table.
2. Atomic radii tend to increase from top to bottom within a periodic table group.

The trends in atomic radii values can be explained on the basis of the number of electrons present in the atom and their energies. In traversing a period in the periodic

FIGURE 6.20 Atomic radii, in picometers, of the representative elements.

table (from left to right), we note that (1) nuclear charge (atomic number) increases and (2) the added electrons enter the same shell, the outermost shell. The increased nuclear charge draws the electrons in this outermost shell closer to the nucleus, resulting in smaller atoms. Down a group in the periodic table, atomic radii increase because the electrons are added to shells with higher numbers. Recall, from Section 6.4, that the higher the shell number the greater the average distances between the electrons and the nucleus.

Key Terms

The new terms or concepts defined in this chapter are

Aufbau diagram (Sec. 6.7)   A listing of subshells arranged in the order in which electrons occupy them.

Aufbau principle (Sec. 6.7)   Electrons normally occupy the lowest energy subshell available.

chemical periodicity (Sec. 6.12)   The variation in properties of elements with their positions in the periodic table.

diamagnetic atom (Sec. 6.8)   An atom having an electron arrangement in which all electrons are paired.

distinguishing electron (Sec. 6.10)   The last electron added to an element's electron configuration when the configuration is written according to the Aufbau principle.

electron configuration (Sec. 6.7)   A statement of how many electrons an atom has in each of its subshells.

electron orbital (Sec. 6.6)   A region of space around a nucleus where an electron with a specific energy is most likely to be found.

electron shell (Sec. 6.4)   An energy level in which the electrons have approximately the same energy and spend most of their time at approximately the same distance from the nucleus.

electron subshell (Sec. 6.5)   An energy sublevel within an electron shell in which the electrons all have the same energy.

group (Sec. 6.2)   A vertical column of elements in the periodic table.

Hund's rule (Sec. 6.8)   When electrons are placed in a set of orbitals of equal energy (the orbitals of a subshell), the order of filling for the orbitals is such that each orbital will be occupied by one electron before any orbital receives a second electron.

inner transition element (Sec. 6.11)   An element of the f area of the periodic table.

metal (Sec. 6.11)   An element that has the characteristic properties of luster, thermal conductivity, electrical conductivity, and malleability.

metalloid (Sec. 6.12)   An element with properties intermediate between those of metals and nonmetals.

noble gas (Sec. 6.11)   An element of the far right column of the periodic table.

nonmetal (Sec. 6.11)   An element characterized by the absence of the properties of luster, thermal conductivity, electrical conductivity, and malleability.

orbital diagram (Sec. 6.8)   A diagram that shows the electron occupancy of each of the orbitals about a nucleus.

paramagnetic atom (Sec. 6.8)   An atom having an electron arrangement containing one or more unpaired electrons.

period (Sec. 6.2)   A horizontal row of elements in the periodic table.

periodic law (Sec. 6.1)   When elements are arranged in order of increasing atomic number, elements with similar properties occur at periodic (regularly recurring) intervals.

periodic table (Sec. 6.2)   A graphical representation of the behavior described by the periodic law.

quantized property (Sec. 6.3)   A property that can have only certain values, that is, not all values are allowed.

representative element (Sec. 6.11)   An element of the s or p area of the periodic table with the exception of the noble gases.

semiconductor (Sec. 6.12)   An element that does not conduct electrical current at room temperature but does so at higher temperatures.

transition element (Sec. 6.11)   An element of the d area of the periodic table.

Practice Problems

Periodic Law and Periodic Table (Sec. 6.1 and 6.2)

6.1 Give a modern-day statement of the periodic law.

6.2 How does the original formulation of the periodic law differ from the current formulation of the law?

6.3 How does the term *group* relate to the periodic table?

6.4 How does the term *period* relate to the periodic table?

6.5 Using the periodic table, indicate the period and group number for each of the following elements.
(a) $_3$Li (b) $_{16}$S
(c) $_{47}$Ag (d) $_{106}$Unh

6.6 Using the periodic table, indicate the period and group number for each of the following elements.
(a) $_9$F (b) $_{31}$Ga
(c) $_{79}$Au (d) $_{89}$Ac

6.7 Give the symbol of the element that occupies each of the following positions in the periodic table.
(a) period 4, group IIIA (b) period 5, group IVB
(c) group IA, period 2 (d) group VIIA, period 3

6.8 Give the symbol of the element that occupies each of the following positions in the periodic table.
(a) period 1, group IA (b) period 6, group IB
(c) group IIIB, period 4 (d) group IVA, period 5

6.9 The following statements either define or are closely related to the terms *periodic law*, *period*, or *group*. Match the terms to the appropriate statements.
(a) This is a vertical arrangement of elements in the periodic table.
(b) The properties of the elements repeat in a regular way as the atomic numbers increase.
(c) The chemical properties of elements 12, 20, and 38 demonstrate this principle.
(d) Elements 24 and 33 belong to this arrangement.

6.10 The following statements either define or are closely related to the terms *periodic law*, *period*, and *group*. Match the terms to the appropriate statements.
(a) This is a horizontal arrangement of elements in the periodic table.
(b) Element 19 begins this arrangement in the periodic table.
(c) The element carbon is the first member of this arrangement.
(d) Elements 10, 18, 36, and 54 belong to this arrangement.

6.11 What three pieces of information about each element are found in the periodic table in Figure 6.2?

6.12 What is the difference between the "long" and "short" forms of the periodic table?

Terminology Associated with Electron Arrangements (Secs. 6.3–6.6)

6.13 What does the term *quantization* mean?

6.14 What property of an electron is quantized?

6.15 The following statements define or are closely related to the terms *shell*, *subshell*, and *orbital*. Match the terms to the appropriate statements.
(a) In terms of electron capacity, this unit is the smallest of the three.
(b) This unit can contain a maximum of two electrons.
(c) This unit can contain as many electrons as, or more electrons than, either of the other two.
(d) This unit is designated by just a number.

6.16 The following statements define or are closely related to the terms *shell*, *subshell*, and *orbital*. Match the terms to the appropriate statements.
(a) The term *energy sublevel* is closely associated with this unit.
(b) The formula $2n^2$ gives the maximum number of electrons that can occupy this unit.
(c) This unit is designated in the same way as the orbitals contained within it.
(d) The term *energy level* is closely associated with this unit.

6.17 Determine the following for the fifth electron shell (the fifth main energy level) of an atom.
(a) the number of subshells it contains
(b) the designations used to describe each of the first four subshells
(c) the number of orbitals in each of the first four subshells
(d) the maximum number of electrons that can occupy this shell
(e) the maximum number of electrons that can occupy each of the first four subshells

6.18 Determine the following for the sixth electron shell (the sixth main energy level) of an atom.
(a) the number of subshells it contains
(b) the designations used to describe each of the first three subshells
(c) the number of orbitals in each of the first three subshells
(d) the maximum number of electrons that can occupy this shell
(e) the maximum number of electrons that can occupy each of the first three subshells

6.19 Indicate whether each of the following statements is *true* or *false*.

(a) An orbital has a definite size and shape, which are related to the energy of the electrons it could contain.
(b) The M and $n = 3$ shells are one and the same.
(c) All of the orbitals in a subshell have the same energy.
(d) A $2p$ and a $3p$ subshell would contain the same number of orbitals.
(e) The fourth shell is made up of six subshells.

6.20 Indicate whether each of the following statements is *true* or *false*.
(a) All of the subshells in a shell have the same energy.
(b) A d subshell always contains five orbitals.
(c) An s orbital is shaped something like a four-leaf clover.
(d) The $n = 3$ shell can accommodate a maximum of 18 electrons.
(e) All subshells accommodate the same number of electrons.

6.21 Describe the general shape of each of the following orbitals.
(a) $2s$ (b) $3s$ (c) $4p$ (d) $3d$

6.22 Describe the general shape of each of the following orbitals.
(a) $4s$ (b) $3p$ (c) $6d$ (d) $4f$

6.23 Give the maximum number of electrons that can occupy each of the following units.
(a) $3d$ subshell (b) $2s$ orbital
(c) second shell (d) shell with $n = 4$

6.24 Give the maximum number of electrons that can occupy each of the following units.
(a) $4p$ subshell (b) $4p$ orbital
(c) third shell (d) shell with $n = 1$

6.25 Which of the following atomic orbital designations is impossible?
(a) $4s$ (b) $2d$ (c) $1p$ (d) $5f$

6.26 Which of the following atomic orbital designations is impossible?
(a) $3p$ (b) $3f$ (c) $7s$ (d) $4d$

Writing Electron Configurations (Sec. 6.7)

6.27 Explain the difference between the Aufbau principle and an Aufbau diagram.

6.28 Within any given shell, how do the energies of the s, p, d, and f subshells compare?

6.29 For each of the following sets of subshells, determine the one of lowest energy and the one of highest energy.
(a) $1s, 2s, 3s$ (b) $3s, 3p, 3d$
(c) $4d, 4f, 5p$ (d) $3d, 5p, 7s$

6.30 For each of the following sets of subshells, determine the one of lowest energy and the one of highest energy.

(a) $4s, 4p, 4d$ (b) $4d, 5d, 6d$
(c) $6s, 4f, 5d$ (d) $3p, 3d, 4s$

6.31 Explain what each number and letter means in the notation $4p^6$.

6.32 Explain what each number and letter means in the notation $3d^7$.

6.33 With the help of an Aufbau diagram, write the complete electron configuration for each of the following atoms.
(a) $_8$O (b) $_{12}$Mg (c) $_{15}$P (d) $_{20}$Ca

6.34 With the help of an Aufbau diagram, write the complete electron configuration for each of the following atoms.
(a) $_6$C (b) $_{13}$Al (c) $_{16}$S (d) $_{18}$Ar

6.35 With the help of an Aufbau diagram, write the complete electron configuration for each of the following atoms.
(a) $_{23}$V (b) $_{34}$Se (c) $_{55}$Cs (d) $_{88}$Ra

6.36 With the help of an Aufbau diagram, write the complete electron configuration for each of the following atoms.
(a) $_{26}$Fe (b) $_{36}$Kr (c) $_{50}$Sn (d) $_{87}$Fr

6.37 Identify the element represented by each of the following electron configurations.
(a) $1s^2 2s^2 2p^4$ (b) $1s^2 2s^2 2p^6 3s^2 3p^6 4s^1$
(c) $1s^2 2s^2 2p^6 3s^2 3p^6 4s^2 3d^6$
(d) $1s^2 2s^2 2p^6 3s^2 3p^6 4s^2 3d^{10}$

6.38 Identify the element represented by each of the following electron configurations.
(a) $1s^2 2s^2 2p^1$ (b) $1s^2 2s^2 2p^6 3s^1$
(c) $1s^2 2s^2 2p^6 3s^2 3p^6$ (d) $1s^2 2s^2 2p^6 3s^2 3p^6 4s^2$

6.39 Write abbreviated electron configurations that give only shell occupancy rather than subshell occupancy for each of the following.
(a) the electron configuration $1s^2 2s^2 2p^6 3s^2 3p^1$
(b) the electron configuration $1s^2 2s^2 2p^6 3s^2 3p^6$
(c) the element $_{14}$Si
(d) the element $_4$Be

6.40 Write abbreviated electron configurations that give only shell occupancy rather than subshell occupancy for each of the following.
(a) the electron configuration $1s^2 2s^2 2p^3$
(b) the electron configuration $1s^2 2s^2 2p^6 3s^1$
(c) the element $_{19}$K
(d) the element $_6$C

Orbital Diagrams (Sec. 6.8)

6.41 Draw the electron orbital diagram associated with each of the following electron configurations.

(a) $1s^2 2s^2 2p^2$ (b) $1s^2 2s^2 2p^6$
(c) $1s^2 2s^2 2p^6 3s^1$ (d) $1s^2 2s^2 2p^6 3s^2 3p^3$

6.42 Draw the electron orbital diagram associated with each of the following electron configurations.
(a) $1s^2 2s^2 2p^3$ (b) $1s^2 2s^2 2p^5$
(c) $1s^2 2s^2 2p^6 3s^2$ (d) $1s^2 2s^2 2p^6 3s^2 3p^4$

6.43 Draw electron orbital diagrams for the following elements.
(a) $_3$Li (b) $_9$F (c) $_{15}$P (d) $_{28}$Ni

6.44 Draw electron orbital diagrams for the following elements.
(a) $_7$N (b) $_{12}$Mg (c) $_{17}$Cl (d) $_{27}$Co

6.45 How many unpaired electrons are there in atoms of the following elements?
(a) $_8$O (b) $_{14}$Si (c) $_{20}$Ca (d) $_{23}$V

6.46 How many unpaired electrons are there in atoms of the following elements?
(a) $_4$Be (b) $_{13}$Al (c) $_{22}$Ti (d) $_{35}$Br

6.47 Indicate whether each of the following atoms is paramagnetic or diamagnetic.
(a) $_6$C (b) $_{11}$Na (c) $_{18}$Ar (d) $_{30}$Zn

6.48 Indicate whether each of the following atoms is paramagnetic or diamagnetic.
(a) $_5$B (b) $_{10}$Ne (c) $_{19}$K (d) $_{26}$Fe

Electron Configurations and the Periodic Law (Sec. 6.9)

6.49 Indicate whether or not the elements represented by the given pairs of electron configurations have similar chemical properties.
(a) $1s^2 2s^1$ and $1s^2 2s^2$
(b) $1s^2 2s^2 2p^6$ and $1s^2 2s^2 2p^6 3s^2 3p^6$
(c) $1s^2 2s^2 2p^3$ and $1s^2 2s^2 2p^6 3s^2 3p^6 4s^2 3d^3$
(d) $1s^2 2s^2 2p^6 3s^2 3p^6$ and $1s^2 2s^2 2p^6 3s^2 3p^6 4s^2 3d^{10} 4p^6$

6.50 Indicate whether or not the elements represented by the given pairs of electron configurations have similar chemical properties.
(a) $1s^2 2s^2 2p^4$ and $1s^2 2s^2 2p^5$
(b) $1s^2 2s^2$ and $1s^2 2s^2 2p^2$
(c) $1s^2 2s^1$ and $1s^2 2s^2 2p^6 3s^2 3p^6 4s^1$
(d) $1s^2 2s^2 2p^6$ and $1s^2 2s^2 2p^6 3s^2 3p^6 4s^2 3d^6$

Electron Configurations and the Periodic Table (Sec. 6.10)

6.51 Indicate the position in the periodic table of each of the following elements in terms of s, p, d, or f area.
(a) $_{13}$Al (b) $_{29}$Cu (c) $_{20}$Ca (d) $_{10}$Ne

6.52 Indicate the position in the periodic table of each of the following elements in terms of s, p, d, or f area.
(a) $_{92}$U (b) $_{21}$Sc (c) $_8$O (d) $_{11}$Na

6.53 For each of the following elements, identify the subshell ($2s$, $3p$, $4f$, etc) that contains the distinguishing electron.
(a) $_{14}$Si (b) $_{50}$Sn (c) $_{76}$Os (d) $_{38}$Sr

6.54 For each of the following elements, identify the subshell ($2s$, $3p$, $4f$, etc.) that contains the distinguishing electron.
(a) $_{35}$Br (b) $_{19}$K (c) $_{61}$Pm (d) $_{80}$Hg

6.55 For each of the following elements, indicate the extent to which the subshell containing the distinguishing electron is filled.
(a) $_{25}$Mn (b) $_3$Li (c) $_{86}$Rn (d) $_{31}$Ga

6.56 For each of the following elements, indicate the extent to which the subshell containing the distinguishing electron is filled.
(a) $_7$N (b) $_{17}$Cl (c) $_{27}$Co (d) $_{37}$Rb

6.57 Indicate the position in the periodic table where each of the following occurs by giving the symbol of the element.
(a) The $3p$ subshell begins filling.
(b) The $2s$ subshell begins filling.
(c) The $5d$ subshell begins filling.
(d) The $3d$ subshell begins filling.

6.58 Indicate the position in the periodic table where each of the following occurs by giving the symbol of the element.
(a) The $4p$ subshell begins filling.
(b) The $5f$ subshell begins filling.
(c) The $3s$ subshell begins filling.
(d) The $7s$ subshell begins filling.

6.59 Indicate the position in the periodic table where each of the following occurs by giving the symbol of the element.
(a) The $4p$ subshell becomes completely filled.
(b) The $2s$ subshell becomes half-filled.
(c) The fourth shell begins filling.
(d) The fourth shell becomes completely filled.

6.60 Indicate the position in the periodic table where each of the following occurs by giving the symbol of the element.
(a) The $3d$ subshell becomes completely filled.
(b) The $4p$ subshell becomes half-filled.
(c) The third shell becomes half-filled.
(d) The third shell becomes completely filled.

6.61 Identify the element whose electron configuration, after the distinguishing electron has been added, ends in
(a) $3d^6$ (b) $5s^1$ (c) $4p^4$ (d) $6d^2$

6.62 Identify the element whose electron configuration, after the distinguishing electron has been added, ends in
(a) $6s^1$ (b) $4f^{10}$ (c) $5d^{10}$ (d) $2p^1$

6.63 Using only the periodic table, determine the complete electron configuration for each of the following elements.
(a) $_{20}$Ca (b) $_{33}$As (c) $_{14}$Si (d) $_{41}$Nb

6.64 Using only the periodic table, determine the complete electron configuration for each of the following elements.
(a) $_{57}$La (b) $_{86}$Rn (c) $_{35}$Br (d) $_{23}$V

6.65 Using the periodic table as a guide, indicate the number of
(a) 3p electrons in a $_{16}$S atom
(b) 3d electrons in a $_{29}$Cu atom
(c) 4s electrons in a $_{37}$Rb atom
(d) 4d electrons in a $_{30}$Zn atom

6.66 Using the periodic table as a guide, indicate the number of
(a) 3s electrons in a $_{12}$Mg atom
(b) 4p electrons in a $_{32}$Ge atom
(c) 3d electrons in a $_{47}$Ag atom
(d) 4p electrons in a $_{15}$P atom

Classification Systems for the Elements (Sec. 6.11)

6.67 Classify each of the following elements as a noble gas, representative element, transition element, or inner transition element.
(a) $_{54}$Xe (b) $_{45}$Rh (c) $_{16}$S (d) $_{95}$Am

6.68 Classify each of the following elements as a noble gas, representative element, transition element, or inner transition element.
(a) $_{3}$Li (b) $_{106}$Unh (c) $_{60}$Nd (d) $_{2}$He

6.69 Classify each of the following elements as a metal or nonmetal.
(a) $_{15}$P (b) $_{4}$Be (c) $_{78}$Pt (d) $_{10}$Ne

6.70 Classify each of the following elements as a metal or nonmetal.
(a) $_{1}$H (b) $_{47}$Ag (c) $_{34}$Se (d) $_{64}$Gd

6.71 Select from the following list of six elements

$_{30}$Zn $_{32}$Ge $_{33}$As $_{36}$Kr $_{38}$Sr $_{40}$Zr

those elements that are
(a) transition metals
(b) inner transition elements
(c) representative metals
(d) representative nonmetals

6.72 Select from the following list of six elements

$_{6}$C $_{14}$Si $_{32}$Ge $_{50}$Sn $_{55}$Cs $_{61}$Pm

those elements that are
(a) inner transition metals
(b) noble gases
(c) representative metals
(d) representative nonmetals

Chemical Periodicity (Sec. 6.12)

6.73 Using the periodic table, indicate which member of each pair is more metallic.
(a) $_{12}$Mg or $_{14}$Si (b) $_{47}$Ag or $_{79}$Au
(c) $_{16}$S or $_{17}$Cl (d) $_{4}$Be or $_{37}$Rb

6.74 Using the periodic table, indicate which member of each pair is more metallic.
(a) $_{11}$Na or $_{19}$K (b) $_{22}$Ti or $_{29}$Cu
(c) $_{32}$Ge or $_{34}$Se (d) $_{30}$Zn or $_{56}$Ba

6.75 Using the periodic table, indicate which member of each pair is more nonmetallic.
(a) $_{9}$F or $_{35}$Br (b) $_{14}$Si or $_{15}$P
(c) $_{25}$Mn or $_{30}$Zn (d) $_{17}$Cl or $_{33}$As

6.76 Using the periodic table, indicate which member of each pair is more nonmetallic.
(a) $_{6}$C or $_{8}$O (b) $_{12}$Mg or $_{20}$Ca
(c) $_{50}$Sn or $_{52}$Te (d) $_{53}$I or $_{81}$Ti

6.77 Indicate which member of each of the following pairs of elements has the larger atomic radius.
(a) $_{7}$N or $_{8}$O (b) $_{17}$Cl or $_{31}$Ga
(c) $_{19}$K or $_{35}$Br (d) $_{19}$K or $_{37}$Rb

6.78 Indicate which member of each of the following pairs of elements has the larger atomic radius.
(a) $_{15}$P or $_{16}$S (b) $_{15}$P or $_{33}$As
(c) $_{35}$Br or $_{52}$Te (d) $_{37}$Rb or $_{53}$I

Additional Problems

6.79 For each of the following sentences decide whether the phrase "dependent on" or "independent of" should be placed in the blank to make the statement a correct statement.
(a) The energy of a subshell is _____ the shell in which it is located.
(b) The maximum number of electrons a subshell can accommodate is _____ the shell in which it is found.
(c) The shape of an orbital is _____ the subshell in which it is found.

6.80 For each of the following sentences, decide whether the phrase "dependent on" or "independent of" should

be placed in the blank to make the statement a correct statement.
(a) The size (volume) of a given type of orbital (s. p, etc.) is _____ the shell in which it is found.
(b) The maximum number of electrons an orbital can accommodate is _____ the subshell in which it is found.
(c) The maximum number of electrons an orbital can accommodate is _____ the shell in which it is found.

6.81 How many orbitals are there in each of the following?
(a) the $3d$ subshell
(b) the $4f$ subshell
(c) the second shell
(d) the sixth shell

6.82 How many orbitals are there in each of the following?
(a) the $4p$ subshell
(b) the $5f$ subshell
(c) the third shell
(d) the fifth shell

6.83 How many electrons are in each of the following?
(a) the outermost p subshell of a group IVA element
(b) the outermost p subshell of a group VIA element
(c) the outermost s subshell of a group IA element
(d) the outermost s subshell of a group IVB element

6.84 How many electrons are in each of the following?
(a) the outermost s subshell of a group IIA element
(b) the outermost d subshell of a group IVB element
(c) the outermost p subshell of a group VA element
(d) the outermost s subshell of a group VIIA element

6.85 What is wrong with each of the following attempts to write an electron configuration?
(a) $1s^2 2s^3$
(b) $1s^2 2s^2 2p^2 3s^2$
(c) $1s^2 2s^2 3s^2$
(d) $1s^2 2s^2 2p^6 3s^2 3d^{10}$

6.86 What is wrong with each of the following attempts to write an electron configuration?
(a) $1s^2 1p^6$
(b) $1s^2 2s^4$
(c) $1s^2 2s^2 2p^4 3s^2$
(d) $1s^2 2s^2 2p^6 3s^2 3p^6 3d^{10}$

6.87 Write electron configurations for the following elements.
(a) the group IVA element in the same period as $_{15}P$
(b) the period 2 element in the same group as $_{50}Sn$
(c) the lowest-atomic-numbered nonmetal in period VA
(d) the period 2 element that has three unpaired electrons

6.88 Write electron configurations for the following elements.
(a) the group IIIA element in the same period as $_4Be$
(b) the period 3 element in the same group as $_5B$
(c) the lowest-atomic-numbered metal in group IA
(d) the period 3 element that has three unpaired electrons

6.89 Referring only to the periodic table, determine the element of lowest atomic number whose electron configuration contains the following.
(a) two completely filled orbitals
(b) two completely filled subshells
(c) two completely filled shells
(d) two completely filled p subshells

6.90 Referring only to the periodic table, determine the element of lowest atomic number whose electron configuration contains the following.
(a) three completely filled orbitals
(b) three completely filled subshells
(c) three completely filled shells
(d) three completely filled s subshells

6.91 In what period and group is an element with each of the following electron configurations located?
(a) $1s^2 2s^2 2p^6 3s^1$
(b) $1s^2 2s^2 2p^6 3s^2 3p^1$
(c) $1s^2 2s^2 2p^6 3s^2 3p^6 4s^2 3d^1$
(d) $1s^2 2s^2 2p^6 3s^2 3p^6 4s^2 3d^{10} 4p^5$

6.92 In what period and group is an element with each of the following electron configurations located?
(a) $1s^2 2s^2 2p^2$
(b) $1s^2 2s^2 2p^6 3s^2$
(c) $1s^2 2s^2 2p^6 3s^2 3p^6 4s^2 3d^2$
(d) $1s^2 2s^2 2p^6 3s^2 3p^6 4s^2 3d^{10} 4p^6 5s^2$

6.93 Identify the element of lowest atomic number for which Hund's rule must be used in determining its electron orbital diagram. Explain how Hund's rule applies to this situation.

6.94 Identify the first element with an atomic number greater than 36 for which Hund's rule must be used in determining its electron orbital diagram. Explain how Hund's rule applies to this situation.

6.95 In what group(s) in the periodic table would you expect to find each of the following?
(a) a representative element with two unpaired electrons
(b) a transition element with three unpaired electrons
(c) an element with two unpaired d electrons
(d) an element with one unpaired s electron

6.96 In what group(s) in the periodic table would you expect to find each of the following?
(a) a transition element with no unpaired electrons
(b) a nonmetal with no unpaired electrons
(c) an element with three unpaired p electrons
(d) an element with one unpaired d electron

6.97 Indicate which element or elements have the electron characteristics below. In those cases where a series of elements have the indicated characteristics, do not write all the symbols but rather write the atomic numbers of the first and last elements in the series, for example, elements 70–83.
(a) a total of 84 electrons
(b) only four $3d$ electrons
(c) two $7s$ electrons (note that more than one element qualifies)
(d) a total of six s electrons (note that more than one element qualifies)

6.98 Indicate which element or elements have the electron characteristics below. In those cases where a series of elements have the indicated characteristics, do not write all the symbols but rather write the atomic numbers of the first and last elements in the series, for example, elements 70–83.
(a) only one electron in shell 7
(b) a total of fifteen p electrons (they are not all in the same subshell)
(c) twelve electrons in shell 3
(d) six $5p$ electrons (note that more than one element qualifies)

6.99 Referring only to the periodic table, determine the element of lowest atomic number whose electron configuration contains the following.
(a) more p electrons than s electrons
(b) more d electrons than p electrons

6.100 Referring only to the periodic table, determine the element of lowest atomic number whose electron configuration contains the following.
(a) more f electrons than s electrons
(b) more d electrons than s electrons

Cumulative Problems

6.101 Write out the complete electron configuration for an atom with $Z = 20$ and $A = 44$.

6.102 Write out the complete electron configuration for an atom with $Z = 43$ and $A = 99$.

6.103 The electron configuration of the isotope $^{12}_{6}C$ is $1s^22s^22p^2$. What is the electron configuration for the isotope $^{13}_{6}C$?

6.104 The electron configuration of the isotope $^{16}_{8}O$ is $1s^22s^22p^4$. What is the electron configuration for the isotope $^{18}_{8}O$?

6.105 How many subatomic particles are present in an atom whose isotopic mass is 36.96590 amu and whose electron configuration is $1s^22s^22p^63s^23p^5$?

6.106 How many subatomic particles are present in an atom whose isotopic mass is 29.97376 amu and whose electron configuration is $1s^22s^22p^63s^23p^2$?

6.107 An element that is a member of periodic table group IIA has atoms that possess a total of eighteen p electrons. What is the atomic number of this element?

6.108 An element that is a member of periodic table group VIA has atoms that possess a total of ten s electrons. What is the atomic number of this element?

Grid Problems

6.109 Select from the grid *all* correct responses for each situation.

1. calcium	2. carbon	3. tin
4. iron	5. germanium	6. phosphorus
7. bismuth	8. silver	9. aluminum

(a) elements that are metalloids
(b) elements whose group number and period number are numerically equal
(c) transition elements
(d) nonmetallic group IVA elements
(e) metallic representative elements
(f) elements whose periodic table group contains at least one metal, metalloid, and nonmetal

6.110 Select from the grid *all* correct responses for each situation.

1. F	2. Ti	3. As
4. P	5. Ca	6. I
7. Si	8. Mg	9. O

(a) is chemically similar to the element Br
(b) has an electron configuration that ends in p^3
(c) has an electron configuration that contains the notation $3p^6$
(d) possesses two unpaired electrons
(e) has an electron configuration in which all occupied subshells are completely filled
(f) has an electron configuration in which the number of s electrons is equal to the number of p electrons

6.111 Select from the grid *all* correct responses for each situation.

1. $1s^2 2s^2$	2. $1s^2 2s^2 2p^2$	3. $1s^2 2s^2 2p^3$
4. $1s^2 2s^2 2p^6$	5. $1s^2 2s^2 2p^6 3s^2 3p^1$	6. $1s^2 2s^2 2p^6 3s^2 3p^4$
7. $1s^2 2s^2 2p^6 3s^2 3p^6$	8. $1s^2 2s^2 2p^6 3s^2 3p^6 4s^1$	9. $1s^2 2s^2 2p^6 3s^2 3p^6 4s^2 3d^3$

(a) electron configurations that represent nonmetallic elements
(b) electron configurations that represent diamagnetic elements
(c) electron configurations that represent elements in period 3 of the periodic table
(d) pairs of electron configurations that represent elements in the same periodic table group
(e) pairs of electron configurations in which the same number of orbitals possess electrons
(f) pairs of electron configurations that represent elements with the same number of unpaired electrons and are in different periodic table groups

6.112 Select from the grid *all* correct responses for each situation.

1. nitrogen	2. silicon	3. fluorine
4. barium	5. strontium	6. aluminum
7. calcium	8. oxygen	9. carbon

(a) elements whose metallic character is greater than that of magnesium
(b) elements whose atomic size is less than that of magnesium
(c) elements in which electrons are found in five or more orbitals
(d) elements in which electrons are found in four or more subshells
(e) elements in which three or more electron shells are completely filled with electrons
(f) elements that possess more unpaired electrons than the elements on either side of them in the periodic table

CHAPTER SEVEN

Chemical Bonding

7.1 Chemical Bonds

In Section 5.2 we dealt with the fact that two types of compounds exist: ionic compounds and molecular compounds. Ionic compounds, we recall, involve an extended three-dimensional assembly of positively and negatively charged particles called ions, and molecular compounds contain separate fixed groupings of atoms called molecules. Why is it that some compounds contain ions, whereas others have molecular building blocks? What is it that determines whether the interaction of two elements produces ions or molecules? In this chapter we consider answers to these questions. The answers are found in a consideration of the subject of *chemical bonding*.

Chemical bonds are the attractive forces that hold atoms or ions together in more complex aggregates. The way in which atoms interact with each other to form chemical bonds is dictated by electron configuration. Thus, many of the concepts of Chapter 6 will be used in treating the subject of chemical bonding.

It is useful to classify chemical attractive forces (chemical bonds) into two categories: ionic bonds and covalent bonds. An *ionic bond* is formed when one or more electrons are *transferred* from one atom or group of atoms to another. As suggested by its name, the ionic bond model is particularly useful in describing the attractive forces in ionic compounds. A *covalent bond* is formed when two atoms *share* one or more electron pairs between them. The covalent bond model is particularly useful in describing attractions between atoms in molecular compounds.

Before we consider the details of these two bond models it is important to emphasize that the notions of ionic and covalent bonds are merely convenient concepts. Most bonds are not 100% ionic or 100% covalent. Instead, most bonds have at least some degree of both ionic and covalent character, that is, some degree of both the transfer and the sharing of electrons. But it is easiest to understand these intermediate bonds (the real bonds) by relating them to the pure or ideal bond types called ionic and covalent.

7.2 Valence Electrons and Electron-Dot Structures

Chemical bonds form as the result of the interaction (transfer or sharing) of electrons found in the combining atoms. Certain electrons, called valence electrons, are particularly important in determining the bonding characteristics of a given atom. For representative elements (Sec. 6.11) **valence electrons** are those electrons in the outermost electron shell, that is, in the shell with the highest shell number (n). These electrons will always be found in either s or p subshells. Note the restriction on the use of this definition; it applies only for representative elements. Most of the common elements are representative elements; hence the definition still finds much use. (We will not consider in this text the more complicated valence electron definitions for transition or inner transition elements (Sec. 6.11); the presence of incompletely filled *inner d* or *f* subshells is the complicating factor in definitions for these elements.)

EXAMPLE 7.1

Determine the number of valence electrons present in atoms of each of the following elements.

(a) $_{20}$Ca (b) $_{16}$S (c) $_{35}$Br

Solution

(a) The element calcium has two valence electrons, as can be seen by examining its electron configuration.

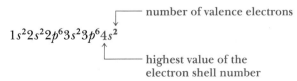

The highest value of the electron shell number is $n = 4$. Only two electrons are found in shell 4, two electrons in the $4s$ subshell.

(b) The element sulfur has six valence electrons.

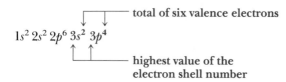

Electrons in two different subshells can simultaneously be valence electrons. The highest shell number is 3, and both the $3s$ and $3p$ subshells belong to shell number 3. Hence, all electrons in both subshells are valence electrons.

(c) The element bromine has seven valence elctrons.

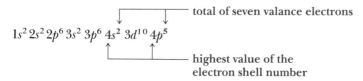

The $3d$ electrons are not counted as valence electrons because the $3d$ subshell is in shell 3 and shell 3 is not the shell with maximum n value. Shell 4 is the outermost shell, the shell with maximum n value.

PRACTICE EXERCISE 7.1

Determine the number of valence electrons present in atoms of each of the following elements.

(a) $_{13}$Al   (b) $_{19}$K   (c) $_{53}$I

Ans. (a) 3; (b) 1; (c) 7.

Similar exercises: Problems 7.3 and 7.4

The fact that the outermost electrons of atoms are those involved in bonding seems reasonable when it is remembered that the outermost electrons will be the first to come into close proximity when atoms collide—an event necessary before atoms can combine. Also, since these electrons are located the farthest from the nucleus, they are therefore the least tightly bound (attraction to the nucleus decreases with distance) and thus the most susceptible to change (transfer or sharing).

A shorthand system for designating numbers of valence electrons, which uses electron-dot structures, has been developed. Use of this system will make it easier, later in this chapter, to picture the role that valence electrons play in chemical bonding. An **electron-dot structure** consists of an element's symbol with one dot for each valence electron placed about the elemental symbol. Electron-dot structures for the first 20 elements, arranged as in the periodic table, are given in Figure 7.1. Note that the location of the dots is not critical. The following all have the same meaning.

$$\overset{\cdot}{\text{Mg}}\cdot \qquad \underset{\cdot}{\text{Mg}}\cdot \qquad \cdot\underset{\cdot}{\text{Mg}} \qquad \cdot\overset{\cdot}{\text{Mg}} \qquad \overset{\cdot}{\underset{\cdot}{\text{Mg}}} \qquad \cdot\text{Mg}\cdot$$

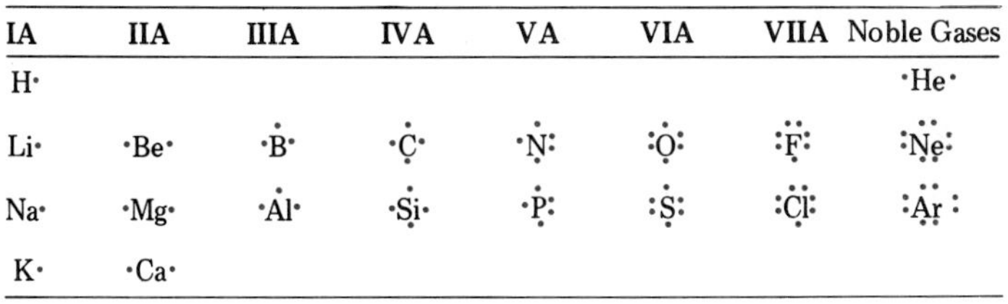

IA	IIA	IIIA	IVA	VA	VIA	VIIA	Noble Gases
H·							·He·
Li·	·Be·	·B̈·	·C̈·	·N̈:	:Ö:	:F̈:	:N̈ë:
Na·	·Mg·	·Äl·	·S̈i·	·P̈:	:S̈:	:C̈l:	:Är:
K·	·Ca·						

FIGURE 7.1 Electron-dot structures of first 20 elements.

Electron-dot structures are also often called Lewis structures. The American chemist Gilbert Newton Lewis (1875–1946), an early contributor to chemical bonding theory, was the first to use such structures.

Three important generalizations about valence electrons can be drawn from a study of the structures in Figure 7.1.

1. *Representative elements in the same group of the periodic table have the same number of valence electrons.* This should not be surprising to you. Elements in the same group in the periodic table have similar chemical properties as a result of having similar outer shell electron configurations (Sec. 6.9). The electrons in the outermost shell are the valence electrons.

2. *The number of valence electrons for representative elements in a group is the same as the periodic table group number.* For example, the electron-dot structures for O and S, both members of group VIA, show six dots. Similarly, the electron-dot structures of H, Li, Na, and K, all members of group IA, show one dot.

3. *The maximum number of valence electrons for any element is eight.* Only the noble gases (Sec. 6.11), beginning with Ne, have the maximum number of eight electrons. Helium, with only two valence electrons, is the exception in the noble gas family; obviously, an element with a grand total of two electrons cannot have eight valence electrons. Although shells with n greater than 2 are capable of holding more than eight electrons, they do so only when they are no longer the outermost shell and thus not the valence shell. For example, bromine (Example 7.1) has 18 electrons in its third shell; however, shell 4 is the valence shell in bromine.

7.3 The Octet Rule

A key concept in modern bonding theory is that certain arrangements of valence electrons are more stable than others. The term *stable* as used here refers to the idea that a system (in this case an arrangement of electrons) does not easily undergo spontaneous change.

The outer shell electron configurations possessed by the noble gases (He, Ne, Ar, Kr, Xe, and Rn; see Sec. 6.11) are considered to be the *most stable of all outer shell electron configurations.* For helium, this most stable electron configuration involves two outer shell electrons. The rest of the noble gases have eight outer shell electrons.

He: $1s^2$
Ne: $1s^2 2s^2 2p^6$
Ar: $1s^2 2s^2 2p^6 3s^2 3p^6$
Kr: $1s^2 2s^2 2p^6 3s^2 3p^6 4s^2 3d^{10} 4p^6$
Xe: $1s^2 2s^2 2p^6 3s^2 3p^6 4s^2 3d^{10} 4p^6 5s^2 4d^{10} 5p^6$
Rn: $1s^2 2s^2 2p^6 3s^2 3p^6 4s^2 3d^{10} 4p^6 5s^2 4d^{10} 5p^6 6s^2 4f^{14} 5d^{10} 6p^6$

These noble gas electron configurations have in common that the outermost s and p subshells are *completely filled*.

The conclusion that a noble gas configuration is the most stable of all outer shell electron configurations is based on the chemical properties of the noble gases. These elements are the *most unreactive* of all the elements. They are the only elemental gases found in nature in the form of individual uncombined atoms. There are no known compounds of He, Ne, and Ar and only a very few compounds of Kr, Xe, and Rn. The noble gases appear to be "happy" the way they are. They have little or no "desire" to form bonds to other atoms.

Atoms of many elements that lack the very stable outer shell electron configuration of the noble gases tend to attain it in chemical reactions that result in compound formation. This observation has become known as the *octet rule* because of the eight outer shell electrons possessed by five of the six noble gases. A formal statement of the **octet rule** is *In forming compounds, atoms of elements lose, gain, or share electrons in such a way as to produce a noble gas electron configuration for each of the atoms involved.*

Application of the octet rule to many different systems has shown that it has value in predicting correctly the observed combining ratios of atoms. For example, it explains why two hydrogen atoms rather than some other number are bonded to one oxygen atom in the molecular compound water. It explains why the formula of the ionic compound sodium chloride is NaCl rather than $NaCl_2$, $NaCl_3$, or Na_2Cl.

There are exceptions to the octet rule, but it is still used because of the large amount of information that it is able to correlate. It is particularly effective in explaining compound formation involving only representative elements. Often complications arise with transition and inner transition elements because of the involvement of d and f electrons in the bonding.

7.4 Ions and Ionic Bonds

The concept of transferring one or more electrons between two or more atoms is central to the ionic bond model. This electron-transfer process produces charged particles called ions. An **ion** is an atom (or group of atoms) that is electrically charged as the result of the loss or gain of electrons. Neutrality of atoms is the result of the number of protons (positive charges) being equal to the number of electrons (negative charges). Loss or gain of electrons destroys this proton–electron balance and leaves a net charge on the atom.

Section 7.4 · Ions and Ionic Bonds

If one or more electrons are gained by an atom, a negatively charged ion is produced; excess negative charge is present because electrons now outnumber protons. The loss of one or more electrons by an atom results in the formation of a positively charged ion; more protons than electrons are now present, resulting in excess positive charge. Note that the excess positive charge associated with a positive ion is never caused by proton gain but always by electron loss. If the number of protons remains constant and the number of electrons decreases, the result is net positive charge. The number of protons, which determines the identity of the element (Sec. 5.7), never changes during ion formation.

The charge on an ion is directly correlated with the number of electrons lost or gained. Loss of one, two, or three electrons gives ions with $+1$, $+2$, and $+3$ charges, respectively. Similarly, a gain of one, two, or three electrons gives ions with -1, -2, and -3 charges, respectively. (Atoms that have lost or gained more than three electrons are very seldom encountered.)

The notation for charges on ions is a superscript placed to the right of the elemental symbol. Some examples of ion symbols are

Positive ions: $Na^+, K^+, Ca^{2+}, Mg^{2+}, Al^{3+}$
Negative ions: $Cl^-, Br^-, O^{2-}, S^{2-}, N^{3-}$

Note that a single plus or minus sign is used to denote a charge of one, instead of using the notation $1+$ or $1-$. Also note that in multicharged ions the number precedes the charge sign; that is, notation for a charge of plus two is $2+$ rather than $+2$.

A final point about ions is that their chemical properties are very different from the neutral atoms from which they are derived. For example, water solutions containing Na^+ ion are very stable even though the element sodium (neutral Na) reacts explosively with water.

EXAMPLE 7.2

Give the symbol for each of the following ions.

(a) the ion formed when a barium atom loses two electrons
(b) the ion formed when a phosphorus atom gains three electrons

Solution

(a) A neutral barium atom contains 56 protons and 56 electrons, since the atomic number of barium is 56 (obtained from the periodic table). The barium ion formed by the loss of two electrons would still contain 56 protons, but would have only 54 electrons because two electrons were lost.

$$\begin{array}{r} 56 \text{ protons} = 56+ \text{ charges} \\ 54 \text{ electrons} = \underline{54- \text{ charges}} \\ \text{Net charge} = 2+ \end{array}$$

The symbol for the barium ion is thus Ba^{2+}

(b) The atomic number of phosphorus is 15. Thus, 15 protons and 15 electrons are present in a neutral phosphorus atom. A gain of three electrons raises the electron count to 18.

$$\begin{aligned} 15 \text{ protons} &= 15 + \text{charges} \\ 18 \text{ electrons} &= 18 - \text{charges} \\ \hline \text{Net charge} &= 3- \end{aligned}$$

The symbol for the phosphorus ion is thus P^{3-}

PRACTICE EXERCISE 7.2

Give the symbol for each of the following ions.

(a) the ion formed when a calcium atom loses two electrons
(b) the ion formed when a nitrogen atom gains three electrons

Ans. (a) Ca^{2+}; (b) N^{3-}.

Similar exercises: Problems 7.15 and 7.16

EXAMPLE 7.3

Determine the number of protons and electrons present in the following ions.

(a) O^{2-} (b) Al^{3+}

Solution

(a) The number of protons is the same as in a neutral atom and is therefore given by the atomic number, which is 8 for oxygen.

$$\text{Number of protons} = \text{atomic number} = 8$$

The number of electrons in a neutral O atom is 8. The charge of -2 on the ion indicates the gain of two electrons.

$$\text{Number of electrons} = 8 + 2 = 10$$

(b) The atomic number of aluminum is 13.

$$\text{Number of protons} = \text{atomic number} = 13$$

The number of electrons in a neutral Al atom is 13. The charge of $+3$ on the ion indicates the loss of three electrons.

$$\text{Number of electrons} = 13 - 3 = 10$$

PRACTICE EXERCISE 7.3

Determine the number of protons and electrons present in the following ions.

(a) K^+ (b) S^{2-} Ans. (a) 19 protons and 18 electrons; (b) 16 protons and 18 electrons.

Similar exercises: Problems 7.19 and 7.20

So far our discussion about electron transfer and ion formation has focused on the loss or gain of electrons by isolated individual atoms. During ionic bond formation, loss and gain of electrons are partner processes; that is, one does not occur without the other. Ion formation occurs only when atoms of two elements are present—an element to donate electrons (electron loss) and an element to accept electrons (electron gain). The number of electrons lost by the one element is the same as the number gained by the other element. Thus, positive and negative ions must always be formed at the same time.

The mutual attraction between the positive and negative ions that results from electron transfer constitutes the force that holds the ions together as an ionic compound. This force is referred to as an ionic bond. An **ionic bond** is the attractive force between positive and negative ions that causes them to remain together as a group.

7.5 Ionic Compound Formation

A simple example of ionic bonding occurs between the elements sodium and chlorine in the compound NaCl. Sodium atoms lose (transfer) one electron to chlorine atoms, producing Na^+ and Cl^- ions. The ions combine in a one-to-one ratio to give the compound NaCl.

Why do sodium atoms form Na^+ and not Na^{2+} or Na^- ions? Why do chlorine atoms form Cl^- ions rather than Cl^{2-} or Cl^+ ions? In general, what determines the specific number of electrons lost or gained in electron-transfer processes?

The octet rule (Sec. 7.3) provides very simple and straightforward answers to these questions. *Atoms tend to gain or lose electrons until they have obtained an electron configuration that is the same as that of a noble gas.*

Consider the element sodium, which has the electron configuration

$$1s^2 2s^2 2p^6 3s^1$$

It can attain a noble gas configuration by losing one electron (to give it the electron configuration of neon) or by gaining seven electrons (to give it the electron configuration of argon).

$$Na\ (1s^2 2s^2 2p^6 3s^1) \begin{array}{c} \xrightarrow{\text{loss of 1 e}^-} Na^+\ (1s^2 2s^2 2p^6)\ \text{electron configuration of neon} \\ \xrightarrow{\text{gain of 7 e}^-} Na^{7-}\ (1s^2 2s^2 2p^6 3s^2 3p^6)\ \text{electron configuration of argon} \end{array}$$

The first process, the loss of one electron, is more energetically favorable than the gain of seven electrons and is the process that occurs. The process that involves the least number of electrons will always be the more energetically favorable process and will be the process that occurs.

Consider the element chlorine, which has the electron configuration

$$1s^2 2s^2 2p^6 3s^2 3p^5$$

It can attain a noble gas configuration by losing seven electrons (to give it the electron configuration of neon) or by gaining one electron (to give it the electron configuration of argon). The latter occurs for the reason cited previously.

$$\text{Cl } (1s^22s^22p^63s^23p^5) \xrightarrow[\text{gain of 1 e}^-]{\text{loss of 7 e}^-} \begin{array}{l} \text{Cl}^{7+} \quad (1s^22s^22p^6) \\ \text{electron configuration of neon} \\ \\ \text{Cl}^- \quad (1s^22s^22p^63s^23p^6) \\ \text{electron configuration of argon} \end{array}$$

The type of consideration we have just used for the elements sodium and chlorine lead to the following generalizations.

1. Metal atoms containing one, two, or three valence electrons (the metals in groups IA, IIA, and IIIA of the periodic table) tend to lose electrons to acquire a noble gas electron configuration.

    Group IA metals form $+1$ ions.
    Group IIA metals form $+2$ ions.
    Group IIIA metals form $+3$ ions.

    Group IA metals are all located one periodic table position past a noble gas. Thus, they will each have one more electron than the preceding noble gas. This electron must be lost if a noble gas configuration is to be obtained. Group IIA and IIIA metals are two and three periodic table positions, respectively, beyond a noble gas. Consequently, two and three electrons, respectively, must be lost for these metals to attain a noble gas electron configuration.

2. Nonmetal atoms containing five, six, or seven valence electrons (the nonmetals in groups VA, VIA, and VIIA of the periodic table) tend to gain electrons to acquire a noble gas configuration.

    Group VIIA nonmetals form -1 ions.
    Group VIA nonmetals form -2 ions.
    Group VA nonmetals form -3 ions.

    The nonmetal ionic charge guidelines can be explained by reasoning similar to that for the metals, only this time the periodic table positions are those immediately preceding the noble gases. Consequently, electrons must be gained to attain noble gas configurations.

Elements in group IVA occupy unique positions relative to the noble gases. They are located equidistant between two noble gases. For example, the element carbon is four positions beyond helium and four positions before neon. Theoretically, ions with charges of $+4$ or -4 could be formed by elements in this group, but in most cases the bonding that results is more adequately described by the covalent bond model to be discussed in Section 7.9.

An ion formed in the preceding manner with an electronic configuration the same as that of a noble gas is said to be *isoelectronic* with the noble gas. **Isoelectronic species** contain the same number of electrons. An atom and an ion or two ions may be isoelectronic. Numerous ions that are isoelectronic with a given noble gas exist, as can be seen from the entries in Table 7.1.

Section 7.5 · Ionic Compound Formation

TABLE 7.1 Ions Isoelectronic with Selected Noble Gases

Helium Structure	Neon Structure	Argon Structure
H⁻ $1s^2$	N³⁻ $1s^2 2s^2 2p^6$	P³⁻ $1s^2 2s^2 2p^6 3s^2 3p^6$
	O²⁻	S²⁻
Li⁺	F⁻	Cl⁻
Be²⁺		
	Na⁺	K⁺
	Mg²⁺	Ca²⁺
	Al³⁺	

It should be emphasized that an ion that is isoelectronic with a noble gas does not have the properties of the noble gas. It has not been converted into the noble gas. The number of protons in the nucleus of the isoelectronic ion is different from that in the noble gas. These points are emphasized by the comparison in Table 7.2 between Mg²⁺ and Ne, the noble gas with which Mg²⁺ is isoelectronic.

The use of electron-dot structures helps in visualizing the formation of ionic compounds through electron transfer. Let us consider, again, the reaction between sodium, which has one valence electron, and chlorine, which has seven valence electrons, to give NaCl. This reaction can be represented as follows with electron-dot structures.

$$\text{Na·} + \text{·}\ddot{\underset{..}{\text{Cl}}}\text{:} \longrightarrow \text{Na}^+[:\ddot{\underset{..}{\text{Cl}}}:]^- \longrightarrow \text{NaCl}$$

The loss of an electron by sodium empties its valence shell. The next inner shell, which contains eight electrons (a noble gas configuration), then becomes the valence shell. After the valence shell of chlorine gains one electron, it then has the "desired" eight valence electrons.

When sodium, which has one valence electron, combines with oxygen, which has six valence electrons, each oxygen atom requires two sodium atoms to meet its need of two additional electrons.

$$\begin{array}{c}\text{Na·} \\ \\ \text{Na·}\end{array} \quad \text{·}\ddot{\underset{..}{\text{O}}}\text{:} \longrightarrow \begin{array}{c}\text{Na}^+ \\ \text{Na}^+\end{array}[:\ddot{\underset{..}{\text{O}}}:]^{2-} \longrightarrow \text{Na}_2\text{O}$$

TABLE 7.2 A Comparison of the Structure of a Mg²⁺ Ion and a Ne Atom, the Noble Gas Atom Isoelectronic with the Ion

	Ne Atom	Mg²⁺ Ion
Protons (in the nucleus)	10	12
Electrons (around the nucleus)	10	10
Atomic number	10	12
Charge	0	2+

Note how oxygen's need for two additional electrons dictates that two sodium atoms are required for each oxygen atom; hence the formula is Na_2O.

An opposite situation to that for Na_2O occurs in the reaction between calcium, which has two valence electrons, and chlorine, which has seven valence electrons. Here, two chlorine atoms are required to accommodate electrons transferred from one calcium atom because a chlorine atom can accept only one electron. (It has seven valence electrons and needs only eight.)

$$Ca \cdot \begin{matrix} \cdot \ddot{\underset{..}{Cl}}: \\ \cdot \ddot{\underset{..}{Cl}}: \end{matrix} \longrightarrow Ca^{2+} \begin{matrix} [:\ddot{\underset{..}{Cl}}:]^- \\ [:\ddot{\underset{..}{Cl}}:]^- \end{matrix} \longrightarrow CaCl_2$$

EXAMPLE 7.4

Show the formation of the following ionic compounds using electron-dot structures.

(a) Na_3N (b) MgO (c) Al_2S_3

Solution

(a) Sodium (a group IA element) has one valence electron, which it would "like" to lose. Nitrogen (a group VA element) has five valence electrons and would thus "like" to acquire three more. Three sodium atoms will be required to supply enough electrons for one nitrogen atom.

$$\begin{matrix} Na \cdot \\ Na \cdot \\ Na \cdot \end{matrix} \; \dot{\underset{\cdot}{N}}: \longrightarrow \begin{matrix} Na^+ \\ Na^+[:\ddot{\underset{..}{N}}:]^{3-} \\ Na^+ \end{matrix} \longrightarrow Na_3N$$

(b) Magnesium (a group IIA element) has two valence electrons, and oxygen (a group VIA element) has six valence electrons. The transfer of the two magnesium valence electrons to an oxygen atom will result in each atom having a noble gas electron configuration. Thus, these two elements combine in a one-to-one ratio.

$$Mg \cdot\cdot + \cdot \ddot{\underset{..}{O}}: \longrightarrow Mg^{2+}[:\ddot{\underset{..}{O}}:]^{2-} \longrightarrow MgO$$

(c) Aluminum (a group IIIA element) has three valence electrons, all of which need to be lost through electron transfer. Sulfur (a group VIA element) has six valence electrons and thus needs to acquire two more. Three sulfur atoms are needed to accommodate the electrons given up by two aluminum atoms.

$$\begin{matrix} Al \cdot \\ \\ Al \cdot \end{matrix} \begin{matrix} \cdot \ddot{S}: \\ \cdot \ddot{S}: \\ \cdot \ddot{S}: \end{matrix} \longrightarrow \begin{matrix} [:\ddot{S}:]^{2-} \\ Al^{3+} \\ [:\ddot{S}:]^{2-} \longrightarrow Al_2S_3 \\ Al^{3+} \\ [:\ddot{S}:]^{2-} \end{matrix}$$

PRACTICE EXERCISE 7.4

Show the formation of the following ionic compounds using electron-dot structures.

(a) KF (b) Li$_2$O (c) Ca$_3$P$_2$

Ans. (a) K· + ·F̈: (c) Ca·⟶ ·P̈:
(b) Li· ⟶ :Ö: ⟵ Li· Ca· ⟶ ·P̈: ⟵ Ca·

Similar exercises: Problems 7.37 and 7.38

7.6 Formulas for Ionic Compounds

It is not always necessary or convenient to write electron-dot structures when determining the formula for an ionic compound. Formulas for ionic compounds can be written directly by using the charges associated with ions being combined and the fact that the total amount of positive and negative charge must add up to zero. Electron loss always equals electron gain in an electron-transfer process. Consequently, ionic compounds are always neutral; no net charge is present. The total positive charge present on the ions that have lost electrons is always exactly counterbalanced by the total negative charge on the ions that have gained electrons. Thus, *the ratio in which positive and negative ions combine is the ratio that achieves charge neutrality for the resulting compound.*

The correct combining ratio when K^+ and S^{2-} ions combine is two to one. Two K^+ ions (each of +1 charge) will be required to balance the charge on a single S^{2-} ion.

$$2(K^+): \quad (2 \text{ ions}) \times (\text{charge of } +1) = +2$$
$$S^{2-}: \quad (1 \text{ ion}) \times (\text{charge of } -2) = -2$$
$$\text{Net charge} = 0$$

Example 7.5 gives further illustration of the procedures needed to determine correct combining ratios between ions and write correct ionic formulas from the combining ratios. There are three items to note about all ionic formulas.

1. The symbol for the positive ion is always written first.
2. The charges on the ions that are present are *not* shown in the formula. Knowledge of charges is necessary to determine the formula, but once it is determined, the charges are not explicitly written.
3. The numbers in the formula (the subscripts) give the combining ratio for the ions.

EXAMPLE 7.5

Determine the formula for the compound that is formed when each of the following pairs of ions interact.

(a) Na^+ and P^{3-} (b) Be^{2+} and P^{3-} (c) Al^{3+} and P^{3-}

Solution

(a) Na^+ and P^{3-} ions will combine in a three-to-one ratio because this combination will cause the total charge to add up to zero. Three Na^+ ions give a total positive charge of 3. One P^{3-} ion results in a total negative charge of 3. Thus, the formula for the compound is **Na_3P**.

(b) The numbers in the charges for these ions are 2 and 3. The lowest common multiple of 2 and 3 is 6 ($2 \times 3 = 6$). Thus, we will need six units of positive charge and six units of negative charge. Three Be^{2+} ions are needed to give the six units of positive charge, and two P^{3-} ions are needed to give the six units of negative charge. The combining ratio of ions is three to two, and the formula is **Be_3P_2**. The strategy used in determining this formula, finding the lowest common multiple in the charges of the ions, will always work.

(c) The formula of this compound is simply **AlP** (a one-to-one ratio between ions). One Al^{3+} ion contributes three units of positive charge, and that is counterbalanced by three units of negative charge from one P^{3-} ion.

PRACTICE EXERCISE 7.5

Determine the formula for the compound that is formed when each of the following pairs of ions interact.

(a) Ba^{2+} and Cl^- (b) Ba^{2+} and S^{2-} (c) Ba^{2+} and N^{3-}

Ans. (a) $BaCl_2$; (b) BaS; (c) Ba_3N_2.   Similar exercises: Problems 7.41–7.44

Before leaving the subject of ions, ionic bonds, and formulas for ionic compounds, let us quickly review the key principles about ionic bonding.

1. Ionic compounds usually contain both a metallic and a nonmetallic element.
2. The metallic element atoms lose electrons to produce positive ions, and the nonmetallic element atoms gain electrons to produce negative ions.

TABLE 7.3 General Formulas for Ionic Compounds as a Function of Periodic Table Position

Metal Group Number	Nonmetal Group Number	Charge on Metal (M) Ion	Charge on Nonmetal (X) Ion	Formula of Compound
IA	VIIA	+1	−1	MX
IA	VIA	+1	−2	M_2X
IA	VA	+1	−3	M_3X
IIA	VIIA	+2	−1	MX_2
IIA	VIA	+2	−2	MX
IIA	VA	+2	−3	M_3X_2
IIIA	VIIA	+3	−1	MX_3
IIIA	VIA	+3	−2	M_2X_3
IIIA	VA	+3	−3	MX

3. The electrons lost by the metal atoms are the same ones that are gained by the nonmetal atoms. Electron loss must always equal electron gain.
4. The ratio in which positive metal ions and negative nonmetal ions combine is the ratio that achieves charge neutrality for the resulting compound.
5. Metals from groups IA, IIA, and IIIA of the periodic table form ions with charges of $+1$, $+2$, and $+3$, respectively. Nonmetals of groups VIIA, VIA, and VA of the periodic table form ions with charges of -1, -2, and -3, respectively. Table 7.3 lists, in general terms, all of the possible metal–nonmetal combinations from these periodic table groups that result in the formation of ionic compounds.

7.7 Structure of Ionic Compounds

The term *molecule* is not appropriate for describing the smallest unit of an ionic compound (Sec. 5.2). In the solid state, ionic compounds consist of an extended array of alternating positive and negative ions. **Ionic solids** consist of positive and negative ions arranged in such a way that each ion is surrounded by nearest neighbors of the opposite charge. Any given ion is bonded by electrostatic (positive–negative) attractions to all of the other ions of opposite charge immediately surrounding it. Figure 7.2 gives two three-dimensional depictions of the arrangement of ions for the ionic compound NaCl (table salt).

We can see in Figure 7.2 that discrete molecules do not exist in an ionic solid. Therefore, the formulas for these solids (Sec. 7.6) cannot represent the composition of a molecule of the substance. Instead, formulas for this type of solid represent the simplest ratio in which the atoms combine. For example, in NaCl (Fig. 7.2) there is

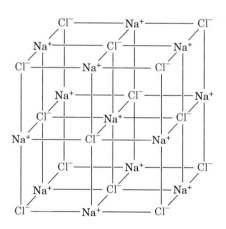
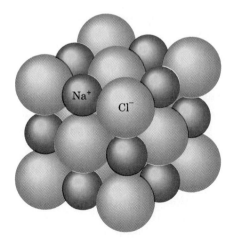

Figure 7.2 The arrangement of ions in sodium chloride. Each of the Na$^+$ ions is surrounded on all sides by Cl$^-$ ions and each of the Cl$^-$ ions is surrounded on all sides by Na$^+$ ions.

no single partner for a sodium ion; there are six immediate neighbors (chloride ions) that are equidistant from it. A chloride ion in turn has six immediate sodium neighbors. The formula NaCl represents the fact that sodium and chloride ions are present in this solid in a one-to-one ratio.

Although the formulas for ionic solids represent only ratios, they are used in equations and chemical calculations in the same way as the formulas for molecular species. Remember, however, that they cannot be interpreted as indicating that molecules exist for these substances.

7.8 Polyatomic Ions

To this point in this chapter all references to and comments about ions have involved monoatomic ions. **Monoatomic ions** are single atoms that have lost or gained electrons. Such ions are very common and very important. Another large and important category of ions, called polyatomic ions, exists. Numerous ionic compounds exist in which the positive or negative ion (sometimes both) is polyatomic. Polyatomic ions are very stable species, generally maintaining their identity during chemical reactions.

An example of a polyatomic ion is the sulfate ion, SO_4^{2-}. This ion contains four oxygen atoms and one sulfur atom, and the whole group of five atoms has acquired a -2 charge. The whole sulfate group is the ion rather than any one atom within the group. Covalent bonding, discussed in Section 7.9, holds the sulfur and oxygen atoms together. A **polyatomic ion** is a group of atoms, held together by covalent bonds, that has acquired a charge.

Polyatomic ions are not molecules. They never occur alone as molecules do. Instead, they are always found associated with ions of opposite charge. Polyatomic ions are *pieces* of compounds, not compounds. Ionic compounds require the presence of both positive and negative ions and are neutral overall. Polyatomic ions are always charged species.

Formulas for ionic compounds containing polyatomic ions are determined in the same way as those for ionic compounds containing monoatomic ions (Sec. 7.6). The basic rule is the same: the total positive and negative charge present must add up to zero.

Two conventions not encountered previously in formula writing often arise when writing formulas with polyatomic ions. They are

1. When more than one polyatomic ion of a given kind is required in a formula, the polyatomic ion is enclosed in parentheses and a subscript is placed outside the parentheses to indicate the number of polyatomic ions needed.
2. To preserve the identity of polyatomic ions, the same elemental symbol may be used more than once in a formula.

Example 7.6 contains examples illustrating the use of both of these new conventions. Besides the sulfate ion, four other polyatomic ions are involved in this example: OH^- (hydroxide ion), NO_3^- (nitrate ion), NH_4^+ (ammonium ion), and CN^- (cyanide ion). The formulas for numerous other polyatomic ions are considered in Section 8.4.

EXAMPLE 7.6

Determine the formulas for the ionic compounds containing the following pairs of ions.

(a) K^+ and OH^-
(b) Na^+ and SO_4^{2-}
(c) Ca^{2+} and NO_3^-
(d) NH_4^+ and CN^-

Solution

(a) Since both of these ions possess a charge of one, combining them in a one-to-one ratio will balance the charge. The formula of the compound is **KOH**.

(b) In order to equalize the total positive and negative charges, we need two Na^+ ions for each SO_4^{2-} ion. We indicate the presence of two Na^+ ions with the subscript 2 following the symbol of this ion. The formula of the compound is **Na_2SO_4**. The convention that the positive ion is always written first in the formula still holds when polyatomic ions are present.

(c) Two NO_3^- ions are needed to balance the charge on one Ca^{2+} ion. Since more than one polyatomic ion is needed, the formula will contain parentheses: **$Ca(NO_3)_2$**. The subscript 2 outside the parentheses indicates two of what is inside the parentheses. If parentheses were not used, the formula would appear to be $CaNO_{32}$, which is not intended and which actually conveys false information. The formula $Ca(NO_3)_2$ indicates a formula unit containing one Ca atom, two N atoms, and six O atoms (Sec. 5.4); the formula $CaNO_{32}$ would indicate a formula unit containing one Ca atom, one N atom, and thirty-two O atoms. Verbally the correct formula, $Ca(NO_3)_2$, would be read as "C-A" (pause) "N-O-three-taken-twice."

(d) In this compound both ions are polyatomic, a perfectly legal situation. Since the ions have equal but opposite charges, they will combine in a one-to-one ratio. The formula is thus **NH_4CN**. No parentheses are needed because we need only one polyatomic ion of each type in a formula unit. Parentheses are used only when there are two or more polyatomic ions of a given kind in a formula unit. What is different about this formula is the appearance of the symbol for the element nitrogen (N) at two locations in the formula. This could be prevented by combining the two nitrogens, giving the formula N_2H_4C. However, combining is not done in situations like this because the identity of the polyatomic ions present is lost in the resulting combined formula. The formula N_2H_4C does not convey the message that NH_4^+ and CN^- ions are present; the formula NH_4CN does. Thus, in writing formulas that contain polyatomic ions we always maintain the identities of these ions even if it means having the same elemental symbol at more than one location in the formula.

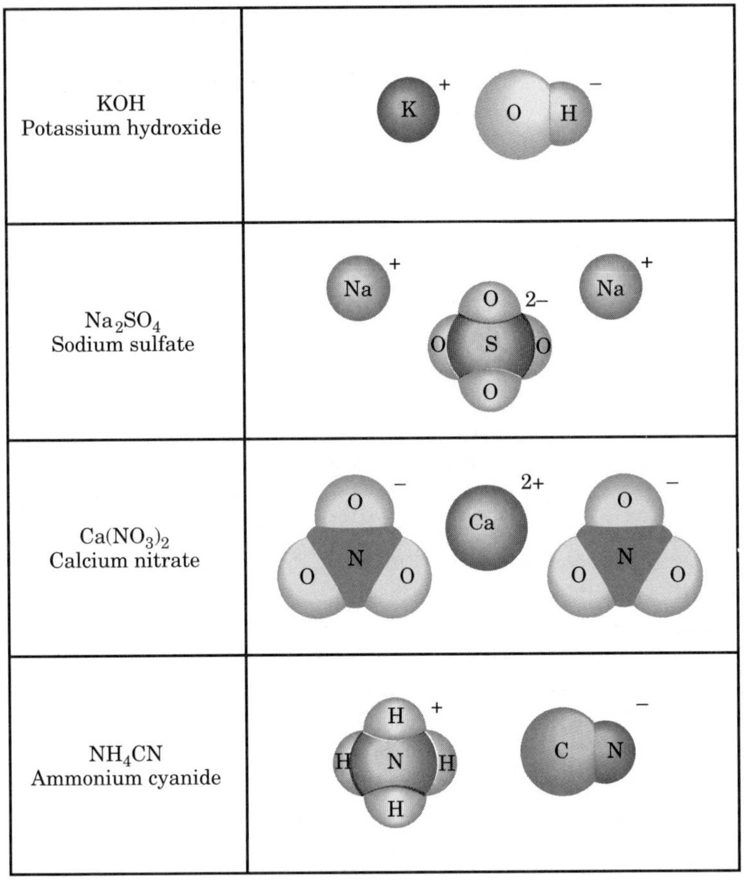
KOH
Potassium hydroxide
K
+
O
H
–
Na2SO4
Sodium sulfate
Na
+
O
2–
O
S
O
O
Na
+
Ca(NO3)2
Calcium nitrate
O
–
N
O
O
Ca
2+
O
–
N
O
O
NH4CN
Ammonium cyanide
H
+
H
N
H
H
C
N
–

FIGURE 7.3 Ionic makeup of some compounds in which polyatomic ions are present.

PRACTICE EXERCISE 7.6

Determine the formulas for the ionic compounds containing the following pairs of ions.

(a) K^+ and $SO_4^{\,2-}$ (b) Na^+ and $NO_3^{\,-}$
(c) Ca^{2+} and CN^- (d) $NH_4^{\,+}$ and OH^-

Ans. (a) K_2SO_4; (b) $NaNO_3$; (c) $Ca(CN)_2$; (d) NH_4OH.

Similar exercises: Problems 7.51 and 7.52

Figure 7.3 gives pictorial representations of the ionic makeup of the four compounds whose formulas were determined in Example 7.6.

7.9 The Nature of a Covalent Bond

In ionic compounds containing monoatomic ions, the two atoms involved in a given ionic bond are a metal and a nonmetal—atoms that are quite *dissimilar*. These dissimilar atoms are complementary to each other; one atom (the metal) "likes" to lose electrons, and the other atom (the nonmetal) "likes" to gain electrons. The net result is electron transfer (Sec. 7.6).

Covalent bonds (Sec. 7.1) are formed between *similar* or even *identical* atoms. Most often the atoms involved are two nonmetal atoms. It is not reasonable to suppose that one atom would give up electrons to another atom when the atoms are identical or very similar. The concept of *electron sharing* rather than *electron transfer* explains bonding between similar or identical atoms. In electron sharing, two nuclei attract the same electrons, and the resulting attractive forces (covalent bond) hold the two nuclei together. A **covalent bond** is the attractive force between two atoms that results from the sharing of one or more electron pairs between the two atoms. (The prefix *co-*, as in covalent, indicates entities that are joined or equal; consider the words cohere, coexist, coalesce, cooperate, coordinate, and so on.)

The hydrogen molecule, H_2, furnishes the simplest possible example of a covalent bond. Hydrogen, with just one $1s$ electron, needs one more electron to obtain a noble gas configuration, that of helium ($1s^2$). A hydrogen atom accomplishes this by sharing its lone electron with another hydrogen atom, which reciprocates by sharing its electron with the first hydrogen atom. The net result is the formation of an H_2 molecule. The two shared electrons in an H_2 molecule do "double duty," helping each of the H atoms achieve a noble gas configuration.

Covalent bonds are represented using electron-dot structures in much the same way as we used them for ionic bonds. A pair of dots placed between the symbols of the bonded atoms indicates the shared pair of electrons. The electron-dot structure for H_2 is

$$H \cdot \;\; \cdot H \longrightarrow H\!:\!H \quad \text{— shared electron pair}$$

An alternative way of representing the shared electron pair in a covalent bond is to draw a dash between the symbols of the bonded atoms.

$$H\!-\!H$$

Both of the atoms in H_2 have access to the two electrons of the shared electron pair. The concept of overlap of orbitals helps visualize this. As shown in Figure 7.4, as the atoms get close together, the orbitals containing the electrons overlap and create an orbital common to both atoms. When this happens, the two electrons move throughout the overlap region between the nuclei and are *shared* by both nuclei.

Two hydrogen atoms in an H_2 molecule sharing two electrons between them is a more stable situation than two separate H atoms, each with one electron. Thus, H atoms are always found in pairs (as H_2 molecules) in samples of elemental hydrogen.

The stability of the H_2 molecule results from each of the hydrogen nuclei attracting the two shared electrons. The following analogy involving "pot-bellied stoves" is useful in understanding this enhanced stability situation. Consider the two hydrogen nuclei to be two pot-bellied stoves and the two electrons to be entities trying to keep warm by moving about the nuclei (stoves). For separated hydrogen atoms (no covalent bond) each electron moves about its own nucleus. If the two nuclei are brought together (H_2 molecule) the two electrons can position themselves between the two nuclei and keep all parts, back and front, warm. Receiving simultaneous

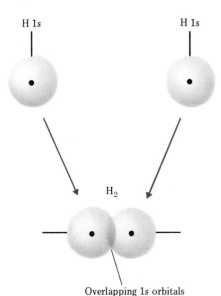

FIGURE 7.4 In the H_2 molecule, the two electrons present are shared by the two atoms as a result of $1s$ orbital overlap.

warmth from two stoves is a more desirable situation than having warmth from only one stove.

Using the octet rule and electron-dot structures, let us consider some other simple molecular compounds where covalent bonding is present.

The element chlorine, located in group VIIA of the periodic table, has seven valence electrons. Its electron-dot structure is

$$\cdot \ddot{\underset{..}{Cl}} :$$

Chlorine needs one additional electron to achieve the octet of electrons that makes it isoelectronic with the noble gas argon. In ionic compounds, where it bonds to metals, the Cl receives the needed electron via electron transfer. When Cl combines with another nonmetal, a common situation, the octet of electrons is completed via electron sharing. Representative of the situation where chlorine obtains its eighth valence electron through an electron-sharing process are the molecules HCl, Cl_2, and BrCl, whose electron-dot structures are as follows.

$$H \cdot \;\; \cdot \ddot{\underset{..}{Cl}} : \longrightarrow H : \ddot{\underset{..}{Cl}} : \;\; \text{or} \;\; H - \ddot{\underset{..}{Cl}} :$$

$$: \ddot{\underset{..}{Cl}} \cdot \;\; \cdot \ddot{\underset{..}{Cl}} : \longrightarrow : \ddot{\underset{..}{Cl}} : \ddot{\underset{..}{Cl}} : \;\; \text{or} \;\; : \ddot{\underset{..}{Cl}} - \ddot{\underset{..}{Cl}} :$$

$$: \ddot{\underset{..}{Br}} \cdot \;\; \cdot \ddot{\underset{..}{Cl}} : \longrightarrow : \ddot{\underset{..}{Br}} : \ddot{\underset{..}{Cl}} : \;\; \text{or} \;\; : \ddot{\underset{..}{Br}} - \ddot{\underset{..}{Cl}} :$$

Note that in each of these molecules Cl atoms have four pairs of electrons about them (an octet) and only one of the four pairs is involved in electron sharing. The three pairs of electrons on each Cl atom not taking part in the bonding are called *nonbonding* electron pairs or *unshared* electron pairs.

The number of covalent bonds that an atom forms is equal to the number of electrons it needs to achieve a noble gas configuration. Note that the chlorine atoms in HCl, Cl_2, and BrCl all formed one covalent bond. For chlorine, seven valence electrons plus one electron acquired by electron sharing (one bond) gives the eight valence electrons needed for a noble gas electronic configuration. The elements oxygen, nitrogen, and carbon have, respectively, six, five, and four valence electrons. Therefore, these elements form, respectively, two, three, and four covalent bonds. The number of covalent bonds these three elements form is reflected in the formulas of their simplest hydrogen compounds H_2O, NH_3, and CH_4. Electron-dot structures for these three molecules are as follows.

$$H\!:\!\ddot{\underset{\cdot\cdot}{O}}\!:\ \longrightarrow\ H\!:\!\ddot{\underset{\cdot\cdot}{O}}\!:\ \text{or}\ H\!-\!\underset{\cdot\cdot}{\overset{H}{\underset{|}{O}}}\!:$$

$$H\!:\!\ddot{\underset{\cdot\cdot}{N}}\!:\ \longrightarrow\ H\!:\!\underset{H}{\overset{H}{N}}\!:\ \text{or}\ H\!-\!\underset{H}{\overset{H}{\underset{|}{N}}}\!:$$

$$H\!:\!\ddot{C}\!:\ \longrightarrow\ H\!:\!\underset{H}{\overset{H}{C}}\!:\!H\ \text{or}\ H\!-\!\underset{H}{\overset{H}{\underset{|}{C}}}\!-\!H$$

Thus, we see that just as the octet rule was useful in determining the ratio of ions in ionic compounds, we can use it to predict formulas in molecular compounds. Example 7.7 gives additional illustrations of the use of the octet rule to determine formulas for molecular compounds.

EXAMPLE 7.7

Write electron-dot structures for the simplest compound formed from the following pairs of elements.

(a) phosphorus and bromine (b) hydrogen and sulfur
(c) oxygen and fluorine

Solution

(a) Phosphorus is in group VA of the periodic table and thus has five valence electrons. It will therefore want to form three covalent bonds. Bromine, in group VIIA of the periodic table, has seven valence electrons and will want to form only one covalent bond. Therefore, we have

$$:\!\ddot{B}r\!-\!\ddot{\underset{\ddot{B}r:}{\overset{\ddot{B}r:}{P}}}\!-\!\ddot{B}r\!:$$

Each atom in PBr$_3$ has an octet of electrons, which is circled in color in the following diagram.

(b) Sulfur has six valence electrons and hydrogen has one valence electron. Thus sulfur will form two covalent bonds $(6 + 2 = 8)$, and hydrogen will form one covalent bond $(1 + 1 = 2)$. Remember that for H an "octet" is two electrons; the noble gas that hydrogen "mimics" is helium, which has only two valence electrons.

(c) Oxygen, with six valence electrons, will form two covalent bonds, and fluorine, with seven valence electrons, will form only one covalent bond. The formula of the compound is thus OF$_2$, which has the following electron-dot structure.

PRACTICE EXERCISE 7.7

Write electron-dot structures for the simplest compound formed from the following pairs of elements.

(a) nitrogen and fluorine (b) carbon and chlorine
(c) sulfur and bromine

Ans. (a) :F̈:N̈:F̈: (b) :C̈l:C:C̈l: (c) :B̈r:S̈:   Similiar exercises:
           :F̈:   :C̈l: :B̈r: Problems 7.57 and 7.58

7.10 Electronegativities and Bond Polarities

Most chemical bonds are neither 100% ionic nor 100% covalent. Instead, most bonds have some characteristics of both types of bonding. We mentioned this fact in Section 7.1 as we started our discussion of chemical bonding.

Ionic and covalent bonding represent the two extremes of a broad continuum of bonding types. Between the extremes of complete transfer of electrons (ionic bonding) and equal sharing of electrons (covalent bonding) we find many intermediate cases where electrons are shared, but shared *unequally*.

Unequal sharing of electrons results from some atoms possessing a greater ability to attract bonding electrons to themselves than others. The greater the difference in electron-attracting ability between atoms involved in a bond, the more unequal the sharing of bonding electrons becomes. The ultimate in unequal sharing is reached when one atom assumes complete control of the bonding electrons; electron transfer has occurred.

A pictorial representation of the continuum of bonding types possible is obtained by considering the distribution of bonding electrons about the nuclei in a diatomic molecule as a function of an increasing difference in the electron-attracting ability of the two nuclei. Such a representation is shown in Figure 7.5. Figure 7.5a shows the distribution of bonding electron density expected when the two atoms of the diatomic molecule are identical. (Note that an "electron cloud" is used to depict the bonding electrons, as in Sec. 5.5). The sharing of electrons must be equal, since identical nuclei must affect the bonding electrons in the same way; hence the symmetrical electron density distribution. The situation depicted in Figure 7.5a corresponds to a 100% covalent bond. Note that this diagram is very similar to Figure

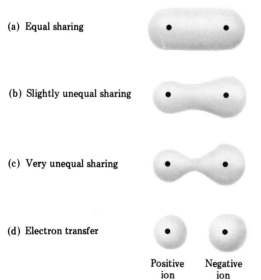

FIGURE 7.5 The continuum of bonding types.

7.3, which describes electron sharing in H_2, a situation where both nuclei are the same.

Any time two nuclei differ in their ability to attract a pair of bonding electrons, unequal sharing results. Figure 7.5b shows the situation where such a difference is very small. Electron sharing will be "close" to being equal, but not exactly equal. Note that the electron distribution is no longer symmetrical. This means that the bonding electrons spend more time associated with the nucleus that has the stronger electron-attracting ability. This will be the nucleus on the right in Figure 7.5b, the side where the electron density is greater.

Figure 7.5c depicts electron density distribution when a relatively large difference in electron-attracting ability exists between nuclei. The sharing of electrons here can be described as being "very unequal."

Finally, when the electron-attracting ability difference becomes very large, one atom "wins the battle." Electron transfer occurs, and the situation depicted in Figure 7.5d results. This situation corresponds to a 100% ionic bond.

An estimate of the degree of inequality of electron sharing in a chemical bond can be obtained from the *electronegativity values* of the bonded atoms. What are electronegativity values? What is electronegativity? **Electronegativity** is a measure of the relative attraction that an atom has for the shared electrons in a bond. Electronegativity values are unitless numbers on a relative scale that are obtained from bond energies and other related experimental data. Electronegativities cannot be directly measured in the laboratory. A number of electronegativity scales exist. The most widely used one, a scale developed by the American chemist Linus Pauling (1901–), is given in the format of a periodic table in Figure 7.6. On this scale, fluorine, the most electronegative of all elements, has arbitrarily been assigned a value of 4.0 (the maximum on the scale) and serves as the reference element. *The higher the electronegativity of an element, the greater the electron-attracting ability of atoms of that element.*

Patterns and trends in the electronegativity values given in Figure 7.6 include the following.

H 2.1																	He —
Li 1.0	Be 1.5											B 2.0	C 2.5	N 3.0	O 3.5	F 4.0	Ne —
Na 0.9	Mg 1.2											Al 1.5	Si 1.8	P 2.1	S 2.5	Cl 3.0	Ar —
K 0.8	Ca 1.0	Sc 1.3	Ti 1.5	V 1.6	Cr 1.6	Mn 1.5	Fe 1.8	Co 1.8	Ni 1.8	Cu 1.8	Zn 1.6	Ga 1.6	Ge 1.8	As 2.0	Se 2.4	Br 2.8	Kr —
Rb 0.8	Sr 1.0	Y 1.2	Zr 1.4	Nb 1.6	Mo 1.8	Tc 1.9	Ru 2.2	Rh 2.2	Pd 2.2	Ag 1.9	Cd 1.7	In 1.7	Sn 1.8	Sb 1.9	Te 2.1	I 2.5	Xe —
Cs 0.7	Ba 0.9	57-71 1.1-1.2	Hf 1.3	Ta 1.5	W 1.7	Re 1.9	Os 2.2	Ir 2.2	Pt 2.2	Au 2.4	Hg 1.9	Tl 1.8	Pb 1.8	Bi 1.9	Po 2.0	At 2.2	Rn —
Fr 0.7	Ra 0.9																

FIGURE 7.6 Electronegativities of the elements.

1. Electronegativity values generally increase from left to right across a period of the periodic table.
2. For the period 2 elements this left-to-right increase is regular, being 0.5 units.

Li	Be	B	C	N	O	F
1.0	1.5	2.0	2.5	3.0	3.5	4.0

3. It is important to note how the electronegativity of hydrogen (period 1) compares with that of the period 2 elements. Hydrogen's value of electronegativity, 2.1, is between that of B and C
4. For the period 3 elements the left-to-right increase is 0.3 units until the last two elements are reached where it is 0.4 and 0.5.

Na	Mg	Al	Si	P	S	Cl
0.9	1.2	1.5	1.8	2.1	2.5	3.0

5. Electronegativity values generally decrease from top to bottom in a periodic table group.

The group and period trends in electronegativity values result in nonmetals generally having higher electronegativities than metals. This fact is consistent with our previous generalization (Sec. 7.5) that metals tend to lose electrons and nonmetals tend to gain electrons when an ionic bond is formed. Metals (low electronegativities, poor electron attractors) will give up electrons to nonmetals (high electronegativities, good electron attractors).

The difference in the electronegativity values of the atoms in a bond is the key to predicting the *polarity* of that bond. **Bond polarity** is a measure of inequality in the sharing of bonding electrons. The terms nonpolar and polar are used in describing bonds. A **nonpolar covalent bond** is one where there is equal sharing of bonding electrons. A **polar covalent bond** is one in which sharing of bonding electrons is unequal. The larger the difference in electronegativity between the atoms involved in a bond the more polar the bond. The "ultimate" in bond polarity is the ionic bond.

In a polar covalent bond, the more electronegative atom, because of its greater electron-attracting ability, pulls the bonding electrons closer to it. The net effect of this is an unbalanced (nonsymmetrical) distribution of electron density in the bond (Fig. 7.5b and c). This results in a slight buildup of negative charge about the more electronegative atom. This additional electron density about the more electronegative atom comes from the other element in the bond (the less electronegative element), which has acquired a slightly positive charge. This means that one end of the bond is negative *with respect to* the other end.

The unequal sharing of electrons in a covalent bond is often indicated by a notation that uses the lowercase Greek letter δ (delta). A $\delta-$ symbol, meaning a "partial negative charge," is placed above the relatively negative atom of the bond, and a $\delta+$ symbol, meaning a "partial positive charge," is placed above the relatively positive atom.

With "delta notation," the bond in hydrogen fluoride (HF) would be depicted as

$$\overset{\delta+}{\text{H}}-\overset{\delta-}{\text{F}}$$

Fluorine has an electronegativity of 4.0, and hydrogen's electronegativity is 2.1 (see

Fig. 7.6). Since F is the more electronegative of the two elements, it dominates the electron-sharing process and draws the electrons closer to it. Hence, the F end of the bond has the $\delta-$ designation. Again, the $\delta-$ over the F atom indicates a partial negative charge—the fluorine end of the molecule is negative with respect to the hydrogen end. The meaning of the $\delta+$ over the H is that the H end of the molecule is positive with respect to the F end. Partial charges are always charges less than $+1$ or -1. Charges of $+1$ and -1, full charges, would result when an electron is transferred from one atom to another. With partial charges we are talking about an intermediate charge state between 0 and 1.

The message of this section is that there is no natural boundary between ionic and covalent bonding. Most bonds are a mixture of pure ionic and pure covalent bonds; that is, unequal sharing of electrons occurs. Most bonds have both ionic and covalent character. Nevertheless, it is still convenient to use the terms ionic and covalent in describing chemical bonds, based on the following guidelines, which relate to electronegativity differences.

1. Bonds between identical atoms (or nonidentical atoms of equal electronegativity), where there is zero difference in electronegativity between atoms, are called *nonpolar covalent bonds*.
2. Bonds where the electronegativity difference between atoms is greater than zero but less than 1.7 are called *polar covalent bonds*.
3. Bonds where the difference in electronegativity between atoms is 1.7 or greater are called *ionic bonds*.

EXAMPLE 7.8

Indicate whether each of the following bonds is nonpolar covalent, polar covalent, or ionic. Also indicate the direction of polarity for bonds that involve a polar covalent designation, using delta notation ($\delta+$ and $\delta-$).
(a) ClF   (b) MgO   (c) HI

Solution

(a) The electronegativities of Cl and F are 3.0 and 4.0, respectively. The electronegativity difference, obtained by subtracting the smaller electronegativity value from the larger, is

$$\text{Electronegativity difference} = 4.0 - 3.0 = 1.0$$

The bond is thus polar covalent. Bonds where the electronegativity difference is greater than zero but less that 1.7 are classified as polar covalent.

The fluorine end of the bond will be negative relative to the chlorine end, since F is the more electronegative of the two elements. Thus, the polarity direction is

$$\overset{\delta+}{\text{Cl}}—\overset{\delta-}{\text{F}}$$

(b) The electronegativity of oxygen is 3.5, and that of magnesium is 1.2. The electronegativity difference is

$$\text{Electronegativity difference} = 3.5 - 1.2 = 2.3$$

The term ionic is used to describe bonds when the electronegativity difference is 1.7 or greater. The polarity of the Mg—O bond can best be described in terms of ions—complete transfer of electrons.

$$[Mg^{2+}][O^{2-}]$$

(c) The electronegativity of iodine is 2.5, and that of hydrogen is 2.1. The electronegativity difference is 0.4. Since the electronegativity difference is greater than zero but less than 1.7, the bond is called a polar covalent bond.

In the polar covalent bond of concern here, the I atom will have acquired a partial negative charge relative to the H atom since the I is the more electronegative of the two atoms. Thus, the polarity is designated as

$$\overset{\delta+\delta-}{\text{H—I}}$$

PRACTICE EXERCISE 7.8

Indicate whether each of the following bonds is nonpolar covalent, polar covalent, or ionic. Also indicate the direction of polarity for bonds that involve a polar covalent designation, using delta notation ($\delta+$ and $\delta-$).

(a) N—S   (b) Na—F   (c) Al—C

Ans. (a) polar covalent, $\overset{\delta-\delta+}{\text{N—S}}$; (b) ionic; (c) polar covalent, $\overset{\delta+\delta-}{\text{Al—C}}$.

Similar exercises: Problems 7.65, 7.66, 7.69, and 7.70

We see, then, that the terms *ionic* and *covalent*, when applied to a particular bond, describe the predominant character of the bond. Almost all bonds have both ionic and covalent character; we describe the bond in terms of the dominant type. Figure 7.7 summarizes the relationships among electronegativity differences, bond types, and bond terminology.

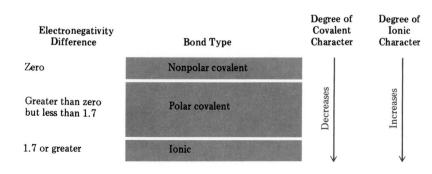

FIGURE 7.7 The relationships among bonding type, electronegativity difference, degree of covalent character, and degree of ionic character.

TABLE 7.4 Relationship Between Electronegativity Difference and Percent Ionic Character of a Bond

Electronegativity Difference	Percent Ionic Character	Electronegativity Difference	Percent Ionic Character
0.0	0.0	1.7	51.0
0.1	0.5	1.8	55.0
0.2	1.0	1.9	59.0
0.3	2.0	2.0	63.0
0.4	4.0	2.1	67.0
0.5	6.0	2.2	70.0
0.6	9.0	2.3	74.0
0.7	12.0	2.4	76.0
0.8	15.0	2.5	79.0
0.9	19.0	2.6	82.0
1.0	22.0	2.7	84.0
1.1	26.0	2.8	86.0
1.2	30.0	2.9	88.0
1.3	34.0	3.0	89.0
1.4	39.0	3.1	91.0
1.5	43.0	3.2	92.0
1.6	47.0		

The justification for using the value 1.7 as the "dividing line" between polar covalent bonds and ionic bonds can be seen by looking at Table 7.4, which gives numerical percents of ionic and covalent character for bonds of various electronegativity differences. An electronegativity difference of 1.7 is the first value for which the ionic character of a bond exceeds the covalent character.

Compounds containing bonds that have greater than 50% ionic character usually produce ions when melted or dissolved in aqueous solution. Hence, such compounds are appropriately characterized as having ionic bonds.

Knowing bond type enables one to predict many general properties of compounds. Covalent compounds are not good conductors of heat or electricity. Their melting points are relatively low—many are below 100 °C. In contrast, ionic compounds have relatively high melting points and are good conductors of electricity in solution or in the liquid state.

Information concerning ionic and covalent character for bonds is useful in deciding whether to formulate the bonding description for a compound in terms of an ionic electron-dot structure (Sec. 7.5) or a covalent electron-dot structure (Sec. 7.9).

EXAMPLE 7.9

Determine the percent ionic character and the percent covalent character for each of the following bonds.

(a) Al—P   (b) Na—F   (c) H—F

Solution

(a) The electronegativities of Al and P are 1.5 and 2.1, respectively. The electronegativity difference is 0.6. From Table 7.4, a 0.6 difference corresponds to a bond with only **9.0% ionic character and 91.0% covalent character.**

(b) The difference in electronegativity between Na (0.9) and F (4.0) is 3.1. This large difference gives a bond that is predominantly ionic: **91.0% ionic character and 9.0% covalent character** (see Table 7.4).

(c) Hydrogen has an electronegativity of 2.1, and fluorine a value of 4.0. The difference between these two values, 1.9, corresponds to a bond that can be characterized as **59.0% ionic and 41.0% covalent** (see Table 7.4).

PRACTICE EXERCISE 7.9

Determine the percent ionic character and the percent covalent character for each of the following bonds.
(a) Al—F   (b) N—S   (c) K—O

Ans. (a) 79.0% ionic, 21.0% covalent; (b) 6.0% ionic, 94.0% covalent; (c) 84.0% ionic, 16.0% covalent.   Similar exercises: Problems 7.71 and 7.72

7.11 Multiple Covalent Bonds

In our discussion of covalent bonding to this point (Secs. 7.9 and 7.10), all of the molecules chosen as examples to illustrate various aspects of such bonding have contained only *single covalent bonds*. A **single covalent bond** is a bond where a single pair of electrons is shared (equally or unequally) between two atoms.

Many molecules exist where two or three pairs of electrons must be shared between the same two atoms in order to provide a complete octet of electrons for each atom involved in the bonding. Such bonds are called multiple covalent bonds. **Multiple covalent bond** is a collective term used to designate covalent bonds where two or three pairs of electrons are shared (equally or unequally) between the same two atoms. Multiple covalent bonds can be further classified as double covalent bonds or triple covalent bonds. A **double covalent bond** is a bond where two pairs of electrons are shared (equally or unequally) between the same two atoms. A double covalent bond is stronger than a single covalent bond, but not twice as strong, because the two electron pairs repel each other and cannot become fully concentrated between the two atoms. (Bond strength, the energy it takes to break a bond, can be measured experimentally.) A **triple covalent bond** is a bond where three pairs of electrons are shared (equally or unequally) between the same two atoms. A triple covalent bond is stronger than a single covalent bond or a double covalent bond, but not three times as strong as a single bond, again because the electron pairs repel each other. Now let us consider some molecules where multiple bonding is present.

A diatomic N_2 molecule, the form in which nitrogen occurs in the atmosphere, contains a triple covalent bond. It is the simplest known triple covalent bond. A nitrogen atom has five valence electrons and thus needs three additional electrons to complete its octet.

$$\cdot \ddot{\underset{\cdot}{N}} \cdot$$

In an N_2 molecule the only sharing that can take place is between the two nitrogen atoms. They are the only atoms present. Thus, to acquire a noble gas electron configuration each nitrogen atom must share three of its electrons with the other nitrogen atom.

$$:\!N\!\cdot \;\; \cdot\!N\!: \longrightarrow \;:\!N\!:\!:\!:\!N\!: \;\; \text{or} \;\; :\!N\!\equiv\!N\!:$$

Notice how all three shared electron pairs are placed in the area between the two nitrogen atoms in the above bonding diagrams. Note also that three lines are used to denote a triple covalent bond, paralleling the use of one line to denote a single covalent bond.

In "bookkeeping" electrons in an electron-dot structure, to make sure that all atoms in the molecule have achieved their octet of electrons, *all* electrons in a multiple covalent bond are considered "to belong" to *both* of the atoms involved in that bond. The "bookkeeping" for the N_2 molecule would be

Each of the circles about an N atom contains eight valence electrons. Again, all of the electrons in a multiple covalent bond are considered to belong to each of the atoms in the bond. Circles are never drawn to include just some of the electrons in a multiple covalent bond.

A slightly more complicated molecule containing a triple covalent bond is the molecule C_2H_2 (acetylene). A carbon–carbon triple covalent bond is present as well as two carbon–hydrogen single covalent bonds. The arrangement of valence electrons in C_2H_2 is as follows.

$$H\!\cdot \; :\!\!C\!\cdot \;\; \cdot\!C\!\!: \; \cdot\!H \longrightarrow H\!:\!C\!:\!:\!:\!C\!:\!H \;\; \text{or} \;\; H\!-\!C\!\equiv\!C\!-\!H$$

The two atoms in a triple covalent bond are commonly the same element. However, they do not have to be. The molecule HCN contains a heteroatomic triple covalent bond.

$$H\!:\!C\!:\!:\!:\!N\!: \;\;\; \text{or} \;\;\; H\!-\!C\!\equiv\!N\!:$$

As might be surmised from our discussion so far, C and N are the two representative elements most frequently encountered in triple covalent bonds.

Double covalent bonds are found in numerous molecules. A very common molecule that contains bonding of this type is carbon dioxide (CO_2). In fact, there are two carbon–oxygen double covalent bonds present in CO_2.

$$:\!\ddot{O}\!\cdot \;\; \cdot\!\ddot{C}\!\cdot \;\; \cdot\!\ddot{O}\!: \longrightarrow \;:\!\ddot{O}\!:\!:\!C\!:\!:\!\ddot{O}\!: \;\; \text{or} \;\; :\!\ddot{O}\!=\!C\!=\!\ddot{O}\!:$$

Note in the following diagram how the circles are drawn for the octet of electrons about each of the atoms in CO_2.

Not all elements can form multiple covalent bonds. There must be at least two vacancies in an atom's valence electron shell prior to bond formation if it is to participate in a multiple covalent bond. This requirement eliminates group VIIA elements (F, Cl, Br, I) and hydrogen from participating in such bonds. The group VIIA elements have seven valence electrons and one vacancy, and hydrogen has one valence electron and one vacancy. All bonds formed by these elements are single covalent bonds.

Double bonding becomes possible for elements needing two electrons to complete their octet, and triple bonding becomes possible when three or more electrons are needed to complete an octet. Note that the word *possible* was used twice in the previous sentence. Multiple bonding does not have to occur when an element has two or three or four vacancies in its octet; single covalent bonds can be formed instead. The "bonding behavior" of an element, when more than one behavior is possible, is determined by what other element or elements it is bonded to.

Let us consider the possible "bonding behaviors" for O (six valence electrons; two octet vacancies), N (five valence electrons; three octet vacancies), and C (four valence electrons; four octet vacancies).

To complete its octet by electron sharing, an oxygen atom can form either two single bonds or one double bond.

$$\overset{|}{\underset{..}{:\text{O}}}- \qquad \underset{..}{:\text{O}}=$$

two single bonds   one double bond

Nitrogen is a very versatile element with respect to bonding. It can form single, double, or triple covalent bonds as dictated by the other atoms present in a molecule.

$$-\overset{..}{\underset{|}{\text{N}}}- \qquad -\overset{..}{\text{N}}= \qquad :\text{N}\equiv$$

three single bonds   one single and   one triple bond
                     one double bond

Note that in each of these bonding situations a nitrogen atom forms three bonds. A double bond counts as two bonds, and a triple bond as three bonds. Since nitrogen has only five valence electrons, it must form three covalent bonds to complete its octet.

Carbon is an even more versatile element than nitrogen with respect to variety of types of bonding as illustrated by the following possibilities for bonding.

$$-\overset{|}{\underset{|}{\text{C}}}- \qquad -\overset{|}{\text{C}}= \qquad =\text{C}= \qquad -\text{C}\equiv$$

four single bonds   two single bonds and   two double bonds   one single bond and
                    one double bond   one triple bond

7.12 Coordinate Covalent Bonds

In the examples of covalent bonding encountered so far, whether the bonds are single, double, or triple, each of the bonded atoms has contributed an equal number of electrons to the bond—one each for a single covalent bond, two each for a double covalent bond, and three each for a triple covalent bond. A few molecules exist for which the covalent bonding within the molecule cannot all be explained in this manner; instead the concept of coordinate covalency must be invoked.

A **coordinate covalent bond** is a bond in which both electrons of a shared pair come from one of the two atoms. Coordinate covalent bonding allows an atom that has two (or more) vacancies in its valence shell to share a pair of nonbonding electrons located on another atom.

The ammonium ion, NH_4^+, is an example of a species containing a coordinate covalent bond. The formation of an NH_4^+ ion can be viewed as resulting from the reaction of a hydrogen ion, H^+, with an ammonia molecule, NH_3. Doing the "bookkeeping" on all of the valence electrons involved in this reaction, using X for nitrogen electrons and dots for hydrogen electrons, we get

$$H \underset{H}{\overset{H}{:\!N\!:}} + H^+ \longrightarrow \left[H \underset{H}{\overset{H}{:\!N\!:}} H \right]^+$$

— coordinate covalent bond

An H^+ ion has no electrons, hydrogen having lost its only electron when it became an ion. The H^+ ion has two vacancies in its valence shell; that is, it needs two electrons to become isoelectronic with the noble gas helium. The nitrogen in NH_3 possesses a pair of nonbonding electrons. These electrons are used in forming the new nitrogen—hydrogen bond. The new species formed, the NH_4^+ ion, is charged, since the H^+ was charged. The $+1$ charge on the NH_4^+ ion is dispersed over the *entire* molecule; it is not localized on the "new" hydrogen atom.

Once a coordinate covalent bond is formed, there is no way to distinguish it from any of the other covalent bonds in a molecule; all electrons are identical regardless of their source. The main use of the concept of coordinate covalency is in helping to rationalize the existence of certain molecules and ions whose bonding-electron arrangement would otherwise present problems.

The molecule N_2O contains a single coordinate covalent bond.

$$:N:::N:O:$$

The nitrogen–nitrogen triple bond in N_2O is a normal covalent bond; the nitrogen–oxygen bond is a coordinate covalent bond where both electrons are supplied by the nitrogen atom.

The concept of coordinate covalency must also be invoked in explaining certain multiple covalent-bonding situations. The triple covalent bond joining C and O in carbon monoxide (CO) is of the coordinate covalent type.

$$\overset{x}{\underset{x}{\text{C}}}\overset{x}{\underset{x}{:}}\text{:O:}$$

Four of the six electrons in the triple bond can be considered to have come from the oxygen atom. Since C has only four valence electrons before bonding, it must share four electrons (from O) in order to achieve an octet of electrons. On the other hand, O has six valence electrons before bonding and thus needs to share only two electrons (from C).

Again, once a coordinate covalent bond is formed, it is no different from any other covalent bond. Electrons are electrons; they are all identical.

7.13 Resonance Structures

In Section 7.11 it was noted that, in general, triple covalent bonds are stronger than double covalent bonds, which in turn are stronger than single covalent bonds. **Bond strength** is measured by the energy it takes to break a bond, that is, to separate bonded atoms to give neutral particles. It can be determined experimentally.

Another experimentally determinable parameter of bonds is bond length. **Bond length** is the distance between the nuclei of bonded atoms (see Fig. 7.8). A direct relationship exists between bond strength and bond length. It is found that as bond strength increases, bond length decreases; that is, the stronger the bond, the shorter the distance between the nuclei of the atoms of the bond. Thus, in general, triple covalent bonds are shorter than double covalent bonds, which are shorter than single covalent bonds.

Most electron-dot structures for molecules give bonding pictures, that are consistent with available experimental information on bond strength and bond length. However, there are some molecules for which no single electron-dot structure that is consistent with such information can be written.

The molecule SO_2 is an example of a situation in which a single electron-dot structure does not adequately describe bonding. A plausible electron-dot structure for SO_2, in which the octet rule is satisfied for all three atoms, is

$$:\overset{..}{\underset{..}{\text{O}}}:\overset{..}{\text{S}}::\overset{..}{\underset{..}{\text{O}}}: \quad \text{or} \quad :\overset{..}{\underset{..}{\text{O}}}-\overset{..}{\text{S}}=\overset{..}{\underset{..}{\text{O}}}:$$

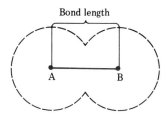

Figure 7.8 The length of a bond between two atoms is the distance separating the nuclei of the atoms.

However, this structure suggests that one sulfur–oxygen bond, the double bond, should be stronger and shorter than the other sulfur–oxygen bond, the single bond. Experiment shows that this is not the case; both sulfur–oxygen bonds are equivalent, with both bond length and bond strength characteristics intermediate between those for known sulfur–oxygen single and double bonds. An electron-dot structure depicting this intermediate situation cannot be written.

The solution to the phenomenon in which no single electron-dot structure adequately describes bonding involves the use of two or more electron-dot structures, known as resonance structures, to represent the bonding in the molecule. **Resonance structures** are two or more electron-dot structures for a molecule or ion that have the same arrangement of atoms, contain the same number of electrons, and differ only in the location of the electrons.

Two resonance structures exist for an SO_2 molecule.

$$:\ddot{O}:\!:\!\ddot{S}:\!:\!\ddot{O}: \longleftrightarrow :\ddot{O}:\!:\!\ddot{S}:\ddot{O}: \quad \text{or} \quad :\ddot{O}\!-\!\ddot{S}\!=\!\ddot{O}: \longleftrightarrow :\ddot{O}\!=\!\ddot{S}\!-\!\ddot{O}:$$

A double-headed arrow is used to connect resonance structures. The only difference between the two SO_2 resonance structures is in the location of one pair of electrons. The positioning of this pair of electrons determines whether the oxygen atom on the right or the left is the oxygen atom involved in the double bond.

The actual bonding in a SO_2 molecule is said to be a *resonance hybrid* of the two contributing resonance structures. Beginning chemistry students frequently misinterpret the concept of a resonance hybrid. They incorrectly envision that a molecule—SO_2 in this case—is constantly changing (resonating) between various resonance structure forms. This is not the case. For example, SO_2 is not a mixture of two kinds of molecules, nor does a single type of molecule flip-flop back and forth between the two resonance forms. There is only one kind of SO_2 molecule, and the bonding in it is an average of those depicted by the resonance structures. SO_2 molecules exist "full time" in this average state. A mule, the offspring of a donkey and a horse, can be considered a hybrid of a donkey and a horse. However, it is not a horse at one instant and a donkey at another; it is always a mule. Likewise, a molecule has only one real structure, which is different from any of the resonance structures; it has characteristics of each one but does not match any one of them exactly.

Sometimes three, four, or even more resonance structures can be drawn for a molecule. Again, such resonance structures must all contain the same number of electrons and have the same arrangement of atoms; the structures may differ only in the location of the electrons about the atoms.

7.14 A Systematic Method for Determining Electron-Dot Structures

The task of constructing electron-dot structures for molecules and polyatomic ions containing many electrons or for which several resonance hybrids must be used to describe the bonding can become quite frustrating if it is approached in a nonsystematic trial-and-error manner. Use of a systematic approach to writing electron-dot structures will enable a student to avoid most of this frustration. The following guidelines make the drawing of an electron-dot structure for any molecule or polyatomic ion that obeys the octet rule, even a very complicated structure, a straightforward procedure.

STEP 1   Determine the arrangement of the atoms in the molecule or polyatomic ion.

Determining which atom is the *central atom*, that is, which atom has the most other atoms bonded to it, is the key to determining the arrangement of atoms in a molecule or ion. Most other atoms present will be bonded to the central atom. For most molecular compounds containing just two elements, the molecular formula is of help in deciding the identity of the central atom. The cental atom is the atom that appears only once in the formula; for example, S is the central atom in SO_3, O is the central atom in H_2O, and P is the central atom in PF_3. In molecular compounds containing hydrogen, oxygen, and an additional element, it is the additional element that is the central atom; for example, N is the central atom in HNO_3, and S is the central atom in H_2SO_4. In compounds of this type the oxygen atoms are bonded to the central atom and the hydrogen atoms are bonded to the oxygens. Carbon is the central atom in almost carbon-containing compounds. Hydrogen and fluorine are never the central atom.

STEP 2   Determine the total number of valence electrons present, that is, the total number of dots that must appear in the electron-dot structure.

The total number of valence electrons is found by adding up the number of valence electrons each atom in the molecule or ion possesses. If the species is a polyatomic ion, add one electron for each unit of negative charge present or subtract one electron for each positive charge. Do not worry about keeping track of which electrons come from which atoms. Only their total number is important.

STEP 3   Draw a single covalent bond between the central atom and each of the surrounding atoms.

Either a pair of dots or a dash can be used in denoting the single bond(s).

STEP 4   Complete the octets of the atoms bonded to the central atom.

Electrons placed on the surrounding atoms are always added in pairs. Remember, also, that hydrogen needs only two electrons.

STEP 5   Place any remaining electrons on the central atom.

The number of remaining electrons is obtained by subtracting from the total number of valence electrons (step 2) the number of electrons used in steps 3 and 4.

STEP 6   If there are not enough electrons to give the central atom an octet, form multiple covalent bonds by shifting nonbonding electron pairs from surrounding atoms into bonding locations.

The elements C, N, and O are the elements most frequently involved in multiple bonding situations. The elements H, F, Cl, Br, and I in terminal atom positions do not participate in multiple bond formation.

The three examples that follow illustrate the use of the procedures for drawing electron-dot structures. In each part of each example the outlined procedure is followed step by step so that you will become familiar with it.

EXAMPLE 7.10

Write the electron-dot structure for the molecule $HClO_3$.

Solution

STEP 1   In compounds containing H, O, and an additional element, the additional element (Cl in this case) is the central atom. The oxygen atoms are attached to this central chlorine atom, and the hydrogen atom is attached to an oxygen atom. The arrangement of atoms is thus

$$O \quad Cl \quad O \quad H$$
$$O$$

STEP 2   Chlorine atoms have seven valence electrons, oxygen has six valence electrons, and hydrogen has one valence electron. The total number of valence electrons present in this molecule is 26.

$$\begin{array}{lll} 1 \text{ Cl:} & 1 \times 7 = & 7 \text{ valence electrons} \\ 3 \text{ O:} & 3 \times 6 = & 18 \text{ valence electrons} \\ 1 \text{ H:} & 1 \times 1 = & 1 \text{ valence electron} \\ \hline & & 26 \text{ valence electrons} \end{array}$$

STEP 3   Four single bonds are placed within the molecule

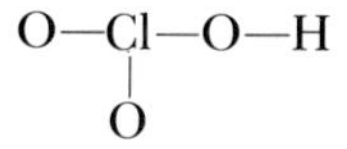

$$\text{O—Cl—O—H}$$
$$|$$
$$\text{O}$$

STEP 4   The two terminal oxygen atoms require six nonbonding electrons, the oxygen bonded to the hydrogen four additional electrons, and the hydrogen atom no additional electrons since it already has two electrons (those in the single bond).

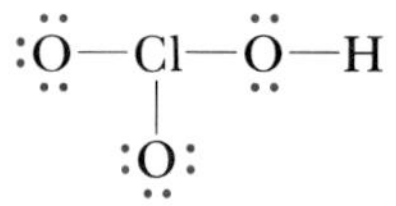

$$:\!\ddot{\text{O}}\!-\!\text{Cl}\!-\!\ddot{\text{O}}\!-\!\text{H}$$
$$|$$
$$:\!\ddot{\text{O}}\!:$$

STEP 5   We started out with 26 electrons (step 2). Eight were used in step 3 (single bonds) and sixteen in step 4 (nonbonding electron pairs). Thus, only two electrons are available for placement on the chlorine atom.

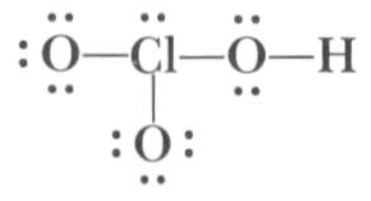

$$:\!\ddot{\text{O}}\!-\!\ddot{\text{Cl}}\!-\!\ddot{\text{O}}\!-\!\text{H}$$
$$|$$
$$:\!\ddot{\text{O}}\!:$$

STEP 6   No double or triple bonds are needed since the action of step 5 causes the central atom (chlorine) to have an octet of electrons.

PRACTICE EXERCISE 7.10

Write the electron-dot structure for the molecule H_2SO_4.

Ans.

$$H-\overset{..}{\underset{..}{O}}-\overset{\overset{\displaystyle :\overset{..}{O}:}{|}}{\underset{\underset{\displaystyle :\overset{..}{O}:}{|}}{S}}-\overset{..}{\underset{..}{O}}-H$$

Similar exercises: Problems 7.89 and 7.90

EXAMPLE 7.11

Write the electron-dot structure for the polyatomic ion SO_4^{2-}, the sulfate ion.

Solution

STEP 1 Sulfur is the central atom according to the guidelines for determining central atoms.

$$\begin{array}{c} O \\ O \quad S \quad O \\ O \end{array}$$

STEP 2 Sulfur has six valence electrons as does each oxygen. There are two additional valence electrons present, since we are dealing with an ion of -2 charge.

$$\begin{array}{rl} 1 \text{ S:} & 1 \times 6 = 6 \text{ valence electrons} \\ 4 \text{ O:} & 4 \times 6 = 24 \text{ valence electrons} \\ \text{charge of } -2 = & \underline{2 \text{ valence electrons}} \\ & 32 \text{ valence electrons} \end{array}$$

If the polyatomic ion had had a positive charge instead of a negative one, we would have had to subtract valence electrons from the total instead of adding. A positive charge denotes loss of electrons, and the electrons lost are valence electrons.

STEP 3 Four single bonds are placed within the molecule.

$$\begin{array}{c} O \\ | \\ O-S-O \\ | \\ O \end{array}$$

STEP 4 Each oxygen atom needs six nonbonding electrons in order to complete its octet.

$$\begin{array}{c} :\overset{..}{O}: \\ | \\ :\overset{..}{\underset{..}{O}}-S-\overset{..}{\underset{..}{O}}: \\ | \\ :\overset{..}{\underset{..}{O}}: \end{array}$$

STEP 5   There are no additional electrons available for placement on the central atom. We started out with 32 electrons, used 8 for the single bonds, and then 24 for the nonbonding electrons on the oxygen atoms.

STEP 6   The fact that no additional electrons are available for sulfur creates no problem. The four single bonds that the sulfur atom participates in gives it an octet of electrons.

The electron-dot structure we have just generated is that of a polyatomic ion. We therefore need to enclose it in large brackets and place the ionic charge outside the right bracket in a superscript position.

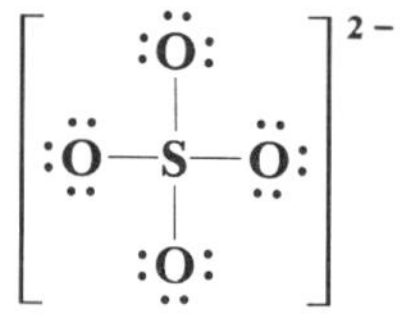

$$\left[\begin{array}{c} ::\!\ddot{O}\!:: \\ | \\ :\!\ddot{O}\!-\!S\!-\!\ddot{O}\!: \\ | \\ :\!\ddot{O}\!: \end{array} \right]^{2-}$$

In this example we have shown the bonding within a polyatomic ion. This polyatomic ion, or any other one for that matter, is not a stable entity that exists alone. Polyatomic ions are parts of ionic compounds. The ion SO_4^{2-} would be found in ionic compounds such as Na_2SO_4, $CaSO_4$, $(NH_4)_2SO_4$, and $Pb(SO_4)_2$. Ionic compounds containing polyatomic ions thus offer an interesting combination of both ionic and covalent bonds: covalent bonding *within* the polyatomic ion and ionic bonding *between* it and ions of opposite charge.

PRACTICE EXERCISE 7.11

Write the electron-dot structure for the polyatomic ion PO_4^{3-}, the phosphate ion.

Ans. $\left[\begin{array}{c} :\!\ddot{O}\!: \\ | \\ :\!\ddot{O}\!-\!P\!-\!\ddot{O}\!: \\ | \\ :\!\ddot{O}\!: \end{array} \right]^{3-}$

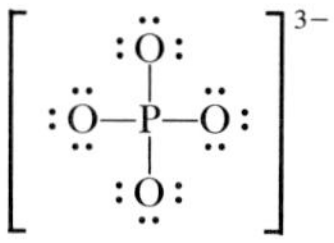

Similar exercises: Problems 7.91 and 7.92

EXAMPLE 7.12

Write the electron-dot structure for the molecule SO_3.

Solution

STEP 1   The central atom is S; all of the O atoms will be attached to it. The atom arrangement is

O   S   O

O

STEP 2   Both S and O atoms have six valence electrons. The total number of valence electrons present is therefore 24.

Section 7.14 · A Systematic Method for Determining Electron-Dot Structures

$$1 \text{ S:} \quad 1 \times 6 = 6 \text{ valence electrons}$$
$$3 \text{ O:} \quad 3 \times 6 = \underline{18 \text{ valence electrons}}$$
$$\phantom{3 \text{ O:} \quad 3 \times 6 = }24 \text{ valence electrons}$$

STEP 3 Three single bonds are placed within the molecule.

$$\text{O}-\text{S}-\text{O}$$
$$\phantom{\text{O}-}|$$
$$\phantom{\text{O}-}\text{O}$$

STEP 4 Each oxygen atom needs six nonbonding electrons in order to complete its octet.

STEP 5 There are no additional electrons available for placement on the sulfur atom. Initially 24 electrons were available for use (step 2). Six electrons were used in step 3 and the remaining 18 were used in step 4.

STEP 6 The central sulfur atom lacks an octet of electrons, having only six. This problem is solved by moving a nonbonding pair of electrons from one of the oxygen atoms into the sulfur–oxygen bonding region, producing a double bond. This action gives sulfur an octet of electrons.

There are three choices for the oxygen atom that is involved in double bond formation. The ramification of this situation is that resonance structures exist. The availability of choices for multiple bond formation (to give a central atom an octet of electrons) is always a signal for the existence of resonance structures. The resonance structures for our current situation, SO_3, are

PRACTICE EXERCISE 7.12

Write the electron-dot structure for the molecule SeO_2.

Ans. $\;\;:\ddot{\text{O}}-\ddot{\text{Se}}=\ddot{\text{O}}:\;\longleftrightarrow\;:\ddot{\text{O}}=\text{Se}-\ddot{\text{O}}:$

Similar exercises: Problems 7.93 and 7.94

The systematic approach to drawing electron-dot structures for molecules and polyatomic ions illustrated in Examples 7.10–7.12 does not take into account the origin of the electrons in a chemical bond, that is, which atoms contribute which electrons to the bond. Thus, no distinction between normal covalent bonds and coordinate covalent bonds is made by the procedures of this system. This is acceptable. Each electron in a bond belongs to the bond as a whole. Electrons do not have labels of genealogy.

7.15 Molecular Polarity

Molecules as well as bonds can have polarity. As we shall see shortly, the mere fact that a molecule contains polar bonds does not mean that the molecule as a whole is polar. Molecular polarity depends on (1) the polarity of the bonds within a molecule and (2) the geometry of the molecule (when three or more atoms are present).

Molecular geometry describes the way in which atoms in a molecule are arranged in space relative to each other. All molecules containing three or more atoms have characteristic three-dimensional shapes. For example, the triatomic CO_2 molecule is linear: that is, its three atoms lie in a straight line (Fig. 7.9a). On the other hand, the H_2O molecule, also a triatomic molecule, has a **nonlinear or bent geometry** (Fig. 7.9b). An NH_3 molecule has a trigonal pyramidal geometry, with the nitrogen atom at the apex and the hydrogen atoms at the base of the pyramid (Fig. 7.9c).

Determining the molecular polarity of a *diatomic* molecule is simple because only one bond is present. If that bond is nonpolar, the molecule is nonpolar; if the bond is polar, the molecule is polar. The collective effect of individual bond polarities must be considered to determine molecular polarity for molecules containing more than one bond (triatomic molecules, tetratomic molecules, and so on). Molecular geometry plays an important role in determining this collective effect. In instances of symmetrical geometry, the effects of polar bonds cancel and a nonpolar molecule results. Let us consider the polarities of three specific triatomic molecules—CO_2, H_2O, and HCN—to illustrate collective effects of bond polarities.

In the linear CO_2 molecule (Fig. 7.9a) both bonds are polar (oxygen is more electronegative than carbon). Despite the presence of these polar bonds, CO_2 molecules are *nonpolar*. The effects of the two polar bonds are cancelled out as a result of the oxygen atoms being arranged symmetrically about the carbon atom. The shift of electronic charge toward one oxygen atom is exactly compensated by the shift of electronic charge toward the other. Thus, one end of the molecule is not negatively charged relative to the other end (a requirement for polarity); hence the molecule is nonpolar. This cancellation of individual bond polarities is diagrammed as follows.

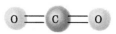

(a) CO_2—a linear molecule

(b) H_2O—a nonlinear or bent molecule

(c) NH_3—a trigonal pyramidal molecule

FIGURE 7.9 Molecular geometries of selected molecules.

$$\overset{\longleftarrow\quad\longrightarrow}{O=C=O}$$

The two individual bond polarities (denoted by arrows), being of equal magnitude (each oxygen affects the carbon atom in the same way) but opposite in direction, cancel.

The nonlinear (angular) triatomic H_2O molecule (Fig. 7.9b) is polar. The bond polarities associated with the two hydrogen–oxygen bonds do not cancel each other because of the nonlinearity of the molecule.

$$\overset{O}{\underset{H\quad H}{\nearrow\ \nwarrow}}$$

As a result of their orientation, both bonds contribute to an accumulation of negative charge on the oxygen atom. The two bond polarities are equal in magnitude but not opposite in their direction.

The generalization that linear triatomic molecules are nonpolar and nonlinear triatomic molecules are polar, a generalization you might be tempted to make on the basis of our discussion of CO_2 and H_2O molecular polarities, is not valid. The linear molecule HCN, which is polar, invalidates this statement. Both bond polarities contribute to nitrogen's acquiring a partial negative charge relative to hydrogen in HCN.

$$\overset{\longrightarrow\quad\longrightarrow}{H-C\equiv N}$$

(Note that the polarity arrows were drawn in the direction they were because nitrogen is more electronegative than carbon and carbon is more electronegative than hydrogen.)

As an analogy to the cancellation or noncancellation of bond polarities in a triatomic molecule, consider the effect that two forces operating on an object have on its position. In Figure 7.10a, the object does not move because both forces are of equal magnitude and are exactly opposite in direction. In Figure 7.10b, the block moves because the two forces are not opposite in direction, even though they are equal in magnitude. In Figure 7.10c, obviously, the block moves because both forces are in the same direction. The first situation is somewhat analogous to the polarity situation in CO_2; the second is analogous to the situation in H_2O; the third is analogous to the situation in HCN. Only when there is "no movement" does nonpolarity result.

Molecules that contain four and five atoms commonly have, respectively, trigonal

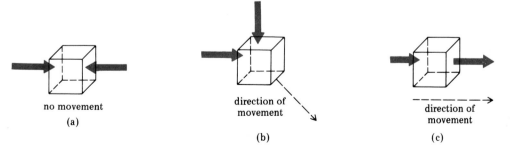

FIGURE 7.10 The effects of forces of various magnitudes and directions on the position of an object.

planar and tetrahedral geometries. The arrangements of atoms associated with these two geometries are

Trigonal planar   Tetrahedral

Trigonal planar and tetrahedral molecules in which all of the atoms attached to the central atom are identical, that is, molecules such as BF_3 (trigonal planar) and CH_4 (tetrahedral), are *nonpolar*. The individual bond polarities cancel as a result of the highly symmetrical arrangement of atoms about the central atom. (Proof that cancellation does occur involves some trigonometric considerations: it is not an obvious situation.)

If two or more kinds of atoms are attached to the central atom in a trigonal planar or tetrahedral molecule, the molecule will be polar. The high symmetry required for cancellation of the individual bond polarities is no longer present. For example, if one of the hydrogen atoms in CH_4 (a nonpolar molecule) is replaced by a chlorine atom, a polar molecule results, even though the resulting CH_3Cl is still a tetrahedral molecule. A C—Cl bond has a greater polarity than a C—H bond; Cl has an electronegativity of 3.0, and H has an electronegativity of only 2.1. Figure 7.11 contrasts the polar CH_3Cl and nonpolar CH_4 molecules. Note that the direction of polarity of the C—Cl bond is opposite to that of the C—H bonds.

EXAMPLE 7.13

Predict the polarity of each of the following molecules.

(a) NH_3 (trigonal pyramidal)

(b) H_2S (angular)

(c) N_2O (linear)

(d) SO_3 (trigonal planar)

Solution

Knowledge of molecular geometry, which is a given quantity for each molecule in this example, is a prerequisite for predicting molecular polarity.

(a) Noncancellation of the individual bond polarities in the trigonal pyramidal NH₃ molecule results in it being a polar molecule.

The bond polarity arrows all point toward the N atom because N is more electronegative than H.

(b) For the angular H₂S molecule, the shift in electron density in the polar sulfur–hydrogen bonds will be toward the S atom because S is more electronegative than H. The H₂S molecule as a whole is polar due to the noncancellation of the individual S—H bond polarities.

(c) The structure of the linear N₂O molecule is unsymmetrical; a nitrogen atom, rather than the O atom, is the central atom.

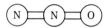

The N—N bond is nonpolar; the N—O bond is polar. The molecule as a whole is polar due to the polarity of the N—O bond.

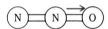

(d) SO₃ is a trigonal planar molecule. Three resonance structures (previously given in Example 7.11) are required to depict the bonding in this molecule. All the S—O bonds, which are equivalent, are polar because S and O differ in electronegativity. However, the molecule as a whole is nonpolar because

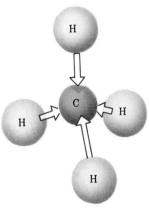
Nonpolar molecule

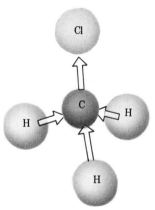
Polar molecule

FIGURE 7.11 CH₄ and CH₃Cl, nonpolar and polar tetrahedral molecules.

of the symmetrical arrangement of the O atoms about the central S atom, which causes the individual bond polarities to cancel.

PRACTICE EXERCISE 7.13

Predict the polarity of each of the following molecules or ions.

(a) SO_2 (angular)
(b) SCN^- (linear)
(c) CCl_4 (tetrahedral)
(d) CCl_3H (tetrahedral)

Ans. (a) polar; (b) polar; (c) nonpolar; (d) polar.

Similar exercises: Problems 7.101 and 7.102

In later chapters we will see that molecular polarity is an important factor in determining the magnitude of some physical properties, for example, the boiling point, as well as some chemical properties for molecular substances.

7.16 Predicting Molecular Geometries (VSEPR Theory)

VSEPR theory is a set of procedures for predicting molecular geometry based on the concept that valence-shell electron pairs about an atom orient themselves so as to minimize the repulsions between them. The acronym VSEPR, pronounced "vesper," stands for valence-shell electron-pair repulsion.

The basic principles behind VSEPR theory are encompassed in the following set of five rules.

RULE 1   Electron pairs in the valence shell of an atom are arranged in space about the nucleus in a manner that minimizes the forces of repulsion between pairs of electrons.

Valence-shell electron pairs about an atom (whether bonding or nonbonding) tend to orient themselves so as to minimize the repulsions between them. Remember that all electrons carry a negative charge and like charges repel each other.

RULE 2   The distribution of valence-shell electron pairs can be predicted from the number of locations in the valence shell of the atom where electrons are found.

A nonchemical analogy involving identical balloons tied together by their ends is helpful in illustrating the concepts germane to rule 2. Two identical balloons tied together at their ends, as shown in Figure 7.12a, naturally adopt an arrangement

(a) (b) (c)

FIGURE 7.12 Balloons tied together at their ends naturally position themselves in particular arrangements: (a) two balloons adopt a linear arrangement, (b) three balloons adopt a trigonal planar arrangement, and (c) four balloons adopt a tetrahedral arrangement.

in which each balloon points away from the other. The word *linear* is used to describe this type of arrangement. Three identical balloons tied together by their ends, as shown in Figure 7.12b, adopt an arrangement in which the balloons orient themselves toward the corners of an equilateral triangle. The phrase *trigonal planar* describes this particular arrangement. Continuing our balloon example one step further, four equivalent balloons tied together at their ends also have a preferred arrangement. This arrangement is not, however, square planar, as some might predict. The preferred arrangement, shown in Figure 7.12c, is that known as *tetrahedral*.

The behavior of electron pairs bound (tied) to a nucleus parallels that of our balloons (Fig. 7.12). The preferred arrangements of the electron pairs match those of the balloons.

Two electron pairs, to be as far apart as possible from each other, will be found on opposite sides of a nucleus, that is, at $180°$ angles to each other. The electron-pair arrangement is linear.

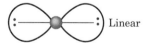 Linear

Three electron pairs are as far apart as possible when they are found at the corners of an equilateral triangle. In such an arrangement, they are separated by angles of $120°$, giving a trigonal planar arrangement of the electron pairs.

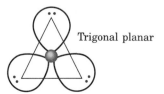 Trigonal planar

A tetrahedral arrangement of electron pairs minimizes repulsions between four sets of electron pairs. A tetrahedron is a four-sided solid, all four sides being identical equivalent triangles. The angle between any two electron pairs is 109°.

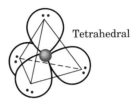
Tetrahedral

The application of rule 2 to actual molecular situations proceeds through electron-dot structures, with the arrangement of electron pairs about the molecule's central atom that predicted by rule 2. The electron-dot structure for the molecule CH_4 is

$$\begin{array}{c} H \\ \cdot\cdot \\ H\!:\!\underset{\cdot\cdot}{C}\!:\!H \\ H \end{array}$$

The central carbon atom has four electron pairs. The arrangement of these electron pairs will be tetrahedral because that is the arrangement that minimizes repulsions between four sets of electron pairs.

RULE 3 Double and triple bonds are considered equivalent to a single pair of bonding electrons when counting the number of locations (places) in the valence shell of an atom where electrons can be found.

The electron-dot structure for CO_2 is

$$:\!\ddot{O}\!::\!C\!::\!\ddot{O}\!:$$

There are four pairs of bonding electrons about the central C atom, but there are only two locations where these electrons can be found. The forces of repulsion between these electrons are minimized when they are located on opposite sides of the carbon atom—a linear arrangement of electron pairs.

The electron-dot structure for the $CO_3{}^{2-}$ ion is

$$\left[\begin{array}{c} :\!\ddot{O}\!:\!C\!::\!\ddot{O}\!: \\ :\!\ddot{O}\!: \end{array} \right]^{2-}$$

As with CO_2, there are four pairs of valence electrons on the carbon atom. This time the electrons are found in three locations. Repulsions between these electrons are minimized when they are arranged toward the corners of an equilateral triangle—a trigonal planar arrangement.

RULE 4 Both bonding and nonbonding electron pairs are counted in determining the number of locations where electrons are found in the valence shell of an atom.

The electron-dot structures for NH_3 and H_2O are, respectively,

$$H:\overset{..}{\underset{H}{N}}:H \quad \text{and} \quad :\overset{..}{\underset{H}{O}}:H$$

Both of these molecules possess electron pairs in four locations about the central atom. Hence, in both cases the arrangement of electron pairs will be tetrahedral. The arrangement of electron pairs is determined by the number of electron locations. Whether the electron pairs are bonding or nonbonding does not affect the basic arrangement.

RULE 5   The molecular geometry (shape) of a molecule is determined by the positions of the atoms within the molecule, which are the same as the positions of the *bonding* electron pairs.

The molecular geometry may or may not be the same as the electron-pair arrangement. If no nonbonding electron pairs are present the two are the same. If nonbonding pairs are present, molecular geometry differs from electron-pair arrangement. Molecular geometry describes the arrangement of atoms, not the arrangement of electron pairs.

The distinction between molecular geometry and electron-pair arrangement can best be illustrated by example. In an H_2O molecule there are four sets of electron pairs about the central O atom—two sets involved in single bonds to hydrogen atoms and two sets present as nonbonding electron pairs. The four pairs of electrons are arranged tetrahedrally.

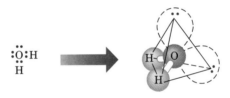

The geometry of the molecule is not, however, tetrahedral. The geometry describes only the arrangement of the atoms. The three atoms are positioned in a nonlinear or angular arrangement, which is the geometry. Note, however, that it is the presence of the nonbonding electron pairs that "forced" the atoms into the nonlinear arrangement. If there were no nonbonding electron pairs present, the atoms would be in a linear arrangement.

Table 7.5 summarizes the relationships among number of electron pairs, number of nonbonding electron pairs, electron-pair arrangement, and molecular geometry. In this table note particularly the three cases where electron-pair arrangement and molecular geometry are different, and the terminology used to describe molecular geometry in these cases.

Examples 7.14 and 7.15, which follow, illustrate the use of the VSEPR theory concepts (rules 1–5) in predicting molecular geometry. Procedurally, the steps followed in applying the VSEPR rules are

1. Draw the electron-dot structure of the molecule or ion.
2. Determine the total number of electron-pair locations around the central atom and arrange them in the way that minimizes electron-pair repulsions.

TABLE 7.5 The Relations Between Electron-Pair Arrangements About a Central Atom and Molecular Geometries

Number of Electron Pairs	Number of Nonbonding Electron Pairs	Number of Bonds	Electron-Pair Arrangement	Molecular Geometry
4	0	4	tetrahedral	tetrahedral
4	1	3	tetrahedral	trigonal pyramidal
4	2	2	tetrahedral	nonlinear or angular
3	0	3	trigonal planar	trigonal planar
3	1	2	trigonal planar	nonlinear or angular
2	0	2	linear	linear

Section 7.16 · Predicting Molecular Geometries (VSEPR Theory) 251

3. Describe the geometry of the molecule or ion in terms of the arrangement of atoms. This atomic arrangement correlates with the positions of the *bonding* electron pairs.

Example 7.14 deals with molecules and polyatomic ions whose central atoms possess no nonbonding valence-shell electron pairs. Molecular geometry and electron-pair arrangement are the same in such situations. Example 7.15 deals with molecules whose central atoms possess nonbonding electron pairs. For these molecules molecular geometry and electron-pair arrangement will always be different.

EXAMPLE 7.14

What is the molecular geometry of each of the following molecules or polyatomic ions?

(a) CCl_4 (b) CO_2 (c) SO_4^{2-}

Solution

(a) The electron-dot structure for CCl_4 is

$$\begin{array}{c} :\ddot{Cl}: \\ :\ddot{Cl}:C:\ddot{Cl}: \\ :\ddot{Cl}: \end{array}$$

The central atom, C, has electron pairs present in four locations; at each location is a pair involved in a single covalent bond. No nonbonding valence shell electron pairs are present on the carbon atom.

The arrangement of the four electron pairs about the central atom is tetrahedral.

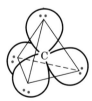

Since each electron pair is a bonding pair, the molecular shape will also be tetrahedral.

(b) The electron-dot structure for CO_2 is

$$:\ddot{O}::C::\ddot{O}:$$

The central atom, C, has electron pairs in two locations; at each location four electrons (a double bond) is present. No nonbonding electron pairs are present on the carbon atom.

A linear geometry is associated with such an electron arrangement.

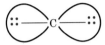

Since all electron locations involve bonding, the molecular shape will also be linear.

(c) The electron-dot structure for SO_4^{2-} is

$$\left[\begin{array}{c} ..\\ :\overset{..}{O}: \\ :\overset{..}{O}:\overset{..}{S}:\overset{..}{O}: \\ :\overset{..}{O}: \end{array} \right]^{2-}$$

The central atom, S, has electron pairs in four locations, all of which are involved in bonding.

The arrangement of four electron pairs about the central atom is tetrahedral.

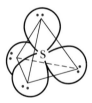

Since all electron locations involve bonding, the molecular geometry will be the same as the electron-pair geometry: tetrahedral.

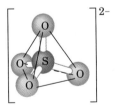

PRACTICE EXERCISE 7.14

What is the molecular geometry of each of the following molecules?

(a) SiH_4 (b) HCN

Ans. (a) tetrahedral; (b) linear.

Similar exercises: Problems 7.109 and 7.110

EXAMPLE 7.15

What is the molecular geometry of each of the following molecules?

(a) PF_3 (b) SO_2

Solution

(a) The electron-dot structure for PF_3 is

$$:\!\overset{..}{\underset{..}{F}}\!:\!\overset{..}{\underset{..}{P}}\!:\!\overset{..}{\underset{..}{F}}\!:$$
$$:\!\overset{..}{\underset{..}{F}}\!:$$

Electrons are present in four locations about the central atom—three electron pairs involved in single bonds and one electron pair that is nonbonding.
 The arrangement of four electron pairs about a central atom is always tetrahedral. It does not matter whether the pairs are bonding or nonbonding. Arrangement depends only on the number of pairs and not on how the pairs function.

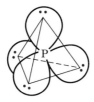

However, the molecular shape will be different from the electron-pair geometry since a nonbonding electron pair is present. The molecular shape is trigonal pyramidal.

Note that the vacant corner of the tetrahedron (the corner that has a nonbonding electron pair instead of an atom) is not considered when coining a word to describe the molecular geometry. Molecular geometry describes the arrangement of *atoms only*.

(b) The electron-dot structure for SO_2 is

$$:\!\overset{..}{\underset{..}{O}}\!:\!\overset{..}{\underset{..}{S}}\!::\!\overset{..}{\underset{..}{O}}\!: \longleftrightarrow :\!\overset{..}{\underset{..}{O}}\!::\!\overset{..}{\underset{..}{S}}\!:\!\overset{..}{\underset{..}{O}}\!:$$

The two resonance structures are equivalent except for the location of the double bond. Because of their equivalency, either one may be used to determine molecular shape; they will both give the same answer. The number of electron locations present about the central S atom is three—a nonbonding pair, two electrons involved in a single bond, and four electrons involved in a double bond.

The arrangement of three electron groupings about a central atom will be trigonal planar.

The molecular shape is angular.

Even though the nonbonding electron pair is ignored when coining a word to describe the molecular shape, its presence is very important. It is the nonbonding pair that causes the molecule to be angular; without it the molecule would be linear. Nonbonding electron pairs are just as important as bonding electron pairs in determining the geometry of the molecule.

PRACTICE EXERCISE 7.15

What is the molecular geometry of each of the following molecules?

(a) OF_2 (b) SeO_2

Ans. (a) angular; (b) angular.

Similar exercises: Problems 7.111 and 7.112

Key Terms

The new terms or concepts defined in this chapter are

bond length (Sec. 7.13) The distance between the nuclei of bonded atoms.

bond polarity (Sec. 7.10) A measure of inequality in the sharing of bonding electrons.

bond strength (Sec. 7.13) A measure of the energy it takes to break a bond.

chemical bond (Sec. 7.1) Attractive forces that hold atoms or ions together in more complex aggregates.

coordinate covalent bond (Sec. 7.12) A covalent bond in which both electrons of a shared pair come from one of the two atoms.

covalent bond (Sec. 7.9) An attractive force between two atoms resulting from the sharing of one or more electron pairs between the two atoms.

double covalent bond (Sec. 7.11) A covalent bond where two pairs of electrons are shared between the same two atoms.

electron-dot structure (Sec. 7.2) An element's symbol with one dot for each valence electron placed about the elemental symbol.

electronegativity (Sec. 7.10) A measure of the relative attraction that an atom has for the shared electrons in a bond.

ion (Sec. 7.4) An atom (or group of atoms) that is electrically charged as a result of the loss or gain of electrons.

ionic bond (Sec. 7.4) An attractive force between positive and negative ions that causes them to remain together as a group.

ionic solid (Sec. 7.7) An array of positive and negative ions arranged in such a way that each ion is surrounded by nearest neighbors of the opposite charge.

isoelectronic species (Sec. 7.5) Species that contain the same number of electrons.

molecular geometry (Sec. 7.15) The way in which atoms in a molecule are arranged in space relative to each other.

monoatomic ion (Sec. 7.8) A single atom that has lost or gained electrons.

multiple covalent bond (Sec. 7.11) A collective term used to designate covalent bonds where two or three pairs of electrons are shared between the same two atoms.

nonpolar covalent bond (Sec. 7.10) A covalent bond in which there is equal sharing of bonding electrons.

octet rule (Sec. 7.3) In forming compounds, atoms of elements lose, gain, or share electrons in such a way as to produce a noble gas electron configuration for each of the atoms involved.

polar covalent bond (Sec. 7.10) A covalent bond in which there is unequal sharing of bonding electrons.

polyatomic ion (Sec. 7.8) A group of atoms, held together by covalent bonds, that has acquired a charge.

resonance structures (Sec. 7.13) Two or more electron-dot structures for a molecule or ion that have the same arrangement of atoms, contain the same number of electrons, and differ only in the location of the electrons.

single covalent bond (Sec. 7.11) A covalent bond where a single pair of electrons is shared between two atoms.

triple covalent bond (Sec. 7.11) A covalent bond where three pairs of electrons are shared between two atoms.

valence electron (Sec. 7.2) An electron in the outermost shell of a representative element.

VSEPR theory (Sec. 7.16) A set of procedures for predicting molecular geometry based on the concept that valence-shell electron pairs about an atom orient themselves so as to minimize the repulsions between them.

Practice Problems

Valence Electrons (Sec. 7.2)

7.1 How many valence electrons do atoms with the following electron configurations have?
(a) $1s^2 2s^2$
(b) $1s^2 2s^2 2p^5$
(c) $1s^2 2s^2 2p^6 3s^2 3p^1$
(d) $1s^2 2s^2 2p^6 3s^2 3p^6 4s^2 3d^{10} 4p^3$

7.2 How many valence electrons do atoms with the following electron configurations have?
(a) $1s^2 2s^2 2p^2$
(b) $1s^2 2s^2 2p^6 3s^2$
(c) $1s^2 2s^2 2p^6 3s^2 3p^4$
(d) $1s^2 2s^2 2p^6 3s^2 3p^6 4s^2 3d^{10} 4p^6 5s^1$

7.3 How many valence electrons do atoms of each of the following elements have?
(a) $_3$Li (b) $_{10}$Ne (c) $_{20}$Ca (d) $_{53}$I

7.4 How many valence electrons do atoms of each of the following elements have?
(a) $_4$Be (b) $_8$O (c) $_{31}$Ga (d) $_{55}$Cs

7.5 For a representative element, what is the relationship between its periodic table group number and the number of valence electrons it contains?

7.6 If Q is the symbol of a representative element that has three valence electrons, in what group is Q in the periodic table?

7.7 Write the complete electron configuration for each of the following elements.
(a) period 2 element with 4 valence electrons
(b) period 2 element with 7 valence electrons
(c) period 3 element with 2 valence electrons
(d) period 3 element with 5 valence electrons

7.8 Write the complete electron configuration for each of the following elements.
(a) period 2 element with 1 valence electron
(b) period 2 element with 3 valence electrons
(c) period 3 element with 3 valence electrons
(d) period 3 element with 6 valence electrons

Electron-Dot Structures for Atoms (Sec. 7.2)

7.9 Draw electron-dot structures for atoms of the following elements.
(a) $_{12}$Mg (b) $_{15}$P (c) $_{19}$K (d) $_{35}$Br

7.10 Draw electron-dot structures for atoms of the following elements.
(a) $_3$Li (b) $_7$N (c) $_{16}$S (d) $_{20}$Ca

7.11 Each of the following electron-dot structures represents a period 2 representative element. Determine the element's identity in each case.
(a) ·Ẋ· (b) ·Ẍ· (c) ·Ẍ: (d) X·

7.12 Each of the following electron-dot structures represents a period 3 representative element. Determine the element's identity in each case.
(a) X· (b) ·Ẍ: (c) ·Ẋ· (d) ·Ẍ·

Octet Rule (Sec. 7.3)

7.13 What is unique about the electron configurations of the noble gases?

7.14 State the octet rule.

Notation for Ions (Sec. 7.4)

7.15 Give the symbol for each of the following ions.
(a) an oxygen atom that has gained two electrons
(b) a magnesium atom that has lost two electrons
(c) a fluorine atom that has gained one electron
(d) an aluminum atom that has lost three electrons

7.16 Give the symbol for each of the following ions.
(a) a chlorine atom that has gained one electron
(b) a sulfur atom that has gained two electrons
(c) a potassium atom that has lost one electron
(d) a beryllium atom that has lost two electrons

7.17 What would be the symbol for an ion with each of the following numbers of protons and electrons?
(a) 20 protons and 18 electrons
(b) 8 protons and 10 electrons
(c) 11 protons and 10 electrons
(d) 13 protons and 10 electrons

7.18 What would be the symbol for an ion with each of the following numbers of protons and electrons?
(a) 15 protons and 18 electrons
(b) 17 protons and 18 electrons
(c) 12 protons and 10 electrons
(d) 19 protons and 18 electrons

7.19 Calculate the number of protons and electrons in each of the following ions.
(a) P^{3-} (b) N^{3-} (c) Mg^{2+} (d) Li$^+$

7.20 Calculate the number of protons and electrons in each of the following ions.
(a) S^{2-} (b) F$^-$ (c) K$^+$ (d) H$^+$

7.21 What is the difference in meaning associated with the notations Li and Li$^+$?

7.22 What is the difference in meaning associated with the notations Cl and Cl$^-$?

Ionic Charge (Sec. 7.5)

7.23 Predict the charge on the ion formed by each of the following elements.
(a) $_{12}$Mg (b) $_9$F (c) $_7$N (d) $_3$Li

7.24 Predict the charge on the ion formed by each of the following elements.
(a) $_{11}$Na (b) $_{35}$Br (c) $_{13}$Al (d) $_{16}$S

7.25 Indicate the number of electrons lost or gained when each of the following atoms forms an ion.
(a) $_4$Be (b) $_{34}$Se (c) $_{37}$Rb (d) $_{38}$Sr

7.26 Indicate the number of electrons lost or gained when each of the following atoms forms an ion.
(a) $_7$N (b) $_{12}$Mg (c) $_{17}$Cl (d) $_{53}$I

7.27 Predict the general kind of behavior, that is, loss or gain of electrons, you would expect from atoms with the following electron configurations.
(a) $1s^2 2s^2$
(b) $1s^2 2s^2 2p^6 3s^2$
(c) $1s^2 2s^2 2p^6 3s^2 3p^1$
(d) $1s^2 2s^2 2p^6 3s^2 3p^5$

7.28 Predict the general kind of behavior, that is, loss or gain of electrons, you would expect from atoms with the following electron configurations.
(a) $1s^2 2s^2 2p^4$
(b) $1s^2 2s^2 2p^3$
(c) $1s^2 2s^1$
(d) $1s^2 2s^2 2p^6 3s^2 3p^6 4s^2 3d^{10} 4p^1$

7.29 Identify the period 3 element that produces each of the following ions.
(a) X$^-$ (b) X$^+$ (c) X^{3+} (d) X^{2-}

7.30 Identify the period 2 element that produces each of the following ions.
(a) X^{2-} (b) X^{2+} (c) X^{3-} (d) X^{3+}

7.31 With which noble gas is each of the following ions isoelectronic?
(a) Mg^{2+} (b) Ca^{2+} (c) Cl$^-$ (d) F$^-$

7.32 With which noble gas is each of the following ions isoelectronic?
(a) K$^+$ (b) Na$^+$ (c) O^{2-} (d) N^{3-}

7.33 For each of the following sets of ions, indicate those ions that are isoelectronic with each other.

(a) Al^{3+}, F^-, Cl^-, Li^+ (b) I^-, Ba^{2+}, Se^{2-}, Cs^+
(c) S^{2-}, P^{3-}, Rb^+, Sr^{2+} (d) Ga^{3+}, Cl^-, P^{3-}, K^+

7.34 For each of the following sets of ions, indicate those ions that are isoelectronic with each other.
(a) As^{3-}, S^{2-}, O^{2-}, Rb^+ (b) Al^{3+}, Mg^{2+}, F^-, N^{3-}
(c) Be^{2+}, S^{2-}, K^+, Li^+ (d) Se^{2-}, Sr^{2+}, Br^-, Rb^+

7.35 The ion Mg^{3+} is unknown. Using the octet rule, explain why.

7.36 The ion Cl^{2-} is unknown. Using the octet rule, explain why.

Ionic Compound Formation and Electron-Dot Structures (Sec. 7.5)

7.37 Show the formation of the following ionic compounds using electron-dot structures.
(a) Na_2O (b) Li_3N (c) CaS (d) $MgCl_2$

7.38 Show the formation of the following ionic compounds using electron-dot structures.
(a) BeF_2 (b) K_3P (c) Al_2O_3 (d) KBr

7.39 Using electron-dot structures, show how ionic compounds are formed by atoms of
(a) Li and Cl   (b) Mg and S
(c) Ga and N   (d) K and O

7.40 Using electron-dot structures, show how ionic compounds are formed by atoms of
(a) Na and F   (b) Be and O
(c) Al and Cl   (d) Ca and N

Formulas for Ionic Compounds (Sec. 7.6)

7.41 Write the formula for an ionic compound formed from Ca^{2+} ion and each of the following ions.
(a) Cl^- (b) O^{2-} (c) P^{3-} (d) S^{2-}

7.42 Write the formula for an ionic compound formed from Mg^{2+} ion and each of the following ions.
(a) F^- (b) S^{2-} (c) Br^- (d) N^{3-}

7.43 Write the formula for an ionic compound formed from S^{2-} ion and each of the following ions.
(a) K^+ (b) Li^+ (c) Al^{3+} (d) Be^{2+}

7.44 Write the formula for an ionic compound formed from N^{3-} ion and each of the following ions.
(a) Na^+ (b) Be^{2+} (c) Al^{3+} (d) Rb^+

7.45 Write the formula of the ionic compound that could form from the elements X and Y if
(a) X has 2 valence electrons and Y has 4 valence electrons
(b) X has 3 valence electrons and Y has 7 valence electrons
(c) X has 1 valence electron and Y has 5 valence electrons
(d) X has 7 valence electrons and Y has 2 valence electrons

7.46 Write the formula of the ionic compound that could form from the elements X and Y if
(a) X has 2 valence electrons and Y has 7 valence electrons
(b) X has 1 valence electron and Y has 6 valence electrons
(c) X has 3 valence electrons and Y has 5 valence electrons
(d) X has 6 valence electrons and Y has 2 valence electrons

7.47 Identify the ions present (by symbol and charge) in each of the following ionic compounds.
(a) Al_2O_3 (b) CaS (c) $LiBr$ (d) Na_3N

7.48 Identify the ions present (by symbol and charge) in each of the following ionic compounds.
(a) Ca_3N_2 (b) Al_2S_3 (c) BeO (d) K_3P

Structure of Ionic Compounds (Sec. 7.7)

7.49 What does the formula for an ionic compound actually represent?

7.50 Explain why it is inappropriate to talk about molecules of an ionic compound.

Polyatomic Ions (Sec. 7.8)

7.51 Determine the formula for the ionic compound formed from each of the following pairs of ions.
(a) Na^+ and OH^- (b) NH_4^+ and NO_3^-
(c) Ca^{2+} and SO_4^{2-} (d) Mg^{2+} and CN^-

7.52 Determine the formula for the ionic compound formed from each of the following pairs of ions.
(a) Na^+ and CN^- (b) NH_4^+ and SO_4^{2-}
(c) Ca^{2+} and OH^- (d) K^+ and NO_3^-

7.53 Determine the formula for the ionic compound in which the Al^{3+} ion is combined with each of the following polyatomic ions.
(a) PO_4^{3-} (b) CO_3^{2-}
(c) ClO_3^- (d) $C_2H_3O_2^-$

7.54 Determine the formula for the ionic compound in which the Mg^{2+} ion is combined with each of the following polyatomic ions.
(a) PO_4^{3-} (b) CO_3^{2-}
(c) ClO_3^- (d) $C_2H_3O_2^-$

Covalent Bond Formation and Electron-Dot Structures (Sec. 7.9)

7.55 Write an electron-dot structure to illustrate the covalent bonding found in each of the following molecules.
(a) Br_2 (b) PH_3 (c) NCl_3 (d) CF_4

7.56 Write an electron-dot structure to illustrate the covalent bonding found in each of the following molecules.
(a) H_2Se (b) SCl_2 (c) F_2 (d) PCl_3

7.57 Write an electron-dot structure for the simplest compound most likely to be formed between each of these pairs of elements.

(a) Si and F (b) H and Br
(c) H and Te (d) Cl and F

7.58 Write an electron-dot structure for the simplest compound most likely to be formed between each of these pairs of elements.
(a) Si and H (b) H and F
(c) S and F (d) O and Cl

7.59 Write an electron-dot structure to illustrate the covalent bonding found in each of the following molecules. All the bonds within the molecules are single bonds.
(a) H_3CCH_3 (C_2H_6) (b) H_2NNH_2 (N_2H_4)
(c) F_2CH_2 (CH_2F_2) (d) H_3CCCl_3 ($C_2H_3Cl_3$)

The above formulas are written in a form that specifies atomic arrangement. When an elemental symbol carries a subscript, these atoms are directly and separately bonded to the atom immediately following or immediately preceding the subscripted symbol.

7.60 Write an electron-dot structure to illustrate the covalent bonding found in each of the following molecules. All the bonds within the molecules are single bonds.
(a) H_2PPH_2 (P_2H_4) (b) H_3CCBr_3 ($C_2H_3Br_3$)
(c) F_2CCl_2 (CF_2Cl_2) (d) H_3SiSiH_3 (Si_2H_6)

The formulas are written in a way that specifies atomic arrangement, as explained in Problem 7.59.

Electronegativities and Bond Polarities (Sec. 7.10)

7.61 Using chemical periodicity trends, select the more electronegative element in each of the following pairs of elements.
(a) H and O (b) C and O
(c) P and S (d) Na and Mg

7.62 Using chemical periodicity trends, select the more electronegative element in each of the following pairs of elements.
(a) N and As (b) Na and Al
(c) H and C (d) Cl and Br

7.63 Using chemical periodicity trends, arrange each of the following sets of atoms in order of increasing electronegativity.
(a) Na, Al, P, Mg (b) Cl, Br, I, F
(c) S, P, O, As (d) Ca, Ge, O, C

7.64 Using chemical periodicity trends, arrange each of the following sets of atoms in order of increasing electronegativity.
(a) Be, N, C, B (b) Te, S, O, Se
(c) B, Na, Al, Mg (d) F, Mg, Al, N

7.65 For each of the following bonds, indicate which end of the bond should carry the $\delta+$ designation.
(a) O—F (b) O—N (c) H—O (d) Br—F

7.66 For each of the following bonds, indicate which end of the bond should carry the $\delta+$ designation.
(a) N—C (b) O—Se (c) B—C (d) H—C

7.67 Arrange each of the following sets of bonds in order of increasing polarity.
(a) H—Cl, H—O, H—Br
(b) O—F, O—P, O—Al
(c) H—Cl, Br—Br, B—N
(d) Al—Cl, C—N, Cl—F

7.68 Arrange each of the following sets of bonds in order of increasing polarity.
(a) H—C, H—N, H—B
(b) O—N, O—S, O—Br
(c) P—N, S—O, Br—F
(d) H—F, Cl—Cl, Si—O

7.69 Classify each of the following bonds as nonpolar covalent, polar covalent, or ionic.
(a) carbon–nitrogen (b) beryllium–oxygen
(c) phosphorus–chlorine (d) silicon–silicon

7.70 Classify each of the following bonds as nonpolar covalent, polar covalent, or ionic.
(a) cesium–fluorine (b) potassium–chlorine
(c) hydrogen–hydrogen (d) silicon–phosphorus

7.71 Using information found in Table 7.4, determine the percent ionic character for each of the following bonds.
(a) N—F (b) N—C (c) N—K (d) N—Si

7.72 Using information found in Table 7.4, determine the percent covalent character for each of the following bonds.
(a) S—N (b) S—F (c) S—Na (d) S—Al

Multiple Covalent Bonds (Sec. 7.11)

7.73 Write electron-dot structures for the following molecules, each of which contains at least one double bond. The formulas are written in a way that specifies atomic arrangement, as explained in Problem 7.59.
(a) H_2CCCH_2 (C_3H_4) (b) FNNF (N_2F_2)
(c) H_2CO (d) H_2CCBr_2 ($C_2H_2Br_2$)

7.74 Write electron-dot structures for the following molecules, each of which contains at least one double bond. The formulas are written in a way that specifies atomic arrangement, as explained in Problem 7.59.
(a) Cl_2CCCl_2 (C_2Cl_4) (b) Cl_2CO
(c) H_3CONO (CH_3O_2N) (d) H_2CCO (C_2H_2O)

7.75 Write electron-dot structures for the following molecules, each of which contains at least one triple bond. The formulas are written in a way that specifies atomic arrangement, as explained in Problem 7.59.

(a) H₃CCN (C₂H₃N) (b) HCCCH₃ (C₃H₄)
(c) NCCN (C₂N₂) (d) H₂NCN (CH₂N₂)

7.76 Write electron-dot structures for the following molecules, each of which contains at least one triple bond. The formulas are written in a way that specifies atomic arrangement, as explained in Problem 7.59.
(a) ClCCCl (C₂Cl₂)
(b) HCCCl (HC₂Cl)
(c) HCCCCH (C₄H₂)
(d) (H₂N)₂CNCN (C₂N₄H₄)

Coordinate Covalent Bonds (Sec. 7.12)

7.77 What is a coordinate covalent bond?
7.78 Once formed, how (if at all) does a coordinate covalent bond differ from an ordinary covalent bond?
7.79 Write electron-dot structures for the following molecules, each of which contains at least one coordinate covalent bond. The formulas are written in a way that specifies atomic arrangement, as explained in Problem 7.59.
(a) CO (b) HClO
(c) OP(OH)₃ (H₃PO₄) (d) O₂BrOH (HBrO₃)
7.80 Write electron-dot structures for the following molecules, each of which contains at least one coordinate covalent bond. The formulas are written in a way that specifies atomic arrangement, as explained in Problem 7.59.
(a) SSO (S₂O) (b) Cl₂SO (SOCl₂)
(c) O₂Se(OH)₂ (H₂SeO₄) (d) OClOH (HClO₂)

Resonance Structures (Sec. 7.13)

7.81 What are resonance structures?
7.82 Explain why H—C≡N: and H—N≡C: are not resonance structures.
7.83 Draw electron-dot structures for the following molecules. At least two resonance structures will be needed in each case. The formulas are written in a way that specifies atomic arrangement, as explained in Problem 7.59.
(a) OOO (O₃) (b) FNO₂
7.84 Write electron-dot structures for the following molecules. At least two resonance structures will be needed in each case. The formulas are written in a way that specifies atomic arrangement, as explained in Problem 7.59.
(a) F₂NNO (b) H₂CNN

A Systematic Method For Determining Electron-Dot Structures (Sec. 7.14)

7.85 Which of each pair of arrangements of atoms is more likely?

(a) O—C—O—O or O—C—O
                        |
                        O

(b) O—P—O or O—O—P
    (with =O above and =O below)

7.86 Which of each pair of arrangements of atoms is more likely?
(a) H—H—O—O or H—O—O—H

(b) O—S—O—H or O—S—O—O
    | |
    O   H
    |
    H

7.87 How many electron dots should appear in the electron-dot structures for each of the following molecules or ions?
(a) HNO₃ (b) PF₃ (c) BF₄⁺ (d) PO₄³⁻
7.88 How many electron dots should appear in the electron-dot structures for each of the following molecules or ions?
(a) O₂F₂ (b) C₂H₂Br₂ (c) S₂²⁻ (d) NH₄⁺

7.89 Write electron-dot structures for the following molecules. Some formulas are written in a way that specifies atomic arrangements, as explained in Problem 7.59.
(a) SiCl₄ (b) PF₃ (c) HOOH (d) H₃COH
7.90 Write electron-dot structures for the following molecules. Some formulas are written in a way that specifies atomic arrangements, as explained in Problem 7.59.
(a) AsH₃ (b) GeBr₄ (c) Cl₃PO (d) O₂IOH

7.91 Write electron-dot structures for the following polyatomic ions.
(a) NF₄⁺ (b) BeH₄²⁻ (c) ClO₃⁻ (d) IO₄⁻
7.92 Write electron-dot structures for the following polyatomic ions.
(a) BH₄⁻ (b) AlCl₄⁻ (c) PF₄⁺ (d) ClO₂⁻

7.93 Write the resonance structure for the following molecules or polyatomic ions. Some formulas are written in a way that specifies atomic arrangements, as explained in Problem 7.59.
(a) H₃CNO₂ (b) NNO (c) CO₃²⁻ (d) SCN⁻
7.94 Write the resonance structures for the following molecules or polyatomic ions. Some formulas are written in a way that specifies atomic arrangements, as explained in Problem 7.59.
(a) HONO (b) H₂NNO₂ (c) NO₃⁻ (d) OCN⁻

7.95 How many bonding and nonbonding electron pairs are present in each of the following diatomic species?
(a) HCl (b) CO (c) OH⁻ (d) CN⁻

7.96 How many bonding and nonbonding electron pairs are present in each of the following diatomic species?
(a) BrCl (b) N_2 (c) HS⁻ (d) NS⁻

7.97 The electron-dot structure for the polyatomic ion $ClO_4{}^{n-}$ is

$$\left[\begin{array}{c} :\!\ddot{\text{O}}\!: \\ | \\ :\!\ddot{\text{O}}\!-\!\text{Cl}\!-\!\ddot{\text{O}}\!: \\ | \\ :\!\ddot{\text{O}}\!: \end{array}\right]^{n-}$$

What is the value of n, the magnitude of the ionic charge, in the formula $ClO_4{}^{n-}$?

7.98 The electron-dot structure for the polyatomic ion $BeCl_4{}^{n-}$ is

$$\left[\begin{array}{c} :\!\ddot{\text{Cl}}\!: \\ | \\ :\!\ddot{\text{Cl}}\!-\!\text{Be}\!-\!\ddot{\text{Cl}}\!: \\ | \\ :\!\ddot{\text{Cl}}\!: \end{array}\right]^{n-}$$

What is the value of n, the magnitude of the ionic charge, in the formula $BeCl_4{}^{n-}$?

Molecular Polarity (Sec. 7.15)

7.99 Is it possible for a nonpolar molecule to contain polar bonds? Explain.

7.100 Is it possible for a polar molecule to contain nonpolar bonds? Explain.

7.101 Determine the polarity of each of the following molecules.
(a) NCl_3 (trigonal pyramid)
(b) H_2S (angular)
(c) H_3CCl (tetrahedral)
(d) CS_2 (linear)

7.102 Determine the polarity of each of the following molecules.
(a) BF_3 (trigonal planar)
(b) $OSCl_2$ (trigonal pyramid)
(c) NO_2 (angular)
(d) $SiCl_4$ (tetrahedral)

7.103 For each of the following hypothetical triatomic molecules, indicate whether the *bonds* are polar or nonpolar and whether the *molecule* is polar or nonpolar. Assume that A and X have different electronegativities.
(a) X—A—X   (b) A—X—X
(c)   (d)

7.104 For each of the following hypothetical square planar molecules, indicate whether the *bonds* are polar or nonpolar and whether the *molecule* is polar or nonpolar. Assume that A, X, and Y have different electronegativities.

Molecular Geometry (VSEPR Theory) (Sec. 7.16)

7.105 Using VSEPR theory predict the geometry of the following triatomic molecules or ions.
(a) H:S:H   (b) H:C:::N:
(c) [:Ö::N::O:]⁺   (d) :Ö:Ö::Ö:

7.106 Using VSEPR theory predict the geometry of the following triatomic molecules or ions.
(a) :N::S:F: (b) :N::N::Ö:
(c) [H:N:H]⁻   (d) [:S:C:::N:]⁻

7.107 Using VSEPR theory predict the geometry of the following sulfur-containing molecules or ions.

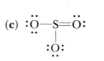

(c) :Ö—S=Ö: (d) [structure with central S, O's, S]²⁻
     |
    :Ö:

7.108 Using VSEPR theory predict the geometry of the following nitrogen-containing molecules or ions.

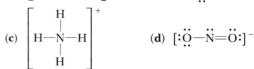

7.109 Using VSEPR theory predict the geometry of the following molecules or polyatomic ions. Some formulas are written in a way that specifies atomic arrangements, as explained in Problem 7.59.
(a) $SiCl_4$ (b) H_2CO (c) $AlCl_4{}^-$ (d) $PO_4{}^{3-}$

7.110 Using VSEPR theory predict the geometry of the following molecules or polyatomic ions. Some formulas are written in a way that specifies atomic arrangements, as explained in Problem 7.59.
(a) CS_2 (b) H_3CBr (c) $BeCl_4{}^{2-}$ (d) $PCl_4{}^+$

7.111 Using VSEPR theory predict the geometry of the following molecules or polyatomic ions.
(a) OF_2 (b) NCl_3 (c) SO_3^{2-} (d) NO_2^-

7.112 Using VSEPR theory predict the geometry of the following molecules or polyatomic ions.
(a) Cl_2O (b) $OSCl_2$ (c) H_3O^+ (d) PO_3^{3-}

7.113 What is the molecular geometry that VSEPR theory predicts for a molecule whose central atom has the environment of
(a) 2 single bonds and 2 nonbonding electron pairs
(b) 1 single bond and 1 triple bond
(c) 1 single bond, 1 double bond, and 1 nonbonding electron pair
(d) 1 double bond and 2 nonbonding electron pairs

7.114 What is the molecular geometry that VSEPR theory predicts for a molecule whose central atom has the environment of
(a) 4 single bonds
(b) 2 double bonds
(c) 3 single bonds and 1 nonbonding electron pair
(d) 1 triple bond and 1 nonbonding electron pair

Additional Problems

7.115 Write the electrons configuration for each of the following ions.
(a) F^- (b) Na^+ (c) S^{2-} (d) Ca^{2+}

7.116 Write the electron configuration for each of the following ions.
(a) O^{2-} (b) N^{3-} (c) Al^{3+} (d) K^+

7.117 Write the electron configuration for only the valence shell of each ion.
(a) F^- (b) Na^+ (c) S^{2-} (d) Ca^{2+}

7.118 Write the electron configuration for only the valence shell of each ion.
(a) O^{2-} (b) N^{3-} (c) Al^{3+} (d) K^+

7.119 Which of the following chemical formulas do not represent stable ionic compounds? Explain your answer.
(a) $AlCl_2$ (b) Ca_2O (c) Na_3N (d) KS

7.120 Which of the following chemical formulas do not represent stable ionic compounds? Explain your answer.
(a) KCl_2 (b) Na_2O (c) BeN (d) CaS

7.121 Write the correct formula for the ionic compound formed by combining
(a) lithium and bromine
(b) beryllium and phosphorus
(c) magnesium and oxygen
(d) calcium and sulfur

7.122 Write the correct formula for the ionic compound formed by combining
(a) sodium and fluorine
(b) potassium and nitrogen
(c) aluminum and oxygen
(d) barium and bromine

7.123 Identify the period 2 element that X represents in each of the following ionic compound formulas.
(a) Li_3X (b) BeX (c) KX (d) AlX_3

7.124 Identify the period 3 element that X represents in each of the following ionic compound formulas.
(a) CaX (b) AlX (c) NaX (d) Mg_3X_2

7.125 The formula P_2Br_4 is correct, but the formula Ca_2Br_4 is incorrect. Explain.

7.126 The formula C_2F_4 is correct, but the formula Mg_2F_4 is incorrect. Explain.

7.127 In each of the following electron-dot structures, X represents a period 3 nonmetal. Identify the nonmetal in each case.
(a) H—Ö—X—Ö: with :Ö: above
(b) [:Ö—X: with :Ö: below]

7.128 In each of the following electron-dot structures, X represents a period 3 nonmetal. Identify the nonmetal in each case.
(a) :Cl—X—Cl: with Ö above (double bond)
(b) [:Ö—X—Ö: with :Ö: above and :Ö: below]$^{2-}$

7.129 What is wrong with each of the following electron-dot structures?
(a) :Br::Cl:
(b) [:H::Ö:]$^-$

7.130 What is wrong with each of the following electron-dot structures?
(a) H:Ö:::Ö:H
(b) [O:N::Ö: with O below]$^-$

7.131 Classify the bonding in each of the following compounds as ionic or covalent.
(a) HN_3 (b) K_3N (c) CCl_4 (d) SF_2

7.132 Classify the bonding in each of the following compounds as ionic or covalent.
(a) NH_3 (b) Na_2O (c) SiH_4 (d) C_2H_4

7.133 Using electron-dot structures, discuss the bonding in the following compounds.
(a) Cs_2SO_4 (b) NH_4NO_3

7.134 Using electron-dot structures, discuss the bonding in the following compounds.
(a) Ca(OH)$_2$ (b) NH$_4$CN

7.135 Four hypothetical elements, A, B, C, and D, have electronegativities A = 3.8, B = 3.3, C = 2.8, and D = 1.3. These elements form the compounds AB, AD, BD, and AC. Arrange these compounds in order of increasing *ionic* bond character.

7.136 Four hypothetical elements, A, B, C, and D, have electronegativities A = 3.6, B = 3.0, C = 2.7 and D = 0.9. These elements form the compounds AC, BC, AD, and CD. Arrange these compounds in order of increasing *covalent* bond character.

7.137 In which of the following pairs of diatomic species do both members of the pair have bonds of the same multiplicity (single, double, triple)?
(a) BrCl and ClF   (b) F$_2$ and N$_2$
(c) NO$^+$ and NO$^-$ (d) CN$^-$ and SN$^-$

7.138 In which of the following pairs of diatomic species do both members of the pair have bonds of the same multiplicity (single, double, triple)?
(a) HCl and HF   (b) S$_2$ and Cl$_2$
(c) CO and NO$^+$ (d) OH$^-$ and HS$^-$

7.139 Specify the electron-pair geometry about the central atom and the molecular geometry for each of the following species.
(a) CF$_4$ (b) NH$_3$ (c) SCN$^-$ (d) PH$_4^+$

7.140 Specify the electron-pair geometry about the central atom and the molecular geometry for each of the following species.
(a) SiH$_4$ (b) NH$_4^+$ (c) ClNO   (d) NO$_3^-$

7.141 Give an approximate value for the indicated bond angle in each of the following molecules.
(a) H—Ö—H   (b) :Cl—C=C—Cl:
                      | |
                      H H

7.142 Give an approximate value for the indicated bond angle in each of the following molecules.
(a) H—N—H   (b) H—C—Ö—H
      | |
      H   H

7.143 Indicate whether each of the following molecules is polar or nonpolar.
(a) PH$_3$
(b) NH$_4^+$
(c) SCN$^-$
(d) N$_2$O (NNO)

7.144 Indicate whether each of the following molecules is polar or nonpolar.
(a) NI$_3$ (b) NH$_2^-$ (c) SO$_2$ (d) SiO$_4^{4-}$

Cumulative Problems

7.145 Given the following information about two elements, A and D, determine their identities and the formula of the binary ionic compound that they form.
(1) A is located in period 3 of the periodic table.
(2) The occupied electron subshell of highest energy for element D contains 2 electrons.
(3) The complete electron configuration for element D contains four fewer electrons than that for element A.

7.146 Given the following information about two elements, A and D, determine their identities and the formula of the binary ionic compound that they form.
(1) Both A and D are in period 3 of the periodic table.
(2) The occupied electron subshell of highest energy for element A contains one electron, and the one for element D contains four electrons.
(3) The atom of the element that is a nonmetal contains three more total electrons than the atom of the element that is a metal.

7.147 Determine the identities of the elements A and D in the ionic compound AD given that
(1) The atomic number of element A is greater than that of element D.
(2) Three electrons are transferred from A to D during compound formation.
(3) The sum of the atomic numbers of A and D is 20.

7.148 Determine the identities of the elements A and D in the ionic compound AD given that
(1) The atomic number of element A is less than that of element D.
(2) Two electrons are transferred from A to D during compound formation.
(3) The sum of the atomic numbers of A and D is 20.

7.149 Determine the electron-dot structure for the covalent compound formed between the elements A and D given that

(1) Both elements A and D are located in the $2p$ area of the periodic table.
(2) Element A has an electronegativity greater than 3.2.
(3) Element D's periodic table position is next to that of a noble gas.

7.150 Determine the electron-dot structure for the covalent compound formed between the elements A and D given that
(1) Both elements A and D are located in the $3p$ area of the periodic table.
(2) Element A has an electronegativity greater than 2.4, and element D has an electronegativity greater than 2.0.
(3) Element A's periodic table position is two positions beyond that of element D.

7.151 A covalent compound containing the elements A and D has the formula A_xD_y. Determine the identities of the elements A and D and the values for x and y in the formula for the compound given the following information about A and D.
(1) A and D are representative elements.
(2) D atoms have electron configurations ending in p^4.
(3) The isotope of D with a mass number of 16 has a percentage abundance greater than 95%.
(4) The period number for element A is one less than that for element D.

7.152 A covalent compound containing the elements A and D has the formula A_xD_y. Determine the identities of the elements A and D and the values for x and y in the formula for the compound given the following information about A and D.
(1) A and D are representative elements in the same period of the periodic table.
(2) A atoms have electron configurations ending in p^2.
(3) The isotope of A with a mass number of 12 has a percentage abundance greater than 95%.
(4) The group number for element A is three less than the group number for element D.

7.153 A polyatomic ion containing the elements A and D has the formula AD_4^{2-}. Determine the identities of the elements A and D, and then draw the electron-dot structure for the polyatomic ion given that
(1) A, the central atom in the polyatomic ion, possesses only s electrons, and exactly one-half of these s electrons are valence electrons.
(2) The D atoms present collectively possess 28 valence electrons.
(3) The atomic number of D is less than 15.

7.154 A polyatomic ion containing the elements A and D has the formula AD_4^+. Determine the identities of the elements A and D, and then draw the electron-dot structure for the polyatomic ion given that
(1) A, the central atom, possesses nine p electrons.
(2) For D atoms, period number in the periodic table and number of valence electrons are the same.
(3) For D atoms, the total number of electrons present is the same as the number of valence electrons present.

Grid Problems

7.155 Select from the grid *all* correct responses for each situation.

1. Na$_2$S	2. NaCl	3. LiCl
4. K$_3$N	5. K$_2$O	6. Ca$_3$P$_2$
7. BeS	8. Al$_2$O$_3$	9. AlF$_3$

(a) compounds in which $+1$ ions are present
(b) compounds in which -2 ions are present
(c) compounds in which either the positive ion or the negative ion (but not both) is isoelectronic with Ar
(d) compounds in which both types of ions present have the same electron configuration
(e) compounds in which the ionic character of the bonds is at least 60%
(f) compounds in which the number of electrons transferred per formula unit is two.

7.156 Select from the grid *all* correct responses for each situation.

1. S_2	2. ClF	3. HI
4. CO	5. HCl	6. N_2
7. I_2	8. F_2	9. BrF

(a) molecules in which a total of 14 valence electrons are present
(b) molecules in which a double bond is present
(c) molecules in which a triple bond is present
(d) molecules in which more bonding than nonbonding valence electrons are present
(e) molecules in which a coordinate covalent bond is present
(f) molecules in which the bond is both multiple and polar

7.157 Select from the grid *all* correct responses for each situation.

1. K	2. Mg	3. Cl
4. Na	5. Ca	6. S
7. Al	8. F	9. N

(a) pairs of elements that combine to produce a compound in which there is a 3 to 1 ratio between atoms or ions
(b) pairs of elements that combine to produce diatomic *molecules*
(c) pairs of elements that combine to produce a compound in which ions with a charge of 2 (plus or minus) are present
(d) pairs of elements that combine through an electron-sharing process
(e) pairs of elements that combine to produce molecules that have a trigonal pyramidal geometry
(f) pairs of elements that combine to produce a compound that has five ions per formula unit

7.158 Select from the grid *all* correct responses for each situation.

1. H_2S	2. N_3^-	3. HOCl
4. SO_2	5. HCN	6. O_3
7. CO_2	8. SF_2	9. HCO^-

(a) molecules that possess an equal number of valence bonding and nonbonding electrons
(b) molecules in which no multiple bonds are present
(c) molecules that have a linear geometry
(d) molecules that are nonpolar
(e) molecules for which the electron-pair arrangement about the central atom is tetrahedral
(f) pairs of molecules in which the total number of valence electrons is identical

CHAPTER EIGHT

Chemical Nomenclature

8.1 Nomenclature Classifications for Compounds

Just as it is important to be able to write formulas for compounds, it is necessary to be able to name those compounds.

Chemical nomenclature is the system of names used to distinguish compounds from each other and the rules needed to devise these names. In the early history of chemistry there was no system for naming compounds. Early names included quicksilver, blue vitriol, Glauber's salt, gypsum, sal ammoniac, and laughing gas. As chemistry grew, it became clear that the "anything goes" system was not acceptable. Without a system for naming compounds, coping with the multitude of known substances would be a hopeless task.

In this chapter we consider the rules now followed for naming compounds. The rules presented here follow the recommendations of the nomenclature committees of the International Union of Pure and Applied Chemistry (IUPAC). **IUPAC rules** are a set of compound-naming rules produced by committees of the International Union of Pure and Applied Chemistry (IUPAC). The committees of this international scientific organization meet periodically to revise and update the nomenclature rules to accommodate any newly discovered types of compounds.

For nomenclatural purposes it is convenient to classify compounds into categories based on the number of elements present in a compound. A **binary compound** is a compound containing just two different elements. The compounds NH_3, H_2O, CO_2, $NaCl$, and P_4O_{10} are all binary compounds. Any number of atoms of the two elements may be present in a molecule or formula unit of a binary compound, but only two elements may be present. A **ternary compound** is a compound containing three different elements. Many of the compounds that students encounter in chemistry laboratory situations are ternary compounds. Examples include nitric acid (HNO_3), sulfuric acid (H_2SO_4), and sodium hydroxide (NaOH). Compounds containing more than three kinds of elements do exist, and we will encounter a few, but will not worry about specific classification terms for such compounds.

Distinguishing between ionic and molecular compounds (Secs. 5.2 and 7.1) is an important key to becoming successful at chemical nomenclature. Although there is no sharp dividing line between ionic and covalent bonds (Sec. 7.10) and thus between ionic and molecular compounds, we will "create" such a line for nomenclatural purposes with the following simple generalizations.

1. Compounds resulting from the combination of a metal and one or more nonmetals are considered *ionic*.
2. Compounds resulting from combinations of nonmetals are considered *molecular*.

We will also treat compounds containing the positive polyatomic NH_4^+ ion (the ammonium ion) as ionic. In such ionic compounds, the NH_4^+ ion (a combination of two nonmetals) is considered to have replaced the metallic ion ordinarily present and to be functioning as a metal.

Metalloid elements (Sec. 6.12) are considered to be nonmetals for purposes of nomenclature. Thus, a compound resulting from the combination of a metalloid and nonmetals is named as a molecular compound.

8.2 Types of Binary Ionic Compounds

Binary ionic compounds are the simplest type of ionic compound; only monoatomic ions (Sec. 7.8) are present. A **binary ionic compound** is an ionic compound in which monoatomic positively charged metallic ions and monoatomic negatively charged nonmetallic ions are present. Binary ionic compounds may be divided into two categories.

1. Binary ionic compounds containing a *fixed-charge* metal.
2. Binary ionic compounds containing a *variable-charge* metal.

In general, all metals lose electrons when forming ions. Fixed-charge metals always exhibit the same behavior in ion formation; that is, they always lose the same number of electrons. A **fixed-charge metal** forms only one type of ion, which always has the

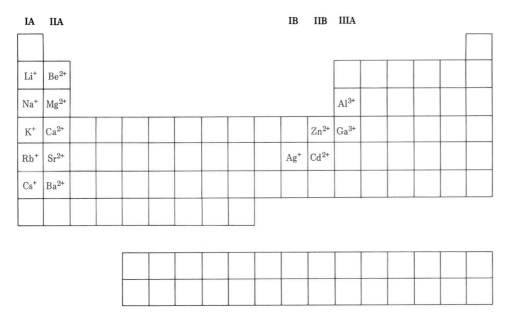

FIGURE 8.1 Periodic table showing the fixed-charge metallic ions. Note the correlation between ionic charge magnitude and periodic-table group number.

same charge magnitude. In demonstrating how to write formulas for ionic compounds (Secs. 7.5 and 7.6), all of the examples involved fixed-charge metals.

Variable-charge metals do not always lose the same number of electrons upon ion formation. A **variable-charge metal** forms more than one type of ion, with the ion types differing in charge magnitude. For example, the variable-charge metal iron sometimes forms an Fe^{2+} ion and sometimes an Fe^{3+} ion.

Which metals are fixed-charge metals? Which metals are variable-charge metals? This information is a prerequisite for using the rules governing binary ionic compound nomenclature. Only a few fixed-charge metals exist. Thus, we learn which metals these are, and then all other metals will be variable-charge metals.

The *fixed-charge metals* are the group IA and IIA metals plus Al, Ga, Zn, Cd, and Ag. Figure 8.1 shows the periodic table positions for these fifteen fixed-charge metals. The charge magnitude for these metals can be related to their position in the periodic table. Looking at Figure 8.1, we see that all group IA elements form +1 ions. The charges for ions of elements in groups IIA and IIIA are +2 and +3, respectively. Group numbers and charge also directly correlate for Zn, Cd, and Ag, the other fixed-charge metallic ions. The reason for this charge–periodic-table correlation, the octet rule, was considered in Section 7.3.

The vast majority of metals, all except the fifteen fixed-charge metals, are variable-charge metals. Charge magnitude for ions of such metals cannot be easily related to periodic-table position. (The presence of *d* or *f* electrons in most of these metals complicates octet rule considerations in a manner beyond what we consider in this

TABLE 8.1 Common Variable-Charge Metallic Element Ions and Their Charges

Element	Ions Formed
Chromium	Cr^{2+} and Cr^{3+}
Cobalt	Co^{2+} and Co^{3+}
Copper	Cu^{+} and Cu^{2+}
Gold	Au^{+} and Au^{3+}
Iron	Fe^{2+} and Fe^{3+}
Lead	Pb^{2+} and Pb^{4+}
Manganese	Mn^{2+} and Mn^{3+}
Tin	Sn^{2+} and Sn^{4+}

book.) Table 8.1 gives the charges for selected commonly encountered variable-charge metal ions. Note that the charges of +2 and +3 are a common combination. However, there are also other combinations: two +2 and +4 pairings (Pb and Sn), a +1 and +2 pairing (Cu), and a +1 and +3 pairing (Au) are listed in Table 8.1.

8.3 Nomenclature for Binary Ionic Compounds

The basic rule for naming binary ionic compounds of both fixed-charge and variable-charge metals is the same. However, naming compounds in the second category requires the use of a subrule in addition to the basic rule.

Compounds Containing a Fixed-Charge Metal

Names for binary ionic compounds in which the metal is a fixed-charge metal are assigned using the following rule.

The full name of the metallic element is given first, followed by a separate word consisting of the stem of the nonmetallic element name and the suffix -ide.

Thus, to name the compound NaF we start with the name of the metal (sodium), follow it with the stem of the name of the nonmetal (fluor-), and then add the suffix -ide. The name becomes *sodium fluoride*.

The stem of the name of the nonmetal is always the first few letters of the nonmetal's name, that is, the name of the nonmetal with its ending chopped off. Table 8.2 gives the stem part of the name for some common nonmetallic elements. The name of the metal ion present is always exactly the same as the name of the metal itself. The metal's name is never shortened. Example 8.1 illustrates further the use of the rule for naming binary ionic compounds containing a fixed-charge metal.

Section 8.3 · Nomenclature for Binary Ionic Compounds

TABLE 8.2 Names of Some Common Nonmetal Ions

Element	Stem	Name of Ion	Formula
Bromine	brom-	bromide ion	Br^-
Carbon	carb-	carbide ion	C^{4-}
Chlorine	chlor-	chloride ion	Cl^-
Fluorine	fluor-	fluoride ion	F^-
Hydrogen	hydr-	hydride ion	H^-
Iodine	iod-	iodide ion	I^-
Nitrogen	nitr-	nitride ion	N^{3-}
Oxygen	ox-	oxide ion	O^{2-}
Phosphorus	phosph-	phosphide ion	P^{3-}
Sulfur	sulf-	sulfide ion	S^{2-}

EXAMPLE 8.1

Name the following binary ionic compounds, each of which contains a fixed-charge metal.

(a) BeO **(b)** AlF_3 **(c)** K_3N **(d)** Na_2S

Solution

The general pattern for naming compounds of this type is

Name of metal + stem of name of nonmetal + -ide

(a) The metal is beryllium, and the nonmetal is oxygen. Thus, the compound name is **beryllium oxide** (the stem of the nonmetal name is underlined).
(b) The metal is aluminum, and the nonmetal is fluorine. Hence, the name is **aluminum fluoride**. Note that no mention is made in the name of the subscript 3 found after the symbol for fluorine in the formula. *The name of an ionic compound never contains any reference to formula subscript numbers.* Since aluminum is a fixed-charge metal, there is only one ratio in which aluminum and fluorine may combine. Thus, just telling the names of the elements present in the compound is adequate nomenclature.
(c) Potassium (K) and nitrogen (N) are present in the compound. Its name is **potassium nitride**.
(d) This compound is named **sodium sulfide**.

PRACTICE EXERCISE 8.1

Name the following binary ionic compounds, each of which contains a fixed-charge metal.

(a) $BaCl_2$ **(b)** AlP **(c)** K_2O **(d)** Ca_3N_2

Ans. (a) barium chloride; (b) aluminum phosphide; (c) potassium oxide; (d) calcium nitride.

Similar exercises: Problems 8.17 and 8.18

Compounds Containing a Variable-Charge Metal

In naming binary ionic compounds containing a variable-charge metal, the charge on the metal ion must be incorporated into the name of the compound. The magnitude of the charge on the metal ion is indicated by using a Roman numeral, inside parentheses, placed immediately after the name of the metal. This Roman numeral is considered to be part of the metal's name, as shown by the following examples.

$$Fe^{2+}: \quad iron(II) \text{ ion}$$
$$Fe^{3+}: \quad iron(III) \text{ ion}$$
$$Au^{+}: \quad gold(I) \text{ ion}$$

Note that the magnitude of the Roman numeral is always the same as the magnitude of the charge on the metal ion.

The chlorides of Fe^{2+} and Fe^{3+} ($FeCl_2$ and $FeCl_3$, respectively), are named iron(II) chloride and iron(III) chloride. Without the use of Roman numerals (or some equivalent system) these two compounds would have the same name, an unacceptable situation.

Sometimes, when given the formula of an ionic compound containing a variable-charge metal, a student is uncertain about the charge on the metal ion. When this is the case, the charge on the nonmetal ion is used to calculate the charge on the metal ion. This technique, as well as the use of Roman numerals in compound names, is illustrated in Example 8.2.

As you "work through" Example 8.2, be sure to note that Roman numerals are not part of the *formula* of a compound, but they are part of the *name* of the compound.

EXAMPLE 8.2

Name the following binary ionic compounds, each of which contains a variable-charge metal ion.

(a) FeO   (b) Fe_2O_3 (c) PbO_2 (d) AuCl

Solution

We will need to indicate the magnitude of the charge on the metal ion, using Roman numerals, in the names of each of these compounds.

(a) Let us assume you are uncertain as to the metal ion charge in this compound, a quantity you need to know to determine the Roman numeral to be used in the name. The metal ion charge is easily calculated using the following procedure.

$$\text{Iron charge} + \text{oxygen charge} = 0$$

The oxide ion has a -2 charge (Sec. 7.5). Therefore,

$$\text{Iron charge} + (-2) = 0$$

Solving, we get

$$\text{Iron charge} = +2$$

Therefore, the iron ions present are Fe^{2+}, and the name of the compound is **iron(II) oxide**.

(b) For charge balance in this compound we have the equation

$$2(\text{Iron charge}) + 3(\text{oxygen charge}) = 0$$

Note that we have to take into account the number of each kind of ion present (2 and 3 in this case). Oxide ions carry a -2 charge (Sec. 7.5). Therefore,

$$2(\text{Iron charge}) + 3(-2) = 0$$
$$2(\text{Iron charge}) = +6$$
$$\text{Iron charge} = +3$$

Here we note that we are interested in the charge on a single iron ion $(+3)$ and not in the total positive charge present $(+6)$. Since Fe^{3+} ions are present, the compound is named **iron(III) oxide**. As is the case for all ionic compounds, the name does not contain any reference to the numerical subscripts in the compound's formula.

(c) The charge balance equation is

$$\text{Lead charge} + 2(\text{oxygen charge}) = 0$$

Substituting a charge of -2 into the equation for oxygen, we get

$$\text{Lead charge} + 2(-2) = 0$$
$$\text{Lead charge} = +4$$

The compound is thus named **lead(IV) oxide**.

(d) This compound is **gold(I) chloride**. Chloride ions carry a -1 charge (Sec. 7.5). Since the compound contains ions in a one-to-one ratio, the gold ions must have a $+1$ charge to counterbalance the -1 charge of the chloride ions. Hence, Au^+ ions are present.

PRACTICE EXERCISE 8.2

Name the following binary ionic compounds, each of which contains a variable-charge metal ion.

(a) CuO (b) Cu_2O (c) Mn_2S_3 (d) $AuCl_3$

Ans. (a) copper(II) oxide; (b) copper(I) oxide; (c) manganese(III) sulfide; (d) gold(III) chloride.

Similar exercises: Problems 8.23 and 8.24

An older method for indicating the charge on metal ions uses suffixes rather than Roman numerals. This system is more complicated and less precise than the Roman numeral system and fortunately is being abandoned. It is mentioned here because it

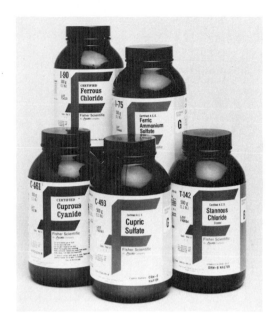

Figure 8.2 The older method of naming ionic compounds containing a variable-charge metal is still used on the labels of many laboratory chemicals. [Fisher Scientific]

is still encountered (especially on the labels of old bottles of chemicals; see Fig. 8.2). In this system, when a metal has two common ionic charges, the suffix **-ous** is used for the ion of lower charge and the suffix **-ic** for the ion of higher charge. Table 8.3 compares the two systems for the metals where the old system is most often encountered. Note that in the old system the Latin names for metals are used (Sec. 4.9).

Table 8.3 Comparison of IUPAC and Old System Names for Selected Metal Ions

Element	Ions	Preferred Name	Old System
Copper	Cu^+	copper(I) ion	cuprous ion
	Cu^{2+}	copper(II) ion	cupric ion
Iron	Fe^{2+}	iron(II) ion	ferrous ion
	Fe^{3+}	iron(III) ion	ferric ion
Tin	Sn^{2+}	tin(II) ion	stannous ion
	Sn^{4+}	tin(IV) ion	stannic ion
Lead	Pb^{2+}	lead(II) ion	plumbous ion
	Pb^{4+}	lead(IV) ion	plumbic ion
Gold	Au^+	gold(I) ion	aurous ion
	Au^{3+}	gold(III) ion	auric ion

EXAMPLE 8.3

Two systems exist for indicating the charge on common variable-charge metal ions: (1) a system that uses Roman numerals and (2) a system that uses the suffixes -ic and -ous. Change the following compound names from one system to the other system.

(a) copper(II) chloride (b) tin(II) fluoride
(c) ferrous bromide   (d) auric oxide

Solution

The information found in Table 8.3 is needed to effect the desired name changes.

(a) The given name indicates that Cu^{2+} ion is present. The alternate name for Cu^{2+} ion, from Table 8.3, is cupric ion. The name copper(II) chloride thus becomes **cupric chloride**.
(b) The alternate name for the tin(II) ion, Sn^{2+}, is stannous ion. Therefore, tin(II) fluoride becomes **stannous fluoride**.
(c) The ferrous ion, from Table 8.3, is Fe^{2+} ion; the alternate name for this ion is iron(II). Therefore, the changed (and preferred) name for this compound is **iron(II) bromide**.
(d) Since auric ion and gold(III) ion are equivalent (Table 8.3), the names auric oxide and **gold(III) oxide** are equivalent.

PRACTICE EXERCISE 8.3

Change each of the following compound names from the "Roman numeral" to the "-ic, -ous" system, or vice versa.

(a) iron(III) oxide   (b) copper(I) chloride
(c) stannic bromide   (d) plumbous fluoride

Ans. (a) ferric oxide; (b) cuprous chloride; (c) tin(IV) bromide; (d) lead(II) fluoride.

Similar exercises: Problems 8.33 and 8.34

8.4 Nomenclature for Ionic Compounds Containing Polyatomic Ions

In Section 7.8 the topic of polyatomic ions was considered, and at that time a limited number of such ions were introduced in the context of formula writing. We now consider some additional facts about polyatomic ions.

Hundreds of different polyatomic ions exist. Table 8.4 gives the names and formulas for the most common ones. Note that almost all the polyatomic ions listed in Table 8.4 contain oxygen atoms. The names, but not necessarily the formulas, of some of these common polyatomic ions should be familiar to you. Many of these ions are

TABLE 8.4 Formulas and Names of Some Common Polyatomic Ions

Key Element Present	Formula	Name of Ion
Nitrogen	NO_3^-	nitrate ion
	NO_2^-	nitrite ion
	NH_4^+	ammonium ion
	N_3^-	azide ion
Sulfur	SO_4^{2-}	sulfate ion
	HSO_4^-	hydrogen sulfate or bisulfate ion
	SO_3^{2-}	sulfite ion
	HSO_3^-	hydrogen sulfite or bisulfite ion
	$S_2O_3^{2-}$	thiosulfate ion
Phosphorus	PO_4^{3-}	phosphate ion
	HPO_4^{2-}	hydrogen phosphate ion
	$H_2PO_4^-$	dihydrogen phosphate ion
	PO_3^{3-}	phosphite ion
Carbon	CO_3^{2-}	carbonate ion
	HCO_3^-	hydrogen carbonate or bicarbonate ion
	$C_2O_4^{2-}$	oxalate ion
	$C_2H_3O_2^-$	acetate ion
	CN^-	cyanide ion
	OCN^-	cyanate ion
	SCN^-	thiocyanate ion
Chlorine	ClO_4^-	perchlorate ion
	ClO_3^-	chlorate ion
	ClO_2^-	chlorite ion
	ClO^-	hypochlorite ion
Oxygen	O_2^{2-}	peroxide ion
Boron	BO_3^{3-}	borate ion
Hydrogen	H_3O^+	hydronium ion
	OH^-	hydroxide ion
Metals	MnO_4^-	permanganate ion
	CrO_4^{2-}	chromate ion
	$Cr_2O_7^{2-}$	dichromate ion

found in commercial products. Examples are fertilizers (phosphates, sulfates, nitrates), baking soda and baking powder (bicarbonates), and building materials (carbonates, sulfates).

There is no easy way to learn the formulas and names for all the common polyatomic ions. Memorization is required. The charges and formulas for the various polyatomic ions cannot be related easily to the periodic table as was the case for many of the monoatomic ions. In Table 8.4 the most frequently encountered polyatomic ions are in color. Their formulas and names should definitely be memorized. Your instructor may want you to memorize others, too. The inability to recognize the presence of polyatomic ions (both by name and by formula) in a compound is a major stumbling block for many chemistry students. It requires some effort to overcome this obstacle.

Note from Table 8.4 the following facts concerning the polyatomic ions.

1. Most of the ions have a negative charge, which can vary from -1 to -3. Only two positive ions are listed in the table: NH_4^+ (ammonium) and H_3O^+ (hydronium).
2. Four of the polyatomic ions have names ending in **-ide**: OH^- (hydroxide), CN^- (cyanide), N_3^- (azide), and O_2^{2-} (peroxide). These names present exceptions to the rule that the suffix -ide be reserved for use in naming binary ionic compounds.
3. A number of **-ate**, **-ite** pairs of ions exist—for example, SO_4^{2-} (sulfate) and SO_3^{2-} (sulfite). The ion in the pair with the higher number of oxygens is always the *-ate* ion. The *-ite* ion always contains one less oxygen then the *-ate* ion.
4. A number of pairs of ions exist where one member of the pair differs from the other by having a hydrogen atom present, for example, CO_3^{2-} (carbonate) and HCO_3^- (hydrogen carbonate or bicarbonate). In such pairs, the charge on the hydrogen-containing ion is always one less than the charge on the other ion.
5. Two pairs of ions exist in which the difference between pair members is that a sulfur atom has replaced an oxygen atom in one member of the pair: (SO_4^{2-}, $S_2O_3^{2-}$) and (OCN^-, SCN^-). The prefix **thio-** is used to denote this replacement of oxygen by sulfur. The names of these pairs of ions, respectively, are sulfate–thiosulfate and cyanate–thiocyanate.

The names of ionic compounds containing polyatomic ions are derived in the same way as those of binary ionic compounds (Sec. 8.3). Recall that the rule for naming binary ionic compounds is: Give the name of the metallic element first (including, when needed, a Roman numeral indicating ion charge), and then as a separate word give the stem of the nonmetallic element name to which the suffix *-ide* is appended.

For our present situation, *if the polyatomic ion is positive, its name is substituted for that of the metal. If the polyatomic ion is negative, its name is substituted for the nonmetal stem plus -ide.* In the case where both positive and negative ions are polyatomic, dual substitution occurs and the resulting name includes just the names of the polyatomic ions. Example 8.4 illustrates the use of these rules.

EXAMPLE 8.4

Name the following compounds, which contain one or more polyatomic ions.

(a) K_2CO_3 (b) $Co(NO_3)_3$ (c) $Fe_2(SO_4)_3$ (d) $(NH_4)_3PO_4$

Solution

(a) The positive ion present is the potassium ion (K^+). The negative ion is the polyatomic carbonate ion (CO_3^{2-}). The name of the compound is potassium carbonate. As in naming binary ionic compounds, subscripts in the formula are not incorporated into the name.

(b) The positive ion present is cobalt, and the negative ion is the nitrate ion. Since cobalt is a variable-charge metal, a Roman numeral must be used to indicate ionic charge. In this case, the Roman numeral is III. The fact that cobalt ions carry a $+3$ charge is deduced by noting that there are three nitrate ions present, each of which carries a -1 charge. The charge on the single cobalt ion present must be a $+3$ in order to counterbalance the total negative charge of -3. The name of the compound, therefore, is **cobalt(III) nitrate**.

(c) The positive ion present is iron(III). The negative ion is the polyatomic sulfate ion (SO_4^{2-}). The name of the compound is **iron(III) sulfate**. The determination that iron is present as iron(III) involves the following calculation dealing with charge balance.

$$2(\text{Iron charge}) + 3(\text{sulfate charge}) = 0$$

The sulfate charge is -2. Therefore,

$$2(\text{Iron charge}) + 3(-2) = 0$$
$$2(\text{Iron charge}) = +6$$
$$\text{Iron charge} = +3$$

(d) Both the positive and negative ions in this compound are polyatomic—the ammonium ion (NH_4^+) and the phosphate ion (PO_4^{3-}). The name of the compound is simply the combination of the names of the two polyatomic ions: **ammonium phosphate**.

PRACTICE EXERCISE 8.4

Name the following compounds, which contain one or more polyatomic ions.

(a) $NaClO_3$ (b) $Ca(C_2H_3O_2)_2$ (c) Cu_2SO_4 (d) NH_4CN

Ans. (a) sodium chlorate; (b) calcium acetate; (c) copper(I) sulfate; (d) ammonium cyanide.

Similar exercises: Problems 8.57–8.60

8.5 Nomenclature for Binary Molecular Compounds

A **binary molecular compound** is a covalently bonded compound in which just two nonmetallic elements are present. Such compounds are named in the following manner. The two nonmetals present are named in the order in which they appear in the formula. [The least electronegative nonmetal (Sec. 7.10) is usually written first in the formula.] *The name of the first nonmetal is used in full. The name of the second nonmetal is treated as was the nonmetal in binary ionic compounds: that is, the stem of the name is given and the suffix -ide is added.*

Section 8.5 · Nomenclature for Binary Molecular Compounds

TABLE 8.5 Greek Numerical Prefixes from 1 to 10

Greek Prefix	Number
Mono-	1
Di-	2
Tri-	3
Tetra-	4
Penta-	5
Hexa-	6
Hepta-	7
Octa-	8
Ennea-[a]	9
Deca-	10

[a] The prefix ennea- is preferred to the Latin nona- by IUPAC, but nona- is still frequently used.

In addition, the number of atoms of each element present in a molecule of the compound is explicitly incorporated into the name of the compound through the use of Greek numerical prefixes. A prefix precedes the name of each nonmetal. The use of these prefixes is in direct contrast to the procedures used for naming ionic compounds. For ionic compounds, formula subscripts are not mentioned in the name.

Prefixes are needed in naming binary molecular compounds because numerous different compounds exist for many pairs of nonmetallic elements. For example, all of the following nitrogen–oxygen molecular compounds exist: NO, NO_2, N_2O, N_2O_3, N_2O_4, and N_2O_5. The prefixes used are the standard Greek numerical prefixes, which are given in Table 8.5 for the numbers 1 through 10, and which are used in words such as mononucleosis, dichromatic, triangle, tetragram, pentagon, hexapod, heptathlon, octave, enneahedron, and decade. Example 8.5 shows how these prefixes are used in naming binary molecular compounds.

EXAMPLE 8.5

Name the following binary molecular compounds.

(a) N_2O (b) N_2O_3 (c) PCl_5 (d) P_4S_{10} (e) CCl_4

Solution

The names of each of these compounds will consist of two words with the following general formats.

First word: numerical prefix + full name of the first element
Second word: numerical prefix + stem of name of second element + -ide

(a) The elements present are nitrogen and oxygen. The two portions of the name, before adding Greek numerical prefixes, are *nitrogen* and *oxide*. Adding the prefixes gives *dinitrogen* (two nitrogen atoms are present) and *monoxide* (one oxygen atom is present). (When an element name begins with an *a* or *o*, the *a* or *o* at the end of the Greek prefix is dropped for ease of pronunciation—monoxide instead of monooxide.) The name of this compound is **dinitrogen monoxide**.

(b) The elements present are again nitrogen and oxygen. This time the two portions of the name are *dinitrogen* and *trioxide*, which are combined to give the name **dinitrogen trioxide**.

(c) When there is only one atom of the first element present, it is standard procedure to omit the prefix *mono-* for that element. Following this guideline, we have for the name of this compound **phosphorus pentachloride**.

(d) The prefix for four atoms is *tetra-* and for ten atoms *deca-*. This compound therefore, has the name **tetraphosphorus decasulfide**.

(e) Omitting the initial *mono-* (see part c), we name this compound **carbon tetrachloride**.

PRACTICE EXERCISE 8.5

Name the following binary molecular compounds.

(a) PF_3 (b) Cl_2O (c) N_2O_5 (d) S_4N_4

Ans. (a) phosphorus trifluoride; (b) dichlorine monoxide; (c) dinitrogen pentoxide; (d) tetrasulfur tetranitride.

Similar exercises: Problems 8.65 and 8.66

There is one exception to the use of Greek numerical prefixes in naming binary molecular compounds. Binary compounds with hydrogen listed as the first element in the formula are named without prefix use. Thus, the compounds H_2S and HCl are named hydrogen sulfide and hydrogen chloride, respectively.

All procedures for naming compounds so far presented in this chapter have been based on IUPAC rules (Sec. 8.1) and have produced systematic names. A **systematic name** for a compound is a name derived from IUPAC rules that conveys information about the composition of the compound. A few binary molecular compounds have names completely unrelated to the IUPAC rules for naming such compounds. They have "common" or "trivial" names, which are coined before the development of the systematic rules. A **common name** for a compound is a name not based on IUPAC rules. Common names usually do not convey any information about the composition of compounds. At one time, in the early history of chemistry, all compounds had common names. With the advent of systematic nomenclature, most common names were discontinued. A few, however, have persisted and are now officially accepted. The most "famous" example of this is the compound H_2O, which has the systematic name hydrogen oxide. This name is never used; H_2O is known as *water*, a name that

TABLE 8.6 Some Binary Molecular Compounds That Have Common Names

Compound Formula	Accepted Common Name
H_2O	water
H_2O_2	hydrogen peroxide
NH_3	ammonia
N_2H_4	hydrazine
CH_4	methane
C_2H_6	ethane
PH_3	phosphine
AsH_3	arsine

is not going to be changed. Another very common example of common nomenclature involves the compound NH_3, which is *ammonia*. Table 8.6 gives additional examples of compounds for which common names are used in preference to systematic names. Such exceptions are actually very few compared to the total number of com-compounds named by systematic rules.

Writing formulas for binary molecular compounds given their names is a very easy task. The Greek prefixes in the names of such compounds tell you exactly how many atoms of each kind are present. For example, the compounds dinitrogen pentoxide and carbon dioxide have, respectively, the formulas N_2O_5 (di- and penta-) and CO_2 (mono- and di-). Remember, the prefix mono- is always dropped at the start of a name, as in carbon dioxide.

8.6 Nomenclature for Acids

Many hydrogen-containing molecular compounds dissolve in water to give solutions with properties markedly different from those of the compounds that were dissolved. These solutions, which we will discuss in detail in Chapter 14, are called *acids*.

Not all hydrogen-containing molecular compounds are acids. Those that are can be recognized by their formulas, which have hydrogen written as the first element in the formula.

Acids: $HCl, H_2S, H_2SO_4, HNO_3$
Nonacids: NH_3, CH_4, PH_3, SiH_4

Water, which is the solvent for most acidic solutions, is generally not considered to be an acid. Despite this, H is written first in its formula (H_2O).

Because of differences in the properties of acids and the anhydrous (without water) compounds from which they are produced, the acids and the anhydrous compounds are given different names. Acid nomenclature is derived from the names of the parent anhydrous compounds (which we have learned to name in this chapter). Because of this close relationship, we will also learn to name acids at this time, even though a detailed discussion of acids and their properties does not come until Chapter 14.

For nomenclatural purposes, acids will be classified into two categories: nonoxyacids and oxyacids. A **nonoxyacid** is a molecular compound composed of hydrogen and one or more nonmetals other than oxygen that produces an acidic aqueous solution. All common nonoxyacids, with one exception, HCN, are binary compounds. An **oxyacid** is a molecular compound composed of hydrogen, oxygen, and another nonmetal that produces an acidic aqueous solution.

Nonoxyacids are named by modifying the names of the parent anhydrous compounds as follows.

1. The word *hydrogen* is replaced with the prefix **hydro-**.
2. The suffix *-ide* on the stem of the name of the nonmetal is replaced with the suffix **-ic**.
3. The word **acid** is added to the end of the name (as a separate word).

By these rules, the compound hydrogen chloride (HCl) in a water solution is called **hydro**chloric **acid**. A water solution of hydrogen selenide (H_2Se) is called **hydro**selenic **acid**, and HCN (hydrogen cyanide) dissolved in water becomes **hydro**cyanic **acid**. Binary molecular compounds containing hydrogen and one or more nonmetals other than oxygen may thus be named in two ways: as a pure compound and as a water solution (acid). Table 8.7 illustrates this dual naming system for some other compounds.

The notation for distinguishing between the chemical formulas for a hydrogen-containing compound in its "pure" and "acid" states involves the use of (g) for gaseous and (aq) for aqueous, as is illustrated for the compound HF.

$$\text{Hydrofluoric acid} \quad \text{HF(aq)}$$
$$\text{Hydrogen fluoride} \quad \text{HF(g)}$$

The elements hydrogen and oxygen are common to all oxyacids. Formulas for common oxyacids include H_2SO_4, HNO_3, H_3PO_4, and H_2CO_3.

When oxyacid molecules are dissolved in water, they break apart to give ions. The positive ions produced are H^+ ions, the species characteristic of all acids. The negative species produced are polyatomic ions, the same polyatomic ions we considered in

TABLE 8.7 The Dual Naming System for Molecular Compounds Containing Hydrogen and a Nonmetal Other Than Oxygen

Formula	Name of Pure Compound	Name of Water Solution
HF	hydrogen fluoride	hydrofluoric acid
HBr	hydrogen bromide	hydrobromic acid
HI	hydrogen iodide	hydroiodic acid
H_2S	hydrogen sulfide	hydrosulfuric acid[a]

[a] For acids involving sulfur, "ur" from sulfur is reinserted in the acid name for pronunciation reasons.

Section 8.4. Even though ions are produced in solution, oxyacids are not ionic compounds. They are molecular compounds that interact with water to produce ions.

Names for oxyacids are derived from the names of the polyatomic ions produced when the acid molecules break into ions in solution.

1. When the polyatomic ion produced in solution has a name ending in *-ate*, the *-ate* ending is dropped, the suffix **-ic** is added in its place, and then the word **acid** is added.
2. When the polyatomic ion produced in solution has a name ending in *-ite*, the *-ite* ending is replaced by the suffix **-ous**, and then the word **acid** is added.

Examples 8.6 and 8.7 illustrate the use of these rules.

EXAMPLE 8.6

Name the following ternary molecular compounds as acids.

(a) HNO_3 (b) $HClO_3$ (c) HNO_2 (d) $HClO$

Solution

(a) The polyatomic ion present in solutions of this acid is the nitrate ion (NO_3^-). Removing the *-ate* ending from the word nitrate, replacing it with the suffix *-ic*, then adding the word *acid* gives the name **nitric acid** for the name of this oxyacid.
(b) The name base this time is the polyatomic ion chlorate (ClO_3^-). Removing the *-ate* suffix and adding an *-ic* suffix and the word *acid* produces the oxyacid name of **chloric acid**.
(c) The polyatomic ion produced when this compound dissolves in water is nitrite (NO_2^-). Note that two closely related (and often confused) polyatomic ions exist: nitr*ite* (NO_2^-) and nitr*ate* (NO_3^-). We are dealing here with nitr*ite*. For polyatomic ions whose names end in *-ite*, the *-ite* suffix is replaced by an *-ous* suffix and then the word *acid* is added. This compound is therefore **nitrous acid**.
(d) The name for the polyatomic ion ClO^-, hypochlorite, is our starting point this time. The *-ite* ending becomes *-ous*, and with the addition of the word *acid* we have the compound's name, **hypochlorous acid**. Note that if a polyatomic ion name contains the prefix *per-* or *hypo-* it carries over to the acid name, as is the case here.

PRACTICE EXERCISE 8.6

Name the following ternary molecular compounds as acids.

(a) H_2SO_3 (b) $H_2C_2O_4$ (c) H_3PO_4 (d) $HC_2H_3O_2$

Ans. (a) sulfurous acid; (b) oxalic acid; (c) phosphoric acid; (d) acetic acid.

Similar exercises: Problems 8.77–8.78

EXAMPLE 8.7

Give the formulas for the following oxyacids.

(a) sulfuric acid   (b) perchloric acid   (c) chlorous acid

Solution

(a) The *-ic* ending of the acid name indicates that the formula will contain a polyatomic ion whose name ends in *-ate*. (We are going through a reasoning process that is the reverse of that used in Example 8.6). The sulfur-containing polyatomic ion whose name ends in *-ate* is sulfate (SO_4^{2-}). The hydrogen present, for formula-writing purposes, is considered to be H^+ ions. Combining these ions in the correct ratio to give a neutral compound (Sec. 7.6) gives the formula H_2SO_4 for sulfuric acid.

(b) The *-ic* ending again indicates the presence of an *-ate* polyatomic ion. This time it will be a *per-.... -ate*—indeed, the perchlorate ion. This ion has the formula ClO_4^- (see Table 8.4). Combining H^+ and ClO_4^- ions in the correct ratio, a 1:1 ratio, gives the formula $HClO_4$ for perchloric acid.

(c) The *-ous* ending in the acid name indicates that an *-ite* polyatomic ion is present, the chlorite ion with the formula ClO_2^- (Table 8.4). Combining H^+ and ClO_2^- ions in a 1:1 ratio gives the formula $HClO_2$ for chlorous acid.

PRACTICE EXERCISE 8.7

Give the formulas for the following oxyacids.

(a) carbonic acid   (b) phosphorous acid   (c) boric acid

Ans. (a) H_2CO_3; (b) H_3PO_3; (c) H_3BO_3.

Similar exercises: Problems 8.81–8.84

For many nonmetals a series of oxyacids exists. The formulas of the members of such a series differ from each other only in oxygen content. For example, there are four chlorine-containing oxyacids whose formulas are $HClO_4$, $HClO_3$, $HClO_2$, and $HClO$. Some useful interrelationships exist between the names of the acids in such a series.

In any series of oxyacids, one of the acids, usually the most common one, will be the "-ic" acid, that is, the acid whose name has no prefix and ends in *-ic*. In the chlorine oxyacid series, $HClO_3$, chloric acid, is the "-ic" acid. The formula of this "-ic" acid is the key to remembering the names for the other members of the series.

The names of the other acids in a series of oxyacids are related, in a constant manner, to the "-ic" acid by prefixes and suffixes as follows.

1. An oxyacid containing *one less oxygen atom* than the "-ic" acid has a name obtained by changing the *-ic* suffix to *-ous*. The oxyacid $HClO_2$, with one less oxygen than chloric acid, $HClO_3$, is chlor**ous** acid.
2. An oxyacid containing *two less oxygen atoms* than the "-ic" acid has a name con-

taining the prefix *hypo-* and the suffix *-ous*. The oxyacid HClO, with two less oxygens than chloric acid, is **hypo**chlor**ous** acid. [The prefix hypo- means below or under (e.g., hypodermic means under the skin).]

3. An oxyacid containing *one more oxygen atom* than the "-ic" acid has a name containing the prefix *per-* and the suffix *-ic*. Thus, the oxyacid HClO$_4$ is **per**chlor**ic** acid. [The prefix per- comes from hyper, meaning above or elevated (e.g., hyperactive).]

Schematically, we can summarize these oxyacid name relationships as follows, where *n* represents the number of oxygen atoms in the "-ic" acid.

Number of Oxygen Atoms	Name
$n + 1$	per____ic acid
n	____ic acid
$n - 1$	____ous acid
$n - 2$	hypo____ous acid

To use these interrelationships you *must know* the "starting point"—the formula of the reference "-ic" acid. Table 8.8 gives the formulas of the common "-ic" acids. There are no common oxyacids for the omitted nonmetals.

Most nonmetals do not form a "complete" series of four simple oxyacids as does chlorine. Bromine is the only other nonmetal for which such a series exists. For both sulfur and nitrogen only the "-ic" and "-ous" acids are known. In general, "-ic" and "-ous" acids are relatively common and "per____ic" and "hypo____ous" acids are relatively rare.

TABLE 8.8 Formulas for the Common Oxy "-ic" Acids

Periodic Table Group of Nonmetal				
IIIA	IVA	VA	VIA	VIIA
H_3BO_3 Boric acid	H_2CO_3 Carbonic acid	HNO_3 Nitric acid		
		H_3PO_4 Phosphoric acid	H_2SO_4 Sulfuric acid	$HClO_3$ Chloric acid
		H_3AsO_4 Arsenic acid	H_2SeO_4 Selenic acid	$HBrO_3$ Bromic acid
			H_6TeO_6 Telluric acid	HIO_3 Iodic acid

EXAMPLE 8.8

Name the following oxyacids using the relationships between oxygen content and the formula of the "-ic" acid.

(a) H_3PO_2 (b) HNO_2 (c) HIO_4

Solution

(a) For phosphorus, the "-ic" acid is H_3PO_4 (see Table 8.8). Since this acid has two less oxygen atoms, it is named by using the prefix *hypo-* and the suffix *-ous*—**hypophosphorous acid**.
(b) For nitrogen, the "-ic" acid is HNO_3. This acid has one less oxygen atom than HNO_3. Thus, the suffix *-ous* is used in naming the acid—**nitrous acid**.
(c) The acid HIO_4 contains one more oxygen atom than HIO_3, the "-ic" acid. Therefore, its name contains the prefix *per-* and the suffix *-ic*—**periodic acid**.

PRACTICE EXERCISE 8.8

Name the following oxyacids using the relationships between oxygen content and the formula of the "-ic" acid.

(a) H_2SeO_3 (b) $HBrO_4$ (c) H_3PO_3

Ans. (a) selenous acid; (b) perbromic acid; (c) phosphorous acid.

Similar exercises: Problems 8.77 and 8.78

The polyatomic ion derived from any oxyacid whose name ends in *-ic* will have a name ending in **-ate**. Polyatomic ions derived from oxyacids whose names end in *-ous* will have names ending in **-ite**. The prefixes **per-** and **hypo-**, when present in the acid name, are retained in the polyatomic ion name. For the chlorine oxyacid system, we thus have

$HClO_4$ (perchloric acid)   and ClO_4^- (perchlorate ion)
$HClO_3$ (chloric acid)   and ClO_3^- (chlorate ion)
$HClO_2$ (chlorous acid)   and ClO_2^- (chlorite ion)
$HClO$ (hypochlorous acid)   and ClO^- (hypochlorite ion)

We have now gone "full circle"; these relationships between acid name and polyatomic ion name are the same ones introduced at the start of this section when the rules for naming oxyacids were given.

8.7 Nomenclature Rules—A Summary

Within this chapter various sets of nomenclature rules have been presented. Students sometimes have problems deciding which set of rules to use in a given situation. This

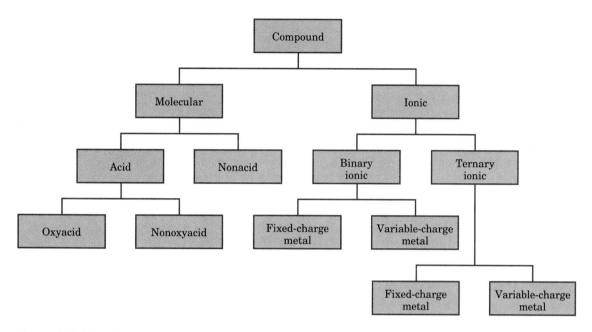

FIGURE 8.3 Classifying a compound for the purpose of naming it.

dilemma can be avoided if, when confronted with the request to name a compound, the student goes through the following reasoning pattern.

1. Decide, first, whether the compound is ionic (metal + nonmetal) or molecular (nonmetals only).
2. If the compound is ionic, then classify it as binary or polyatomic-ion-containing (ternary) and use the rules appropriate for that classification, taking into account whether the metal ions are fixed-charge or variable-charge. If the compound is ternary ionic, it will always contain polyatomic ions.
3. If the compound is molecular, classify it as an acid or nonacid. An acid must have the element hydrogen present (written first in the formula), and the compound must be in water solution. If the compound is a nonacid, name it according to the rules for binary molecular compounds.
4. If the compound is an acid, further classify it as a nonoxyacid or an oxyacid and then use the appropriate nomenclature rules.

Figure 8.3 summarizes in diagrammatic form the steps involved in deciding how to name a compound.

EXAMPLE 8.9

Name the following compounds.

(a) K_3P
(b) N_2O_4
(c) $FeSO_4$
(d) $HClO_4$ (in solution)
(e) H_2S (in solution)

Solution

(a) This is a binary compound containing a metal (K) and a nonmetal (P). The presence of both a metal and a nonmetal indicates that the compound is an ionic one. Its name is thus **potassium phosphide**. Remember that Greek numerical prefixes are never used in naming ionic compounds. Their use is limited to molecular compounds. The name tripotassium phosphide is incorrect. Note that no Roman numeral is needed for the K since it is a fixed-charge metal (Fig. 8.1).

(b) This is a binary compound containing two nonmetals. Therefore, it is a molecular compound. It cannot be an acid since it does not contain hydrogen. Using the rules for binary molecular compounds, we name the compound **dinitrogen tetroxide**.

(c) Iron (Fe) is a metal. Sulfur (S) and oxygen (O) are nonmetals. This is a ternary ionic compound; it is ionic since both a metal and nonmetals are present. Since it is a ternary compound, polyatomic ions must be present. The polyatomic ion is the sulfate ion (SO_4^{2-}). The metal ion is iron(II). The $+2$ charge on the iron counterbalances the -2 charge on the sulfate ion. The compound's name is **iron(II) sulfate**.

(d) The compound is an oxyacid. The polyatomic ion on which the name is based is ClO_4^- (perchlorate). Changing the -*ate* ending to -*ic* yields the name **perchloric acid**.

(e) This compound is a nonoxyacid. The pure compound would be named hydrogen sulfide. Using the prefix *hydro*- and the suffix -*ic* transforms this name into **hydrosulfuric acid**.

PRACTICE EXERCISE 8.9

Name the following compounds.

(a) NO_2 (b) $Cu(NO_3)_2$ (c) HBr (in solution)
(d) KCN   (e) Na_3N

Ans. (a) nitrogen dioxide; (b) copper(II) nitrate; (c) hydrobromic acid; (d) potassium cyanide; (e) sodium nitride.

Similar exercises: Problems 8.93–8.96

Example 8.9, as well as many other examples in this chapter, involved naming compounds given their formulas. Let us now consider the reverse process, deriving formulas when given names. Often students who can go "forward" without problems have difficulty with this "reverse" process.

EXAMPLE 8.10

Write formulas for the following compounds.

(a) calcium borate   (b) iron(II) nitrate
(c) dinitrogen trioxide   (d) aluminum oxide

Solution

(a) This is a ternary ionic compound containing the polyatomic borate ion. As the name indicates, the metal present is calcium. The calcium ion is Ca^{2+} (Fig. 8.1), and the formula of the borate ion (Table 8.4) is BO_3^{3-}. Combining these ions in the ratio that will cause their charges to add to zero gives **$Ca_3(BO_3)_2$**. Recall that any time more than one polyatomic ion of a given kind is needed in a formula, parentheses must be used about the polyatomic ion. (Review Examples 7.5 and 7.6 if you are still having trouble determining the ratio in which ions combine.)

(b) This is also an ionic compound containing a polyatomic ion. The Roman numeral associated with the word iron indicates that the iron ions present are Fe^{2+} ions. Combining Fe^{2+} ions and nitrate ions, NO_3^-, in a one-to-two ratio gives **$Fe(NO_3)_2$**. Note, again, that a Roman numeral in the name of a compound never becomes part of the formula of the compound. The Roman numeral's purpose is to give information about ionic charge.

(c) The presence of only nonmetals and the presence of Greek numerical prefixes both indicate that this is a molecular compound. The Greek numerical prefixes give directly the number of each type of atom present—**N_2O_3**.

(d) This is a binary ionic compound since a metal and a nonmetal are present. The lack of a Roman numeral in the name indicates that the metal present is a fixed-charge metal. Aluminum forms a +3 ion (Fig. 8.1). The oxide ion has a charge of −2 (Sec. 7.5). Combining the ions Al^{3+} and O^{2-} in the ratio that will cause their charges to add to zero gives the formula **Al_2O_3**.

PRACTICE EXERCISE 8.10

Write formulas for the following compounds.

(a) heptasulfur dioxide (b) copper(II) carbonate
(c) magnesium sulfide (d) beryllium hydroxide

Ans. (a) S_7O_2; (b) $CuCO_3$; (c) MgS; (d) $Be(OH)_2$

Similar exercises: Problems 8.89–8.92

Key Terms

The new terms or concepts defined in this chapter are

binary compound (Sec. 8.1)   A compound containing just two different elements.

binary ionic compound (Sec. 8.2)   A binary compound in which monoatomic positively charged metallic ions and monoatomic negatively charged nonmetallic ions are present.

binary molecular compound (Sec. 8.5)   A binary compound in which two nonmetallic elements are covalently bonded to each other.

chemical nomenclature (Sec. 8.1)   A system of names used to distinguish compounds from each other and the rules needed to devise these names.

common name (Sec. 8.5)   A name for a compound not based on IUPAC rules.

fixed-charge metal (Sec. 8.2)   A metal that forms only

one type of ion, which always has the same charge magnitude.

IUPAC rules (Sec. 8.1) A set of rules for naming compounds produced by committees of the International Union of Pure and Applied Chemistry (IUPAC).

nonoxyacid (Sec. 8.6) A molecular compound composed of hydrogen and one or more nonmetals other than oxygen that produces an acidic aqueous solution.

oxyacid (Sec. 8.6) A molecular compound composed of hydrogen, oxygen, and another nonmetal that produces an acidic aqueous solution.

systematic name (Sec. 8.5) A name for a compound, based on IUPAC rules, that conveys information about the composition of the compound.

ternary compound (Sec. 8.1) A compound containing three different elements.

variable-charge metal (Sec. 8.2) A metal that forms more than one type of ion, with the ion types differing in charge magnitude.

Practice Problems

Nomenclature Classifications for Compounds (Sec. 8.1)

8.1 Classify each of the following compounds as binary or ternary.
(a) N_2O_5 (b) $(NH_4)_2S$ (c) Na_2SO_4 (d) SO_3

8.2 Classify each of the following compounds as binary or ternary.
(a) CO_2 (b) SiF_4 (c) $Fe(CN)_2$ (d) $AlPO_4$

8.3 Classify each of the following compounds, for nomenclatural purposes, as an ionic or molecular compound.
(a) CBr_4 (b) Li_2O (c) K_2SO_4 (d) NH_4Cl

8.4 Classify each of the following compounds, for nomenclatural purposes, as an ionic or molecular compound.
(a) NaH (b) Fe_2S_3 (c) ClF_3 (d) NH_4CN

8.5 Classify each of the following compounds, for nomenclatural purposes, as an ionic or molecular compound.
(a) $CaCl_2$ (b) SCl_2 (c) NCl_3 (d) $AsCl_3$

8.6 Classify each of the following compounds, for nomenclatural purposes, as an ionic or molecular compound.
(a) CO_2 (b) Al_2O_3 (c) OF_2 (d) $GeCl_4$

Types of Binary Ionic Compounds (Sec. 8.2)

8.7 Classify each of the following metals as a fixed-charge metal or a variable-charge metal.
(a) Na (b) Fe (c) Cu (d) Al

8.8 Classify each of the following metals as a fixed-charge metal or a variable-charge metal.
(a) K (b) Ag (c) Au (d) Zn

8.9 Classify the metal in each of the following compounds as a fixed-charge metal or a variable-charge metal.
(a) $BeCl_2$ (b) SnO_2 (c) Pb_3S_2 (d) GaN

8.10 Classify the metal in each of the following compounds as a fixed-charge metal or a variable-charge metal.
(a) Mg_3N_2 (b) $CdBr_2$ (c) NiO (d) BaI_2

8.11 What is the magnitude of the charge associated with the ion of each of the following fixed-charge metals?
(a) K (b) Ag (c) Al (d) Ga

8.12 What is the magnitude of the charge associated with the ion of each of the following fixed-charge metals?
(a) Mg (b) Li (c) Zn (d) Cd

Nomenclature of Binary Ionic Compounds (Sec. 8.3)

8.13 Name the following common nonmetal ions.
(a) N^{3-} (b) Cl^- (c) S^{2-} (d) Br^-

8.14 Name the following common nonmetal ions.
(a) O^{2-} (b) P^{3-} (c) F^- (d) I^-

8.15 Name the following common metal ions.
(a) Mg^{2+} (b) K^+ (c) Co^{2+} (d) Co^{3+}

8.16 Name the following common metal ions.
(a) Ca^{2+} (b) Al^{3+} (c) Fe^{3+} (d) Fe^{2+}

8.17 Name each of the following binary ionic compounds, each of which contains a fixed-charge metal.
(a) Li_2O (b) MgS (c) $ZnCl_2$ (d) $AgBr$

8.18 Name each of the following binary ionic compounds, each of which contains a fixed-charge metal.
(a) $BeCl_2$ (b) Ca_3P_2 (c) CdO (d) Ga_2S_3

8.19 Explain why calcium bromide is a satisfactory name for $CaBr_2$ but iron bromide is not a satisfactory name for $FeBr_2$.

8.20 Explain why magnesium oxide is a satisfactory name for MgO but tin oxide is not a satisfactory name for SnO.

8.21 What is the charge on the metal ion in each of the following variable-charge metal binary ionic compounds?
(a) $SnCl_2$ (b) $SnCl_4$ (c) SnO (d) SnO_2

8.22 What is the charge on the metal ion in each of the following variable-charge metal binary ionic compounds?
(a) $CrCl_2$ (b) $CrCl_3$ (c) CrO (d) Cr_2O_3

8.23 Name each of the following binary ionic compounds, each of which contains a variable-charge metal.
(a) Au_2S_3 (b) Cu_2S (c) PbS_2 (d) NiS

8.24 Name each of the following binary ionic compounds, each of which contains a variable-charge metal.
(a) Fe_2S_3 (b) PbS (c) Au_2S (d) CuS

8.25 Indicate whether or not a Roman numeral is required in the name of each of the following binary ionic compounds.
(a) Au_2O_3 (b) $CuCl_2$ (c) Ag_2O (d) AlN

8.26 Indicate whether or not a Roman numeral is required in the name of each of the following binary ionic compounds.
(a) $ZnCl_2$ (b) $PbCl_2$ (c) $MnCl_2$ (d) $BaCl_2$

8.27 Name each of the following binary ionic compounds.
(a) AgCl (b) $CoCl_3$ (c) $AlCl_3$ (d) $PbCl_4$

8.28 Name each of the following binary ionic compounds.
(a) Au_2O (b) Ag_2O (c) Na_2O (d) PbO

8.29 Name each compound in the following pairs of binary ionic compounds using the Roman numeral system.
(a) CuO and Cu_2O (b) MnO and Mn_2O_3
(c) SnS and SnS_2 (d) $PbBr_2$ and $PbBr_4$

8.30 Name each compound in the following pairs of binary ionic compounds using the Roman numeral system.
(a) $FeCl_2$ and $FeCl_3$ (b) SnO and SnO_2
(c) CoO and Co_2O_3 (d) Cr_3N_2 and CrN

8.31 Name each compound in the following pairs of binary ionic compounds using the Roman numeral system.
(a) FeO and Fe_2O_3 (b) MnS and Mn_2S_3
(c) $CoBr_2$ and $CoBr_3$ (d) CuI and CuI_2

8.32 Name each compound in the following pairs of binary ionic compounds using the Roman numeral system.
(a) $CuCl_2$ and CuCl (b) Ni_2O_3 and NiO
(c) MnN and Mn_3N_2 (d) SnI_2 and SnI_4

8.33 Change each of the following compound names from the Roman numeral to the "-ic, -ous" system or vice versa.
(a) lead(IV) oxide (b) gold(III) chloride
(c) ferric iodide (d) stannous bromide

8.34 Change each of the following compound names from the Roman numeral to the "-ic, -ous" system or vice versa.
(a) cuprous chloride (b) ferrous sulfide
(c) tin(IV) nitride (d) lead(II) oxide

8.35 Write formulas for the following binary ionic compounds.
(a) iron(II) sulfide (b) tin(IV) sulfide
(c) lithium sulfide (d) zinc sulfide

8.36 Write formulas for the following binary ionic compounds.
(a) cobalt(III) sulfide (b) manganese(II) sulfide
(c) calcium sulfide (d) aluminum sulfide

8.37 Write formulas for the following binary ionic compounds.
(a) potassium iodide (b) cobalt(II) bromide
(c) ferrous iodide (d) barium phosphide

8.38 Write formulas for the following binary ionic compounds.
(a) manganese(II) fluoride (b) cupric chloride
(c) aluminum nitride (d) gallium oxide

Nomenclature for Ionic Compounds Containing Polyatomic Ions (Sec. 8.4)

8.39 Give the names for the following polyatomic ions.
(a) NO_3^- (b) SO_4^{2-} (c) OH^- (d) CN^-

8.40 Give the names for the following polyatomic ions.
(a) CO_3^{2-} (b) H_3O^+ (c) $C_2H_3O_2^-$ (d) NH_4^+

8.41 Give the names for the following polyatomic ions.
(a) MnO_4^- (b) BO_3^{3-} (c) $S_2O_3^{2-}$ (d) N_3^-

8.42 Give the names for the following polyatomic ions.
(a) SCN^- (b) O_2^{2-} (c) $Cr_2O_7^{2-}$ (d) $C_2O_4^{2-}$

8.43 Write formulas (including charge) for the following series of polyatomic ions: phosphate, hydrogen phosphate, and dihydrogen phosphate.

8.44 Write formulas (including charge) for the following series of polyatomic ions: perchlorate, chlorate, chlorite, and hypochlorite.

8.45 What is the difference in chemical composition (including charge) between the following ions?
(a) sulfate and sulfite
(b) hydrogen carbonate and bicarbonate
(c) cyanate and thiocyanate
(d) chromate and dichromate

8.46 What is the difference in chemical composition (including charge) between the following ions?
(a) nitrate and nitrite
(b) sulfate and thiosulfate
(c) hydroxide and peroxide
(d) cyanide and azide

8.47 Indicate which of the following compounds contain polyatomic ions, and identify the polyatomic ion by name if present.
(a) Al_2S_3 (b) $Ca_3(PO_4)_2$
(c) $ZnSO_4$ (d) KNO_3

8.48 Indicate which of the following compounds contain polyatomic ions, and identify the polyatomic ion by name if present.

(a) NH_4Cl (b) KCN
(c) K_3PO_4 (d) P_4O_{10}

8.49 Name the following ionic compounds containing polyatomic ions and fixed-charge metals.
(a) $MgCO_3$ (b) $ZnSO_4$
(c) $Be(NO_3)_2$ (d) Ag_3PO_4

8.50 Name the following ionic compounds containing polyatomic ions and fixed-charge metals.
(a) $LiOH$ (b) $Al(CN)_3$
(c) $Ba(ClO_3)_2$ (d) $NaNO_3$

8.51 What is the charge on the variable-charge metal in each of the following polyatomic-ion-containing compounds?
(a) $Fe(OH)_2$ (b) $CuCO_3$
(c) $AuCN$ (d) $Mn_3(PO_4)_2$

8.52 What is the charge on the variable-charge metal in each of the following polyatomic-ion-containing compounds?
(a) $Fe(NO_3)_3$ (b) $Co_2(CO_3)_3$
(c) Cu_3PO_4 (d) $Pb(SO_4)_2$

8.53 Name the following ionic compounds containing polyatomic ions and variable-charge metals.
(a) $FeSO_4$ (b) $Mn(NO_3)_2$
(c) $Ni(CN)_2$ (d) $Cu(OH)_2$

8.54 Name the following ionic compounds containing polyatomic ions and variable-charge metals.
(a) $FePO_4$ (b) $Co(CN)_2$
(c) $Cu(C_2H_3O_2)_2$ (d) $CrSO_4$

8.55 Name each compound in the following pairs of polyatomic-ion-containing compounds.
(a) $Fe_2(CO_3)_3$ and $FeCO_3$
(b) Au_2SO_4 and $Au_2(SO_4)_3$
(c) $Sn(OH)_2$ and $Sn(OH)_4$
(d) $Cr(C_2H_3O_2)_3$ and $Cr(C_2H_3O_2)_2$

8.56 Name each compound in the following pairs of polyatomic-ion-containing compounds.
(a) $CuNO_3$ and $Cu(NO_3)_2$
(b) $Pb_3(PO_4)_2$ and $Pb_3(PO_4)_4$
(c) $Mn(CN)_3$ and $Mn(CN)_2$
(d) $Co(ClO_3)_2$ and $Co(ClO_3)_3$

8.57 Name each of the following polyatomic-ion-containing compounds.
(a) $Au(NO_3)_3$ (b) Ag_2SO_4
(c) $Al_2(CO_3)_3$ (d) $Pb(OH)_2$

8.58 Name each of the following polyatomic-ion-containing compounds.
(a) $Cd(NO_3)_2$ (b) $SnSO_4$
(c) $AgClO_3$ (d) $CrPO_4$

8.59 Name each of the following polyatomic-ion-containing compounds.
(a) NH_4Cl (b) $(NH_4)_2SO_4$
(c) $Cu_3(PO_4)_2$ (d) Na_3PO_4

8.60 Name each of the following polyatomic-ion-containing compounds.
(a) $Ni_2(SO_4)_3$ (b) NH_4ClO_3
(c) NH_4Br (d) $BeSO_4$

8.61 Write formulas for the following compounds containing polyatomic ions.
(a) silver carbonate
(b) gold(I) nitrate
(c) chromium(III) sulfate
(d) ammonium acetate

8.62 Write formulas for the following compounds containing polyatomic ions.
(a) copper(II) sulfate
(b) manganese(III) hydroxide
(c) ammonium nitrate
(d) magnesium phosphate

Nomenclature for Binary Molecular Compounds (Sec. 8.5)

8.63 Write the number that corresponds to each of the following prefixes.
(a) hepta- (b) penta- (c) tri- (d) deca-

8.64 Write the number that corresponds to each of the following prefixes.
(a) tetra- (b) octa- (c) hexa- (d) ennea-

8.65 Name the following binary molecular compounds.
(a) P_4O_{10} (b) SF_4 (c) CBr_4 (d) ClO_2

8.66 Name the following binary molecular compounds.
(a) S_4N_2 (b) SO_3 (c) IF_7 (d) N_2O_4

8.67 Name the following binary molecular compounds.
(a) H_2S (b) HF (c) NH_3 (d) CH_4

8.68 Name the following binary molecular compounds.
(a) HCl (b) H_2Se (c) N_2H_4 (d) H_2O_2

8.69 Write formulas for the following binary molecular compounds.
(a) iodine monochloride
(b) nitrogen trichloride
(c) sulfur hexafluoride
(d) oxygen difluoride

8.70 Write formulas for the following binary molecular compounds.
(a) disulfur monoxide
(b) tetraphosphorus hexoxide
(c) carbon dioxide
(d) silicon tetrachloride

8.71 Write formulas for the following binary molecular compounds.
(a) phosphine (b) hydrogen bromide
(c) ethane (d) hydrogen telluride

8.72 Write formulas for the following binary molecular compounds.
(a) hydrogen iodide (b) hydrazine
(c) hydrogen peroxide (d) arsine

Nomenclature for Acids (Sec. 8.6)

8.73 Indicate whether or not each of the following hydrogen-containing compounds forms an acid in aqueous solution.
(a) CH_4 (b) H_2S (c) HCN (d) NH_3

8.74 Indicate whether or not each of the following hydrogen-containing compounds forms an acid in aqueous solution.
(a) HCl (b) HClO (c) SiH_4 (d) CH_4

8.75 Classify each of the following compounds as an oxyacid or nonoxyacid.
(a) H_2SO_4 (b) HCN (c) $HClO_3$ (d) HF

8.76 Classify each of the following compounds as an oxyacid or nonoxyacid.
(a) HCl (b) HClO (c) $HC_2H_3O_2$ (d) H_2S

8.77 Name each of the following compounds as acids.
(a) HCl (b) HCN (c) $HClO_2$ (d) H_2SO_4.

8.78 Name each of the following compounds as acids.
(a) H_2S (b) H_2CO_3 (c) $HClO_4$ (d) HI

8.79 Write names and formulas for the "ic" oxyacid of each of the following nonmetals
(a) C (b) N (c) Br (d) Cl

8.80 Write names and formulas for the "ic" oxyacid of each of the following nonmetals.
(a) P (b) S (c) I (d) As

8.81 Write formulas for the following oxyacids.
(a) perbromic acid (b) iodous acid
(c) nitric acid (d) hypobromous acid

8.82 Write formulas for the following oxyacids.
(a) sulfurous acid (b) nitrous acid
(c) iodic acid (d) arsenous acid

8.83 Write formulas for the following acids.
(a) hydrocyanic acid (b) hydrochloric acid
(c) chloric acid (d) bromous acid

8.84 Write formulas for the following acids.
(a) hydrosulfuric acid (b) hydrobromic acid
(c) phosphoric acid (d) bromic acid

8.85 Given that the name of the hypothetical acid H_2ZO_4 is zooic acid, how would you name the following acids?
(a) H_2ZO_2 (b) H_2ZO_5 (c) H_2ZO_3

8.86 Given that the name of the hypothetical acid H_2XO_3 is exic acid, how would you name the following acids?
(a) H_2XO_4 (b) H_2XO (c) H_2XO_2

Additional Problems

8.87 Classify the following compounds as binary ionic, ternary ionic, binary molecular, nonoxyacid, or oxyacid.
(a) Na_3P (b) NF_3 (c) K_3PO_4 (d) HClO

8.88 Classify the following compounds as binary ionic, ternary ionic, binary molecular, nonoxyacid, or oxyacid.
(a) LiOH (b) H_2SO_4 (c) NH_4Cl (d) P_4O_{10}

8.89 Write formulas for the following compounds.
(a) calcium nitride (b) calcium nitrate
(c) calcium nitrite (d) calcium cyanide

8.90 Write formulas for the following compounds.
(a) sodium sulfide (b) sodium sulfate
(c) sodium sulfite (d) sodium thiosulfate

8.91 Write formulas for the following compounds.
(a) potassium phosphide
(b) potassium phosphate
(c) potassium hydrogen phosphate
(d) potassium dihydrogen phosphate

8.92 Write formulas for the following compounds.
(a) magnesium carbide
(b) magnesium bicarbonate
(c) magnesium carbonate
(d) magnesium hydrogen carbonate

8.93 Indicate which compounds in each of the following groups have names that contain the prefix *di-*.
(a) N_2O, Li_2O, CO_2, K_2S
(b) K_2O, K_2CO_3, NO_2, SO_2
(c) BeF_2, $BeCl_2$, SF_2, SCl_2
(d) Au_2O_3, Fe_2O_3, Al_2O_3, N_2O_3

8.94 Indicate which compounds in each of the following groups have names that contain the suffix *-ide*.
(a) CaS, CO, SO_3, Be_3N_2
(b) K_3N, KNO_3, KNO_2, KClO
(c) MgO, AlN, KF, KOH
(d) NaCN, NaOH, Na_2CO_3, NaF

8.95 Indicate which compounds in each of the following groups have names that contain the suffix *-ous* or *-ate*.
(a) $CaCO_3$, H_2CO_3, $Ca(NO_2)_2$, HNO_2
(b) $NaClO_4$, $NaClO_3$, $NaClO_2$, $NaClO$
(c) $HClO_4$, $HClO_3$, $HClO_2$, $HClO$
(d) $LiOH$, $LiCN$, Li_2CO_3, Li_3PO_4

8.96 Indicate which compounds in each of the following groups have names that contain the suffix *-ic* or *-ite*.
(a) $HBr(aq)$, $HBrO$, $HBrO_2$, $HBrO_3$
(b) K_2SO_4, KCN, $KMnO_4$, K_3PO_3
(c) $HI(g)$, $HI(aq)$, HIO_2, KIO_2
(d) NH_4CN, $NaSCN$, Na_2SO_3, $NaNO_2$

8.97 Indicate which compounds in each of the following groups possess molecules or formula units that are pentatomic.
(a) sodium cyanide, sodium thiocyanate, sodium hypochlorite, sodium nitrate
(b) aluminum sulfide, magnesium nitride, beryllium phosphide, potassium hydroxide
(c) beryllium oxide, iron(II) oxide, iron(III) oxide, sulfur dioxide.
(d) gold(I) cyanide, gold(III) cyanide, gold(I) chlorate, gold(III) chlorate

8.98 Indicate which compounds in each of the following groups possess molecules or formula units that are heptatomic.
(a) magnesium cyanide, magnesium oxalate, magnesium chlorite, magnesium perchlorate
(b) aluminum oxide, calcium thiosulfate, beryllium thiocyanate, dinitrogen pentoxide
(c) iron(III) sulfide, iron(III) nitride, iron(III) sulfate, iron(III) hypochlorite
(d) dichlorine heptoxide, sulfur trioxide, sulfur hexafluoride, ammonium sulfide

8.99 The formula of a hydroxide of nickel is $Ni(OH)_3$. What are the formulas of the following nickel compounds in which nickel has the same ionic charge as in $Ni(OH)_3$?
(a) nickel sulfate
(b) nickel oxide
(c) nickel oxalate
(d) nickel nitrate

8.100 The formula of a phosphate of vanadium is VPO_4. What are the formulas of the following vanadium compounds in which vanadium has the same ionic charge as in VPO_4?
(a) vanadium sulfate
(b) vanadium hydroxide
(c) vanadium oxide
(d) vanadium nitrate

8.101 The formula of the ionic compound potassium superoxide is KO_2. The formula of the ionic compound nitronium perchlorate is NO_2ClO_4. What is the formula for the ionic compound nitronium superoxide?

8.102 The formula of the ionic compound calcium perrhenate is $Ca(ReO_4)_2$. The formula of the ionic compound nitrosonium hydrogen sulfate is $NOHSO_4$. What is the formula for the ionic compound nitrosonium perrhenate?

Cumulative Problems

8.103 Give the name of the simplest binary compound that forms between elements with the following electron configurations.
(a) $1s^22s^22p^63s^23p^5$ and $1s^22s^22p^63s^2$
(b) $1s^22s^22p^4$ and $1s^22s^22p^5$

8.104 Give the name of the simplest binary compound that forms between elements with the following electron configurations.
(a) $1s^22s^22p^3$ and $1s^22s^22p^63s^23p^64s^1$
(b) $1s^22s^22p^63s^23p^4$ and $1s^22s^22p^63s^23p^5$

8.105 After determining the value of x in each of the following formulas, name the compounds.
(a) $SiCl_x$ (b) $MgCl_x$ (c) K_xN (d) NCl_x

8.106 After determining the value of x in each of the following formulas, name the compounds.
(a) CCl_x (b) BeO_x (c) NF_x (d) $AlBr_x$

8.107 A compound has the formula $M(XO_3)_2$, where M is a metal and X is a nonmetal. Assign a name to this compound given the following information.
(1) M is a metal with two valence electrons.
(2) X forms a monoatomic ion with a charge of -1.
(3) X has an electronegativity between 2.6 and 2.9.
(4) The sum of the periodic table period numbers for X and M is 6.

8.108 A compound has the formula M_3XO_3, where M is a metal and X is a nonmetal. Assign a name to this compound given the following information.
(1) M is a metal whose electron configuration "ends" in s^1.
(2) Both M and X are found in period 2 of the periodic table.
(3) There are 38 protons in one formula unit of M_3XO_3.

8.109 Determine the name for a binary ionic compound that has the following characteristics.

(1) Positive and negative ions are present in a one to one ratio.
(2) All ions present have the electron configuration $1s^2 2s^2 2p^6$.
(3) One of the elements present has an atomic number that is six less than the atomic number of the other element.

8.110 Determine the name for a binary ionic compound that has the following characteristics.
(1) Positive and negative ions are present in a one-to-two ratio.
(2) All ions present have the electron configuration $1s^2 2s^2 2p^6 3s^2 3p^6$.
(3) Neutral atoms of the more electronegative element present have 17 electrons.

8.111 Determine the name for a binary covalent compound that has the following characteristics.
(1) Its molecules are triatomic.
(2) There are twice as many atoms of the more electronegative element per molecule as there are of the less electronegative element.
(3) Both elements are in period 2 of the periodic table.
(4) Atoms of one element contain four valence electrons, and atoms of the other element contain six valence electrons.

8.112 Determine the name for a binary covalent compound that has the following characteristics.
(1) Its molecules are hexatomic.
(2) There are twice as many atoms of the more electronegative element per molecule as there are of the less electronegative element.
(3) The two elements occupy adjacent positions in period 2 of the periodic table.
(4) The sum of the valence electrons for an atom of each element is 11.

8.113 Determine the name for a compound that has the following characteristics.
(1) Monoatomic and polyatomic ions are present.
(2) The ratio between positive and negative ions is one to two.
(3) A group IIA element that loses one half of its total electrons upon ion formation is present.
(4) The polyatomic ion contains equal numbers of atoms of two nonmetallic elements.
(5) The sum of the atomic numbers for the two elements involved in the polyatomic ion is 13.
(6) One of the elements present in the compound forms a monoatomic ion with a charge of -3 that is isoelectronic with Ne.

8.114 Determine the name for a compound that has the following characteristics.
(1) Monoatomic and polyatomic ions are present.
(2) The ratio between positive and negative ions is one to three.
(3) A group IIIA element whose ion electron configuration is isoelectronic with Ne is present.
(4) The polyatomic ion contains equal numbers of atoms of two nonmetallic elements.
(5) The sum of the atomic numbers for the two elements involved in the polyatomic ion is 9.

Grid Problems

8.115 Select from the grid *all* correct responses for each situation.
(a) compounds in which polyatomic ions are present
(b) compounds in which electron sharing is present
(c) compounds in which ions are present in a one-to-one ratio
(d) compounds in which hexatomic *molecules* are present
(e) compounds in which $+2$ ions are present
(f) compounds in which two elements are present in a one-to-one ratio

1. potassium sulfide	2. nitrogen trifluoride	3. hydrazine
4. copper(II) hydroxide	5. aluminum nitrate	6. ammonium chloride
7. hydrogen sulfide	8. perchloric acid	9. beryllium cyanide

8.116 Select from the grid *all* correct responses for each situation.

1. sulfate	2. nitrate	3. carbonate
4. phosphate	5. cyanide	6. hydroxide
7. ammonium	8. chlorate	9. acetate

(a) polyatomic ions with the same charge as the azide ion
(b) polyatomic ions that are pentatomic
(c) polyatomic ions that contain no oxygen atoms
(d) polyatomic ions that are ternary
(e) polyatomic ions that combine with +2 ions in a one-to-one ratio
(f) polyatomic ions in which covalent bonding is present

8.117 Select from the grid *all* correct responses for each situation.

1. NaCN, BeO	2. $AlCl_3$, NCl_3	3. $NaClO_4$, $NaClO_3$
4. NaClO, $Na_2S_2O_3$	5. H_2S, SCl_2	6. CO_2, Cl_2O
7. $NaClO_4$, $HClO_4$	8. HNO_2, H_2SO_3	9. NaN_3, Na_2O_2

(a) paired compounds that have names ending in -ide
(b) paired compounds that have names containing the word acid
(c) paired compounds that have names containing both a suffix and a prefix
(d) paired compounds that have names ending in -ate
(e) paired compounds that have names containing the suffix -ous
(f) paired compounds that have names containing a numerical prefix

8.118 Select from the grid *all* correct responses for each situation. Assume variable-charge metal ions are named using Roman numerals rather than the "-ic, -ous" system.

1. per____ic	2. ____ide	3. hydro____ic
4. ____ous	5. hypo____ite	6. per____ate
7. hypo____ous	8. tri____ide	9. ____ic

(a) possible prefix and/or suffix usage in the name of a nonoxyacid in aqueous solution
(b) possible prefix and/or suffix usage in the name of an oxyacid in aqueous solution
(c) possible prefix and/or suffix usage in the name of a polyatomic-ion-containing compound
(d) possible prefix and/or suffix usage in the name of a binary ionic compound
(e) possible prefix and/or suffix usage in the name of a molecular compound
(f) possible prefix and/or suffix usage in the name of a compound containing a variable-charge metal ion

CHAPTER NINE

Chemical Calculations: The Mole Concept and Chemical Formulas

9.1 The Law of Definite Proportions

This chapter is the first of two chapters dealing with "chemical arithmetic," that is, with the quantitative relationships between elements and compounds. The emphasis in this chapter will be on quantitative relationships that involve chemical formulas. Emphasis in Chapter 10 will be on quantitative relationships that involve chemical equations.

In Section 4.7 we saw that compounds are pure substances with the following characteristics.

1. They are chemical combinations of two or more elements.
2. They can be broken down into constituent elements by chemical, but not physical, means.
3. They have a definite, constant elemental composition.

In this section we consider further the fact that compounds have a definite composition.

The composition of a compound can be determined by decomposing a weighed amount of the compound into its elements and then determining the masses of the individual elements. Alternatively, determining the mass of a compound formed by the combination of known masses of elements will also lead to its composition.

Studies of composition data for many compounds have led to the conclusion that the percentage of each element present in a given compound does not vary. This conclusion has been formalized into a statement known as the **law of definite proportions**: In a pure compound, the elements are always present in the same definite proportion by mass. The French chemist Joseph Louis Proust (1754–1826) is responsible for the work that established this law as one of the fundamentals of chemistry.

Let us consider how composition data obtained from decomposing a compound are used to illustrate the law of definite proportions. Samples of the compound of *known masses* are decomposed. The masses of the constituent elements present in each sample are then obtained. This experimental information (elemental masses and the original sample mass) is then used to calculate the percent composition of the samples. Within the limits of experimental error, the calculated percentages for any given element present turn out to be the same, validating the law of definite proportions. Example 9.1 gives actual decomposition data for a compound and shows how they are used to verify the law of definite proportions.

EXAMPLE 9.1

Two samples of ammonia (NH_3) of differing mass and from different sources are individually decomposed to yield ammonia's constituent elements (nitrogen and hydrogen). The results of the decomposition experiments are as follows.

	Sample Mass Before Decomposition (g)	Mass of Nitrogen Produced (g)	Mass of Hydrogen Produced (g)
Sample 1	1.840	1.513	0.327
Sample 2	2.000	1.644	0.356

Show that these data are consistent with the law of definite proportions.

Solution

Calculating the percent nitrogen in each sample will be sufficient to show whether the data are consistent with the law. Both nitrogen percentages should come out the same if the law is obeyed.

$$\text{Percent nitrogen} = \frac{\text{mass of nitrogen obtained}}{\text{total sample mass}} \times 100$$

Sample 1

$$\% \, N = \frac{1.513 \, \cancel{g}}{1.840 \, \cancel{g}} \times 100 = 82.22826\% \quad \text{(calculator answer)}$$

$$= 82.23\% \quad \text{(correct answer)}$$

Sample 2

$$\% \text{ N} = \frac{1.644 \text{ g}}{2.000 \text{ g}} \times 100 = 82.2\% \quad \text{(calculator answer)}$$

$$= 82.20\% \quad \text{(correct answer)}$$

Note that the percentages are close to being equal but are not identical. This is due to measuring uncertainties and rounding in the original experimental data. The two percentages are the same to three significant figures. The difference lies in the fourth significant digit. Recall from Section 2.2 that the last of the significant digits in a number (the fourth one here) has uncertainty in it. The two percentages above are considered to be the same within experimental error.

An alternative way of treating the given data to illustrate the law of definite proportions involves calculating the mass ratio between nitrogen and hydrogen This ratio should be the same for each sample; otherwise the two samples are not the same compound.

Sample 1

$$\frac{\text{Mass of N}}{\text{Mass of H}} = \frac{1.513 \text{ g}}{0.327 \text{ g}} = 4.6269113 \quad \text{(calculator answer)}$$

$$= 4.63 \quad \text{(correct answer)}$$

Sample 2

$$\frac{\text{Mass of N}}{\text{Mass of H}} = \frac{1.644 \text{ g}}{0.356 \text{ g}} = 4.6179775 \quad \text{(calculator answer)}$$

$$= 4.62 \quad \text{(correct answer)}$$

Again, slight differences in the ratios are caused by measuring errors and rounding in the original data

PRACTICE EXERCISE 9.1

Two samples of chlorine dioxide (ClO_2) of differing mass and from different sources are individually decomposed to yield the elements chlorine and oxygen. The results of the decomposition experiments are as follows.

	Sample Mass Before Decomposition (g)	Mass of Chlorine Produced (g)	Mass of Oxygen Produced (g)
Sample 1	2.724	1.056	1.668
Sample 2	8.367	3.243	5.124

Show that these data are consistent with the law of definite proportions.

Ans. The percents by mass of chlorine in the two samples are 38.77% and 38.76%.

Similar exercises: Problems 9.1 and 9.2

Chapter 9 · Chemical Calculations: The Mole Concept and Chemical Formulas

TABLE 9.1 Data Illustrating the Law of Definite Proportions

Mass of Ca Used (g)	Mass of S Used (g)	Mass of CaS Formed (g)	Mass of Excess Unreacted Sulfur (g)	Ratio in Which Substances React
55.6	44.4	100.0	none	1.25
55.6	50.0	100.0	5.6	1.25
55.6	100.0	100.0	55.6	1.25
55.6	200.0	100.0	155.6	1.25
111.2	88.8	200.0	none	1.25

An alternative way of examining the constancy of composition found for compounds involves looking at the mass ratios in which elements combine to form compounds. Let us consider the reaction between the elements calcium and sulfur to produce the compound calcium sulfide. Suppose an attempt is made to combine various masses of sulfur with a fixed mass of calcium. A set of possible experimental data for this attempt is given in the first four lines of Table 9.1. Note that, regardless of the mass of S present, only a certain amount, 44.4 g, reacts with the 55.6 g of Ca. The excess S is left over in an unreacted form. The data therefore illustrate that Ca and S will react in only one fixed mass ratio (55.6/44.4 = 1.25) to form CaS. This fact is consistent with the law of definite proportions. Note also that if the amount of Ca used is doubled (line 5 of Table 9.1), the amount of S with which it reacts also doubles (compare lines 1 and 5 of the table). Nevertheless, the ratio in which the substances react (111.2/88.8) still remains 1.25.

9.2 Calculation of Formula Masses

Formula masses play a role in almost all chemical calculations and will be used extensively in later sections of this chapter and in succeeding chapters. The **formula mass** of a substance is the sum of the atomic masses of the atoms present in one formula unit of the substance. Formula masses, like the atomic masses from which they are calculated, are relative masses based on the $^{12}_{6}C$ relative mass scale (Sec. 5.9). They can be calculated for compounds (both molecular and ionic) and for elements that exist in molecular form.

The term *molecular mass* is used interchangeably with formula mass by many chemists in referring to substances that contain discrete molecules. It is incorrect, however, to use the term molecular mass when talking about ionic substances, since molecules are not their basic structural unit (Sec. 7.7).

Some chemistry books use the terms *formula weight* and *molecular weight* rather than *formula mass* and *molecular mass*. This practice is not followed in this book because mass is technically more correct (Sec. 3.3)

Once the formula of a substance has been established, its formula mass is calculated by adding together the atomic masses of all the atoms in the formula. If more than one atom of any element is present, that element's atomic mass must be added as many times as there are atoms of the element present.

Example 9.2 contains sample formula mass calculations. You will note as you follow through this example that significant figures are back to haunt us again. Significant figures (Sec 2.4) are a consideration in any calculation that involves experimental data or numbers obtained from such data. Atomic masses (Sec. 5.9), the basis for formula mass calculations, fall into this latter category.

EXAMPLE 9.2

Calculate the formula masses for the following substances.

(a) H_2O_2 (hydrogen peroxide, a bleaching agent)
(b) $Na_2S_2O_3$ (sodium thiosulfate, a photographic chemical)
(c) $(NH_2)_2CO$ (urea, a chemical fertilizer for crops)
(d) O_3 (ozone, a stratosphere component that absorbs ultraviolet radiation)

Solution

Formula masses are obtained simply by adding the atomic masses of the constituent elements, counting each atomic mass as many times as the symbol for the element occurs in the formula.

(a) A molecule of H_2O_2 contains four atoms: two atoms of H and two atoms of O. The formula mass, the collective mass of these four atoms, is calculated as follows.

$$2 \text{ atoms H} \times \frac{1.00794 \text{ amu}}{1 \text{ atom H}} = 2.01588 \text{ amu}$$

$$2 \text{ atoms O} \times \frac{15.9994 \text{ amu}}{1 \text{ atom O}} = 31.9988 \text{ amu}$$

$$\text{Formula mass} = 34.01468 \text{ amu} \quad \text{(calculator answer)}$$
$$= 34.0147 \text{ amu} \quad \text{(correct answer)}$$

The conversion factors in the calculation were derived from the atomic masses listed on the inside front or back cover of the text. The atomic mass, the mass of an average atom, is 1.00794 amu for H and 15.9994 amu for O. The calculator answer, 34.01468 amu, had to be rounded down one significant figure, since the atomic mass of O is known only to the fourth decimal place. (Recall the rules for significant figures in addition; Sec. 2.4.)

Often conversion factors are not explicitly shown in a formula mass calculation as just shown; the calculation is simplified as follows.

H: 2 × 1.00794 amu = 2.01588 amu
O: 2 × 15.9994 amu = 31.9988 amu
Formula mass = 34.01468 amu (calculator answer)
= 34.0147 amu (correct answer)

(b) Using the simplified calculation method, we calculate the formula mass for $Na_2S_2O_3$ as

$$
\begin{aligned}
\text{Na:} &\quad 2 \times 22.98977 \text{ amu} = 45.97954 \text{ amu} \\
\text{S:} &\quad 2 \times 32.066 \text{ amu} = 64.132 \text{ amu} \\
\text{O:} &\quad 3 \times 15.9994 \text{ amu} = 47.9982 \text{ amu} \\
&\quad \text{Formula mass} = 158.10974 \text{ amu} \quad \text{(calculator answer)} \\
&\qquad\qquad\qquad\quad = 158.110 \text{ amu} \quad \text{(correct answer)}
\end{aligned}
$$

Note how the atomic mass of S limits the precision of the formula mass to thousandths. Why does the periodic table not give a more precise atomic mass for S, you may ask. A more fundamental question is "Why does the preciseness of atomic masses of the elements vary as much as it does?" Some elements have atomic masses known to 0.00001 amu; others are known to only 0.1 amu. A major factor in atomic mass precision is the constancy of the abundance percentages for the isotopes of an element. Abundance percentages fluctuate more for some elements than for others; the less the fluctuation, the greater the atomic mass precision. (The fluctuation is a relative matter; none of the fluctuations is large on an absolute scale.) The atomic masses given in the periodic table are stated to current precision limits.

(c) The formula for this compound contains parentheses. Improper interpretation of parentheses (see Sec. 5.4) is a common error made by students doing formula mass calculations. In this formula, $(NH_2)_2CO$, the subscript 2 outside the parentheses effects all symbols inside the parentheses. Thus we have

$$
\begin{aligned}
\text{N:} &\quad 2 \times 14.0067 \text{ amu} = 28.0134 \text{ amu} \\
\text{H:} &\quad 4 \times 1.00794 \text{ amu} = 4.03176 \text{ amu} \\
\text{C:} &\quad 1 \times 12.011 \text{ amu} = 12.011 \text{ amu} \\
\text{O:} &\quad 1 \times 15.9994 \text{ amu} = 15.9994 \text{ amu} \\
&\quad \text{Formula mass} = 60.05556 \text{ amu} \quad \text{(calculator answer)} \\
&\qquad\qquad\qquad\quad = 60.056 \text{ amu} \quad \text{(correct answer)}
\end{aligned}
$$

(d) This is a very simple problem. We multiply the atomic mass of O by 3 to get the formula mass.

$$\text{O:} \quad 3 \times 15.9994 \text{ amu} = 47.9982 \text{ amu} \quad \text{(calculator and correct answer)}$$

The point of this simple calculation is that elements that exist in molecular form have both a formula mass and an atomic mass. Numerically, these two masses are not the same.

PRACTICE EXERCISE 9.2

Calculate the formula masses for the following substances.

(a) H_2SO_4 (sulfuric acid, an industrial acid)
(b) $C_{16}H_{18}N_2O_5S$ (Penicillin V, an important antibiotic)

Ans. (a) 98.082 amu; (b) 350.395 amu.

Similar exercises: Problems 9.7–9.10

In Example 9.2, all the formula masses were calculated purposely to the maximum number of significant figures possible. This was done to emphasize the fact that the number of significant figures in "input" atomic masses determines the precision of calculated formula masses.

In most chemical calculations you will not need maximum formula mass precision. Other numbers that enter into the calculation will usually restrict the answer to three or four significant figures. Thus, in most problems, formula masses rounded to tenths or hundredths may be used. Occasionally, however, formula masses of maximum precision will be required.

Formula masses, relative masses on the $^{12}_{6}C$ relative mass scale, can be used for mass comparisons. For example, using the formula masses calculated in Example 9.2, we can make the statement that one molecule of $(NH_2)_2CO$ is 1.7656 times as heavy as one molecule of H_2O_2 (60.056 amu/34.0147 amu = 1.7656).

9.3 Percent Composition

A useful piece of information about a compound is its percent composition. **Percent composition** specifies the percent by mass of each element present in a compound. For instance, the percent composition of water is 88.81% oxygen and 11.19% hydrogen.

Percent compositions are frequently used to compare compound compositions. For example, the compounds gold(III) iodide (AuI_3), gold(III) nitrate [$Au(NO_3)_3$], and gold(I) cyanide (AuCN) contain, respectively, 34.10%, 51.43%, and 88.33% gold by mass. If you were given the choice of receiving a gift of 1 lb of one of these three gold compounds, which one would you choose?

The percent composition of a compound can be calculated from experimental decomposition data as was done in Example 9.1. Such a calculation can be carried out even if the formula or the identity of the compound is unknown. Another way to calculate percent composition for a compound is from its chemical formula. There is sufficient information just in the formula for such a calculation, as is illustrated in Example 9.3.

EXAMPLE 9.3

The characteristic odor of pineapple is due to ethyl butyrate, a compound with the formula $C_6H_{12}O_2$. What is the percentage composition (to two decimal places) of ethyl butyrate?

Solution

First, we calculate the formula mass of $C_6H_{12}O_2$, using atomic masses rounded to the hundredths decimal place.

$$
\begin{align*}
C: & \quad 6 \times 12.01 \text{ amu} = 72.06 \text{ amu} \\
H: & \quad 12 \times 1.01 \text{ amu} = 12.12 \text{ amu} \\
O: & \quad 2 \times 16.00 \text{ amu} = 32.00 \text{ amu} \\
& \quad \text{Formula mass} = 116.18 \text{ amu}
\end{align*}
$$

The mass percent of each element in the compound is found by dividing the mass contribution of each element, in amu, by the total mass (formula mass), in amu, and multiplying by 100.

$$\% \text{ element} = \frac{\text{mass of element in one formula unit}}{\text{formula mass}} \times 100$$

Finding percentages, we have

% C: $\dfrac{72.06 \text{ amu}}{116.18 \text{ amu}} \times 100 = 62.024445\%$ (calculator answer)

$= 62.02\%$ (correct answer)

% H: $\dfrac{12.12 \text{ amu}}{116.18 \text{ amu}} \times 100 = 10.432088\%$ (calculator answer)

$= 10.43\%$ (correct answer)

% O: $\dfrac{32.00 \text{ amu}}{116.18 \text{ amu}} \times 100 = 27.543467\%$ (calculator answer)

$= 27.54\%$ (correct answer)

To check our work we can add the percentages of all the parts. They, of course, have to total 100. (On occasion, round-off errors may not cancel, and totals such as 99.99% or 100.01% may be obtained.)

$$62.02\% + 10.43\% + 27.54\% = 99.99\%$$

PRACTICE EXERCISE 9.3

What is the percent composition (to two decimal places) of vitamin C, a compound necessary in small amounts for the normal growth of humans, whose formula is $C_6H_8O_6$?

Ans. 40.91% C; 4.59% H; 54.50% O.

Similar exercises: Problems 9.15 and 9.16

Percent compositions can also be calculated from mass data obtained from compound synthesis or compound decomposition experiments. Example 9.4 shows how synthesis data are treated to yield percent composition.

EXAMPLE 9.4

To produce a 13.50-g sample of adrenaline, a hormone secreted into the bloodstream in times of stress, requires 7.96 g of C, 0.96 g of H, 3.54 g of O, and 1.04 g of N. What is the percent composition of this compound?

Solution

The total mass of the compound sample is given as 13.50 g. We divide the mass of each element present by this total mass (13.50 g) and multiply by 100 to give percentage.

$$\% \text{ element} = \frac{\text{mass of element}}{\text{total sample mass}} \times 100$$

Finding the percentages, we have

% C: $\dfrac{7.96 \text{ g}}{13.50 \text{ g}} = 100 = 58.962963\%$ (calculator answer)

$\phantom{\dfrac{7.96 \text{ g}}{13.50 \text{ g}} = 100} = 59.0\%$ (correct answer)

% H: $\dfrac{0.96 \text{ g}}{13.50 \text{ g}} \times 100 = 7.1111111\%$ (calculator answer)

$\phantom{\dfrac{0.96 \text{ g}}{13.50 \text{ g}} \times 100} = 7.1\%$ (correct answer)

% O: $\dfrac{3.54 \text{ g}}{13.50 \text{ g}} \times 100 = 26.222222\%$ (calculator answer)

$\phantom{\dfrac{3.54 \text{ g}}{13.50 \text{ g}} \times 100} = 26.2\%$ (correct answer)

% N: $\dfrac{1.04 \text{ g}}{13.50 \text{ g}} \times 100 = 7.7037037\%$ (calculator answer)

$\phantom{\dfrac{1.04 \text{ g}}{13.50 \text{ g}} \times 100} = 7.70\%$ (correct answer)

Checking our work, we see that the percentages do add up correctly.

$$59.0\% + 7.1\% + 26.2\% + 7.70\% = 100.0\%$$

PRACTICE EXERCISE 9.4

The sour taste of vinegar is caused by the compound acetic acid. Calculate the percent composition of acetic acid, knowing that the synthesis of 24.03 g of this compound requires 9.61 g of C, 1.62 g of H, and 12.80 g of O.

Ans. 40.0% C, 6.74% H, 53.27% O.

Similar exercises: Problems 9.19 and 9.20

9.4 The Mole: The Chemist's Counting Unit

Two common methods exist for specifying the quantity of material in a sample of a substance: (1) in terms of units of *mass* and (2) in terms of units of *amount*. We measure *mass* by using a balance (Sec. 3.3). Common mass units are gram, kilogram, and pound. For substances that consist of discrete units, we can specify the *amount* of substance present by indicating the number of units present—12, 27, 113, and so on.

We all use both units of mass and units of amount on a daily basis. We work well with this dual system. Sometimes it does not matter which type of unit is used; at other times one system is preferred over the other. When buying potatoes at the grocery store we can decide on quantity in either mass units (10-lb bag, 20-lb bag, etc.) or amount units (9 potatoes, 15 potatoes, etc.). When buying eggs, amount units are used almost exclusively—12 eggs (1 dozen), 24 eggs (2 dozen), and so on. On the other hand, peanuts and grapes are almost always purchased in weighed quantities. It is impractical to count the number of grapes in a bunch. Very few people go to the store with the idea of buying 117 grapes.

In chemistry, as in everyday life, both the mass and amount methods of specifying quantity find use. Again, the specific situation dictates the method used. In laboratory work, practicality dictates working with quantities of known mass (12.3 g, 0.1365 g, etc.). (Counting out a given number of atoms for a laboratory experiment is somewhat impractical, since we cannot see individual atoms.)

In performing chemical calculations, after the laboratory work has been done, it is often useful (even necessary) to think of quantities of substances present in terms of atoms or formula units. A problem exists when this is done—very, very large numbers are always encountered. Any macroscopic sample of a chemical substance contains many trillions of atoms or formula units.

In order to cope with this "large number problem" chemists have found it convenient to use a special counting unit. Employment of such a unit should not surprise you, as specialized counting units are used in many areas. The two most common counting units are *dozen* and *pair*. Other more specialized counting units exist. For example, at an office supply store, paper is sold by the *ream* (500 sheets), and pencils by the *gross* (144 pencils). (See Fig. 9.1.)

The chemist's counting unit is called a *mole*. What is unusual about the mole is its magnitude. A **mole** is 6.02×10^{23} objects. This extremely large number is necessitated by the extremely small size of atoms, molecules, and ions. The use of a tradi-

2 gloves—1 pair
(a)

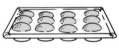

12 rolls—1 dozen
(b)

144 pencils—1 gross
(c)

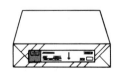

500 sheets of paper—1 ream
(d)

6.02×10^{23} iron atoms—1 mole
(e)

FIGURE 9.1 Some counting units used to denote quantities in terms of groups.

tional counting unit, such as a dozen, would be, at best, only a slight improvement over counting atoms singly.

$$6.02 \times 10^{23} \text{ atoms} = 5.02 \times 10^{22} \text{ dozen atoms} = 1 \text{ mole atoms}$$

Note how the use of the mole counting unit decreases very significantly the magnitude of numbers encountered. The number 1 represents 6.02×10^{23} objects, the number 2, double that number of objects. (Why the number 6.02×10^{23} was chosen as the counting unit rather than some other number will be discussed in Sec. 9.5. A more formal definition of the mole will also be presented in that section.)

The number 6.02×10^{23} also has a special name. **Avogadro's number** is the name given to the numerical value 6.02×10^{23}, the number of particles in a mole. This designation honors the Italian physicist Lorenzo Romano Amedeo Carlo Avogadro (1776–1856), whose pioneering work on gases later proved to be valuable in determining the number of particles present in a given volume of a substance.

In solving mathematical problems dealing with the number of atoms, molecules, or ions present in a given amount of material, Avogadro's number becomes part of the conversion factor used to relate number of particles present to moles present.

$$\boxed{\text{Moles of substance}} \xleftrightarrow{\text{Avogadro's number}} \boxed{\text{Particles of substance}}$$

From the definition

$$1 \text{ mole} = 6.02 \times 10^{23} \text{ objects}$$

two conversion factors can be derived.

$$\frac{1 \text{ mole}}{6.02 \times 10^{23} \text{ objects}} \quad \text{and} \quad \frac{6.02 \times 10^{23} \text{ objects}}{1 \text{ mole}}$$

Example 9.5 illustrates the use of particle-to-mole conversion factors.

EXAMPLE 9.5

How many objects are there in each of the following quantities?

(a) 3.20 moles of water (H_2O) molecules
(b) 1.11 moles of copper (Cu) atoms
(c) 0.782 mole of calcium oxide (CaO) formula units
(d) 3.00 moles of elephants

Solution

We will use dimensional analysis (Sec. 3.7) in solving each part of this problem. All of the parts are similar in that we are given a certain number of moles of substance and want to find the number of particles contained in the given number of moles. All parts can be classified as moles-to-particles problems, and each solution will involve the use of Avogadro's number.

$$\boxed{\text{Moles of substance}} \xrightarrow{\text{Avogadro's number}} \boxed{\text{Particles of substance}}$$

(a) The given quantity is 3.20 moles of H_2O molecules, and the desired quantity is number of H_2O molecules.

$$3.20 \text{ moles } H_2O = ?\ H_2O \text{ molecules}$$

The setup, by dimensional analysis, involves only one conversion factor.

$$3.20 \text{ moles } H_2O \times \frac{6.02 \times 10^{23}\ H_2O \text{ molecules}}{1 \text{ mole } H_2O} = 1.9264 \times 10^{24}\ H_2O \text{ molecules}$$
(calculator answer)
$$= 1.93 \times 10^{24}\ H_2O \text{ molecules}$$
(correct answer)

(b) The given quantity is 1.11 moles of copper atoms, and the desired quantity is the actual number of copper atoms present.

$$1.11 \text{ moles } Cu = ?\ Cu \text{ atoms}$$

The setup, with the same conversion factor as in part (a), is

$$1.11 \text{ moles } Cu \times \frac{6.02 \times 10^{23}\ Cu \text{ atoms}}{1 \text{ mole } Cu} = 6.6822 \times 10^{23}\ Cu \text{ atoms} \quad \text{(calculator answer)}$$
$$= 6.68 \times 10^{23}\ Cu \text{ atoms} \quad \text{(correct answer)}$$

(c) The fact that we are dealing with formula units here, rather than atoms or molecules, does not change the way the problem is solved.

$$0.782 \text{ mole } CaO = ?\ \text{formula units}$$

The conversion factor setup is

$$0.782 \text{ mole } CaO \times \frac{6.02 \times 10^{23}\ CaO \text{ formula units}}{1 \text{ mole } CaO} = 4.70764 \times 10^{23}\ CaO \text{ formula units}$$
(calculator answer)
$$= 4.71 \times 10^{23}\ CaO \text{ formula units}$$
(correct answer)

(d) Use of the mole as a counting unit is usually found only in a chemical context. Technically, however, any type of object can be counted in units of moles. One mole denotes 6.02×10^{23} objects; it does not matter what the objects are—even elephants. Just as we can talk about 3 dozen elephants, we can talk about 3 moles of elephants, although the latter is an above-average-sized herd of elephants.

$$3.00 \text{ moles elephants} \times \frac{6.02 \times 10^{23} \text{ elephants}}{1 \text{ mole elephants}} = 1.806 \times 10^{24} \text{ elephants}$$
(calculator answer)
$$= 1.81 \times 10^{24} \text{ elephants}$$
(correct answer)

PRACTICE EXERCISE 9.5

How many objects (molecules or formula units) are present in each of the following amounts of substance?

(a) 2.67 moles of carbon dioxide (CO_2) molecules
(b) 1.45 moles of sodium chloride (NaCl) formula units

Ans. (a) 1.61×10^{24} molecules of CO_2; (b) 8.73×10^{23} formula units of NaCl.

Similar exercises: Problems 9.29–9.34

It is somewhat unfortunate, because of its similarity to the word molecule, that the name mole was selected as the name for the chemist's counting unit. Students often think that *mole* is an abbreviated form of the word *molecule*. That is not the case. The word mole comes from the Latin *moles*, which means "heap or pile." The word molecule is a diminutive form of *moles* meaning "the smallest piece" of that heap or pile. A mole is a macroscopic amount, a heap or pile of objects, that can easily be seen. A molecule is a particle too small to be seen with the naked eye.

In Example 9.5 we calculated the number of objects present in samples ranging in size from 0.782 mole to 3.20 moles. Our answers were numbers carrying the exponents 10^{23} or 10^{24}. Numbers with these exponents are inconceivably large. The magnitude of Avogadro's number itself is so large that it is almost incomprehensible. There is nothing in our experience to relate to it. (When chemists count, they really count in "big" jumps.) Many attempts have been made to create word pictures of the vast size of Avogadro's number. Such pictures, however, really only hint at its magnitude, since other large numbers must be used in the word pictures. Three such pictures are as follows.

1. Suppose a fraternity decided to throw a large pizza party and ordered 1 mole of pizza pies. Two thousand students showed up at the party. It would take these 2000 students, each downing one pizza every 3 minutes, 2×10^{15} years to finish the stack of pizzas. If everyone living on this planet attended the party (5 billion pizza eaters), it would still take 1×10^9 years to eat the pizzas—a billion years of nonstop pizza eating.
2. If each one of the 5 billion people on Earth were made a millionaire (receiving 1 million dollar bills), we would still need 120 million other worlds, each inhabited with the same number of millionaires, in order to have Avogadro's number of dollar bills in circulation. (Where would we put all of the dollar bills?) One mole of dollar bills is a lot of money, enough to pay all the expenses of the United States government for the next billion years or so (in inflated dollars).
3. It would take an ultramodern computer that can count 100 million times a second 190 million years to count 6.02×10^{23} times (Avogadro's number).

9.5 The Mass of a Mole

How much does a mole weigh; that is, what is its mass? Are you uncertain about the answer to that question? Let us, then, consider a similar (but more familiar) question first: "How much does a dozen weigh?" Your response is now immediate. "A dozen what?" you reply. The mass of a dozen identical objects obviously depends on the identity of the object. For example, the mass of a dozen elephants will be "somewhat greater" than the mass of a dozen marshmallows. The mole, like the dozen, is a counting unit. Similarly, the mass of a mole of objects will depend on the identity of the object. Thus, the mass of a mole, *molar mass*, is not one set number; it varies, being different for each different chemical substance. This is in direct contrast to the *molar number*, Avogadro's number, which is the same for all chemical substances.

The **molar mass of an element**, when the element is in atomic form, is a mass in grams that is numerically equal to the atomic mass of the element. Thus, if we know the atomic mass of an element, we also know the mass of 1 mole of atoms of the element. The two quantities are numerically the same, differing only in units. For the elements carbon, oxygen, and sodium we can write the following mass–number relationships.

$$\text{Mass of 1 carbon atom} = 12.011 \text{ amu} \quad \text{(atomic mass)}$$
$$\text{Mass of 1 mole of carbon atoms} = 12.011 \text{ g} \quad \text{(molar mass)}$$

$$\text{Mass of 1 oxygen atom} = 15.9994 \text{ amu} \quad \text{(atomic mass)}$$
$$\text{Mass of 1 mole of oxygen atoms} = 15.9994 \text{ g} \quad \text{(molar mass)}$$

$$\text{Mass of 1 sodium atom} = 22.98977 \text{ amu} \quad \text{(atomic mass)}$$
$$\text{Mass of 1 mole of sodium atoms} = 22.98977 \text{ g} \quad \text{(molar mass)}$$

It is not a coincidence that the mass in grams of a mole of atoms of an element and the element's atomic mass are numerically equal. Avogadro's number has the value that it has in order to cause this relationship to exist. Experimentally it was determined that when 6.02×10^{23} atoms of an element are present, molar masses (in grams) and atomic masses (in amu) are numerically equal. Again, only when the chemist's counting unit has the value 6.02×10^{23} does this relationship hold. Avogadro's number is thus the "connecting link" between the chemist's microscopic mass scale (amu) and macroscopic mass scale (grams). The use of the mass unit grams in specifying the molar mass of an element is a must. The use of other mass units would require a different counting unit than 6.02×10^{23} for a numerical match between atomic mass and molar mass.

The molecular form of an element will have a different molar mass than its atomic form. Consider the element chlorine, which is found in nature in the form of diatomic molecules (Cl_2). The mass of 1 mole of chlorine atoms (Cl) is different from the mass of 1 mole of chlorine molecules (Cl_2). Since there are two atoms in each molecule of chlorine, the molar mass of molecular chlorine is twice the molar mass of atomic chlorine. The following relationships hold for chlorine, whose atomic mass is 35.453 amu.

$$6.02 \times 10^{23} \text{ Cl atoms} = 1 \text{ mole Cl atoms} = 35.453 \text{ g Cl}$$

$$6.02 \times 10^{23} \text{ Cl}_2 \text{ molecules} = 1 \text{ mole Cl}_2 \text{ molecules} = 70.906 \text{ g Cl}_2$$

Note that 1 mole of molecular chlorine contains twice as many atoms as 1 mole of atomic chlorine; however, the number of discrete particles present is the same (Avogadro's number) in both cases. For atomic chlorine, atoms are considered to be the object counted; for molecular chlorine, molecules are considered to be the discrete particle counted. There is the same number of atoms in the former case as there is of molecules in the latter case. This atomic–molecular chlorine situation is analogous to the difference between a dozen shoes and a dozen pairs of shoes. Both the mass and the actual number of shoes for the dozen pairs of shoes are double those for the dozen shoes.

The existence of some elements in molecular form becomes a source of error in chemical calculations if care is not taken to distinguish properly between atomic and molecular forms of the element. The phrase "1 mole of chlorine" is an ambiguous term. Does it mean 1 mole of chlorine atoms (Cl), or does it mean 1 mole of chlorine molecules (Cl_2)?

The **molar mass of a compound** is a mass in grams that is numerically equal to the formula mass of the compound. Thus, for compounds a numerical equivalence exists between molar mass and formula mass if the molar mass is specified in grams. When we add atomic masses to get the formula mass (in amu) of a compound, we are simultaneously finding the mass of 1 mole of compound (in grams). The molar mass–formula mass relationships for the compounds water (H_2O), ammonia (NH_3), and barium chloride ($BaCl_2$) are

Mass of 1 H_2O molecule = 18.0152 amu (formula mass)
Mass of 1 mole of H_2O molecules = 18.0152 g (molar mass)

Mass of 1 NH_3 molecule = 17.0304 amu (formula mass)
Mass of 1 mole of NH_3 molecules = 17.0304 g (molar mass)

Mass of 1 $BaCl_2$ formula unit = 208.25 amu (formula mass)
Mass of 1 mole of $BaCl_2$ formula units = 208.25 g (molar mass)

Figure 9.2 pictures molar quantities of a number of common substances. Note

Figure 9.2 One mole of a substance is that amount of substance with a mass, in grams, numerically equal to the atomic mass, molecular mass, or formula mass.

again how the mass of a mole varies in numerical value. On the other hand, all of the pictured amounts of substances contain the same number of units—6.02×10^{23} atoms, molecules, or formula units.

It should now be very evident to you why the chemist's counting unit, the mole, has the value it does—6.02×10^{23}. *Avogadro's number represents the experimentally determined number of atoms, molecules, or formula units contained in a sample of a pure substance with a mass in grams numerically equal to the atomic mass or formula mass of the pure substance.*

The numerical match between molar mass and atomic or formula mass makes the calculation of the mass of any given number of moles of a substance a very simple procedure. In solving problems of this type, the numerical value of the molar mass becomes part of the conversion factor used to convert from moles to grams.

$$\boxed{\text{Moles of substance}} \xrightarrow{\text{molar mass}} \boxed{\text{Grams of substance}}$$

For example, for the compound CO_2, which has a formula mass of 44.0 amu, we can write the equality

$$44.0 \text{ g } CO_2 = 1 \text{ mole } CO_2$$

From this statement two conversion factors can be written.

$$\frac{44.0 \text{ g } CO_2}{1 \text{ mole } CO_2} \quad \text{and} \quad \frac{1 \text{ mole } CO_2}{44.0 \text{ g } CO_2}$$

Example 9.6 illustrates the use of gram-to-mole conversion factors like these.

EXAMPLE 9.6

Calculate the mass, in grams, of each of the following molar quantities of matter.

(a) 2.05 moles of IF_3
(b) 6.50 moles of Na_2SO_4
(c) 2.22 moles of O_2 molecules
(d) 2.22 moles of O atoms

Solution

We will use dimensional analysis to solve each of these problems. The relationship between molar mass and atomic or formula mass will serve as a conversion factor in the setup of each problem.

(a) The given quantity is 2.05 moles of IF_3, and the desired quantity is grams of IF_3. Thus,

$$2.05 \text{ moles } IF_3 = ? \text{ g } IF_3$$

The formula mass of IF_3 is calculated to be 184 amu.

$$\begin{aligned}
\text{I: } & 1 \times 127 \text{ amu} = 127 \text{ amu} \\
3 \text{ F: } & 3 \times 19 \text{ amu} = \underline{57 \text{ amu}} \\
& \phantom{3 \times 19 \text{ amu} = } 184 \text{ amu}
\end{aligned}$$

Section 9.5 · The Mass of a Mole

(The formula mass need contain only three significant figures since the given quantity, 2.05 moles, contains only that number of significant figures). Using the formula mass, we can write the equality

$$184 \text{ g IF}_3 = 1 \text{ mole IF}_3$$

The dimensional analysis setup for the problem, with the gram-to-mole equation as a conversion factor, is

$$2.05 \text{ moles IF}_3 \times \frac{184 \text{ g IF}_3}{1 \text{ mole IF}_3} = 377.2 \text{ g IF}_3 \quad \text{(calculator answer)}$$

$$= 377 \text{ g IF}_3 \quad \text{(correct answer)}$$

(b) The given quantity is 6.50 moles Na_2SO_4, and the desired quantity is grams of Na_2SO_4.

$$6.50 \text{ moles } Na_2SO_4 = ? \text{ g } Na_2SO_4$$

The calculated formula mass of Na_2SO_4 is 142 amu. Thus,

$$142 \text{ g } Na_2SO_4 = 1 \text{ mole } Na_2SO_4$$

With this relationship as a conversion factor, the setup for the problem becomes

$$6.50 \text{ moles } Na_2SO_4 \times \frac{142 \text{ g } Na_2SO_4}{1 \text{ mole } Na_2SO_4} = 923 \text{ g } Na_2SO_4 \quad \text{(calculator and correct answer)}$$

(c) The given quantity is 2.22 moles of O_2 molecules. The desired quantity is grams of O_2 molecules. Thus

$$2.22 \text{ moles } O_2 = ? \text{ g } O_2$$

We are dealing here with diatomic oxygen molecules (O_2) and not oxygen atoms. Thus, 32.0 amu, twice the atomic mass of oxygen, is the formula mass used in the mole-to-gram equality statement.

$$32.0 \text{ g } O_2 = 1 \text{ mole } O_2$$

With this relationship as a conversion factor, the setup becomes

$$2.22 \text{ moles } O_2 \times \frac{32.0 \text{ g } O_2}{1 \text{ mole } O_2} = 71.04 \text{ g } O_2 \quad \text{(calculator answer)}$$

$$= 71.0 \text{ g } O_2 \quad \text{(correct answer)}$$

(d) The given quantity is 2.22 moles of O atoms, and the desired quantity is grams of O atoms. Thus,

$$2.22 \text{ moles O} = ? \text{ g O}$$

This problem differs from the previous one in that atoms, rather than molecules, of oxygen are being counted. The atomic mass of oxygen is 16.0 amu, and the mole-to-gram equality statement is

$$16.0 \text{ g O} = 1 \text{ mole O}$$

With this relationship as a conversion factor, the setup becomes

$$2.22 \text{ moles O} \times \frac{16.0 \text{ g O}}{1 \text{ mole O}} = 35.52 \text{ g O} \quad \text{(calculator answer)}$$

$$= 35.5 \text{ g O} \quad \text{(correct answer)}$$

The answer to part (c) is double the answer here. That is the way it should be. An oxygen molecule (O_2) has a mass twice that of an oxygen atom (O).

PRACTICE EXERCISE 9.6

Calculate the mass, in grams, of each of the following molar quantities of matter.

(a) 1.81 moles of ClF_5 molecules   (b) 0.622 mole of O_3 molecules

Ans. (a) 236 g ClF_5; (b) 29.9 g O_3.

Similar exercises: Problems 9.39 and 9.40

The atomic mass unit (amu) and the grams (g) unit are related to each other through Avogadro's number.

$$6.02 \times 10^{23} \text{ amu} = 1.00 \text{ g}$$

That the above is the case can be deduced from the following equalities.

$$\text{Atomic mass of N} = \text{mass of 1 atom} = 14.0 \text{ amu}$$

$$\text{Molar mass of N} = \text{mass of } 6.02 \times 10^{23} \text{ atoms} = 14.0 \text{ g}$$

Since the second line of the equalities involves 6.02×10^{23} times as many atoms and the masses come out numerically equal, the gram unit must be 6.02×10^{23} times larger than the amu unit.

EXAMPLE 9.7

What is the mass in grams of an atom whose mass on the amu scale is 28.1 amu?

Solution

This is a one-step problem based on the conversion factor

$$6.02 \times 10^{23} \text{ amu} = 1.00 \text{ g}$$

The dimensional analysis setup is

$$28.1 \text{ amu} \times \frac{1.00 \text{ g}}{6.02 \times 10^{23} \text{ amu}} = 4.6677741 \times 10^{-23} \text{ g} \quad \text{(calculator answer)}$$

$$= 4.67 \times 10^{-23} \text{ g} \quad \text{(correct answer)}$$

PRACTICE EXERCISE 9.7

What is the mass in grams of an atom whose mass on the amu scale is 40.1 amu?

Ans. 6.66×10^{-23} g.

Similar exercises: Problems 9.45 and 9.46

In Section 9.4 we defined the mole simply as

$$1 \text{ mole} = 6.02 \times 10^{23} \text{ objects}$$

Although this statement conveys correct information (the value of Avogadro's number to three significant figures is 6.02×10^{23}), it is not the officially accepted definition for Avogadro's number. The official definition, which is mass based, is: The **mole** is the amount of substance in a system that contains as many elementary units (atoms, molecules, or formula units) as there are $^{12}_{6}\text{C}$ atoms in exactly 12.00000 grams of $^{12}_{6}\text{C}$. The value of Avogadro's number is an experimentally determined quantity (the number of atoms in exactly 12.00000 g of $^{12}_{6}\text{C}$ atoms) rather than an exactly defined quantity. Its value is not even mentioned in the definition. The most up-to-date experimental value for Avogadro's number is 6.022137×10^{23}, which is consistent with our previous definition. In calculations we will never need such a precise value as the experimentally determined one. Most often three significant figures will suffice; occasionally four significant figures (6.022×10^{23}) will be needed for a calculation. But remember, more significant figures are available if the need ever arises for their use.

9.6 Counting Particles by Weighing

In a laboratory situation it is often necessary to work with equal numbers of atoms of two different substances or twice as many atoms of one type than of another, and so forth. Atoms are so small that is impossible to count them one by one. How does the chemist resolve this requirement of "equal or proportional numbers of atoms"? The problem is solved by means of the concept of "counting by weighing," a concept closely related to molar mass (Sec. 9.5).

To illustrate the "counting-by-weighing" concept, let us compare the masses of two kinds of atoms—oxygen and nitrogen—when varying but equal numbers of atoms are present. (The choice of elements in the comparison is arbitrary; any two could be used.) From a table of atomic masses we determine that the mass of a single oxygen atom (on the average) is 16.0 amu and that of a single nitrogen atom (on the average) is 14.0 amu. Using these single atom masses and varying the equal numbers of atoms present from 1 to 2 to 5 to 100 to 1 billion to Avogadro's number generates the mass comparisons given in Table 9.2. Note that the mass ratio is identical in every case; also note that the numerical value of this ratio is the same as the ratio of the atomic masses of oxygen and nitrogen. What Table 9.2 establishes is that *the mass ratio of equal numbers of atoms of two elements is the same as the ratio of the atomic masses of the two elements.* This statement is true regardless of the elements involved in the comparison. The ratio between two other elements will not be 1.143 (as was the case for O and N); each pair of elements will have its own unique mass ratio determined by the atomic masses of the elements involved.

Turning the statement of the conclusion derived from Table 9.2 around (reversing it) gives an even more useful generalization (from a laboratory viewpoint): *If samples of two or more elements have mass ratios equal to their atomic mass ratios, they must contain equal numbers of atoms.* Thus, for example, since the atomic mass ratio between Au and Cu is 3.100 (197.0/63.55), any time samples of these elements are weighed out

TABLE 9.2 Mass Ratios for Varying Equal Numbers of Oxygen and Nitrogen Atoms

Atomic Ratio	Mass Ratio of Equal Numbers of Atoms
$\dfrac{1 \text{ atom O}}{1 \text{ atom N}}$	$\dfrac{1 \times 16.00 \text{ amu}}{1 \times 14.00 \text{ amu}} = \dfrac{16.00}{14.00} = 1.143$
$\dfrac{2 \text{ atoms O}}{2 \text{ atoms N}}$	$\dfrac{2 \times 16.00 \text{ amu}}{2 \times 14.00 \text{ amu}} = \dfrac{32.00}{28.00} = 1.143$
$\dfrac{5 \text{ atoms O}}{5 \text{ atoms N}}$	$\dfrac{5 \times 16.00 \text{ amu}}{5 \times 14.00 \text{ amu}} = \dfrac{80.00}{70.00} = 1.143$
$\dfrac{100 \text{ atoms O}}{100 \text{ atoms N}}$	$\dfrac{100 \times 16.00 \text{ amu}}{100 \times 14.00 \text{ amu}} = \dfrac{1600.}{1400.} = 1.143$
$\dfrac{10^9 \text{ atoms O}}{10^9 \text{ atoms N}}$	$\dfrac{10^9 \times 16.00 \text{ amu}}{10^9 \times 14.00 \text{ amu}} = \dfrac{1.600 \times 10^{10}}{1.400 \times 10^{10}} = 1.143$
$\dfrac{6.022 \times 10^{23} \text{ atoms O}}{6.022 \times 10^{23} \text{ atoms N}}$	$\dfrac{6.022 \times 10^{23} \times 16.00 \text{ amu}}{6.022 \times 10^{23} \times 14.00 \text{ amu}} = \dfrac{9.635 \times 10^{24}}{8.431 \times 10^{24}} = 1.143$

in a 3.100-to-1.000 ratio the samples will contain equal numbers of atoms. A 3.100-g sample of Au contains the same number of atoms as a 1.000-g sample of Cu. Similarly, 3.100-lb and 3.100-ton samples of Au contain, respectively, the same numbers of atoms as 1.000-lb and 1.000-ton samples of Cu. Any time, regardless of units (g, lb, kg, ton, etc.), Au and Cu are present in a 3.100-to-1.000 mass ratio, equal numbers of atoms are present.

Our comparisons (and generalizations) so far have involved atoms and elements. Parallel comparisons (and generalizations) exist for molecules (or formula units) and compounds. The generalizations for compounds differ from those for elements only in that "formula mass" has replaced atomic mass and "molecules" (or formula units) has replaced atoms. The substitution of formula masses for atomic masses is valid because both are based on the same $^{12}_{6}C$ scale (Sec. 9.2). Therefore, the following samples all contain the same number of molecules.

$$18.0 \text{ g H}_2\text{O} \quad \text{(formula mass of H}_2\text{O} = 18.0 \text{ amu)}$$
$$44.0 \text{ g CO}_2 \quad \text{(formula mass of CO}_2 = 44.0 \text{ amu)}$$
$$64.1 \text{ g SO}_2 \quad \text{(formula mass of SO}_2 = 64.1 \text{ amu)}$$

We can now make the following generalization, which is applicable to both elements and compounds for **counting by weighing:** samples of two or more pure substances (elements or compounds) found to have mass ratios equal to the ratios of their atomic or formula masses must contain identical numbers of particles (atoms, molecules, or formula units). The following samples therefore contain the same numbers of particles.

$$27.0 \text{ g Al} \quad \text{(atomic mass} = 27.0 \text{ amu)}$$
$$18.0 \text{ g H}_2\text{O} \quad \text{(formula mass} = 18.0 \text{ amu)}$$

The particles for water are molecules, and the particles for aluminum are atoms.

We see, therefore, that to obtain samples of elements or compounds containing equal numbers of particles, we merely weigh out quantities (in any units) whose mass ratio is numerically equal to the ratio of the substance's atomic or formula masses.

EXAMPLE 9.8

Carbon monoxide (CO) and nitrogen monoxide (NO) are air pollutants that enter the atmosphere in automobile exhaust. Using the concept of "counting particles by weighing," determine whether equal numbers of molecules of these two pollutants are present in the following pairs of samples.

(a) 10.00 g CO and 11.00 g NO   (b) 7.011 g CO and 7.511 g NO

Solution

The formula mass of CO is 28.01 amu, and that of NO 30.01 amu. The ratio of formula masses for these two compounds is

$$\frac{\text{Formula mass CO}}{\text{Formula mass NO}} = \frac{28.01 \text{ amu}}{30.01 \text{ amu}} = 0.93335554 \quad \text{(calculator answer)}$$

$$= 0.9334 \quad \text{(correct answer)}$$

The ratio of masses for the two samples in each pair must equal this number, 0.9334, if equal numbers of molecules are present.

(a) The ratio of the masses of the two samples is

$$\frac{\text{Mass CO}}{\text{Mass NO}} = \frac{10.00 \text{ g}}{11.00 \text{ g}} = 0.9090909 \quad \text{(calculator answer)}$$

$$= 0.9091 \quad \text{(correct answer)}$$

Since this ratio is not equal to 0.9334, equal numbers of molecules are not present. The fact that the ratio is less than 0.9934 indicates that there are fewer CO molecules (the numerator of the ratio) than NO molecules.

(b) The ratio of the masses of the two samples is

$$\frac{\text{Mass CO}}{\text{Mass NO}} = \frac{7.011 \text{ g}}{7.511 \text{ g}} = 0.93343096 \quad \text{(calculator answer)}$$

$$= 0.9334 \quad \text{(correct answer)}$$

This time the atomic mass ratio and the mass ratio are the same (to four significant figures). Thus, equal numbers of molecules are present (to four significant figures).

PRACTICE EXERCISE 9.8

Using the concept of "counting particles by weighing," determine whether each of the following pairs of samples contains the same number of particles.

(a) 160.3 g of Ca and 107.9 g Al   (b) 60.07 g of H_2O and 48.66 g NH_3

Ans. (a) same number of particles present in each sample;
(b) different numbers of particles present in the two samples.

Similar exercises: Problems 9.49 and 9.50

9.7 The Mole and Chemical Formulas

A chemical formula has two meanings or interpretations: (1) a microscopic level interpretation and (2) a macroscopic level interpretation.

The first of these two interpretations was discussed in Section 5.4. At a **microscopic level** a chemical formula indicates the number of atoms of each element present in one molecule or formula unit of a substance. *The subscripts in the formula are interpreted to mean the numbers of atoms of the various elements present in one formula unit of the substance.* The formula C_2H_6, interpreted at the microscopic level, conveys the information that two atoms of C and six atoms of H are present in one molecule of C_2H_6.

Now that the mole concept has been introduced, a macroscopic interpretation of formulas is possible. At a **macroscopic level** a chemical formula indicates the number of moles of atoms of each element present in 1 mole of a substance. *The subscripts in the formula are interpreted to mean the numbers of moles of atoms of the various elements present in 1 mole of the substance.* The designation "macroscopic" is given to this molar interpretation, since moles are "laboratory-sized" quantities of atoms. The formula C_2H_6, interpreted at the macroscopic level, conveys the information that 2 moles of C atoms and 6 moles of H atoms are present in 1 mole of C_2H_6.

It is now evident, then, that the subscripts in a formula always carry a dual meaning: "atom" at the microscopic level and "moles of atoms" at the macroscopic level.

The validity of the molar interpretation for subscripts in a formula derives from the following line of reasoning. In x molecules of C_2H_6, where x is any number, there are $2x$ atoms of C and $6x$ atoms of H. Regardless of the value of x, there must always be 2 times as many C atoms as molecules and six times as many H atoms as molecules; that is,

$$\text{Number of } C_2H_6 \text{ molecules} = x$$
$$\text{Number of C atoms} = 2x$$
$$\text{Number of H atoms} = 6x$$

Now let x equal 6.02×10^{23}, the value of Avogadro's number. With this x value, the following statements are true.

Number of C_2H_6 molecules = 6.02×10^{23}

Number of C atoms = $2 \times 6.02 \times 10^{23}$ = 1.204×10^{24} (calculator answer)
= 1.20×10^{24} (correct answer)

Number of H atoms = $6 \times 6.02 \times 10^{23}$ = 3.612×10^{24} (calculator answer)
= 3.61×10^{24} (correct answer)

Since 6.02×10^{23} is equal to 1 mole, 1.20×10^{24} to 2 moles, and 3.61×10^{24} to 6 moles, these statements may be changed to read

$$\text{Number of } C_2H_6 \text{ molecules} = 1 \text{ mole}$$
$$\text{Number of C atoms} = 2 \text{ moles}$$
$$\text{Number of H atoms} = 6 \text{ moles}$$

Thus, the mole ratio is the same as the subscript ratio: 2 to 6.

In calculations where the moles of a particular element within a compound are asked for, the subscript of that particular element in the chemical formula of the compound becomes part of the conversion factor used to convert from moles of compound to moles of element within the compound.

$$\boxed{\text{Moles of compound}} \xrightarrow{\text{formula subscript}} \boxed{\text{Moles of element in compound}}$$

For example, again using C_2H_6 as our chemical formula, we can write the following conversion factors.

For C: $\dfrac{2 \text{ moles C atoms}}{1 \text{ mole } C_2H_6 \text{ molecules}}$ or $\dfrac{1 \text{ mole } C_2H_6 \text{ molecules}}{2 \text{ moles C atoms}}$

For H: $\dfrac{6 \text{ moles H atoms}}{1 \text{ mole } C_2H_6 \text{ molecules}}$ or $\dfrac{1 \text{ mole } C_2H_6 \text{ molecules}}{6 \text{ moles H atoms}}$

Example 9.9 illustrates the use of conversion factors of this type in a problem-solving context.

EXAMPLE 9.9

The compound aspirin, the most widely used of all pain relievers, has the formula $C_9H_8O_4$. How many moles of each type of atom present in $C_9H_8O_4$ are contained in a 1.75-mole sample of this compound?

Solution

The formula $C_9H_8O_4$ specifies that 1 mole of this substance will contain 9 moles of carbon atoms, 8 moles of hydrogen atoms, and 4 moles of oxygen atoms. This information leads to the following conversion factors.

$$\dfrac{9 \text{ moles C atoms}}{1 \text{ mole } C_9H_8O_4}, \quad \dfrac{8 \text{ moles H atoms}}{1 \text{ mole } C_9H_8O_4}, \quad \text{and} \quad \dfrac{4 \text{ moles O atoms}}{1 \text{ mole } C_9H_8O_4}$$

Using the first of these conversion factors, the moles of C atoms present are calculated as follows.

$$1.75 \text{ moles } C_9H_8O_4 \times \dfrac{9 \text{ moles C atoms}}{1 \text{ mole } C_9H_8O_4} = 15.75 \text{ moles C atoms} \quad \text{(calculator answer)}$$

$$= 15.8 \text{ moles C atoms} \quad \text{(correct answer)}$$

Similarly, using the second conversion factor, the moles of H atoms present are calculated.

$$1.75 \text{ moles } C_9H_8O_4 \times \frac{8 \text{ moles H atoms}}{1 \text{ mole } C_9H_8O_4} = 14 \text{ moles H atoms} \quad \text{(calculator answer)}$$
$$= 14.0 \text{ moles H atoms} \quad \text{(correct answer)}$$

Finally, using the third conversion factor, we obtain the moles of O atoms present.

$$1.75 \text{ moles } C_9H_8O_4 \times \frac{4 \text{ moles O atoms}}{1 \text{ mole } C_9H_8O_4} = 7 \text{ moles O atoms} \quad \text{(calculator answer)}$$
$$= 7.00 \text{ moles O atoms} \quad \text{(correct answer)}$$

If the question "How many total moles of atoms are present?" were asked, we could obtain the answer by adding the moles of C, H, and O atoms just calculated.

15.8 moles C atoms + 14.0 moles H atoms + 7.00 moles O atoms = 36.8 moles total atoms

(calculator and correct answer)

Alternatively, by noting that there are a total of 21 moles of atoms present in 1 mole of $C_9H_8O_4$, we could calculate the total moles of atoms present using the following setup.

$$1.75 \text{ moles } C_9H_8O_4 \times \frac{21 \text{ moles atoms}}{1 \text{ mole } C_9H_8O_4} = 36.75 \text{ moles atoms} \quad \text{(calculator answer)}$$
$$= 36.8 \text{ moles atoms} \quad \text{(correct answer)}$$

PRACTICE EXERCISE 9.9

How many moles of each type of atom are present in each of the following molar quantities?

(a) 0.753 mole of CO_2 molecules   (b) 1.31 moles of P_4O_{10} molecules

Ans. (a) 0.753 moles of C atoms and 1.51 moles of O atoms; (b) 5.24 moles of P atoms and 13.1 moles of O atoms.

Similar exercises: Problems 9.57 and 9.58

9.8 The Mole and Chemical Calculations

In this section we combine the major points we've learned about moles in previous sections to produce a general approach to problem solving that is applicable to a variety of types of chemical calculations.

The three quantities most often calculated in chemical problems are

1. The number of *particles* of a substance, that is, the number of atoms, molecules, or formula units.

2. The number of *moles* of a substance.
3. The number of *grams* (mass) of a substance.

These quantities are interrelated. The conversion factors dealing with these relationships, as previously noted, involve the concepts of (1) Avogadro's number, (2) molar mass, and (3) molar interpretation of chemical formula subscripts.

1. Avogadro's number (Sec. 9.4) provides a relationship between the number of particles of a substance and the number of moles of the same substance.

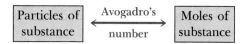

2. Molar mass (Sec. 9.5) provides a relationship between the number of grams of a substance and the number of moles of the same substance.

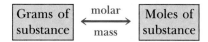

3. Molar interpretation of chemical formula subscripts (Sec. 9.7) provides a relationship between the number of moles of a substance and the number of moles of its component parts.

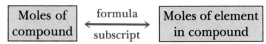

The preceding three concepts can be combined into a single diagram that is very useful in problem solving. This diagram, Figure 9.3, can be viewed as a "road map" from which conversion factor sequences (pathways) can be obtained. It gives all the needed relationships for solving two general types of problems.

1. Calculations for which information (moles, grams, particles) is given about a particular substance, and additional information (moles, grams, particles) is needed concerning the *same* substance.

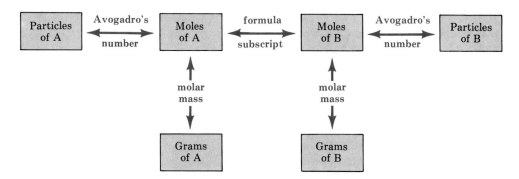

FIGURE 9.3 Useful relationships for solving chemical-formula-based problems.

2. Calculations for which information (moles, grams, particles) is given about a particular substance, and information (moles, grams, particles) is needed concerning a *component* of that same substance.

For the first type of problem, only the left side of Figure 9.3 (the A boxes) is needed. For problems of the second type, both sides of the diagram (both A and B boxes) are used.

The thinking pattern needed to use Figure 9.3 is very simple.

1. Determine which box in the diagram represents the *given* quantity in the problem.
2. Next, locate the box that represents the *desired* quantity.
3. Finally, follow the indicated pathway that takes you from the *given* quantity to the *desired* quantity. This involves simply following the arrows. There will always be only one pathway possible for the needed transition.

Examples 9.10–9.13 illustrate a few of the types of problems that can be solved using the relationships shown in Figure 9.3. In the first two examples we will need only the A side of the diagram; the following two examples make use of both the A and B sides of the diagram.

EXAMPLE 9.10

The koala bear feeds exclusively on eucalyptus leaves. Its digestive system detoxifies the eucalyptus oil, a poison to other animals. The predominant compound in eucalyptus oil is eucalyptol, which has the formula $C_{10}H_{18}O$. How many eucalyptol molecules are present in a 2.00-g sample of this compound?

Solution

We will solve this problem by using the three steps of dimensional analysis (Sec. 3.7) and Figure 9.3.

STEP 1 The given quantity is 2.00 g of $C_{10}H_{18}O$, and the desired quantity is molecules of $C_{10}H_{18}O$.

$$2.00 \text{ g } C_{10}H_{18}O = ? \text{ molecules } C_{10}H_{18}O$$

In terms of Figure 9.3, this is a "grams of A" to "particles of A" problem. We are given grams of substance A and desire to find particles molecules) of that same substance.

STEP 2 Figure 9.3 gives us the "pathway" (sequence of conversion factors) needed to work the problem. We want to start in the "grams of A" box and end up in the "particles of A" box. The pathway is

$$\boxed{\text{Grams of A}} \xrightarrow{\text{molar mass}} \boxed{\text{Moles of A}} \xrightarrow{\text{Avogadro's number}} \boxed{\text{Particles of A}}$$

Using dimensional analysis, the setup for this sequence of conversion factors is

$$2.00 \text{ g } \cancel{C_{10}H_{18}O} \times \frac{1 \text{ mole } \cancel{C_{10}H_{18}O}}{154 \text{ g } \cancel{C_{10}H_{18}O}} \times \frac{6.02 \times 10^{23} \text{ molecules } C_{10}H_{18}O}{1 \text{ mole } \cancel{C_{10}H_{18}O}}$$

grams A $\longrightarrow$ moles A $\longrightarrow$ particles A

The number 154 that is used in the first conversion factor is the formula mass of $C_{10}H_{18}O$. It was not given in the problem, but had to be calculated using atomic masses.

STEP 3 The solution to the problem, obtained by doing the arithmetic, is

$$\frac{2.00 \times 1 \times 6.02 \times 10^{23}}{154 \times 1} \text{ molecules } C_{10}H_{18}O = 7.8181818 \times 10^{21} \text{ molecules } C_{10}H_{18}O$$
(calculator answer)

$$= 7.82 \times 10^{21} \text{ molecules } C_{10}H_{18}O$$
(correct answer)

PRACTICE EXERCISE 9.10

Sucrose, common table sugar, has the formula $C_{12}H_{22}O_{11}$. How many sucrose molecules are present in a 0.100-g sample of this compound?

Ans. 1.76×10^{20} molecules $C_{12}H_{22}O_{11}$.

Similar Exercises: Problems 9.67 and 9.68

EXAMPLE 9.11

The artifical sweetener aspartame (Nutra-Sweet), with the chemical formula $C_{14}H_{18}N_2O_5$, is the sugar substitute used in most diet soft drinks. What would be the mass in grams of a pure sample of this sweetener that contains 1 billion (1.000×10^9) molecules?

Solution

STEP 1 The given quantity is 1.000×10^9 aspartame molecules. The desired quantity is grams of aspartame.

$$1.000 \times 10^9 \text{ } C_{14}H_{18}N_2O_5 \text{ molecules } = ? \text{ g } C_{14}H_{18}N_2O_5$$

In terms of Figure 9.3, this is a "particles of A" to "grams of A" problem.

STEP 2 The pathway for this problem is the exact reverse of the one used in the previous example. We are given particles and asked to find grams of the same substance.

Particles of A $\xrightarrow{\text{Avogadro's number}}$ Moles of A $\xrightarrow{\text{molar mass}}$ Grams of A

Using dimensional analysis, the setup is

$$1.000 \times 10^9 \text{ molecules } \cancel{C_{14}H_{18}N_2O_5} \times \frac{1 \text{ mole } \cancel{C_{14}H_{18}N_2O_5}}{6.022 \times 10^{23} \text{ molecules } \cancel{C_{14}H_{18}N_2O_5}} \times \frac{294.3 \text{ g } C_{14}H_{18}N_2O_5}{1 \text{ mole } \cancel{C_{14}H_{18}N_2O_5}}$$

particles A $\longrightarrow$ moles A $\longrightarrow$ grams A

The molar mass for aspartame, which was used in the second conversion factor, was calculated using the atomic masses of carbon, hydrogen, nitrogen, and oxygen and the formula for aspartame. Avogadro's number to four significant figures was used since the given number, 1.000×10^9, has four significant figures.

STEP 3 The final answer is obtained by doing the arithmetic.

$$\frac{1.000 \times 10^9 \times 1 \times 294.3}{6.022 \times 10^{23} \times 1} \text{ g } C_{14}H_{18}N_2O_5 = 4.8870807 \times 10^{-13} \text{ g } C_{14}H_{18}N_2O_5$$
(calculator answer)

$$= 4.887 \times 10^{-13} \text{ g } C_{14}H_{18}N_2O_5$$
(correct answer)

PRACTICE EXERCISE 9.11

Cocaine, a powerful central nervous system stimulant, has the chemical formula $C_{17}H_{21}NO_4$. What would be the mass in grams of a pure sample of this drug that contains 1 billion (1.000×10^9) molecules?   *Ans.* 5.038×10^{-13} g $C_{17}H_{21}NO_4$.

Similar exercises: Problems 9.69 and 9.70

EXAMPLE 9.12

Caffeine, the stimulant in coffee and tea, has the formula $C_8H_{10}N_4O_2$. How many grams of nitrogen are present in a 50.0-g sample of caffeine?

Solution

STEP 1 There is an important difference between this problem and the preceding two; here we are dealing with not one but two substances, caffeine and nitrogen. The given quantity is grams of caffeine (substance A) and we are asked to find the grams of nitrogen (substance B). This is a "grams of A" to "grams of B" problem.

$$50.0 \text{ g } C_8H_{10}N_4O_2 = ? \text{ g N}$$

STEP 2 The appropriate set of conversions for a "grams of A" to "grams of B" problem, from Figure 9.3, is

$$\boxed{\text{Grams of A}} \xrightarrow{\text{molar mass}} \boxed{\text{Moles of A}} \xrightarrow{\text{formula subscript}} \boxed{\text{Moles of B}} \xrightarrow{\text{molar mass}} \boxed{\text{Grams of B}}$$

The mathematical setup involving conversion factors is

$$50.0 \text{ g } C_8H_{10}N_4O_2 \times \frac{1 \text{ mole } C_8H_{10}N_4O_2}{194 \text{ g } C_8H_{10}N_4O_2} \times \frac{4 \text{ moles N}}{1 \text{ mole } C_8H_{10}N_4O_2} \times \frac{14.0 \text{ g N}}{1 \text{ mole N}}$$

grams A ⟶ moles A ⟶ moles B ⟶ grams B

The number 194 that is used in the first conversion factor is the formula mass for caffeine. The conversion from "moles of A" to "moles of B" (the second conversion factor) is derived using the information contained in the formula for caffeine. One mole of caffeine contains 4 moles of N because the subscript for N in the formula is 4. The number 14.0 in the final conversion factor is the atomic mass of nitrogen.

STEP 3 Collecting the numbers from the various conversion factors together and doing the arithmetic gives us our answer.

$$\frac{50.0 \times 1 \times 4 \times 14.0}{194 \times 1 \times 1} \text{ g N} = 14.432989 \text{ g N} \quad \text{(calculator answer)}$$

$$= 14.4 \text{ g N} \quad \text{(correct answer)}$$

PRACTICE EXERCISE 9.12

The industrial explosive TNT (trinitrotoluene) has the chemical formula $C_7H_5O_6N_3$. How many grams of nitrogen are present in 50.0 g of TNT?

Ans: 9.25 g N.

Similar exercises: Problems 9.73 and 9.74

EXAMPLE 9.13

Ethylene glycol, the major ingredient in most automobile antifreezes, has the formula $C_2H_6O_2$. How many atoms of oxygen are present in a 2.000-g sample of ethylene glycol?

Solution

STEP 1 The given quantity is 2.000 g of $C_2H_6O_2$, the desired quantity is atoms of oxygen.

$$2.000 \text{ g } C_2H_6O_2 = ? \text{ atoms O}$$

In the jargon of Figure 9.3, this is a "grams of A" to "particles of B" problem.

STEP 2 From Figure 9.3, the appropriate pathway for solving this problem is

$$\boxed{\text{Grams of A}} \xrightarrow{\text{molar mass}} \boxed{\text{Moles of A}} \xrightarrow{\text{formula subscript}} \boxed{\text{Moles of B}} \xrightarrow{\text{Avogadro's number}} \boxed{\text{Particles of B}}$$

Using dimensional analysis, the conversion factor setup becomes

$$2.000 \text{ g } C_2H_6O_2 \times \frac{1 \text{ mole } C_2H_6O_2}{62.08 \text{ g } C_2H_6O_2} \times \frac{2 \text{ moles O}}{1 \text{ mole } C_2H_6O_2} \times \frac{6.022 \times 10^{23} \text{ atoms O}}{1 \text{ mole O}}$$

$$\text{grams A} \longrightarrow \text{moles A} \longrightarrow \text{moles B} \longrightarrow \text{particles B}$$

The numbers in the middle conversion factor (2 and 1) were obtained from the chemical formula of ethylene glycol. One mole of $C_2H_6O_2$ contains 2 moles of oxygen; the oxygen subscript is 2. Other numbers in the set-up must be specified to four significant figures.

STEP 3 The solution to the problem, obtained by doing the arithmetic, is

$$\frac{2.000 \times 1 \times 2 \times 6.022 \times 10^{23}}{62.08 \times 1 \times 1} \text{ atoms O} = 3.8801546 \times 10^{22} \text{ atoms O}$$
(calculator answer)

$$= 3.880 \times 10^{22} \text{ atoms O}$$
(correct answer)

PRACTICE EXERCISE 9.13

Glucose, $C_6H_{12}O_6$, is one of the main sources of energy used by living organisms. How many atoms of carbon are present in 10.0 g of glucose?

Ans. 2.01×10^{23} atoms of C.

Similar exercises: Problems 9.75 and 9.76

9.9 Determination of Empirical and Molecular Formulas

Chemical formulas provide a great deal of useful information about the substances they represent and, as seen in previous sections of this chapter, are key entities in many types of calculations. How are formulas themselves determined? They are determined by calculation from experimentally obtained information.

Depending on the amount of experimental information available, two types of chemical formulas may be obtained: an empirical formula or a molecular formula. The **empirical formula** (or simplest formula) gives the *smallest* whole number ratio of atoms present in a formula unit of a compound. In empirical formulas the subscripts in the formula cannot be reduced to a simpler set of numbers by division with a small integer. The **molecular formula** (or true formula) gives the *actual* number of atoms present in a formula unit of a compound.

For ionic compounds the empirical and molecular formulas are almost always the same; that is, the actual ratio of atoms present in a formula unit and the smallest ratio of atoms present are one and the same. For molecular compounds the two types of formulas may be the same, but frequently they are not. When they are not the same, the molecular formula is a multiple of the empirical formula.

Molecular formula = whole number multiplier × empirical formula

Table 9.3 contrasts the differences between the two types of formulas for selected compounds.

Elemental composition data for compounds can be used to determine empirical formulas. Such data may be in the form of percent composition or simply the mass of each element present in a known mass of compound. Example 9.14 shows the steps used in obtaining an empirical formula from percent composition data. Example 9.15 uses the mass of the elements present in a compound sample of known mass to obtain an empirical formula.

Section 9.9 · Determination of Empirical and Molecular Formulas

TABLE 9.3 A Comparison of Empirical and Molecular Formulas for Selected Compounds

Compound	Empirical Formula	Molecular Formula	Whole Number Multiplier
Dinitrogen tetrafluoride	NF_2	N_2F_4	2
Hydrogen peroxide	HO	H_2O_2	2
Sodium chloride	NaCl	NaCl	1
Benzene	CH	C_6H_6	6

EXAMPLE 9.14

Acetone is the chemical solvent present in fingernail polish remover. Its percent composition is 62.0% carbon, 10.4% hydrogen, and 27.5% oxygen. What is the empirical formula for the compound?

Solution

The problem of calculating the empirical formula of a compound from percent composition data can be broken down into three steps.

1. Express the composition of each element in the compound in grams.
2. Convert the grams of each element to moles of element.
3. Express the mole ratio between the elements in terms of *small whole numbers*.

STEP 1   In working with mass percentages in chemical calculations, it is convenient to assume for the purposes of the calculation a 100.0-g sample size; mass percentages can be easily converted to grams with this sample size. The mass of each element in this case becomes the same as the percentage figure.

$$C: \quad 62.0\% \text{ of } 100.0 \text{ g} = 62.0 \text{ g}$$
$$H: \quad 10.4\% \text{ of } 100.0 \text{ g} = 10.4 \text{ g}$$
$$O: \quad 27.5\% \text{ of } 100.0 \text{ g} = 27.5 \text{ g}$$

STEP 2   The subscripts in a formula give the ratio of the number of moles of each element present in the compound (Sec. 9.7). We next convert the grams (from step 1) to moles.

$$62.0 \text{ g C} \times \frac{1 \text{ mole C}}{12.0 \text{ g C}} = 5.1666666 \text{ moles C} \quad \text{(calculator answer)}$$
$$= 5.17 \text{ moles C} \quad \text{(correct answer)}$$

$$10.4 \text{ g H} \times \frac{1 \text{ mole H}}{1.01 \text{ g H}} = 10.297029 \text{ moles H} \quad \text{(calculator answer)}$$
$$= 10.3 \text{ moles H} \quad \text{(correct answer)}$$

$$27.5 \text{ g O} \times \frac{1 \text{ mole O}}{16.0 \text{ g O}} = 1.71875 \text{ moles O} \quad \text{(calculator answer)}$$

$$= 1.72 \text{ moles O} \quad \text{(correct answer)}$$

Thus, for every 5.17 moles of C there are 10.3 moles of H and 1.72 moles of O.

STEP 3   The subscripts in a formula are expressed as whole numbers, not as decimals. To obtain whole numbers from the decimals of step 2, we proceed as follows. Each of the numbers is divided by the smallest of the numbers.

$$\text{C:} \quad \frac{5.17 \text{ moles}}{1.72 \text{ moles}} = 3.0058139 \quad \text{(calculator answer)}$$

$$= 3.01 \quad \text{(correct answer)}$$

$$\text{H:} \quad \frac{10.3 \text{ moles}}{1.72 \text{ moles}} = 5.988372 \quad \text{(calculator answer)}$$

$$= 5.99 \quad \text{(correct answer)}$$

$$\text{O:} \quad \frac{1.72 \text{ moles}}{1.72 \text{ moles}} = 1 \quad \text{(calculator answer)}$$

$$= 1.00 \quad \text{(correct answer)}$$

Thus the ratio of carbon to hydrogen to oxygen in this compound is 3 to 6 to 1, and the empirical formula is C_3H_6O.

The calculated values for two of the three formula subscripts just obtained (3.01 and 5.99) are not *exactly* whole numbers. This is because of experimental error in the original mass percentages. These calculated values are, however, sufficiently close to whole numbers that we easily recognize what the whole numbers are. If experimental data having four significant figures instead of three had been used, agreement with whole numbers would have been closer.

PRACTICE EXERCISE 9.14

The compound Freon-12 was widely used as an aerosol can propellant until it was found that it can cause damage to the ozone layer in the upper atmosphere. The percent composition for Freon-12 is 9.90% carbon, 58.7% chlorine, and 31.4% fluorine. What is the empirical formula for the compound?

Ans. CCl_2F_2.

Similar exercises: Problems 9.87, 9.88, 9.91, and 9.92

EXAMPLE 9.15

The tip of the head of a strike-anywhere match contains a phosphorus–sulfur compound that readily ignites when drawn over a rough surface. What is the empirical formula of this ignitable compound given that a 1.0000 g sample of the compound contains 0.5629 g of P and 0.4371 g of S?

Solution

Having data given in grams instead of as percent composition (Example 9.14) simplifies the calculation of an empirical formula. The grams are changed to moles, and we are ready to find the smallest whole number ratio between elements.

STEP 1   Each of the given gram amounts is converted to moles using dimensional analysis.

$$0.5629 \text{ g P} \times \frac{1 \text{ mole P}}{30.97 \text{ g P}} = 0.018175654 \text{ mole P} \quad \text{(calculator answer)}$$

$$= 0.01818 \text{ mole P} \quad \text{(correct answer)}$$

$$0.4371 \text{ g S} \times \frac{1 \text{ mole S}}{32.07 \text{ g S}} = 0.01362956 \text{ mole S} \quad \text{(calculator answer)}$$

$$= 0.01363 \text{ mole S} \quad \text{(correct answer)}$$

STEP 2   Dividing the mole quantities from step 1 by the smaller of the two numbers gives

$$\text{P:} \quad \frac{0.01818 \text{ mole}}{0.01363 \text{ mole}} = 1.3338225 \quad \text{(calculator answer)}$$

$$= 1.334 \quad \text{(correct answer)}$$

$$\text{S:} \quad \frac{0.01363 \text{ mole}}{0.01363 \text{ mole}} = 1 \quad \text{(calculator answer)}$$

$$= 1.000 \quad \text{(correct answer)}$$

Frequently all whole numbers will result at this point (as was the case in Example 9.14) and the calculation is finished. Sometimes, however, as in this case, we obtain small decimals that can be recognized as simple fractions. Such decimals can be cleared by multiplying *all* numbers by a small whole number. Since 1.334, the decimal occurring in this problem, equals four-thirds ($\frac{4}{3}$), it can be cleared by multiplying all numbers by 3.

$$\text{P:} \quad 1.334 \times 3 = 4.002$$
$$\text{S:} \quad 1.000 \times 3 = 3.000$$

We now have a whole number ratio and the desired empirical formula is

$$P_4S_3$$

The decimals that most commonly occur in empirical formula calculations are fourths, 0.25 ($\frac{1}{4}$) and 0.75 ($\frac{3}{4}$), which are multiplied by 4 to clear; thirds, 0.33 ($\frac{1}{3}$) and 0.67 ($\frac{2}{3}$), which are multiplied by 3 to clear; and half, 0.50 ($\frac{1}{2}$), which is multiplied by 2 to clear.

PRACTICE EXERCISE 9.15

Iron and structural steel, when exposed to a moist atmosphere, rust. The reddish brown "rust" which forms is an iron–oxygen compound. Analysis of a 1.386 g sample of dry rust indicates that 0.9694 g of Fe and 0.4166 g of O are present. What is the empirical formula of "rust"?

Ans. Fe_2O_3.

Similar exercises: Problems 9.95 and 9.96

Examples 9.14 and 9.15 suggest that compounds are always broken down completely into their elements when they are analyzed. This is not always the case. Elemental analysis often involves changing the compound to be analyzed into other compounds that contain the elements present in the original compound. A common procedure of this type is *combustion analysis*. This procedure is used extensively for compounds containing carbon, hydrogen, and oxygen (or just carbon and hydrogen). The basic idea in combustion analysis is as follows. A sample of the compound is burned completely in pure oxygen. The products of the reaction are CO_2 and H_2O. All carbon atoms originally present end up in CO_2 molecules, and all hydrogen atoms originally present, in H_2O molecules (see Fig. 9.4). From the mass of CO_2 and H_2O produced, it is possible to calculate the empirical formula of the original compound. Example 9.16 shows how this is done.

EXAMPLE 9.16

The food preservative butylated hydroxytoluene (BHT) is known to contain only the elements C, H, and O. A sample of this compound is burned in a combustion analysis apparatus. The sample size is 2.204 g, and 6.601 g of CO_2 and 2.162 g of H_2O are produced. What is the empirical formula of BHT?

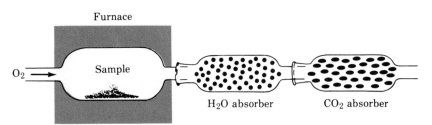

FIGURE 9.4 Combustion method for determining the percentages of carbon and hydrogen in a compound. A weighed sample of the compound is burned in a stream of oxygen, producing gaseous CO_2 and H_2O. These gases then pass through a series of tubes. One tube contains a substance that absorbs H_2O; the substance in another tube absorbs CO_2. By comparing the masses of these tubes before and after the reaction, the analyst can determine the masses of hydrogen and carbon present in the compound that was burned.

Section 9.9 · Determination of Empirical and Molecular Formulas

Solution

STEP 1 To calculate the moles of C, H, and O present will require a number of manipulations. We first convert the grams of CO_2 and grams of H_2O to moles of C and moles of H, respectively. (Both of these calculations are "grams of A" to "moles of B" problems.)

$$6.601 \text{ g } CO_2 \times \frac{1 \text{ mole } CO_2}{44.01 \text{ g } CO_2} \times \frac{1 \text{ mole C}}{1 \text{ mole } CO_2} = 0.14998863 \text{ mole C} \quad \text{(calculator answer)}$$
$$= 0.1500 \text{ mole C} \quad \text{(correct answer)}$$

$$2.162 \text{ g } H_2O \times \frac{1 \text{ mole } H_2O}{18.02 \text{ g } H_2O} \times \frac{2 \text{ moles H}}{1 \text{ mole } H_2O} = 0.2399556 \text{ mole H} \quad \text{(calculator answer)}$$
$$= 0.2400 \text{ mole H} \quad \text{(correct answer)}$$

The last conversion factor in the CO_2 calculation comes from the formula CO_2. One molecule contains one carbon atom. Likewise, the formula H_2O is the key to obtaining information about hydrogen; one water molecule contains two hydrogen atoms.

Information concerning the amount of oxygen present can be calculated by difference: the difference between the original sample mass and the masses of carbon and hydrogen. The masses of carbon and hydrogen present are obtained from the moles of these substances, which were just calculated.

$$0.1500 \text{ mole C} \times \frac{12.01 \text{ g C}}{1 \text{ mole C}} = 1.8015 \text{ g C} \quad \text{(calculator answer)}$$
$$= 1.802 \text{ g C} \quad \text{(correct answer)}$$

$$0.2400 \text{ mole H} \times \frac{1.008 \text{ g H}}{1 \text{ mole H}} = 0.24192 \text{ g H} \quad \text{(calculator answer)}$$
$$= 0.2419 \text{ g H} \quad \text{(correct answer)}$$

The mass of oxygen present can now be calculated by difference.

$$\text{Mass O} = \text{mass sample} - \text{mass C} - \text{mass H}$$
$$= 2.204 \text{ g} - 1.802 \text{ g} - 0.2419 \text{ g}$$
$$= 0.1601 \text{ g} \quad \text{(calculator answer)}$$
$$= 0.160 \text{ g} \quad \text{(correct answer)}$$

From the grams of oxygen the moles of oxygen can be calculated. This will give us the final number needed to determine the empirical formula. (The moles of C and H were calculated previously.)

$$0.160 \text{ g O} \times \frac{1 \text{ mole O}}{16.0 \text{ g O}} = 0.01 \text{ mole O} \quad \text{(calculator answer)}$$
$$= 0.0100 \text{ mole O} \quad \text{(correct answer)}$$

STEP 2   To express the mole ratio between the three elements in terms of small whole numbers, we divide each mole amount by the smallest of these amounts.

C: $\dfrac{0.1500 \text{ mole}}{0.0100 \text{ mole}} = 15$ (calculator answer)

$= 15.0$ (correct answer)

H: $\dfrac{0.2400 \text{ mole}}{0.0100 \text{ mole}} = 24$ (calculator answer)

$= 24.0$ (correct answer)

O: $\dfrac{0.0100 \text{ mole}}{0.0100 \text{ mole}} = 1$ (calculator answer)

$= 1.00$ (correct answer)

Since the C-to-H-to-O ratio is 15 to 24 to 1, the empirical formula will be **$C_{15}H_{24}O$**.

PRACTICE EXERCISE 9.16

A 1.000 g sample of a compound containing only C, H, and O is burned in oxygen and produces 1.913 g of CO_2 and 1.174 g of H_2O. What is the empirical formula of the compound?

Ans. C_2H_6O.

Similar exercises: Problems 9.99 and 9.100

All of the calculations in this section so far (Examples 9.14–9.16) have dealt with empirical formulas. What about molecular formulas, the other type of chemical formula defined at the beginning of this section? Calculation of the molecular formula of a compound requires *both* an elemental analysis (percent composition or its equivalent) and information concerning the molecular mass of the compound. A number of experimental methods (not discussed in this text) exist for determining this latter quantity. The procedure followed for a molecular formula calculation is identical to that for an empirical formula calculation with an additional step required at the end. Composition data are used to calculate the empirical formula (as in Examples 9.14–9.16.), and then the molecular mass and empirical formula are used to calculate the molecular formula, the new last step. Example 9.17 shows how this last step is accomplished.

EXAMPLE 9.17

The compound deoxyribose is an important component of DNA molecules, the molecules responsible for the transfer of genetic information from one generation to the next in living organisms. Deoxyribose has a molecular mass of 150.0 amu and the empirical formula CH_2O. What is deoxyribose's molecular formula?

Solution

The molecular mass is always a whole number multiple of the formula mass for the empirical formula. This whole number relating the two quantities is obtained by dividing the molecular mass by the formula mass of the empirical formula.

We know the molecular mass is 150.0 amu (it was given). We will need to calculate the formula mass of the empirical formula (CH_2O).

$$\begin{aligned} \text{C:} & \quad 1 \times 12.0 \text{ amu} = 12.0 \text{ amu} \\ \text{H:} & \quad 2 \times 1.0 \text{ amu} = 2.0 \text{ amu} \\ \text{O:} & \quad 1 \times 16.0 \text{ amu} = \underline{16.0 \text{ amu}} \\ & \qquad\qquad\qquad\qquad\; 30.0 \text{ amu} \end{aligned}$$

Dividing the formula mass of the empirical formula into the molecular mass for the molecular formula gives

$$\frac{150.0 \text{ amu}}{30.0 \text{ amu}} = 5.00 \quad \text{(whole number multiplier)}$$

The molecular formula is determined by multiplying each of the subscripts in the empirical formula by 5.

$$(CH_2O)_5 = C_{1 \times 5} H_{2 \times 5} O_{1 \times 5} = C_5H_{10}O_5$$

Therefore, $C_5H_{10}O_5$ is the molecular formula of deoxyribose.

PRACTICE EXERCISE 9.17

A compound known to have a molecular mass of 176.0 amu is found to have an empirical formula of $C_3H_4O_3$. What is its molecular formula?   *Ans.* $C_6H_8O_6$.

Similar exercises: Problems 9.101 and 9.102

Figure 9.5 summarizes in a diagrammatic fashion the various problem-solving procedures used in calculating empirical and molecular formulas from experimental data.

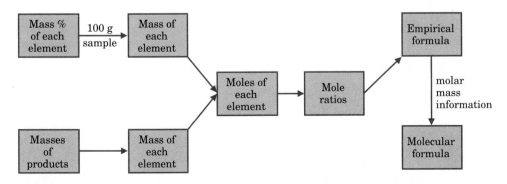

FIGURE 9.5 Outline of the procedures used in calculating empirical and molecular formulas from experimental data.

Key Terms

The new terms or concepts defined in this chapter are

Avogadro's number (Sec. 9.4) The name given to the numerical value 6.02×10^{23}, the number of particles in a mole.

counting by weighing (Sec. 9.6) Samples of two or more pure substances (elements or compounds) found to have mass ratios equal to the ratios of their atomic or formula masses must contain identical numbers of particles (atoms, molecules, or formula units).

empirical formula (Sec. 9.9) A formula that gives the smallest whole number ratio of atoms present in a formula unit of a compound.

formula mass (Sec. 9.2) A number that is the sum of the atomic masses of the atoms in one formula unit.

law of definite proportions (Sec. 9.1) In a pure compound, the elements are always present in the same definite proportion by mass.

macroscopic level interpretation of a chemical formula (Sec. 9.7) The subscripts in the formula of a substance are interpreted to mean the numbers of moles of atoms of the various elements present in one mole of the substance.

microscopic level interpretation of a chemical formula (Sec. 9.7) The subscripts in the formula of a substance are interpreted to mean the numbers of atoms of the various elements present in one formula unit of the substance.

molar mass of a compound (Sec. 9.5) A mass, in grams, that is numerically equal to the formula mass of the compound.

molar mass of an element (Sec. 9.5) For an element in atomic form, a mass, in grams, that is numerically equal to the atomic mass of the element.

mole (Secs. 9.4 and 9.5) 6.02×10^{23} objects or an amount of substance that contains as many elementary units (atoms, molecules, or formula units) as there are $^{12}_{6}C$ atoms in exactly 12.00000 g of $^{12}_{6}C$.

molecular formula (Sec. 9.9) A formula that gives the actual number of atoms present in a formula unit of a compound.

percent composition (Sec. 9.3) The percent by mass of each element present in a compound.

Practice Problems

Law of Definite Proportions (Sec. 9.1)

9.1 Two different samples of a pure compound (containing elements A and B) were analyzed with the following results.
Sample I: 17.35 g of compound yielded 10.03 g of A and 7.32 g of B.
Sample II: 22.78 g of compound yielded 13.17 g of A and 9.61 g of B.
Show that these data are consistent with the law of definite proportions.

9.2 Two different samples of a pure compound (containing elements C and D) were analyzed with the following results.
Sample I: 11.21 g of compound yielded 5.965 g of C and 5.245 g of D.
Sample II: 13.65 g of compound yielded 7.263 g of C and 6.387 g of D.
Show that these data are consistent with the law of definite proportions.

9.3 It has been found experimentally that the elements X and Y react to produce two different compounds, depending upon conditions. Some sample experimental data are as follows.

Experiment	Grams of X	Grams of Y	Grams of Compound
1	3.37	8.90	12.27
2	0.561	1.711	2.272
3	26.0	71.0	97.9

Which two of the three experiments produced the same compound?

9.4 It has been found experimentally that the elements S and T react to produce two different compounds, depending upon conditions. Some sample experimental data are as follows.

Experiment	Grams of S	Grams of T	Grams of Compound
1	1.926	1.074	3.000
2	3.440	1.560	5.000
3	4.494	2.506	7.000

Which two of the three experiments produced the same compound?

9.5 The elemental composition, by mass, of ammonia (NH_3) is 82.2% nitrogen (N) and 17.8% hydrogen (H). Suppose you mixed 82.2 g of N with 25.0 g of H. What is the maximum amount of NH_3 that can be formed from this mixture? Which element is in excess, and how much of it will remain after completion of the chemical reaction?

9.6 The elemental composition, by mass, of water (H_2O) is 88.9% oxygen (O) and 11.1% hydrogen (H). Suppose you mixed 102.6 g of O with 11.1 g of H. What is the maximum amount of H_2O that can be formed from this mixture? Which element is in excess, and how much of it will remain after completion of the chemical reaction?

Formula Masses (Sec. 9.2)

9.7 Calculate the formula mass of each of the following substances. Round off all atomic masses to the hundredths digit before using.
(a) SnF_2 [tin(II) fluoride, a toothpaste additive]
(b) $FeSO_4$ [iron(II) sulfate, used in treatment of iron deficiency]
(c) $C_{12}H_{22}O_{11}$ (sucrose, table sugar)
(d) $C_{14}H_9Cl_5$ (DDT, an insecticide)

9.8 Calculate the formula mass of each of the following substances. Round off all atomic masses to the hundredths digit before using.
(a) $C_9H_8O_4$ (naphthalene, ingredient in some mothballs)
(b) NaCN (sodium cyanide, used to extract gold from ores)
(c) C_7H_{16} (heptane, a component of gasoline)
(d) $C_{55}H_{72}MgN_4O_5$ (chlorophyll, present in all green plants)

9.9 Calculate the formula mass of each of the following substances. Round off all atomic masses to the tenths digit before using.
(a) $Ca(NO_3)_2$ (calcium nitrate, gives red color to fireworks)
(b) $Pb(C_2H_5)_4$ (tetraethyllead, a gasoline additive)
(c) $Al_2(SO_4)_3$ (aluminum sulfate, used in tanning animal hides)
(d) $C_7H_5(NO_2)_3$ (TNT, an explosive)

9.10 Calculate the formula mass of each of the following substances. Round off all atomic masses to the tenths digit before using.
(a) $Ca(OH)_2$ (calcium hydroxide, slaked lime)
(b) $Ni(NO_3)_2$ [nickel(II) nitrate, a pigment in ceramics]
(c) $(CH_3)_2O$ (dimethyl ether, an inhalation anesthetic)
(d) $Mg_3(Si_2O_5)_2(OH)_2$ (talc, a mineral used to make talcum powder)

9.11 The compound 2-butene-1-thiol, which is responsible in part for the characteristic odor of skunks, has a formula mass of 88.19 amu and the formula C_yH_8S. What number does y stand for in this formula?

9.12 The compound 1-propanethiol, which is the eye irritant that is released when fresh onions are chopped up, has a formula mass of 76.18 amu and the formula C_3H_yS. What number does y stand for in this formula?

9.13 A compound that is 51.5% oxygen by mass contains two oxygen atoms per molecule. What is the formula mass of the compound?

9.14 A compound that is 52.1% oxygen by mass contains three oxygen atoms per molecule. What is the formula mass of the compound?

Percent Composition (Sec. 9.3)

9.15 Calculate the percent composition for each of the following compounds. (Round off all atomic masses to 0.01 amu before using them.)
(a) NaCl (sodium chloride, table salt)
(b) H_3BO_3 (boric acid, a mild antiseptic)
(c) $C_{10}H_{14}N_2$ (nicotine, a chemical stimulant)
(d) $C_{20}H_{25}N_3O$ (LSD, a hallucinogenic drug)

9.16 Calculate the percent composition for each of the following compounds. (Round off all atomic masses to 0.01 amu before using them.)
(a) CH_4 (methane, major component of natural gas)
(b) $NaHCO_3$ (sodium bicarbonate, baking soda)
(c) $C_3H_5N_3O_9$ (nitroglycerin, an explosive)
(d) $C_{10}H_{19}O_6PS_2$ (malathion, an insecticide)

9.17 Arrange the following compounds in order of increasing mass percent of calcium: $CaCO_3$, $CaSO_4$, CaC_2O_4, and $Ca_3(PO_4)_2$.

9.18 Arrange the following compounds in order of increasing mass percent of nitrogen: $(NH_4)_2SO_4$, KNO_3, N_2F_4, $(H_2N)_2CO$.

9.19 An iron–oxygen compound is formed by combining 2.495 g of iron with 1.072 g of oxygen. What is the percent composition of the compound?

9.20 An iron–sulfur compound is formed by combining 2.233 g of iron with 1.926 g of sulfur. What is the percent composition of the compound?

9.21 A 6.48-g sample of N is reacted with oxygen to give 25.00 g of a compound. What is the percent composition of the compound?

9.22 A 5.76-g sample of P is reacted with oxygen to give 13.20 g of a compound. What is the percent composition of the compound?

9.23 When 49.31 g of the compound sodium sulfate is decomposed, 15.96 g of Na, 11.13 g of S, and 22.22 g of O are produced. Using this information, calculate the percent composition of sodium sulfate.

9.24 When 18.03 g of the compound potassium dichromate is decomposed, 4.79 g of K, 6.38 g of Cr, and 6.86 g of O are produced. Using this information, calculate the percent composition of potassium dichromate.

9.25 A sample of the compound hydrazine, N_2H_4, contains 52.34 g of H. How many grams of N does it contain?

9.26 A sample of the compound sodium azide, NaN_3, contains 57.08 g of Na. How many grams of N does it contain?

The Mole as a Counting Unit (Sec. 9.4)

9.27 How many atoms are present in 2.50 moles of each of the following elements?
(a) zinc (b) gold
(c) silver (d) manganese

9.28 How many atoms are present in 1.75 moles of each of the following elements?
(a) sulfur (b) boron
(c) phosphorus (d) silicon

9.29 Calculate the number of atoms present in each of the following samples of elements.
(a) 2.00 moles B (b) 1.45 moles Na
(c) 0.100 mole C (d) 0.875 mole U

9.30 Calculate the number of atoms present in each of the following samples of elements.
(a) 3.00 moles Li (b) 6.75 moles S
(c) 0.00100 mole As (d) 1.33 moles Se

9.31 Calculate the number of molecules present in each of the following samples of molecular compounds.
(a) 1.50 moles CO_2 (b) 0.500 mole NH_3
(c) 2.33 moles PF_3 (d) 1.115 moles N_2H_4

9.32 Calculate the number of molecules present in each of the following samples of molecular compounds.
(a) 4.69 moles CO (b) 0.433 mole SO_3
(c) 1.44 moles P_2H_4 (d) 2.307 moles H_2O_2

9.33 Calculate the number of formula units present in each of the following samples of ionic compounds.
(a) 2.00 moles CaO (b) 2.00 moles MgO
(c) 5.20 moles Na_2SO_4 (d) 0.344 mole KOH

9.34 Calculate the number of formula units present in each of the following samples of ionic compounds.
(a) 3.50 moles Na_3N (b) 3.50 moles AlF_3
(c) 0.333 mole KNO_3 (d) 1.19 moles NaOH

Molar Mass (Sec. 9.5)

9.35 Calculate the mass, in grams, of each of the following samples of matter.
(a) 1.00 mole of Cu atoms
(b) 2.00 moles of As atoms
(c) 3.00 moles of Be atoms
(d) 4.00 moles of S atoms

9.36 Calculate the mass, in grams, of each of the following samples of matter.
(a) 1.000 mole of Cl atoms
(b) 2.000 moles of Br atoms
(c) 3.000 moles of Rb atoms
(d) 4.000 moles of Si atoms

9.37 What is the molar mass, to four significant figures, of each of the following compounds?
(a) $Al(OH)_3$ (b) Mg_3N_2
(c) $Cu(NO_3)_2$ (d) La_2O_3

9.38 What is the molar mass, to four significant figures, of each of the following compounds?
(a) NH_4Cl (b) $HClO_4$
(c) $Ca(ClO_4)_2$ (d) HfO_2

9.39 Calculate the mass, in grams, of each of the following samples of lithium-containing compounds.
(a) 2.33 moles of Li_2CO_3
(b) 1.20 moles of LiF
(c) 3.25 moles of Li_3N
(d) 0.725 mole of LiOH

9.40 Calculate the mass, in grams, of each of the following samples of sulfur-containing compounds.
(a) 1.47 moles of SCl_4
(b) 2.04 moles of SO_3
(c) 4.50 moles of S_2F_2
(d) 0.0312 mole of SO_2Cl_2

9.41 Calculate the mass, in grams, of 1.727 moles of
(a) UF_6 (b) B_2O_3
(c) C_2H_6O (d) $Pb(C_2H_5)_4$

9.42 Calculate the mass, in grams, of 3.527 moles of
(a) Cr_2O_3 (b) CH_3Br
(c) $NaClO_4$ (d) $Ca_3(PO_4)_2$

9.43 A sample of a pure substance with a mass of 0.1116 g contains 1.397×10^{-3} mole of molecules. What is the molar mass of the substance comprising the sample?

9.44 A sample of a pure substance with a mass of 0.3215 g contains 1.191×10^{-2} mole of molecules. What is the molar mass of the substance comprising the sample?

Practice Problems

9.45 What is the mass, in grams, of an atom whose mass on the amu scale is 19.0 amu?

9.46 What is the mass, in grams, of an atom whose mass on the amu scale is 23.0 amu?

9.47 Identify the element that contains atoms with an average mass of 5.143×10^{-23} g.

9.48 Identify the element that contains atoms with an average mass of 2.326×10^{-23} g.

Counting Particles by Weighing (Sec. 9.6)

9.49 Using the concept of counting by weighing, determine whether each of the following pairs of samples contains the same number of particles (atoms or molecules).
(a) 28.086 g of Si and 35.453 g of Cl
(b) 72.59 g of Ge and 74.72 g of Ga
(c) 1.427 g of NH_3 and 2.685 g of N_2H_4
(d) 64.14 g of S and 128.14 g of SO_2

9.50 Using the concept of counting by weighing, determine whether each of the following pairs of samples contains the same number of particles (atoms or molecules).
(a) 3.813 g of B and 2.314 g of Li
(b) 2.0020 g of C and 2.3347 g of N
(c) 9.01 g of H_2O and 15.78 g of H_2O_2
(d) 41.4 g of Au and 44.6 g of AuCN

9.51 Using the concept of counting by weighing, indicate whether each of the following samples contains more atoms, the same number of atoms, or fewer atoms than 23.6 g of copper (Cu).
(a) 24.3 g of Fe   (b) 19.6 g of Ni
(c) 11.5 g of P   (d) 11.2 g of Al

9.52 Using the concept of counting by weighing, indicate whether each of the following samples contains more atoms, the same number of atoms, or fewer atoms than 14.0 g of sodium (Na).
(a) 77.3 g of I   (b) 126 g of Pb
(c) 20.2 g of K   (d) 9.00 g of N

9.53 Using the concept of counting by weighing, determine the number of grams of each of the following elements that will contain three times as many atoms as 10.00 g of P.
(a) S   (b) Be   (c) U   (d) Si

9.54 Using the concept of counting by weighing, determine the number of grams of each of the following elements that will contain twice as many atoms as 20.00 g of Zn.
(a) Li   (b) Al   (c) Cu   (d) Ba

The Mole and Chemical Formulas (Sec. 9.7)

9.55 Write the six mole-to-mole conversion factors that can be derived from the formula H_3PO_4.

9.56 Write the six mole-to-mole conversion factors that can be derived from the formula H_2SO_3.

9.57 How many moles of each type of atom are present in each of the following chemical quantities?
(a) 2.00 moles of N_2O_3
(b) 0.100 mole of $SOCl_2$
(c) 3.50 moles of $Ca_3(PO_4)_2$
(d) Avogadro's number of CO_2 molecules

9.58 How many moles of each type of atom are present in each of the following chemical quantities?
(a) 3.00 moles of P_4S_3
(b) 0.200 mole of C_2H_6O
(c) 2.25 moles of $Al_2(SO_4)_3$
(d) Avogadro's number of N_2O molecules

9.59 How many moles of sulfur atoms are present in 2.50 moles of each of the following sulfur-containing compounds?
(a) Na_2SO_4 (b) S_3Cl_2
(c) $H_2S_2O_5$ (d) $Na_3Ag(S_2O_3)_2$

9.60 How many moles of gold atoms are present in 1.50 moles of each of the following gold-containing compounds?
(a) $NaAuBr_4$ (b) Au_2Cl_6
(c) Au_2Te_3 (d) $Au_4(SCH_3)_4$

9.61 How many *total* moles of atoms are present in 2.00 moles of each of the following substances?
(a) KN_3 (b) $C_2H_2Cl_4$
(c) $Ba(NO_3)_2$ (d) $(NH_4)_3PO_4$

9.62 How many *total* moles of atoms are present in 3.00 moles of each of the following substances?
(a) CH_4 (b) K_2SO_4
(c) Cl_2O_7 (d) $(NH_4)_2SO_4$

9.63 Which member of each of the following pairs contains the larger number of moles of nitrogen atoms?
(a) 0.1 mole of N_2O_5 or 0.2 mole of N_2O_3
(b) 0.2 mole of N_2O_3 or 0.3 mole of NO_2
(c) 0.3 mole of NO_2 or 0.2 mole of N_2O_4
(d) 0.1 mole of N_2O_4 or 0.2 mole of N_2O

9.64 Which member of each of the following pairs contains the larger number of moles of oxygen atoms?
(a) 0.2 mole of Cl_2O or 0.1 mole of Cl_2O_3
(b) 0.3 mole of Cl_2O_3 or 0.1 mole of Cl_2O_6
(c) 0.1 mole of Cl_2O_6 or 0.1 mole of Cl_2O_3
(d) 0.1 mole of Cl_2O_7 or 0.5 mole of Cl_2O

The Mole and Chemical Calculations (Sec. 9.8)

9.65 Calculate the number of atoms present in each of the following samples of an element.
(a) 7.35 g of Au   (b) 23.0 g of Ba
(c) 0.000312 g of H   (d) 22.22 g of Be

9.66 Calculate the number of atoms present in each of the following samples of an element.

(a) 32.0 g of S (b) 1.22 g of He
(c) 55.3 g of Fe (d) 0.0000250 g of U

9.67 Determine the number of molecules present in each of the following samples of a molecular compound.
(a) 10.0 g of CO_2 (b) 20.0 g of H_2O_2
(c) 32.0 g of NH_3 (d) 120.0 g of Cl_2O_7

9.68 Determine the number of molecules present in each of the following samples of a molecular compound.
(a) 10.0 g of CO (b) 20.0 g of C_2H_6
(c) 32.0 g of HN_3 (d) 120.0 g of IF_7

9.69 Determine the mass, in grams, of each of the following quantities of chemical substance.
(a) 6.022×10^{23} atoms of Zn
(b) 4.000×10^{23} atoms of S
(c) 1.000×10^{23} molecules of NH_3
(d) 2457 molecules of SO_2

9.70 Determine the mass, in grams, of each of the following quantities of chemical substance.
(a) 6.022×10^{23} atoms of Be
(b) 2.000×10^{23} atoms of F
(c) 9.000×10^{23} molecules of CO
(d) 3368 molecules of N_2O

9.71 What is the mass, in grams (to four significant figures), of a single atom (for elements) or molecule (for compounds) of the following substances?
(a) S (b) Ti (c) C_2H_6 (d) C_7H_{16}

9.72 What is the mass, in grams (to four significant figures), of a single atom (for elements) or molecule (for compounds) of the following substances?
(a) P (b) Fe (c) C_3H_8 (d) $C_{10}H_{22}$

9.73 Determine the number of grams of Cl present in each of the following amounts of chlorine-containing compound.
(a) 100.0 g of NaCl (b) 1000.0 g of CCl_4
(c) 10.0 g of HCl (d) 50.0 g of $BaCl_2$

9.74 Determine the number of grams of N present in each of the following amounts of nitrogen-containing compound.
(a) 100.0 g of $NaNO_3$ (b) 1000.0 g of N_2H_4
(c) 10.0 g of Na_3N (d) 50.0 g of KCN

9.75 Determine the number of phosphorus atoms present in 25.0 g of each of the following substances.
(a) PF_3 (b) Be_3P_2
(c) $POCl_3$ (d) $Na_5P_3O_{10}$

9.76 Determine the number of sulfur atoms present in 35.0 g of each of the following substances.
(a) CS_2 (b) S_4N_4
(c) SF_6 (d) $Al_2(SO_4)_3$

9.77 Determine the number of grams of C present in each of the following chemical samples.

(a) 4.00×10^{20} molecules of C_9H_{20}
(b) 2.45 moles of $NaHCO_3$
(c) 10.00 g of $H_2C_2O_4$
(d) 1000.0 g $C_8H_{11}NO_3$

9.78 Determine the number of grams of O present in each of the following chemical samples.
(a) 2.50×10^{25} molecules of $C_6H_{12}O_6$
(b) 0.500 mole of Na_2SO_4
(c) 15.00 g of CO
(d) 1000.0 g of $K_6Si_3O_9$

9.79 The active ingredient in the substance marijuana is the compound tetrahydrocannabinol, which is commonly called THC. The formula for THC is $C_{21}H_{36}O_2$. Determine the number of grams of THC that you need to obtain one billion (1.00×10^9)
(a) THC molecules (b) C atoms
(c) H atoms (d) total atoms

9.80 Enflurane, which has the formula $C_3H_2ClF_5O$, is widely used as a surgical anesthetic. Determine the number of grams of enflurane that you need to obtain one million (1.00×10^6)
(a) enflurane molecules (b) F atoms
(c) Cl atoms (d) total atoms

9.81 Treflan is a herbicide used to control weeds. Its formula is $C_{13}H_{16}N_3O_4F_3$. In a 15.00-g sample of this compound,
(a) How many moles of atoms are present?
(b) How many atoms of carbon are present?
(c) How many grams of oxygen are present?
(d) How many molecules of Treflan are present?

9.82 Acetaminophen is the painkilling ingredient in over-the-counter Tylenol products. Its formula is $C_8H_9O_2N$. In a 25.00-g sample of this compound,
(a) How many moles of atoms are present?
(b) How many atoms of hydrogen are present?
(c) How many grams of nitrogen are present?
(d) How many acetaminophen molecules are present?

9.83 What amount or mass of each of the following substances would be needed to obtain 1.000 g of S?
(a) moles of H_2S (b) molecules of S_8
(c) grams of $S_3N_3O_3Cl_3$ (d) atoms of S

9.84 What amount or mass of each of the following substances would be needed to obtain 1.000 g of Si?
(a) moles of SiH_4 (b) molecules of SiO_2
(c) grams of $(CH_3)_3SiCl$ (d) atoms of Si

Empirical and Molecular Formulas (Sec. 9.9)

9.85 Each of the following is a correctly written molecular formula. In each case write the empirical formula for the substance.

(a) $C_{10}H_{22}$ (b) C_7H_{16} (c) Cl_2O_6 (d) $P_3N_3Cl_6$

9.86 Each of the following is a correctly written molecular formula. In each case write the empirical formula for the substance.
(a) N_2H_4 (b) $C_2H_6O_2$ (c) H_2S_6 (d) Cl_2O_7

9.87 Given the following percent compositions, determine the empirical formula.
(a) 58.91% Na and 41.09% S
(b) 24.74% K, 34.76% Mn, and 40.50% O
(c) 2.057% H, 32.69% S, and 65.25% O
(d) 19.84% C, 2.503% H, 66.08% O, and 11.57% N

9.88 Given the following percent compositions, determine the empirical formula.
(a) 47.26% Cu and 52.74% Cl
(b) 40.27% K, 26.78% Cr, and 32.96% O
(c) 40.04% Ca, 12.00% C, and 47.96% O
(d) 28.03% Na, 29.28% C, 3.690% H, and 39.01% O

9.89 Convert each of the following molar ratios between elements in a compound to a whole number molar ratio.
(a) 1.00:1.67
(b) 3.00:1.50
(c) 1.50:2.50:3.25
(d) 1.33:2.67:1.00

9.90 Convert each of the following molar ratios between elements in a compound to a whole number molar ratio.
(a) 2.00:1.25
(b) 1.33:1.00
(c) 1.50:2.25:2.00
(d) 2.67:1.25:1.33

9.91 Determine the empirical formula for substances with each of the following percent compositions.
(a) 43.64% P and 56.36% O
(b) 72.24% Mg and 27.76% N
(c) 29.08% Na, 40.56% S, and 30.36% O
(d) 21.85% Mg, 27.83% P, and 50.32% O

9.92 Determine the empirical formula for substances with each of the following percent compositions.
(a) 54.88% Cr and 45.12% S
(b) 38.76% Cl and 61.24% O
(c) 59.99% C, 4.485% H, and 35.52% O
(d) 26.58% K, 35.35% Cr, and 38.06% O

9.93 Cobalt forms two oxides. One has 21.4% O by mass, and the other 28.9% O. What are the empirical formulas of these two oxides?

9.94 Gold forms two chlorides. One has 84.75% Au by mass, and the other 64.94% Au. What are the empirical formulas of these two chlorides?

9.95 Ethyl mercaptan is an odorous substance added to natural gas to make leaks easily detectable. Analysis of a 15.00-g sample of ethyl mercaptan indicates that 5.798 g of C, 1.462 g of H, and 7.740 g of S are present. What is the empirical formula of this compound?

9.96 Hydroquinone is a compound used in developing photographic film. Analysis of a 25.00-g sample of hydroquinone indicates that 16.36 g of C, 1.375 g of H, and 7.265 g of O are present. What is the empirical formula of this compound?

9.97 A 4.00-g sample of bromine was reacted completely with excess fluorine. The mass of the compound formed was 8.75 g. Determine the compound's empirical formula.

9.98 A 1.83-g sample of aluminum metal is burned in an oxygen atmosphere to produce 3.46 g of aluminum oxide. Determine the empirical formula for aluminum oxide.

9.99 A 3.750-g sample of the compound responsible for the odor of cloves (containing only C, H, and O) is burned in a combustion analysis apparatus. The mass of CO_2 produced is 10.05 g, and the mass of H_2O produced is 2.470 g. What is the empirical formula of the compound?

9.100 A 0.8640-g sample of the compound responsible for the pungent odor of rancid butter (containing only C, H, and O) is burned in a combustion analysis apparatus. The mass of CO_2 produced is 1.727 g, and the mass of H_2O produced is 0.7068 g. What is the empirical formula of the compound?

9.101 Determine the molecular formulas of compounds with the following empirical formulas and molecular masses.
(a) BH_3, 27.6 amu
(b) SO_2F_2, 102 amu
(c) C_4H_9, 114 amu
(d) SNF, 26$\overline{0}$ amu

9.102 Determine the molecular formulas of compounds with the following empirical formulas and molecular masses.
(a) NO_2, 92 amu
(b) CH_2, 56 amu
(c) IO_3F, 194 amu
(d) C_5H_7N, 162 amu

9.103 The compound isopropyl alcohol (rubbing alcohol) has a molecular mass of 60.1 amu, and its percent composition by mass is 60.0% C, 13.4% H, and 26.6% O. What is the molecular formula of isopropyl alcohol?

9.104 The compound oxalic acid, which is used in commercial laundries to remove rust stains, has a molecular mass of 90.0 amu, and its percent composition by mass is 26.7% C, 2.2% H, and 71.1% O. What is the molecular formula of oxalic acid?

9.105 Lactic acid, the substance that builds up in muscles and causes them to "hurt" when they are worked hard, has a molecular mass of 90.0 amu and a percent composition of 40.0% C, 6.71% H, and 53.3% O. Determine the empirical and molecular formulas for lactic acid.

9.106 Citric acid, a flavoring agent in many carbonated beverages, has a molecular mass of 192 amu and a percent composition of 37.50% C, 4.21% H, and 58.29% O. Determine the empirical and molecular formulas for citric acid.

Additional Problems

9.107 A chemist prepared a sample by mixing 10.0 g of Na_3PO_4 with 1.00 mole of K_3PO_4. How many oxygen atoms are present in the sample?

9.108 A chemist prepared a sample by mixing 20.0 g of H_2O with 1.50 moles of $C_{12}H_{22}O_{11}$ (table sugar). How many molecules are present in the sample?

9.109 For the compound $(CH_3)_3SiCl$ calculate the
  (a) mass percent of H present
  (b) atom percent of H present
  (c) mole percent of H present

9.110 For the compound $(CH_3)_2SiCl_2$ calculate the
  (a) mass percent of H present
  (b) atom percent of H present
  (c) mole percent of H present

9.111 Calculate the ratio of the number of SO_2 molecules in 1.00 g of SO_2 to the number of U atoms in 1.00 g of U.

9.112 Calculate the ratio of the number of H_2O_2 molecules in 25.00 g of H_2O_2 to the number of Pb atoms in 25.00 g of Pb.

9.113 How many grams of Be would contain the same number of atoms as there are in 2.23 moles of N?

9.114 How many grams of Si would contain the same number of atoms as there are in 3.41 moles of Ca?

9.115 Individual samples of the three compounds Cr_2O_3, CrO_2, and CrO_3 are found to contain identical masses of Cr.
  (a) Which sample has the largest total mass?
  (b) Which sample contains the greatest number of oxygen atoms?

9.116 Individual samples of the three compounds V_2O_5, V_2O_3, and VO_2 are found to contain identical masses of V.
  (a) Which sample has the largest total mass?
  (b) Which sample contains the greatest number of oxygen atoms?

9.117 A certain alloy of Au, Cu, and Ni contains these elements in the atomic proportions 3:2:1, respectively. What is the mass, in grams, of a sample of this alloy containing a total of 1.00×10^{24} atoms?

9.118 A certain alloy of Sn, Pb, and Bi contains these elements in the atomic proportions 2:4:3, respectively. What is the mass, in grams, of a sample of this alloy containing Avogadro's number of atoms?

9.119 If a molecule of X has a mass of 1.792×10^{-22} g, what is the molar mass of X?

9.120 If a molecule of Y has a mass of 6.647×10^{-24} g, what is the molar mass of Y?

9.121 A pure sample of an element with a mass of 0.05232 g contains 4.840×10^{-3} mole of atoms. What is the atomic mass of the element, and which element is it?

9.122 A pure sample of an element with a mass of 0.2346 g contains 3.380×10^{-2} mole of atoms. What is the atomic mass of the element, and which element is it?

9.123 What are the empirical formulas for the compounds that contain each of the following?
  (a) 9.0×10^{23} atoms of Na, 3.0×10^{23} atoms of Al, and 1.8×10^{24} atoms of F
  (b) 3.2 g of S and 1.20×10^{23} atoms of O
  (c) 0.36 mole of Ba, 0.36 mole of C, and 17.2 g of O
  (d) 1.81×10^{23} atoms of H, 0.30 mole of O atoms, and 10.65 g of Cl

9.124 What are the empirical formulas for the compounds that contain each of the following?
  (a) 3.0×10^{30} atoms of Fe, 3.0×10^{30} atoms of Cr, and 1.2×10^{31} atoms of O
  (b) 0.0023 g of N and 2.0×10^{20} atoms of O
  (c) 0.40 mole of Li, 6.4 g of S, and 0.80 mole of O
  (d) 0.15 mole of S, 1.8×10^{23} atoms of O, and 5.7 g of F

9.125 A sample of a compound containing only C, H, and S was burned in oxygen, and 6.601 g of CO_2, 5.406 g of H_2O, and 9.615 g of SO_2 were obtained.
  (a) What is the empirical formula of the compound?
  (b) What was the mass, in grams, of the sample that was burned?

9.126 A sample of a compound containing only C, H, and N was burned in oxygen, and 6.601 g of CO_2, 6.758 g of H_2O, and 4.502 g of NO were obtained.
  (a) What is the empirical formula of the compound?
  (b) What was the mass, in grams, of the sample that was burned?

9.127 By analysis, a compound with the formula $KClO_x$ is found to contain 28.9% by mass of chlorine. What is the value of the integer x?

9.128 By analysis, a compound with the formula H_3AsO_x is found to contain 52.78% by mass of arsenic. What is the value of the integer x?

9.129 A 7.503-g sample of metal is reacted with excess oxygen to yield 10.498 g of the oxide MO. Calculate the molar mass of the element M.

9.130 A 11.17-g sample of metal is reacted with excess oxygen to yield 15.97 g of the oxide M$_2$O$_3$. Calculate the molar mass of the element M.

9.131 A certain compound contains only carbon, hydrogen, and oxygen. If it contains 47.4% carbon by mass, and if there is one oxygen atom present for every four hydrogen atoms, what is its empirical formula?

9.132 A certain compound contains only lead, carbon, and hydrogen. If it contains 64.07% lead by mass, and if there are two carbon atoms present for every five hydrogen atoms, what is its empirical formula?

Cumulative Problems

9.133 Calculate the density of the metal lead, in g/cm^3, given that 0.422 mole of lead occupies a volume of 7.74 cm^3.

9.134 Calculate the density of the metal potassium, in g/cm^3, given that 0.674 mole of potassium occupies a volume of 30.8 cm^3.

9.135 Calculate the volume of 2.50 moles of zinc, assuming the density of the metal to be 7.14 g/cm^3.

9.136 Calculate the volume of 1.25 moles of magnesium, assuming the density of the metal to be 1.74 g/cm^3.

9.137 Calculate the molar mass, to four significant figures, of each of the following compounds.
  (a) magnesium nitrate
  (b) sodium azide
  (c) nickel(II) fluoride
  (d) ammonium perchlorate

9.138 Calculate the molar mass, to four significant figures, of each of the following compounds.
  (a) sodium thiosulfate
  (b) copper(II) hydroxide
  (c) ammonium chlorate
  (d) aluminum hydrogen phosphate

9.139 The specific heat of the metal silver is 0.24 J/(g · °C). How many joules of energy are needed to raise the temperature of a block of silver containing 2.50 moles of silver atoms by 10.0 °C?

9.140 The specific heat of the metal gold is 0.13 J/(g · °C). How many joules of energy are needed to raise the temperature of a block of gold containing 1.40 moles of gold atoms by 15.0°C?

9.141 A 2.33-mole sample of the ionic compound Al$_2$(SO$_4$)$_3$ contains how many of the following?
  (a) Al$_2$(SO$_4$)$_3$ formula units
  (b) Al^{3+} ions
  (c) SO$_4^{2-}$ ions
  (d) total ions

9.142 A 3.50-mole sample of the ionic compound (NH$_4$)$_3$PO$_4$ contains how many of the following?

  (a) (NH$_4$)$_3$PO$_4$ formula units
  (b) NH$_4^+$ ions
  (c) PO$_4^{3-}$ ions
  (d) total ions

9.143 A mixture consists of 42.0% NaCl and 58.0% CaCl$_2$ by mass. What is the total number of chloride ions (Cl$^-$) in 425 g of mixture?

9.144 A mixture consists of 22.0% Cu(NO$_3$)$_2$ and 78.0% Fe(NO$_3$)$_3$ by mass. What is the total number of nitrate ions (NO$_3^-$) in 25.00 g of mixture?

9.145 The mass percent composition for an alloy that has a density of 8.31 g/cm^3 is 56.0% copper, 43.0% nickel, and 1.0% manganese. How many nickel atoms are in a block of this alloy measuring 10.0 cm × 30.0 cm × 62.0 cm?

9.146 The mass percent composition for an alloy that has a density of 8.28 g/cm^3 is 64.0% iron, 12.0% cobalt, and 24.0% molybdenum. How many molybdenum atoms are in a block of this alloy measuring 2.0 cm × 1.5 cm × 6.7 cm?

9.147 What volume, in milliliters, of a NaOH solution that is 12.0% NaOH by mass contains 0.275 mole of NaOH? The density of the solution is 1.131 g/ml.

9.148 What volume, in milliliters, of a H$_3$PO$_4$ solution that is 85.5% H$_3$PO$_4$ by mass contains 0.100 mole of H$_3$PO$_4$? The density of the solution is 1.70 g/ml.

9.149 How many total atoms are present in 15.0 mL of an ethyl alcohol–water solution that is 85.0% ethanol by mass? The formula of ethyl alcohol is C$_2$H$_6$O, and the density of the solution is 0.807 g/mL.

9.150 How many total atoms are present in 500.0 mL of an ethylene glycol–water solution that is 56.0% ethylene glycol by mass? The formula for ethylene glycol is C$_2$H$_6$O$_2$, and the density of the solution is 1.072 g/mL.

9.151 When copper is selling for $0.645 per pound, how many copper atoms could you buy for 1 cent?

9.152 When gold is selling for $390 per ounce, how many gold atoms could you buy for 1 cent?

9.153 What would be the value of Avogadro's number if it were defined as the number of particles in a sample of a pure substance with a mass in pounds (rather than grams) numerically equal to the atomic or molecular mass of a substance?

9.154 What would be the value of Avogadro's number if it were defined as the number of particles in a sample of a pure substance with a mass in kilograms (rather than grams) numerically equal to the atomic or molecular mass of a substance?

Grid Problems

9.155 Select from the grid *all* correct responses for each situation. Use *whole number* formula masses in any calculations you do.

1. N_2H_4	2. C_2H_6	3. NO
4. C_3H_6	5. CO_2	6. N_2O
7. NH_3	8. H_2O_2	9. O_2

(a) pairs of substances that have the same molecular mass
(b) pairs of substances that have the same molar mass
(c) pairs of substances in which there is the same number of atoms per molecule
(d) pairs of substances in which there is the same number of atoms per mole
(e) pairs of substances in which there is the same number of molecules per mole
(f) pairs of substances for which 50.0-g samples of each substance contain the same number of moles

9.156 Select from the grid *all* correct responses for each situation.

1. 2 moles CH_4	2. 1 mole C_2H_6	3. 2 moles C_3H_8
4. 1 mole C_4H_{10}	5. $\frac{1}{2}$ mole C_2H_4	6. 1 mole C_3H_6
7. $\frac{1}{2}$ mole C_6H_6	8. $2\frac{1}{2}$ moles C_2H_2	9. 2 moles C_3H_4

(a) pairs of quantities in which the molar ratio between carbon and hydrogen is the same
(b) pairs of quantities in which the mass percent carbon is the same
(c) pairs of quantities in which the number of total atoms is the same
(d) pairs of quantities in which the moles of hydrogen are the same
(e) pairs of quantities in which the mass of hydrogen is the same
(f) pairs of quantities in which the mass of carbon is the same

9.157 Select from the grid *all* correct responses for each situation. Use *whole number* formula masses in any calculations you do.

1. 34 g NH_3	2. 48 g N_2H_4	3. $\frac{1}{2}$ mole O_3
4. $1\frac{1}{2}$ moles O_2	5. $\frac{1}{2}$ mole H_2O	6. 28 g N_2
7. 28 g CO	8. 3 moles CO_2	9. 68 g H_2O_2

(a) compounds that have the same empirical and molecular formulas
(b) quantities for which the number of molecules present exceeds Avogadro's number
(c) pairs of quantities that have the same mass
(d) pairs of quantities that have the same molar mass
(e) pairs of quantities in which the same number of molecules is present
(f) pairs of quantities in which the same number of moles of atoms is present

9.158 Select from the grid *all* correct responses for each situation. Use *whole number* formula masses in any calculations you do. Assume the two components of each mixture do not react with each other.

Grid Problems

1. 3 moles He $\frac{1}{2}$ mole O_2	2. 1 mole C $\frac{1}{2}$ mole O_2	3. 2 moles C 1 mole He
4. 2 moles C 2 moles H_2	5. 1 mole Ne $\frac{1}{4}$ mole O_2	6. 1 mole Ne 2 moles He
7. 1 mole Be $\frac{1}{2}$ mole F_2	8. $\frac{1}{2}$ mole Si $\frac{1}{2}$ mole N_2	9. $\frac{1}{4}$ mole Si $\frac{3}{4}$ mole N_2

(a) mixtures that contain equal masses of the two components
(b) mixtures that contain an equal number of moles of two kinds of atoms
(c) pairs of mixtures in which the total mass is the same
(d) pairs of mixtures in which the mass ratio between components is the same
(e) pairs of mixtures in which the total moles of substances present is the same
(f) pairs of mixtures in which the total moles of atoms present is the same

CHAPTER TEN

Chemical Calculations Involving Chemical Equations

10.1 The Law of Conversation of Mass

In a previous consideration of chemical change (Sec. 4.4), it was noted that chemical changes are referred to as chemical reactions. As we learned there, a *chemical reaction is a process in which at least one new substance is produced as a result of chemical change.* It is usually easy to see that a chemical reaction has occurred. Color change, emission of heat and/or light, gas evolution, and solid formation are all indications that a chemical reaction has taken place.

The starting materials for a chemical reaction are known as reactants. **Reactants** are all substances present prior to the start of a chemical reaction. As a chemical reaction proceeds, reactants are consumed (used up) and new materials with new chemical properties are produced. **Products** are the substances produced as a result of a chemical reaction.

From a molecular viewpoint, a chemical reaction (chemical change) involves the

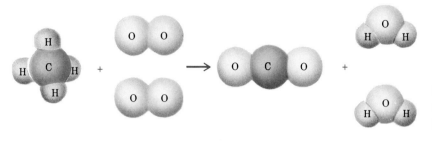

FIGURE 10.1 Rearrangement of atoms that occurs when methane (CH_4) reacts with oxygen (O_2).

union, separation, or rearrangement of atoms to produce new substances. Figure 10.1 shows the rearrangement of atoms that occurs when methane (CH_4) reacts with oxygen (O_2) to produce carbon dioxide (CO_2) and water (H_2O). Hydrogen atoms originally associated with carbon atoms (CH_4) become associated with oxygen atoms (H_2O) as the result of the chemical reaction.

Studies of countless chemical reactions over a period of 200 years have shown that there is no detectable change in the quantity of matter present during an ordinary chemical reaction. This generalization concerning chemical reactions has been formalized into a statement known as the **law of conservation of mass**: mass is neither created nor destroyed in any ordinary chemical reaction. To demonstrate the validity of this law, the masses of all reactants (substances that react together) and all products (substances formed) in a chemical reaction are carefully determined. It is found that the sum of the masses of the products is always the same as the sum of the masses of the reactants. The French chemist Antoine Laurent Lavoisier (1743–1794) is given credit for being the first to state this important relationship between the reactants and products of a chemical reaction.

Consider, as an illustrative example of this law, the reaction of known masses of the elements beryllium (Be) and oxygen (O) to form the compound beryllium oxide (BeO). Experimentally it is found that 36.03 g of Be will react with *exactly* 63.97 g of O. After the reaction no Be or O remains in elemental form; the only substance present is the product BeO, combined Be and O. When this product is weighed, its mass is found to be 100.00 g, which is the sum of the masses of the reactants (36.03 g + 63.97 g = 100.00 g). It is also found that when the 100.00 g of product BeO is heated to a high temperature in the absence of air, the BeO decomposes into Be and O, producing 36.03 g of Be and 63.97 g of O. Once again, no detectable mass change is observed; the mass of the reactants is equal to the mass of the products.

The law of conservation of mass is consistent with the statements of atomic theory (Sec. 5.1). Since all reacting chemical substances are made up of atoms (statement 1), each with its unique identity (statement 2), and these atoms can be neither created nor destroyed in a chemical reaction but merely rearranged (statement 4), it follows that the total mass after the reaction must equal the total mass before the reaction. We have the same number of atoms of each kind after the reaction as we started out with. An alternative way of stating the law of conservation of mass is: *The total mass of reactants and the total mass of products in a chemical reaction are always equal.*

The law of conservation of mass applies to all ordinary chemical reactions, and there is no known case of a *measurable* change in total mass during an *ordinary* chemical reaction.* This law will be a "guiding principle" for the discussion that follows about chemical equations and their use.

10.2 Writing Chemical Equations

A **chemical equation** is a written statement that uses symbols and formulas instead of words to describe the changes that occur in a chemical reaction. The following example shows the contrast between a word description of a chemical reaction and a chemical equation for the same reaction.

Word description: Magnesium oxide reacts with carbon to produce carbon monoxide and magnesium.

Chemical equation:

$$MgO + C \longrightarrow CO + Mg$$

In the same way that chemical symbols are considered the *letters* of chemical language, and formulas the *words* of the language, chemical equations can be considered the *sentences* of chemical language.

The conventions used in writing chemical equations are

1. The correct formulas of the *reactants* are always written on the *left* side of the equation.

$$MgO + C \longrightarrow CO + Mg$$

2. The correct formulas of the *products* are always written on the *right* side of the equation.

$$MgO + C \longrightarrow CO + Mg$$

3. The reactants and products are separated by an arrow pointing toward the products.

$$MgO + C \longrightarrow CO + Mg$$

4. Plus signs are used to separate different reactants or different products from each other.

$$MgO + C \longrightarrow CO + Mg$$

In reading chemical equations, plus signs on the reactant side of the equation are

---

* In the last half century the law of conservation of mass has had to be qualified. Certain types of reactions that involve radioactive processes have been found to deviate from this law. In these processes, there is a conversion of a small amount of matter into energy rather than into another form of matter. A more general law incorporates this apparent discrepancy—the law of conservation of mass and energy. This law takes into account the fact that matter and energy are interconvertible. Note that the statement of the law of conservation of mass as given at the start of this discussion contains the phrase "ordinary chemical reaction." Radioactive processes are not considered to be ordinary chemical reactions.

taken to mean "reacts with"; the arrow, "to produce"; and plus signs on the product side, "and."

A catchy, informal way of defining a chemical equation is to say that it gives the "before and after" picture of a chemical reaction. *Before* the reaction starts, only reactants are present—the left side of the equation. *After* the reaction is completed, products are present—the right side of the equation.

In order for a chemical equation to be *valid* it must satisfy two conditions.

1. *It must be consistent with experimental facts.* Only the reactants and products actually involved in a reaction are shown in an equation. An accurate formula must be used for each of these substances. For compounds, molecular rather than empirical formulas (Sec. 9.9) are always used. Elements in the solid and liquid states are represented in equations by the chemical symbol for the element. Elements that are gases at room temperatures are represented by the molecular form in which they actually occur in nature. Monoatomic, diatomic, and tetraatomic elemental gases are known.

   Monoatomic: He, Ne, Ar, Kr, Xe, Rn
   Diatomic: H_2, O_2, N_2, F_2, Cl_2, Br_2 (vapor), I_2 (vapor)
   Tetraatomic: P_4 (vapor), As_4 (vapor)*

2. *It must be consistent with the law of conservation of mass* (Sec. 10.1). There must be the same number of product atoms of each kind as there are reactant atoms of each kind, since atoms are neither created nor destroyed in an ordinary chemical reaction. Equations that satisfy the conditions of this law are said to be *balanced*. Using the four conventions previously listed for writing equations does not guarantee a balanced equation. Section 10.3 considers the steps that must be taken to ensure that an equation is balanced.

10.3 Balancing Chemical Equations

A **balanced chemical equation** has the same number of atoms of each element involved in the reaction on each side of the equation. It is therefore an equation consistent with the law of conservation of mass (Sec. 10.1).

An unbalanced equation is brought into balance by adding coefficients to the equation; such coefficients adjust the number of reactant and/or product molecules (or formula units) present. A **coefficient** is a number placed to the left of the formula of a substance that denotes the amount of the substance. In the notation 2 H_2O, the 2 on the left is a coefficient; 2 H_2O means two molecules of H_2O, and 3 H_2O means three molecules of H_2O. Coefficients tell how many formula units of a given substance are present.

---

* The four elements listed as vapors are not gases at room temperature but vaporize at slightly higher temperatures. The resultant vapors contain molecules with the formulas indicated. Even if these elements do not vaporize, they can still be represented with these formulas.

The following is a balanced chemical equation with the coefficients shown in color.

$$3\ Cu + 8\ HNO_3 \longrightarrow 3\ Cu(NO_3)_2 + 2\ NO + 4\ H_2O$$

The message of this balanced equation is "three Cu atoms react with eight HNO_3 molecules to produce three $Cu(NO_3)_2$ formula units, two NO molecules, and four H_2O molecules." A coefficient of 1 in a balanced equation is not explicitly written; it is considered to be understood. Both PCl_3 and H_3PO_3 have understood coefficients of 1 in the following balanced equation.

$$PCl_3 + 3\ H_2O \longrightarrow H_3PO_3 + 3\ HCl$$

A coefficient placed in front of a formula applies to the whole formula. In contrast, subscripts, also present in formulas, affect only parts of a formula.

$2\ H_2O$ — coefficient (affects both the H and O)

— subscript (affects only H)

The above notation denotes two molecules of H_2O; it also denotes a total of four H atoms and two O atoms.

We now proceed to the method of determining the proper coefficients to balance a given equation. It is introduced in the context of actually balancing two equations (Examples 10.1 and 10.2). Both of these examples should be studied carefully, since each includes detailed commentary concerning the "ins and outs" of balancing equations.

EXAMPLE 10.1

Balance the equation $Fe_2O_3 + C \longrightarrow Fe + CO_2$.

Solution

STEP 1 *Examine the equation, and pick one element to balance first.* It is often convenient to identify the most complex substance first, that is, the substance with the greatest number of atoms per formula unit. For this most complex substance, whether a reactant or product, "key in" on the element within it that is present in the greatest amount (greatest number of atoms). Using this guideline, we select Fe_2O_3 and the element oxygen.

We note that there are three oxygen atoms on the left side of the equation (in Fe_2O_3) and two oxygen atoms on the right side (in CO_2). For the oxygen atoms to balance we will need six on each side; six is the lowest number that both two and three will divide into evenly. To obtain six atoms on each side of the equation we place the coefficient 2 in front of Fe_2O_3 and the coefficient 3 in front of CO_2.

$$2\ Fe_2O_3 + C \longrightarrow Fe + 3\ CO_2$$

We now have six oxygen atoms on each side of the equation.

Section 10.3 · Balancing Chemical Equations

$$2\ Fe_2O_3: \quad 2 \times 3 = 6$$
$$3\ CO_2: \quad 3 \times 2 = 6$$

STEP 2   *Now pick a second element to balance.* We will balance the element Fe next. (In this particular equation it does not matter whether we balance Fe or C second.) The number of Fe atoms on the left side of the equation is four; the coefficient 2 in front of Fe_2O_3 sets the Fe atom number at four. We will need four Fe atoms on the product side, instead of the one Fe atom now present. This is accomplished by placing the coefficient 4 in front of Fe.

$$2\ Fe_2O_3 + C \longrightarrow 4Fe + 3\ CO_2$$

Now there are four Fe atoms on each side of the equation.

STEP 3   *Now pick a third element to balance.* The only element left to balance is C. There is one C atom on the left, but three carbon atoms on the right. Placing the coefficient 3 in front of C balances the carbon atoms at three.

$$2\ Fe_2O_3 + 3C \longrightarrow 4\ Fe + 3\ CO_2$$

STEP 4   *As a final check on the correctness of the balancing procedure, count atoms on each side of the equation.* The following table can be constructed from our balanced equation.

$$2\ Fe_2O_3 + 3\ C \longrightarrow 4\ Fe + 3\ CO_2$$

Atom	Left Side	Right Side
Fe	$2 \times 2 = 4$	$4 \times 1 = 4$
O	$2 \times 3 = 6$	$3 \times 2 = 6$
C	$3 \times 1 = 3$	$3 \times 1 = 3$

All elements are in balance: four Fe atoms on each side, six O atoms on each side, and three C atoms on each side.

PRACTICE EXERCISE 10.1

Balance the equation $KClO_3 \longrightarrow KCl + O_2$.   Ans. $2\ KClO_3 \rightarrow 2\ KCl + 3\ O_2$.

Similar exercises: Problems 10.15 and 10.16

EXAMPLE 10.2

Balance the equation $C_2H_6 + O_2 \longrightarrow CO_2 + H_2O$.

Solution

STEP 1   *Examine the equation, and pick one element to balance first.* The formula containing the most atoms is C_2H_6. We will balance the element H first. We have six H atoms on the left and two H atoms on the right. The two sides are brought into balance by placing the coefficient 3 in front of H$_2$O on the right side. We now have six H atoms on each side.

$$3 \text{ H}_2\text{O}: \quad 3 \times 2 = 6$$

Our equation now has the following appearance.

$$1 \text{ C}_2\text{H}_6 + \text{O}_2 \longrightarrow \text{CO}_2 + 3 \text{ H}_2\text{O}$$

In setting the H balance at six atoms we are setting the coefficient in front of C$_2$H$_6$ at 1. The 1 has been explicitly shown in the above equation to remind us that the C$_2$H$_6$ coefficient has been determined. (In the final balanced equation the 1 need not be shown.)

STEP 2 *Now pick a second element to balance.* We will balance C next. It is always better to balance the elements that appear in only one reactant and one product before trying to balance any elements appearing in several formulas on one side of the equation. Oxygen, our other choice for an element to balance at this stage, appears in two places on the product side of the equation. The number of carbon atoms is already set at two on the left side of the equation.

$$1 \text{ C}_2\text{H}_6: \quad 1 \times 2 = 2$$

We obtain a balance of two carbon atoms on each side of the equation by placing the coefficient 2 in front of CO$_2$.

$$1 \text{ C}_2\text{H}_6 + \text{O}_2 \longrightarrow 2 \text{ CO}_2 + 3 \text{ H}_2\text{O}$$

STEP 3 *Now pick a third element to balance.* Only one element is left to balance—oxygen. The number of oxygen atoms on the right side of the equation is already set at seven—four O atoms from the CO$_2$ and three O atoms from the H$_2$O.

$$2 \text{ CO}_2: \quad 2 \times 2 = 4$$
$$3 \text{ H}_2\text{O}: \quad 3 \times 1 = 3$$

To obtain seven O atoms on the left side of the equation we need a fractional coefficient, $3\frac{1}{2}$.

$$3\frac{1}{2} \text{ O}_2: \quad 3\frac{1}{2} \times 2 = 7$$

The coefficient 3 gives six atoms, and the coefficient 4 gives eight atoms. The only way we can get seven atoms is by using $3\frac{1}{2}$.

All of the coefficients in the equation have now been determined.

$$1 \text{ C}_2\text{H}_6 + 3\frac{1}{2} \text{ O}_2 \longrightarrow 2 \text{ CO}_2 + 3 \text{ H}_2\text{O}$$

Equations containing fractional coefficients are not considered to be written in their most conventional form. Although such equations are "mathematically" correct, they have some problems "chemically." The above equation indicates the need for $3\frac{1}{2}$ O$_2$ molecules among the reactants, and half an O$_2$ molecule does not exist as such. Step 4 shows how to take care of this "problem."

STEP 4   *After all coefficients have been determined, clear any fractional coefficients that are present.* We can clear the fraction present in this equation, $3\frac{1}{2}$, by multiplying *each* of the coefficients in the equation by the factor 2.

$$2\,C_2H_6 + 7\,O_2 \longrightarrow 4\,CO_2 + 6\,H_2O$$

Now we have the equation in its conventional form. Note that *all of the coefficients* had to be multiplied by 2, not just the fractional one. It will always be the case that whatever is done to a fractional coefficient to make it a whole number must also be carried out on all of the other coefficients.

If a coefficient involving $\frac{1}{3}$ had been present in the equation, we would have multiplied by 3 instead of by 2.

STEP 5   *As a final check on the correctness of the balancing procedure, count atoms on each side of the equation.* The following table can be constructed from our balanced equation.

$$2\,C_2H_6 + 7\,O_2 \longrightarrow 4\,CO_2 + 6\,H_2O$$

Atom	Left Side	Right Side
C	$2 \times 2 = 4$	$4 \times 1 = 4$
H	$2 \times 6 = 12$	$6 \times 2 = 12$
O	$7 \times 2 = 14$	$(4 \times 2) + (6 \times 1) = 14$

All atom counts balance. We have accomplished our task.

PRACTICE EXERCISE 10.2

Balance the equation $C_4H_{10} + O_2 \longrightarrow CO_2 + H_2O$.

Ans. $2\,C_4H_{10} + 13\,O_2 \rightarrow 8\,CO_2 + 10\,H_2O$.

Similar exercises: Problems 10.17 and 10.18

Some additional comments and guidelines concerning equations in general and the process of balancing in particular are

1. The coefficients in a balanced equation are always the *smallest set of whole numbers* that will balance the equation. We mention this because more than one set of coefficients will balance an equation. Consider the following three equations.

$$2\,H_2 + O_2 \longrightarrow 2\,H_2O$$
$$4\,H_2 + 2\,O_2 \longrightarrow 4\,H_2O$$
$$8\,H_2 + 4\,O_2 \longrightarrow 8\,H_2O$$

All three of these equations are mathematically correct; there are equal numbers of H and O atoms on each side of the equation. The first equation, however, is considered the conventional form, because the coefficients used there are the smallest set of whole numbers that will balance the equation. The coefficients in the second equation are double those in the first, and the third equation has coefficients four times those of the first equation.

2. It is helpful to consider polyatomic ions as single entities in balancing an equation, provided they maintain their identity in the chemical reaction, that is, provided they appear on both sides of the equation in the same form. For example, in an equation where sulfate units are present (SO_4^{2-}), balance them as a unit rather than trying to balance S and O separately. The reasoning would be: "We have two sulfates on this side, so we need two sulfates on that side, etc."

3. Subscripts in a formula may never be altered during the balancing process. A student might try to balance the atoms of K in KCl at two by using the notation K_2Cl_2 instead of 2 KCl. This is incorrect. The notation K_2Cl_2 denotes a formula unit containing four atoms, whereas the notation 2 KCl denotes two formula units, each containing two atoms. The experimental fact is that a formula unit of KCl contains two rather than four atoms. *The coefficient deals with the number of formula units of a substance, and the subscript deals with the composition of the substance.* Subscripts illustrate the law of definite proportions (Sec. 9.1); coefficients relate to the law of conservation of mass (Sec. 10.1).

4. You are not expected, at this point, to be able to write down the products for a chemical reaction given what the reactants are. After learning how to balance equations, students sometimes get the mistaken idea that they ought to be able to write down equations from "scratch." This is not so. At this stage, you should be able to balance simple equations given *all* of the reactants and products. In Section 10.5 guidelines are given for predicting the products for selected types of reaction.

5. Some equations are much more difficult to balance than those you encounter in this chapter's examples and problem exercises. The procedures discussed here simply are not adequate for these more difficult equations. In Chapter 15 a more systematic method for balancing equations, specifically designed for these more difficult situations, will be presented.

6. The ultimate source of any chemical equation is experimental information. The products formed in a chemical reaction are facts learned by experiment; we cannot discover them simply by writing an equation. The products are first identified experimentally, and then they can be represented by an equation.

7. Finally, the only way to learn to balance equations is through practice. The problem set at the end of this chapter contains numerous equations for you to practice on.

10.4 Special Symbols Used in Equations

In addition to the essential plus sign and arrow notation used in chemical equations, a number of optional symbols convey more information about a chemical reaction than just the chemical species involved. In particular, it is often useful to know the

TABLE 10.1 Symbols Used in Equations

Symbol	Meaning
Essential	
→	"to produce"
+	"reacts with" or "and"
Optional	
(s)	solid
(l)	liquid
(g)	gas
(aq)	aqueous solution (a substance dissolved in water)

physical state of the substances involved in a chemical reaction. The optional symbols listed in Table 10.1 are used to specify the physical state of reactants and products.

The equations we balanced in Section 10.3 (Examples 10.1 and 10.2) are written as follows when the optional symbols are included.

$$2\ Fe_2O_3(s) + 3\ C(s) \longrightarrow 4\ Fe(s) + 3\ CO_2(g)$$
$$2\ C_2H_6(g) + 7\ O_2(g) \longrightarrow 4\ CO_2(g) + 6\ H_2O(g)$$

Two more examples of the use of optional symbols are

$$NaCl(aq) + AgNO_3(aq) \longrightarrow AgCl(s) + NaNO_3(aq)$$
$$NaOH(aq) + HCl(aq) \longrightarrow NaCl(aq) + H_2O(l)$$

The optional symbols in these two equations indicate that both reactions take place in aqueous solution. In the first reaction, one of the products, AgCl, is insoluble, being present in the solution as a solid. In the second reaction, the product NaCl is soluble and thus remains in solution.

Also, note that in the last equation the reactant HCl is functioning as an acid and must be named as such (Sec. 8.6). The notation HCl(g) indicates hydrogen chloride in its gaseous state; the notation HCl(aq), hydrogen chloride dissolved in water, denotes hydrochloric acid.

10.5 Patterns in Chemical Reactivity

To this point in our discussion of chemical equations the focus has been on balancing them given all of the reactants and all of the products. We now consider, for selected situations, the prediction of what the products of a chemical reaction will be.

Combustion reactions are a most common type of reaction. A **combustion reaction** involves the reaction of a substance with oxygen (usually from air) that proceeds with evolution of heat and usually also a flame. Hydrocarbons, binary compounds of carbon and hydrogen (of which many exist), are the most common type of compound that undergoes combustion. In hydrocarbon combustion, the carbon of the hydrocarbon combines with oxygen to produce carbon dioxide (CO_2). The hydrocarbon hydrogen also interacts with oxygen of air to give water (H_2O) as a product. The relative amounts of CO_2 and H_2O produced depends on the composition of the hydrocarbon.

$$2\ C_2H_2(g) + 5\ O_2(g) \longrightarrow 4\ CO_2(g) + 2\ H_2O(g)$$
$$C_3H_8(g) + 5\ O_2(g) \longrightarrow 3\ CO_2(g) + 4\ H_2O(g)$$
$$C_4H_8(g) + 6\ O_2(g) \longrightarrow 4\ CO_2(g) + 4\ H_2O(g)$$

Note that the equation that was balanced in Example 10.2 was of the type we have just considered.

Combustion of compounds containing oxygen as well as carbon and hydrogen (for example CH_4O or C_3H_8O) also produce CO_2 and H_2O as products.

$$2\ CH_4O(l) + 3\ O_2(g) \longrightarrow 2\ CO_2(g) + 4\ H_2O(g)$$
$$2\ C_3H_8O(l) + 9\ O_2(g) \longrightarrow 6\ CO_2(g) + 8\ H_2O(g)$$

Example 10.3 is an equation-balancing exercise involving a reaction of this type.

EXAMPLE 10.3

Write the balanced chemical equation for the reaction that occurs when ethyl alcohol, C_2H_6O, is burned in air.

Solution

The C atoms in the alcohol will end up in product CO_2 and the H atoms of the alcohol in product H_2O. There are two sources for the oxygen present in the products CO_2 and H_2O. Most of it comes from the O_2 of air; however, the reactant C_2H_6O is also an oxygen source. The unbalanced equation for this reaction is

$$C_2H_6O + O_2 \longrightarrow CO_2 + H_2O$$

The equation is balanced using the procedures of Section 10.3.

STEP 1 *Balancing of C atoms:* There are two C atoms on the left and only one C atom on the right. Placing the coefficient 2 in front of CO_2 balances the C atoms at two on each side.

$$1\ C_2H_6O + O_2 \longrightarrow 2\ CO_2 + H_2O$$

STEP 2 *Balancing of H atoms:* An effect of balancing the C atoms at two (step 1) is the setting of the H atoms on the left side of the equation at six; the coefficient in front of C_2H_6O is 1. Placing the coefficient 3 in front of H_2O causes the hydrogen atoms to balance at six on each side of the equation.

$$1\ C_2H_6O + O_2 \longrightarrow 2\ CO_2 + 3\ H_2O$$

STEP 3 *Balancing of O atoms:* The oxygen content of the right side of the equation is set at seven atoms; four oxygen atoms from 2 CO_2 and three oxygen atoms from 3 H_2O. To obtain seven oxygen atoms on the left side of the equation the coefficient 3 is placed in front of O_2; 3 O_2 gives six oxygen atoms and there is an additional O in 1 C_2H_6O. Note that the element oxygen is present in all four formulas in the equation.

$$1\ C_2H_6O + 3\ O_2 \longrightarrow 2\ CO_2 + 3\ H_2O$$

STEP 4 *Final check:* The equation is balanced. There are two carbon atoms, six hydrogen atoms and seven oxygen atoms on each side of the equation.

$$C_2H_6O + 3\ O_2 \longrightarrow 2\ CO_2 + 3\ H_2O$$

PRACTICE EXERCISE 10.3

Write the balanced chemical equation for the reaction that occurs when diethyl ether, $C_4H_{10}O$, is burned in air.   Ans. $C_4H_{10}O + 6\ O_2 \rightarrow 4\ CO_2 + 5\ H_2O$.

Similar exercises: Problems 10.25 and 10.26

Thermal decomposition of metal carbonates is another common type of reaction. Metal carbonates, when heated to a high temperature (the temperature needed varies with the metal), break down, releasing carbon dioxide gas and producing the metal oxide. Typical carbonate decomposition equations include the following.

$$Na_2CO_3(s) \longrightarrow Na_2O(s) + CO_2(g)$$
$$MgCO_3(s) \longrightarrow MgO(s) + CO_2(g)$$
$$Al_2(CO_3)_3(s) \longrightarrow Al_2O_3(s) + 3\ CO_2(g)$$

10.6 Classes of Chemical Reactions

An almost inconceivable number of chemical reactions is possible. The problems associated with organizing our knowledge about them are diminished considerably by grouping the reactions into classes. We consider in this section an important classification system for reactions based on the *form* of the equation for the reaction. In this system four general types of reactions are recognized: synthesis, decomposition, single replacement, and double replacement.

The first of the four categories of reactions is the synthesis reaction. A **synthesis reaction** is one in which a single product is produced from two (or more) reactants. The general equation for a synthesis reaction is

$$X + Y \longrightarrow XY$$

This is an extremely simple type of reaction, one in which two substances join together to form a more complicated product. The reactants X and Y may be elements or compounds or an element and a compound. The product of the reaction, XY, is always a compound.

Some representative synthesis reactions with elements as the reactants are

$$H_2 + Cl_2 \longrightarrow 2\ HCl$$
$$S + O_2 \longrightarrow SO_2$$
$$Ni + S \longrightarrow NiS$$

Some examples of synthesis reactions in which compounds are involved as reactants are

$$SO_3 + H_2O \longrightarrow H_2SO_4$$
$$K_2O + H_2O \longrightarrow 2\ KOH$$
$$2\ NO + O_2 \longrightarrow 2\ NO_2$$
$$2\ NO_2 + H_2O_2 \longrightarrow 2\ HNO_3$$

A **decomposition reaction** is one in which a single reactant is converted into two or more simpler substances. It is thus the exact opposite of a synthesis reaction. The general equation for a decomposition reaction is

$$XY \longrightarrow X + Y$$

At sufficiently high temperatures all compounds can be broken down (decomposed) into their constituent elements. Such reactions include

$$2\ CuO \longrightarrow 2\ Cu + O_2$$
$$2\ H_2O \longrightarrow 2\ H_2 + O_2$$

Examples of decomposition reactions that result in at least one compound as a product are

$$CaCO_3 \longrightarrow CaO + CO_2$$
$$2\ KClO_3 \longrightarrow 2\ KCl + 3\ O_2$$

A **single-replacement reaction** is one in which one element within a compound is replaced by another element. In this type of reaction there are always two reactants, an element and a compound, and two products, also an element and a compound. The general equation for a single-replacement reaction is

$$X + YZ \longrightarrow Y + XZ$$

Both metals and nonmetals can be replaced in this manner. Such reactions usually involve aqueous solutions. Examples include

$$Zn + H_2SO_4 \longrightarrow H_2 + ZnSO_4$$
$$Ni + 2\ HCl \longrightarrow H_2 + NiCl_2$$
$$Fe + CuSO_4 \longrightarrow Cu + FeSO_4$$
$$Mg + Ni(NO_3)_2 \longrightarrow Ni + Mg(NO_3)_2$$

A **double-replacement reaction** is one in which two compounds exchange parts with each other and form two different compounds. The general equation for such a reaction is

$$AX + BY \longrightarrow AY + BX$$

Double-replacement reactions generally involve ionic compounds in aqueous solution. Most often the positive ion from one compound exchanges with the positive

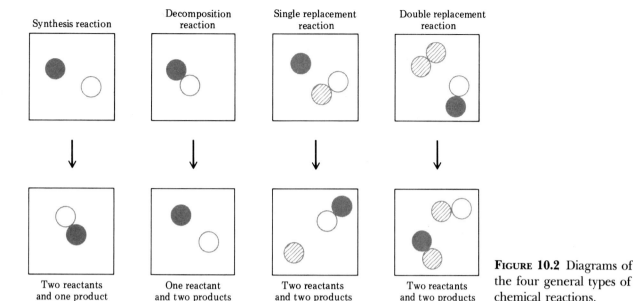

FIGURE 10.2 Diagrams of the four general types of chemical reactions.

ion of the other. The process may be thought of as "partner swapping," since each negative ion ends up paired with a new partner (positive ion). Such reactions include

$$AgNO_3 + NaCl \longrightarrow NaNO_3 + AgCl$$
$$NaF + HCl \longrightarrow NaCl + HF$$
$$AgNO_3 + HCl \longrightarrow AgCl + HNO_3$$

Figure 10.2 summarizes in pictorial form the four general types of chemical reactions.

EXAMPLE 10.4

Classify each of the following reactions as synthesis, decomposition, single replacement, or double replacement.

(a) $2\,C + O_2 \longrightarrow 2\,CO$
(b) $2\,KNO_3 \longrightarrow 2\,KNO_2 + O_2$
(c) $Zn + 2\,AgNO_3 \longrightarrow Zn(NO_3)_2 + 2\,Ag$
(d) $Ni(NO_3)_2 + 2\,NaOH \longrightarrow Ni(OH)_2 + 2\,NaNO_3$

Solution

(a) Two substances combine to form a single substance; hence, this reaction is classified as a synthesis reaction.
(b) Since two substances are produced from a single substance, this reaction is a decomposition reaction.
(c) Having an element and a compound as reactants and an element and compound as products is a characteristic of a single-replacement reaction. That is the type of reaction we have here.
(d) This reaction is a double-replacement reaction. Nickel and sodium are changing places, that is, "swapping partners."

PRACTICE EXERCISE 10.4

Classify each of the following reactions as synthesis, decomposition, single replacement, or double replacement.

(a) $CuCO_3 \longrightarrow CuO + CO_2$
(b) $Fe_2O_3 + 3C \longrightarrow 2Fe + 3CO$
(c) $3Mg + N_2 \longrightarrow Mg_3N_2$
(d) $NH_4Cl + AgNO_3 \longrightarrow NH_4NO_3 + AgCl$

Ans. (a) decomposition; (b) single replacement; (c) synthesis; (d) double replacement.

Similar exercises: Problems 10.33 and 10.34

Many, but not all, reactions fall into the four categories we have just discussed. For example, combustion reactions (Sec. 10.5) do not fit any of the four general patterns. Despite the system not being all inclusive, it is still very useful because of the many reactions it does help correlate.

We should note, before leaving this topic, that other ways of classifying chemical reactions exist. Two important additional categories of reactions are *acid–base reactions* and *oxidation–reduction reactions*. Acid–base reactions are the topic of Chapter 14 and oxidation–reduction reactions are considered in Chapter 15. Combustion reactions are examples of oxidation–reduction reactions.

10.7 Chemical Equations and the Mole Concept

The coefficients in a balanced chemical equation, like the subscripts in a chemical formula (Sec. 9.7), have two levels of interpretation—a microscopic level of meaning and a macroscopic level of meaning.

The first of these two interpretations, the microscopic level, has been used in the previous sections of this chapter. At this level, a balanced chemical equation gives the relative numbers of formula units of the various reactants and products involved in a chemical reaction. *The coefficients in the equation give directly the numerical relationships among formula units consumed (used up) and/or produced in the chemical reaction.* Interpreted at the microscopic level, the equation

$$4NH_3 + 5O_2 \longrightarrow 4NO + 6H_2O$$

conveys the information that four molecules of NH_3 react with five molecules of O_2 to produce four molecules of NO and six molecules of H_2O.

At the macroscopic level of interpretation, chemical equations are used to relate mole-sized quantities of reactants and products to each other. At this molar level of interpretation, the coefficients in an equation give the fixed molar ratios between substances consumed and/or produced in the chemical reaction. Thus the equation

$$4NH_3 + 5O_2 \longrightarrow 4NO + 6H_2O$$

conveys the information that four moles of NH_3 react with five moles of O_2 to produce four moles of NO and six moles of H_2O.

The validity of the molar interpretation of coefficients in an equation can be derived very straightforwardly from the microscopic level of interpretation. A balanced chemical equation remains valid (mathematically correct) when all of its coefficients are multiplied by the same number. (If molecules react in a 3-to-1 ratio, they will also react in a 6-to-2 or 9-to-3 ratio.) Multiplying the previous equation by y, where y is any number, we have

$$4y\ NH_3 + 5y\ O_2 \longrightarrow 4y\ NO + 6y\ H_2O$$

The situation where $y = 6.02 \times 10^{23}$ is of particular interest because $6.02 \times 10^{23} = 1$ mole. Using $y = 1$ mole, we have by substitution

$$4 \text{ moles } NH_3 + 5 \text{ moles } O_2 \longrightarrow 4 \text{ moles } NO + 6 \text{ moles } H_2O$$

Thus, as with the subscripts in formulas, the coefficients in equations carry a dual meaning: "number of formula units" at the microscopic level and "moles of formula units" at the macroscopic level.

In Section 10.3 it was noted that fractional equation coefficients are often needed in the equation-balancing process. We can now further note that such fractional coefficients do have valid meaning for the macroscopic level interpretation of a chemical equation ($3\frac{1}{3}$ moles, and so on), whereas they are totally unacceptable for the microscopic level of interpretation ($3\frac{1}{2}$ molecules, and so on).

The coefficients in an equation may be used to generate conversion factors used in problem solving. Numerous conversion factors are obtainable from a single balanced equation. Consider the balanced equation

$$P_4O_{10} + 6\ H_2O \longrightarrow 4\ H_3PO_4$$

Three mole-to-mole relationships can be obtained from this equation.

$$1 \text{ mole of } P_4O_{10} \text{ produces } 4 \text{ moles of } H_3PO_4$$
$$6 \text{ moles of } H_2O \text{ produces } 4 \text{ moles of } H_3PO_4$$
$$1 \text{ mole of } P_4O_{10} \text{ reacts with } 6 \text{ moles of } H_2O$$

From these three macroscopic-level relationships, six conversion factors can be written.

From the first relationship:

$$\frac{1 \text{ mole } P_4O_{10}}{4 \text{ moles } H_3PO_4} \quad \text{and} \quad \frac{4 \text{ moles } H_3PO_4}{1 \text{ mole } P_4O_{10}}$$

From the second relationship:

$$\frac{6 \text{ moles } H_2O}{4 \text{ moles } H_3PO_4} \quad \text{and} \quad \frac{4 \text{ moles } H_3PO_4}{6 \text{ moles } H_2O}$$

From the third relationship:

$$\frac{1 \text{ mole } P_4O_{10}}{6 \text{ moles } H_2O} \quad \text{and} \quad \frac{6 \text{ moles } H_2O}{1 \text{ mole } P_4O_{10}}$$

Any chemical equation can be the source of numerous conversion factors. The more reactants and products there are in the equation, the greater the number of derivable conversion factors.

Conversion factors obtained from equations are used in many different types of calculations. Example 10.5 illustrates some very simple applications of their use. In Section 10.8 we explore their use in more complicated problem-solving situations.

EXAMPLE 10.5

The natural gas burned to provide heat in many homes is predominantly methane (CH_4). Methane burns (reacts with O_2) as shown by the equation.

$$CH_4 + 2\,O_2 \longrightarrow CO_2 + 2\,H_2O$$

(a) How many moles of H_2O are produced when 1.23 moles of CH_4 burns?
(b) How many moles of O_2 are needed to react with 3.61 moles of CH_4?

Solution

Both parts of this problem are one-step mole-to-mole calculations. In each case the needed conversion factor is derived from the coefficients of the chemical equation.

$$\boxed{\text{Moles of substance A}} \xrightarrow{\text{equation coefficients}} \boxed{\text{Moles of substance B}}$$

(a)
STEP 1 The given quantity is 1.23 moles of CH_4, and the desired quantity is moles of H_2O.

$$1.23 \text{ moles } CH_4 = ?\text{ moles } H_2O$$

STEP 2 The conversion factor needed to convert from moles of CH_4 to moles of H_2O is derived from the coefficients of CH_4 and H_2O in the balanced equation. The equation tells us that 1 mole of CH_4 produces 2 moles of H_2O. From this relationship two conversion factors are obtainable.

$$\frac{1 \text{ mole } CH_4}{2 \text{ moles } H_2O} \quad \text{and} \quad \frac{2 \text{ moles } H_2O}{1 \text{ mole } CH_4}$$

We will use the second of these conversion factors in solving the problem. The setup is

$$1.23 \text{ moles } CH_4 \times \frac{2 \text{ moles } H_2O}{1 \text{ mole } CH_4}$$

We used the second of the two conversion factors because it had moles of CH_4 in the denominator, a requirement for the units moles of CH_4 to cancel.

STEP 3 Collecting numerical terms, after cancellation of units, gives

$$\frac{1.23 \times 2}{1} \text{ moles } H_2O = 2.46 \text{ moles } H_2O \quad \text{(calculator and correct answer)}$$

Note that the coefficients in the equation enter directly into the numerical calculation. Having a correctly balanced equation is therefore of vital importance. Using an unbalanced or misbalanced equation as a source of a conversion factor will lead to a wrong numerical answer.

(b)

STEP 1 Both the given species, CH_4, and the desired species, O_2, are reactants.

$$3.61 \text{ moles } CH_4 = ? \text{ moles } O_2$$

STEP 2 Molar relationships obtained from an equation are not required to always involve one reactant and one product, as was the case in part (a). Molar relationships involving only reactants or only products are often needed and used. In this problem we will need the molar relationship between the two reactants, O_2 and CH_4, which is two to one. With this ratio the conversion factor

$$\frac{2 \text{ moles } O_2}{1 \text{ mole } CH_4}$$

can be constructed, which is used in the setup of the problem as follows.

$$3.61 \text{ moles } CH_4 \times \frac{2 \text{ moles } O_2}{1 \text{ mole } CH_4}$$

STEP 3 Collecting numerical terms, after cancellation of units, gives

$$\frac{3.61 \times 2}{1} \text{ moles } O_2 = 7.22 \text{ moles } O_2 \quad \text{(calculator and correct answer)}$$

In terms of significant figures, the numbers in conversion factors obtained from equation coefficients are considered exact numbers. Thus, since 3.61 contains three significant figures, the answer to this problem should also contain three significant figures.

PRACTICE EXERCISE 10.5

The rusting of iron can be represented by the equation

$$4 \text{ Fe} + 3 \text{ O}_2 \longrightarrow 2 \text{ Fe}_2\text{O}_3$$

(a) How many moles of Fe_2O_3 are produced when 2.68 moles of O_2 are consumed?
(b) How many moles of Fe are needed to react with 7.21 moles of O_2?

Ans. (a) 1.79 moles Fe_2O_3; (b) 9.61 moles Fe.

Similar exercises: Problems 10.41–10.46

10.8 Calculations Based on Chemical Equations—Stoichiometry

A major area of concern for chemists is the quantities of materials consumed and produced in chemical reactions. This area of study is called chemical stoichiometry. **Chemical stoichiometry** is the study of the quantitative relationships among reactants

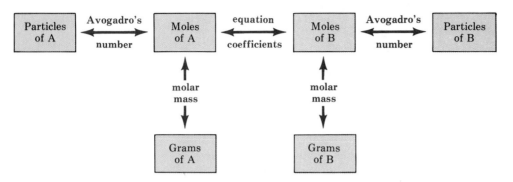

FIGURE 10.3 Conversion factor relationships needed for solving chemical-equation-based problems.

and products in a chemical reaction. The word *stoichiometry,* pronounced stoy-kee-om-eh-tree, is derived from the Greek *stoicheion* (element) and *metron* (measure). The stoichiometry of a chemical reaction always involves the *molar relationships* between reactants and products (Sec. 10.5) and thus is given by the coefficients in the balanced equation for the chemical reaction.

In a typical stoichiometric (stoy-kee-oh-met-rik) calculation, information is given about one reactant or product of a reaction (number of grams, moles, or particles), and information is requested concerning another reactant or product of the same reaction. The substances involved in such a calculation may both be reactants, may both be products, or may be one of each.

The conversion factor relationships needed to solve problems of the above general type are given in Figure 10.3. This diagram should seem very familiar to you, for it is almost identical to Figure 9.3, with which you have worked repeatedly. There is only one difference between the two. In the Chapter 9 diagram the subscripts in a chemical formula were listed as the basis for relating moles of given and desired substances to each other. In this new diagram, these same two quantities are related using the coefficients of a balanced chemical equation.

The most common type of stoichiometric calculation is a mass-to-mass (gram-to-gram) problem. In such problems the mass of one substance involved in a chemical reaction (either reactant or product) is given, and information is requested about the mass of another of the substances involved in the reaction (either a reactant or product). Situations requiring the solution of problems of this type are frequently encountered in laboratory settings. For example, a chemist has available so many grams of a certain chemical and wants to know how many grams of another substance can be produced from it, or how many grams of a third substance are needed to react with it. Examples 10.6 and 10.7 are both problem-solving situations of the gram-to-gram type.

EXAMPLE 10.6

Silicon carbide, SiC, used as an abrasive on sandpaper, is prepared using the following chemical reaction

$$SiO_2 + 3\,C \longrightarrow SiC + 2\,CO$$

How many grams of C are needed to completely react with 4.50 g of SiO_2?

Solution

STEP 1   Here we are given information about one reactant (4.50 g of SiO_2) and asked to calculate information about C, the other reactant.

$$4.50 \text{ g } SiO_2 = ? \text{ g } C$$

STEP 2   This problem is of the "grams of A" to "grams of B" type. The pathway to be used in solving this type of problem, in terms of Figure 10.3, is

$$\boxed{\text{Grams of A}} \xrightarrow{\text{molar mass}} \boxed{\text{Moles of A}} \xrightarrow{\text{equation coefficients}} \boxed{\text{Moles of B}} \xrightarrow{\text{molar mass}} \boxed{\text{Grams of B}}$$

STEP 3   The dimensional analysis setup for this pathway is

$$4.50 \text{ g } \cancel{SiO_2} \times \frac{1 \text{ mole } \cancel{SiO_2}}{60.1 \text{ g } \cancel{SiO_2}} \times \frac{3 \text{ moles } \cancel{C}}{1 \text{ mole } \cancel{SiO_2}} \times \frac{12.0 \text{ g C}}{1 \text{ mole } \cancel{C}}$$

grams A $\longrightarrow$ moles A $\longrightarrow$ moles B $\longrightarrow$ grams B

The number 60.1 in the first conversion factor is the molar mass of SiO_2; the 3 and 1 in the second conversion factor are the coefficients, respectively, of C and SiO_2 in the balanced chemical equation; and the 12.0 g in the third conversion factor is the molar mass of C.

STEP 4   The solution, obtained from combining all of the numerical factors, is

$$\frac{4.50 \times 1 \times 3 \times 12.0}{60.1 \times 1 \times 1} \text{ g C} = 2.6955074 \text{ g C} \quad \text{(calculator answer)}$$

$$= 2.70 \text{ g C} \quad \text{(correct answer)}$$

PRACTICE EXERCISE 10.6

The active ingredient in many antacid formulations is magnesium hydroxide [$Mg(OH)_2$], which reacts with stomach acid (HCl) to produce magnesium chloride ($MgCl_2$) and water. The equation for the reaction is

$$Mg(OH)_2(s) + 2\,HCl(aq) \longrightarrow MgCl_2(aq) + 2\,H_2O(l)$$

How many grams of $Mg(OH)_2$ are needed to react with 1.00 g of HCl?

Ans.   0.799 g $Mg(OH)_2$.

Similar exercises: Problems 10.49 and 10.50

EXAMPLE 10.7

The reusable booster rockets of the U.S. space shuttle use a mixture of aluminum, Al, and ammonium perchlorate, NH_4ClO_4, for fuel. The reaction between these substances is

$$3\,Al + 3\,NH_4ClO_4 \longrightarrow Al_2O_3 + AlCl_3 + 3\,NO + 6\,H_2O$$

How many grams of H_2O are produced as 15.23 g of Al is consumed?

Solution

STEP 1 Here we are given information about a reactant (15.23 g Al) and asked to calculate information about one of the products (H_2O).

$$15.23 \text{ g Al} = ? \text{ g } H_2O$$

STEP 2 This problem, like Example 10.6, is a "grams of A" to "grams of B" problem. The pathway used in solving it will be the same, which, in terms of Figure 10.3, is

$$\boxed{\text{Grams of A}} \xrightarrow{\text{molar mass}} \boxed{\text{Moles of A}} \xrightarrow{\text{equation coefficients}} \boxed{\text{Moles of B}} \xrightarrow{\text{molar mass}} \boxed{\text{Grams of B}}$$

STEP 3 The dimensional analysis setup is

$$15.23 \text{ g Al} \times \frac{1 \text{ mole Al}}{26.98 \text{ g Al}} \times \frac{6 \text{ moles } H_2O}{3 \text{ moles Al}} \times \frac{18.02 \text{ g } H_2O}{1 \text{ mole } H_2O}$$

grams A ⟶ moles A ⟶ moles B ⟶ grams B

The chemical equation is the "bridge" that enables us to go from Al to H_2O. The numbers in the second conversion factor, the "bridge factor," are coefficients from this equation.

STEP 4 The solution, obtained from combining all of the numerical factors in the setup, is

$$\frac{15.23 \times 1 \times 6 \times 18.02}{26.98 \times 3 \times 1} \text{ g } H_2O = 20.344299 \text{ g } H_2O \quad \text{(calculator answer)}$$

$$= 20.34 \text{ g } H_2O \quad \text{(correct answer)}$$

PRACTICE EXERCISE 10.7

The chemical equation for the photosynthesis reaction in plants is

$$6 \, CO_2 + 6 \, H_2O \longrightarrow C_6H_{12}O_6 + 6 \, O_2$$

How many grams of oxygen (O_2) are produced by a plant from the consumption of 25.0 g of carbon dioxide (CO_2)?

Ans. 18.2 g O_2.

Similar exercises: Problems 10.55 and 10.56

"Grams of A" to "grams of B" problems (Examples 10.6 and 10.7) are not the only type of problem for which the coefficients in a balanced equation can be used to relate quantities of two substances. As further examples of the use of equation coefficients in problem solving, consider Example 10.8 (a "grams of A" to "moles of B" problem) and Example 10.9 (a "particles of A" to "grams of B" problem).

EXAMPLE 10.8

Automotive airbags inflate when sodium azide, NaN_3, rapidly decomposes to its constituent elements. The equation for the chemical reaction is

$$2\,NaN_3(s) \longrightarrow 2\,Na(s) + 3\,N_2(g)$$

How many moles of N_2 are produced when 1.000 g of NaN_3 decomposes?

Solution

STEP 1 The given quantity is 1.000 g of NaN_3 and the desired quantity is moles of N_2.

$$1.000\text{ g }NaN_3 = ?\text{ moles }N_2$$

STEP 2 This is a "grams of A" to "moles of B" problem. The pathway used to solve such a problem is, according to Figure 10.3,

$$\boxed{\text{Grams of A}} \xrightarrow{\text{molar mass}} \boxed{\text{Moles of A}} \xrightarrow{\text{equation coefficients}} \boxed{\text{Moles of B}}$$

STEP 3 The dimensional analysis setup is

$$1.000\text{ g }NaN_3 \times \frac{1\text{ mole }NaN_3}{65.02\text{ g }NaN_3} \times \frac{3\text{ moles }N_2}{2\text{ moles }NaN_3}$$

$$\text{grams A} \longrightarrow \text{moles A} \longrightarrow \text{moles B}$$

The number 65.02 in the first conversion factor is the molar mass of NaN_3.

STEP 4 The solution, obtained from combining all of the numbers in the manner indicated in the setup, is

$$\frac{1.000 \times 1 \times 3}{65.02 \times 2}\text{ moles }N_2 = 0.023069824\text{ mole }N_2 \quad \text{(calculator answer)}$$

$$= 0.02307\text{ mole }N_2 \quad \text{(correct answer)}$$

PRACTICE EXERCISE 10.8

Decomposition of $KClO_3$ serves as a convenient laboratory source of small amounts of oxygen gas. The reaction is

$$2\,KClO_3 \longrightarrow 2\,KCl + 3\,O_2$$

How many moles of $KClO_3$ must be decomposed to produce 5.00 g of O_2?

Ans. 0.104 mole $KClO_3$.

Similar exercises: Problems 10.51 and 10.52

EXAMPLE 10.9

The reaction of sulfuric acid (H_2SO_4) with elemental copper (Cu) produces three products—sulfur dioxide (SO_2), water (H_2O), and copper(II) sulfate ($CuSO_4$). How many grams of water will be produced at the same time that 5 billion (5.00×10^9) sulfur dioxide molecules are produced?

Solution

Although a calculation of this type will not have a lot of practical significance, it will test your understanding of the problem-solving relationships under discussion in this section of the text.

The specifics of the chemical reaction of concern to us in this problem were given in "word" rather than "equation" form in the problem statement. These "words" must be translated into an equation before we can proceed with the problem solving. The equation is

$$H_2SO_4 + Cu \longrightarrow SO_2 + H_2O + CuSO_4$$

Having a chemical equation is not enough. It must be a *balanced* chemical equation. Using the balancing procedures of Section 10.3, the above equation, in balanced form, becomes

$$2\,H_2SO_4 + Cu \longrightarrow SO_2 + 2\,H_2O + CuSO_4$$

Now we are ready to proceed with the solving of our problem.

STEP 1 We are given a certain number of particles (molecules) and asked to find the number of grams of a related substance.

$$5.00 \times 10^9 \text{ molecules } SO_2 = ?\text{ g } H_2O$$

STEP 2 This is a "particles of A" to "grams of B" problem. Even though SO_2 and H_2O are both products, we can still work this problem in a manner similar to previous problems. The coefficients in a balanced equation relate reactants to products, reactants to reactants, and *products to products*. The pathway for this problem (see Fig. 10.3) is

Particles of A $\xrightarrow{\text{Avogadro's number}}$ Moles of A $\xrightarrow{\text{equation coefficients}}$ Moles of B $\xrightarrow{\text{molar mass}}$ Grams of B

STEP 3 The dimensional analysis setup is

$$5.00 \times 10^9 \text{ molecules } SO_2 \times \frac{1 \text{ mole } SO_2}{6.02 \times 10^{23} \text{ molecules } SO_2} \times \frac{2 \text{ moles } H_2O}{1 \text{ mole } SO_2} \times \frac{18.0 \text{ g } H_2O}{1 \text{ mole } H_2O}$$

particles A $\longrightarrow$ moles A $\longrightarrow$ moles B $\longrightarrow$ grams B

STEP 4 The solution, obtained by combining all of the numerical factors in the setup, is

$$\frac{5.00 \times 10^9 \times 1 \times 2 \times 18.0}{6.02 \times 10^{23} \times 1 \times 1} \text{ g } H_2O = 2.9900332 \times 10^{-13} \text{ g } H_2O$$
(calculator answer)
$$= 2.99 \times 10^{-13} \text{ g } H_2O$$
(correct answer)

PRACTICE EXERCISE 10.9

When silver carbonate (Ag_2CO_3) is decomposed by heating, three products are produced—metallic silver (Ag), carbon dioxide (CO_2), and oxygen (O_2).

How many grams of O_2 will be produced when 100 billion (1.00×10^{11}) Ag_2CO_3 formula units decompose?

Ans. 2.66×10^{-12} g O_2.

Similar exercises: Problems 10.57 and 10.58

10.9 The Limiting Reactant Concept

When a chemical reaction is carried out in a laboratory or industrial setting, the reactants are not usually present in the exact molar ratios specified in the balanced chemical equation for the reaction. Most often, on purpose, excess quantities of one or more of the reactants are present.

Numerous reasons exist for having some reactants present in excess. Sometimes such a procedure will cause a reaction to occur more rapidly. For example, large amounts of oxygen make combustible materials burn faster. Sometimes an excess of one reactant will ensure that another reactant, perhaps a very expensive one, is completely consumed. (Reactions do not always go to completion in the way that theory predicts they should—that is, the reactants are not completely converted to products.)

When one or more reactants are present in excess, the excess will not react because there is not enough of the other reactant to react with it. The reactant *not* in excess thus limits the amount of product(s) formed and is called *the limiting reactant*. The **limiting reactant** is the reactant in a chemical reaction that determines how much product(s) can be formed.

The concept of a limiting reactant plays a major role in chemical calculations of certain types. It must be thoroughly understood. Let us consider some simple but analogous nonchemical examples of a "limiting reactant" before we go on to limiting reactant calculations.

Suppose we have a vending machine that contains forty 50¢ candy bars and we have 30 quarters. In this case we can purchase only 15 candy bars. The quarters are the limiting reactant. The candy bars are present in excess. Suppose we have ten slices of cheese and eighteen slices of bread and we want to make as many cheese sandwiches as possible using one slice of cheese and two slices of bread per sandwich. The eighteen slices of bread limit us to nine sandwiches; one slice of cheese is left over. The bread is the limiting reactant in this case even though initially there was more bread (18 slices) than cheese (10 slices) present. The bread is still limiting because it is used up first.

An additional limiting reactant analogy that comes closer to the realm of molecules and atoms involves nuts and bolts. Assume we have 10 identical nuts and 10 identical bolts. From this collection we can make 10 "one nut–one bolt" entities by screwing a nut on each bolt. This situation is depicted in Figure 10.4a.

Next, let us make "two nut–one bolt" entities from our same collection of nuts and bolts. This time we can make only 5 combinations and we will have 5 bolts left over, as is shown in Figure 10.4b. We run out of nuts before all of the bolts are used up. In chemical jargon, we say that the nuts are the limiting reactant.

Finally, let us consider making "one nut–two bolt" combinations. As is shown in Figure 10.4c, this time we do not have enough bolts; we can make 5 combinations and we will have 5 nuts left over. The bolts are the limiting reactant.

Now let us consider, as is presented in Example 10.10, an extension of our nut–bolt

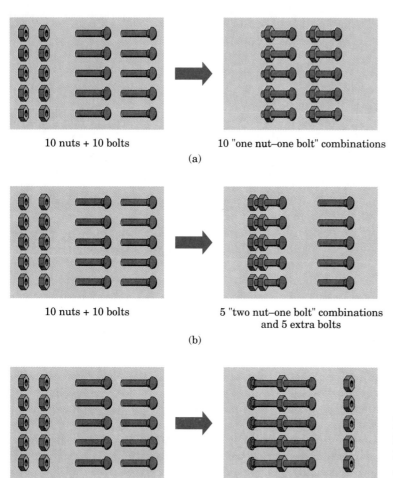

FIGURE 10.4 Starting with 10 nuts and 10 bolts, we can make (a) 10 "one nut–one bolt" combination, (b) 5 "two nut–one bolt combinations with 5 bolts left over; the nuts are the limiting reactant, and (c) 5 one nut–two bolt combinations with 5 nuts left over; the bolts are the limiting reactant.

discussion to a situation that cannot easily be reasoned out in one's head. Instead, a calculation must be performed.

EXAMPLE 10.10

What will be the limiting reactant in the production of "three nut–two bolt" combinations from a collection of 261 nuts and 176 bolts?

Solution

There are three possible answers to this problem.

1. We will run out of bolts first; bolts are the limiting reactant.
2. We will run out of nuts first; nuts are the limiting reactant.
3. The ratio of nuts to bolts is such that we run out of both at the same time; there is no limiting reactant.

The approach we will use in deciding which of the possible answers is correct is to test the possible answers for validity. (We will see shortly that this is the approach we actually use in "real" chemical situations.)

Let us assume that we will run out of bolts first. We have 176 bolts which will require 264 nuts.

$$176 \text{ bolts} \times \frac{3 \text{ nuts}}{2 \text{ bolts}} = 264 \text{ nuts} \quad \text{(calculator and correct answer)}$$

We need 264 nuts and we have only 261. Thus our supposition that we would run out of bolts first is not correct; we will run out of nuts first.

Let us now assume we will run out of nuts first (which we now know to be the case) and look at the calculation for this case.

$$261 \text{ nuts} \times \frac{2 \text{ bolts}}{3 \text{ nuts}} = 174 \text{ bolts} \quad \text{(calculator and correct answer)}$$

Do we have 174 bolts; yes, we have 176 bolts. Thus we can make sufficient combinations to use up all the nuts and we will have 2 bolts left over. Nuts are the limiting reactant.

PRACTICE EXERCISE 10.10

What will be the limiting reactant in the production of "two nut–three bolt" combinations from a collection of 363 nuts and 236 bolts?

Ans. Bolts are the limiting reactant.

Similar exercises: Problems 10.65 and 10.66

Now let us proceed to chemical calculations that involve a limiting reactant. *Whenever you are given the quantities of two or more reactants in a chemical reaction, it is necessary for you to determine which of the given quantities is the limiting reactant.*

Determining the limiting reactant can be accomplished by the following procedure.

1. Determine the number of *moles* of each of the reactants present.
2. Calculate the number of moles of product *each* of the molar amounts of reactant would produce if it were the only reactant amount given. If more than one product is formed in the reaction, you need to do this mole calculation for only one of the products.
3. The reactant that produces the *least number* of moles of product is the limiting reactant.

EXAMPLE 10.11

Silver and silver-plated objects tarnish in the presence of hydrogen sulfide (H_2S), a gas that originates from the decay of food, because of the reaction

$$4 \text{ Ag} + 2 \text{ H}_2\text{S} + \text{O}_2 \longrightarrow 2 \text{ Ag}_2\text{S} + 2 \text{ H}_2\text{O}$$

The black product Ag_2S is the tarnish. If 25.00 g of Ag, 5.00 g of H_2S, and 4.00 g of O_2 are present in a reaction mixture, which is the limiting reactant for tarnish formation?

Solution

To determine the limiting reactant, we determine how many moles of product each of the reactants can form. In this particular problem there are two products: Ag_2S and H_2O. It is sufficient to calculate how many moles of either Ag_2S or H_2O are formed. The decision as to which one is arbitrary; we choose H_2O.

The calculation type will be "grams of A" to "moles of B." We start with a given number of grams of reactant and desire to calculate moles of product. The pathway for the calculation, in terms of Figure 10.3, is

$$\boxed{\text{Grams of A}} \xrightarrow{\text{molar mass}} \boxed{\text{Moles of A}} \xrightarrow{\text{equation coefficients}} \boxed{\text{Moles of B}}$$

Note that we will have to go through this type of calculation three times because we have three reactants: once for Ag, once for H_2S, and once for O_2.

For Ag:

$$25.00 \text{ g Ag} \times \frac{1 \text{ mole Ag}}{107.9 \text{ g Ag}} \times \frac{2 \text{ moles } H_2O}{4 \text{ moles Ag}} = 0.11584801 \text{ mole } H_2O$$
$$\text{(calculator answer)}$$
$$= 0.1158 \text{ mole } H_2O$$
$$\text{(correct answer)}$$

For H_2S:

$$5.00 \text{ g } H_2S \times \frac{1 \text{ mole } H_2S}{34.1 \text{ g } H_2S} \times \frac{2 \text{ moles } H_2O}{2 \text{ moles } H_2S} = 0.14662757 \text{ mole } H_2O$$
$$\text{(calculator answer)}$$
$$= 0.147 \text{ mole } H_2O$$
$$\text{(correct answer)}$$

For O_2:

$$4.00 \text{ g } O_2 \times \frac{1 \text{ mole } O_2}{32.0 \text{ g } O_2} \times \frac{2 \text{ moles } H_2O}{1 \text{ mole } O_2} = 0.25 \text{ mole } H_2O$$
$$\text{(calculator answer)}$$
$$= 0.250 \text{ mole } H_2O$$
$$\text{(correct answer)}$$

The limiting reactant is the reactant that will produce the least number of moles of H_2O. Looking at the numbers just calculated we see that **Ag will be the limiting reactant**. Once the limiting reactant has been determined, the amount of that reactant present becomes the starting point for any further calculations about the chemical reaction under consideration. Example 10.12 illustrates this point.

PRACTICE EXERCISE 10.11

Old oil paintings are darkened by PbS, which forms by the reaction of the Pb in paint (a coloring agent) with H_2S in air.

$$Pb + H_2S \longrightarrow PbS + H_2$$

If 5.00 g of Pb and 0.500 g of H_2S are present in a reaction mixture, which is the limiting reactant for the darkening reaction?   Ans. H_2S.

Similar exercises: Problems 10.69 and 10.70

EXAMPLE 10.12

Ammonia (NH_3) reacts with oxygen as shown in the following equation.

$$4\, NH_3 + 3\, O_2 \longrightarrow 2\, N_2 + 6\, H_2O$$

How many grams of H_2O can be formed from a reaction mixture containing 35.0 g of NH_3 and 50.0 g of O_2?

Solution

First, we must determine the limiting reactant since specific amounts of both reactants are given in the problem. This determination involves two "grams of A" to "moles of B" calculations—one for the reactant NH_3 and one for the reactant O_2. The product we "key in" on is H_2O because our final goal is the mass of water produced.

$$\boxed{\text{Grams of A}} \xrightarrow{\text{molar mass}} \boxed{\text{Moles of A}} \xrightarrow{\text{equation coefficients}} \boxed{\text{Moles of B}}$$

For NH_3:

$$35.0 \text{ g } NH_3 \times \frac{1 \text{ mole } NH_3}{17.0 \text{ g } NH_3} \times \frac{6 \text{ moles } H_2O}{4 \text{ moles } NH_3} = 3.0882353 \text{ moles } H_2O$$
(calculator answer)
$$= 3.09 \text{ moles } H_2O$$
(correct answer)

For O_2:

$$50.0 \text{ g } O_2 \times \frac{1 \text{ mole } O_2}{32.0 \text{ g } O_2} \times \frac{6 \text{ moles } H_2O}{3 \text{ moles } O_2} = 3.125 \text{ moles } H_2O \quad \text{(calculator answer)}$$
$$= 3.12 \text{ moles } H_2O \quad \text{(correct answer)}$$

Thus, NH_3 is the limiting reactant since fewer moles of H_2O can be produced from it (3.09 moles) than from the O_2 (3.12 moles).

We can now calculate the grams of H_2O formed in the reaction using the 3.09 moles of H_2O (formed from our limiting reactant) as our starting factor. This calculation will be a simple one-step "moles of A" to "grams of A" conversion.

$$\boxed{\text{Moles of } H_2O} \xrightarrow{\text{molar mass}} \boxed{\text{Grams of } H_2O}$$

$$3.09 \text{ moles } H_2O \times \frac{18.0 \text{ g } H_2O}{1 \text{ mole } H_2O} = 55.62 \text{ g } H_2O \quad \text{(calculator answer)}$$
$$= 55.6 \text{ g } H_2O \quad \text{(correct answer)}$$

PRACTICE EXERCISE 10.12

Chloroform ($CHCl_3$), an important solvent, can be made by the reaction of chlorine and methane (CH_4) according to the equation

$$3\ Cl_2 + CH_4 \longrightarrow CHCl_3 + 3\ HCl$$

How many grams of $CHCl_3$ can be made from a reaction mixture containing 150.0 g of Cl_2 and 50.0 g of CH_4?   *Ans.* 84.27 g $CHCl_3$.

Similar exercises: Problems 10.71–10.74

10.10 Yields: Theoretical, Actual, and Percent

When stoichiometric relationships (Secs. 10.8 and 10.9) are used to calculate the amount of a product that will be produced in a chemical reaction from given amounts of reactants, the answer obtained represents a theoretical yield. A **theoretical yield** is the maximum amount of a product that can be obtained from given amounts of reactants in a chemical reaction that proceeds completely in the manner described by its chemical equation. Examples 10.7 and 10.12 are theoretical yield calculations, although this fact was not noted at the time they were presented.

In most chemical reactions the amount of a given product isolated from the reaction mixture is less than that theoretically possible. Why is this so? Two major factors contribute to this situation.

1. Some product is almost always lost in the process of its isolation and purification and in such mechanical operations as transferring materials from one container to another.
2. Often a particular set of reactants undergoes two or more reactions simultaneously, forming undesired products (in small amounts) as well as the desired products. Reactants consumed in these side reactions obviously will not end up in the form of the desired products.

The net effect of these factors is that the actual quantities of product isolated—that is, the actual yield—are less, sometimes far less, than the theoretically possible amount. An **actual yield** is the amount of a product actually obtained from a chemical reaction. Actual yield is always an experimentally determined number; it cannot be calculated.

Product loss is specified in terms of percent yield. **Percent yield** is the ratio of the actual yield to the theoretical yield multiplied by 100 (to give percent). The mathematical equation for percent yield is

$$\text{Percent yield} = \frac{\text{actual yield}}{\text{theoretical yield}} \times 100$$

If the theoretical yield of a product for a reaction is calculated to be 17.9 g and the amount of product actually obtained (the actual yield) is 15.8 g, the percent yield is 88.3%.

$$\text{Percent yield} = \frac{15.8\ \text{g}}{17.9\ \text{g}} \times 100 = 88.3\%$$

EXAMPLE 10.13

A mixture of 80.0 g of chromium(III) oxide (Cr_2O_3) and 8.00 g of carbon (C) is used to produce elemental chromium (Cr) by the reaction

$$Cr_2O_3 + 3\ C \longrightarrow 2\ Cr + 3\ CO$$

(a) What is the theoretical yield of Cr that can be obtained from the reaction mixture?
(b) The actual yield is 21.7 g Cr. What is the percent yield for the reaction?

Solution

(a) The limiting reactant must be determined before the theoretical yield can be calculated. Recalling the procedures of Examples 10.11 and 10.12 for determining the limiting reactant, we calculate the number of moles of Cr that can be produced from each individual reactant amount using the "grams of A" to "moles of B" type of calculations.

$$\boxed{\text{Grams of A}} \xrightarrow{\text{molar mass}} \boxed{\text{Moles of A}} \xrightarrow{\text{equation coefficients}} \boxed{\text{Moles of B}}$$

For Cr_2O_3:

$$80.0\ g\ Cr_2O_3 \times \frac{1\ \text{mole}\ Cr_2O_3}{152.0\ g\ Cr_2O_3} \times \frac{2\ \text{moles Cr}}{1\ \text{mole}\ Cr_2O_3} = 1.0526316\ \text{moles Cr}\quad \text{(calculator answer)}$$

$$= 1.05\ \text{moles Cr}\quad \text{(correct answer)}$$

For C:

$$8.00\ g\ C \times \frac{1\ \text{mole C}}{12.0\ g\ C} \times \frac{2\ \text{moles Cr}}{3\ \text{moles C}} = 0.44444444\ \text{mole Cr}\quad \text{(calculator answer)}$$

$$= 0.444\ \text{mole Cr}\quad \text{(correct answer)}$$

Our calculations show that C is the limiting reactant.
    The maximum number of grams of Cr obtainable from the limiting reactant, that is, the theoretical yield, can now be calculated. It is done using a one-step "moles of A" to "grams of A" setup.

$$\boxed{\text{Moles of Cr}} \xrightarrow{\text{molar mass}} \boxed{\text{Grams of Cr}}$$

$$0.444\ \text{mole Cr} \times \frac{52.0\ g\ Cr}{1\ \text{mole Cr}} = 23.088\ g\ Cr\quad \text{(calculator answer)}$$

$$= 23.1\ g\ Cr\quad \text{(correct answer)}$$

(b) The percent yield is obtained by dividing the actual yield by the theoretical yield and multiplying by 100.

$$\text{Percent yield} = \frac{\text{actual yield}}{\text{theoretical yield}} \times 100 = \frac{21.7 \text{ g Cr}}{23.1 \text{ g Cr}} \times 100 = 93.939393\%$$
(calculator answer)
$$= 93.9\%$$
(correct answer)

PRACTICE EXERCISE 10.13

Solutions of sodium hypochlorite (NaClO) are sold as laundry bleach. This bleaching agent can be produced by the reaction

$$2 \text{ NaOH} + \text{Cl}_2 \longrightarrow \text{NaCl} + \text{NaClO} + \text{H}_2\text{O}$$

(a) What is the theoretical yield of NaClO that can be obtained from a reaction mixture containing 75.0 g of NaOH and 50.0 g of Cl_2?
(b) The actual yield is 43.2 g of NaClO. What is the percent yield of NaClO for this reaction? *Ans.* (a) 52.4 g NaClO; (b) 82.4%.

Similar exercises: Problems 10.85 and 10.86

10.11 Simultaneous and Consecutive Reactions

The concepts presented so far in this chapter can easily be adapted to problem-solving situations that involve two or more chemical reactions. In some cases the two or more chemical reactions occur simultaneously, and in other cases they occur consecutively (one right after the other). Example 10.14 deals with a pair of simultaneous reactions, and Example 10.15 deals with three consecutive reactions.

EXAMPLE 10.14

A mixture of gaseous fuel has the composition 83.1% methane (CH_4) and 16.9% ethane (C_2H_6). The combustion of this mixture produces CO_2 and H_2O as the only products. The combustion equations are

$$CH_4 + 2 O_2 \longrightarrow CO_2 + 2 H_2O$$
$$2 C_2H_6 + 7 O_2 \longrightarrow 4 CO_2 + 6 H_2O$$

How many moles of CO_2 would be produced from the combustion of 72.0 g of this gaseous fuel mixture?

Solution

In solving this problem we will need to carry out two parallel calculations. In the one calculation we will determine the moles of CO_2 produced from the CH_4 component of the mixture and in the other the moles of CO_2 produced from the C_2H_6 component. Then we will add together the answers from the two parallel calculations to get our final answer, the total moles of CO_2 produced. The sequence of conversion factors for each setup is derived from the following pathway.

Grams of mixture $\xrightarrow{\text{percent composition}}$ Grams of fuel $\xrightarrow{\text{molar mass}}$ Moles of fuel $\xrightarrow{\text{equation coefficients}}$ Moles of CO_2

The moles of CO_2 produced from the CH_4 is given by the following setup.

$$72.0 \text{ g mixture} \times \frac{83.1 \text{ g } CH_4}{100 \text{ g mixture}} \times \frac{1 \text{ mole } CH_4}{16.0 \text{ g } CH_4} \times \frac{1 \text{ mole } CO_2}{1 \text{ mole } CH_4}$$

$$= 3.7395 \text{ moles } CO_2 \quad \text{(calculator answer)}$$
$$= 3.74 \text{ moles } CO_2 \quad \text{(correct answer)}$$

The moles of CO_2 produced from the C_2H_6 is given by the following parallel setup.

$$72.0 \text{ g mixture} \times \frac{16.9 \text{ g } C_2H_6}{100 \text{ g mixture}} \times \frac{1 \text{ mole } C_2H_6}{30.0 \text{ g } C_2H_6} \times \frac{4 \text{ moles } CO_2}{2 \text{ moles } C_2H_6}$$

$$= 0.8112 \text{ mole } CO_2 \quad \text{(calculator answer)}$$
$$= 0.811 \text{ mole } CO_2 \quad \text{(correct answer)}$$

The first conversion factor in each setup is derived from the given percentage of that compound in the fuel mixture. The use of percentages as conversion factors was covered in Section 3.9.

The total moles of CO_2 produced is the sum of the moles of CO_2 in the individual reactions.

$$3.74 \text{ moles } CO_2 + 0.811 \text{ mole } CO_2 = 4.551 \text{ moles } CO_2 \quad \text{(calculator answer)}$$
$$= 4.55 \text{ moles } CO_2 \quad \text{(correct answer)}$$

PRACTICE EXERCISE 10.14

A mixture of composition 47.3% magnesium carbonate ($MgCO_3$) and 52.7% calcium carbonate ($CaCO_3$) is heated until the carbonates are decomposed to oxides as shown by the following equations.

$$MgCO_3 \longrightarrow MgO + CO_2$$
$$CaCO_3 \longrightarrow CaO + CO_2$$

How many grams of CO_2 are produced from decomposition of 78.3 g of mixture?

Ans. 37.5 g CO_2.

Similar exercises: Problems 10.93 and 10.94

EXAMPLE 10.15

An older method for preparation of nitric acid (HNO_3) involves the following sequence of three reactions.

Reaction (1): $4 NH_3 + 5 O_2 \longrightarrow 4 NO + 6 H_2O$
Reaction (2): $2 NO + O_2 \longrightarrow 2 NO_2$
Reaction (3): $3 NO_2 + H_2O \longrightarrow 2 HNO_3 + NO$

Assuming an excess of O_2 and H_2O as reactants, how many grams of HNO_3 can be produced from 244 g of NH_3?

Solution

The key substances in this set of reactions, from a calculational point of view, are the nitrogen-containing species: NH_3, NO, NO_2, and HNO_3. Note that the nitrogen-containing species produced in the first and second reactions (NO and NO_2, respectively) are the reactants for the second and third reactions, respectively.

$$NH_3 \xrightarrow{\text{reaction (1)}} NO \xrightarrow{\text{reaction (2)}} NO_2 \xrightarrow{\text{reaction (3)}} HNO_3$$

We can solve this problem using a single multiple-step setup. The sequence of conversion factors needed is that for a gram-to-gram problem with two additional intermediate mole-to-mole steps added.

$$\boxed{\text{Grams of } NH_3} \xrightarrow{\text{molar mass}} \boxed{\text{Moles of } NH_3} \xrightarrow{\text{eq. (1) coeff.}} \boxed{\text{Moles of } NO} \xrightarrow{\text{eq. (2) coeff.}} \boxed{\text{Moles of } NO_2} \xrightarrow{\text{eq. (3) coeff.}} \boxed{\text{Moles of } HNO_3} \xrightarrow{\text{molar mass}} \boxed{\text{Grams of } HNO_3}$$

The dimensional analysis setup is

$$244 \text{ g } NH_3 \times \frac{1 \text{ mole } NH_3}{17.0 \text{ g } NH_3} \times \frac{4 \text{ moles } NO}{4 \text{ moles } NH_3} \times \frac{2 \text{ moles } NO_2}{2 \text{ moles } NO} \times \frac{2 \text{ moles } HNO_3}{3 \text{ moles } NO_2} \times \frac{63.0 \text{ g } HNO_3}{1 \text{ mole } HNO_3}$$

grams $NH_3 \longrightarrow$ moles $NH_3 \longrightarrow$ moles $NO \longrightarrow$ moles $NO_2 \longrightarrow$ moles $HNO_3 \longrightarrow$ grams HNO_3

$$= 602.82353 \text{ g } HNO_3 \quad \text{(calculator answer)}$$
$$= 603 \text{ g } HNO_3 \quad \text{(correct answer)}$$

An alternative approach to solving this problem would involve setting up a separate calculation for each equation. As a first step, the number of moles of NO produced in the first reaction would be calculated. In the second step, one would determine the moles of NO_2 obtained if all of the NO produced in the first reaction entered into the second reaction. In the final step, one would determine the grams of HNO_3 derivable from the NO_2 produced in the second reaction. The answer obtained from this three-step method is the same as that obtained from the multiple-step method.

PRACTICE EXERCISE 10.15

In steelmaking a three-step process leads to the conversion of Fe_2O_3 (the iron-containing component of iron ore) to molten iron.

$$3\ Fe_2O_3 + CO \longrightarrow 2\ Fe_3O_4 + CO_2$$
$$Fe_3O_4 + CO \longrightarrow 3\ FeO + CO_2$$
$$FeO + CO \longrightarrow Fe + CO_2$$

Assuming that the reactant CO is present in excess, how many grams of Fe can be produced from 125 g of Fe_2O_3?

Ans. 87.2 g Fe.

Similar exercises: Problems 10.97–10.98

Key Terms

The new terms or concepts defined in this chapter are

actual yield (Sec. 10.10) The amount of product actually obtained from a chemical reaction.

balanced chemical equation (Sec. 10.3) A chemical equation in which the same number of atoms of each element involved in the reaction appears on each side of the equation.

chemical equation (Sec. 10.2) A written statement that uses symbols and formulas instead of words to describe the changes that occur in a chemical reaction.

chemical stoichiometry (Sec. 10.8) The study of the quantitative relationships among reactants and products in a chemical reaction.

coefficient (Sec. 10.3) A number placed to the left of the formula of a substance that denotes the amount of the substance.

combustion reaction (Sec. 10.5) The reaction of a substance with oxygen (usually from air) that proceeds with evolution of heat and usually also a flame.

decomposition reaction (Sec. 10.6) A reaction in which a single reactant is converted into two or more simpler substances.

double-replacement reaction (Sec. 10.6) A reaction in which two compounds exchange parts with each other and form two different compounds.

law of conservation of mass (Sec. 10.1) Mass is neither created nor destroyed in any ordinary chemical reaction.

limiting reactant (Sec. 10.9) The reactant in a chemical reaction that determines how much product(s) can be formed.

molar (macroscopic level) interpretation of coefficients (Sec. 10.7) The coefficients in a balanced chemical equation are interpreted to mean moles of reactants and moles of products.

percent yield (Sec. 10.10) The ratio of the actual yield to the theoretical yield (of a product from a chemical reaction) multiplied by 100.

products (Sec. 10.1) Substances produced as a result of a chemical reaction.

reactants (Sec. 10.1) Substances present prior to the start of a chemical reaction.

single-replacement reaction (Sec. 10.6) A reaction in which one element within a compound is replaced by another element.

synthesis reaction (Sec. 10.6) A reaction in which a single product is produced from two or more reactants.

theoretical yield (Sec. 10.10) The maximum amount of product that can be obtained from given amounts of reactants in a chemical reaction if no losses or inefficiencies of any kind occur.

Practice Problems

The Law of Conservation of Mass (Sec. 10.1)

10.1 Consider a hypothetical reaction in which A and B are reactants and C and D are products. If 10.0 g of A completely reacts with 7.0 g of B to produce 14.0 g of C, how many grams of D will be produced?

10.2 Consider a hypothetical reaction in which A and B are reactants and C and D are products. If 5.0 g of A completely reacts with 15.0 g of B to produce 12.0 g of C, how many grams of D will be produced?

10.3 A 4.2-g sample of sodium hydrogen carbonate is added to a solution of acetic acid weighing 10.0 g. The two substances react, releasing carbon dioxide gas to the atmosphere. After the reaction, the contents of the reaction vessel weigh 12.0 g. What is the mass of carbon dioxide gas given off during the reaction?

10.4 A 1.00-g sample of solid calcium carbonate is added to a reaction flask containing 10.00 g of hydrochloric acid solution. The calcium carbonate slowly dissolves in the acid solution as evidenced by the generation of carbon dioxide gas. After 5 min of reaction, 0.21 g of carbon dioxide gas has been produced. At that time, what is the mass, in grams, of the reaction flask contents?

Chemical Equations (Secs. 10.2 and 10.4)

10.5 The formulas used in equations for gaseous elements should reflect their molecular makeup. Which gases (or vapors) should be written as diatomic molecules?

10.6 The formulas used in equations for gaseous elements should reflect their molecular makeup. Which gases (or vapors) should be written as tetraatomic molecules?

10.7 What do the symbols in parentheses stand for in the following equations?
(a) $CaCO_3(s) \rightarrow CaO(s) + CO_2(g)$
(b) $SO_2(g) + H_2O(l) \rightarrow H_2SO_3(aq)$

10.8 What do the symbols in parentheses stand for in the following equations?
(a) $PCl_3(l) + Cl_2(g) \rightarrow PCl_5(s)$
(b) $NaCl(aq) + AgNO_3(aq) \rightarrow AgCl(s) + NaNO_3(aq)$

10.9 What symbol is used in a chemical equation to represent the phrase "to produce"?

10.10 What symbol is used in a chemical equation to represent the phrase "reacts with"?

Balancing Chemical Equations (Sec. 10.3)

10.11 What fundamental law is the basis for balancing a chemical equation?

10.12 What is meant by a *balanced* chemical equation?

10.13 Classify each of the following equations as *balanced* or *unbalanced*.
(a) $CaCO_3 \rightarrow CaO + CO_2$
(b) $KClO_3 \rightarrow KCl + O_2$
(c) $2 Na + 2 H_2O \rightarrow 2 NaOH + H_2$
(d) $PbO_2 + 2 H_2 \rightarrow 2 Pb + H_2O$

10.14 Classify each of the following equations as *balanced* or *unbalanced*.
(a) $TiCl_4 + 2 Ti \rightarrow 2 TiCl_3$
(b) $SO_2 + H_2O \rightarrow H_2SO_3$
(c) $4 NH_3 + 3 O_2 \rightarrow 2 N_2 + 6 H_2O$
(d) $CS_2 + 2 O_2 \rightarrow CO_2 + 2 SO_2$

10.15 Balance the following equations.
(a) $N_2 + O_2 \rightarrow NO$
(b) $SO_2 + O_2 \rightarrow SO_3$
(c) $U + F_2 \rightarrow UF_6$
(d) $Al + O_2 \rightarrow Al_2O_3$

10.16 Balance the following equations.
(a) $H_2O \rightarrow H_2 + O_2$
(b) $Li + N_2 \rightarrow Li_3N$
(c) $Al + N_2 \rightarrow AlN$
(d) $H_2O + O_2 \rightarrow H_2O_2$

10.17 Balance the following equations.
(a) $Mg + Fe_3O_4 \rightarrow MgO + Fe$
(b) $Na_2NH + H_2O \rightarrow NH_3 + NaOH$
(c) $NH_3 + O_2 \rightarrow NO + H_2O$
(d) $Cl_2 + H_2O \rightarrow HCl + HClO$

10.18 Balance the following equations.
(a) $Na_2O_2 + H_2O \rightarrow O_2 + NaOH$
(b) $Sb_2S_3 + O_2 \rightarrow Sb + SO_2$
(c) $PCl_3 + H_2O \rightarrow H_3PO_3 + HCl$
(d) $Zn + HCl \rightarrow ZnCl_2 + H_2$

10.19 Balance the following equations.
(a) $PbO + NH_3 \rightarrow Pb + N_2 + H_2O$
(b) $NaHCO_3 + H_2SO_4 \rightarrow Na_2SO_4 + H_2O + CO_2$
(c) $TiO_2 + C + Cl_2 \rightarrow TiCl_4 + CO_2$
(d) $NBr_3 + NaOH \rightarrow N_2 + NaBr + HOBr$

10.20 Balance the following equations.
(a) $NH_3 + O_2 + CH_4 \rightarrow HCN + H_2O$
(b) $KClO_3 + HCl \rightarrow KCl + ClO_2 + Cl_2 + H_2O$
(c) $SO_2Cl_2 + HI \rightarrow H_2S + H_2O + HCl + I_2$
(d) $NO + CH_4 \rightarrow HCN + H_2O + H_2$

10.21 Balance the following equations.
(a) $Ca(OH)_2 + HNO_3 \rightarrow Ca(NO_3)_2 + H_2O$
(b) $BaCl_2 + (NH_4)_2SO_4 \rightarrow BaSO_4 + NH_4Cl$
(c) $Fe(OH)_3 + H_2SO_4 \rightarrow Fe_2(SO_4)_3 + H_2O$
(d) $Na_3PO_4 + AgNO_3 \rightarrow NaNO_3 + Ag_3PO_4$

10.22 Balance the following equations.
(a) $Al + Sn(NO_3)_2 \rightarrow Al(NO_3)_3 + Sn$
(b) $Na_2CO_3 + Mg(NO_3)_2 \rightarrow MgCO_3 + NaNO_3$
(c) $Al(NO_3)_3 + H_2SO_4 \rightarrow Al_2(SO_4)_3 + HNO_3$
(d) $Ba(C_2H_3O_2)_2 + (NH_4)_3PO_4 \rightarrow Ba_3(PO_4)_2 + NH_4C_2H_3O_2$

10.23 Write balanced chemical equations to represent the reactions described by the following statements.
(a) Steam (gaseous water) reacts with carbon at high temperatures to produce carbon monoxide (CO) and hydrogen gases.
(b) Ethyl alcohol (C_2H_6O) burns in oxygen to produce carbon dioxide (CO_2) and water.

(c) Aluminum and sulfur react at elevated temperatures to produce aluminum sulfide (Al_2S_3).
(d) Sulfuric acid (H_2SO_4) and calcium hydroxide [$Ca(OH)_2$] react to produce water and calcium sulfate ($CaSO_4$).

10.24 Write balanced chemical equations to represent the reactions described by the following statements.
(a) Limestone ($CaCO_3$) decomposes when heated to produce lime (CaO) and gaseous carbon dioxide (CO_2).
(b) At high temperatures the gases chlorine and water react to produce hydrogen chloride (HCl) and oxygen gases.
(c) During photosynthesis in plants carbon dioxide and water are converted into glucose ($C_6H_{12}O_6$) and oxygen.
(d) Silver nitrate ($AgNO_3$) and calcium chloride ($CaCl_2$) react to produce silver chloride (AgCl) and calcium nitrate [$Ca(NO_3)_2$].

Patterns in Chemical Reactivity (Sec. 10.5)

10.25 Write a balanced chemical equation for the combustion of each of the following hydrocarbons in air.
(a) C_2H_4 (b) C_3H_8 (c) C_3H_6 (d) C_4H_6

10.26 Write a balanced chemical equation for the combustion of each of the following hydrocarbons in air.
(a) C_3H_4 (b) C_5H_{12} (c) C_5H_{10} (d) C_6H_{10}

10.27 Write a balanced chemical equation when each of the following substances is combusted in an atmosphere of pure O_2.
(a) $C_4H_8O_2$ (b) C_3H_6O (c) $C_5H_{10}O_2$ (d) $C_6H_{13}O$

10.28 Write a balanced chemical equation when each of the following substances is combusted in an atmosphere of pure oxygen.
(a) $C_2H_4O_2$ (b) $C_5H_{10}O$ (c) $C_3H_6O_2$ (d) C_4H_8O

10.29 Write complete balanced equations for the following reactions.
(a) $PbCO_3$ is heated to a high enough temperature for decomposition to occur.
(b) $Co_2(CO_3)_3$ is heated to a high enough temperature for decomposition to occur.

10.30 Write complete balanced equations for the following reactions.
(a) $MgCO_3$ is heated to a high enough temperature for decomposition to occur.
(b) $Fe_2(CO_3)_3$ is heated to a high enough temperature for decomposition to occur.

10.31 Write a complete balanced equation for the combustion in air of each of the following compounds.
(a) C_2H_7N, where NO_2 is one of the products
(b) CH_4S, where SO_2 is one of the products

10.32 Write a complete balanced equation for the combustion in air of each of the following compounds.
(a) C_2H_6S, where SO_2 is one of the products
(b) CH_5N, where NO_2 is one of the products

Classes of Chemical Reactions (Sec. 10.6)

10.33 Classify each of the following reactions as synthesis, decomposition, single replacement, or double replacement.
(a) $SO_3 + H_2O \rightarrow H_2SO_4$
(b) $2\,H_2 + O_2 \rightarrow 2\,H_2O$
(c) $Na_2CO_3 + Ca(OH)_2 \rightarrow CaCO_3 + 2\,NaOH$
(d) $Cu(NO_3)_2 + Fe \rightarrow Cu + Fe(NO_3)_2$

10.34 Classify each of the following reactions as synthesis, decomposition, single replacement, or double replacement.
(a) $3\,CuSO_4 + 2\,Al \rightarrow Al_2(SO_4)_3 + 3\,Cu$
(b) $K_2CO_3 \rightarrow K_2O + CO_2$
(c) $2\,AgNO_3 + K_2SO_4 \rightarrow Ag_2SO_4 + 2\,KNO_3$
(d) $2\,SO_2 + O_2 \rightarrow 2\,SO_3$

10.35 Classify each of the following reactions as synthesis, decomposition, single replacement, or double replacement.
(a) $Al(OH)_3 + 3\,HCl \rightarrow AlCl_3 + 3\,H_2O$
(b) $2\,Fe + 3\,Cl_2 \rightarrow 2\,FeCl_3$
(c) $Pb(NO_3)_2 + H_2SO_4 \rightarrow PbSO_4 + 2\,HNO_3$
(d) $2\,NaHCO_3 \rightarrow Na_2CO_3 + CO_2 + H_2O$

10.36 Classify each of the following reactions as synthesis, decomposition, single replacement, or double replacement.
(a) $2\,H_2O_2 \rightarrow 2\,H_2O + O_2$
(b) $Mg + 2\,HCl \rightarrow MgCl_2 + H_2$
(c) $2\,CO + O_2 \rightarrow 2\,CO_2$
(d) $2\,Ag_2CO_3 \rightarrow 4\,Ag + 2\,CO_2 + O_2$

Chemical Equations and the Mole Concept (Sec. 10.7)

10.37 Give a word interpretation of the balanced equation

$$4\,NH_3 + 3\,O_2 \rightarrow 2\,N_2 + 6\,H_2O$$

in terms of (a) molecules and (b) moles.

10.38 Give a word interpretation of the balanced equation

$$CS_2 + 3\,O_2 \rightarrow CO_2 + 2\,SO_2$$

in terms of (a) molecules and (b) moles.

10.39 Write the twelve mole-to-mole conversion factors that can be derived from the balanced equation

$$3 \, HNO_2 \rightarrow 2 \, NO + HNO_3 + H_2O$$

10.40 Write the twelve mole-to-mole conversion factors that can be derived from the balanced equation

$$N_2H_4 + 2 \, H_2O_2 \rightarrow N_2 + 4 \, H_2O$$

10.41 How many moles of the first-listed product in each of the following equations could be obtained by reacting 0.750 mole of the first-listed reactant with an excess of the other reactant?
(a) $H_2O_2 + H_2S \rightarrow 2 \, H_2O + S$
(b) $4 \, NH_3 + 3 \, O_2 \rightarrow 2 \, N_2 + 6 \, H_2O$
(c) $Mg + 2 \, HCl \rightarrow MgCl_2 + H_2$
(d) $6 \, HCl + 2 \, Al \rightarrow 3 \, H_2 + 2 \, AlCl_3$

10.42 How many moles of the first-listed product in each of the following equations could be obtained by reacting 0.750 mole of the first-listed reactant with an excess of the other reactant?
(a) $SiO_2 + 3 \, C \rightarrow 2 \, CO + SiC$
(b) $5 \, O_2 + C_3H_8 \rightarrow 3 \, CO_2 + 4 \, H_2O$
(c) $CH_4 + 4 \, Cl_2 \rightarrow 4 \, HCl + CCl_4$
(d) $3 \, NO_2 + H_2O \rightarrow 2 \, HNO_3 + NO$

10.43 Using each of the following equations, calculate the number of moles of the first-listed reactant that are needed to produce 3.00 moles of N_2.
(a) $2 \, NaN_3 \rightarrow 2 \, Na + 3 \, N_2$
(b) $3 \, CO + 2 \, NaCN \rightarrow Na_2CO_3 + 4 \, C + N_2$
(c) $2 \, NH_2Cl + N_2H_4 \rightarrow 2 \, NH_4Cl + N_2$
(d) $4 \, C_3H_5O_9N_3 \rightarrow 12 \, CO_2 + 6 \, N_2 + O_2 + 10 \, H_2O$

10.44 Using each of the following equations, calculate the number of moles of the first-listed reactant that are needed to produce 4.00 moles of N_2.
(a) $4 \, NH_3 + 3 \, O_2 \rightarrow 2 \, N_2 + 6 \, H_2O$
(b) $(NH_4)_2Cr_2O_7 \rightarrow Cr_2O_3 + N_2 + 4 \, H_2O$
(c) $N_2H_4 + 2 \, H_2O_2 \rightarrow N_2 + 4 \, H_2O$
(d) $2 \, Li_3N \rightarrow 6 \, Li + N_2$

10.45 Given the equation

$$4 \, NH_3(g) + 5 \, O_2(g) \rightarrow 4 \, NO(g) + 6 \, H_2O(g)$$

(a) How many moles of O_2 are needed to produce 1.34 mole of NO?
(b) How many moles of H_2O will be produced from 0.789 mole of NH_3?
(c) How many moles of NH_3 are needed to react with 3.22 moles of O_2?
(d) How many moles of NO are produced when 0.763 mole of H_2O is produced?

10.46 Given the equation

$$C_3H_8(g) + 5 \, O_2(g) \rightarrow 3 \, CO_2(g) + 4 \, H_2O(g)$$

(a) How many moles of O_2 are needed to produce 2.38 moles of H_2O?
(b) How many moles of CO_2 will be produced from 0.57 mole of C_3H_8?
(c) How many moles of C_3H_8 are needed to react with 1.45 moles of O_2?
(d) How many moles of CO_2 are produced when 1.11 moles of H_2O are produced?

10.47 Using each of the following equations, calculate the total number of moles of products that can be obtained from the decomposition of 1.75 moles of the reactant.
(a) $2 \, NH_4NO_3 \rightarrow 2 \, N_2 + O_2 + 4 \, H_2O$
(b) $2 \, NaClO_3 \rightarrow 2 \, NaCl + 3 \, O_2$
(c) $2 \, KNO_3 \rightarrow 2 \, KNO_2 + O_2$
(d) $4 \, I_4O_9 \rightarrow 6 \, I_2O_5 + 2 \, I_2 + 3 \, O_2$

10.48 Using each of the following equations, calculate the total number of moles of products that can be obtained from the decomposition of 2.25 moles of the reactant.
(a) $2 \, Ag_2CO_3 \rightarrow 4 \, Ag + 2 \, CO_2 + O_2$
(b) $2 \, KClO_3 \rightarrow 2 \, KCl + 3 \, O_2$
(c) $4 \, HNO_3 \rightarrow 4 \, NO_2 + 2 \, H_2O + O_2$
(d) $2 \, H_2O_2 \rightarrow 2 \, H_2O + O_2$

Stoichiometry (Sec. 10.8)

10.49 How many grams of the first reactant in each of the following reactions would be needed to completely react with 2.025 g of the second reactant?
(a) $C_7H_{16} + 11 \, O_2 \rightarrow 7 \, CO_2 + 8 \, H_2O$
(b) $2 \, HCl + CaCO_3 \rightarrow CaCl_2 + CO_2 + H_2O$
(c) $Na_2SO_4 + 2 \, C \rightarrow Na_2S + 2 \, CO_2$
(d) $4 \, Na_2CO_3 + Fe_3Br_8 \rightarrow 8 \, NaBr + 4 \, CO_2 + Fe_3O_4$

10.50 How many grams of the first reactant in each of the following reactions would be needed to completely react with 2.725 g of the second reactant?
(a) $3 \, O_2 + CS_2 \rightarrow CO_2 + 2 \, SO_2$
(b) $FeO + CO \rightarrow Fe + CO_2$
(c) $2 \, C_8H_{18} + 25 \, O_2 \rightarrow 16 \, CO_2 + 18 \, H_2O$
(d) $Fe_3O_4 + CO \rightarrow 3 \, FeO + CO_2$

10.51 A mixture of hydrazine (N_2H_4) and hydrogen peroxide (H_2O_2) is used as a fuel for rocket engines. These two substances react as shown by the equation

$$N_2H_4(l) + 2 \, H_2O_2(l) \rightarrow N_2(g) + 4 \, H_2O(g)$$

(a) How many grams of N_2H_4 are needed to react with 0.453 mole of H_2O_2?
(b) How many grams of N_2 are obtained when 1.37 moles of N_2H_4 react?

(c) How many grams of H_2O are produced when 0.317 mole of N_2 is produced?
(d) How many grams of H_2O_2 must react in order to produce 19.6 moles of H_2O?

10.52 One way to remove gaseous carbon dioxide (CO_2) from the air in a spacecraft is to let cannisters of solid lithium hydroxide (LiOH) absorb it according to the reaction

$$CO_2(g) + 2\ LiOH(s) \rightarrow Li_2CO_3(s) + H_2O(l)$$

(a) How many grams of LiOH are needed to react with 1.30 moles of CO_2?
(b) How many grams of H_2O are obtained when 2.50 moles of CO_2 react?
(c) How many grams of Li_2CO_3 are produced when 0.500 mole of H_2O is produced?
(d) How many grams of LiOH must react in order to produce 12.0 moles of Li_2CO_3?

10.53 Determine what mass, in grams, of K_3AsO_4 can be prepared by the reaction of 4.511 g of KOH with an excess of H_3AsO_4 according to the equation

$$H_3AsO_4 + 3\ KOH \rightarrow K_3AsO_4 + 3\ H_2O$$

10.54 Silver is often extracted from ores as $KAg(CN)_2$ and then recovered by the reaction

$$2\ KAg(CN)_2 + Zn \rightarrow 2\ Ag + Zn(CN)_2 + 2\ KCN$$

What mass of zinc, in grams, is required to produce 28.35 g (1.00 oz) of silver metal?

10.55 When chromium metal is reacted with chlorine gas, a violet solid with the formula $CrCl_3$ is formed.

$$2\ Cr + 3\ Cl_2 \rightarrow 2\ CrCl_3$$

How many grams of Cr are needed to produce 100.0 g of $CrCl_3$?

10.56 Tungsten (W) metal, used to make incandescent bulb filaments, is produced by the reaction

$$WO_3 + 3\ H_2 \rightarrow W + 3\ H_2O$$

How many grams of H_2 are needed to produce 1.00 g of W?

10.57 A common method for producing lime (CaO) is the thermal decomposition of limestone ($CaCO_3$).

$$CaCO_3 \rightarrow CaO + CO_2$$

Based on this equation, how many grams of $CaCO_3$ must be decomposed to produce each of the following?
(a) 3.00 moles of CO_2
(b) 3.00×10^{20} molecules of CO_2
(c) 0.627 g of CaO
(d) 125 formula units of CaO

10.58 Pure silver metal results when silver carbonate (Ag_2CO_3) is decomposed by heating as shown by the equation

$$2\ Ag_2CO_3 \rightarrow 4\ Ag + 2\ CO_2 + O_2$$

Based on this equation, how many grams of Ag_2CO_3 must be decomposed to produce each of the following?
(a) 1.50 moles of O_2
(b) 1 billion (1.00×10^9) molecules of CO_2
(c) 100.0 g of O_2
(d) 115,000 atoms of Ag

10.59 Hydrofluoric acid, HF, cannot be stored in glass bottles because it attacks silicate compounds present in the glass. For example, sodium silicate, Na_2SiO_3, reacts with HF in the following way:

$$Na_2SiO_3 + 8\ HF \rightarrow H_2SiF_6 + 2\ NaF + 3\ H_2O$$

(a) How many moles of Na_2SiO_3 must react to produce 25.00 g of NaF?
(b) How many grams of HF must react to produce 27.00 g of H_2O?
(c) How many molecules of H_2SiF_6 are produced from the reaction of 2.000 g of Na_2SiO_3?
(d) How many grams of HF are needed to react with 50.00 g of Na_2SiO_3?

10.60 Potassium thiosulfate, $K_2S_2O_3$, is used to remove any excess chlorine from fibers and fabrics that have been bleached with that gas.

$$K_2S_2O_3 + 4\ Cl_2 + 5\ H_2O \rightarrow 2\ KHSO_4 + 8\ HCl$$

(a) How many moles of $K_2S_2O_3$ must react to produce 2.500 g of HCl?
(b) How many grams of Cl_2 must react to produce 20.00 g of $KHSO_4$?
(c) How many molecules of HCl are produced at the same time that 2.000 g of $KHSO_4$ is produced?
(d) How many grams of H_2O are consumed as 12.50 g of Cl_2 reacts?

10.61 How many grams of potassium (K) are needed to react completely with 27.3 g of bromine (Br_2) to produce KBr?

10.62 How many grams of aluminum (Al) are needed to react completely with 35.2 g of sulfur (S) to produce Al_2S_3?

10.63 A sample of magnesium carbonate ($MgCO_3$) was decomposed by heating to give magnesium oxide (MgO) and carbon dioxide (CO_2). If 7.89 g of CO_2 was produced, how many grams of MgO were produced?

10.64 A sample of ammonium nitrate (NH_4NO_3) was decomposed by careful heating to give dinitrogen monoxide (N_2O) and water (H_2O). If 11.33 g of H_2O was produced, how many grams of N_2O were produced?

Limiting Reactant (Sec. 10.9)

10.65 What will be the limiting reactant in the production of "four nut–three bolt" combinations from a collection of 348 nuts and 267 bolts?

10.66 What will be the limiting reactant in the production of "five nut–three bolt" combinations from a collection of 215 nuts and 123 bolts?

10.67 A model airplane kit is designed to contain two wings, one fuselage, four engines, and six wheels. How many model airplane kits can a manufacturer produce from a parts inventory of 426 wings, 224 fuselages, 860 engines, and 1578 wheels?

10.68 A model car kit is designed to contain one body, four wheels, two bumpers, and one steering wheel. How many model car kits can a manufacturer produce from a parts inventory of 137 bodies, 532 wheels, 246 bumpers, and 139 steering wheels?

10.69 Magnesium nitride can be prepared by the direct reaction of the elements as shown by the equation

$$3\ Mg + N_2 \rightarrow Mg_3N_2$$

For each of the following combinations of reactants, decide which is the limiting reactant.
(a) 2.00 moles of Mg and 0.500 mole of N_2
(b) 3.00 moles of Mg and 1.25 moles of N_2
(c) 3.00 g of Mg and 0.100 mole of N_2
(d) 20.00 g of Mg and 7.00 g of N_2

10.70 Under appropriate conditions water can be produced from the reaction of the elements hydrogen and oxygen as shown by the equation

$$2\ H_2 + O_2 \rightarrow 2\ H_2O$$

For each of the following combinations of reactants, decide which is the limiting reactant.
(a) 1.75 moles of H_2 and 1.00 mole of O_2
(b) 2.50 moles of H_2 and 2.00 moles of O_2
(c) 6.00 g of H_2 and 1.25 moles of O_2
(d) 1.00 g of H_2 and 7.00 g of O_2

10.71 At high temperatures and pressures nitrogen will react with hydrogen to produce ammonia as shown by the equation

$$N_2 + 3\ H_2 \rightarrow 2\ NH_3$$

How many grams of ammonia can be produced from the following amounts of reactants?
(a) 3.0 g of N_2 and 5.0 g of H_2
(b) 30.0 g of N_2 and 10.0 g of H_2
(c) 50.0 g of N_2 and 8.00 g of H_2
(d) 56 g of N_2 and 12 g of H_2

10.72 Aluminum oxide can be prepared by the direct reaction of the elements as shown by the equation

$$4\ Al + 3\ O_2 \rightarrow 2\ Al_2O_3$$

How many grams of aluminum oxide can be produced from the following amounts of reactants?
(a) 11.0 g of Al and 9.40 g of O_2
(b) 10.0 g of Al and 10.0 g of O_2
(c) 50.0 g of Al and 100.0 g of O_2
(d) 6.20 g of Al and 5.51 g of O_2

10.73 How many grams of Fe_2O_3 can be produced according to the equation

$$4\ Fe_3O_4 + O_2 \rightarrow 6\ Fe_2O_3$$

from a reaction mixture containing 4.00 g of Fe_3O_4 and 13.3 g of O_2?

10.74 How many grams of $TiCl_3$ can be produced according to the equation

$$3\ TiCl_4 + Ti \rightarrow 4\ TiCl_3$$

from a reaction mixture containing 40.0 g of $TiCl_4$ and 4.00 g of Ti?

10.75 How many grams of $NiCl_2$ can be produced from a reaction mixture containing 10.0 g of Ni and 0.750 mole of HCl according to the following equation

$$Ni + 2\ HCl \rightarrow NiCl_2 + H_2$$

10.76 How many grams of $CoCl_3$ can be produced from a reaction mixture containing 20.0 g of Co and 0.850 mole of HCl according to the following equation

$$2\ Co + 6\ HCl \rightarrow 2\ CoCl_3 + 3\ H_2$$

10.77 Determine the number of grams of each of the products that can be made from 8.00 g of SCl_2 and 4.00 g of NaF by the reaction

$$3\ SCl_2 + 4\ NaF \rightarrow SF_4 + S_2Cl_2 + 4\ NaCl$$

10.78 Determine the number of grams of each of the products that can be made from 100.0 g of Na_2CO_3 and 300.0 g of Fe_3Br_8 by the reaction

$$4\ Na_2CO_3 + Fe_3Br_8 \rightarrow 8\ NaBr + 4\ CO_2 + Fe_3O_4$$

10.79 If 70.0 g of H_2S and 125 g of O_2 are present in a reaction mixture, how many grams of each reactant will be left unreacted upon completion of this reaction?

$$2\ H_2S + 3\ O_2 \rightarrow 2\ SO_2 + 2\ H_2O$$

10.80 If 100.0 g of Al and 200.0 g of HCl are present in a reaction mixture, how many grams of each reactant will be left unreacted upon completion of this reaction?

$$2\,Al + 6\,HCl \rightarrow 2\,AlCl_3 + 3\,H_2$$

10.81 Criticize the statement "In every reaction involving more than one reactant, one of the reactants must be a limiting reactant."

10.82 Criticize the statement "The limiting reactant is the reactant present in the smallest mass amount."

Theoretical and Percent Yield (Sec. 10.10)

10.83 Because of "sloppiness" in his procedures, a student was able to isolate only 16.0 g of a desired product from a chemical reaction rather than the 52.0 g that was theoretically possible. What was the percent yield of product that the student obtained?

10.84 The theoretical yield of product for a particular reaction is 25.31 g. A very "meticulous" student isolates 24.79 g of product when the reaction is run. What is the percent yield that this student obtained?

10.85 Aluminum and sulfur react at elevated temperatures to form aluminum sulfide as shown by the equation.

$$2\,Al + 3\,S \rightarrow Al_2S_3$$

In a certain experiment, 125 g of Al_2S_3 was produced from 75.0 g of Al and 300.0 g of S.
(a) What is the theoretical yield of Al_2S_3?
(b) What is the percent yield of Al_2S_3?

10.86 Aluminum and oxygen react at room temperature to form aluminum oxide by the equation

$$4\,Al + 3\,O_2 \rightarrow 2\,Al_2O_3$$

In a certain experiment, 125 g of Al_2O_3 was produced from 75.0 g of Al and 200.0 g of O_2.
(a) What is the theoretical yield of Al_2O_3?
(b) What is the percent yield of Al_2O_3?

10.87 3.000 g of $TiCl_3$ was produced from 3.513 g of $TiCl_4$ and an excess of Ti according to the reaction

$$3\,TiCl_4 + Ti \rightarrow 4\,TiCl_3$$

What is the percent yield of $TiCl_3$?

10.88 5.000 g of Ag_2S was produced from 5.000 g of Ag and an excess of S according to the reaction

$$2\,Ag + S \rightarrow Ag_2S$$

What is the percent yield of Ag_2S?

10.89 Under appropriate reaction conditions Zn and S produce ZnS according to the equation

$$Zn + S \rightarrow ZnS$$

In a certain experiment with 40.7 g of Zn and an excess of S, a percent yield of 83.7% ZnS was obtained. What is the actual yield of ZnS in grams for this experiment?

10.90 Under appropriate reaction conditions Cu and O produce CuO according to the equation

$$2\,Cu + O_2 \rightarrow 2\,CuO$$

In a certain experiment with 55.3 g of Cu and an excess of O_2, a percent yield of 85.7% CuO was obtained. What is the actual yield of CuO in grams for this experiment?

10.91 The percent yield for the reaction

$$PCl_3 + Cl_2 \rightarrow PCl_5$$

is 74.3%. What mass of product, in grams, can be produced from the reaction of 20.0 g of Cl_2 and an excess of PCl_3?

10.92 The percent yield for the reaction

$$2\,NO + O_2 \rightarrow 2\,NO_2$$

is 35.6%. What mass of product, in grams, can be produced from the reaction of 35.0 g of NO and an excess of O_2?

Simultaneous Reactions (Sec. 10.11)

10.93 A mixture of composition 60.0% ZnS and 40.0% CuS is heated in air until the sulfides are completely converted to oxides as shown by the following equations.

$$2\,ZnS + 3\,O_2 \rightarrow 2\,ZnO + 2\,SO_2$$
$$2\,CuS + 3\,O_2 \rightarrow 2\,CuO + 2\,SO_2$$

How many grams of SO_2 are produced from the reaction of 82.5 g of the sulfide mixture?

10.94 A mixture of composition 50.0% H_2S and 50.0% CH_4 is reacted with oxygen, producing SO_2, CO_2, and H_2O. The equations for the reactions are

$$2\,H_2S + 3\,O_2 \rightarrow 2\,SO_2 + 2\,H_2O$$
$$CH_4 + 2\,O_2 \rightarrow CO_2 + 2\,H_2O$$

How many grams of H_2O are produced from the reaction of 65.0 g of mixture?

10.95 A mixture of magnesium carbonate ($MgCO_3$) and magnesium hydroxide [$Mg(OH)_2$] has the composition 10.0% $MgCO_3$ and 90.0% $Mg(OH)_2$. Both components of the mixture dissolve in hydrochloric acid (HCl), the reactions being

$$MgCO_3 + 2\,HCl \rightarrow MgCl_2 + H_2O + CO_2$$
$$Mg(OH)_2 + 2\,HCl \rightarrow MgCl_2 + 2\,H_2O$$

(a) How many grams of HCl are required to dissolve a 20.0 g sample of the mixture?
(b) How many grams of $MgCl_2$ are produced from dissolving a 50.0 g sample of the mixture?

10.96 A mixture of composition 75.0% carbon disulfide (CS_2) and 25.0% hydrogen sulfide (H_2S) is burned in oxygen to produce SO_2, CO_2, and H_2O. The reactions occurring are

$$CS_2 + 3\,O_2 \rightarrow CO_2 + 2\,SO_2$$
$$2\,H_2S + 3\,O_2 \rightarrow 2\,SO_2 + 2\,H_2O$$

(a) How many grams of O_2 are required to completely react with 45.3 g of the mixture?
(b) How many grams of SO_2 are produced from the reaction of 75.0 g of mixture?

Consecutive Reactions (Sec. 10.11)

10.97 How many grams of SO_2 can be obtained from 50.0 g of $KClO_3$ by the following two-step chemical process?

$$2\,KClO_3 \rightarrow 2\,KCl + 3\,O_2$$
$$S + O_2 \rightarrow SO_2$$

10.98 NO_2 is a reddish brown gas that is a component of smog. It is formed in the following two-step process.

$$N_2 + O_2 \rightarrow 2\,NO$$
$$2\,NO + O_2 \rightarrow 2\,NO_2$$

How many grams of NO_2 result from the reaction of 2.00 g of N_2?

10.99 What mass, in grams, of potassium chlorate ($KClO_3$) would be needed to supply the proper amount of oxygen to burn 33.5 g of methane (CH_4)? The two reactions are

$$2\,KClO_3 \rightarrow 2\,KCl + 3\,O_2$$
$$CH_4 + 2\,O_2 \rightarrow CO_2 + 2\,H_2O$$

10.100 What mass, in grams, of sodium chloride (NaCl) would be needed to produce the proper amount of chlorine necessary to react with 50.0 g of CH_4? The two reactions are

$$2\,NaCl \rightarrow 2\,Na + Cl_2$$
$$CH_4 + 4\,Cl_2 \rightarrow 4\,HCl + CCl_4$$

10.101 The following process has been used for obtaining iodine from oilfield brines.

$$NaI + AgNO_3 \rightarrow AgI + NaNO_3$$
$$2\,AgI + Fe \rightarrow FeI_2 + 2\,Ag$$
$$2\,FeI_2 + 3\,Cl_2 \rightarrow 2\,FeCl_3 + 2\,I_2$$

How much $AgNO_3$, in grams, is required in the first step for every 5.00 g of I_2 produced in the third step?

10.102 Sodium bicarbonate, $NaHCO_3$, can be prepared from sodium sulfate, Na_2SO_4, using the following three-step process.

$$Na_2SO_4 + 4\,C \rightarrow Na_2S + 4\,CO$$
$$Na_2S + CaCO_3 \rightarrow CaS + Na_2CO_3$$
$$Na_2CO_3 + H_2O + CO_2 \rightarrow 2\,NaHCO_3$$

How much carbon, C, in grams, is required in the first step for every 10.00 g of $NaHCO_3$ produced in the third step?

Additional Problems

10.103 Ammonium dichromate decomposes according to the following reaction.

$$(NH_4)_2Cr_2O_7 \rightarrow N_2 + 4\,H_2O + Cr_2O_3$$

How many grams of each of the products can be formed from the decomposition of 75.0 g of $(NH_4)_2Cr_2O_7$?

10.104 Nitrous acid decomposes according to the following reaction.

$$3\,HNO_2 \rightarrow 2\,NO + HNO_3 + H_2O$$

How many grams of each of the products can be formed from the decomposition of 63.5 g of HNO_2?

10.105 Pure, dry NO gas can be made by the following reaction.

$$3\,KNO_2 + KNO_3 + Cr_2O_3 \rightarrow 4\,NO + 2\,K_2CrO_4$$

How many grams of NO can be produced from a reaction mixture containing 2.00 moles each of KNO_2, KNO_3, and Cr_2O_3?

10.106 Sodium cyanide, NaCN, can be prepared by the reaction

$$Na_2CO_3 + 4\,C + N_2 \rightarrow 2\,NaCN + 3\,CO$$

How many grams of CO can be produced from a reaction mixture containing 3.00 moles each of Na_2CO_3, C, and N_2?

10.107 Terephthalic acid, $C_8H_6O_4$, an important raw material for making the synthetic fiber Dacron, is produced from xylene, C_8H_{10}, by the reaction

$$C_8H_{10} + 3\,O_2 \rightarrow C_8H_6O_4 + 2\,H_2O$$

How many grams of C_8H_{10} must react in order to produce a total of 100.0 g of products?

10.108 Adipic acid, $C_6H_{10}O_4$, an important raw material for making synthetic nylon fibers, is produced from cyclohexane, C_6H_{12}, by the reaction

$$2\,C_6H_{12} + 5\,O_2 \rightarrow 2\,C_6H_{10}O_4 + 2\,H_2O$$

How many grams of C_6H_{12} must react in order to produce a total of 50.00 g of products?

10.109 An *impure* sample of $CuSO_4$ weighing 7.53 g was dissolved in water. The dissolved $CuSO_4$, but not the impurities, then reacted with excess zinc.

$$CuSO_4 + Zn \rightarrow ZnSO_4 + Cu$$

What was the mass percent $CuSO_4$ in the sample if 1.33 g of Cu was produced?

10.110 An *impure* sample of $Hg(NO_3)_2$ weighing 64.5 g was dissolved in water. The dissolved $Hg(NO_3)_2$, but not the impurities, then reacted with excess Mg metal.

$$Hg(NO_3)_2 + Mg \rightarrow Mg(NO_3)_2 + Hg$$

What was the mass percent $Hg(NO_3)_2$ in the sample if 23.6 g of Hg was produced?

10.111 Silver oxide (Ag_2O) decomposes completely at high temperatures to produce metallic silver and oxygen gas. A 1.80-g sample of *impure* silver oxide yielded 0.115 g of O_2. Assuming that Ag_2O was the only source of O_2, what was the percent by mass of Ag_2O in the sample?

10.112 Gold(III) oxide (Au_2O_3) decomposes completely at high temperatures to produce metallic gold and oxygen gas. A 2.21 g sample of *impure* gold(III) oxide yielded 0.233 g of O_2. Assuming that Au_2O_3 was the only source of O_2, what was the percent by mass of Au_2O_3 in the sample?

10.113 The reaction of 113.4 g of I_2O_5 with 132.2 g of BrF_3 was found to produce 97.0 g of IF_5. The equation for the reaction is

$$6\,I_2O_5 + 20\,BrF_3 \rightarrow 12\,IF_5 + 15\,O_2 + 10\,Br_2$$

(a) What is the percent yield of IF_5?
(b) What is the percent yield of Br_2?

10.114 The reaction of 20.0 g of NH_3 with 20.0 g of CH_4 and an excess of oxygen was found to produce 15.0 g of HCN. The equation for the reaction is

$$2\,NH_3 + 3\,O_2 + 2\,CH_4 \rightarrow 2\,HCN + 6\,H_2O$$

(a) What is the precent yield of HCN?
(b) What is the percent yield of H_2O?

10.115 Copper metal can be recovered from an ore containing $CuCO_3$ by the decomposition reaction

$$2\,CuCO_3 \rightarrow 2\,Cu + 2\,CO_2 + O_2$$

What mass of copper ore, in tons, is needed to produce 500.0 lb of Cu if the ore is 13.22% by mass $CuCO_3$? Assume complete decomposition of the $CuCO_3$.

10.116 Silver metal can be recovered from an ore containing Ag_2CO_3 by the decomposition reaction

$$2\,Ag_2CO_3 \rightarrow 4\,Ag + 2\,CO_2 + O_2$$

What mass of silver ore, in tons, is needed to produce 500.0 lb of Ag if the ore is 24.21% by mass Ag_2CO_3? Assume complete decomposition of the Ag_2CO_3.

10.117 A 13.20-g sample of a mixture of $CaCO_3$ and $NaHCO_3$ is heated, and the compounds decompose as follows.

$$CaCO_3 \rightarrow CaO + CO_2$$
$$2\,NaHCO_3 \rightarrow Na_2CO_3 + CO_2 + H_2O$$

The decomposition of the sample yields 4.35 g of CO_2 and 0.873 g of H_2O. What percentage, by mass, of the original sample is $CaCO_3$?

10.118 A 4.00-g sample of a mixture of H_2S and CS_2 was burned in oxygen. The equations for the reactions are

$$2\,H_2S + 3\,O_2 \rightarrow 2\,H_2O + 2\,SO_2$$
$$CS_2 + 3\,O_2 \rightarrow CO_2 + 2\,SO_2$$

7.32 g of SO_2 and 0.577 g of CO_2 were produced along with some H_2O. What percentage, by mass, of the original sample was H_2S?

Cumulative Problems

10.119 Write a balanced chemical equation for each of the following chemical reactions.
(a) calcium oxide + sulfur trioxide → calcium sulfate
(b) ammonium nitrate → dinitrogen monoxide + water
(c) aluminum + silver nitrate →
  aluminum nitrate + silver
(d) hydrobromic acid + potassium hydroxide →
  potassium bromide + water

10.120 Write a balanced chemical equation for each of the following chemical reactions.
(a) sulfur dioxide + oxygen → sulfur trioxide
(b) magnesium carbonate →
magnesium oxide + carbon dioxide
(c) copper(II) sulfate + zinc → zinc sulfate + copper
(d) sulfuric acid + barium hydroxide →
barium sulfate + water

10.121 After the following equation was balanced, the name of one of the reactants was substituted for its formula.

$$2 \text{ Cyclopropane} + 9 \text{ O}_2 \rightarrow 6 \text{ CO}_2 + 6 \text{ H}_2\text{O}$$

Using only the information found within this equation, determine the molecular formula of cyclopropane.

10.122 After the following equation was balanced, the name of one of the reactants was substituted for its formula.

$$2 \text{ Butyne} + 11 \text{ O}_2 \rightarrow 8 \text{ CO}_2 + 6 \text{ H}_2\text{O}$$

Using only the information found within this equation, determine the molecular formula of butyne.

10.123 Write a balanced chemical equation for the reaction in which Cu_2S and O_2 are reactants and SO_2 and a copper oxide containing 88.82% copper by mass are products. The molecular and empirical formulas of the copper oxide are the same.

10.124 Write a balanced chemical equation for the reaction in which NH_3 and O_2 are reactants and H_2O and a nitrogen oxide containing 46.68% nitrogen by mass are products. The molecular and empirical formulas of the nitrogen oxide are the same.

10.125 Write a balanced chemical equation for the reaction in which CO_2 and H_2O are the products, O_2 is one of the reactants, and a compound with an empirical formula of CH and a formula mass of 78.12 amu is the other.

10.126 Write a balanced chemical equation for the reaction in which CO_2 and H_2O are the products, O_2 is one of the reactants, and a compound with an empirical formula of C_3H_5 and a formula mass of 82.16 amu is the other.

10.127 A particular coal contains 4.3% sulfur by mass as an impurity. When the coal is burned the S is converted to gaseous SO_2. The SO_2 enters the exhaust gases where it is removed by reaction with powdered CaO to produce solid $CaSO_3$. How much by-product $CaSO_3$, in tons, is produced by the burning of 1.0 ton of coal?

10.128 A particular coal contains 3.5% sulfur by mass as an impurity. When the coal is burned the S is converted to gaseous SO_2. The SO_2 enters the exhaust gases where it is removed by reaction with powdered CaO to produce solid $CaSO_3$. How much coal would have to be burned, in tons, to produce 2.0 tons of by-product $CaSO_3$?

10.129 The reusable booster rockets of the U.S. space shuttle employ a mixture of aluminum and ammonium perchlorate for fuel. The chemical reaction that occurs is

$$3 \text{ Al} + 3 \text{ NH}_4\text{ClO}_4 \rightarrow \text{Al}_2\text{O}_3 + \text{AlCl}_3 + 3 \text{ NO} + 6 \text{ H}_2\text{O}$$

How many moles of electrons are present in the products produced when 10.0 g of Al reacts?

10.130 The fluoride in many toothpastes is tin(II) fluoride produced by the reaction of Sn metal with gaseous HF.

$$\text{Sn} + 2 \text{ HF} \rightarrow \text{SnF}_2 + \text{H}_2$$

How many moles of electrons are present in the reactants consumed when 25.0 g of H_2 is produced?

10.131 If 100.0 g of $KClO_3$ and 200.0 g of HCl are allowed to react according to the equation

$$2 \text{ KClO}_3 + 4 \text{ HCl} \rightarrow 2 \text{ KCl} + 2 \text{ ClO}_2 + \text{Cl}_2 + 2 \text{ H}_2\text{O}$$

what is the combined total number of moles of chlorine-containing products produced?

10.132 If 50.0 g of SO_2Cl_2 and 200.0 g of HI are allowed to react according to the equation

$$\text{SO}_2\text{Cl}_2 + 8 \text{ HI} \rightarrow \text{H}_2\text{S} + 2 \text{ H}_2\text{O} + 2 \text{ HCl} + 4 \text{ I}_2$$

what is the combined total number of moles of hydrogen-containing products produced?

10.133 If the products produced by the reaction of 500.0 g of $CaCl_2$ with excess Na_2CO_3 according to the equation

$$\text{CaCl}_2 + \text{Na}_2\text{CO}_3 \rightarrow 2 \text{ NaCl} + \text{CaCO}_3$$

were broken up into ions, how many positive ions would result?

10.134 If the products produced by the reaction of 220.0 g of $AgNO_3$ with excess K_3PO_4 according to the equation

$$\text{K}_3\text{PO}_4 + 3 \text{ AgNO}_3 \rightarrow \text{Ag}_3\text{PO}_4 + 3 \text{ KNO}_3$$

were broken up into ions, how many negative ions would result?

10.135 The concentration of an aqueous NaBr solution, whose density is 1.046 g/mL, is 6.00% by mass. Determine the volume, in milliliters, of NaBr solution needed to prepare 10.0 g of AgBr by the reaction

$$\text{NaBr} + \text{AgNO}_3 \rightarrow \text{AgBr} + \text{NaNO}_3$$

10.136 The concentration of an aqueous NH_3 solution, whose density is 0.979 g/mL, is 4.50% by mass. Determine the volume, in milliliters, of NH_3 solution needed to prepare 10.0 g of NH_4NO_3 by the reaction

$$NH_3 + HNO_3 \longrightarrow NH_4NO_3$$

10.137 A three-step process for producing nitric acid, HNO_3, from gaseous ammonia, NH_3, is

$$4 NH_3 + 5 O_2 \longrightarrow 4 NO + 6 H_2O$$
$$2 NO + O_2 \longrightarrow 2 NO_2$$
$$3 NO_2 + H_2O \longrightarrow 2 HNO_3 + NO$$

Assuming yields, respectively, of 85.2%, 82.7%, and 87.0% for the three steps, how many grams of nitric acid can be produced from 75.0 mL of ammonia with a density of 0.695 g/L?

10.138 A three-step process for producing sulfuric acid, H_2SO_4, from gaseous sulfur dioxide, SO_2, is

$$2 SO_2 + O_2 \longrightarrow 2 SO_3$$
$$SO_3 + H_2SO_4 \longrightarrow H_2S_2O_7$$
$$H_2S_2O_7 + H_2O \longrightarrow 2 H_2SO_4$$

Assuming yields, respectively, of 63.1%, 87.5%, and 73.8% for the three steps, how many grams of sulfuric acid can be produced from 125 mL of sulfur dioxide with a density of 0.773 g/L?

Grid Problems

10.139 Select from the grid *all* correct responses for each situation. In each response box are found the reactants for a specified type of reaction.

1. Al, Cl_2 synthesis	2. Na_2S decomposition	3. Zn, H_2SO_4 single replacement
4. $Mg, Ni(NO_3)_2$ single replacement	5. NaF, HCl double replacement	6. $NaOH, HNO_3$ double replacement
7. C_3H_6, O_2 combustion	8. C_3H_6O, O_2 combustion	9. C_3H_6N, O_2 combustion

(a) reactions in which all products are elements
(b) reactions in which all products are compounds
(c) reactions in which there are three products
(d) reactions in which there is only one product
(e) reactions in which H_2O is one of the products
(f) reactions in which the total moles of reactants consumed and the total moles of products produced are equal

10.140 Select from the grid *all* correct responses for each situation.

1. $X + Y \rightarrow X_2Y_3$	2. $X_2 + Y_2 \rightarrow XY$	3. $X_2 + Y \rightarrow XY$
4. $X_2 + Y_4 \rightarrow XY_2$	5. $X_2 + Y \rightarrow X_2Y$	6. $X_2 + Y_2 \rightarrow X_3Y$
7. $X_2 + Y \rightarrow X_3Y$	8. $X + Y_2 \rightarrow X_2Y$	9. $X_2 + Y_2 \rightarrow X_2Y_5$

(a) balanced equations where all coefficients are the same
(b) balanced equations where the coefficient sum is five
(c) reactions for which the total moles of reactants and total moles of products are equal
(d) pairs of equations where the sum of the coefficients is equal
(e) pairs of reactions where the molar ratio of reactants to products is the same
(f) pairs of reactions where the product formulas have the same atomic ratio

10.141 Select from the grid *all* correct responses for each situation.

1. $Na + N_2 \rightarrow NaN_3$	2. $C + H_2 \rightarrow CH_4$	3. $NO + O_2 \rightarrow NO_2$
4. $Al + S \rightarrow Al_2S_3$	5. $N_2 + H_2 \rightarrow NH_3$	6. $Mg + N_2 \rightarrow Mg_3N_2$
7. $SO_3 + H_2O \rightarrow H_2SO_4$	8. $K_2O + H_2O \rightarrow KOH$	9. $H_2 + Cl_2 \rightarrow HCl$

(a) reactions for which the three coefficients in the balanced equations have different values
(b) reactions for which the coefficient set (6, 9, 3) would *mathematically* balance the equation
(c) reactions where the needed molar amount of the first reactant exceeds that for the second reactant
(d) reactions where the needed mass amount of the first reactant exceeds that for the second reactant
(e) reactions in which the molar amounts of reactants and products are the same
(f) reactions in which the mass amounts of reactants and products are the same

10.142 Select from the grid *all* correct responses for each situation.

1. 2 moles X, 2 moles Y	2. 3 moles X, 4 moles Y	3. $\frac{1}{4}$ mole X, $\frac{1}{4}$ mole Y
4. 3 moles X, 2 moles Y	5. 2 moles X, 3 moles Y	6. 1 mole X, 3 moles Y
7. 6 moles X, 4 moles Y	8. 4 moles X, 3 moles Y	9. 2 moles X, 8 moles Y

(a) reactant mixtures in which X is the limiting reactant for the reaction $4X + 3Y \rightarrow$ products
(b) reactant mixtures in which Y is the limiting reactant for the reaction $4X + 3Y \rightarrow$ products
(c) reactant mixtures in which there is no limiting reactant for the reaction $3X + 2Y \rightarrow$ products
(d) reactant mixtures in which X is the limiting reactant for the reaction $2X + Y \rightarrow$ products
(e) reactant mixtures in which both X and Y may be considered limiting reactants for the reaction $X + 4Y \rightarrow$ products
(f) reactant mixtures in which all X and all Y will be consumed for the reaction $2X + 3Y \rightarrow$ products

CHAPTER ELEVEN

States of Matter

11.1 Physical States of Matter

In this chapter we consider the physical states of matter—the solid state, the liquid state, and the gaseous state.

From everyday experience we know that the physical state of a substance is determined both by what it is—its chemical identity—and by the temperature and pressure it is under. At room temperature and pressure some substances are solids (gold, sodium chloride, etc.), others are liquids (water, mercury, etc.), and still others are gases (oxygen, carbon dioxide, etc.). Thus chemical identity must be a determining factor for physical state, since all three states are observed at room temperature and pressure. On the other hand, when the physical state of a single substance is considered, temperature and pressure are determining variables. Liquid water can be changed to a solid by lowering the temperature or to a gas by raising the temperature.

We tend to characterize a substance almost exclusively in terms of its most common physical state, that is, the state in which it is found at room temperature and pressure. Oxygen is almost always thought of as a gas, its most common state; gold

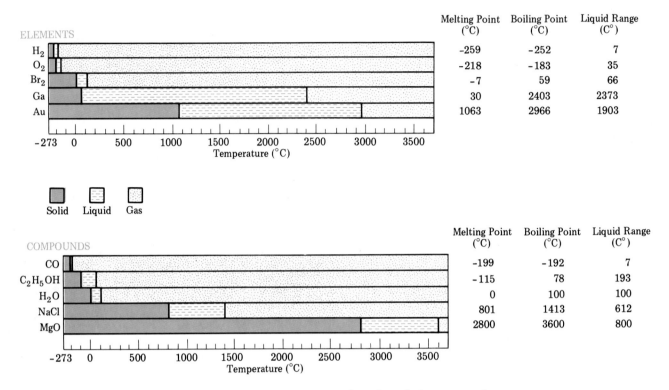

FIGURE 11.1 Solid, liquid, and gaseous temperature ranges for selected elements and compounds.

is almost always thought of as a solid, its most common state. A major reason for such "single state" characterization is the narrow range of temperatures encountered on this planet. Most substances are never encountered in more than one state under "natural" conditions. We must be careful not to fall into the error of assuming that the commonly observed state of a substance is the *only* state in which it can exist. Under laboratory conditions, states other than the "natural" one can be obtained for almost all substances. Figure 11.1 shows the temperature ranges for the solid, liquid, and gaseous states of a few elements and compounds. As can be seen in this figure, the size and location of the physical-state temperature ranges vary widely among chemical substances. Extremely high temperatures are required to obtain some substances in the gaseous state; other substances are gases at temperatures below room temperature. The size of a given physical-state range also varies dramatically. For example, the elements H_2 and O_2 are liquids over a very narrow temperature range, whereas the elements Ga and Au remain liquids over ranges of hundreds of degrees.

Among the elements, at room temperature and pressure, the solid state is the dominant physical state. Ninety-six of the 109 known elements are solids at room temperature, as shown in Figure 11.2. It is usually stated that only two elements, bromine and mercury, are liquids at room temperature. However, there are three other elements, cesium and francium of group IA and gallium of group IIIA, that have melting

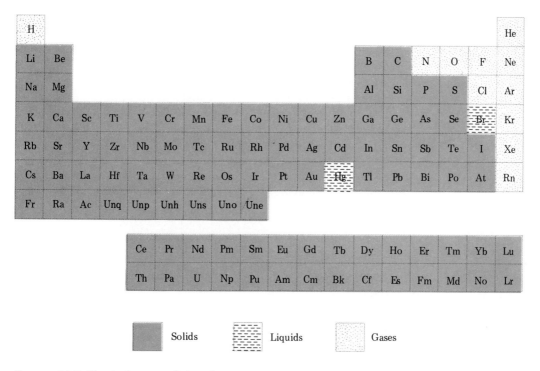

FIGURE 11.2 Physical states of the elements at room temperature and pressure.

points between 25 and 30 °C. Temperatures in this range are reached in a "hot" room. Thus, on "hot" days, there could be five elements that are liquids. Eleven of the elements are gases under "normal" conditions.

Explanations for the experimentally observed physical state variations among substances can be derived from the concepts of atomic structure and bonding we have considered in previous chapters. Such explanations are found in later sections of this chapter.

11.2 Property Differences of Physical States

The differences among solids, liquids, and gases are so great that only a few gross distinguishing features need to be mentioned in order to differentiate them clearly. Certain obvious differences among the three states of matter are apparent to even the most casual observer—differences related to (1) volume and shape, (2) density, (3) compressibility, and (4) thermal expansion. These distinguishing properties are compared in Table 11.1 for the three states of matter. The properties of volume and density have been discussed in detail previously (Secs. 3.4 and 3.8, respectively). **Compressibility** is a measure of the change in volume resulting from a pressure change. **Thermal expansion** is a measure of the volume change resulting from a temperature change.

Table 11.1 Distinguishing Properties of Solids, Liquids, and Gases

Property	Solid State	Liquid State	Gaseous State
Volume and shape	definite volume and definite shape	definite volume and indefinite shape; takes the shape of container to the extent it is filled	indefinite volume and indefinite shape; takes the volume and shape of container that it fills
Density	high	high, but usually lower than corresponding solid	low
Compressibility	small	small, but usually greater than corresponding solid	large
Thermal expansion	very small: about 0.01% per °C	small: about 0.10% per °C	moderate: about 0.30% per °C

The contents of Table 11.1 should be studied in detail; they will serve as the starting point for further discussions about the states of matter.

11.3 The Kinetic Molecular Theory

The physical characteristics of the solid, liquid, and gaseous states can be explained by kinetic molecular theory, one of the fundamental theories of chemistry. A basic idea of this theory is that the particles (atoms or molecules) present in a substance, independent of the physical state of the substance, have *motion* associated with them. The word *kinetic* comes from the Greek word *kinesis*, which means movement, hence the name kinetic molecular theory.

The specific statements of **kinetic molecular theory** are

1. Matter is ultimately composed of tiny particles (atoms, molecules, or ions) with definite and characteristic sizes that never change.
2. The particles are in constant random motion and therefore possess kinetic energy.
3. The particles interact with each other through attractions and repulsions and therefore possess potential energy.
4. The velocity of the particles increases as the temperature is increased. The average kinetic energy of all particles in a system depends on the temperature, increasing as the temperature increases.
5. The particles in a system transfer energy from one to another during collisions in which no net energy is lost from the system. The energy of any given particle is thus continually changing.

The statements of kinetic molecular theory refer to kinetic and potential energy—the two basic types of energy. Definitions for these types of energy were previously given in Section 3.11. Recall that kinetic energy is energy associated with motion and potential energy is stored energy. We now consider additional concepts relative to these types of energy.

The amount of kinetic energy a particle possesses depends on both its mass and its velocity. The exact mathematical relationship between the kinetic energy and the mass and velocity of a particle is

$$\text{Kinetic energy} = \tfrac{1}{2}mv^2$$

where m is the mass of the particle and v is its velocity. From this expression we see that any differences in kinetic energy between particles of the same mass must be caused by differences in their velocities. Similarly, differences in kinetic energy between particles moving at the same velocity must be due to mass differences.

In any system of particles (atoms, molecules, or ions), potential energy resulting from particle interactions is present. Electrostatic interactions are the potential energy interactions of most importance for atomic-sized particles. **Electrostatic interactions** occur between charged particles; objects of opposite electric charge (one positive and one negative) attract each other, and objects of identical charge (both positive or both negative) repel each other. The magnitude of an electrostatic interaction depends on the sizes of the charges associated with the particles and their separation distance. Potential energy of attraction increases as the separation between particles increases, while that of repulsion increases with decreasing separation, as shown graphically in Figure 11.3. An analogy to the dependence of electrostatic potential energy on distance is found in two particles connected by a spring. When the spring is either stretched or compressed, the system has more potential energy than when the spring is in a nonextended position. The stretched spring represents attractive potential energy (the particles will come together if the spring is released), and the compressed spring represents repulsive potential energy (the particles will move apart when the spring is released). The greater the stretching of the spring (the greater the distance between particles), the greater the potential energy of attraction. The greater the compression of the spring (the smaller the distance between particles), the greater the potential energy of repulsion.

The relative influence of kinetic energy and potential energy in a chemical system is the major consideration in uses of kinetic molecular theory to explain the general properties of the solid, liquid, and gaseous states of matter. The important question is whether the kinetic energy or the potential energy dominates the energetics of the chemical system under study.

Kinetic energy may be considered a *disruptive force* within the chemical system, tending to make the particles of the system increasingly independent of each other.

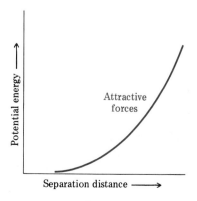
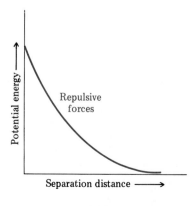

FIGURE 11.3 Effect of separation distance on potential energy.

As the result of energy of motion the particles will tend to move away from each other. Potential energy may be considered a *cohesive force* tending to cause order and stability among the particles of the system.

The role that temperature plays in determining the state of a system is related to kinetic energy magnitude. Kinetic energy increases as temperature increases (statement 4 of the kinetic molecular theory). Thus, the higher the temperature the greater the magnitude of disruptive influences within a chemical system. The magnitude of potential energy is essentially independent of temperature change. Neither charge nor separation distance, the two factors on which the magnitude of potential energy depends, is affected by temperature change.

Sections 11.4, 11.5, and 11.6 deal, respectively, with kinetic molecular theory explanations for the general properties of the solid, liquid, and gaseous states.

11.4 The Solid State

The solid state is characterized by a dominance of potential energy (cohesive forces) over kinetic energy (disruptive forces). The particles in a solid are drawn close together in a regular pattern by the strong cohesive forces present. Each particle occupies a fixed position about which it vibrates, owing to disruptive kinetic energy. An explanation of the characteristic properties of solids is obtained from this model.

1. *Definite volume and definite shape.* The strong cohesive forces hold the particles in essentially fixed positions, resulting in definite volume and definite shape.
2. *High density.* The constituent particles of solids are located as close together as possible. Therefore, large numbers of particles are contained in a unit volume, resulting in a high density.
3. *Small compressibility.* Since there is very little space between particles, increased pressure cannot push them any closer together and therefore has little effect on the solid's volume.
4. *Very small thermal expansion.* An increase in temperature increases the kinetic energy (disruptive forces), thereby causing more vibrational motion of the particles. Each particle "occupies" a slightly larger volume. The result is a slight expansion of the solid. The strong cohesive forces prevent this effect from becoming very large.

11.5 The Liquid State

The liquid state consists of particles randomly packed relatively close to each other. The molecules are in constant random motion, freely sliding over one another but without sufficient energy to separate from each other. *The liquid state is a situation in which neither potential energy (cohesive forces) nor kinetic energy (disruptive forces) dominates.* The fact that the particles freely slide over each other indicates the influence of disruptive forces, but the fact that the particles do not separate indicates a fairly strong influence from cohesive forces. The characteristic properties of liquids are explained by this model.

1. *Definite volume and indefinite shape.* Attractive forces are strong enough to restrict particles to movement within a definite volume. They are not strong enough, however, to prevent the particles from moving over each other in a random manner, limited only by the container walls. Thus liquids have no definite shape, with the exception that they maintain a horizontal upper surface in containers that are not completely filled.
2. *High density.* The particles in a liquid are not widely separated; they essentially touch each other. Therefore, there will be a large number of particles per unit volume and a resultant high density.
3. *Small compressibility.* Since the particles in a liquid essentially touch each other, there is very little empty space. Therefore, a pressure increase cannot squeeze the particles much closer together.
4. *Small thermal expansion.* Most of the particle movement in a liquid is vibrational, because a particle can move only a short distance before colliding with a neighbor. The increased particle velocity that accompanies a temperature increase results only in increased vibrational amplitudes. The net effect is an increase in the effective volume a particle "occupies," which causes a slight volume increase in the liquid.

11.6 The Gaseous State

Kinetic energy (disruptive forces) completely dominates potential energy (cohesive forces) in the gaseous state. As a result, the particles of a gas move essentially independently of one another in a totally random manner. Under ordinary pressure, the particles are relatively far apart except, of course, when they collide with each other. Between collisions with each other or with the container walls, gas particles travel in straight lines. The particle velocities and resultant collision frequencies are extremely high; at room temperature the collisions experienced by one molecule in 1 sec at 1 atm pressure are of the order of 10^{10}.

The kinetic theory explanation of gaseous state properties follows the same pattern we saw earlier for solids and liquids.

1. *Indefinite volume and indefinite shape.* The attractive (cohesive) forces between particles have been overcome by kinetic energy, and the particles are free to travel in all directions. Therefore, the particles completely fill the container the gas is in and assume its shape.
2. *Low density.* The particles of a gas are widely separated. There are relatively few of them in a given volume, which means little mass per unit volume.
3. *Large compressibility.* Particles in a gas are widely separated; a gas is essentially mostly empty space. When pressure is applied, the particles are easily pushed closer together, decreasing the amount of empty space and the volume of the gas (see Fig. 11.4).

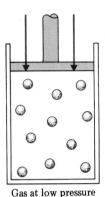
Gas at low pressure

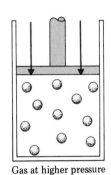
Gas at higher pressure

FIGURE 11.4 The compression of a gas—decreasing the amount of empty space in the container.

4. *Moderate thermal expansion.* An increase in temperature means an increase in particle velocity. The increased kinetic energy of the particles enables them to push back whatever barrier is confining them into a given volume. Hence, the volume increases.

It must be understood that the size of the particles is not changed during expansion or compression of gases, solids, or liquids. The particles merely move farther apart or closer together; the space between them is what changes.

11.7 A Comparison of Solids, Liquids, and Gases

Two obvious conclusions about the similarities and differences between the various states of matter may be drawn from a comparison of the descriptive materials in Sections 11.4–11.6.

1. One of the states of matter, the gaseous state, is markedly different from the other two states.
2. Two of the states of matter, the solid and the liquid states, have many similar characteristics.

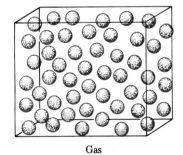
Gas
Molecules far apart and disordered
Negligible interactions between molecules

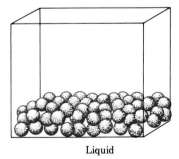
Liquid
Intermediate situation

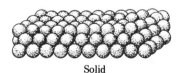
Solid
Molecules close together and ordered
Strong interactions between molecules

FIGURE 11.5 Similarities and differences among the states of matter.

These two conclusions are illustrated diagrammatically in Figure 11.5.

The average distance between particles is only slightly different in the solid and liquid states but markedly different in the gaseous state. Roughly speaking, at ordinary temperatures and pressures, particles in a liquid are about 10% and particles in a gas about 1000% farther apart than those in the solid state. The distance ratio between particles in the three states (solid to liquid to gas) is thus 1 to 1.1 to 10.

11.8 Endothermic and Exothermic Changes of State

Physical changes of state were discussed in Section 4.4. The terminology associated with such changes—evaporation, condensation, sublimation, etc.—was introduced at that time (Fig. 4.4).

Changes of state are usually accomplished through heating or cooling a substance. (Pressure change is also a factor in some systems.) Changes of state may be classified according to whether heat (thermal energy) is given up or absorbed. An **endothermic change** of state is a change that requires the input (absorption) of heat energy. The endothermic changes of state are melting, sublimation, and evaporation. An **exothermic change** of state is a change that requires heat energy to be given up (released). Exothermic changes of state are the reverse of endothermic changes of state and include deposition, condensation, and freezing. Figure 11.6 summarizes the classification of changes of state as endothermic or exothermic.

11.9 Temperature Change as a Substance Is Heated

For a given pure substance, molecules in the gaseous state contain more energy than molecules in the liquid state, which in turn contain more energy than molecules in the solid state. This fact is obvious; we know that it takes energy (heat) to melt a

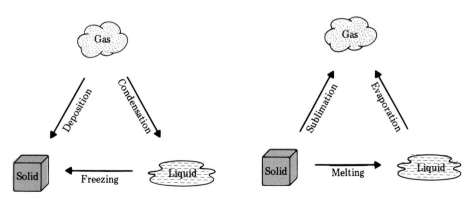

FIGURE 11.6 Endothermic and exothermic changes of state.

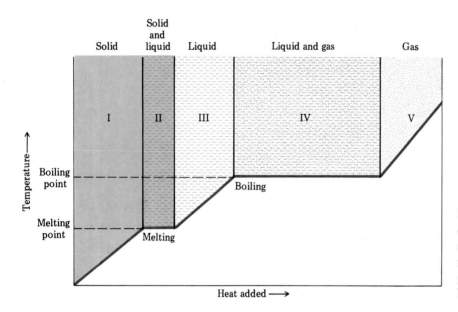

FIGURE 11.7 A heating curve depicting the addition of heat to a solid at a temperature below its melting point until it becomes a gas at a temperature above its boiling point.

solid and still more energy (heat) to change the resulting liquid to a gas. Additional information concerning the relationship of energy to the states of matter can be obtained by a closer examination of what happens, step by step, to a solid (which is below its melting point) as heat is continuously supplied, causing it to melt and ultimately to change to a gas. The heating curve shown in Figure 11.7 gives the steps involved in changing a solid to a gas. As the solid is heated (region I of the graph), its temperature rises until the melting point is reached. The temperature increase indicates that the added heat causes an increase in the kinetic energy of the particles (recall Sec. 11.3, kinetic molecular theory). Once the melting point is reached, the temperature remains constant while the solid melts (region II of the graph). The constant temperature during melting indicates that the added heat has increased the potential energy of the particles without increasing their kinetic energy—the interparticle attractions are being weakened by the increase in potential energy. The addition of more heat to the system, which is now in the liquid state, increases the temperature until the boiling point is reached (region III of the graph). During this stage the molecules in the system are again gaining kinetic energy. At the boiling point another state change occurs as heat is added with the temperature remaining constant (region IV of the graph). The constant temperature again indicates an increase in potential energy. Once the system is completely vaporized, the temperature of the gas will again increase with further heating (region V of the graph).

The actual amount of energy that must be added to a system to cause it to undergo the series of changes just described depends on values of three properties of the substance. These properties are (1) specific heat, (2) heat of fusion, and (3) heat of vaporization. Specific heat was discussed in Section 3.12. Heats of fusion and vaporization are the subject of Section 11.10.

11.10 Energy and Changes of State

When heat energy is added to a solid, its temperature rises until the melting point is reached, at a rate that is governed by the specific heat (Sec. 3.12) of the solid. Once the melting point is reached, the temperature then remains constant while the solid changes to a liquid (Sec. 11.9). The "energetics" of the system during this transition from the solid state to the liquid state depend on the value of the substance's heat of fusion. (The term *fusion* means melting.)

The **heat of fusion** is the amount of heat energy needed for the conversion of 1 g of a solid to a liquid at its melting point. Units for heat of fusion are joules per gram (J/g). Note that these units do not involve temperature (degrees) as was the case for specific heat [J/(g · °C)]. No temperature units are needed since temperature remains constant during a change of state.

The reverse of the fusion process is solidification (or freezing). The **heat of solidification** is the amount of heat energy evolved in the conversion of 1 g of a liquid to a solid at its freezing point. The heat of solidification always has the *same numerical value* as the heat of fusion. The only difference between these two entities is in the *direction* of heat flow (in or out). Heat of solidification is associated with an exothermic process, and heat of fusion with an endothermic process. The amount of heat required to melt 50.0 g of ice at its melting point is the same as the amount of heat that must be removed to freeze 50.0 g of water at its freezing point. The heats of fusion (or solidification) for selected substances are given in Table 11.2 in the units joules per gram.

Table 11.2 also gives a second heat of fusion (solidification) value for each substance, the *molar* heat of fusion (solidification), which has the units kJ/mole. In some types of calculations kJ/mole are more convenient units to use.

The magnitude of the heat of fusion for a given solid depends on the intermolecular forces of attraction in the solid state. The strength of such forces is the subject of Section 11.17.

TABLE 11.2 Heats of Fusion (or Solidification) for Various Substances at Their Melting (or Freezing) Points

Solid	Melting (or Freezing) Point (°C)	Heat of Fusion (or Solidification)	
		J/g	kJ/mole
Methane	−182	59	0.94
Ethyl alcohol	−117	109	5.01
Carbon tetrachloride	−23	16.3	2.51
Water	0	334	6.01
Benzene	6	126	9.87
Aluminum	658	393	10.6
Copper	1083	205	13.0

The general equation for calculating the amount of heat absorbed as a substance changes from a solid to a liquid is

$$\text{Heat absorbed (J)} = \text{heat of fusion (J/g)} \times \text{mass (g)}$$

Similarly, for the amount of heat released as a liquid freezes to a solid we have

$$\text{Heat released (J)} = \text{heat of solidification (J/g)} \times \text{mass (g)}$$

Example 11.1 illustrates the use of the first of these two equations.

EXAMPLE 11.1

How much heat energy, in joules, is required to melt 35.2 g of ice at its melting point of 0 °C?

Solution
The heat of fusion for ice (water), from Table 11.2, is 334 J/g. Therefore, the total amount of heat energy required is

$$\text{Heat absorbed} = 35.2 \text{ g} \times \frac{334 \text{ J}}{\text{g}} = 11{,}756.8 \text{ J} \quad \text{(calculator answer)}$$

$$= 11{,}800 \text{ J} \quad \text{(correct answer)}$$

PRACTICE EXERCISE 11.1

How much heat energy, in joules, is required to melt 25.1 g of aluminum at its melting point of 658 °C?

Ans. 9860 J

Similar exercises:
Problems 11.17 and 11.18

Principles similar to those just considered for solid–liquid or liquid–solid changes apply to changes between the liquid and gaseous states. Here the specific energy quantities involved are heats of vaporization and heats of condensation. The **heat of vaporization** is the amount of heat energy needed for the conversion of 1 g of a liquid to a gas at its boiling point. Heats of vaporization are usually measured at the normal boiling point (Sec. 11.14) of the liquid.

The reverse of the vaporization (evaporation) process is condensation (liquefaction). The **heat of condensation** is the amount of heat energy evolved in the conversion of 1 g of a gas to a liquid at the liquid's boiling point. The heat of vaporization and the heat of condensation will always have the same numerical value because these two quantities characterize processes that are the "reverse" of each other. The only difference is direction of heat flow; one process is endothermic (vaporization), and the other is exothermic (condensation). Table 11.3 gives heats of vaporization (or condensation) for selected substances.

The general equation for calculating the amount of heat absorbed as a substance changes from a liquid to a gas is

$$\text{Heat absorbed (J)} = \text{heat of vaporization (J/g)} \times \text{mass (g)}$$

TABLE 11.3 Heats of Vaporization (or Condensation) for Various Substances at Their Normal Boiling Points

Liquid	Normal Boiling Point (°C)	Heat of Vaporization (or Condensation)	
		J/g	kJ/mole
Methane	−161	65.0	10.4
Ammonia	−33	1380	23.4
Diethyl ether	34.6	375	27.8
Carbon tetrachloride	77	195	30.0
Ethyl alcohol	78.3	837	38.6
Benzene	80.1	394	30.8
Water	100	2260	40.7

Similarly, for the amount of heat released as a gas condenses to a liquid we have

$$\text{Heat released (J)} = \text{heat of condensation (J/g)} \times \text{mass (g)}$$

EXAMPLE 11.2

The vaporization of 20.0 g of ethyl alcohol at its boiling point requires 16,740 J of heat energy. Using this information, calculate the heat of vaporization, in joules per gram, of ethyl alcohol.

Solution

Rearranging the equation

$$\text{Heat absorbed} = \text{heat of vaporization} \times \text{mass}$$

to isolate "heat of vaporization," which is our desired quantity, gives

$$\text{Heat of vaporization} = \frac{\text{heat absorbed}}{\text{mass}}$$

Substituting the given quantities into this equation gives

$$\text{Heat of vaporization} = \frac{16{,}740 \text{ J}}{20.0 \text{ g}}$$
$$= 837 \text{ J/g} \quad \text{(calculator and correct answer)}$$

PRACTICE EXERCISE 11.2

5744 joules of heat energy are required to vaporize 15.0 g of an unknown liquid at its boiling point. What is the heat of vaporization of this substance, in joules per gram? *Ans.* 383 J/g.

Similar exercises: Problems 11.25 and 11.26

11.11 Heat Energy Calculations

In "bookkeeping" on heat energy added to or removed from a chemical system, the following two generalizations always apply.

1. The amount of energy added to or removed from a chemical system that undergoes a temperature change *but no change of state* is governed by the substance's specific heat. Examples 3.20 and 3.21 in Section 3.12 show how specific heat is used in calculations involving this situation.
2. The amount of energy added to or removed from a chemical system that undergoes a change of state at *constant temperature* is governed by heat of fusion or solidification (solid-to-liquid changes) and heat of vaporization or condensation (liquid-to-gas changes). Examples 11.1 and 11.2 in Section 11.10 show how these entities are used in change of state situations.

Figure 11.8 is a useful summary diagram of all of the possible "energy–change-of-temperature" and "energy–change-of-state" situations that occur. Also, at the bottom

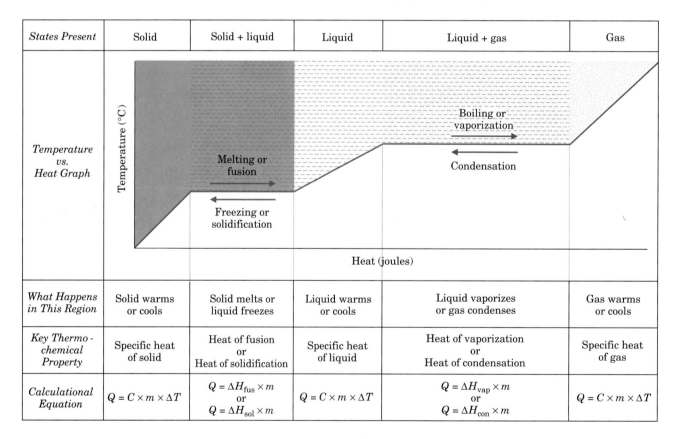

FIGURE 11.8 A temperature–heat-energy graph showing what happens as a solid, below its melting point, is heated to higher and higher temperatures.

Section 11.10 · Energy and Changes of State 401

of this diagram are given the key equations needed for doing heat energy calculations. The three equations that involve specific heat were originally introduced in Section 3.12, a section that it would be appropriate to review at this time. Those equations involving changes of state are the equations of Section 11.10.

The equations, as given in Figure 11.8, have new symbolism (abbreviations) associated with them. This symbolism involves the following relationships.

Q = heat absorbed or heat released
C = specific heat of the substance in a given physical state
m = mass of the substance
ΔT = change in temperature of the substance
ΔH_{fus} = heat of fusion of the substance
ΔH_{sol} = heat of solidification of the substance
ΔH_{vap} = heat of vaporization of the substance
ΔH_{con} = heat of condensation of the substance

In problem-solving situations where a pure substance experiences both a temperature change and one or more changes of state (a common situation) the calculational equations of Figure 11.8 are used in an additive manner. Examples 11.3 and 11.4 illustrate problem solving in such "combination" situations.

EXAMPLE 11.3

Calculate the heat released when 32.0 g of steam at 122 °C condenses to water at 100 °C in a radiator of a steam heating system.

Solution

We may consider this process to occur in two steps.

1. Cooling the steam from 122 to 100 °C (the condensation temperature for steam).
2. Condensing the steam to liquid water at 100 °C.

We will calculate the amount of heat released in each of the steps and then add these amounts together to get our final answer.

STEP 1 $Gas_{122°C} \longrightarrow gas_{100°C}$

From the last column of Figure 11–8 we note that the heat energy equation for the change in temperature of a gaseous substance is

Heat released = specific heat × mass × temperature change

From the specific heats table in Chapter 3 (Table 3.4) we find that the specific heat of water (as steam) is 2.0 J/g · °C. The temperature change is 22 °C and the mass is given as 32.0 g. Plugging these values into our equation gives

$$\text{Heat released} = \frac{2.0 \text{ J}}{\text{g} \cdot {}^\circ\text{C}} \times 32.0 \text{ g} \times 22 \, {}^\circ\text{C}$$

$$= 1408 \text{ J} \quad \text{(calculator answer)}$$

$$= 1400 \text{ J} \quad \text{(correct answer)}$$

(The answer may contain only two significant figures.)

STEP 2 $\text{Gas}_{100\,{}^\circ\text{C}} \longrightarrow \text{liquid}_{100\,{}^\circ\text{C}}$

This step involves a change of state, from the gaseous to the liquid state (the fourth column of the heating curve of Figure 11.8). The heat energy equation for this change is

$$\text{Heat released} = \text{heat of condensation} \times \text{mass}$$

The heat of condensation of H_2O, from Table 11.3, is 2260 J/g. Therefore, we have

$$\text{Heat released} = \frac{2260 \text{ J}}{\text{g}} \times 32.0 \text{ g} = 72{,}320 \text{ J} \quad \text{(calculator answer)}$$

$$= 72{,}300 \text{ J} \quad \text{(correct answer)}$$

(The correct answer may contain only three significant figures.)

The total heat released is the sum of that released in each step.

From step 1:   1,400 J
From step 2: 72,300 J
              ─────────
              73,700 J   (calculator and correct answer)

PRACTICE EXERCISE 11.3

Calculate the heat released when 40.0 g of steam at 115 °C condenses to water at 100 °C in a radiator of a steam heating system.

Ans. 91,600 J.

Similar exercises:
Problems 11.29 and 11.30

EXAMPLE 11.4

Calculate the number of joules of heat needed to change 12.8 g of ice at −18 °C to 12.8 g of steam at 121 °C.

Solution

We may consider the change process to occur in five steps.

1. Heating the ice from −18 to 0 °C, its melting point. (This change corresponds to the first column of the heating curve of Figure 11.8.)
2. Melting the ice, while the temperature remains at 0 °C. (This change corresponds to the second column of the heating curve of Figure 11.8.)
3. Heating the liquid water from 0 to 100 °C, its boiling point. (This change corresponds to the third column of the heating curve of Figure 11.8.)
4. Evaporating the water, while the temperature remains at 100 °C. (This change corresponds to the fourth column of the heating curve of Figure 11.8.)

5. Heating the steam from 100 to 121 °C. (This change corresponds to the last column of the heating curve of Figure 11.8.)

We will calculate the amount of heat required in each of the steps and then add these amounts together to get our desired answer. In doing this we will need the following heat energy values associated with water.

Step 1: specific heat of ice, 2.1 J/g · °C (Table 3.4)
Step 2: heat of fusion, 334 J/g (Table 11.2)
Step 3: specific heat of water, 4.18 J/g · °C (Table 3.4)
Step 4: heat of vaporization, 2260 J/g (Table 11.3)
Step 5: specific heat of steam, 2.0 J/g · °C (Table 3.4)

STEP 1 Solid$_{-18\,°C}$ ⟶ solid$_{0\,°C}$

Heat required = specific heat × mass × temperature change

$$= \frac{2.1 \text{ J}}{\text{g} \cdot °C} \times 12.8 \text{ g} \times 18\,°C = 483.84 \text{ J} \quad \text{(calculator answer)}$$

$$= 480 \text{ J} \quad \text{(correct answer)}$$

STEP 2 Solid$_{0\,°C}$ ⟶ liquid$_{0\,°C}$

Heat required = heat of fusion × mass

$$= \frac{334 \text{ J}}{\text{g}} \times 12.8 \text{ g} = 4275.2 \text{ J} \quad \text{(calculator answer)}$$

$$= 4280 \text{ J} \quad \text{(correct answer)}$$

STEP 3 Liquid$_{0\,°C}$ ⟶ liquid$_{100\,°C}$

Heat required = specific heat × mass × temperature change

$$= \frac{4.18 \text{ J}}{\text{g} \cdot °C} \times 12.8 \text{ g} \times 100\,°C = 5350.4 \text{ J} \quad \text{(calculator answer)}$$

$$= 5350 \text{ J} \quad \text{(correct answer)}$$

STEP 4 Liquid$_{100\,°C}$ ⟶ gas$_{100\,°C}$

Heat required = heat of vaporization × mass

$$= \frac{2260 \text{ J}}{\text{g}} \times 12.8 \text{ g} = 28,928 \text{ J} \quad \text{(calculator answer)}$$

$$= 28,900 \text{ J} \quad \text{(correct answer)}$$

STEP 5 Gas$_{100\,°C}$ ⟶ gas$_{121\,°C}$

Heat required = specific heat × mass × temperature change

$$= \frac{2.0 \text{ J}}{\text{g} \cdot °C} \times 12.8 \text{ g} \times 21\,°C = 537.6 \text{ J} \quad \text{(calculator answer)}$$

$$= 540 \text{ J} \quad \text{(correct answer)}$$

The total heat required is the sum of the results of the five steps.

$(480 + 4280 + 5350 + 28{,}900 + 540)\text{ J} = 39{,}550\text{ J}$ (calculator answer)

$= 39{,}600\text{ J}$ (correct answer)

(The answer must be rounded to the hundreds place. In adding, the number 28,900 is the limiting number, with uncertainty in the hundreds place.)

PRACTICE EXERCISE 11.4

Calculate the number of joules of heat needed to change 35.0 g of ice at $-10\,°C$ to 35.0 g of steam at 110 °C.

Ans. 106,800 J.

Similar exercises: Problems 11.31–11.34

11.12 Evaporation of Liquids

Evaporation is the process by which molecules escape from a liquid phase to the gas phase. It is a familiar process. We are all aware that water left in an open container at room temperature will slowly disappear by evaporation.

The phenomenon of evaporation can readily be explained using kinetic molecular theory. Statement 5 of this theory (Sec. 11.3) indicates that the molecules in a liquid (or solid or gas) do not all possess the same kinetic energy. At any given instant some molecules will have above average kinetic energies and others below average kinetic energies as a result of collisions between molecules. A given molecule's energy constantly changes as a result of collisions with neighboring molecules. Molecules considerably above average in kinetic energy can overcome the attractive forces (potential energy) that are holding them in the liquid and escape if they are near the liquid surface and are moving in a favorable direction relative to the surface.

Note that evaporation is a surface phenomenon. Molecules within the interior of a liquid are surrounded on all sides by other molecules, making escape very improbable. Surface molecules are subject to fewer attractive forces since they are not completely surrounded by other molecules; escape is much more probable. Liquid surface area is an important factor in determining the rate at which evaporation occurs. Increased surface area results in an increased evaporation rate; a greater fraction of molecules occupy "surface" locations.

Water evaporates faster from a glass of hot water than from a glass of cold water. Why is this so? A certain minimum kinetic energy is required for molecules to escape from the attractions of neighboring molecules. As the temperature of a liquid increases, a larger fraction of the molecules present acquire this needed minimum kinetic energy. Consequently, the rate of evaporation always increases as liquid temperature increases. Figure 11.9 contrasts the fraction of molecules possessing the needed minimum kinetic energy for escape at two temperatures. Note that at both the lower and higher temperatures a broad distribution of kinetic energies is present and that at each temperature some molecules possess the needed minimum kinetic energy. However, at the higher temperature a larger fraction of molecules present have the requisite kinetic energy.

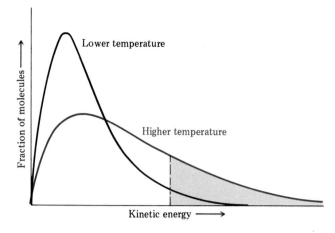

FIGURE 11.9 Kinetic energy distributions of molecules of a liquid at two different temperatures. The dashed line represents the minimum kinetic energy required for molecules of the liquid to overcome attractive forces and escape into the gas phase. Molecules in the shaded area have the necessary energy to overcome attractions.

The escape of high-energy molecules from a liquid during evaporation affects the liquid in two ways: the amount of liquid decreases, and the liquid temperature is lowered. The temperature lowering reflects the fact that the average kinetic energy of the remaining molecules is lower than the pre-evaporation value due to the loss of the most energetic molecules. (Analogously, if all the tall people are removed from a classroom of students, the average height of the remaining students decreases.) A lower average kinetic energy corresponds to a lower temperature (statement 4 of kinetic molecular theory); hence a cooling effect is produced.

Evaporative cooling is important in many processes. Our own bodies use evaporation to maintain a constant temperature. We perspire in hot weather because evaporation of the perspiration cools our skin. The cooling effect of evaporation is quite noticeable when someone first comes out of a swimming pool on a hot day (especially if a breeze is blowing). A canvas water bag keeps water cool because some of the water seeps through the canvas and evaporates, a process that removes heat from the remaining water. In medicine, the local skin anesthetic ethyl chloride (C_2H_5Cl) exerts its effect through evaporative cooling (freezing). The minimum kinetic energy molecules of this substance need to acquire to escape (evaporate) is very low, resulting in a very rapid evaporation rate. Evaporation is so fast that the cooling "freezes" tissue near the surface of the skin, with temporary loss of feeling in the region of application. Certain alcohols also evaporate quite rapidly. An alcohol "rub" is sometimes used to reduce body temperature when a high fever is present.

For a liquid in a container, the decrease in temperature that occurs as a result of evaporation can be actually measured only if the container is an *insulated* one. When a liquid evaporates from a noninsulated container, there is sufficient heat flow from the surroundings into the container to counterbalance the loss of energy in the escaping molecules and thus prevent any cooling effect. The Thermos bottle is an insulated container that minimizes heat transfer, making it useful for maintaining liquids at a cool temperature for a short period of time. Figure 11.10 contrasts the evaporation process occuring in an uninsulated container with that which occurs within an insulated one. The temperature of liquid and surroundings is the same in an uninsulated container; in an insulated one, liquid temperature drops below that of the surroundings because of evaporative cooling.

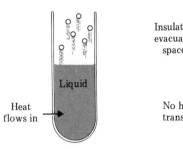
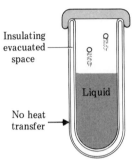

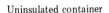

FIGURE 11.10 In order for an evaporative cooling effect to be measured, a container that minimizes heat flow into the container from its surroundings must be used.

Collectively, the molecules that escape from an evaporating liquid are often referred to as vapor rather than gas. The term **vapor** describes the gaseous state of a substance at a temperature and pressure at which the substance is normally a liquid or solid. For example, at room temperature and atmospheric pressure the normal state for water is the liquid state. Molecules that escape (evaporate) from liquid water at these conditions are frequently called water vapor.

The evaporative behavior of a liquid in a *closed* container is quite different from that in an *open* container. In a closed container we observe that some liquid evaporation occurs, as indicated by a drop in liquid level. However, unlike the open container system, the liquid level, with time, ceases to drop (becomes constant), an indication that not all of the liquid will evaporate.

Kinetic molecular theory explains these observations in the following way. The molecules that do evaporate are unable to move completely away from the liquid as they did in the open container. They find themselves confined in a fixed space immediately above the liquid (see Fig. 11.11a). These "trapped" vapor molecules undergo many random collisions with the container walls, other vapor molecules and the liquid surface. Molecules colliding with the liquid surface may be recaptured by the liquid. Thus, two processes—evaporation (escape) and condensation (recapture)—take place in the closed container.

In a closed container, for a short time, the rate of evaporation exceeds the rate of condensation and the liquid level drops. However, as more and more of the liquid evaporates, the number of vapor molecules increases and the chance of their recap-

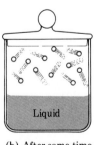
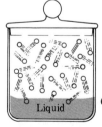

FIGURE 11.11 Evaporation of a liquid in a closed container.

ture through striking the liquid surface also increases. Eventually the rate of condensation becomes equal to the rate of evaporation and the liquid level stops dropping (see Fig. 11.11c). At this point the number of molecules that escape in a given time is the same as the number recaptured; a steady-state condition has been reached. The amounts of liquid and vapor in the container are not changing, even though both evaporation and condensation are still occurring.

This steady-state situation, which will continue as long as the temperature of the system remains constant, is an example of a state of equilibrium. A **state of equilibrium** is a situation in which two opposite processes take place at equal rates. For systems in a state of equilibrium, no net macroscopic changes can be detected. However, the system is dynamic; both forward and reverse processes are still occurring but in a manner such that they balance each other.

11.13 Vapor Pressure of Liquids

For a liquid–vapor equilibrium in a closed container, the vapor in the fixed space immediately above the liquid exerts a constant pressure on the liquid surface and the walls of the container. This pressure is called the liquid's vapor pressure. **Vapor pressure** is the pressure exerted by a vapor above a liquid when the liquid and vapor are in equilibrium.

The magnitude of a vapor pressure depends on the nature and temperature of the liquid. Liquids with strong attractive forces between molecules will have lower vapor pressures than liquids in which only weak attractive forces exist between particles. Substances with high vapor pressures evaporate readily; that is, they are volatile. A **volatile substance** is a substance that readily evaporates at room temperature because of a high vapor pressure.

The vapor pressures of all liquids increase with temperature. Why? An increase in temperature results in more molecules having the minimum energy required for evaporation. Hence, at equilibrium the pressure of the vapor is greater. Table 11.4 shows the variation in vapor pressure with increasing temperature for water.

TABLE 11.4 Vapor Pressure of Water at Various Temperatures

Temperature (°C)	Vapor Pressure (mm Hg)[a]	Temperature (°C)	Vapor Pressure (mm Hg)[a]
0	4.6	60	149.4
10	9.2	70	233.7
20	17.5	80	355.1
30	31.8	90	525.8
40	55.3	100	760.0
50	92.5		

[a] The units used to specify vapor pressure in this table will be discussed in detail in Sec. 12.2.

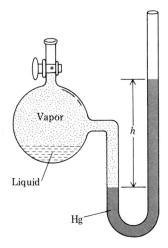

Figure 11.12 Apparatus for determining the vapor pressure of a liquid.

When the temperature of a liquid–vapor system in equilibrium is changed, the equilibrium is upset. The system immediately begins the process of establishing a new equilibrium. Let us consider a case where the temperature is increased. The higher temperature signifies that energy has been added to the system. More molecules will have the minimum energy needed to escape. Thus, immediately after the temperature is increased, molecules begin to escape at a rate greater than that at which they are recaptured. With time, however, the rates of escape and recapture will again become equal. The new rates will, however, be different from those for the previous equilibrium. Since energy has been added to the system, the rates will be higher, resulting in a higher vapor pressure.

The size (volume) of the space that the vapor occupies does not affect the magnitude of the vapor pressure. A larger fixed space will enable more molecules to be present in the vapor at equilibrium. However, the larger number of molecules spread over a larger volume results in the same pressure as a small number of molecules in a small volume.

Vapor pressures for liquids are commonly measured in an apparatus similar to that pictured in Figure 11.12. At the moment liquid is added to the vessel, the space above the liquid is filled only with air and the levels of mercury in the U-tube are equal. With time, liquid evaporates and equilibrium is established. The vapor pressure of the liquid is proportional to the difference between the heights of the mercury columns. The larger the vapor pressure, the greater the extent to which the mercury column is "pushed up."

11.14 Boiling and Boiling Point

Usually, for a molecule to escape from the liquid state, it must be on the surface of the liquid. **Boiling** is a special form of evaporation in which conversion from the liquid to the vapor state occurs within the body of a liquid through bubble formation. This phenomenon begins to occur when the vapor pressure of a liquid, which is steadily increasing as a liquid is heated, reaches a value equal to that of the prevailing external pressure on the liquid; for liquids in open containers this value is atmospheric pressure. When these two pressures become equal, bubbles of vapor form around any speck of dust or any rough surface of the container. Being less dense than the liquid itself, these vapor bubbles quickly rise to the surface and escape. The quick ascent of the bubbles causes the agitation associated with a boiling liquid.

Let us consider this "bubble phenomenon" in more detail. As the heating of a liquid begins, the first small bubbles that form on the bottom of the container are bubbles of dissolved air (oxygen and nitrogen), which have been driven out of solution by the rising liquid temperature. (The solubilities of oxygen and nitrogen in liquids decrease with increasing temperature.)

As the liquid is heated further, larger bubbles form and begin to rise. These bubbles, usually also formed on the bottom of the container (where the liquid is hottest), are liquid vapor bubbles rather than air bubbles. Initially, these vapor bubbles "disappear" as they rise, never reaching the liquid surface. Their disappearance is related to vapor pressure. In the hotter lower portions of the liquid, the liquid's vapor pres-

Section 11.14 · Boiling and Boiling Point

FIGURE 11.13 Bubble formation associated with a liquid that is boiling.

sure is high enough to sustain bubble formation. [For a bubble to exist, the pressure within it (vapor pressure) must equal external pressure (atmospheric pressure).] In the cooler, higher portions of the liquid, where the vapor pressure is lower, the bubbles are collapsed by external pressure. Finally, with further heating, the temperature throughout the liquid becomes high enough to sustain bubble formation. Then bubbles rise all the way to the surface and escape (see Fig. 11.13). At this point we say the liquid is boiling.

Like evaporation, boiling is actually a cooling process. When heat is taken away from a boiling liquid, boiling ceases almost immediately. It is the highest energy molecules that are escaping. Quickly the temperature of the remaining molecules drops below the boiling point of the liquid.

The **boiling point** of a liquid is the temperature at which the vapor pressure of the liquid becomes equal to the external (atmospheric) pressure exerted on the liquid. Since atmospheric pressure fluctuates from day to day, so does the boiling point of a liquid. To compare the boiling points of different liquids the external pressure must be the same. The boiling point of a liquid most often used for comparison and tabulation purposes (reference books, for example) is the normal boiling point. A liquid's **normal boiling point** is the temperature at which a liquid boils under a pressure of 760 mm Hg.

At any given location the changes in the boiling point of liquids due to *natural variation* in atmospheric pressure seldom exceed a few degrees; in the case of water the maximum is about 2 °C. However, variations in boiling points *between* locations at different elevations can be quite striking, as shown by the data in Table 11.5.

The boiling point of a liquid can be increased by increasing the external pressure. Use is made of this principle in the operation of a pressure cooker. Foods cook faster in pressure cookers because the elevated pressure causes water to boil above 100 °C. An increase in temperature of only 10 °C will cause food to cook in approximately half the normal time. (Cooking involves chemical reactions, and the rate of a chemical reaction generally doubles with every 10 °C increase in temperature.) Table 11.6 gives the boiling temperatures reached by water in normal household pressure cookers. Hospitals use the same principle in sterilizing instruments and laundry in autoclaves; sufficiently high temperatures are reached to destroy bacteria.

Liquids that have high normal boiling points or that undergo undesirable chemical reactions at boiling temperatures can be made to boil at low temperatures by re-

TABLE 11.5 Variation of the Boiling Point of Water with Elevation

Location	Elevation (ft above sea level)	Boiling Point of Water (°C)
San Francisco, CA	0	100.0
Salt Lake City, UT	4,390	95.6
Denver, CO	5,280	95.0
La Paz, Bolivia	12,795	91.4
Mount Everest	28,028	76.5

TABLE 11.6 Boiling Point of Water in a Pressure Cooker

Pressure Above Atmospheric		Boiling Point of Water (°C)
lb/in.^{2}	mm Hg	
5	259	108
10	517	116
15	776	121

ducing the external pressure. This principle is used in the preparation of numerous food products including frozen fruit juice concentrates. At a reduced pressure some of the water in a fruit juice is boiled away, concentrating the juice without having to heat it to a high temperature. Heating to a high temperature would cause changes that would spoil the taste of the juice and/or reduce its nutritional value.

11.15 Intermolecular Forces in Liquids

In order for a liquid in an open container to boil, its vapor pressure must reach atmospheric pressure. For some substances this occurs at temperatures well below zero; for example, oxygen has a boiling point of $-183\,°C$. Other substances do not boil until the temperature is much higher. Mercury, for example, has a boiling point of $357\,°C$, which is $540\,°C$ higher than that of oxygen. An explanation for this variation involves a consideration of the nature of the intermolecular forces that must be overcome in order for molecules to escape from the liquid state into the vapor state. **Intermolecular forces** are forces that act between a molecule and another molecule.

Intermolecular forces are similar in one way to the previously discussed *intra*molecular forces (*within* molecules) involved in covalent bonding (Sec. 7.9). They are electrostatic in origin. A major difference between inter- and intramolecular forces is their magnitude; the former are much weaker. However, intermolecular forces, despite their relative weakness, are sufficiently strong to influence the behavior of liquids, often in a very dramatic way. There are three principal types of intermolecular forces: dipole–dipole interactions, hydrogen bonds, and London forces.

Dipole–dipole interactions are electrostatic attractions between polar molecules. Polar molecules (which are often called dipoles), it should be recalled, are electrically nonsymmetrical (Sec. 7.10.). Therefore, when polar molecules approach each other, they tend to line up so that the relatively positive end of one molecule is directed toward the relatively negative end of the other molecule. As a result, there is an electrostatic attraction between the molecules. The greater the polarity of the molecules, the greater the strength of the dipole–dipole interaction. Figure 11.14 shows the many dipole–dipole interactions possible for a random arrangement of polar ClF molecules.

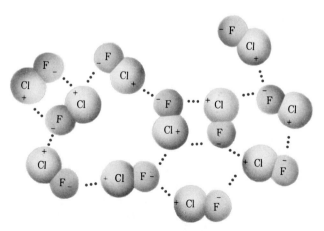

FIGURE 11.14 Dipole–dipole interactions between randomly arranged ClF molecules.

A **hydrogen bond** is a special type of dipole–dipole interaction that occurs between polar molecules when one molecule contains hydrogen bonded to a very electronegative element. When hydrogen is bonded to a very electronegative element (F, O, and N are the main elements involved), the electronegativity difference is sufficient essentially to strip the H atom of its electron, leaving a nearly exposed nucleus. (Remember, hydrogen has only one electron.) The small size of the hydrogen nucleus allows it to approach the F, O, or N atoms of other molecules very closely, resulting in a much stronger than normal dipole–dipole interaction. It is significant that hydrogen bonding appears to be limited to compounds containing the three elements F, O, and N, all of which are very small in addition to being very electronegative. The larger Cl and S atoms, even though Cl has an electronegativity similar to that of N, show little tendency to hydrogen bond. The strongest hydrogen bond is about one-tenth as strong as a covalent bond and is the strongest of all intermolecular forces. Figure 11.15 shows hydrogen bonding in liquid water.

It is not an overstatement to say that hydrogen bonding makes life possible. Were it not for hydrogen bonding, water, the most abundant compound in the human body, would be a gas at room temperature. Life as we know it could not exist under such a condition. Section 11.16 will consider in detail the effects that hydrogen bonding has on the properties of water. Many other molecules of biological importance, such as DNA and proteins, contain O—H and N—H bonds, and hydrogen bonding plays a role in the behavior of these substances. Certain bonds in such compounds must be capable of breaking and reforming with relative ease. Only hydrogen bonds have just the right energies to permit this.

The third type of intermolecular force, and the weakest, is the London force, named after the German physicist Fritz London (1900–1954), who first postulated its existence. **London forces** are instantaneous dipole–dipole interactions that exist between all atoms and molecules, nonpolar as well as polar. The origin of London forces is more difficult to visualize than that of dipole–dipole interactions.

London forces result from momentary (temporary) uneven electron distributions in molecules. Most of the time the electrons can be visualized as being distributed

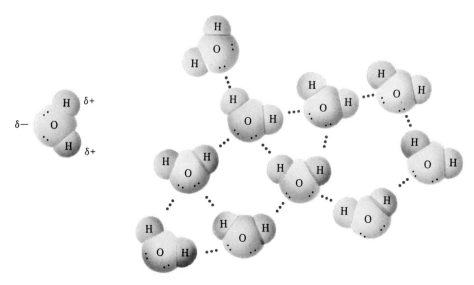

FIGURE 11.15 Hydrogen bonds, indicated by dotted lines, between water molecules in liquid water.

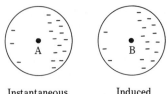

FIGURE 11.16 A London force. The instantaneous (temporary) polarity present in atom A results in induced polarity in atom B.

in a molecule in a definite pattern determined by their energies and the electronegativities of the atoms. However, there is a small statistical chance (probability) that the electrons will deviate from their normal pattern. For example, in the case of a nonpolar diatomic molecule, more electron density may temporarily be located on one side of the molecule than on the other. This condition causes the molecule to become polar for an instant. The negative side of this instantaneously polar molecule will tend to repel electrons of adjoining molecules and cause these molecules to also become polar (an *induced polarity*). The original (statistical) polar molecule and all of the molecules with induced polarity are then attracted to each other. This happens many, many times per second throughout the liquid, resulting in a net attractive force. Figure 11.16 depicts the situation present when London forces exist.

The strength of London forces depends on the ease with which an electron distribution in a molecule can be distorted (polarized) by the polarity present in another molecule. In large molecules the outermost electrons are necessarily located farther from the nucleus than are the outermost electrons in small molecules. The farther electrons are located from the nucleus, the more freedom they have and the more susceptible they are to polarization. This leads to the observation that for *related* molecules boiling points increase with molecular mass, which usually parallels size. This trend is reflected in the boiling points given in Table 11.7 for two series of related substances: the noble gases and the elements of group VIIA.

If two molecules have approximately the same molecular mass and polarity but are of different size, the smaller one will have the lower boiling point, since its electrons are less susceptible to polarization. The nonpolar SF_6 and $C_{10}H_{22}$ molecules (molecular masses 146 and 142 amu, respectively) reflect this trend with boiling points of -64 and $174\,°C$, respectively.

Of the three types of intermolecular forces (dipole–dipole interactions, hydrogen bonds, and London forces), the London forces are the most common and, in the

TABLE 11.7 Boiling Point Trends for Related Series of Nonpolar Molecules

Substance	Molecular Mass (amu)	Boiling Point (°C)
Noble Gases		
He	4.0	−269
Ne	20.2	−246
Ar	39.9	−186
Kr	83.8	−153
Xe	131.3	−107
Rn	222	−62
Group VIIA Elements		
F_2	39.0	−187
Cl_2	70.9	−35
Br_2	159.8	+59
I_2	253.8	+184

majority of cases, the most important. They are the only attractive forces present between *nonpolar* molecules, and it is estimated that London forces contribute 85% of the total intermolecular force in the *polar* HCl molecule. Only where hydrogen bonding is involved do London forces play a minor role. In water, for example, about 80% of the intermolecular attraction is attributed to hydrogen bonding and only 20% to London forces.

In this section it has been assumed that the particles making up the liquids are molecules or atoms. This is a valid assumption for liquids at normal temperatures. Only at extremely high temperatures are liquids encountered where this is not the case. Liquids obtained by melting ionic compounds, which always requires an extremely high temperature, are ionic in the liquid state. The attractive forces in such liquids are those that result from the attraction of positive and negative ions.

11.16 Water—A Most Unusual (Unique) Substance

Water is the most abundant and most essential compound known to human beings. This substance covers approximately 75% of the Earth's surface and constitutes 60–70% of the mass of a human body.

The structure of a water molecule is very simple. Two hydrogen atoms are attached to a central oxygen atom to give a V-shaped (angular) molecule, which is polar. Despite this molecular simplicity, the behavior and properties of water are both complex and unusual. The unusual behavior of water makes it the key substance that it is. Most of water's unusual behavior is a consequence of the extensive hydrogen bonding (Sec. 11.15) that occurs between water molecules, in both the liquid and solid states.

We will consider here four unusual hydrogen-bonding-caused properties of water.

1. A lower than expected vapor pressure.
2. Higher than expected thermal properties.
3. A very rare temperature dependence for density.
4. A higher than expected surface tension.

Vapor Pressure

Liquids in which significant hydrogen bonding occurs have vapor pressures that are significantly lower than those of similar liquids where little or no hydrogen bonding occurs. The presence of hydrogen bonds makes it more difficult for molecules to escape from the condensed state; additional energy is needed to overcome the hydrogen bonds. The greater the hydrogen bond strength, the lower the vapor pressure at any given temperature.

Since the boiling point of a liquid depends upon vapor pressure (Sec. 11.14), liquids with low vapor pressures will have to be heated to higher temperatures to bring their vapor pressures up to the point where boiling occurs (atmospheric pressure). Hence, boiling points are much higher for liquids in which hydrogen bonding occurs.

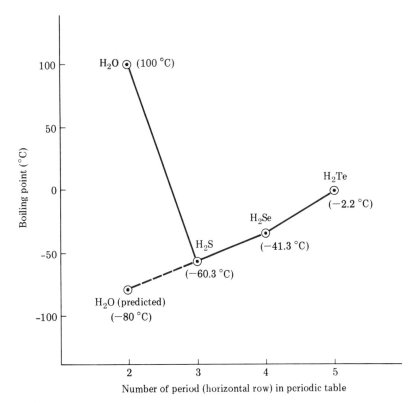

FIGURE 11.17 Boiling points of the hydrogen compounds of group VIA elements. Water is the only one of the four compounds in which significant hydrogen bonding occurs.

The effect that hydrogen bonding has on water's boiling point can be seen by comparing it with the boiling points of other hydrogen compounds of group VIA elements—H_2S, H_2Se, and H_2Te. In this series of compounds—H_2O, H_2S, H_2Se, and H_2Te—water is the only one in which significant hydrogen bonding occurs (Sec. 11.15). Normally, the boiling points of a series of compounds containing elements in the same periodic table group increase with increasing molar mass. Thus, in the hydrogen–group VIA element series, we would expect that H_2Te, the heaviest member of the series, would have the highest boiling point and that water, the compound of lowest molar mass, would have the lowest boiling point. Contrary to expectation, H_2O has the highest boiling point, as can be seen from the boiling-point data shown in Figure 11.17. The data in Figure 11.17 indicate that water "should have" a boiling point of approximately $-80\,°C$, a value obtained by extrapolation of the line connecting the three heavier compounds. The actual boiling point of water, $100\,°C$, is nearly $200\,°C$ higher than predicted. Indeed, in the absence of hydrogen bonding, water would be a gas at room temperature, as mentioned in Section 11.15.

A higher than expected freezing point is also characteristic of water, and a plot of freezing points for similar compounds would have the same general shape as that in Figure 11.17.

Thermal Properties

The presence of extensive hydrogen bonding between water molecules increases significantly the ability of water to absorb heat energy when evaporating (an endothermic process) and to release heat energy when freezing (an exothermic process). Water's thermal properties, together with its abundance, largely account for its widespread use as a coolant.

Large bodies of water exert a temperature-moderating effect on their surroundings, primarily because of water's ability to absorb and release large amounts of heat energy. In the heat of a summer day, extensive water evaporation occurs, and in the process energy is absorbed from the surroundings. The net effect of the evaporation is a lowering of the temperature of the surroundings. In the cool of evening, some of this water vapor condenses back to the liquid state, releasing heat that raises the temperature of the surroundings. In this manner the temperature variation between night and day is reduced. In the winter a similar process occurs; water freezes on cold days and releases heat energy to the surroundings. The hottest and coldest regions on Earth are all inland regions, those distant from the moderating effects of large bodies of water.

The large release of heat from freezing water is the basis for the practice of spraying orange trees with water when freezing temperatures are expected. The freezing water liberates enough heat energy to keep the temperature of the air higher than the freezing point of the fruits.

Water's thermal properties are a factor in the cooling of the human body. The coolant in this case is perspiration, which evaporates, absorbs heat in the process, and lowers skin temperature. For similar reasons, a swimmer upon emerging from the water on a warm but windy day will feel cold and will shiver as excess water rapidly evaporates from his or her body. Additionally, water's ability to absorb large amounts of heat helps keep water loss to a minimum, thus making it easier for humans, animals, and plants to exist in environments where water is scarce.

Density

A very striking and unusual behavior pattern occurs in the variation of the density of water with temperature. For most liquids, density increases with decreasing temperature, reaching a maximum for the liquid at its freezing point. The density pattern for water is different. Maximum density is reached not at its freezing point, but at a temperature a few degrees higher than the freezing point. As shown in Figure 11.18, the maximum density for liquid water occurs at 4 °C. This abnormality, that water at its freezing point is less dense than water at slightly higher temperatures, has tremendous ecological significance. Furthermore, at 0 °C, solid water (ice), is significantly less dense than liquid water—0.9170 g/mL versus 0.9998 g/mL. All of this "strange" density behavior of water is directly related to hydrogen bonding.

Of primary importance is the fact that hydrogen bonding between water molecules is directional. Hydrogen bonds can form only at certain angles between molecules because of the angular geometry of the water molecule. The net result is that water molecules that are hydrogen bonded are farther apart than those that are not.

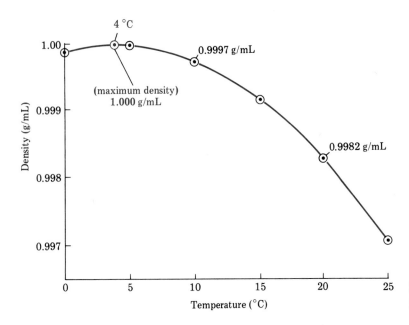

FIGURE 11.18 A plot of density versus temperature for liquid water.

On a molecular scale, let us now consider what happens to water molecules as the temperature of water is lowered. At high temperatures, such as 80 °C, the kinetic energy of the water molecules is sufficiently great to prevent hydrogen bonding from having much of an orientation effect on the molecules. Hydrogen bonds are rapidly and continually being formed and broken. As the temperature is lowered, the accompanying decrease in kinetic energy decreases molecular motion and the molecules move closer together. This results in an increase in density. The kinetic energy is still sufficient to negate most of the orientation effects from hydrogen bonding. When the temperature is lowered still further, the kinetic energy finally becomes insufficient to prevent hydrogen bonding from orienting molecules into definite patterns that require "open spaces" between molecules. At 4 °C, the temperature at which water has its maximum density, the trade-off between random motion from kinetic energy and orientation from hydrogen bonding is such that the molecules are as close together as they will ever be. Cooling below 4 °C causes a decrease in density as hydrogen bonding causes more and more "open spaces" to be present in the liquid. Density decrease continues down to the freezing point, at which temperature the hydrogen bonding causes molecular orientation to the maximum degree, and the solid crystal lattice of ice is formed. This solid crystal lattice of ice (Fig. 11.19) is an open structure, with the result that solid ice has a lower density than liquid water. Water is one of only a few substances known whose solid phase is less dense than the liquid phase at the freezing point.

Two important consequences of ice being less dense than water are that ice floats in liquid water and liquid water expands upon freezing.

When ice floats in liquid water, approximately 8% of its volume is above water. In the case of icebergs found in far northern locations, 92% of the volume of an iceberg is located below the liquid surface. The common expression "it is only the tip

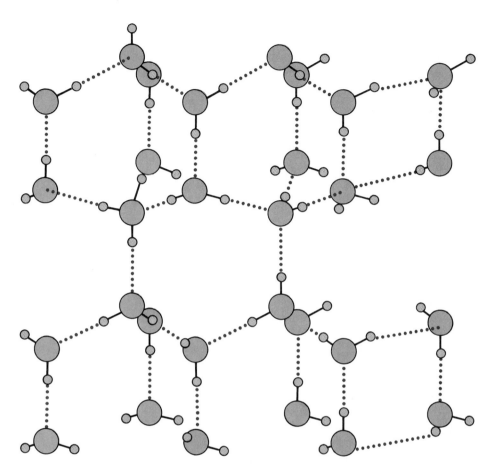

FIGURE 11.19 The crystalline structure of ice. Normal covalent bonds between oxygen and hydrogen, which hold water molecules together, are shown by solid short lines. The weaker hydrogen bonds are shown by dotted lines.

of the iceberg" is literally true; what is seen is only a very small part of what is actually there.

If a container is filled with water and sealed, the force generated from the expansion of the water upon freezing will break the container. Antifreeze is added to car radiators in the winter to prevent the water present from freezing and cracking the engine block; sufficient force is generated from water expansion upon freezing to burst even iron or copper parts. During the winter season, also, the weathering of rocks and concrete and the formation of "potholes" in the streets are hastened by the expansion of freezing water in cracks.

Water's density pattern also explains why lakes freeze from top to bottom and not vice versa, and why aquatic life can continue to exist for extended periods of time in bodies of water that are "frozen over." In the fall of the year, surface water is cooled through contact with cold air. This water becomes denser than the warmer water

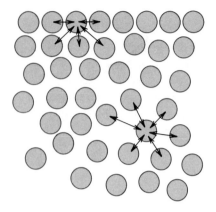

FIGURE 11.20 Molecules on the surface of a liquid experience an imbalance in intermolecular forces. Molecules within the liquid are uniformly attracted in all directions.

underneath and sinks. In this way, cool water is circulated from the top of a lake to the bottom until the entire lake has reached the temperature of water's maximum density, 4 °C. During the circulation process, oxygen and nutrients are distributed throughout the water. Upon further cooling, below 4 °C, a new behavior pattern emerges. Surface water, upon cooling, no longer sinks, since it is less dense than the water underneath. Eventually a thin layer of surface water is cooled to the freezing point and changed to ice, which floats because of its still lower density. Even in the coldest winters, lakes will usually not freeze to a depth of more than a few feet because the ice forms an insulating layer over the water. Thus, aquatic life can live throughout the winter, under the ice, in water that is "thermally insulated" and contains nutrients. If water behaved as "normal" substances do, freezing would occur from the bottom up, and most, if not all, aquatic life would be destroyed.

For a few weeks in the early spring, circulation again occurs in a body of water during the melting and warming of the surface water. This process stops when the surface water is warmed above 4 °C. Above this temperature surface water is less dense and forms a layer over the colder water of higher density. This causes a thermal stratification that persists through the hot summer months. In the fall, circulation again begins, and the cycle is repeated.

Surface Tension

Surface tension is a property of liquids that is directly related to intermolecular forces. Molecules *within* a liquid, because they are completely surrounded by other molecules, experience equal intermolecular attractions in all directions. Molecules *on the surface* of a liquid, because there are no molecules above them, experience an imbalance in intermolecular attractions that pulls them toward the interior of the liquid. Figure 11.20 illustrates this imbalance of force. **Surface tension** is a measure of the inward force on the surface of a liquid caused by unbalanced intermolecular forces.

FIGURE 11.21 Surface tension enables one to fill a glass with water above the rim. [Photo by John Schultz, PAR/NYC]

The strong attraction of water molecules for each other as a result of hydrogen bonding gives water a very high surface tension. This high surface tension causes water surfaces to seem to have a "membrane" or "skin" covering them. A steel needle or a razor blade will float on the surface of water because of surface tension even though each has a density greater than that of water. Water bugs are able to skitter across water because of surface tension.

Because of surface tension it is possible to fill a container with water slightly above the rim (see Fig. 11.21). Surface tension prevents the water from running over. It is also surface tension that causes water to form droplets in many situations. Almost everyone has observed water droplets that have formed on a car windshield or on greasy or waxed surfaces. Surface tension causes water to resist increasing its surface area. For a given volume of any liquid, the geometrical shape having the minimum surface area is a sphere; hence, water tends to form spherical droplets.

The property of surface tension helps water to rise in the narrow vessels in plant stems and roots. High surface tension also helps to hold water in the small spaces between soil particles.

The surface tension of a liquid can be "broken" by dissolved substances. Soaps and detergents are used for this purpose. Water bugs will sink in soapy water. Soapy water spreads out over a glass surface instead of forming droplets as pure water would.

11.17 Types of Solids

Solids may be classified into two categories: crystalline solids and amorphous solids. Most solids are crystalline. **Crystalline solids** are solids characterized by a regular three-dimensional arrangement of the atoms, ions, or molecules present. This regular arrangement of particles is manifested in the outward appearance of the solid; crystalline features are discernible. These features are often very obvious; at other times, however, the crystals may be so tiny as to be visible only with the aid of a microscope. Many times the crystals are imperfect, making it difficult to discern them, but they are there.

Amorphous solids are solids characterized by a random, nonrepetitive three-dimensional arrangement of the atoms, ions, or molecules present. Only a few such solids exist. The word *amorphous* comes from the Greek *amorphos* meaning "without form." Examples of amorphous solids include glass, tar, rubber, and many plastics. In some textbooks amorphous solids are referred to as *supercooled liquids*, since they have the molecular disorder associated with liquids. However, they differ from liquids and resemble solids in being rigid and immobile.

Most amorphous solids consist of very long chainlike molecules that are randomly intertwined. To form a crystalline material from such a solid, the intertwined molecules would have to become regularly arranged in the melted state. When such a liquid is cooled to a solid, molecular motion decreases to the point where "solidification" takes place before the untwining can occur, resulting in a disordered structure.

The remainder of this section deals with the more commonly encountered crystalline solid. Indeed, in this text the word *solid* without a modifier will always mean a *crystalline solid.*

The highly ordered pattern of particles found in a crystalline solid is called a *crystal lattice.* The positions occupied by the particles in the lattice are called *crystal lattice sites.* Crystalline solids may be classified into five groups based on the type of particles

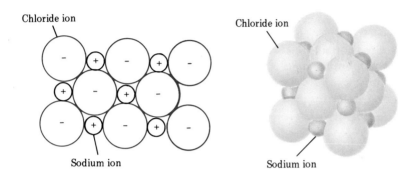

Figure 11.22 Two-dimensional cross section and three-dimensional view of the ionic compound NaCl.

at the crystal lattice sites and the forces that hold the particles together. A consideration of these classifications provides some insight into the reasons some solids have very low and others very high melting points and some solids are more volatile than others. The five classes of crystalline solids are ionic, polar molecular, nonpolar molecular, macromolecular, and metallic.

Ionic solids consist of positive and negative ions arranged in such a way that each ion is surrounded by nearest neighbors of the opposite charge (as was previously discussed in Sec. 7.7). Any given ion is bonded by electrostatic attractions to all the ions of opposite charge immediately surrounding it. Since these interionic attractions are relatively strong, ionic solids have high melting points and negligible vapor pressures. Figure 11.22 shows a two-dimensional cross section and also a three-dimensional view of the arrangement of ions for the compound NaCl.

Polar molecular solids have polar molecules at the crystal lattice sites. The molecules are oriented so that the positive end of each molecule is near the negative end of an adjacent molecule in the lattice. Dipole–dipole interactions and London forces (Sec. 11.15) hold the molecules in the lattice positions. Since these forces are relatively weak compared to other types of electrostatic forces (ionic and covalent bonds), the resulting solids have moderate melting points and moderate vapor pressures.

Nonpolar molecular solids have nonpolar molecules (or atoms in the case of the noble gases) at the crystal lattice sites. They are held together solely by London forces, the weakest of electrostatic forces. Consequently, these solids have low melting points and are volatile.

Macromolecular solids have atoms that are bonded to their nearest neighbors by covalent bonds at the crystal lattice sites. A crystal of such a solid can be visualized as a gigantic molecule. Such a molecule, for the form of carbon called diamond, is shown in Figure 11.23. Each lattice site in diamond is occupied by a carbon atom. Macromolecular solids are not common, but they are noteworthy because of their extremely high melting points and their low volatility. Covalent bonds must be broken to cause physical change in such solids. The atoms at the lattice sites in a macromolecular solid need not all be the same kind, as was the case for diamond. Examples of macromolecular solids where different kinds of atoms are found at the lattice sites include quartz (SiO_2, the main component of sand) and silicon carbide (SiC, an abrasive used in sandpaper and other similar materials).

Metallic solids have metal atoms occupying the crystal lattice sites. The nature of bonding in such solids is not completely understood. However, the solids are often considered to have a lattice of metal ions, with the outer electrons of each metal

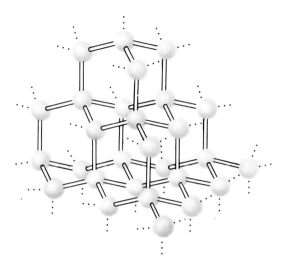

FIGURE 11.23 The diamond crystal lattice—a macromolecular solid.

atom free to move about in the lattice. The bonding results from the interaction of the electrons with the various nuclei. The movement of the free electrons also accounts for the electrical conductivity of metals. The melting points of metallic solids show a wide range. Their volatility is generally low.

Table 11.8 gives a summary of the various types of crystalline solids and their distinguishing characteristics along with examples of each type.

TABLE 11.8 Characteristics and Examples of the Various Types of Crystalline Solids

Type	Particles Occupying Lattice Sites	Forces Between Particles	Properties of Solids	Examples
Ionic	positive and negative ions	electrostatic attraction between oppositely charged ions	high melting points; nonvolatile	NaCl, MgSO$_4$, KBr
Polar molecular	polar molecules	dipole–dipole attractions and London forces	moderate melting points; moderate volatility	H$_2$O, NH$_3$, SO$_2$
Nonpolar molecular	nonpolar molecules (or atoms)	London forces	low melting points; volatile	CO$_2$, CH$_4$, I$_2$, O$_2$, Ar
Macromolecular	atoms	covalent bonds between atoms	extremely high melting points; nonvolatile	C (diamond), SiO$_2$ (sand), SiC
Metallic	metal atoms	attraction between outer electrons and positive atomic centers	variable melting points; low volatility	Cu, Ag, Au, Fe, Al

Key Terms

The new terms or concepts defined in this chapter are

amorphous solid (Sec. 11.17) A solid characterized by a random, nonrepetitive three-dimensional arrangement of atoms or molecules.

boiling (Sec. 11.14) A special form of evaporation in which conversion from the liquid to the vapor state occurs within the body of a liquid through bubble formation.

boiling point (Sec. 11.14) The temperature at which the vapor pressure of a liquid equals the external (atmospheric) pressure exerted on the liquid.

compressibility (Sec. 11.2) A measure of the change in volume resulting from a pressure change.

crystalline solid (Sec. 11.17) A solid characterized by a regular three-dimensional arrangement of atoms, ions, or molecules.

dipole–dipole interaction (Sec. 11.15) Electrostatic attraction between polar molecules.

electrostatic interaction (Sec. 11.13) Interaction between charged particles; objects of opposite electric charge attract each other, and objects of the same electric charge repel each other.

endothermic change (Sec. 11.8) A change that requires the input (absorption) of heat energy.

evaporation (Sec. 11.12) Process in which molecules escape from a liquid phase to a gaseous phase.

exothermic change (Sec. 11.8) A change that requires heat energy to be given up (released).

heat of condensation (Sec. 11.10) Amount of heat energy evolved in the conversion of 1 g of a gas to a liquid at the liquid's boiling point.

heat of fusion (Sec. 11.10) Amount of heat energy needed for the conversion of 1 g of a solid to a liquid at its melting point.

heat of solidification (Sec. 11.10) Amount of heat energy evolved in the conversion of 1 g of a liquid to a solid at its freezing point.

heat of vaporization (Sec. 11.10) Amount of heat energy needed for the conversion of 1 g of a liquid to a gas at its boiling point.

hydrogen bond (Sec. 11.15) A special type of dipole–dipole interaction that occurs between polar molecules when one molecule contains hydrogen bonded to a very electronegative element.

intermolecular force (Sec. 11.15) A force that acts between a molecule and another molecule.

ionic solid (Sec. 11.17) A solid consisting of positive and negative ions arranged in such a way that each ion is surrounded by nearest neighbors of the opposite charge.

kinetic molecular theory (Sec. 11.3) A series of statements useful in explaining the physical characteristics of the solid, liquid, and gaseous states.

London force (Sec. 11.15) Instantaneous dipole–dipole interaction that exists between all atoms and molecules, nonpolar as well as polar.

macromolecular solid (Sec. 11.17) A solid in which atoms are bonded to their nearest neighbors by covalent bonds.

metallic solid (Sec. 11.17) A solid that has metal atoms occupying the crystal lattice sites.

nonpolar molecular solid (Sec. 11.17) A solid in which nonpolar molecules (or atoms in the case of the noble gases) are at the lattice sites.

normal boiling point (Sec. 11.14) The temperature at which a liquid boils under a pressure of 760 mm Hg.

polar molecular solid (Sec. 11.17) A solid in which polar molecules are at the crystal lattice points.

state of equilibrium (Sec. 11.12) A situation in which two opposite processes take place at equal rates.

surface tension (Sec. 11.16) A measure of the inward force on the surface of a liquid caused by unbalanced intermolecular forces.

thermal expansion (Sec. 11.2) A measure of the volume change resulting from a temperature change.

vapor (Sec. 11.12) The gaseous state of a substance at a temperature and pressure at which the substance is normally a liquid or solid.

vapor pressure (Sec. 11.13) The pressure exerted by a vapor above a liquid when the liquid and vapor are in equilibrium.

volatile substance (Sec. 11.13) A substance that readily evaporates at room temperature because of a high vapor pressure.

Practice Problems

States of Matter (Secs. 11.4–11.6)

11.1 The following statements relate to the terms *solid state*, *liquid state*, and *gaseous state*. Match the terms to the appropriate statements.
(a) This state is characterized by the lowest density of the three.
(b) This state is characterized by an indefinite shape and a definite volume.
(c) Temperature changes influence the volume of this state significantly.
(d) In this state constituent particles are more free to move about than in other states.

11.2 The following statements relate to the terms *solid state*, *liquid state*, and *gaseous state*. Match the terms to the appropriate statements.
(a) This state is characterized by an indefinite shape and high density.
(b) Pressure changes influence the volume of this state more than that of the other two.
(c) In this state constituent particles are less free to move about than in other states.
(d) This state is characterized by a definite shape and a definite volume.

11.3 List characteristics that distinguish solids from liquids.

11.4 List characteristics that distinguish liquids from gases.

Kinetic Molecular Theory (Secs. 11.3–11.6)

11.5 According to the statements of the kinetic molecular theory, how do molecules transfer energy from one to another?

11.6 According to the statements of the kinetic molecular theory, what is the relationship between temperature and the average velocity with which particles move?

11.7 What type of energy is related to cohesive forces?

11.8 What type of energy is related to disruptive forces?

11.9 What effect does temperature have on the magnitude of disruptive forces and cohesive forces?

11.10 Contrast the effects of cohesive and disruptive forces on a system of particles.

11.11 In which of the three states of matter are disruptive forces greater than cohesive forces?

11.12 In which of the three states of matter are disruptive forces and cohesive forces of about the same magnitude?

11.13 Explain each of the following observations using the kinetic molecular theory.
(a) Gases have a low density.
(b) Liquids show little change in volume with changes in temperature.
(c) Solids maintain characteristic shapes.

11.14 Explain each of the following observations using the kinetic molecular theory.
(a) Both liquids and solids are practically incompressible.
(b) A gas always exerts a pressure on the object or container with which it is in contact.
(c) An aerosol can heated in an open fire may explode.

Physical Changes of State (Sec. 11.8)

11.15 Indicate whether each of the following is an exothermic or endothermic change of state.
(a) sublimation   (b) melting   (c) freezing

11.16 Indicate whether each of the following is an exothermic or endothermic change of state.
(a) condensation   (b) deposition   (c) evaporation

Energy and Changes of State (Sec. 11.10)

11.17 How much heat energy, in joules, is required to effect each of the following changes of state?
(a) melting of 35.0 g of copper that is at its melting point
(b) melting of 10.0 g of solidified methane that is at its melting point.
(c) vaporizing of 35.0 g of liquid water that is at its boiling point
(d) vaporizing of 10.0 g of liquid carbon tetrachloride that is at its boiling point.

11.18 How much heat energy, in joules, is required to effect each of the following changes of state?
(a) melting of 50.0 g of aluminum that is at its melting point
(b) melting of 5.23 g of solidified benzene that is at its melting point
(c) vaporizing of 125 g of liquid benzene that is at its boiling point
(d) vaporizing of 0.100 g of liquid water that is at its boiling point

11.19 How much heat energy, in joules, would be evolved when each of the following changes occurs?
(a) 50.0 g of liquid water at 0 °C is changed to ice at 0 °C.
(b) 50.0 g of methane gas at −161 °C (its boiling point) is changed to a liquid.

11.20 How much heat energy, in joules, would be evolved when each of the following changes occurs?
(a) 50.0 g of steam at 100 °C is condensed to liquid water at 100 °C.
(b) 50.0 g of molten aluminum at 658 °C (its melting point) is allowed to solidify.

11.21 How many times greater is the energy input needed to melt 50.0 g of Al than is needed to melt 50.0 g of Cu? Assume both metals are at their melting points.

11.22 How many times greater is the energy input needed to vaporize 50.0 g of water than is needed to vaporize 50.0 g of ethyl alcohol? Assume both substances are at their boiling points.

11.23 How many moles of gaseous benzene at its boiling point must be condensed to the liquid state at the same temperature in order to release 125 kJ of heat energy?

11.24 How many moles of gaseous diethyl ether at its boiling point must be condensed to the liquid state at the same temperature in order to release 255 kJ of heat energy?

11.25 Experiments involving small samples of two metals, each at its melting point temperature, are carried out with the following results.
(1) 1016 J of heat energy is required to melt 3.25 g of metal A.
(2) 983 J of heat energy is required to melt 3.20 g of metal B.
Which metal has the higher heat of fusion? By how many joules per gram does this metal's heat of fusion exceed that of the other metal?

11.26 Experiments involving small samples of two liquids, each at its boiling point temperature, are carried out with the following results.
(1) 2238 J of heat energy is required to vaporize 1.79 g of liquid A.
(2) 1933 J of heat energy is required to vaporize 1.66 g of liquid B.
Which liquid has the higher heat of vaporization? By how many joules per gram does this liquid's heat of vaporization exceed that of the other liquid?

11.27 The heat of fusion of Na at its melting point is 2.40 kJ/mole. How much heat, in joules, must be absorbed by 7.00 g of solid Na at its melting point to convert it to molten Na?

11.28 The heat of vaporization of Hg at its boiling point is 58.6 kJ/mole. How much heat, in joules, must be absorbed by 13.0 g of liquid Hg at its boiling point to convert it to gaseous Hg?

Heat Energy Calculations (Sec. 11.11)

11.29 The melting point of iron is 1530° C, its solid state specific heat is 7.78 J/(g · °C), and its heat of fusion is 63.8 J/g. How much heat energy, in joules, is required to melt 35.2 g of iron that is at each of the following initial temperatures?
(a) 20.0 °C below its melting point
(b) 855 °C below its melting point
(c) its melting point
(d) 1235 °C

11.30 The melting point of calcium is 851 °C, its solid state specific heat is 11.4 J/(g · °C), and its heat of fusion is 233 J/g. How much heat energy, in joules, is required to melt 52.0 g of calcium that is at each of the following initial temperatures?
(a) 45.0 °C below its melting point
(b) 685 °C below its melting point
(c) its melting point
(d) 35.0 °C

11.31 Calculate the heat required to convert 15.0 g of ethyl alcohol, C_2H_6O, from a solid at -135 °C into the gaseous state at 95 °C. The normal melting and boiling points of this substance are -117 and 78 °C, respectively. The heat of fusion is 109 J/g, and the heat of vaporization is 837 J/g. The specific heats of the solid, liquid, and gaseous states are 0.97, 2.3, and 0.95 J/(g · °C).

11.32 Calculate the heat required to convert 25.0 g of propyl alcohol, C_3H_8O, from a solid at -150 °C into the gaseous state at 115 °C. The normal melting and boiling points of this substance are -127 and 97 °C, respectively. The heat of fusion is 86.2 J/g, and the heat of vaporization is 694 J/g. The specific heats of the solid, liquid, and gaseous states are 2.36, 2.83, and 1.76 J/g · °C).

11.33 Calculate the amount of heat released, in joules, in converting 50.0 g of steam at 125.0 °C to each of the following.
(a) steam at 100.0 °C   (b) water at 90.0 °C
(c) ice at 0.0 °C   (d) ice at -20.0 °C

11.34 Calculate the amount of heat released, in joules, in converting 75.0 g of steam at 115.0 °C to each of the following.
(a) steam at 110.0 °C   (b) water at 15.0 °C
(c) water at 0.0 °C   (d) ice at -5.0 °C

Evaporation of Liquids (Sec. 11.12)

11.35 Water evaporates faster from a glass of hot water than from a glass of cold water. Why is this so?

11.36 Water evaporates faster from a shallow pan than from a glass even when the temperature is the same in the two containers. Why is this so?

11.37 For a given liquid, what two factors determine the rate of evaporation?

11.38 Liquids A and B are placed in identical containers on a tabletop. Liquid B evaporates at a faster rate than liquid A even though both liquids are at the same temperature. Why is this so?

11.39 The cooling effect produced by evaporation can be measured if the liquid is in an insulated container, but the effect cannot be measured if the liquid is in an uninsulated container. Why is this so?

11.40 What is the difference between a gas and a vapor?

Vapor Pressure of Liquids (Sec. 11.13)

11.41 What two factors affect the magnitude of the vapor pressure a liquid exerts?

11.42 Not all liquids have the same vapor pressure at a given temperature. Why is this so?

11.43 Changing the volume of a container in which there is a liquid–vapor equilibrium will not change the magnitude of the vapor pressure. Why is this so?

11.44 Increasing the temperature of a liquid–vapor equilibrium system causes an increase in the magnitude of the vapor pressure. Why is this so?

11.45 Given that the vapor pressures, at 25 °C, of CS_2 and CCl_4 are, respectively, 309 and 107 mm Hg, which substance is the most volatile? Explain your answer.

11.46 Given that the vapor pressures, at 25 °C, of CS_2 and CCl_4 are, respectively, 309 and 107 mm Hg, which substance has the stronger intermolecular forces? Explain your answer.

Boiling and Boiling Point (Sec. 11.14)

11.47 A given liquid has many boiling points but only one *normal* boiling point. Why is this so?

11.48 The boiling point of a liquid varies with atmospheric pressure. Why is this so?

11.49 It takes more time to cook an egg in water on a mountaintop than at sea level. Why is this so?

11.50 Food will cook just as fast in boiling water with the stove set at "low heat" as in boiling water at "high heat." Why is this so?

11.51 What happens to the boiling point of a liquid in an open container when the temperature of the surroundings increase without a change in atmospheric pressure?

11.52 What happens to the boiling point of a liquid in an open container when the volume of the liquid is doubled?

11.53 What is contained in the bubbles that are observed in boiling liquids? Explain your answer.

11.54 At 30 °C the vapor pressure of substance A is 653 mm Hg, and that of substance B is 227 mm Hg. Which substance would you predict to have the lower boiling point? Explain your answer.

Intermolecular Forces in Liquids (Sec. 11.15)

11.55 What is the difference in meaning between the terms *intermolecular force* and *intramolecular force*?

11.56 What are the three major types of intermolecular forces?

11.57 Describe the molecular conditions necessary for the existence of a dipole–dipole interaction.

11.58 Describe the molecular conditions necessary for the existence of a London force.

11.59 In liquids, what is the relationship between boiling point and the strength of intermolecular forces?

11.60 In liquids, what is the relationship between vapor pressure magnitude and the strength of intermolecular forces?

11.61 Which is harder to break, an ordinary covalent bond or a hydrogen bond? Explain your answer.

11.62 What is the difference between a dipole–dipole interaction and a hydrogen bond?

11.63 Which of the two group VIIA diatomic gases F_2 and Cl_2 would have the stronger intermolecular forces? Explain your answer.

11.64 Which of the two nonpolar molecules CO_2 (molecular mass 44 amu) and F_2 (molecular mass 38 amu) would have the stronger intermolecular forces?

11.65 Explain why the replacement of a hydrogen atom in CH_4 by a fluorine atom, to form CH_3F, causes an increase in boiling point from -164 to -78 °C?

11.66 Assign the boiling points -88 °C, -42 °C, and 0 °C to the following molecules: C_4H_{10}, C_2H_6, and C_3H_8. Explain the basis for your assignment.

11.67 How many hydrogen bonds can one water molecule form to other water molecules?

11.68 How many hydrogen bonds can one ammonia molecule form to other ammonia molecules?

Properties of Water (Sec. 11.16)

11.69 Explain, in terms of hydrogen bonding, why water has a higher than expected boiling point.

11.70 If water were a "normal" compound, what would its boiling point be?

11.71 Explain why the maximum density of water occurs at a temperature that is higher than its freezing point.

11.72 Discuss the "density behavior" of a sample of water as its temperature is lowered from 20 to 0 °C.

11.73 Use the structure of ice to explain its being lower in density than water.

11.74 Discuss the consequences to the Earth and its inhabitants if water, on going from the liquid to the solid state, increased rather than decreased in density.

11.75 Explain why large bodies of water have a moderating effect on the climate of the surrounding area.

11.76 Explain why the formation of potholes in streets is hastened during the winter.

11.77 Define the property of liquids called surface tension.

11.78 What is the relationship between intermolecular forces and the property of surface tension?

Types of Solids (Sec. 11.17)

11.79 Indicate the type of particle that occupies the lattice sites in each of the following type of crystalline solids.
(a) nonpolar molecular
(b) ionic

11.80 Indicate the type of particle that occupies the lattice sites in each of the following types of crystalline solids.
(a) macromolecular
(b) metallic

11.81 Indicate the nature of the forces between particles in each of the following types of crystalline solids.
(a) ionic
(b) polar molecular

11.82 Indicate the nature of the forces between particles in each of the following types of crystalline solids.
(a) macromolecular
(b) nonpolar molecular

11.83 Describe melting point characteristics and volatility characteristics for each of the following types of crystalline solids.
(a) macromolecular
(b) metallic

11.84 Describe melting point characteristics and volatility characteristics for each of the following types of crystalline solids.
(a) polar molecular
(b) ionic

11.85 What is the difference between a *crystalline solid* and an *amorphous solid*?

11.86 Give three examples of materials that are amorphous solids.

Additional Problems

11.87 For each of the following pairs of substances, predict which will have the higher melting point and indicate why.
(a) NaCl and Cl_2
(b) SiO_2 and CO_2
(c) Cu and CO
(d) NaF and MgF_2

11.88 For each of the following pairs of substances, predict which will have the higher melting point and indicate why.
(a) HCl and HBr
(b) Ar, Xe
(c) Na_2O and SiC
(d) CH_4 and CCl_4

11.89 If heat is supplied at an identical, constant rate in each case, which would take longer, heating 50.0 g of water from 0.0 to 80.0 °C or vaporizing 50.0 g of water at 100.0 °C?

11.90 If heat is supplied at an identical, constant rate in each case, which would take longer, heating 80.0 g of water from 20.0 to 100.0 °C or melting 80.0 g of ice at 0.0 °C?

11.91 A quantity of ice at 0.0 °C was added to 40.0 g of water at 19.0 °C in an insulated container. All of the ice melted, and the water temperature decreased to 0.0 °C. How many grams of ice were added?

11.92 A quantity of ice at 0.0 °C was added to 55.0 g of water at 25.3 °C in an insulated container. All of the ice melted, and the water temperature decreased to 0.0 °C. How many grams of ice were added?

11.93 If 10.0 g of ice at −10.0 °C and 30.0 g of liquid water at 80.0 °C are mixed in an insulated container, what will the final temperature of the liquid be?

11.94 If 10.0 g of ice at −20.0 °C and 60.0 g of liquid water at 70.0 °C are mixed in an insulated container, what will the final temperature of the liquid be?

11.95 A 500.0-g piece of metal at 50.0 °C is placed in 100.0 g of water at 10.0 °C in an insulated container. The metal and water come to the same temperature of 23.4 °C. What is the specific heat, in joules per gram per degree Celsius, of the metal?

11.96 A 100.0-g piece of metal at 80.0 °C is placed in 200.0 g of water at 20.0 °C in an insulated container. The metal and water come to the same temperature of 33.6 °C. What is the specific heat, in joules per gram per degree Celsius, of the metal?

11.97 Heat is added to a 25.3-g sample of a metal at the rate of 136 J/sec. After the sample reaches the metal's melting point temperature, the temperature remains constant for 4.3 min. Calculate the heat of vaporization for this metal in joules per gram.

11.98 Heat is added to a 35.2-g sample of a metal at the rate of 189 J/sec. After the sample reaches the metal's melting point temperature, the temperature remains constant for 5.2 min. Calculate the heat of vaporization for this metal in joules per gram.

Cumulative Problems

11.99 To solidify a 10.0-g sample of a liquid at its freezing point, 1485 J of heat must be removed. To solidify a 10.0-mL sample of this same liquid at its freezing point, 1151 J of heat must be removed. What is the density, in grams per milliliter, of this liquid?

11.100 To solidify a 10.0-mL sample of a liquid at its freezing point, 1063 J of heat must be removed. To solidify a 15.0-g sample of this same liquid at its freezing point, 1276 J of heat must be removed. What is the density, in grams per milliliter, of this liquid?

11.101 The heat of vaporization for a substance is 18.1 kJ/mole. Determine the molecular mass of this substance given that 1.00 g, at its boiling point, releases 602 J of heat as it condenses.

11.102 The heat of vaporization for a substance is 36.6 kJ/mole. Determine the molecular mass of this substance given that 1.00 g, at its boiling point, releases 314 J of heat as it condenses.

11.103 How many kilojoules of heat energy are needed to completely melt a pure sample of copper at its melting point that was obtained from the complete decomposition of 52.0 g of Cu_2O?

11.104 How many kilojoules of heat energy are needed to completely melt a pure sample of aluminum at its melting point that was obtained from the complete decomposition of 75.0 g of Al_2O_3?

11.105 How many joules of heat energy are needed to heat a sample of liquid benzene (C_6H_6) from a temperature 10 °C below its boiling point to its boiling point, given that the sample contains 6.32×10^{24} hydrogen atoms and that the specific heat of liquid benzene is 1.74 J/(g · °C)?

11.106 How many joules of heat energy must be removed from a sample of liquid benzene (C_6H_6) to lower its temperature from 10 °C above its melting point to its melting point, given that the sample contains 4.03×10^{23} carbon atoms and the specific heat of liquid benzene is 1.74 J/(g · °C).

11.107 A nuclear power plant produces 3.3×10^6 kJ of waste heat energy every second. If this heat is to be dissipated by evaporating 25 °C water (in cooling towers), how many gallons of water per day would be needed? The density of water is 0.997 g/mL at 25 °C. Atmospheric pressure is 1.00 atm.

11.108 A nuclear power plant produces 3.3×10^6 kJ of waste heat energy every second. If this waste heat is dumped into a river that is 42 ft wide and has an average depth of 5.1 ft and a current of 3.0 mi/hr, what is the temperature difference between the water upstream and the water downstream? Assume that the density of water is 1.0 g/mL.

Grid Problems

11.109 Select from the grid *all* correct responses for each situation.

(a) characterizations that apply to the solid state
(b) characterizations that apply to the liquid state
(c) characterizations that apply to the gaseous state
(d) characterizations that apply to only one of the states of matter
(e) characterizations that apply to both solids and liquids
(f) characterizations that apply to two states of matter

1. definite shape	2. indefinite volume	3. high density
4. small compressibility	5. moderate thermal expansion	6. kinetic energy dominance
7. vibrational particle movement	8. cohesive force dominance	9. large interparticle distance

11.110 Select from the grid *all* correct responses for each situation.

1. evaporation	2. deposition	3. solidification
4. fusion	5. melting	6. liquefaction
7. sublimation	8. condensation	9. freezing

(a) processes in which the gaseous state is involved
(b) processes in which the liquid state is involved
(c) processes that are exothermic
(d) processes that are endothermic
(e) pairs of processes that denote the same change
(f) pairs of processes that denote "opposite" changes

11.111 Select from the grid *all* correct responses for each situation.

1. heat of vaporization	2. boiling point	3. heat of condensation
4. specific heat of solid	5. heat of fusion	6. specific heat of liquid
7. freezing point	8. specific heat of gas	9. heat of solidification

(a) quantities whose definition involves the word liquid
(b) pairs of quantities that have the same numerical value
(c) quantities that might be needed in solving a problem in which there is a temperature change but no change of state
(d) quantities that might be needed in solving a problem in which there is a change of state but no change in temperature
(e) quantities needed to solve the problem "How many joules of energy are released as 25.0 g of 115 °C steam is cooled to 30 °C?"
(f) quantities needed to solve the problem "How many joules of energy must be supplied to heat 25.0 g of −15 °C ice to 115 °C?"

11.112 Select from the grid *all* correct responses for each situation. Each statement relates to crystalline solids.

1. ionic solid	2. polar molecular solid	3. nonpolar molecular solid
4. macromolecular solid	5. metallic solid	6. diamond (C)
7. dry ice (CO_2)	8. ice (H_2O)	9. table salt (NaCl)

(a) ions are present at the lattice sites
(b) single atoms must or can be present at the lattice sites
(c) London forces are the only interparticle forces
(d) volatility is low or zero
(e) covalent bonds between lattice sites are present
(f) covalent bonds *must* be present within the lattice sites

CHAPTER TWELVE

Gas Laws

12.1 Properties of Some Common Gases

The word *gas* is used to refer to a substance that is normally in the gaseous state at ordinary temperatures and pressures. The word *vapor* (Sec. 11.12) describes the gaseous form of any substance that is a liquid or solid at normal temperatures and pressures. Thus, we speak of oxygen gas and water vapor.

The normal state for eleven of the elements is that of a gas (Sec. 11.1). A listing of these elements along with some of their properties is given in Table 12.1. This table also lists some common compounds that are gases at room temperature and pressure.

The three most commonly encountered elemental gases—hydrogen, oxygen, and nitrogen—are colorless and odorless. So also are all of the noble gases (Sec. 6.11). From these observations one should not conclude that all gases, or even all elemental gases, are colorless and odorless. Note from Table 12.1 the pale yellow and greenish yellow colors associated with elemental fluorine and chlorine and the irritating odors of both of them. Many gaseous compounds have pungent odors.

TABLE 12.1 Color, Odor, and Toxicity of Elements and Common Compounds That Are Gases at Ordinary Temperatures and Pressures

Element		Properties
H_2	hydrogen	colorless, odorless
O_2	oxygen	colorless, odorless
N_2	nitrogen	colorless, odorless
Cl_2	chlorine	greenish yellow, choking odor, toxic
F_2	fluorine	pale yellow, pungent odor, toxic
He	helium	colorless, odorless
Ne	neon	colorless, odorless
Ar	argon	colorless, odorless
Kr	krypton	colorless, odorless
Xe	xenon	colorless, odorless
Rn	radon	colorless, odorless

Compound		Properties
CO_2	carbon dioxide	colorless, faintly pungent odor
CO	carbon monoxide	colorless, odorless, toxic
NH_3	ammonia	colorless, pungent odor, toxic
CH_4	methane	colorless, odorless
SO_2	sulfur dioxide	colorless, pungent choking odor, toxic
H_2S	hydrogen sulfide	colorless, rotten egg odor, toxic
HCl	hydrogen chloride	colorless, choking odor, toxic
NO_2	nitrogen dioxide	reddish brown, irritating odor, toxic

Table 12.1 also brings to your attention again (recall Sec. 10.2) that all of the elemental gases, except for the noble gases, exist in the form of diatomic molecules (H_2, O_2, N_2, F_2, and Cl_2).

Many of the gaseous compounds listed in Table 12.1 are colorless, have odors, and are toxic. Note that odor and toxicity do not have to go together. Carbon monoxide, a "deadly" poison, is odorless but toxic.

12.2 Gas Law Variables

The behavior of a gas can be described reasonably well by *simple* quantitative relationships called *gas laws*. **Gas laws** are generalizations that describe in mathematical terms the relationships among the *pressure, temperature,* and *volume* of a specific quantity of a gas.

It is only the gaseous state that is describable by simple mathematical relationships. Laws describing liquid and solid state behavior are mathematically extremely complex. Consequently, quantitative treatments of these behaviors will not be given in this text.

Before we discuss the mathematical form of the various gas laws, some comments concerning the major variables involved in gas law calculations—volume, temperature, and pressure—are in order. Two of these three variables, volume and temperature, have been discussed previously (Secs. 3.4 and 3.10, respectively). The units of *liter* or *milliliter* are usually used in specifying gas volume. Only one of the three temperature scales discussed in Section 3.10, *the Kelvin scale,* can be used in gas law calculations if the results are to be valid. Therefore, you should be thoroughly familiar with the conversion of Celsius and Fahrenheit scale readings to kelvins (Sec. 3.6). We have not yet discussed pressure, the third variable. Comments concerning pressure will occupy the remainder of this section.

Pressure is defined as the force applied per unit area, that is, the total force on a surface divided by the area of that surface.

$$P \text{ (pressure)} = \frac{F \text{ (force)}}{A \text{ (area)}}$$

Note that pressure and force are not the same. Identical forces give rise to different pressures if they are acting on areas of different size. For areas of the same size, the larger the force, the greater the pressure.

Barometers, manometers, and gauges are the instruments commonly used by chemists to measure gas pressure. Barometers and manometers measure pressure in terms of the height of a column of mercury, whereas gauges are usually calibrated in terms of force per area, for example, in pounds per square inch.

The air that surrounds the Earth exerts a pressure on all objects with which it has contact. The **barometer** is the most commonly used device for measuring atmospheric pressure. It was invented by the Italian physicist Evangelista Torricelli (1608–1647) in 1643. The essential components of a barometer are shown in Figure 12.1. A barometer can be constructed by filling a long glass tube, sealed at one end, all the way to the top with mercury and then inverting the tube (without letting any air in) into a dish of mercury. The mercury in the tube falls until the pressure from the mass of the mercury in the tube is just balanced by the pressure of the atmosphere on the mercury in the dish. The pressure of the atmosphere is then expressed in terms of the height of the supported column of mercury.

Mercury is the liquid of choice in a barometer for two reasons: (1) it is a very dense liquid, and therefore only a short glass tube is needed; and (2) it has a very low vapor pressure, so the pressure reading does not have to be corrected for vapor pressure.

The height of the mercury column is most often expressed in millimeters or inches. Millimeters of mercury (mm Hg) are used in laboratory work. The most common use of inches of mercury (in. Hg) is in weather reporting.

The pressure of the atmosphere varies with altitude, decreasing at the rate of approximately 25 mm Hg per 1000 ft increase in altitude. It also fluctuates with weather conditions. Recall the terminology used in a weather report: high pressure front, low pressure front, and so on. At sea level the height of a column of mercury in a barometer fluctuates with weather conditions between about 740 and 770 mm Hg and averages about 760 mm Hg. Another pressure unit, the atmosphere (atm), is defined in terms of this average sea-level pressure.

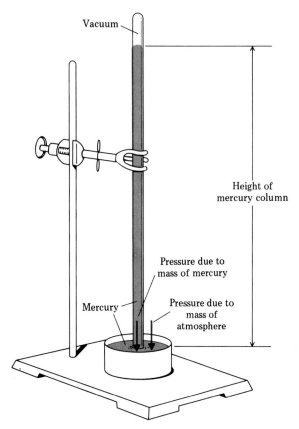

FIGURE 12.1 The essential components of a mercury barometer.

$$1 \text{ atm} = 760 \text{ mm Hg}$$

Because of its size, 760 times larger than 1 mm Hg, the atmosphere unit is frequently used to express the high pressures encountered in many industrial processes and in some experimental work.

It is impractical to measure the pressure of gases other than air with a barometer. One cannot usually introduce a mercury barometer directly into a container of a gas. A **manometer** is the device used to measure gas pressures in a laboratory. It is a U-tube filled with mercury. One side of the U-tube is connected to the container in which the pressure is to be measured, and the other side is connected to a region of known pressure. Such an arrangement is depicted in Figure 12.2. The gas in the container exerts a pressure on the Hg on that side of the tube, while atmospheric pressure pushes on the open end. A difference in the heights of the Hg columns in the two arms of the manometer indicates a pressure difference between the gas and the atmosphere. In Figure 12.2 the gas pressure in the container exceeds atmospheric pressure by an amount, ΔP, that is equal to the difference in the heights of the Hg columns in the two arms.

$$P_{\text{container}} = P_{\text{atm}} + \Delta P$$

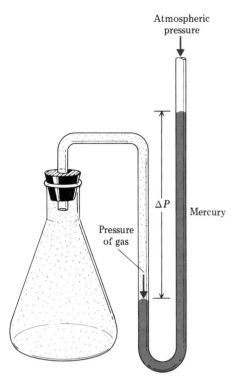

FIGURE 12.2 Pressure measurement by means of a manometer.

Pressures of contained gas samples can also be measured by special gas pressure gauges attached to their containers. Such gauges are commonly found on tanks (cylinders) of gas purchased commercially. These gauges are most often calibrated in terms of pounds per square inch (lb/in.^{2} or psi). The relationship between psi and atmospheres is

$$14.68 \text{ psi} = 1 \text{ atm}$$

Table 12.2 summarizes the relationships among the various pressure units we have discussed and also lists one additional pressure unit, the *pascal*. The pascal is the SI unit (Sec. 3.1) of pressure.

TABLE 12.2 Units of Pressure and Their Relationship to the Unit Atmosphere

Unit	Relationship to Atmosphere	Area of Use
Atmosphere	—	gas law calculations
Millimeters of mercury	760 mm Hg = 1 atm	gas law calculations
Inches of mercury	29.92 in. Hg = 1 atm	weather reports
Pound per square inch	14.68 psi = 1 atm	stored or bottled gases
Pascal	1.013×10^5 Pa = 1 atm	calculations requiring SI units

The values 1 atm and 760 mm Hg are exact numbers since they arise from a definition. Consequently, these two values have an infinite number of significant figures. All other values in Table 12.2 are not exact numbers; they are given to four significant figures in the table.

The equalities of Table 12.2 can be used to construct conversion factors for problem solving in the usual way (Sec. 3.6). Example 12.1 illustrates the use of such conversion factors in the context of change in pressure unit problems.

EXAMPLE 12.1

A weather reporter gives the day's barometric pressure as 30.12 in. of mercury. Express this pressure in the units of

(a) atmospheres   (b) millimeters of mercury
(c) pounds per square inch

Solution

(a) The given quantity is 30.12 in. Hg, and the unit of the desired quantity is atmospheres.

$$30.12 \text{ in. Hg} = ? \text{ atm}$$

The conversion factor relating these two units, found in Table 12.1, is

$$\frac{1 \text{ atm}}{29.92 \text{ in. Hg}}$$

The result of using this conversion factor is

$$30.12 \text{ in. Hg} \times \frac{1 \text{ atm}}{29.92 \text{ in. Hg}} = 1.0066845 \text{ atm} \quad \text{(calculator answer)}$$

$$= 1.007 \text{ atm} \quad \text{(correct answer)}$$

(b) This time, the unit for the desired quantity is mm Hg.

$$30.12 \text{ in. Hg} = ? \text{ mm Hg}$$

A direct conversion factor between in. Hg and mm Hg is not given in Table 12.1. However, the table does give the relationships of both inches and millimeters of mercury to atmospheres. Hence, we can use atmospheres as an intermediate step in the unit conversion sequence.

$$\text{in. Hg} \longrightarrow \text{atm} \longrightarrow \text{mm Hg}$$

The dimensional analysis setup is

$$30.12 \text{ in. Hg} \times \frac{1 \text{ atm}}{29.92 \text{ in. Hg}} \times \frac{760 \text{ mm Hg}}{1 \text{ atm}} = 765.08021 \text{ mm Hg}$$
$$\text{(calculator answer)}$$

$$= 765.1 \text{ mm Hg}$$
$$\text{(correct answer)}$$

(c) The problem to be solved here is

$$30.12 \text{ in. Hg} = ? \text{ psi}$$

The unit conversion sequence will be

$$\text{in. Hg} \longrightarrow \text{atm} \longrightarrow \text{psi}$$

The dimensional analysis sequence of conversion factors, by the above pathway, is

$$30.12 \text{ in. Hg} \times \frac{1 \text{ atm}}{29.92 \text{ in. Hg}} \times \frac{14.68 \text{ psi}}{1 \text{ atm}} = 14.778128 \text{ psi} \quad \text{(calculator answer)}$$

$$= 14.78 \text{ psi} \quad \text{(correct answer)}$$

PRACTICE EXERCISE 12.1

The atmospheric pressure in a chemistry laboratory at Weber State University in Ogden, Utah, on a particular day is found to be 674.3 mm Hg. Express this pressure in the following units.

(a) atmospheres
(b) inches of mercury
(c) pounds per square inch

Ans. (a) 0.8872 atm; (b) 26.55 in. Hg; (c) 13.02 psi.

Similar exercises Problems 12.1 and 12.2

12.3 Boyle's Law: A Pressure–Volume Relationship

Of the several relationships that exist between gas-law variables, the first to be discovered was the one that relates gas pressure to gas volume. It was formulated over 300 years ago, in 1662, by the British chemist and physicist Robert Boyle (1627–1691) and is known as Boyle's law. **Boyle's law** states that the volume of a sample of gas is *inversely proportional* to the pressure applied to the gas if the temperature is kept constant. This means that if the pressure on the gas increases, the volume decreases proportionally; and conversely, if the pressure is decreased, the volume will increase. Doubling the pressure cuts the volume in half; tripling the pressure cuts the volume to one-third its original value; quadrupling the pressure cuts the volume to one-fourth; and so on. Any time two quantities are *inversely proportional*, as pressure and volume are (Boyle's law), one increases as the other decreases. Data illustrating Boyle's law are given in Figure 12.3.

Boyle's law can be illustrated physically quite simply with the J-tube apparatus shown in Figure 12.4. The pressure on the trapped gas is increased by adding mercury to the J-tube. The volume of the trapped gas decreases as the pressure is raised.

Boyle's law can be stated mathematically as

$$P \times V = \text{constant}$$

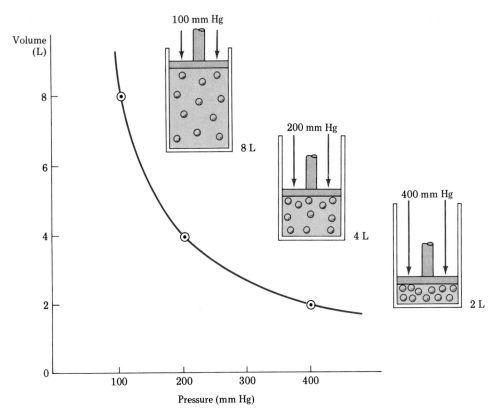

FIGURE 12.3 Data illustrating the "inverse proportionality" associated with Boyle's law.

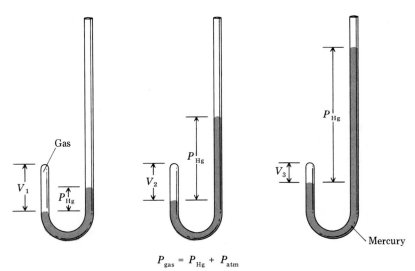

$P_{gas} = P_{Hg} + P_{atm}$

FIGURE 12.4 Boyle's law apparatus. The volume of the trapped gas in the closed end of the tube (V_1, V_2, and V_3) decreases as mercury is added through the open end of the tube.

In this expression V is the volume of the gas at a given temperature and P is the pressure. This expression thus indicates that at constant temperature the product of the pressure times the volume is always the same (or constant). (Note that Boyle's law is valid only if the temperature of the gas does not change.)

An alternative, and more useful, mathematical form of Boyle's law can be derived by considering the following situation. Suppose a gas is at an initial pressure P_1 and has a volume V_1. (We will use the subscript 1 to indicate the initial conditions.) Now imagine that the pressure is changed to some final pressure P_2. The volume will also change, and we will call the final volume V_2. (We will use the subscript 2 to indicate the final conditions.) According to Boyle's law,

$$P_1 \times V_1 = \text{constant}$$

After the change in pressure and volume, we have

$$P_2 \times V_2 = \text{constant}$$

The constant is the same in both cases; we are dealing with the same sample of gas. Thus, we can combine the two PV products to give the equation

$$P_1 \times V_1 = P_2 \times V_2$$

When we know any three of the four quantities in this equation, we can calculate the fourth, which will usually be the final pressure, P_2, or the final volume, V_2, as illustrated in Examples 12.2 and 12.3.

EXAMPLE 12.2

An inflated air mattress has a volume of 114 L at a pressure of 0.925 atm. The mattress, still inflated, is taken into the mountains where the pressure, because of the higher elevation, drops to 0.825 atm. What will be the air-mattress volume, in liters, at this lower pressure, assuming that temperature remains constant?

Solution

A suggested first step in working all gas law problems involving two sets of conditions is to analyze the given data in terms of initial and final conditions. Doing this, we find that

$$P_1 = 0.925 \text{ atm} \qquad P_2 = 0.825 \text{ atm}$$
$$V_1 = 114 \text{ L} \qquad V_2 = ? \text{ L}$$

Next, we arrange Boyle's law to isolate V_2 (the quantity to be calculated) on one side of the equation. This is accomplished by dividing both sides of the Boyle's law equation by P_2.

$$P_1 V_1 = P_2 V_2 \quad \text{(Boyle's law)}$$

$$\frac{P_1 V_1}{P_2} = \frac{\cancel{P_2} V_2}{\cancel{P_2}} \quad \text{(division of each side of the equation by } P_2\text{)}$$

$$V_2 = V_1 \times \frac{P_1}{P_2}$$

Substituting the given data in the rearranged equation and doing the arithmetic gives

$$V_2 = 114 \text{ L} \times \frac{0.925 \text{ atm}}{0.825 \text{ atm}}$$

$$= 127.81818 \text{ L} \quad \text{(calculator answer)}$$

$$= 128 \text{ L} \quad \text{(correct answer)}$$

In gas law problems, when possible, it is always a good idea to check qualitatively the reasonableness of your final answer. This is a double check against two types of errors frequently made: improper substitution of variables into the equation and improper rearrangement of the equation. Decreasing the pressure on a fixed amount of gas at constant temperature, which is the case in this problem, should result in a volume increase for the gas. Our answer is consistent with this conclusion.

PRACTICE EXERCISE 12.2

An air bubble forms at the bottom of a lake, where the total pressure is 2.45 atm. At this pressure, the bubble has a volume of 3.1 mL. What volume will it have when it rises to the surface, where the pressure is that of the atmosphere, 0.93 atm? Assume that the temperature remains constant, as does the amount of gas in the bubble.　　　Ans.　8.2 mL.

Similar exercises: Problems 12.13 and 12.14

EXAMPLE 12.3

On an airliner, an inflated toy has a volume of 0.310 L at 25 °C and 745 mm Hg. During flight, the cabin pressure unexpectedly drops to 0.810 atm with the temperature remaining the same. What is the volume, in liters, of the toy at the new conditions?

Solution
Analyzing the given data in terms of initial and final conditions gives

$$P_1 = 745 \text{ mm Hg} \qquad P_2 = 0.810 \text{ atm}$$
$$V_1 = 0.310 \text{ L} \qquad V_2 = ? \text{ L}$$

The temperature is constant; it will not enter into the calculation.
  A slight complication exists with the pressure values; they are not given in the same units. We must make the units the same before proceeding with the calculation. It does not matter whether they are both millimeters of mercury or both atmospheres. The same answer is obtained with either unit. Let us arbitrarily decide to change atmospheres to mm Hg.

$$0.810 \text{ atm} \times \frac{760 \text{ mm Hg}}{1 \text{ atm}} = 615.6 \text{ mm Hg} \quad \text{(calculator)}$$

$$= 616 \text{ mm Hg} \quad \text{(correct answer)}$$

Our given conditions are now

$$P_1 = 745 \text{ mm Hg} \qquad P_2 = 616 \text{ mm Hg}$$
$$V_1 = 0.310 \text{ L} \qquad V_2 = ? \text{ L}$$

Boyle's law with V_2 isolated on the left side has the form

$$V_2 = V_1 \times \frac{P_1}{P_2}$$

Plugging the given quantities into this equation and doing the arithmetic gives

$$V_2 = 0.310 \text{ L} \times \frac{745 \text{ mm Hg}}{616 \text{ mm Hg}} = 0.37491883 \text{ L} \quad \text{(calculator answer)}$$

$$= 0.375 \text{ L} \quad \text{(correct answer)}$$

The answer is reasonable. Decreased pressure, at constant temperature, should produce a volume increase.

In solving this problem we arbitrarily chose to use mm Hg pressure units. If we had used atmosphere pressure units instead, the answer obtained would still be the same, as we now illustrate.

$$745 \text{ mm Hg} \times \frac{1 \text{ atm}}{760 \text{ mm Hg}} = 0.98026315 \text{ atm} \quad \text{(calculator answer)}$$

$$= 0.980 \text{ atm} \quad \text{(correct answer)}$$

Using $P_1 = 0.980$ atm and $P_2 = 0.810$ atm, we get

$$V_2 = 0.310 \text{ L} \times \frac{0.980 \text{ atm}}{0.810 \text{ atm}} = 0.37506172 \text{ L} \quad \text{(calculator answer)}$$

$$= 0.375 \text{ L} \quad \text{(correct answer)}$$

PRACTICE EXERCISE 12.3

A sample of cyclopropane (C_3H_6), a colorless gas with a pleasant odor that finds use as a general anesthetic, occupies a volume of 2.00 L at 27 °C and 1.00 atm pressure. What volume will this sample occupy at the same temperature but at a pressure of 655 mm Hg?

Ans. 2.32 L.

Similar exercises: Problems 12.13 and 12.14

Boyle's law is consistent with kinetic molecular theory (Sec. 11.3). The theory states that the pressure a gas exerts results from collisions of the gas molecules with the sides of the container. The number of collisions within a given area on the container wall in a given time is proportional to the pressure of the gas at a given temperature. If the volume of a container holding a specific number of gas molecules is increased, the total wall area of the container will also increase and the number of collisions in a given area (the pressure) will decrease due to the greater wall area. Conversely, if the volume of the container is decreased, the wall area will

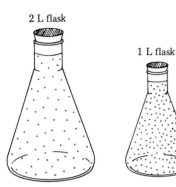
(a) The volume is decreased by one half.

(b) A given molecule hits container walls twice as often.

FIGURE 12.5 The average number of times a molecule hits container walls is doubled when the volume of the gas, at constant temperature, is cut by one half.

be smaller and there will be more collisions in a given wall area. This means an increase in pressure. Figure 12.5 illustrates this idea.

The phenomenon described by Boyle's law has practical importance. Helium-filled research balloons, used to study the upper atmosphere, are only half-filled with helium when launched. As the balloon ascends, it encounters lower and lower pressures. As the pressure decreases, the balloon expands until it reaches full inflation. A balloon launched at full inflation would burst in the upper atmosphere because of the reduced external pressure.

Breathing is an example of Boyle's law in action, as is the operation of a respirator, a machine designed to help patients with respiration difficulties to breathe. A respirator contains a movable diaphragm that works in opposition to the patient's lungs. When the diaphragm is moved out so that the volume inside the respirator increases, the lower pressure in the respirator allows air to expand out of the patient's lungs. When the diaphragm is moved in the opposite direction, the higher pressure inside the respirator compresses the air into the lungs and causes them to increase in volume.

12.4 Charles's Law: A Temperature–Volume Relationship

How does the volume of a fixed quantity of gas respond to a temperature change at constant pressure? In 1787 the French scientist Jacques Alexander Cesar Charles (1746–1823) showed that a simple mathematical relationship exists between the volume and temperature of a gas, at constant pressure, *provided the temperature is expressed in kelvins*. **Charles's law** states that the volume of a sample of gas is *directly proportional* to its Kelvin temperature if the pressure is kept constant. Contained within the wording of this law is the phrase *directly proportional*; this contrasts with Boyle's law, which contains the phrase *inversely proportional*. Any time a *direct* proportion exists between two quantities, one increases when the other increases, and one

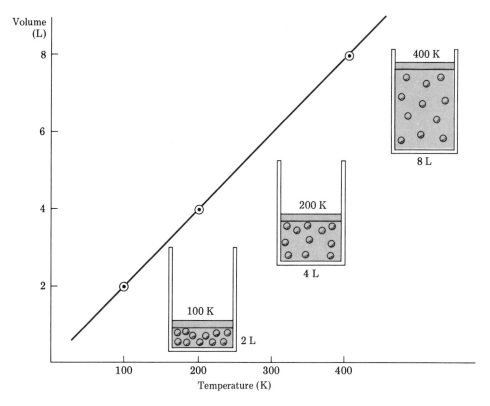

FIGURE 12.6 Data illustrating the "direct proportionality" associated with Charles' law

decreases when the other decreases. Thus a direct proportion and an inverse proportion portray "opposite" behaviors. The direct proportion relationship of Charles's law means that if the temperature increases the volume will also increase and if the temperature decreases the volume will also decrease. Data illustrating Charles's law are given in Figure 12.6.

Charles's law can be qualitatively illustrated by a balloon filled with air. If the balloon is placed near a heat source, such as a light bulb that has been on for some time, the heat will cause the balloon to increase in size (volume). The change in volume is visually noticeable. Putting the same balloon in the refrigerator will cause it to shrink.

Charles's law can be stated mathematically as

$$\frac{V}{T} = \text{constant}$$

In this expression V is the volume of a gas at a given pressure and T is the temperature, expressed on the Kelvin temperature scale. A consideration of two sets of temperature–volume conditions for a gas, in a manner similar to that done for Boyle's law, leads to the following useful form of the law.

$$\frac{V_1}{T_1} = \frac{V_2}{T_2}$$

Again, note that in any of the mathematical expressions of Charles's law, or any other gas law, the symbol T is understood to mean the Kelvin temperature.

EXAMPLE 12.4

A helium-filled birthday balloon has a volume of 223 mL at a temperature of 24 °C (room temperature). The balloon is placed in a car overnight during the winter where the temperature drops to -13 °C. What is the balloon's volume, in milliliters, when it is removed from the car in the morning (assuming no helium has escaped)?

Solution
Writing all the given data in the form of initial and final conditions, we have

$$V_1 = 223 \text{ mL} \qquad V_2 = ? \text{ mL}$$
$$T_1 = 42 \,°\text{C} = 315 \text{ K} \qquad T_2 = -13 \,°\text{C} = 260 \text{ K}$$

Note that both of the given temperatures have been converted to Kelvin scale readings by adding 273 to them.

Rearranging Charles's law to isolate V_2, the desired quantity, on one side of the equation is accomplished by multiplying each side of the equation by T_2.

$$\frac{V_1}{T_1} = \frac{V_2}{T_2} \qquad \text{(Charles's law)}$$

$$\frac{V_1 T_2}{T_1} = \frac{V_2 \cancel{T_2}}{\cancel{T_2}} \qquad \text{(multiplication of each side by } T_2\text{)}$$

$$V_2 = V_1 \times \frac{T_2}{T_1}$$

Substituting the given data into the equation and doing the arithmetic gives

$$V_2 = 223 \text{ mL} \times \frac{260 \,\cancel{K}}{315 \,\cancel{K}}$$

$$= 184.06349 \text{ mL} \quad \text{(calculator answer)}$$

$$= 184 \text{ mL} \quad \text{(correct answer)}$$

Our answer is consistent with what reasoning says it should be. Decreasing the temperature of a gas, at constant pressure, should result in a volume decrease.

PRACTICE EXERCISE 12.4

At constant pressure, a 675 mL sample of N_2 gas is warmed from 23 °C (a comfortable room temperature) to 38 °C (the temperature outdoors on a hot summer day). What is the new volume of the N_2 gas?   *Ans.* 709 mL.

Similar exercises: Problems 12.21 and 12.22

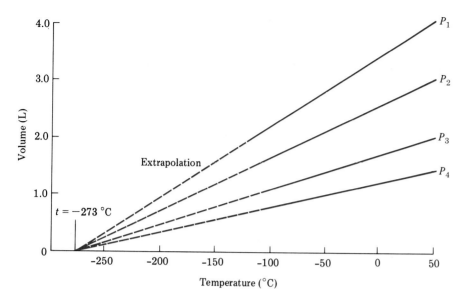

FIGURE 12.7 A graph showing the variation of the volume of a gas with temperature at a constant pressure. Each line in the graph represents a different fixed pressure for the gas. The pressures increase from P_1 to P_4. The solid portion of each line represents the temperature region before the gas condenses to a liquid. The extrapolated (dashed) portion of each line represents the hypothetical volume of the gas if it had not condensed. All extrapolated lines intersect at the same point, a point corresponding to zero volume and a temperature of $-273\ °C$.

Figure 12.7 is a graph showing four sets of volume–temperature data for a gas, each data set being at a different constant pressure. The plot of each data set gives a straight line. These four straight lines can be used to show the basis for the use of the Kelvin temperature scale in gas-law calculations. If each of these lines is extrapolated (extended) to lower temperatures (the dashed portions of the lines in Fig. 12.7), we find that they all intersect at a common point on the temperature axis. This point of intersection, corresponding to a temperature value of $-273\ °C$, is the point at which the volume of the gas "would become" zero. Notice that the words "would become" in the last sentence are in quotation marks. In reality, the volume of a gas never reaches zero; all gases would liquefy before they reach a temperature of $-273\ °C$ and Charles's law would no longer apply. Thus, the extrapolated portions of the four straight lines portray a hypothetical situation. It is not, however, a situation without significance.

The Scottish mathematician and physicist William Thomson (1824–1907), better known as Lord Kelvin, was the first to recognize the importance of the "zero volume" temperature value of $-273\ °C$. It is the lowest temperature that is theoretically attainable, a temperature now referred to as *absolute zero*. It is this temperature that is now used as the starting point for the temperature scale called the Kelvin temperature scale. When temperatures are specified in kelvins, Charles's law assumes the simple mathematical form in which it was previously given in this section. The use of other temperature scales results in more complicated versions of Charles's law.

Charles's law is readily understood in terms of kinetic molecular theory. The theory states that when the temperature of a gas increases, the velocity of the gas particles increases. The speedier particles hit the container walls harder and more often. In order for the pressure of the gas to remain constant, it is necessary for the container volume to increase. Moving in a larger volume, the particles will hit the container walls less often, and thus the pressure can remain the same. A similar argument applies if the temperature of the gas is lowered, only this time the velocity of the molecules decreases and the wall area (volume) must decrease in order to increase the number of collisions in a given area in a given time.

Charles's law is the principle used in the operation of an incubator. The air in contact with the heating element expands (its density becomes less). The hot, less dense air rises, causing continuous circulation of warm air. This same principle has ramifications in closed rooms in which there is not effective air circulation. The warmer and less dense air stays near the top of the room. This is desirable in the summer but not in the winter.

12.5 Gay-Lussac's Law: A Temperature–Pressure Relationship

If a gas is placed in a rigid container, one that cannot expand, the volume of the gas must remain constant. What is the relationship between the pressure and temperature of a fixed amount of gas in such a situation? This question was answered in 1802 by the French scientist Joseph Louis Gay-Lussac (1778–1850). Performing a number of experiments similar to those performed by Charles, he discovered the relationship between pressure and the temperature of a gas when it is held at constant volume. **Gay-Lussac's law** states that the pressure of a sample of gas, at constant volume, is *directly proportional* to its Kelvin temperature. Thus, as the temperature of the gas increases, the pressure increases; conversely, as temperature is decreased, pressure decreases.

The kinetic molecular theory explanation for Gay-Lussac's law is very simple. As the temperature increases, the velocity of the gas molecules increases. This increases the number of times the molecules hit the container walls in a given time, which translates into an increase in pressure.

Gay-Lussac's law explains the observed pressure increase inside an automobile tire after a car has been driven for a period of time on a hot day. It also explains why aerosol cans should not be disposed of in a fire. The aerosol can keeps the gas at a constant volume. The elevated temperatures encountered in the fire increase the pressure of the confined gas to the point that the can may explode.

Mathematical forms of **Gay-Lussac's law** are

$$\frac{P}{T} = \text{constant} \quad \text{and} \quad \frac{P_1}{T_1} = \frac{P_2}{T_2}$$

Again, note that the symbol T is understood to mean the Kelvin temperature. Any pressure unit is acceptable; P_1 and P_2 must, however, be in the same units.

EXAMPLE 12.5

Early in the morning, with the temperature at 17 °C, automobile tire pressure was found to be 30.1 psi. At the end of an 8-hr trip, the combination of tire friction and hot pavement has raised the temperature of the air inside the tires to 110 °C. (A temperature this high, which is equivalent to 230 °F, is definitely attainable within a tire on a hot summer day.) What will be the tire pressure (gauge pressure) at this higher temperature?

Solution

In working this problem we are assuming a negligible change in tire volume—a reasonable assumption.

We also note that the pressure inside a tire is the pressure shown by the gauge plus atmospheric pressure. We will assume that atmospheric pressure for the day was 14.5 psi. Initially, then, the total pressure in the tire was

$$30.1 \text{ psi} + 14.5 \text{ psi} = 44.6 \text{ psi}$$

Writing all the given data in the form of initial and final conditions, we have

$$P_1 = 44.6 \text{ psi} \qquad P_2 = ? \text{ psi}$$
$$T_1 = 17\,°\text{C} = 290 \text{ K} \qquad T_2 = 110\,°\text{C} = 383 \text{ K}$$

Rearrangement of Gay-Lussac's law to isolate P_2, the quantity we desire, on the left side on the equation gives

$$P_2 = P_1 \times \frac{T_2}{T_1}$$

Substituting the given data into the equation and doing the arithmetic gives

$$P_2 = 44.6 \text{ psi} \times \frac{383 \, K}{290 \, K} = 58.902759 \text{ psi} \quad \text{(calculator answer)}$$

$$= 58.9 \text{ psi} \quad \text{(correct answer)}$$

The gauge pressure observed for the "hot" tire will be

$$58.9 \text{ psi} - 14.5 \text{ psi} = 44.4 \text{ psi}$$

The answer is reasonable. Increased temperature, at constant volume, should produce a pressure increase.

PRACTICE EXERCISE 12.5

An aerosol spray can contains gas under a pressure of 2.00 atm at 27 °C. The can itself can only withstand a pressure of 3.00 atm. To what temperature, in degrees Celsius, can the can and its contents be heated before the can explodes?

Similar exercises: Problems 12.33 and 12.34

Ans. 177 °C.

12.6 The Combined Gas Law

The **combined gas law** is the expression obtained from mathematically combining Boyle's, Charles's and Gay-Lussac's laws. Its mathematical form is

$$\frac{P_1 V_1}{T_1} = \frac{P_2 V_2}{T_2}$$

TABLE 12.3 Relationship of the Individual Gas Laws to the Combined Gas Law

Law	Constancy Requirement	Mathematical Form of the Law
Combined gas law	none	$\dfrac{P_1 V_1}{T_1} = \dfrac{P_2 V_2}{T_2}$
Boyle's law	$T_1 = T_2$	Since T_1 and T_2 are equal, substitute T_1 for T_2 in the combined gas law and cancel. $\dfrac{P_1 V_1}{\cancel{T_1}} = \dfrac{P_2 V_2}{\cancel{T_1}}$ or $P_1 V_1 = P_2 V_2$
Charles's law	$P_1 = P_2$	Since P_1 and P_2 are equal, substitute P_1 for P_2 in the combined gas law and cancel. $\dfrac{\cancel{P_1} V_1}{T_1} = \dfrac{\cancel{P_1} V_2}{T_2}$ or $\dfrac{V_1}{T_1} = \dfrac{V_2}{T_2}$
Gay-Lussac's law	$V_1 = V_2$	Since V_1 and V_2 are equal, substitute V_1 for V_2 in the combined gas law and cancel. $\dfrac{P_1 \cancel{V_1}}{T_1} = \dfrac{P_2 \cancel{V_1}}{T_2}$ or $\dfrac{P_1}{T_1} = \dfrac{P_2}{T_2}$

This combined gas law is a much more versatile equation than the individual gas laws. With it, a change in any one of the three gas law variables brought about by changes in *both* of the other two variables can be calculated. Each of the individual gas laws requires that one of the three variables be held constant.

The combined gas law reduces (simplifies) to each of the equations for the individual gas laws when the constancy provision is invoked for the appropriate variable. These reduction relationships are given in Table 12.3. Because of the ease with which they can be derived from the combined gas law, you need not memorize the mathematical forms for the individual gas laws if you know the mathematical form for the combined gas law.

The three most used forms of the combined gas law are those that isolate V_2, P_2, and T_2 on the left side of the equation.

$$V_2 = V_1 \times \frac{P_1}{P_2} \times \frac{T_2}{T_1}$$

$$P_2 = P_1 \times \frac{V_1}{V_2} \times \frac{T_2}{T_1}$$

$$T_2 = T_1 \times \frac{P_2}{P_1} \times \frac{V_2}{V_1}$$

Students frequently have questions about the algebra involved in accomplishing these rearrangements. Example 12.6 should clear up any such questions. Examples 12.7 and 12.8 illustrate the use of the combined gas law equation.

EXAMPLE 12.6

Rearrange the standard form of the combined gas law equation such that the variable T_2 is by itself on the left side of the equation.

Solution

In rearranging the standard form of the combined gas law into various formats, the following rule from algebra is useful.

If two fractions are equal,

$$\frac{a}{b} = \frac{c}{d}$$

then the numerator of the first fraction (a) times the denominator of the second fraction (d) is equal to the numerator of the second fraction (c) times the denominator of the first fraction (b).

$$\text{If } \frac{a}{b} = \frac{c}{d}, \text{ then } a \times d = c \times b.$$

Applying this rule to the standard form of the combined gas law gives

$$\frac{P_2 V_2}{T_2} = \frac{P_1 V_1}{T_1} \longrightarrow P_2 V_2 T_1 = P_1 V_1 T_2$$

With the combined gas law in the form $P_1 V_1 T_2 = P_2 V_2 T_1$, any of the six variables can be isolated by a simple division. To isolate T_2, we divide both sides of the equation by $P_1 V_1$ (the other quantities on the same side of the equation as T_2).

$$\frac{P_1 V_1 T_2}{P_1 V_1} = \frac{P_2 V_2 T_1}{P_1 V_1}$$

$$T_2 = \frac{P_2 V_2 T_1}{P_1 V_1} \quad \text{or} \quad T_2 = T_1 \times \frac{P_2}{P_1} \times \frac{V_2}{V_1}$$

PRACTICE EXERCISE 12.6

Rearrange the standard form of the combined gas law equation such that the variable P_2 is by itself on the left side of the equation.

Ans. $P_2 = P_1 \times \dfrac{V_1}{V_2} \times \dfrac{T_2}{T_1}$. Similar exercises: Problems 12.37 and 12.38

EXAMPLE 12.7

On a winter day, a person takes in a breath of 425 mL of cold air ($-12\,°C$). The atmospheric pressure is 682 mm Hg. What is the volume of this air, in milliliters, in the lungs where the temperature is $37\,°C$ and the pressure is 684 mm Hg?

Solution
Writing all the given data in the form of initial and final conditions, we have

$$P_1 = 682 \text{ mm Hg} \qquad P_2 = 684 \text{ mm Hg}$$
$$V_1 = 425 \text{ mL} \qquad V_2 = ? \text{ mL}$$
$$T_1 = -12\,°C = 261 \text{ K} \qquad T_2 = 37\,°C = 310 \text{ K}$$

Rearrangement of the combined gas law expression to isolate V_2 on the left side gives

$$V_2 = V_1 \times \frac{P_1}{P_2} \times \frac{T_2}{T_1}$$

Substituting numerical values into this equation and doing the arithmetic gives

$$V_2 = 425 \text{ mL} \times \frac{682 \text{ mm Hg}}{684 \text{ mm Hg}} \times \frac{310 \text{ K}}{261 \text{ K}}$$
$$= 503.31328 \text{ mL} \quad \text{(calculator answer)}$$
$$= 503 \text{ mL} \quad \text{(correct answer)}$$

PRACTICE EXERCISE 12.7

A helium-filled weather balloon, when released, has a volume of 10.0 L at $27\,°C$ and a pressure of 663 mm Hg. What volume, in liters, will the balloon occupy at an altitude where the pressure is 96 mm Hg and the temperature is $-30.0\,°C$?

Ans. 56 L.

Similar exercises: Problems 12.39 and 12.40

EXAMPLE 12.8

A sample of argon gas (the gas used in electric light bulbs) occupies a volume of 80.0 mL at a pressure of 1.10 atm and a temperature of $29\,°C$. What will be the temperature of the gas, in degrees Celsius, if the volume of the container is decreased to 40.0 mL and the pressure is increased to 2.20 atm?

Solution
Writing all the given data in the form of initial and final conditions we have

$$P_1 = 1.10 \text{ atm} \qquad P_2 = 2.20 \text{ atm}$$
$$V_1 = 80.0 \text{ mL} \qquad V_2 = 40.0 \text{ mL}$$
$$T_1 = 29\,°C = 302 \text{ K} \qquad T_2 = ?\,°C$$

Rearrangement of the combined gas law to isolate T_2 on the left side of the equation gives

$$T_2 = T_1 \times \frac{P_2}{P_1} \times \frac{V_2}{V_1}$$

Substituting the given data into this equation and doing the arithmetic gives

$$T_2 = 302 \text{ K} \times \frac{2.20 \text{ atm}}{1.10 \text{ atm}} \times \frac{40.0 \text{ mL}}{80.0 \text{ mL}}$$

$$= 302 \text{ K} \quad \text{(calculator and correct answer in kelvins)}$$

Converting the temperature to the Celsius scale by subtracting 273 gives 29 °C as the final answer.

$$302 \text{ K} - 273 = 29 \text{ °C}$$

The temperature did not change! The pressure correction factor, considered by itself, would cause the temperature to increase by a factor of 2; the pressure was doubled. The volume correction factor, considered by itself, would cause the temperature to decrease by a factor of 2; the volume was halved. Considered together, the effects of the two factors cancel each other and result in the temperature not changing.

PRACTICE EXERCISE 12.8

A sample of helium gas occupies a volume of 2.00 L at a pressure of 5.21 atm and a temperature of 41 °C. What will be the temperature, in °C, of the gas if the volume is changed to 3.55 L and the pressure is changed to 7.33 atm?

Ans. 511 °C.

Similar exercises: Problems 12.41 and 12.42

12.7 Standard Conditions for Temperature and Pressure

The volumes of liquids and solids change only slightly with temperature and pressure changes (Sec. 11.2). This is not the case for volumes of gases. As the gas law discussions of previous sections pointedly show, the volume of a gas is very dependent on temperature and pressure.

Gas volumes can be compared only if the gases are at the same temperature and pressure. It is convenient to specify a particular temperature and pressure as standards for comparison purposes. **Standard temperature** is 0 °C (273 K). **Standard pressure** is 1 atm (760 mm Hg). **STP conditions** are those of standard temperature and standard pressure.

Example 12.9 is a problem involving STP conditions.

EXAMPLE 12.9

A sample of Cl_2 gas occupies a volume of 13.6 L at STP. What volume, in liters, will it occupy at a pressure of 0.500 atm and a temperature of $-10\,°C$?

Solution
STP conditions denote a temperature of $0\,°C$ and a pressure of 1.00 atm. Writing all the given data in the form of initial and final conditions, we have

$$P_1 = 1.00 \text{ atm} \qquad P_2 = 0.500 \text{ atm}$$
$$V_1 = 13.6 \text{ L} \qquad V_2 = ? \text{ L}$$
$$T_1 = 0\,°C = 273 \text{ K} \qquad T_2 = -10\,°C = 263 \text{ K}$$

Rearrangement of the combined gas law expression to isolate V_2 on the left side gives

$$V_2 = V_1 \times \frac{P_1}{P_2} \times \frac{T_2}{T_1}$$

Substituting numerical values into the equation gives

$$V_2 = 13.6 \text{ L} \times \frac{1.00 \text{ atm}}{0.500 \text{ atm}} \times \frac{263 \text{ K}}{273 \text{ K}}$$
$$= 26.203663 \text{ L} \quad \text{(calculator answer)}$$
$$= 26.2 \text{ L} \quad \text{(correct answer)}$$

In this problem the pressure correction factor (P_1/P_2) and the temperature correction factor (T_2/T_1) work in opposite directions. The decreased pressure should cause the volume to increase. The decreased temperature should cause the volume to decrease. The fact that, overall, the volume increases indicates that the pressure change is the dominant change.

PRACTICE EXERCISE 12.9

A sample of H_2 gas occupies a volume of 1.37 L at STP. What volume, in liters, will it occupy at a pressure of 4.00 atm and a temperature of $340\,°C$?

Ans. 0.769 L.

Similar exercises: Problems 12.47 and 12.48

12.8 Gay-Lussac's Law of Combining Volumes

Gases are involved as reactants or products in many chemical reactions. In such reactions it is usually easier to determine the volumes rather than the masses of the gases involved.

A very simple relationship exists between the volumes of different gases consumed or produced in chemical reactions, provided the volumes are all determined at the same temperature and pressure. This relationship, known as Gay-Lussac's law of combining volumes, was first formulated in 1808 by Joseph Louis Gay-Lussac (the same Gay-Lussac discussed in Sec. 12.5). **Gay-Lussac's law of combining volumes** states that the volumes of different gases involved in a reaction, measured at the same temperature and pressure, are in the same ratio as the coefficients for these gases in the balanced equation for the reaction. For example, 1 volume of nitrogen and 3 volumes of hydrogen react to give 2 volumes of ammonia (NH_3).

$$N_2(g) + 3 H_2(g) \longrightarrow 2 NH_3(g)$$
$$\text{1 volume} \quad \text{3 volumes} \quad \text{2 volumes}$$

"Volume" is used here in the general sense of relative volume in any units. It could, for example, be 1 L of N_2, 3 L of H_2, and 2 L of NH_3 or 0.1 mL of N_2, 0.3 mL of H_2, and 0.2 mL of NH_3. Note that in making volume comparisons the units must always be the same; if the volume of N_2 is measured in milliliters, the volumes of H_2 and NH_3 must also be in milliliters.

It is important to remember that these volume relationships apply only to gases, and then only when all gaseous volumes are measured at the same temperature and pressure. The volumes of solids and liquids involved in reactions cannot be treated this way.

The previously given statement of the law of combining volumes is a modern version of the law. At the time the law was first formulated by Gay-Lussac, chemists were still struggling with the difference between atoms and molecules. Equations as we now write them were unknown; formulas for substances were still to be determined. The original statement of the law simply noted that the ratio of volumes was always a ratio of small whole numbers. Gay-Lussac's work was based solely on volume measurements and had nothing to do with chemical equations. The explanation for the small whole numbers and the linkup with equations came later.

In Section 10.5 we learned that coefficients in a balanced equation can be interpreted in terms of moles. For example, for the equation

$$4 NH_3(g) + 3 O_2(g) \longrightarrow 2 N_2(g) + 6 H_2O(g)$$

it is correct to say

$$4 \text{ moles } NH_3(g) + 3 \text{ moles } O_2(g) \longrightarrow 2 \text{ moles } N_2(g) + 6 \text{ moles } H_2O(g)$$

As a result of Gay-Lussac's law of combining volumes, all of the mole designations in this equation can be replaced with volume designations.

$$4 \text{ volumes } NH_3(g) + 3 \text{ volumes } O_2(g) \longrightarrow 2 \text{ volumes } N_2(g) + 6 \text{ volumes } H_2O(g)$$

Again, all of the volumes must be measured at the same temperature and pressure for the above relationship to be valid.

In calculations involving chemical reactions, where two or more gases are participants, this volume interpretation of coefficients can be used to generate conversion factors useful in problem solving. Consider the balanced equation

$$2 CO(g) + O_2(g) \longrightarrow 2 CO_2(g)$$

At constant temperature and pressure, three volume–volume relationships are obtainable from this equation.

$$2 \text{ volumes CO} \quad \text{produce} \quad 2 \text{ volumes CO}_2$$
$$1 \text{ volume O}_2 \quad \text{produces} \quad 2 \text{ volumes CO}_2$$
$$2 \text{ volumes CO} \quad \text{react with} \quad 1 \text{ volume O}_2$$

From these three relationships, six conversion factors can be written.

From the first relationship:

$$\frac{2 \text{ volumes CO}}{2 \text{ volumes CO}_2} \quad \text{and} \quad \frac{2 \text{ volumes CO}_2}{2 \text{ volumes CO}}$$

From the second relationship:

$$\frac{1 \text{ volume O}_2}{2 \text{ volumes CO}_2} \quad \text{and} \quad \frac{2 \text{ volumes CO}_2}{1 \text{ volume O}_2}$$

From the third relationship:

$$\frac{2 \text{ volumes CO}}{1 \text{ volume O}_2} \quad \text{and} \quad \frac{1 \text{ volume O}_2}{2 \text{ volumes CO}}$$

The more gaseous reactants and products there are in a chemical reaction, the greater the number of volume–volume conversion factors obtainable from the equation for the chemical reaction. The use of volume–volume conversion factors is illustrated in Example 12.10.

EXAMPLE 12.10

Hydrogen, H_2, reacts with acetylene, C_2H_2, according to the equation

$$2 H_2(g) + C_2H_2(g) \longrightarrow C_2H_6(g)$$

What volume of H_2, in liters, is required to react with 13.7 L of C_2H_2? Assume that both gases are at the same conditions of temperature and pressure.

Solution

STEP 1 The given quantity is 13.7 L of C_2H_2 and the desired quantity is liters of H_2.

$$13.7 \text{ L } C_2H_2 = ? \text{ L } H_2$$

STEP 2 This is a volume-to-volume problem. The conversion factor needed for this one-step problem is derived from the coefficients of H_2 and C_2H_2 in the equation for the chemical reaction. The equation tells us that at constant temperature and pressure two volumes of H_2 react with one volume of C_2H_2. From this relationship the conversion factor

$$\frac{2 \text{ L } H_2}{1 \text{ L } C_2H_2}$$

is obtained.

STEP 3   The dimensional analysis setup for this problem is

$$13.7 \text{ L } C_2H_2 \times \frac{2 \text{ L } H_2}{1 \text{ L } C_2H_2}$$

Note that it makes no difference what the temperature and pressure are, as long as they are the same for the two gases involved in the calculation.

STEP 4   Doing the arithmetic, after cancellation of units, gives

$$\frac{13.7 \times 2}{1} \text{ L } H_2 = 27.4 \text{ L } H_2 \quad \text{(calculator and correct answer)}$$

PRACTICE EXERCISE 12.10

Nitrogen reacts with hydrogen to produce ammonia as shown by the equation

$$N_2(g) + 3 H_2(g) \longrightarrow 2 NH_3(g)$$

What volume of H_2, in liters, at 750 mm Hg and 25 °C is required to produce 1.75 L of NH_3 at the same temperature and pressure? *Ans.* 2.62 L H_2.

Similar exercises: Problems 12.51 and 12.52

12.9 Avogadro's Law: A Volume–Quantity Relationship

In the year 1811, Amedeo Avogadro, the Avogadro involved in Avogadro's number (Sec. 9.4), published work in which he proposed an explanation for Gay-Lussac's observations (Sec. 12.8). His proposal, a hypothesis at the time it was first published, is now known as Avogadro's law. Its validity has been demonstrated in a number of ways.

Avogadro's law states that equal volumes of different gases, measured at the same temperature and pressure, contain equal numbers of molecules. Although Avogadro's law seems very simple in terms of today's scientific knowledge, it was a very astute conclusion at the time it was first proposed. At that time scientists were still struggling with the differences between atoms and molecules, and Avogadro's reasoning eventually led to the realization that many of the common gaseous elements, such as hydrogen, oxygen, nitrogen, and chlorine, occur naturally as diatomic molecules (H_2, O_2, N_2, Cl_2) rather than single atoms.

Equal volumes of gases at the same temperature and pressure, because they contain equal numbers of molecules, must also contain equal numbers of moles of molecules. Thus, Avogadro's law can also be stated in terms of volumes and moles rather than volumes and molecules.

Avogadro's law can also be stated in another alternative form, which uses terminology similar to that used in Boyle's, Charles's, and Gay-Lussac's laws. This alternative form deals with two samples of the same gas rather than two samples of different gases. *The volume of a gas, at constant temperature and pressure, is directly proportional to the number of moles of gas present.* When the original number of moles of gas present is doubled, the volume of the gas increases twofold. Halving the number of moles present halves the volume.

Experimentally, it is easy to show the direct relationship between volume and amount of gas present at constant temperature and pressure. All that is needed is an inflated balloon. Adding more gas to the balloon increases the volume of the balloon. The more gas added to the balloon, the greater the volume increase. Conversely, letting gas out of the balloon decreases its volume. It should be noted that this balloon demonstration is an approximate rather than exact demonstration for the law, since the tension of the rubber material from which the balloon is made exerts an effect on the proportionality between volume and moles.

Mathematical statements of Avogadro's law, where n represents the number of moles, are

$$\frac{V}{n} = \text{constant} \quad \text{and} \quad \frac{V_1}{n_1} = \frac{V_2}{n_2}$$

Note the similarity in mathematical form between this law and the laws of Charles and Gay-Lussac. The similarity results from all three laws being direct proportionality relationships between two variables.

Avogadro's law and the combined gas law can be combined to give the expression

$$\frac{P_1 V_1}{n_1 T_1} = \frac{P_2 V_2}{n_2 T_2}$$

This equation covers the situation where none of the four variables P, T, V, and n, is constant. With it, a change in any one of the four variables brought about by changes in the other three variables can be calculated.

EXAMPLE 12.11

A balloon containing 1.83 moles of He has a volume of 0.673 L at a given temperature and pressure. If an additional 0.50 mole of He gas is introduced into the balloon without changing temperature and pressure conditions, what will be the new volume, in liters, of the balloon?

Solution
We will use the equation

$$\frac{V_1}{n_1} = \frac{V_2}{n_2}$$

in solving the problem.

Writing the given data in terms of initial and final conditions, we have

$$V_1 = 0.673 \text{ L} \qquad V_2 = ? \text{ L}$$
$$n_1 = 1.83 \text{ moles} \qquad n_2 = (1.83 + 0.50) \text{ moles} = 2.33 \text{ moles}$$

Rearrangement of the Avogadro's law expression to isolate V_2 on the left side gives

$$V_2 = V_1 \times \frac{n_2}{n_1}$$

Substituting numerical values into the equation gives

$$V_2 = 0.673 \text{ L} \times \frac{2.33 \text{ moles}}{1.83 \text{ moles}} = 0.85687978 \text{ L} \quad \text{(calculator answer)}$$

$$= 0.857 \text{ L} \quad \text{(correct answer)}$$

Our answer is consistent with reasoning. Increasing the number of moles of gas present in the balloon at constant temperature and pressure should increase the volume of the balloon.

PRACTICE EXERCISE 12.11

A balloon containing 2.00 moles of Ar gas has a volume of 0.500 L at a given temperature and pressure. If 0.480 mole of Ar gas is removed from the balloon without changing temperature and pressure conditions, what will be the new volume, in liters, of the balloon?

Ans. 0.380 L.

Similar exercises: Problems 12.55–12.58

12.10 Molar Volume of a Gas

The **molar volume** of a gas is the volume occupied by 1 mole of the gas at STP conditions. It follows from Avogadro's law (Sec. 12.9) that all gases will have the same molar volume. Experimentally, it is found that the molar volume of a gas is 22.4 L. To visualize a volume this size, 22.4 L, think of standard-sized basketballs. The volume occupied by three standard-sized basketballs is almost exactly equal to 22.4 L.

The fact that all gases have the same molar volume at STP (or any other temperature–pressure combination) is a property unique to the gaseous state. Similar statements cannot be made about the liquid and solid states. The reason for the difference can be understood by considering the relationship between the volume the molecules occupy in the given physical state and the volume of the molecules themselves. In the gaseous state, because the molecules are so far apart—a gas is mostly empty space—the volume of the gas molecules themselves is negligible compared to the total volume. This is not the case in the liquid and solid states, where the molecules are in close contact with each other.

It is informative to contrast the properties of molar volume and molar mass, the latter having been discussed in Section 9.5. The *molar mass* is a constant for a given substance, and is independent of temperature and pressure. Each substance has its own unique molar mass. The *molar volume* of a substance has a temperature and pressure specification (STP conditions). All gases have the same molar volume at the same temperature and pressure (STP conditions).

The molar volume concept is very useful in a variety of types of calculations. When used in calculations, the concept "translates" into a conversion factor having either of two forms

$$\frac{1 \text{ mole gas}}{22.4 \text{ L gas}} \quad \text{or} \quad \frac{22.4 \text{ L gas}}{1 \text{ mole gas}}$$

A most common type of problem involving these conversion factors is one where

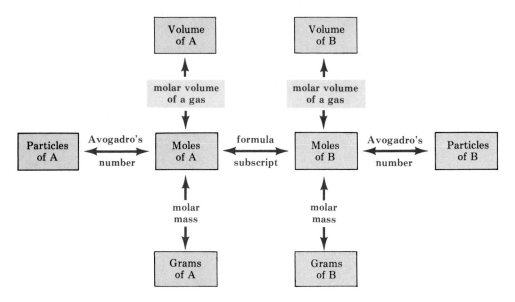

FIGURE 12.8 Quantitative relationships needed for solving mass-to-volume and volume-to-mass chemical-formula-based problems.

the volume of a gas at STP is known and you are asked to calculate from it the moles, grams, or particles of gas present, or vice versa. Figure 12.8 summarizes the relationships needed in performing calculations of this general type.

Perhaps you recognize Figure 12.8 as being very similar to a diagram you have already encountered many times in problem solving, Figure 9.3. This "new diagram" differs from Figure 9.3 in only one way: volume boxes have been added at the top of the diagram. The diagram in Figure 12.8 is used in the same way as the earlier one. The given and desired quantities are determined, and the arrows of the diagram are used to "map out" the pathway to be used in going from the given quantity to the desired quantity. Examples 12.12–12.14 illustrate the usefulness of Figure 12.8.

EXAMPLE 12.12

Sulfur dioxide, SO_2, is a colorless, nonflammable gas that has a pungent and choking odor. What would be the volume, in liters, of a 10.3-g sample of this gas at STP conditions?

Solution

STEP 1 The given quantity is 10.3 g of SO_2, and the unit of the desired quantity is liters of SO_2.

$$10.3 \text{ g } SO_2 = ? \text{ L } SO_2$$

STEP 2 In terms of Figure 12.8, this problem is a "grams of A" to "volume of A" problem. The pathway is

Grams of A $\xrightarrow[\text{mass}]{\text{molar}}$ Moles of A $\xrightarrow[\text{volume}]{\text{molar}}$ Volume of A

Using dimensional analysis, we set up this sequence of conversion factors as

$$10.3\ \cancel{g\ SO_2} \times \frac{1\ \cancel{mole\ SO_2}}{64.1\ \cancel{g\ SO_2}} \times \frac{22.4\ L\ SO_2}{1\ \cancel{mole\ SO_2}}$$

$$\text{grams A} \longrightarrow \text{moles A} \longrightarrow \text{volume A}$$

Note how the molar volume relationship that 1 mole of gas is equal to 22.4 L is used as the last conversion factor.

STEP 3 The solution, obtained by doing the arithmetic, is

$$\frac{10.3 \times 1 \times 22.4}{64.1 \times 1}\ L\ O_2 = 3.5993759\ L\ SO_2 \quad \text{(calculator answer)}$$

$$= 3.60\ L\ SO_2 \quad \text{(correct answer)}$$

PRACTICE EXERCISE 12.12

What is the volume, in liters, occupied by 25.6 g of N_2 gas at STP? *Ans.* 20.5 L.

Similar exercises: Problems 12.65 and 12.66

Equal volumes of different gases at STP do not have equal masses. This is because different kinds of molecules have different masses. Example 12.13 illustrates the validity of this concept.

EXAMPLE 12.13

What is the mass, in grams, of 4.78 L of (**a**) O_2 gas and (**b**) F_2 gas at STP?

Solution

(**a**)

STEP 1 The given quantity is 4.78 L of O_2 at STP, and the desired quantity is grams of O_2.

$$\mathbf{4.78\ L\ O_2 = ?\ g\ O_2}$$

STEP 2 This is a "volume of A" to "grams of A" problem. The pathway appropriate for solving this problem, as indicated in Figure 12.8, is

Volume of A $\xrightarrow[\text{volume}]{\text{molar}}$ Moles of A $\xrightarrow[\text{mass}]{\text{molar}}$ Grams of A

The setup, from dimensional analysis, is

$$4.78\ \cancel{L\ O_2} \times \frac{1\ \cancel{mole\ O_2}}{22.4\ \cancel{L\ O_2}} \times \frac{32.0\ g\ O_2}{1\ \cancel{mole\ O_2}}$$

STEP 3 Collecting the numerical terms after cancellation of units and doing the arithmetic gives

$$\frac{4.78 \times 1 \times 32.0}{22.4 \times 1} \text{ g O}_2 = 6.8285714 \text{ g O}_2 \quad \text{(calculator answer)}$$

$$= 6.83 \text{ g O}_2 \quad \text{(correct answer)}$$

(b) The analysis and setup for this part are identical to those in (a) except that F_2 and its molecular mass replace O_2 and its molecular mass. Thus, the setup is

$$4.78 \text{ L } F_2 \times \frac{1 \text{ mole } F_2}{22.4 \text{ L } F_2} \times \frac{38.0 \text{ g } F_2}{1 \text{ mole } F_2} = 8.1089286 \text{ g } F_2 \quad \text{(calculator answer)}$$

$$= 8.11 \text{ g } F_2 \quad \text{(correct answer)}$$

Equal volumes of different gases at STP do not have equal masses; 4.78 L of O_2 has a mass of 6.83 g, and 4.78 L of F_2 has a mass of 8.11 g.

PRACTICE EXERCISE 12.13

What is the mass, in grams, of 0.375 L of each of the following gases at STP?
(a) CO_2 (b) CO

Ans. (a) 0.737 g CO_2; (b) 0.469 g CO.

Similar exercises: Problems 12.67 and 12.68

The density of a gas at STP conditions can be calculated by dividing the molar mass of the substance by its molar volume.

$$\text{Density (STP)} = \frac{\text{molar mass}}{\text{molar volume (STP)}}$$

The units of density, g/L, are, indeed, obtained from this division.

$$\text{Density (STP)} = \frac{\dfrac{\text{grams}}{\text{mole}}}{\dfrac{\text{liters}}{\text{mole}}} = \frac{\text{grams} \times \text{mole}}{\text{mole} \times \text{liters}} = \frac{\text{grams}}{\text{liters}}$$

Rearrangement of this density equation enables one to calculate molar mass

$$\text{Molar mass} = \text{density (STP)} \times \text{molar volume (STP)}$$

EXAMPLE 12.14

Ozone, a form of oxygen with the formula O_3, is naturally found in the upper atmosphere where it screens out 95–99% of the ultraviolet solar radiation headed for earth. Calculate the density, in grams per liter, of ozone at STP conditions.

Solution

The molar mass and molar volume relationships needed to solve this problem are

Molar mass: $\dfrac{48.0 \text{ g } O_3}{1 \text{ mole } O_3}$ Molar volume: $\dfrac{22.4 \text{ L } O_3}{1 \text{ mole } O_3}$

Substituting into the density equation gives

$$\text{Density} = \frac{\text{molar mass}}{\text{molar volume}}$$

$$= \frac{\dfrac{48.0 \text{ g } O_3}{1 \text{ mole } O_3}}{\dfrac{22.4 \text{ L } O_3}{1 \text{ mole } O_3}} = \frac{48.0 \text{ g } O_3 \times 1 \text{ mole } O_3}{1 \text{ mole } O_3 \times 22.4 \text{ L } O_3} = \frac{48.0 \times 1}{1 \times 22.4} \text{ g/L } O_3$$

$$= 2.1428571 \text{ g/L } O_3 \quad \text{(calculator answer)}$$

$$= 2.14 \text{ g/L } O_3 \quad \text{(correct answer)}$$

PRACTICE EXERCISE 12.14

Hydrogen is the least dense of all substances. Calculate the density of H_2 gas at STP conditions, in grams per liter.

Ans. 0.0902 g/L H_2.

Similar exercises: Problems 12.68 and 12.69

Example 12.15 differs from previous examples in that it involves both the A and B sides of Figure 12.8.

EXAMPLE 12.15

If the Xe atoms present in 10.6 g of XeF_6 are converted to gaseous Xe, what volume, in liters, will they occupy at STP conditions?

Solution

STEP 1 The given quantity is 10.6 g of XeF_6, and the desired quantity is liters of Xe at STP. In the jargon of Figure 12.8, this is a "grams of A" to "volume of B" problem.

STEP 2 The sequence of conversion factors needed for this problem, using Figure 12.8 as a guide, follows the pathway

$$\boxed{\text{Grams of A}} \xrightarrow{\text{molar mass}} \boxed{\text{Moles of A}} \xrightarrow{\text{formula subscript}} \boxed{\text{Moles of B}} \xrightarrow{\text{molar volume}} \boxed{\text{Volume of B}}$$

Translating this pathway into a dimensional analysis setup gives

$$10.6 \text{ g } XeF_6 \times \frac{1 \text{ mole } XeF_6}{245 \text{ g } XeF_6} \times \frac{1 \text{ mole Xe}}{1 \text{ mole } XeF_6} \times \frac{22.4 \text{ L Xe}}{1 \text{ mole Xe}}$$

STEP 3 Collecting numerical terms after cancellation of units and doing the arithmetic gives

$$\frac{10.6 \times 1 \times 1 \times 22.4}{245 \times 1 \times 1} \text{ L Xe} = 0.96914285 \text{ L Xe} \quad \text{(calculator answer)}$$

$$= 0.969 \text{ L Xe} \quad \text{(correct answer)}$$

PRACTICE EXERCISE 12.15

If the Xe atoms present in 14.5 g of XeF_2 are converted to monoatomic Xe gas, what volume, in liters, will they occupy at STP conditions?

Ans. 1.92 L Xe

Similar exercises: Problems 12.73 and 12.74

12.11 The Ideal Gas Law

The **ideal gas law** describes the relationships among the four variables temperature (T), pressure (P), volume (V), and moles of gas (n) for gaseous substances.

In previous sections we have discussed three independent relationships dealing with the volume of a gas. They are

Boyle's law: $\quad V = k \times \dfrac{1}{P} \quad$ (n and T constant)

Charles's law: $\quad V = kT \quad$ (n and P constant)

Avogadro's law: $\quad V = kn \quad$ (P and T constant)

We can combine these three equations into a single expression

$$V = \dfrac{kTn}{P}$$

since we know (from mathematics) that if a quantity is independently proportional to two or more quantities, it is also proportional to their product. This combined equation is a mathematical statement of the ideal gas law. Any gas that obeys the individual laws of Boyle, Charles, and Avogadro will also obey the ideal gas law.

The usual form in which the ideal gas law is written is

$$PV = nRT$$

This form of the ideal gas law, besides being rearranged, differs from the previous form in that the proportionality constant k has been given the symbol R. The constant R is called the *ideal gas constant*. In order to use the ideal gas equation we must know the value of R. This can be determined by substituting STP values for P, V, T, and n into the ideal gas equation and solving for R. For 1 mole of gas at STP, $P = 1$ atm, $V = 22.4$ L, $T = 273$ K, and $n = 1$ mole. Substituting these values into the ideal gas equation arranged to have R isolated on the left side gives

$$R = \dfrac{PV}{nT} = \dfrac{(1 \text{ atm})(22.4 \text{ L})}{(1 \text{ mole})(273 \text{ K})} = 0.0821 \; \dfrac{\text{atm} \cdot \text{L}}{\text{mole} \cdot \text{K}}$$

Notice the complex units associated with R—the four variables temperature, pressure, volume, and moles are all involved. This will always be the case.

The value of R is dependent on the units used to express pressure and volume. If we use 760 mm Hg instead of 1 atm as standard pressure, then R would have the value 62.4 and the units (mm Hg·L)/(mole·K).

$$R = \dfrac{PV}{nT} = \dfrac{(760 \text{ mm Hg})(22.4 \text{ L})}{(1 \text{ mole})(273 \text{ K})} = 62.4 \; \dfrac{\text{mm Hg} \cdot \text{L}}{\text{mole} \cdot \text{K}}$$

The most used values of R are the two values just encountered. When pressure units other than millimeters of mercury or atmosphere and volume units other than liter are encountered, convert them to these units and then use the appropriate known R value.

If three of the four variables in the ideal gas equation are known, then the fourth can be calculated by the equation. The ideal gas equation is used in calculations when *one* set of conditions is given with one missing variable. The combined gas law (Sec. 12.6) is used when *two* sets of conditions are given with one missing variable. Examples 12.16 and 12.17 illustrate the use of the ideal gas equation.

EXAMPLE 12.16

The colorless, odorless, tasteless gas carbon monoxide, CO, is a by-product of incomplete combustion of any material that contains the element carbon. Calculate the volume, in liters, occupied by 1.52 moles of this gas at 0.992 atm pressure and a temperature of 65 °C.

Solution

This problem deals with only one set of conditions, a situation where the ideal gas equation is applicable. Three of the four variables in the ideal gas equation (P, n, and T) are given, and the fourth (V) is to be calculated.

$$P = 0.992 \text{ atm} \qquad n = 1.52 \text{ moles}$$
$$V = ? \text{ L} \qquad T = 65\,°\text{C} = 338 \text{ K}$$

Rearranging the ideal gas equation to isolate V on the left side of the equation gives

$$V = \frac{nRT}{P}$$

Since the pressure is given in atmospheres and the volume unit is liters, the appropriate R value is

$$R = 0.0821 \frac{\text{atm} \cdot \text{L}}{\text{mole} \cdot \text{K}}$$

Substituting the given numerical values into the equation and canceling units gives

$$V = \frac{(1.52 \text{ moles})\left(0.0821 \frac{\text{atm} \cdot \text{L}}{\text{mole} \cdot \text{K}}\right)(338 \text{ K})}{0.992 \text{ atm}}$$

Note how all of the parts of the ideal gas constant unit cancel except for one, the volume part.

Doing the arithmetic, we get as an answer 42.5 L CO.

$$V = \frac{1.52 \times 0.0821 \times 338}{0.992} \text{ L CO}$$

$$= 42.519854 \text{ L CO} \quad \text{(calculator answer)}$$

$$= 42.5 \text{ L CO} \quad \text{(correct answer)}$$

PRACTICE EXERCISE 12.16

Calculate the volume, in liters, occupied by 4.25 moles of Cl_2 gas at 1.54 atm pressure and a temperature of 213 °C.

Ans. $11\overline{0}$ L.

Similar exercises: Problems 12.75 and 12.76

EXAMPLE 12.17

Dinitrogen monoxide (nitrous oxide), N_2O, is a naturally present constituent, in trace amounts, of the atmosphere. Its source is decomposition reactions occurring in soils. What would be the temperature, in degrees Celsius, of a 1.53-mole sample of N_2O gas under a pressure of 5.00 atm in a 7.00-L container?

Solution
Three of the four variables in the ideal gas equation (P, V, and n) are given, and the fourth (T) is to be calculated.

$$P = 5.00 \text{ atm} \qquad n = 1.53 \text{ moles}$$
$$V = 7.00 \text{ L} \qquad T = ? \text{ K}$$

Rearranging the ideal gas equation to isolate T on the left side gives

$$T = \frac{PV}{nR}$$

Since the pressure is given in atmospheres and the volume in liters, the value of R to be used is

$$R = 0.0821 \, \frac{\text{atm} \cdot \text{L}}{\text{mole} \cdot \text{K}}$$

Substituting numerical values into the equation gives

$$T = \frac{(5.00 \text{ atm})(7.00 \text{ L})}{(1.53 \text{ moles})\left(0.0821 \, \frac{\text{atm} \cdot \text{L}}{\text{mole} \cdot \text{K}}\right)}$$

Notice again how the gas constant units, except for K, cancel. After cancellation, the expression $1/(1/K)$ remains. This expression is equivalent to K. That this is the case can be easily shown. All we need to do is multiply both the numerator and denominator of the fraction by K.

$$\frac{1 \times K}{\frac{1}{K} \times K} = K$$

Doing the arithmetic, we get as an answer 279 K for the temperature of the O_2 gas.

$$T = \frac{(5.00)(7.00)}{(1.53)(0.0821)} \text{ K} = 278.63358 \text{ K} \quad \text{(calculator answer)}$$

$$= 279 \text{ K} \quad \text{(correct answer)}$$

The calculated temperature is in kelvins. To convert to degrees Celsius, the unit specified in the problem statement, we subtract 273 from the Kelvin temperature.

$$T(°C) = 279\ K - 273 = 6\ °C$$

PRACTICE EXERCISE 12.17

What is the temperature, in degrees Celsius, of a 1.76-mole sample of NO_2 gas under a pressure of 3.15 atm in a 7.37-L container? *Ans.* $-112°C$.

Similar exercises: Problems 12.77 and 12.78

A word about the phrase "*ideal gas*" in the name ideal gas law is in order. An **ideal gas** is a gas that obeys exactly all of the statements of kinetic molecular theory (Sec. 11.3) and obeys exactly the ideal gas law. Real gases are not ideal gases; that is, real gases do not obey exactly the ideal gas equation. Nonetheless, for real gases under ordinary conditions of temperature and pressure, deviations from ideal gas behavior are small, and the ideal gas law (as well as the other laws discussed in this chapter) gives accurate information about gas behavior. It is only at low temperatures and/or high pressures (near liquefaction conditions) that ideal gas behavior breaks down. At low temperatures, because of the slower motion of the molecules, attractive forces between molecules begin to become important. At high pressures, where the molecules are forced closer together, attractive forces again begin to cause deviation from behavior predicted by the ideal gas law.

12.12 Equations Derived from the Ideal Gas Law

Some of the most useful calculations involving the ideal gas equation are those in which the mass, molar mass, or density of a gas is determined. Such calculations are performed by using modified forms of the ideal gas equation.

The number of moles of any substance is equal to the mass (m) of the substance in grams divided by the substance's molar mass ($\mathcal{M}$).

$$n = \frac{m}{\mathcal{M}}$$

Replacing n in the ideal gas equation with this equivalent expression gives

$$PV = \frac{m}{\mathcal{M}} RT$$

This equation, in the rearranged form

$$m = \frac{PV\mathcal{M}}{RT}$$

is used to calculate the mass, in grams, of a gas. This same equation, in the rearranged form

$$M = \frac{mRT}{PV}$$

is used to calculate the molar mass of a gas.

The density, d, of a gas has the units of mass (grams) per unit volume (liters).

$$d = \frac{m}{V}$$

Solving the modified ideal gas equation for m/V gives

$$\frac{m}{V} = \frac{PM}{RT} \quad \text{or} \quad d = \frac{PM}{RT}$$

Examples 12.18 and 12.19 illustrate the use of these new ideal-gas-equation relationships.

EXAMPLE 12.18

A 0.276 g sample of oxygen gas (O_2) occupies a volume of 0.270 L at 739 mm Hg and 98 °C. Calculate, from these data, the molar mass of gaseous O_2.

Solution

The molar mass of a gas is calculated by the ideal gas equation in the modified form

$$M = \frac{mRT}{PV}$$

All of the quantities on the right side of the equation are known.

$$m = 0.276 \text{ g} \quad R = 62.4 \frac{\text{mm Hg} \cdot \text{L}}{\text{mole} \cdot \text{K}} \quad T = 98 \text{ °C} = 371 \text{ K}$$

$$P = 739 \text{ mm Hg} \quad V = 0.270 \text{ L}$$

Substitution of these values into the equation gives

$$M = \frac{(0.276 \text{ g})\left(62.4 \frac{\cancel{\text{mm Hg}} \cdot \cancel{L}}{\text{mole} \cdot \cancel{K}}\right)(371 \cancel{K})}{(739 \cancel{\text{mm Hg}})(0.270 \cancel{L})}$$

All units cancel except for g/mole, the units of molar mass. Recall that the molar mass of a substance is the mass in grams of 1 mole of the substance.

Doing the arithmetic, we obtain a value of 32.0 g/mole for the molar mass of oxygen (O_2).

$$M = \frac{(0.276)(62.4)(371)}{(739)(0.270)} \frac{\text{g}}{\text{mole}} = 32.022806 \text{ g/mole} \quad \text{(calculator answer)}$$
$$= 32.0 \text{ g/mole} \quad \text{(correct answer)}$$

This experimental molar mass value is in agreement with the molar mass value obtained with a table of atomic masses, 32.0 g/mole.

PRACTICE EXERCISE 12.18

A 1.07-g sample of nitric oxide (NO) occupies a volume of 1.00 L at 979 mm Hg and 167 °C. Calculate, from these data, the molar mass of gaseous NO.

Ans. 30.0 g/mole.

Similar exercises: Problems 12.89 and 12.90

EXAMPLE 12.19

Decomposition processes in wet locations where oxygen is scarce, such as swamps and natural wetlands, produce the gas methane, CH_4. Calculate what the density of this gas would be, in grams per liter, at 0.954 atm pressure and a temperature of 45 °C.

Solution

The ideal gas equation in the modified form

$$d = \frac{P\mathcal{M}}{RT}$$

is used to calculate the density of a gas.

All of the quantities on the right side of this equation are known.

$P = 0.954$ atm

$\mathcal{M} = 16.0$ g/mole
(calculated from a table of atomic masses)

$R = 0.0821 \dfrac{\text{atm} \cdot \text{L}}{\text{mole} \cdot \text{K}}$

$T = 45\,°C = 318$ K

Substitution of these values into the equation gives

$$d = \frac{(0.954\,\text{atm})\left(16.0\,\dfrac{\text{g}}{\text{mole}}\right)}{\left(0.0821\,\dfrac{\text{atm} \cdot \text{L}}{\text{mole} \cdot \text{K}}\right)(318\,\text{K})}$$

All units cancel except for the desired ones, grams per liter.

Doing the arithmetic, we obtain a value of 0.585 g/L for the density of CH_4 gas at the specified temperature and pressure.

$$d = \frac{(0.954)(16.0)}{(0.0821)(318)} \text{ g/L}$$

$= 0.58465286$ g/L (calculator answer)

$= 0.585$ g/L (correct answer)

PRACTICE EXERCISE 12.19

Calculate the density of CO_2 gas, in grams per liter, at 1.21 atm pressure and a temperature of 35 °C.

Ans. 2.11 g/L.

Similar exercises: Problems 12.91–12.94

12.13 Gas Laws and Chemical Equations

In Section 10.8 we learned to calculate the mass of any component (reactant or product) of a chemical reaction, given the balanced equation for the reaction and mass of any other component—a mass-to-mass (or gram-to-gram) problem. A simple extension of the procedures involved in solving such problems will enable us to do mass-to-volume and volume-to-mass calculations for reactions where at least one gas is involved.

If you are given the number of liters of any gaseous component of a reaction and are asked to calculate the mass of any other component (gaseous or otherwise)—a *volume-to-mass problem*—first convert the given volume of gas to moles of gas, using the principles in this chapter. If the reaction is at STP conditions, use molar volume as the conversion factor to go from volume to moles. At other temperatures and pressures, use the ideal gas law (or the combined gas law and molar volume) to make the conversion. Once you obtain moles of gas the rest of the calculation is a standard mole-to-gram conversion (Sec. 10.8).

If you are given the number of grams of any component (gaseous or otherwise) in a chemical reaction and are asked to calculate the volume of any gaseous component (reactant or product)—*a mass-to-volume problem*—use standard procedures to calculate the moles of gas and then the new procedures of this chapter to go from moles to volume.

Figure 12.9 summarizes the relationships between the conversion factors needed to solve mass-to-volume and volume-to-mass types of chemical-equation-based problems.

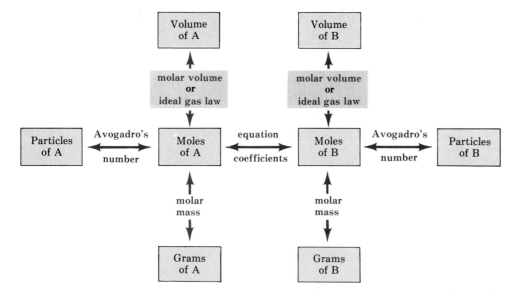

FIGURE 12.9 Quantitative relationships needed for solving mass-to-volume and volume-to-mass chemical-equation-based problems.

EXAMPLE 12.20

Magnesium and nitrogen react to produce magnesium nitride as shown by the equation

$$3\ Mg(s) + N_2(g) \longrightarrow Mg_3N_2(s)$$

What volume, in liters, of N_2 at STP conditions will completely react with 63.0 g of Mg?

Solution

STEP 1   The given quantity is 63.0 g of Mg, and the desired quantity is liters of N_2 at STP conditions.

$$63.0\ g\ Mg = ?\ L\ N_2$$

STEP 2   This is a "grams of A" to "volume of B" type of problem. The pathway used in solving it, in terms of Figure 12.9, is

$$\boxed{\text{Grams of A}} \xrightarrow{\text{molar mass}} \boxed{\text{Moles of A}} \xrightarrow{\text{equation coefficients}} \boxed{\text{Moles of B}} \xrightarrow{\text{molar volume}} \boxed{\text{Volume of B}}$$

The dimensional analysis setup for the calculation is

$$63.0\ \cancel{g\ Mg} \times \frac{1\ \cancel{mole\ Mg}}{24.3\ \cancel{g\ Mg}} \times \frac{1\ \cancel{mole\ N_2}}{3\ \cancel{moles\ Mg}} \times \frac{22.4\ L\ N_2}{1\ \cancel{mole\ N_2}}$$

STEP 3   The solution, obtained from combining all the numerical factors, is

$$\frac{63.0 \times 1 \times 1 \times 22.4}{24.3 \times 3 \times 1}\ L\ N_2 = 19.358024\ L\ N_2 \quad \text{(calculator answer)}$$

$$= 19.4\ L\ N_2 \quad \text{(correct answer)}$$

PRACTICE EXERCISE 12.20

Aluminum and oxygen react to produce aluminum oxide as shown by the equation

$$4\ Al(s) + 3\ O_2(g) \longrightarrow 2\ Al_2O_3(s)$$

What volume, in liters, of O_2 at STP conditions will completely react with 75.0 g of Al?

Ans. $46.7\ L\ O_2$.

Similar exercises: Problems 12.101 and 12.102

EXAMPLE 12.21

Glucose ($C_6H_{12}O_6$), the human body's main energy source, is metabolized according to the overall reaction

$$C_6H_{12}O_6(s) + 6\ O_2(g) \longrightarrow 6\ CO_2(g) + 6\ H_2O(l)$$

How many grams of H_2O can be produced from the reaction of 7.50 L of O_2 at 1.00 atm pressure and a temperature of 37 °C with an excess of glucose?

Solution

STEP 1 The given quantity is 7.50 L of O_2, and the desired quantity is grams of H_2O.

$$7.50 \text{ L } O_2 = ? \text{ g } H_2O$$

STEP 2 This is a "volume of A" to "grams of B" problem. The pathway, in terms of Figure 12.9, is

$$\boxed{\text{Volume of A}} \xrightarrow{\text{ideal gas law}} \boxed{\text{Moles of A}} \xrightarrow{\text{equation coefficients}} \boxed{\text{Moles of B}} \xrightarrow{\text{molar mass}} \boxed{\text{Grams of B}}$$

The first conversion step, from volume (O_2) to moles (O_2), can be made by applying the ideal gas law.

$$n = \frac{PV}{RT} = \frac{(1.00 \cancel{\text{ atm}})(7.50 \cancel{\text{ L}})}{\left(0.0821 \frac{\cancel{\text{atm}} \cdot \cancel{\text{L}}}{\text{mole} \cdot \cancel{\text{K}}}\right)(310 \cancel{\text{ K}})}$$

$$= 0.2946839 \text{ mole } O_2 \quad \text{(calculator answer)}$$

$$= 0.295 \text{ mole } O_2 \quad \text{(correct answer)}$$

The dimensional analysis setup for the remaining conversion steps is

$$0.295 \cancel{\text{ mole } O_2} \times \frac{6 \cancel{\text{ moles } H_2O}}{6 \cancel{\text{ moles } O_2}} \times \frac{18.0 \text{ g } H_2O}{1 \cancel{\text{ mole } H_2O}}$$

STEP 3 The solution, obtained from combining all the numerical factors, is

$$\frac{0.295 \times 6 \times 18.0}{6 \times 1} \text{ g } H_2O = 5.31 \text{ g } H_2O \quad \text{(calculator and correct answer)}$$

An alternative method can be used for converting the liters of O_2 to moles of O_2. The combined gas law can be used to calculate the STP volume of oxygen, and then the molar volume relationship can be used to calculate the moles of oxygen.

The combined gas law, rearranged to isolate V_2 on the left side, is

$$V_2 = V_1 \times \frac{P_1}{P_2} \times \frac{T_2}{T_1}$$

The given values for the variables are

$P_1 = 1.00$ atm $P_2 = 1.00$ atm (STP)
$V_1 = 7.50$ L $V_2 = ?$ L
$T_1 = 37\,°C = 310$ K $T_2 = 0\,°C = 273$ K (STP)

Substituting these values into the combined gas law equation we get

$$V_2 = 7.50 \text{ L} \times \frac{1.00 \text{ atm}}{1.00 \text{ atm}} \times \frac{273 \text{ K}}{310 \text{ K}}$$

$$= 6.6048387 \text{ L} \quad \text{(calculator answer)}$$

$$= 6.60 \text{ L} \quad \text{(correct answer)}$$

Now, changing the STP volume to moles by the molar volume relationship, we get

$$6.60 \text{ L} \times \frac{1 \text{ mole O}_2}{22.4 \text{ L O}_2} = 0.29464285 \text{ mole O}_2 \quad \text{(calculator answer)}$$

$$= 0.295 \text{ mole O}_2 \quad \text{(correct answer)}$$

This is the same value we previously obtained with the ideal gas law.

PRACTICE EXERCISE 12.21

Lead(IV) chloride ($PbCl_4$) can be produced from its constituent elements as shown by the equation

$$Pb(s) + 2 Cl_2(g) \longrightarrow PbCl_4(s)$$

How many grams of $PbCl_4$ can be produced from the reaction of 5.00 L of Cl_2 at 1.33 atm pressure and a temperature of 7 °C with an excess of Pb metal?

Ans. 50.4 g $PbCl_4$.

Similar exercises: Problems 12.103 and 12.104

12.14 Dalton's Law of Partial Pressures

In a mixture of gases that do not react with each other, each type of molecule moves about in the container as if the other kinds were not there. This type of behavior is possible because attractions between molecules in the gaseous state are negligible at most temperatures and pressures and because a gas is mostly empty space (Sec. 11.6). Each gas in the mixture occupies the entire volume of the container, that is, it distributes itself uniformly throughout the container. The molecules of each type strike the walls of the container as frequently and with the same energy as if they were the only gas in the mixture. Consequently, the pressure exerted by a gas in a mixture is the same as it would be if the gas were alone in the same container under the same conditions.

John Dalton—the same John Dalton discussed in Section 5.1—was the first to notice this independent behavior of gases in mixtures. In 1803 he published a summary statement concerning such behavior, which is now known as **Dalton's law of partial pressures:** The total pressure exerted by a mixture of gases is the sum of the partial pressures of the individual gases. A new term, *partial pressure*, is used in stating Dalton's law. A **partial pressure** is the pressure that a gas in a mixture would exert if it were present alone under the same conditions.

470 Chapter 12 · Gas Laws

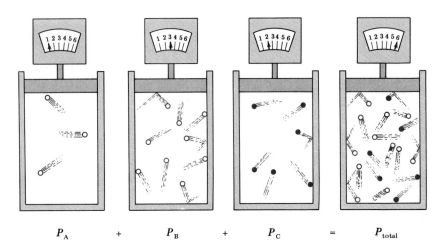

FIGURE 12.10 Dalton's law of partial pressures.

Expressed mathematically, Dalton's law states that

$$P_T = P_1 + P_2 + P_3 + \cdots$$

where P_T is the total pressure of a gaseous mixture and P_1, P_2, P_3, and so on are the partial pressures of the individual gaseous components of the mixture. (When the identity of a gas is known, its molecular formula is used as a subscript in the partial-pressure notation; for example, P_{CO_2} is the partial pressure of carbon dioxide in a mixture.)

To illustrate Dalton's law, consider the four identical gas containers shown in Figure 12.10. Suppose we place amounts of three different gases (represented by A, B, and C) into three of the containers and measure the pressure exerted by each sample. We then place all three samples in the fourth container and measure the pressure exerted by this mixture of gases. It is found that

$$P_{total} = P_A + P_B + P_C$$

Using the pressures given in Figure 12.10, we see that

$$P_{total} = 1 + 3 + 2 = 6$$

EXAMPLE 12.22

Assume that a sample of air, collected when the atmospheric pressure is 742 mm Hg, contains only three gases: nitrogen, oxygen, and water vapor. The partial pressures of nitrogen and oxygen in the sample are found to be 581 mm Hg and 143 mm Hg, respectively. What is the partial pressure of the water vapor present in the air?

Solution
Dalton's law says that

$$P_{total} = P_{N_2} + P_{O_2} + P_{H_2O}$$

The known values for variables in this equation are

$$P_{total} = 742 \text{ mm Hg (atmospheric pressure)}$$
$$P_{N_2} = 581 \text{ mm Hg}$$
$$P_{O_2} = 143 \text{ mm Hg}$$

Rearranging Dalton's law to isolate P_{H_2O} on the left side of the equation gives

$$P_{H_2O} = P_{total} - P_{N_2} - P_{O_2}$$

Substituting the known numerical values into this equation and doing the arithmetic gives

$$P_{H_2O} = 742 \text{ mm Hg} - 581 \text{ mm Hg} - 143 \text{ mm Hg}$$
$$= 18 \text{ mm Hg} \quad \text{(calculator and correct answer)}$$

PRACTICE EXERCISE 12.22

A gaseous mixture contains the three noble gases helium, argon, and krypton. The total pressure exerted by the mixture is 1.57 atm, and the partial pressures of the helium and argon are 0.33 and 0.39 atm, respectively. What is the partial pressure of the krypton present in the mixture?

Ans. 0.85 atm.

Similar exercises: Problems 12.109 and 12.110

The validity of Dalton's law of partial pressures is easily demonstrated by using the ideal gas law (Sec. 12.11). For a mixture of three gases (A, B, and C) the total pressure is given by the expression

$$P_{total} = n_{total} \frac{RT}{V}$$

The total number of moles present is the sum of the moles of A, B, and C; that is,

$$n_{total} = n_A + n_B + n_C$$

Substituting this equation into the previous one gives

$$P_{total} = (n_A + n_B + n_C) \frac{RT}{V}$$

Expanding the right side of this equation results in the expression

$$P_{total} = n_A \frac{RT}{V} + n_B \frac{RT}{V} + n_C \frac{RT}{V}$$

The three individual terms on the right side of this equation are respectively the partial pressures of A, B, and C.

$$P_A = n_A \frac{RT}{V} \quad P_B = n_B \frac{RT}{V} \quad P_C = n_C \frac{RT}{V}$$

Substitution of this information into the previous equation gives

$$P_{total} = P_A + P_B + P_C$$

which is a statement of Dalton's law of partial pressures for a mixture of three gases. In the preceding derivation two of the expressions that were encountered are

$$P_A = n_A \frac{RT}{V} \quad \text{and} \quad P_{total} = n_{total} \frac{RT}{V}$$

If we divide the first of these expressions by the second, we obtain

$$\frac{P_A}{P_{total}} = \frac{n_A \frac{\cancel{RT}}{\cancel{V}}}{n_{total} \frac{\cancel{RT}}{\cancel{V}}} = \frac{n_A}{n_{total}} \quad \text{or} \quad \frac{P_A}{P_{total}} = \frac{n_A}{n_{total}}$$

Rearrangement of this equation gives

$$P_A = P_{total} \times \frac{n_A}{n_{total}}$$

The fraction

$$\frac{n_A}{n_{total}}$$

appearing in this equation is called the mole fraction of A in the mixture. It is the fraction of the total moles that is accounted for by gas A. Expressing the preceding equation in word form gives the following generalization: *the partial pressure of a gas in a mixture is equal to its mole fraction multiplied by the total pressure.* This is a generalization that is very useful in problem solving.

EXAMPLE 12.23

A gaseous mixture under a pressure of 1.20 atm contains 0.40 mole of O_2 and 0.65 mol of N_2. What is the partial pressure of O_2 in the mixture?

Solution

Substituting known quantities into the equation

$$P_A = P_{total} \times \frac{n_A}{n_{total}}$$

gives

$$P_{O_2} = 1.20 \text{ atm} \times \frac{0.40 \;\cancel{\text{mole}}}{(0.40 + 0.65) \;\cancel{\text{mole}}}$$

$$= 0.45714285 \text{ atm} \quad \text{(calculator answer)}$$

$$= 0.46 \text{ atm} \quad \text{(correct answer)}$$

PRACTICE EXERCISE 12.23

A gaseous mixture under a pressure of 0.75 atm contains 0.35 mole of Xe and 1.45 mole of Ar. What is the partial pressure of the Ar in the mixture?

Ans. 0.60 atm.

Similar Exercises: Problems 12.115 and 12.116

The air we breathe is a most important mixture of gases. The composition of clean air from which all water vapor has been removed (dry air) is found to be virtually constant over the entire Earth. Table 12.4 gives the composition of clean dry air in volume percent. All components that make up at least 0.001% of the total volume are listed.

Atmospheric pressure is the sum of the partial pressures of the gaseous components present in air. Table 12.4 also gives the partial pressure of each component of air in a situation where total atmospheric pressure is 760 mm Hg.

The composition of air is not absolutely constant. The variability in composition is caused predominantly by the presence of water vapor, a substance not listed in Table 12.4 because those statistics are for *dry* air. The amount of water vapor in air varies between a few tenths of 1% and 5–6%, depending on weather and temperature.

A common application of Dalton's law of partial pressures is encountered in the laboratory preparation of gases. Such gases are often collected by displacement of water. Figure 12.11 shows O_2, prepared from the decomposition of $KClO_3$, being collected by water displacement. A gas collected by water displacement is never pure. It always contains some water vapor. The total pressure exerted by the gaseous mixture is the sum of the partial pressures of the gas being collected and the water vapor.

$$P_{total} = P_{gas} + P_{H_2O}$$

The pressure exerted by the water vapor in the mixture will be constant at any given temperature if sufficient time has been allowed to establish equilibrium conditions.

For gases collected by water displacement, the partial pressure of the water vapor can be obtained from a table showing the variation of water vapor pressure with temperature (see Table 12.5). Thus the partial pressure of the collected gas is easily determined.

$$P_{gas} = P_{atm} - P_{H_2O}$$

TABLE 12.4 The Major Components of Clean, Dry Air

Gaseous Component	Formula	Volume Percent	Partial Pressure (mm Hg)
Nitrogen	N_2	78.084	593.4
Oxygen	O_2	20.948	159.2
Argon	Ar	0.934	7.1
Carbon dioxide	CO_2	0.031	0.2
Neon	Ne	0.002	—
Helium	He	0.001	—

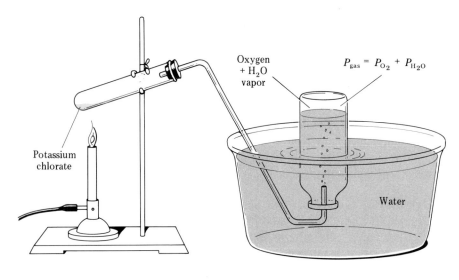

FIGURE 12.11 Collection of oxygen gas by water displacement. Potassium chlorate ($KClO_3$) decomposes to form oxygen (O_2), which is collected over water.

EXAMPLE 12.24

What is the partial pressure of oxygen collected over water at 24 °C when the atmospheric pressure is 752 mm Hg?

Solution
From Table 12.5 we determine that water has a vapor pressure of 22.4 mm Hg at 24 °C.
  Using the equation

$$P_{O_2} = P_{atm} - P_{H_2O}$$

we find the partial pressure of the oxygen to be $73\overline{0}$ mm Hg.

P_{O_2} = 752 mm Hg − 22.4 mm Hg = 729.6 mm Hg   (calculator answer)
       = $73\overline{0}$ mm Hg   (correct answer)

TABLE 12.5 Vapor Pressure of Water at Various Temperatures

T (°C)	Vapor Pressure (mm Hg)	T (°C)	Vapor Pressure (mm Hg)	T (°C)	Vapor Pressure (mm Hg)
15	12.8	22	19.8	29	30.0
16	13.6	23	21.1	30	31.8
17	14.5	24	22.4	31	33.7
18	15.5	25	23.8	32	35.7
19	16.5	26	25.2	33	37.7
20	17.5	27	26.7	34	39.9
21	18.7	28	28.3	35	42.2

PRACTICE EXERCISE 12.24

What is the partial pressure of oxygen collected over water at 33 °C when the atmospheric pressure is 657 mm Hg?

Ans. 619 mm Hg.

Similar exercises: Problems 12.121 and 12.122

Key Terms

The new terms or concepts defined in this chapter are

Avogadro's law (Sec. 12.9) Equal volumes of different gases, measured at the same temperature and pressure, contain equal numbers of molecules.

barometer (Sec. 12.2) Commonly used device for measuring atmospheric pressure.

Boyle's law (Sec. 12.3) The volume of a sample of gas is inversely proportional to the pressure applied to the gas if the temperature is kept constant.

Charles's law (Sec. 12.4) The volume of a sample of gas is directly proportional to its Kelvin temperature if the pressure is kept constant.

combined gas law (Sec. 12.6) The expression obtained from mathematically combining Boyle's, Charles's, and Gay-Lussac's laws.

Dalton's law of partial pressures (Sec. 12.14) The total pressure exerted by a mixture of gases is the sum of the partial pressures of the individual gases.

gas laws (Sec. 12.2) Generalizations that describe in mathematical terms the relationships among the pressure, temperature, and volume of a specific quantity of a gas.

Gay-Lussac's law (Sec. 12.5) The pressure of a sample of gas, at constant volume, is directly proportional to its Kelvin temperature.

Gay-Lussac's law of combining volumes (Sec. 12.8) The volumes of different gases involved in a reaction, if measured at the same temperature and pressure, are in the same ratio as the coefficients for these gases in the balanced equation for the reaction.

ideal gas (Sec. 12.11) A gas that obeys exactly all of the statements of kinetic molecular theory.

ideal gas law (Sec. 12.11) An expression that describes the relationships among the four variables temperature (T), pressure (P), volume (V), and moles of gas (n) for gaseous substances.

manometer (Sec. 12.2) A device used to measure gas pressures in the laboratory.

molar volume (Sec. 12.10) The volume occupied by 1 mole of a gas at STP conditions.

partial pressure (Sec. 12.14) The pressure that a gas in a mixture would exert if it were present alone under the same conditions.

pressure (Sec. 12.2) The force applied per unit area, that is, the total force on a surface divided by the area of that surface.

standard pressure (Sec. 12.7) A pressure of 1 atm (760 mm Hg).

standard temperature (Sec. 12.7) A temperature of 0 °C (273 K).

STP conditions (Sec. 12.7) The conditions of standard temperature and standard pressure.

Practice Problems

Measurement of Pressure (Sec. 12.2)

12.1 Express each of the following pressures in millimeters of mercury (mm Hg).
  (a) 0.456 atm   (b) 20.5 in. Hg   (c) 756 psi

12.2 Express each of the following pressures in millimeters of mercury (mm Hg).
  (a) 1.31 atm   (b) 1.56 in. Hg   (c) 12.2 psi

12.3 A weather report atmospheric pressure is given as

30.31 in. Hg. Express this pressure in the following units.
(a) mm Hg   (b) psi   (c) atm

12.4 A weather report atmospheric pressure is given as 29.76 in. Hg. Express this pressure in the following units.
(a) mm Hg   (b) lb/in.^{2} (c) atm

12.5 Describe what would happen to the level of the mercury column in the barometer of Figure 12.1 in the following situations.
(a) The atmospheric pressure decreases.
(b) More mercury is added to the circular container.

12.6 Describe what would happen to the level of the mercury column in the barometer of Figure 12.1 in the following situations.
(a) A small leak develops in the sealed end of the glass tube.
(b) Some mercury is removed from the circular container.

12.7 The mercury level in the arm of a manometer (see Fig. 12.2) that is open to the atmosphere is found to be 237 mm higher than the mercury level in the arm of the manometer connected to the container of gas. Measured barometric pressure is 762 mm Hg. What is the pressure, in millimeters of mercury, of the gas in the container?

12.8 The mercury level in the arm of a manometer (see Fig. 12.2) that is open to the atmosphere is found to be 35 mm lower than the mercury level in the arm of the manometer connected to the container of gas. Measured barometric pressure is 743 mm Hg. What is the pressure, in millimeters of mercury, of the gas in the container?

Boyle's Law (Sec. 12.3)

12.9 State Boyle's law in terms of a mathematical equation.

12.10 Explain in your own words what it means to say that "the volume of a gas is inversely proportional to its pressure at constant temperature."

12.11 A sample of carbon monoxide (CO) gas in an expandable container has a volume of 3.73 L at a temperature of 23 °C and a pressure of 742 mm Hg. What pressure, in millimeters of mercury, will the CO gas be under if the volume of the container is changed, at constant temperature, to each of the following values?
(a) 3.83 L   (b) 20.00 L
(c) 2650 mL   (d) 975 mL

12.12 A sample of carbon dioxide (CO_2) gas, in an expandable container, has a volume of 1.02 L at a temperature of 25 °C and a pressure of 0.577 atm. What pressure, in atmospheres, will the CO_2 gas be under if the volume of the container is changed, at constant temperature, to each of the following values?
(a) 0.998 L   (b) 7.32 L
(c) 253 mL   (d) 2830 mL

12.13 A sample of ammonia (NH_3), a colorless gas with a pungent odor, occupies a volume of 3.00 L at 27 °C and 2.00 atm pressure. What volume, in liters, will this NH_3 sample occupy at the same temperature, but at each of the following pressures?
(a) 4.23 atm   (b) 0.122 atm
(c) 853 mm Hg   (d) 3350 mm Hg

12.14 A sample of nitrogen dioxide (NO_2), a toxic gas with a reddish brown color, occupies a volume of 76.3 mL at 31 °C and 755 mm Hg pressure. What volume, in milliliters, will this NO_2 sample occupy at the same temperature, but at each of the following pressures?
(a) 252 mm Hg   (b) 1630 mm Hg
(c) 0.911 atm   (d) 4.73 atm

12.15 A sample of O_2 gas under a pressure of 1.0 atm occupies 231 mL in a piston–cylinder arrangement before compression. If the gas is compressed to 0.15 its original volume, what must the new pressure, in atmospheres, be? Assume that the temperature remains constant.

12.16 A sample of N_2 gas under a pressure of 15.4 atm occupies 47.3 mL in a piston–cylinder arrangement before compression. If the gas is compressed to 0.88 its original volume, what must the new pressure, in atmospheres, be? Assume that the temperature remains constant.

12.17 A balloon is inflated to a volume of 11.4 L on a day when the atmospheric pressure is 698 mm Hg. The next day a storm front arrives and the atmospheric pressure drops to 651 mm Hg. Assuming the temperature remains constant, what is the *percent change in the volume* of the balloon?

12.18 On a stormy day, a balloon has a volume of 7.72 L at a barometric pressure of 673 mm Hg. As the storm clears the barometric pressure rises to 712 mm Hg. Assuming that the temperature remains constant, what is the *percent change in the volume* of the balloon.

Charles's Law (Sec. 12.4)

12.19 Explain in your own words what it means to say that "the volume of a gas is directly proportional to its Kelvin temperature at constant pressure."

12.20 State Charles's law in terms of a mathematical equation.

12.21 A sample of H_2 gas has a volume of 2.73 L at 27 °C. What volume, in liters, will the H_2 gas occupy at each of the following temperatures if the pressure is held constant?
(a) 127 °C   (b) 473 °C   (c) −27 °C   (d) 123 °F

12.22 A sample of F_2 gas has a volume of 453 mL at 53 °C. What volume, in liters, will the F_2 gas occupy at each

of the following temperatures if the pressure is held constant?
(a) 55 °C (b) 279 °C (c) −53 °C (d) 53 °F

12.23 A sample of N_2 gas occupies a volume of 355 mL at 25 °C and a pressure of 1.00 atm. At the same pressure, at what temperature, in degrees Celsius, would the volume of the gas be equal to each of the following?
(a) 675 mL (b) 57.2 mL (c) 0.721 L (d) 3.50 L

12.24 A sample of SO_2 gas occupies a volume of 3.53 L at 55 °C and a pressure of 3.58 atm. At the same pressure, at what temperature, in degrees Celsius, would the volume of the gas be equal to each of the following?
(a) 0.275 L (b) 10.3 L (c) 2220 mL (d) 355 mL

12.25 Decomposition of a sample of $KClO_3$ produced 25.2 mL of O_2 gas at a temperature of 27 °C and a pressure of 672 mm Hg. What is the volume, in milliliters, of this O_2 at 45 °C and a pressure of 672 mm Hg?

12.26 A bacterial culture isolated from sewage produced 32.3 mL of methane (CH_4) gas at 29 °C and 741 mm Hg pressure. What is the volume, in milliliters, of this CH_4 at 0 °C and 741 mm Hg pressure?

12.27 It is desired to increase the volume of 75.0 mL of Ne gas by 15.0% while holding the pressure constant. To what temperature, in degrees Celsius, must the gas be heated if the initial temperature is 28 °C?

12.28 It is desired to decrease the volume of 85.0 mL of Ar gas by 35.0% while holding the pressure constant. To what temperature, in degrees Celsius, must the gas be cooled if the initial temperature is 22 °C?

Gay-Lussac's Law (Sec. 12.5)

12.29 State Gay-Lussac's law in terms of a mathematical equation.

12.30 Which gas variables are held constant in Gay-Lussac's law?

12.31 At a room temperature of 27 °C, the pressure exerted by O_2 gas stored in a 20.0 L steel cylinder is 1.50 atm. What will be the pressure of the O_2, in atmospheres, in the cylinder if the temperature is allowed to reach the following levels?
(a) 350 °C (b) −27 °C (c) 1250 °C (d) 1250 °F

12.32 A sample of N_2 gas in a rigid container at −45 °C is at a pressure of 500 mm Hg. What will be the pressure, in millimeters of mercury, of the N_2 in the container if the temperature is allowed to reach the following levels?
(a) −65 °C (b) 45 °C (c) 237 °C (d) 145 °F

12.33 A sample of He gas at 55 °C in a constant volume container exerts a pressure of 1.31 atm. To what temperature, in degrees Celsius, would the gas have to be heated for it to exert each of the following pressures?
(a) 3.93 atm (b) 7.00 atm
(c) 15.0 atm (d) 2450 mm Hg

12.34 A sample of He gas at 122 °C in a constant volume container exerts a pressure of 9.75 atm. To what temperature, in degrees Celsius, would the gas have to be cooled for it to exert each of the following pressures?
(a) 5.00 atm (b) 3.34 atm
(c) 2.00 atm (d) 3550 mm Hg

12.35 A spray can is empty except for the propellant gas, which exerts a pressure of 1.2 atm at 24 °C. If the can is thrown into a fire (485 °C), what will be the pressure, in atmospheres, inside the hot can?

12.36 A spray can is empty, except for the propellant gas, which exerts a pressure of 1.2 atm at 24 °C. If the can is placed in a refrigerator (3 °C), what will be the pressure, in atmospheres, inside the cold can?

The Combined Gas Law (Sec. 12.6)

12.37 Rearrange the standard form of the combined gas law equation such that the variable V_1 is isolated on the left side of the equation.

12.38 Rearrange the standard form of the combined gas law equation such that the variable P_1 is isolated on the left side of the equation.

12.39 A sample of CO_2 gas has a volume of 15.2 L at a pressure of 1.35 atm and a temperature of 33 °C. Determine the following for this gas sample.
(a) volume, in liters, at $T = 35$ °C and $P = 3.50$ atm
(b) volume, in milliliters, at $T = 97$ °C and $P = 6.70$ atm
(c) pressure, in atmospheres, at $T = 42$ °C and $V = 10.0$ L
(d) temperature, in degrees Celsius, at $P = 7.00$ atm and $V = 0.973$ L

12.40 A sample of NO_2 gas has a volume of 37.3 mL at a pressure of 621 mm Hg and a temperature of 52 °C. Determine the following for this gas sample.
(a) volume, in milliliters, at $T = 35$ °C and $P = 650$ mm Hg
(b) volume, in liters, at $T = 43$ °C and $P = 1.11$ atm
(c) pressure, in millimeters of mercury, at $T = 125$ °C and $V = 52.4$ mL
(d) temperature, in degrees Celsius, at $P = 775$ mm Hg and $V = 23.0$ mL

12.41 Gasoline vapor is injected into a cylinder of volume 3.80 mL at a temperature of 22 °C and a pressure of 741 mm Hg. The vapor is compressed under a pressure of 9.25 atm to 1.01 mL. What is the final temperature, in degrees Celsius, of the gasoline vapor?

12.42 A helium-filled weather balloon, when released, has a volume of 10.0 L at 27.0 °C and a pressure of 0.833 atm. In the upper atmosphere at a location where the pressure is 0.329 atm the balloon volume has increased to 20.5 L. What is the temperature, in degrees Celsius, of the balloon at this higher altitude?

12.43 A sample of N_2O gas in a 275-mL container at a pressure of 655 mm Hg and a temperature of 25 °C is transferred to a container with a volume of 382 mL.
(a) What is the new pressure, in millimeters of mercury, if no change in temperature occurs?
(b) What is the new temperature, in degrees Celsius, if no change in pressure occurs?

12.44 A sample of NO gas in a 375-mL container at a pressure of 713 mm Hg and a temperature of 22 °C is transferred to a container with a volume of 282 mL.
(a) What is the new pressure, in millimeters of mercury, if no change in temperature occurs?
(b) What is the new temperature, in degrees Celsius, if no change in pressure occurs?

12.45 A sample of NH_3 gas in a nonrigid container occupies a volume of 4.00 L at a certain temperature and pressure. What will be its volume, in liters, in each of the following situations?
(a) Both pressure and Kelvin temperature are doubled.
(b) Both pressure and Kelvin temperature are cut in half.
(c) The pressure is doubled, and the Kelvin temperature is cut in half.
(d) The pressure is cut in half, and the Kelvin temperature is tripled.

12.46 A sample of NO_2 gas in a nonrigid container, at a temperature of 24 °C, occupies a certain volume at a certain pressure. What will be its temperature, in degrees Celsius, in each of the following situations?
(a) Both pressure and volume are doubled.
(b) Both pressure and volume are cut in half.
(c) The pressure is doubled, and the volume is cut in half.
(d) The pressure is cut in half, and the volume is tripled.

STP Conditions (Sec. 12.7)

12.47 A quantity of CH_4 gas has a volume of 0.835 L at STP. What volume, in liters, will the CH_4 occupy if the pressure and temperature are changed to the following values?
(a) 1.03 atm and 21 °C
(b) 5.71 atm and 325 °C
(c) 2.10 atm and −25 °C
(d) 453 mm Hg and 546 °C

12.48 A quantity of C_2H_2 gas has a volume of 1.41 L at STP. What volume, in liters, will the C_2H_2 occupy if the pressure and temperature are changed to the following values?
(a) 0.0573 atm and 19 °C
(b) 0.567 atm and −43 °C
(c) 2.20 atm and 257 °C
(d) 453 mm Hg and 546 °C

12.49 Change the following volumes of neon gas to STP conditions.
(a) 5.73 L at 30 °C and 2.00 atm
(b) 25.1 mL at 75 °C and 5.35 atm
(c) 0.431 L at 235 °C and 150.3 atm
(d) 275 mL at −198 °C and 285 mm Hg

12.50 Change the following volumes of argon gas to STP conditions.
(a) 4.51 L at 35 °C and 5.54 atm
(b) 357 mL at −10 °C and 1.50 atm
(c) 0.0571 L at 457 °C and 0.571 atm
(d) 3.00 mL at −103 °C and 702 mm Hg

Gay-Lussac's Law of Combining Volumes (Sec. 12.8)

12.51 The equation for the combustion of acetylene, C_2H_2, in an oxyacetylene torch is

$$2\,C_2H_2(g) + 5\,O_2(g) \rightarrow 4\,CO_2(g) + 2\,H_2O(g)$$

(a) What volume, in liters, of O_2 is needed to react with 15.0 L of C_2H_2, if both gaseous volumes are determined at the same temperature and pressure?
(b) What volume, in liters, of CO_2 can be prepared from 2.5 L of C_2H_2 and an excess of O_2, if both volumes are determined at the same temperature and pressure?

12.52 Elemental nitrogen gas can be obtained from ammonia, NH_3, using the following reaction:

$$4\,NH_3(g) + 3\,O_2(g) \rightarrow 2\,N_2(g) + 6\,H_2O(g)$$

(a) What volume, in liters, of O_2 is needed to react with 15.0 L of NH_3, if both gaseous volumes are determined at the same temperature and pressure?
(b) What volume, in liters, of N_2 can be prepared from 2.5 L of NH_3 and an excess of O_2, if both volumes are determined at the same temperature and pressure?

12.53 The equation for the combustion of the fuel propane, C_3H_8, is

$$C_3H_8(g) + 5\,O_2(g) \rightarrow 3\,CO_2(g) + 4\,H_2O(g)$$

(a) How many liters of C_3H_8 must be burned to produce 1.30 L of CO_2 if both volumes are measured at STP?

(b) How many liters of C_3H_8 must be burned to produce 1.30 L of H_2O if both volumes are measured at 2.00 atm and 56 °C?

12.54 The equation for the combustion of the fuel butane, C_4H_{10}, is

$$2\,C_4H_{10}(g) + 13\,O_2(g) \rightarrow 8\,CO_2(g) + 10\,H_2O(g)$$

(a) How many liters of C_4H_{10} must be burned to produce 2.60 L of CO_2 if both volumes are measured at STP?
(b) How many liters of C_4H_{10} must be burned to produce 2.60 L of H_2O if both volumes are measured at 1.75 atm and 43 °C?

Avogadro's Law (Sec. 12.9)

12.55 A 3.67-mole sample of argon gas has a volume of 2.00 L at a certain temperature and pressure. What is the volume of a 1.50-mole sample of the same gas at the same temperature and pressure?

12.56 A 0.513-mole sample of xenon gas has a volume of 10.2 L at a certain temperature and pressure. What is the volume of a 5.13-mole sample of the same gas at the same temperature and pressure?

12.57 A 0.527-mole sample of O_2 gas at 1.00 atm and 32 °C occupies a volume of 1.37 L. What volume, in liters, would a 0.527-mole sample of N_2 gas occupy at the same temperature and pressure?

12.58 A 0.333-mole sample of SO_2 gas at 1.25 atm and 314 °C occupies a volume of 2.54 L. What volume, in liters, would a 0.333-mole sample of CO_2 gas occupy at the same temperature and pressure?

12.59 A balloon containing 1.83 moles of He has a volume of 0.673 L at a certain temperature and pressure. How many *grams* of He would have to be added to the balloon in order for the volume to increase to 0.811 L at the same temperature and pressure?

12.60 A balloon containing 1.83 moles of He has a volume of 0.673 L at a certain temperature and pressure. How many *grams* of He would have to be removed from the balloon in order for the volume to decrease to 0.455 L at the same temperature and pressure?

12.61 A nitrogen–oxygen compound decomposes to form nitrogen and oxygen. Determine the formula of this nitrogen–oxygen compound given that 2.78 L of the compound decomposes to form 2.78 L of N_2 and 1.39 L of O_2, all volumes being measured at the same temperature and pressure.

12.62 A nitrogen–oxygen compound decomposes to form nitrogen and oxygen. Determine the formula of this nitrogen–oxygen compound given that 3.22 L of the compound decomposes to form 1.61 L of N_2 and 3.22 L of O_2, all volumes being measured at the same temperature and pressure.

Molar Volume (Sec. 12.10)

12.63 What is the volume, in liters, at STP occupied by 2.33 moles of each of the following gases?
(a) H_2 (b) CH_4 (c) SO_2 (d) F_2

12.64 What is the volume, in liters, at STP occupied by 4.07 moles of each of the following gases?
(a) O_2 (b) N_2O (c) CO   (d) HCN

12.65 What is the volume, in liters, at STP occupied by 24.5 g of each of the following gases?
(a) N_2 (b) NH_3 (c) CO_2 (d) Cl_2

12.66 What is the volume, in liters, at STP occupied by 42.1 g of each of the following gases?
(a) O_3 (b) O_2 (c) NO_2 (d) C_2H_6

12.67 What is the mass, in grams, of each of the following volumes of gas at STP?
(a) 43.0 L of H_2 (b) 1.33 L of NO
(c) 23.0 L of CO_2 (d) 9.76 L of CH_4

12.68 What is the mass, in grams, of each of the following volumes of gas at STP?
(a) 125 L of N_2 (b) 12.5 L of N_2O
(c) 13.5 L of C_2H_6 (d) 0.555 L of Ar

12.69 Calculate the density at STP conditions, in grams per liter, of each of the following gases.
(a) He   (b) CO   (c) HCN   (d) O_3

12.70 Calculate the density at STP conditions, in grams per liter, of each of the following gases.
(a) Ne   (b) OF_2 (c) NO_2 (d) SO_2

12.71 What is the molar mass of a gas with a density of 1.16 g/L at STP?

12.72 What is the molar mass of a gas with a density of 2.03 g/L at STP?

12.73 If the Xe atoms present in 22.5 g of XeO_2F_2 are converted to gaseous Xe, what volume, in liters, will they occupy at STP conditions?

12.74 If the Xe atoms present in 235 g of $XeOF_4$ are converted to gaseous Xe, what volume, in liters, will they occupy at STP conditions?

The Ideal Gas Law (Sec. 12.11)

12.75 Using the ideal gas law, calculate the volume, in liters, of 0.100 mole of O_2 gas at each of the following sets of conditions.

(a) STP
(b) 98 °C and 1.21 atm
(c) 515 °C and 20.0 atm
(d) −7 °C and 775 mm Hg

12.76 Using the ideal gas law, calculate the volume, in liters, of 0.315 mole of CO_2 gas at each of the following sets of conditions.
(a) STP
(b) −145 °C and 0.0567 atm
(c) 427 °C and 1345 mm Hg
(d) 23 °C and 23 mm Hg

12.77 3.00 moles of Ne gas is introduced into a 6.00-L fixed-volume container. The container is heated until the pressure of the gas becomes 27.5 atm. What is the temperature, in degrees Celsius, of the Ne gas?

12.78 What is the temperature, in degrees Celsius, of 1.23 moles of He gas confined to a volume of 1.23 L at a pressure of 1.23 atm?

12.79 Chlorine gas, Cl_2, is widely used to purify municipal water supplies and to treat swimming pool waters. Calculate the following quantities for the described samples of Cl_2 gas.
(a) the volume of Cl_2, in liters, if 1.14 moles has a pressure of 365 mm Hg at a temperature of 22 °C
(b) the quantity of Cl_2, in moles, if 2.30 L at 35 °C has a pressure of 2.12 atm

12.80 Fluorine gas, F_2, is one of the most reactive of all gases. Calculate the following quantities for the described samples of F_2 gas.
(a) the volume of F_2, in liters, if 2.08 moles has a pressure of 1.23 atm at a temperature of 37 °C
(b) the quantity of F_2, in moles, if 2.30 L at 23 °C has a pressure of 1.45 atm

12.81 1.14 moles of the noble gas argon occupies a volume of 3.00 L at a temperature of 152 °C and a pressure of 13.3 atm. Use this information to calculate the value of the ideal gas constant R in the units atm · L/mole · K.

12.82 0.317 moles of the noble gas helium occupies a volume of 8.67 L at a temperature of 25 °C and a pressure of 679 mm Hg. Use this information to calculate the value of the ideal gas constant R in the units mm Hg · L/mole · K.

12.83 A 1.00-mole sample of liquid water is placed in a flexible sealed container and allowed to evaporate. After complete evaporation, what will be the container volume, in liters, at 27 °C and 0.908 atm pressure?

12.84 A 1.00-mole sample of dry ice (solid CO_2) is placed in a flexible sealed container and allowed to sublime. After all the CO_2 has changed from solid to gas, what will be the container volume, in liters, at 23 °C and 0.983 atm pressure?

12.85 A 25.0-L cylinder contains 87.0 g of N_2 at 17 °C. How many grams of N_2 must be removed to reduce the pressure in the cylinder to 1.65 atm?

12.86 A 25.0-L cylinder contains 87.0 g of N_2 at 17 °C. How many grams of N_2 must be added to increase the pressure in the cylinder to 9.78 atm?

Equations Derived from the Ideal Gas Law (Sec. 12.12)

12.87 Calculate the mass, in grams, of each of the following quantities of gas.
(a) 30.0 L of CH_4 at 1.25 atm and 31 °C
(b) 1.11 L of H_2 at 546 mm Hg and 123 °C
(c) 4.00 L of O_2 at STP
(d) 67.5 mL of N_2 at 382 mm Hg and −100 °C

12.88 Calculate the mass, in grams, of each of the following quantities of gas.
(a) 4.62 L of C_2H_6 at 0.879 atm and 57 °C
(b) 1.07 L of N_2O at 781 mm Hg and 15 °C
(c) 15.0 L of SiH_4 at STP
(d) 154 mL of Cl_2 at 1570 mm Hg and −35 °C

12.89 What is the molar mass of a gas if a 3.00-g sample of it occupies a volume of 6.00 L at 85 °C and 1.11 atm pressure?

12.90 What is the molar mass of a gas if a 1.65-g sample of it occupies a volume of 3.00 L at 75 °C and 1.05 atm pressure?

12.91 Calculate the densities, in grams per liter, of the following gases at 2.00 atm pressure and a temperature of 55 °C.
(a) N_2 (b) He   (c) ClF   (d) N_2O

12.92 Calculate the densities, in grams per liter, of the following gases at 1.50 atm pressure and a temperature of 35 °C.
(a) O_2 (b) O_3 (c) H_2S (d) SCl_2

12.93 Calculate the density of SO_2 gas, in grams per liter, at each of the following temperature–pressure conditions.
(a) 27 °C and 0.889 atm
(b) 27 °C and 8.89 atm
(c) 127 °C and 0.889 atm
(d) 515 °C and 7.50 atm

12.94 Calculate the density of CH_4 gas, in grams per liter, at each of the following temperature–pressure conditions.
(a) 25 °C and 0.952 atm
(b) 253 °C and 0.952 atm
(c) 25 °C and 3.14 atm
(d) 425 °C and 25.0 atm

12.95 Calculate the molar mass of a gas having a density of 2.20 g/L at 31 °C and 745 mm Hg.

12.96 Calculate the molar mass of a gas having a density of 1.80 g/L at 51 °C and 857 mm Hg.

12.97 What pressure, in atm, is required to cause CO gas, at 47 °C, to have a density of 0.500 g/L?

12.98 What pressure, in mm Hg, is required to cause SO_2 gas, at 25 °C, to have a density of 1.08 g/L?

Gas Laws and Chemical Equations (Sec. 12.13)

12.99 A mixture of 50.0 g of N_2 and an excess of O_2 react according to the balanced equation

$$N_2(g) + O_2(g) \rightarrow 2\ NO(g)$$

How many liters of NO, at STP, are produced?

12.100 A mixture of 25.0 g of CO and an excess of O_2 react according to the balanced equation

$$2\ CO(g) + O_2(g) \rightarrow 2\ CO_2(g)$$

How many liters of CO_2, at STP, are produced?

12.101 How many liters of O_2, at STP, can be produced by heating 50.0 g of $KClO_3$? The equation for the reaction is

$$2\ KClO_3(s) \rightarrow 2\ KCl(s) + 3\ O_2(g)$$

12.102 How many liters of CO_2, at STP, can be produced by heating 100.0 g of $CaCO_3$? The equation for the reaction is

$$CaCO_3(s) \rightarrow CaO(s) + CO_2(g)$$

12.103 A sample of O_2 with a volume of 10.3 L at 27 °C and 743 mm Hg pressure is reacted with excess NO to produce NO_2. The equation for the reaction is

$$2\ NO(g) + O_2(g) \rightarrow 2\ NO_2(g)$$

How many grams of NO_2 are produced?

12.104 A sample of NH_3 with a volume of 10.3 L at 27 °C and 1.21 atm pressure is reacted with excess O_2 according to the following reaction.

$$4\ NH_3(g) + 5\ O_2(g) \rightarrow 4\ NO(g) + 6\ H_2O(g)$$

How many grams of H_2O are produced?

12.105 Hydrogen gas can be produced in the laboratory through reaction of magnesium metal with nitric acid:

$$Mg(s) + 2\ HNO_3(l) \rightarrow Mg(NO_3)_2(aq) + H_2(g)$$

What volume, in liters, of H_2 at 23 °C and 0.980 atm pressure can be produced from the reaction of 12.0 g of Mg with an excess of HNO_3?

12.106 A common laboratory preparation for O_2 gas involves the thermal decomposition of potassium nitrate:

$$2\ KNO_3(s) \rightarrow 2\ KNO_2(s) + O_2(g)$$

What volume, in liters, of O_2 at 35 °C and 1.31 atm pressure can be produced from the decomposition of 35.0 g of KNO_3?

12.107 How many liters of NO_2 gas at 21 °C and 2.31 atm must be consumed in producing 75.0 L of NO gas at 38 °C and 645 mm Hg according to the following reaction?

$$3\ NO_2(g) + H_2O(l) \rightarrow 2\ HNO_3(l) + NO(g)$$

12.108 How many liters of $Cl_2(g)$ at 25 °C and 1.50 atm are needed to completely react with 3.42 L of NH_3 gas at 50 °C and 2.50 atm according to the following reaction?

$$2\ NH_3(g) + 3\ Cl_2(g) \rightarrow N_2(g) + 6\ HCl(g)$$

Dalton's Law of Partial Pressures (Sec. 12.14)

12.109 A steel cylinder contains a mixture of nitrogen, oxygen, and carbon dioxide gases. The total pressure in the cylinder is 2235 mm Hg. The pressures exerted by the nitrogen and oxygen are 545 and 685 mm Hg, respectively. What is the partial pressure, in millimeters of mercury, of the carbon dioxide?

12.110 The total pressure exerted by a mixture of O_2, N_2, and He gases is 1.50 atm. What is the partial pressure, in millimeters of mercury, of the O_2, given that the partial pressures of the N_2 and He are 0.75 and 0.33 atm, respectively?

12.111 A gas mixture contains O_2, N_2, and Ar at partial pressures, respectively, of 125, 175, and 225 mm Hg. If CO_2 gas is added to the mixture until the total pressure reaches 623 mm Hg, what is the partial pressure, in millimeters of mercury, of CO_2?

12.112 A gas mixture contains He, Ne, and H_2S at partial pressures, respectively, of 125, 175, and 225 mm Hg. If all of the H_2S is removed from the mixture, what will be the partial pressure, in milliliters of mercury, of Ne?

12.113 Calculate the partial pressure of O_2, in atmospheres, in a gaseous mixture with a volume of 2.50 L at 20 °C given that the mixture composition is
(a) 0.50 mole O_2 and 0.50 mole N_2
(b) 0.50 mole O_2 and 0.75 mole N_2
(c) 0.50 mole O_2, 0.75 mole N_2, and 0.75 mole Ar
(d) 0.50 g O_2 and 0.75 g N_2

12.114 Calculate the partial pressure of Xe, in atmospheres, in a gaseous mixture with a volume of 1.20 L at 32 °C given that the mixture composition is
(a) 0.40 mole Xe and 0.40 mole Ne
(b) 0.40 mole Xe and 0.60 mole Ne

(c) 0.40 mole Xe, 0.60 mole Ne, and 1.25 moles O_2
(d) 0.40 g Xe and 0.60 g Ne

12.115 What is the partial pressure of O_2 in a gaseous mixture whose total pressure is 1.20 atm given that the composition of the mixture is
(a) 0.40 mole O_2 and 0.40 mole Ne
(b) 0.40 mole O_2 and 0.80 mole Ne
(c) an equal number of moles of O_2, N_2, and H_2
(d) an equal number of molecules of O_2, N_2, and H_2

12.116 What is the partial pressure of Xe in a gaseous mixture whose total pressure is 1.55 atm given that the composition of the mixture is
(a) 0.50 mole Xe and 0.50 mole Ne
(b) 0.50 mole Xe and 1.00 mole Ne
(c) an equal number of moles of Xe, Ne, and He
(d) an equal number of atoms of Xe, Ne, and He

12.117 What is the total pressure in a flask that contains 4.0 moles He, 2.0 moles Ne, and 0.50 mole Ar, and in which the partial pressure of Ar is 0.40 atm?

12.118 What is the total pressure in a flask that contains 2.0 moles H_2, 6.0 moles O_2, and 0.50 mole N_2, and in which the partial pressure of H_2 is 0.80 atm?

12.119 The partial pressure of CO_2 in a CO_2–N_2 mixture under a total pressure of 1.20 atm is 0.35 atm. What is the mole fraction of N_2 in the mixture?

12.120 The partial pressure of CO in a CO–He mixture under a total pressure of 4.50 atm is 0.22 atm. What is the mole fraction of He in the mixture?

12.121 What would be the partial pressure, in millimeters of mercury, of O_2 collected over water at the following conditions of temperature and atmospheric pressure?
(a) 19 °C and 743 mm Hg
(b) 28 °C and 645 mm Hg
(c) 34 °C and 762 mm Hg
(d) 21 °C and 0.933 atm

12.122 What would be the partial pressure, in millimeters of mercury, of O_2 collected over water at the following conditions of temperature and atmospheric pressure?
(a) 15 °C and 632 mm Hg
(b) 35 °C and 749 mm Hg
(c) 31 °C and 682 mm Hg
(d) 26 °C and 0.975 atm

Additional Problems

12.123 A large flask (of unknown volume) is filled with air until the pressure reaches 3.6 atm. The flask is then attached to a second evacuated flask of known volume, and the air from the first flask is allowed to expand into the second flask. The final pressure of the air (in both flasks) is 2.6 atm, and the volume of the second flask is 5.21 L. Calculate the volume, in liters, of the first flask.

12.124 A large 4.2-L flask is filled with air to a pressure that is unknown. This flask is then attached to a second evacuated flask of known volume, and the air from the first flask is allowed to expand into the second flask. The final pressure of the air (in both flasks) is 2.6 atm, and the volume of the second flask is 5.21 L. Calculate the original pressure, in atmospheres, in the first flask.

12.125 At a particular temperature and pressure, 8.00 g of N_2 gas occupies 6.00 L. What would be the volume, in liters, occupied by 1.00×10^{23} molecules of NH_3 at the same temperature and pressure?

12.126 At a particular temperature and pressure, 2.00×10^{23} molecules of N_2 gas occupies 5.00 L. What would be the volume, in liters, occupied by 25.7 g of SO_2 at the same temperature and pressure?

12.127 A piece of Al is placed in a 1.00-L container with pure O_2. The O_2 is at a pressure of 1.00 atm and a temperature of 25 °C. One hour later the pressure has dropped to 0.880 atm and the temperature has dropped to 22 °C. Calculate the number of grams of O_2 that reacted with the Al.

12.128 A piece of Ca is placed in a 1.00-L container with pure N_2. The N_2 is at a pressure of 1.12 atm and a temperature of 26 °C. One hour later the pressure has dropped to 0.924 atm and the temperature has dropped to 24 °C. Calculate the number of grams of N_2 that reacted with the Ca.

12.129 The volume of a fixed quantity of gas, at constant temperature, is decreased by 20.0%. What is the resulting percentage increase in the pressure of the gas?

12.130 The volume of a fixed quantity of gas, at constant temperature, is increased by 30.0%. What is the resulting percentage decrease in the pressure of the gas?

12.131 The volume of a fixed quantity of gas, at constant pressure, is increased by 50.0%. What is the resulting percentage increase in the absolute temperature of the gas?

12.132 The volume of a fixed quantity of gas, at constant pressure, is decreased by 10.0%. What is the resulting percentage decrease in the absolute temperature of the gas?

12.133 At constant temperature, a 2.00-L mixture of gases is produced from 1.00 L of N_2 at $35\overline{0}$ mm Hg, 6.00 L of O_2 at $30\overline{0}$ mm Hg, and 1.00 L of H_2 at $25\overline{0}$ mm Hg. What is the pressure, in millimeters of mercury, of the mixture?

12.134 At constant temperature, a 2.00-L mixture of gases is produced from 2.00 L of N_2 at $25\overline{0}$ mm Hg, 4.00 L of O_2 at $25\overline{0}$ mm Hg, and 1.00 L of H_2 at $25\overline{0}$ mm Hg. What is the pressure, in millimeters of mercury, of the mixture?

12.135 A mixture of 15.0 g of Ar and 15.0 g of CH_4 occupies a 4.0-L container at 8.80 atm and 54 °C. What is the partial pressure, in atmospheres, of Ar in the mixture?

12.136 A mixture of 30.0 g of Ar and 15.0 g of CH_4 occupies a 4.0-L container at 11.9 atm and 27 °C. What is the partial pressure, in atmospheres, of CH_4 in the mixture?

12.137 Suppose 30.0 mL of N_2 gas at 27 °C and 645 mm Hg pressure is added to a 40.0-mL container that already contains He at 37 °C and 765 mm Hg. If the resulting mixture is brought to 32 °C, what is the total pressure, in millimeters of mercury, of the mixture?

12.138 Suppose 50.0 mL of Xe gas at 45 °C and 0.998 atm pressure is added to a 100.0-mL container that already contains He at 37 °C and 765 mm Hg. If the resulting mixture is warmed to 75 °C, what is the total pressure, in millimeters of mercury, of the mixture?

12.139 At a temperature of 27 °C, 50.0 mL of O_2 gas is collected by water displacement. Barometric pressure is 638 mm Hg. What would be the volume, in liters, of O_2 gas (dry) at STP?

12.140 At a temperature of 31 °C, 75.0 mL of O_2 gas is collected by water displacement. Barometric pressure is 672 mm Hg. What would be the volume, in liters, of O_2 gas (dry) at 20 °C and $76\overline{0}$ mm Hg?

12.141 Calculate the ratio of the densities of O_2 and N_2 at
(a) STP conditions
(b) 1.25 atm and 25 °C

12.142 Calculate the ratio of the densities of He and Ne at
(a) STP conditions
(b) 2.30 atm and 57 °C

12.143 How many molecules of sulfur dioxide (SO_2) gas are contained in 1.00 L of SO_2 at STP?

12.144 How many molecules of nitrogen dioxide (NO_2) gas are contained in 2.00 L of NO_2 gas at STP?

12.145 A near-vacuum pressure of 0.0010 mm Hg is readily obtained in a laboratory by means of a vacuum pump. Calculate the number of molecules in 1.00 mL of O_2 gas at this pressure and 23 °C.

12.146 A near-vacuum pressure of 0.0010 mm Hg is readily obtained in a laboratory by means of a vacuum pump. Calculate the number of atoms in 1.00 mL of Xe gas at this pressure and 27 °C.

Cumulative Problems

12.147 A 2.24-L sample of a gaseous compound has a mass of 7.5 g at STP. What is the molecular mass of a molecule of this compound in atomic mass units?

12.148 A 4.48-L sample of a gaseous compound has a mass of 20.0 g at STP. What is the molecular mass of a molecule of this compound in atomic mass units?

12.149 What is the value of x in the formula PH_x if the density of PH_x gas is 1.517 g/L at 0 °C and 1.00 atm?

12.150 What is the value of x in the formula P_2H_x if the density of P_2H_x gas is 2.944 g/L at 0 °C and 1.00 atm?

12.151 Calculate the density of SO_2 gas, in kilograms per cubic meter, at 35 °C and 1.07 atm.

12.152 Calculate the density of N_2O gas, in micrograms per cubic centimeter, at 45 °C and 1.53 atm.

12.153 A 0.581-g sample of a gaseous compound containing only carbon and hydrogen contains 0.480 g of carbon and 0.101 g of hydrogen. At STP, 33.6 mL of the gas has a mass of 0.0869 g. What is the molecular formula for the compound?

12.154 A 6.01-g sample of a gaseous carbon–hydrogen compound is found to contain 4.80 g of carbon and 1.21 g of hydrogen. At STP, 762 mL of the gas has a mass of 1.02 g. What is the molecular formula for the compound?

12.155 The elemental analysis of a certain compound is 24.3% C, 4.1% H, and 71.6% Cl by mass. If 0.132 g of compound occupies 41.4 mL at 741 mm Hg pressure and 96 °C, what are the molar mass and molecular formula of the compound?

12.156 The elemental analysis of a certain compound is 88.82% C and 11.18% H by mass. A 20.87-mg sample of compound vapor occupies 11.63 mL at 1.016 atm pressure and $10\overline{0}$ °C. What are the molar mass and the molecular formula of the compound?

12.157 1.75 L of H_2S, measured at 25.0 °C and 625 mm Hg, is mixed with 5.75 L of O_2, measured at 10.0 °C and 715 mm Hg, and the mixture is allowed to react.

$$2\ H_2S(g) + 3\ O_2(g) \rightarrow 2\ SO_2(g) + 2\ H_2O(g)$$

How much H_2O, in grams, is produced?

12.158 2.00 L of N_2, measured at 30.0 °C and 1.08 atm, is mixed with 4.00 L of O_2, measured at 25.0 °C and 0.118 atm, and the mixture is allowed to react.

$$N_2(g) + O_2(g) \rightarrow 2\ NO(g)$$

How much NO, in grams, is produced?

12.159 A 22.0-g sample of NH_3 reacts with an excess of Cl_2 gas according to the equation

$$2\ NH_3(g) + 3\ Cl_2(g) \rightarrow N_2(g) + 6\ HCl(g)$$

What volume, in cubic meters at STP, of HCl gas is produced?

12.160 A 43.0-g sample of NO reacts with an excess of O_2 gas according to the equation

$$2\ NO(g) + O_2(g) \rightarrow 2\ NO_2(g)$$

What volume, in cubic meters at STP, of NO_2 gas is produced?

12.161 Ammonia gas reacts with hydrogen chloride gas according to the equation

$$NH_3(g) + HCl(g) \rightarrow NH_4Cl(s)$$

If 7.00 g of NH_3 is reacted with 12.0 g of HCl in a 1.00-L container at 25 °C, what will the final pressure, in atmospheres, be in the reaction container?

12.162 At elevated temperatures phosphorus reacts with oxygen according to the equation

$$4\ P(g) + 5\ O_2(g) \rightarrow P_4O_{10}(s)$$

If 50.0 g of P is reacted with 25.0 g of O_2 in a 8.00-L container at 175 °C, what will the final pressure, in atmospheres, be in the reaction container?

12.163 An adult person breathes in approximately 0.500 L of air at 1.00 atm pressure with every breath. If a gas storage cylinder contains 48.0 L of air at $\overline{200}$ atm pressure, how many breaths of air will the cylinder supply? The temperature is constant at 36 °C.

12.164 A weather balloon needs to be inflated to a volume of 96,000 L at a pressure of 0.12 atm. The source for inflation is 75-L cylinders of compressed helium gas at a pressure of $1\overline{00}$ atm. Assuming that the temperature is constant, how many cylinders of He are needed to do the job?

Grid Problems

12.165 Select from the grid *all* correct responses for each situation.

1. $P_1V_1 = P_2V_2$	2. $\dfrac{V_1}{T_1} = \dfrac{V_2}{T_2}$	3. $\dfrac{P_1}{T_1} = \dfrac{P_2}{T_2}$
4. $\dfrac{T_1}{V_1} = \dfrac{T_2}{V_2}$	5. $\dfrac{P_1V_1}{n_1T_1} = \dfrac{P_2V_2}{n_2T_2}$	6. $\dfrac{V_1}{n_1} = \dfrac{V_2}{n_2}$
7. $n_1T_1 = n_2T_2$	8. $\dfrac{P_1V_1}{T_1} = \dfrac{P_2V_2}{T_2}$	9. $P_1T_2 = P_2T_1$

(a) pairs of expressions that are equivalent to each other
(b) expressions that require two of the four gas law variables (*n*, *P*, *V*, *T*) to remain constant
(c) expressions that require a fixed quantity of gas to be present
(d) expressions that require both an initial and a final set of conditions
(e) expressions in which an inverse proportion is present
(f) expressions that could be used to solve the problem "A sample of CO gas has a volume of 3.73 L at 23 °C under a pressure of 1.23 atm. What pressure will this CO gas be under if the volume is changed to 1.25 L and the temperature remains at 23 °C?"

12.166 Select from the grid *all* correct responses for each situation. Consider all numbers in the grid to be exact numbers and use whole number formula masses in any calculations you do.

Grid Problems

1. 2 moles CH_4	2. 17 g NH_3	3. $\frac{1}{3}$ mole NO
4. 10 g C_2H_6	5. $\frac{1}{4}$ mole C_2H_2	6. 1 mole CO_2
7. 7 g CO	8. 17 g H_2S	9. 22 g N_2O

(a) quantities for which the volume at STP is 11.2 L
(b) quantities for which the density at STP is 1.34 moles/L
(c) quantities for which the molar volume at STP is 22.4 L
(d) pairs of quantities that have the same volume at STP
(e) pairs of quantities that have the same density at STP
(f) pairs of quantities that have the same number of moles and the same volume at STP

12.167 Select from the grid *all* correct responses for each situation. All responses involve the same gas sample.

1. $T = 23\,°C$ $P = 1.50$ atm	2. $T = 25\,°C$ $P = 25.0$ atm	3. $T = 303$ K $P = 6.22$ atm
4. $T = 321$ K $P = 1.82$ atm	5. $T = 202\,°C$ $P = 9.75$ atm	6. $T = 296$ K $P = 1140$ mm Hg
7. $T = -28\,°C$ $P = 1911$ mm Hg	8. $T = 298$ K $P = 2324$ mm Hg	9. $T = 595$ K $P = 927$ mm Hg

(a) pairs of conditions that have identical T and P values
(b) conditions in which 1.00 mole of gas has a volume of 4.00 L
(c) conditions in which 1.00 mole of gas has a volume greater than 6.00 L
(d) pairs of conditions for which 1 mole of gas has the same volume value
(e) pairs of conditions in which the density, in moles per liter, is the same
(f) conditions in which the number of moles per liter is less than 1.00

12.168 Select from the grid *all* correct responses for each situation. In each case a mixture of gases A, B, and C is under a total pressure of 2 atm.

1. $n_A = 4$ moles $P_A = 1$ atm	2. $n_A = 4$ moles $n_T = 8$ moles	3. $n_T = 8$ moles $P_A = 1$ atm
4. mole fraction A $= 0.5$ $n_T = 6$ moles	5. pressure fraction A $= 0.5$ $n_A = 3$ moles	6. $n_B = 0.5$ mole $n_C = 0.5$ mole $n_T = 2$ moles
7. $n_A = 6$ moles partial pressure A $= 760$ mm Hg	8. $n_B = 0.1$ mole $n_C = 0.1$ mole $P_A = 1.8$ atm	9. $n_A = 2$ moles $n_B = 1$ mole $n_C = 1$ mole

(a) partial pressure of A is 1 atm
(b) mole fraction of A is 0.5
(c) partial pressure of B can be determined from the given information
(d) numerical value of the product $n_T P_A$ is less than 8
(e) total moles of B present is definitely less than 4
(f) mole fraction of A is equal to the pressure fraction of A

ns
CHAPTER THIRTEEN

Solutions

13.1 Types of Solutions

A **solution** is a homogeneous (uniform) mixture of two or more substances. To achieve a homogeneous mixture, the intermingling of components must be on the molecular level; that is, the particles present must be of atomic and molecular size.

In discussing solutions it is often convenient to call one component the *solvent* and the others *solutes*. The **solvent** is the component of the solution present in the greatest amount. The solvent may be thought of as the medium in which the other substances present are *dissolved*. A **solute** is a solution component present in a small amount relative to that of solvent. More than one solute may be present in the same solution. For example, both sugar and salt (two solutes) may be dissolved in a container of water.

In most situations we will encounter, the solutes present in a solution will be of more interest to us than the solvent. The solutes are the "active ingredients" in the solution. They are the substances that undergo reaction when solutions are mixed.

Solutions used in the laboratory are usually liquids, and the solvent is almost always water. However, as we shall see shortly, gaseous solutions and solid solutions of numerous types do exist.

A solution, since it is homogeneous, will have the same properties throughout. No matter where we take a sample from a solution we will obtain material with the same composition as that of any other sample from the same solution. The composition of a solution can be varied, usually within certain limits, by changing the relative amounts of solvent and solute present. (If the composition limits are violated, a heterogeneous mixture is formed.)

Two-component solutions can be classified into nine types according to the physical states of the solvent and solute before mixing. These types, along with an example of each, are listed in Table 13.1. Solutions in which the final state of the solution components is liquid are the most common and are the type that will be emphasized in this book.

The physical state of a solute becomes that of the solvent when a solution is formed. For example, solid naphthalene (moth repellent) must be sublimed (Sec. 4.4) in order for it to dissolve in air. Merely finely pulverizing a solid and dispersing it in air does not produce a solution. (Dust particles in air would be an example of this.) The particles of the solid must be subdivided to the molecular level; the solid must sublime. Similarly, fog is a suspension of water droplets in air; the droplets are large enough to reflect light, a fact that becomes evident when we drive an automobile on a foggy night. Thus, fog is not a solution. Water vapor, however, is present in solution form in air. When hydrogen gas dissolves in platinum metal (a gas-in-solid solution),

TABLE 13.1 Examples of Various Types of Solutions

Solution Type (solute listed first)	Example
Gaseous Solutions	
Gas dissolved in gas	Dry air (oxygen and other gases dissolved in nitrogen)
Liquid dissolved in gas[a]	Wet air (water vapor in air)
Solid dissolved in gas[a]	Moth repellent (or moth balls) sublimed into air
Liquid Solutions	
Gas dissolved in liquid	Carbonated beverage (carbon dioxide in water)
Liquid dissolved in liquid	Vinegar (acetic acid dissolved in water)
Solid dissolved in liquid	Saltwater
Solid Solutions	
Gas dissolved in solid	Hydrogen in platinum
Liquid dissolved in solid	Dental filling (mercury dissolved in silver)
Solid dissolved in solid	Sterling silver (copper dissolved in silver)

[a] An alternative viewpoint is that liquid-in-gas and solid-in-gas solutions do not actually exist as true solutions. From this viewpoint water vapor or moth repellent in air is considered to be a gas-in-gas solution since the water or moth repellent must evaporate or subline first in order to enter the air.

the gas molecules take up fixed positions in the metal lattice. The gas is "solidified" as a result.

13.2 Terminology Used in Describing Solutions

In addition to *solvent* and *solute*, several other terms are useful in describing characteristics of solutions.

The **solubility** of a solute is the maximum amount of solute that will dissolve in a given amount of solvent. Numerous factors affect the numerical value of a solute's solubility in a given solvent, including the nature of the solvent itself, the temperature, and in some cases the pressure and the presence of other solutes.

Common units for expressing solubility are grams of solute per 100 g of solvent. The temperature of the solvent must also be specified. Table 13.2 gives the solubilities of selected solutes in the solvent water at three different temperatures.

The use of specific units for specifying solubility, as in Table 13.2, allows us to compare solubilities quite precisely. Such precision is often unnecessary, and instead *qualitative* statements about solubilities are made by using terms such as *very soluble*, *slightly soluble*, and so forth. The guidelines for the use of such terms are given in Table 13.3.

TABLE 13.2 Solubilities of Various Compounds in Water at 0, 50, and 100 °C

	Solubility (g solute/100 g H_2O)		
Solute	0 °C	50 °C	100 °C
Lead(II) bromide ($PbBr_2$)	0.455	1.94	4.75
Silver sulfate (Ag_2SO_4)	0.573	1.08	1.41
Copper(II) sulfate ($CuSO_4$)	14.3	33.3	75.4
Sodium chloride (NaCl)	35.7	37.0	39.8
Silver nitrate ($AgNO_3$)	122	455	952
Cesium chloride (CsCl)	161.4	218.5	270.5

TABLE 13.3 Qualitative Solubility Terms

Solute Solubility (g solute/100 g solvent)	Qualitative Solubility Description
Less than 0.1	insoluble
0.1–1	slightly soluble
1–10	soluble
Greater than 10	very soluble

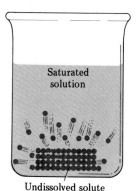

FIGURE 13.1 The dynamic equilibrium process occurring in a saturated solution that contains undissolved excess solute.

A **saturated solution** contains the maximum amount of solute that can be dissolved under the conditions at which the solution exists. A saturated solution together with excess undissolved solute is an equilibrium situation where the rate of dissolution of undissolved solute is equal to the rate of crystallization of dissolved solute. Consider the process of adding table sugar (sucrose) to a container of water. Initially the sugar dissolves as the solution is stirred. Finally, as we add more sugar, a point is reached where no amount of stirring will cause the added sugar to dissolve. Sugar remains as a solid on the bottom of the container; the solution is saturated. Although it appears to the eye that nothing is happening once the saturation point is reached, on the molecular level this is not the case. Solid sugar from the bottom of the container is continuously dissolving in the water, and an equal amount of sugar is coming out of solution. Accordingly, the net number of sugar molecules in the liquid remains the same, and outwardly it appears that the dissolution process has stopped. This equilibrium situation in the saturated solution is somewhat similar to the previously discussed evaporation of a liquid in a closed container (Sec. 11.12). Figure 13.1 illustrates the dynamic equilibrium process occurring in a saturated solution in the presence of undissolved excess solute.

An **unsaturated solution** is a solution where less solute than the maximum amount possible is dissolved in the solution.

The terms *dilute* and *concentrated* are also used to convey qualitative information about the degree of saturation of a solution. A **dilute solution** contains a small amount of solute in solution relative to the amount that could dissolve. On the other hand, a **concentrated solution** contains a large amount of solute relative to the amount that could dissolve. A concentrated solution need not be a saturated solution.

In dealing with liquid-in-liquid solutions, the terms *miscible, partially miscible*, and *immiscible* are frequently used to describe solubility characteristics associated with the liquids. **Miscible substances** dissolve in any amount in each other. For example, methyl alcohol (CH_3OH) and water are miscible—they completely mix with each other in any and all proportions. Always, after these two liquids are mixed, only one phase is present. **Partially miscible substances** have limited solubility in each other. Benzene (C_6H_6) and water are partially miscible. If benzene is added slowly to water, a small amount of benzene initially dissolves; a single phase results. However, as soon as the benzene solubility limit is reached, the excess benzene (saturated with water) forms a separate layer on top of the water (on top because it is less dense). **Immiscible substances** do not dissolve in each other. When such substances are mixed, two layers (phases) immediately form. Very few liquids are totally immiscible in each other; toluene (C_7H_8) and water approach this limiting case. Figure 13.2 illustrates the results obtained by mixing liquids of various miscibilities with each other.

Another term commonly encountered in discussions of solutions is *aqueous solution*. An **aqueous solution** is simply a solution in which water is the solvent.

13.3 Solution Formation

In a solution, solute particles are uniformly dispersed throughout the solvent. Considering what happens at the molecular level during the solution process will help us to understand how this is achieved.

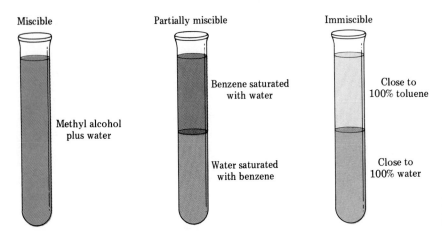

FIGURE 13.2 Miscibility of selected liquids with each other.

For a solute to dissolve in a solvent, two types of interparticle attractions must be overcome: (1) attractions between solute particles (solute–solute attractions) and (2) attractions between solvent particles (solvent–solvent attractions). Only when these attractions are overcome can particles in both pure solute and pure solvent begin to intermingle. A new type of interaction, which exists only in solutions, arises as the result of the mixing of solute and solvent. This new interaction is the attraction between solute and solvent particles (solute–solvent attractions). These new attractions are the primary driving force for solution formation. The extent to which a substance dissolves depends on the degree to which the newly formed solute–solvent attractions are able to compensate for the energy needed to overcome the solute–solute and solvent–solvent interactions. A solute will not dissolve in a solvent if either solute–solute or solvent–solvent interactions are too strong to be offset by the formation of the new solute–solvent interactions.

A most important type of solution process is the dissolution of an ionic solid in water. Let us consider in detail the process of dissolving sodium chloride, a typical ionic solid, in water. We will consider the process to occur in steps. The fact that water molecules are polar (Sec. 7.15) will become very important in our considerations.

Figure 13.3 shows what is thought to happen when sodium chloride is placed in water. The polar water molecules become oriented so that the negative oxygen portion points toward positive sodium ions and the positive hydrogen portion points toward negative chloride ions. As the polar water molecules begin to surround ions on the crystal surface, they exert sufficient attraction to cause these ions to break away from the crystal surface. After leaving the crystal, the ion retains its surrounding group of water molecules; it has become a *hydrated ion*. As each hydrated ion leaves the surface, other ions are exposed to the water, and the crystal is picked apart ion by ion. Once in solution, the hydrated ions are uniformly distributed by stirring or by random collisions with other molecules or ions.

The random motion of solute ions in solution causes them to collide with each other, with solvent molecules, and occasionally with the surface of the undissolved solute. Ions undergoing this last type of collision occasionally stick to the solid surface and thus leave the solution. When the number of ions in solution is low, the chances for collision with the undissolved solute are low. However, as the number of ions in

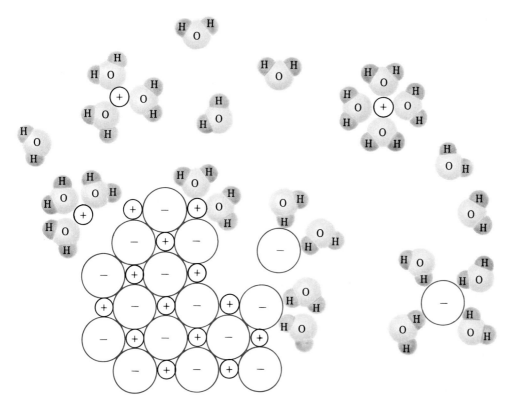

FIGURE 13.3 The solution process for an ionic solid in water.

solution increases, so do the chances for such collisions, and more ions are recaptured by the undissolved solute. Eventually, the number of ions in solution reaches such a level that ions return to the undissolved solute at the same rate as other ions leave. At this point the solution is saturated, and the equilibrium process discussed in the last section is in operation.

13.4 Solubility Rules

In this section we present some rules for qualitatively predicting solute solubilities. These rules summarize in a concise form the results of thousands of experimental solute–solvent solubility determinations.

A very useful generalization that relates polarity to solubility is *substances of like polarity tend to be more soluble in each other than substances that differ in polarity*. This conclusion is often expressed as the simple phrase "*like dissolves like.*" Polar substances, in general, are good solvents for other polar substances but not for nonpolar substances. Similarly, nonpolar substances exhibit greater solubility in nonpolar solvents than they do in polar solvents.

TABLE 13.4 Solubility Guidelines for Ionic Compounds in Water

If ____ Ion Is Present	Compound Is ____	Unless ____ Ion Also Is Present
Group IA (Li$^+$, Na$^+$, K$^+$, etc.)	soluble	
Ammonium (NH$_4^+$)	soluble	
Acetate (C$_2$H$_3$O$_2^-$)	soluble	
Nitrate (NO$_3^-$)	soluble	
Chloride (Cl$^-$), bromide (Br$^-$), and iodide (I$^-$)	soluble	Ag$^+$, Pb^{2+}, Hg$_2^{2+}$
Sulfate (SO$_4^{2-}$)	soluble	Ca^{2+}, Sr^{2+}, Ba^{2+}, Pb^{2+}
Carbonate (CO$_3^{2-}$)	insoluble	group IA and NH$_4^+$
Phosphate (PO$_4^{3-}$)	insoluble	group IA and NH$_4^+$
Sulfide (S^{2-})	insoluble	groups IA and IIA and NH$_4^+$
Hydroxide (OH$^-$)	insoluble	group IA, Ba^{2+}, Sr^{2+}, Ca^{2+}

The generalization "like dissolves like" is a useful tool for predicting solubility behavior in many, but not all, solute–solvent situations. Results that agree with the generalization are almost always obtained in the cases of gas-in-liquid and liquid-in-liquid solutions and for solid-in-liquid solutions in which the solute is not an ionic compound. For example, NH$_3$ gas (a polar gas) is much more soluble in H$_2$O (a polar liquid) than is O$_2$ gas (a nonpolar gas). (The actual solubilities of NH$_3$ and O$_2$ in water at 20 °C are, respectively, 51.8 g/100 g H$_2$O and 0.0043 g/100 g H$_2$O.)

For those solid-in-liquid solutions in which the solute is an ionic compound—a very common situation—the rule "like dissolves like" is not adequate. One would predict that since all ionic compounds are polar, they all would dissolve in polar solvents such as water. This is not the case. The failure of the generalization here is related to the complexity of the factors involved in determining the magnitude of the solute–solute (ion–ion) and solute–solvent (ion–polar solvent molecule) interactions. Among other things, both the charge on and size of the ions in the solute must be considered. Changes in these factors affect both types of interactions, but not to the same extent.

Some guidelines concerning the solubility of ionic compounds in water, which should be used in place of "like dissolves like," are given in Table 13.4.

All ionic compounds, even the most insoluble ones, dissolve to a slight extent in water. The insoluble classification used in Table 13.4 thus really means ionic compounds that have very limited solubility in water.

You should become thoroughly familiar with the rules in Table 13.4. They find extensive use in chemical discussions. We will next encounter them in Section 14.7 when the topic of net ionic equations is presented.

EXAMPLE 13.1

Predict the solubility of each solute-in-solvent combination.

(a) SO$_2$ (a polar gas) in water
(b) paraffin wax (nonpolar) in CCl$_4$ (nonpolar)
(c) AgNO$_3$ (an ionic solid) in water
(d) CaSO$_4$ (an ionic solid) in water
(e) NH$_4$Cl (an ionic solid) in water

Solution

(a) **Soluble.** SO_2 is polar, as is water—like dissolves like.
(b) **Soluble.** Since both substances are nonpolar, they should be relatively soluble in each other—like dissolves like.
(c) **Soluble.** Table 13.4 indicates that all nitrates are soluble.
(d) **Insoluble.** Table 13.4 indicates that calcium is an exception to the rule that all sulfates are soluble.
(e) **Soluble.** Table 13.4 indicates that all compounds containing the ammonium ion are soluble. Alternatively, we could use the rule that all chlorides are soluble except those of Ag^+, Pb^{2+}, and Hg_2^{2+} to predict that NH_4Cl would be soluble.

PRACTICE EXERCISE 13.1

Predict the solubility of each solute-in-solvent combination.

(a) CO_2 (a nonpolar gas) in water
(b) CCl_4 (a nonpolar liquid) in benzene (a nonpolar liquid)
(c) NaBr (an ionic solid) in water
(d) $MgCO_3$ (an ionic solid) in water
(e) $(NH_4)_3PO_4$ (an ionic solid) in water

Ans. (a) soluble; (b) soluble; (c) soluble; (d) insoluble; (e) soluble.

Similar exercises: Problems 13.11 and 13.12

13.5 Solution Concentrations

In Section 13.2 we learned that, in general, there is a limit to the amount of solute that can be dissolved in a specified amount of solvent and also that a solution is said to be saturated when this maximum amount of solute has been dissolved. We can determine the amount of dissolved solute in a saturated solution by the solute's solubility.

Most solutions chemists deal with are *unsaturated* rather than saturated solutions. The amount of solute present in an unsaturated solution is specified by stating the concentration of the solution. The **concentration** of a solution is the amount of solute present in a specified amount of solvent or a specified amount of solution. Thus, concentration is a ratio of two quantities, being either the ratio

$$\frac{\text{Amount of solute}}{\text{Amount of solvent}} \quad \text{or} \quad \frac{\text{Amount of solute}}{\text{Amount of solution}}$$

In specifying a concentration, what are the units used to indicate the amounts of solute and solution present? In practice, a number of different unit combinations are used, with the choice of units depending on the use to be made of the solution. In each of the next four sections we shall discuss a commonly encountered set of units used to express solution concentration. The concentration expressions to be discussed are (1) percentage of solute (Sec. 13.6), (2) parts per million and parts per billion (Sec. 13.7), (3) molarity (Sec. 13.8), and (4) molality (Sec. 13.9). A fifth concentration unit, normality, which is used extensively in situations that involve the reactions of acids and bases, will be discussed in the next chapter (Sec. 14.17) where the topic of acids and bases is considered.

13.6 Concentration: Percentage of Solute

The concentration of a solution is often specified in terms of the percentage of solute in the total amount of solution. Since the amounts of solute and solution present can be stated in terms of either mass or volume, different types of percent units exist. The three most common are

1. Percent by mass (or mass–mass percent).
2. Percent by volume (or volume–volume percent).
3. Mass–volume percent.

The percent unit most frequently used by chemists is percent by mass (or mass–mass percent). **Percent by mass** is equal to the mass of solute divided by the total mass of solution multiplied by 100 (to put the value in terms of percentage). (Percentage is always part of the whole divided by the whole times 100; see Sec. 3.9.)

$$\text{Percent by mass} = \frac{\text{mass solute}}{\text{mass solution}} \times 100$$

The solute and solution masses must be in the same units. The mass of solution is equal to the mass of *solute* plus the mass of *solvent*.

$$\text{Percent by mass} = \frac{\text{mass solute}}{\text{mass solute} + \text{mass solvent}} \times 100$$

A solution of 5.0% by mass concentration would contain 5.0 g of solute in 100.0 g of solution (5.0 g of solute and 95.0 g of solvent). Thus, percent by mass gives directly the number of grams of solute in 100 g of solution. The abbreviation for percent by mass is % (m/m).

EXAMPLE 13.2

What is the percent by mass, % (m/m), concentration of sucrose (table sugar) in a solution made by dissolving 5.4 g of sucrose in enough water to give 87.3 g of solution?

Solution
Both the mass of solute and mass of solution are given directly in the problem statement. Substituting these numbers into the equation

$$\text{Percent by mass} = \frac{\text{mass solute}}{\text{mass solution}} \times 100$$

gives

$$\text{Percent by mass} = \frac{5.4 \text{ g}}{87.3 \text{ g}} \times 100 = 6.185567\% \quad \text{(calculator answer)}$$

$$= 6.2\% \quad \text{(correct answer)}$$

PRACTICE EXERCISE 13.2

What is the percent by mass concentration of K_3PO_4 in a solution made by dissolving 13.4 g of K_3PO_4 in enough water to give 55.4 g of solution?

Ans. 24.2% (m/m).

Similar exercises: Problems 13.17 and 13.18

EXAMPLE 13.3

How many grams of sucrose must be added to 375 g of water to prepare a 2.75% (m/m) solution of sucrose?

Solution

Often, when a solution concentration is given as part of a problem statement, the concentration information is used in the form of a conversion factor in solving the problem. That will be the case in this problem.

The given quantity is 375 g of H_2O (grams of solvent), and the desired quantity is grams of sucrose (grams of solute).

$$375 \text{ g } H_2O = ? \text{ g sucrose}$$

The conversion factor relating these two quantities (solvent and solute) is obtained from the given concentration. In a 2.75% (m/m) sucrose solution there is 2.75 g of sucrose for every 97.25 g of H_2O

$$100.00 \text{ g solution} - 2.75 \text{ g sucrose} = 97.25 \text{ g } H_2O$$

This relationship between grams of solute and grams of solvent (2.75 to 97.25) gives us the needed conversion factor.

$$\frac{2.75 \text{ g sucrose}}{97.25 \text{ g } H_2O}$$

Dimensional analysis gives the problem setup, which is solved in the following manner.

$$375 \text{ g } H_2O \times \frac{2.75 \text{ g sucrose}}{97.25 \text{ g } H_2O} = 10.604113 \text{ g sucrose} \quad \text{(calculator answer)}$$

$$= 10.6 \text{ g sucrose} \quad \text{(correct answer)}$$

PRACTICE EXERCISE 13.3

How many grams of $LiNO_3$ must be added to 25.0 g of water to prepare a 5.00% (m/m) solution of $LiNO_3$?

Ans. 1.32 g $LiNO_3$.

Similar exercises: Problems 13.19 and 13.20

Percent by volume (or volume–volume percent) finds use as a concentration unit when both the solute and solvent are liquids or gases. In such cases it is often more convenient to measure volumes than masses. **Percent by volume** is equal to the volume of solute divided by the total volume of solution multiplied by 100.

$$\text{Percent by volume} = \frac{\text{volume solute}}{\text{volume solution}} \times 100$$

Solute and solution volumes must always be expressed in the same units when this expression is used. The abbreviation for percent by volume is % (v/v).

The numerical value of a concentration expressed as a percent by volume gives directly the number of milliliters of solute in 100 mL of solution. Thus, a 100-mL sample of a 5.0% alcohol-in-water solution contains 5.0 mL of alcohol dissolved in enough water to give 100 mL of solution. Note that such a 5.0% by volume solution could not be made by adding 5 mL of alcohol to 95 mL of water, since volumes of liquids are not usually additive. Differences in the way molecules are packed as well as in the distances between molecules almost always result in the volume of a solution being different from the sum of the volumes of solute and solvent. For example, the final volume resulting from the addition of 50.0 mL of ethyl alcohol to 50.0 mL of water is 96.5 mL of solution.

EXAMPLE 13.4

80.0 mL of methyl alcohol and 80.0 mL of water are mixed to give a solution that has a final volume of 154 mL. What is the concentration of the solution expressed as percent by volume methyl alcohol?

Solution

To calculate a percent by volume, the volumes of solute and solution are needed. Both are given in this problem.

$$\text{Solute volume} = 80.0 \text{ mL}$$
$$\text{Solution volume} = 154 \text{ mL}$$

Note that the solution volume is not the sum of the solute and solvent volumes. As previously mentioned, liquid volumes are generally not additive.

Substituting the given values into the equation

$$\text{Percent by volume} = \frac{\text{volume solute}}{\text{volume solution}} \times 100$$

gives

$$\text{Percent by volume} = \frac{80.0 \text{ mL}}{154 \text{ mL}} \times 100 = 51.948052\% \quad \text{(calculator answer)}$$
$$= 51.9\% \quad \text{(correct answer)}$$

PRACTICE EXERCISE 13.4

An antifreeze solution was prepared by dissolving $25\overline{0}$ mL of methyl alcohol in sufficient water to produce $7\overline{00}$ mL of solution. What is the concentration of this solution expressed as percent by volume methyl alcohol? *Ans.* 35.7% (v/v).

Similar exercises: Problems 13.27 and 13.28

The third type of percentage unit in common use is mass–volume percentage. This unit, which is often encountered in hospital and industrial settings, is particularly convenient to use when working with a solid solute (which is easily weighed) and a liquid solvent. Concentrations are specified using this unit when dealing with physiological fluids such as blood and urine. **Mass–volume percent** is equal to the mass of solute (in grams) divided by the total volume of solution (milliliters) multiplied by 100.

$$\text{Mass–volume percent} = \frac{\text{mass of solute (g)}}{\text{volume of solution (mL)}} \times 100$$

Note that specific mass and volume units are given in the definition of mass–volume percent. This is necessary because the units do not cancel as was the case with mass percent and volume percent. The abbreviation for mass–volume percent is % (m/v).

EXAMPLE 13.5

In the treatment of certain illnesses of the human body, a 0.92% (m/v) sodium chloride (NaCl) solution is administered intravenously. How many grams of sodium chloride are required to prepare 345 mL of this solution?

Solution
The given quantity is 345 mL of solution, and the desired quantity is grams of NaCl.

$$345 \text{ mL solution} = ? \text{ g NaCl}$$

The given concentration of 0.92% (m/v), which means 0.92 g NaCl per 100 mL solution, can be used as a conversion factor to go from milliliters of solution to grams of NaCl. The setup for the conversion is

$$345 \text{ mL solution} \times \frac{0.92 \text{ g NaCl}}{100 \text{ mL solution}}$$

Doing the arithmetic, after cancellation of units, gives

$$\frac{345 \times 0.92}{100} \text{ g NaCl} = 3.174 \text{ g NaCl} \quad \text{(calculator answer)}$$
$$= 3.2 \text{ g NaCl} \quad \text{(correct answer)}$$

PRACTICE EXERCISE 13.5

How many grams of glucose ($C_6H_{12}O_6$) are needed to prepare $5\overline{00}$ mL of a 4.50% (m/v) glucose–water solution? *Ans.* 22.5 g $C_6H_{12}O_6$.

Similar exercises: Problems 13.37 and 13.38

13.7 Concentration: Parts per Million and Parts per Billion

The concentration units parts per million (ppm) and parts per billion (ppb) find use when dealing with extremely dilute solutions. Environmental chemists frequently use such units in specifying the concentrations of the minute amounts of trace pollutants or toxic chemicals in air and water samples.

Parts per million and parts per billion units are closely related to percentage concentration units. Not only are the defining equations very similar, but also various forms of the units exist. Because amounts of solute and solution present may be stated in terms of either mass or volume, there are three different forms for each unit: mass–mass (m/m), volume–volume (v/v), and mass–volume (m/v).

A **part per million** (ppm) is one part of solute per million parts of solution. In terms of defining equations, we can write

$$\text{ppm (m/m)} = \frac{\text{mass solute}}{\text{mass solution}} \times 10^6$$

$$\text{ppm (v/v)} = \frac{\text{volume solute}}{\text{volume solution}} \times 10^6$$

$$\text{ppm (m/v)} = \frac{\text{mass solute (g)}}{\text{volume solution (mL)}} \times 10^6$$

Note that the units of grams and milliliters are specified in the last of the three defining equations, but that no units are given in the first two equations. For the first two equations, the only unit restriction is that the units be the same for both numerator and denominator.

A **part per billion** (ppb) is one part of solute per billion parts of solution. The mathematical defining equations for the three types of part per billion units are identical to those just shown for parts per million except that a multiplicative factor of 10^9 instead of 10^6 is used.

The use of parts per million and parts per billion in specifying concentration often avoids the very small numbers that result when other concentration units are used. For example, a pollutant in water might be present at a level of 0.0013 g per 100 mL of solution. In terms of mass–volume percent, this concentration is 0.0013%. In parts per million, however, the concentration is 13.

$$\text{ppm (m/v)} = \frac{0.0013 \text{ g}}{100 \text{ mL}} \times 10^6 = 13$$

The only difference in the ways in which percent concentrations and parts per million or billion are calculated is in the multiplicative factor used. For percentages it is 10^2, for parts per million 10^6, and for parts per billion 10^9. An alternative name for percentage concentration units would be *parts per hundred*.

EXAMPLE 13.6

A sample of water, upon analysis, is found to contain 6.3×10^{-3} g of lead per 375 mL of solution. What is the lead concentration in (a) ppm (m/v) and (b) ppb (m/v)?

Solution

(a) The defining equation for ppm (m/v) is

$$\text{ppm (m/v)} = \frac{\text{mass solute (g)}}{\text{volume solution (mL)}} \times 10^6$$

Substituting the known values for the two variables into this equation gives

$$\text{ppm (m/v)} = \frac{6.3 \times 10^{-3} \text{ g}}{375 \text{ mL}} \times 10^6$$

$$= 16.8 \quad \text{(calculator answer)}$$

$$= 17 \quad \text{(correct answer)}$$

(b) For parts per billion we have

$$\text{ppb (m/v)} = \frac{6.3 \times 10^{-3} \text{ g}}{375 \text{ mL}} \times 10^9$$

$$= 16,800 \quad \text{(calculator answer)}$$

$$= 17,000 \quad \text{(correct answer)}$$

Any time a concentration is expressed in both parts per million and parts per billion, the parts per billion value will be 1000 times larger than the parts per million value.

PRACTICE EXERCISE 13.6

A sample of air, upon analysis, is found to contain 0.032 g of CO per 875 mL of air. What is the CO concentration in (a) ppm (m/v) and (b) ppb (m/v)?

Ans. (a) 37 ppm (m/v); (b) 37,000 ppb (m/v).

Similar exercises: Problems 13.43 and 13.44

EXAMPLE 13.7

The carbon monoxide, CO, content of the tobacco smoke that reaches a smoker's lungs is estimated to be $2\overline{0}0$ ppm (v/v). At this concentration, how much CO, in milliliters, would be present in a sample of air the size of standard-sized basketball (7.5 L)?

Solution

The defining equation for ppm (v/v) is

$$\text{ppm (v/v)} = \frac{\text{volume solute}}{\text{volume solution}} \times 10^6$$

The solute is CO. Rearranging this equation to isolate volume of solute on the left gives

$$\text{Volume solute (mL)} = \frac{\text{ppm (v/v)} \times \text{volume solution (mL)}}{10^6}$$

Substituting the known values into the equation, remembering to change 7.5 L to 7500 mL so that the answer will have the unit of milliliters, gives

$$\text{Volume CO} = \frac{2\overline{0}0 \times 7500 \text{ mL}}{10^6}$$

$$= 1.5 \text{ mL} \quad \text{(calculator and correct answer)}$$

PRACTICE EXERCISE 13.7

The ozone, O_3, content of smog is approximately 0.5 ppm (v/v). At this concentration, how much O_3, in milliliters, would be present in a sample of smog the size of a 2-L soda pop container?

Ans. 0.001 mL O_3

Similar exercises: Problems 13.47 and 13.48

13.8 Concentration: Molarity

The **molarity** of a solution, abbreviated M, is a ratio giving the number of moles of solute per liter of solution.

$$\text{Molarity (M)} = \frac{\text{moles of solute}}{\text{liters of solution}}$$

A solution containing 1 mole of KBr in 1 L of solution has a molarity of 1 and is said to be a 1 M (1 *molar*) solution.

When a solution is to be used for a chemical reaction, concentration is almost always expressed in units of molarity. A major reason for this is the fact that the amount of solute is expressed in moles, a most convenient unit for dealing with chemical reactions. Because chemical reactions occur between molecules and atoms, a unit that counts particles, as the mole does, is desirable.

To find the molarity of a solution we need to know the solution volume in liters and the number of moles of solute present. An alternative to knowing the number of moles of solute is knowledge about the grams of solute present and the solute's molar mass.

EXAMPLE 13.8

Determine the molarities of the following solutions.

(a) 2.37 moles of KNO_3 dissolved in enough water to give $65\overline{0}$ mL of solution.
(b) 25.0 g of NaOH dissolved in enough water to give 2.50 L of solution.

Solution

(a) The number of moles of solute is given in the problem statement.

$$\text{Moles of solute} = 2.37 \text{ moles KNO}_3$$

The volume of the solution is also given in the problem statement, but not in the right units. Molarity requires liters for volume units. Making the unit change gives

$$650 \text{ mL} \times \frac{10^{-3} \text{ L}}{1 \text{ mL}} = 0.65 \text{ L} \quad \text{(calculator answer)}$$

$$= 0.650 \text{ L} \quad \text{(correct answer)}$$

The molarity of the solution is obtained by substituting the known quantities into the equation

$$\text{Molarity} = \frac{\text{moles of solute}}{\text{L of solution}}$$

which gives

$$\text{Molarity} = \frac{2.37 \text{ moles KNO}_3}{0.650 \text{ L solution}} = 3.6461538 \frac{\text{moles KNO}_3}{\text{L solution}} \quad \text{(calculator answer)}$$

$$= 3.65 \frac{\text{moles KNO}_3}{\text{L solution}} \quad \text{(correct answer)}$$

Note that the units of molarity are always moles per liter.

(b) This time the volume of solution is given in the right units, liters.

$$\text{Volume of solution} = 2.50 \text{ L}$$

The moles of solute must be calculated from the grams of solute (given) and the solute's formula mass, which is 40.0 amu (calculated from a table of atomic masses).

$$25.0 \text{ g NaOH} \times \frac{1 \text{ mole NaOH}}{40.0 \text{ g NaOH}} = 0.625 \text{ mole NaOH}$$
$$\text{(calculator and correct answer)}$$

Substituting the known quantities into the defining equation for molarity gives

$$\text{Molarity} = \frac{0.625 \text{ mole NaOH}}{2.50 \text{ L solution}} = 0.25 \frac{\text{mole NaOH}}{\text{L solution}} \quad \text{(calculator answer)}$$

$$= 0.250 \frac{\text{mole NaOH}}{\text{L solution}} \quad \text{(correct answer)}$$

PRACTICE EXERCISE 13.8

What is the molarity of a solution prepared by dissolving 40.0 g of KCl in enough water to give $85\overline{0}$ mL of solution?

Ans. 0.631 M KCl.

Similar exercises: Problems 13.55 and 13.56

As the previous example indicates, when you perform a molarity calculation the *identity* of the solute is always needed. You cannot calculate moles of solute without knowing the chemical identity of the solute. In contrast, when you perform percent concentration calculations (or parts per million or billion)—Sections 13.6 and 13.7—the *identity* of the solute is not used in the calculation.

The mass of solute present in a known volume of solution is an easily calculated quantity if the molarity of the solution is known. In doing such a calculation, molarity serves as a conversion factor relating liters of solution to moles of solute.

$$\boxed{\text{Volume of solution (liters)}} \xleftrightarrow{\text{molarity}} \boxed{\text{Moles of solute}}$$

EXAMPLE 13.9

How many grams of ascorbic acid (vitamin C), $C_6H_8O_6$, are present in 125 mL of a 0.400 M vitamin C solution?

Solution

The given quantity is 125 mL of solution, and the desired quantity is grams of $C_6H_8O_6$.

$$125 \text{ mL solution} = ? \text{ g } C_6H_8O_6$$

The pathway to be used in solving this problem is

$$\text{mL solution} \longrightarrow \text{L solution} \longrightarrow \text{moles } C_6H_8O_6 \longrightarrow \text{g } C_6H_8O_6$$

The given molarity (0.400 M) will serve as the conversion factor for the second unit change; the molecular mass of vitamin C (which must be calculated as it is not given) is used in accomplishing the third unit change.

The dimensional analysis setup from this pathway is

$$125 \text{ mL solution} \times \frac{10^{-3} \text{ L solution}}{1 \text{ mL solution}} \times \frac{0.400 \text{ moles } C_6H_8O_6}{1 \text{ L solution}} \times \frac{176 \text{ g } C_6H_8O_6}{1 \text{ mole } C_6H_8O_6}$$

Canceling units and doing the arithmetic gives

$$\frac{125 \times 10^{-3} \times 0.400 \times 176}{1 \times 1 \times 1} \text{ g } C_6H_8O_6 = 8.8 \text{ g } C_6H_8O_6 \quad \text{(calculator answer)}$$

$$= 8.80 \text{ g } C_6H_8O_6 \quad \text{(correct answer)}$$

PRACTICE EXERCISE 13.9

How many grams of silver nitrate ($AgNO_3$) are present in 375 mL of a 1.50 M silver nitrate solution?

Ans. 95.6 g $AgNO_3$.

Similar exercises: Problems 13.57 and 13.58

EXAMPLE 13.10

Caffeine, the central-nervous-system stimulant found in coffee and tea, has the molecular formula $C_8H_{10}N_4O_2$. How many liters of 0.100 M $C_8H_{10}N_4O_2$ solution can be prepared from 50.0 g of $C_8H_{10}N_4O_2$?

Solution
The given quantity is 50.0 g of $C_8H_{10}N_4O_2$, and the desired quantity is liters of solution.

$$50.0 \text{ g } C_8H_{10}N_4O_2 = ? \text{ L } C_8H_{10}N_4O_2 \text{ solution}$$

The pathway to be used in solving this problem will involve the following steps.

$$\text{g } C_8H_{10}N_4O_2 \longrightarrow \text{moles } C_8H_{10}N_4O_2 \longrightarrow \text{L } C_8H_{10}N_4O_2 \text{ solution}$$

The first unit conversion will be accomplished by using the molar mass of $C_8H_{10}N_4O_2$ (which must be calculated, since it is not given) as a conversion factor. The second unit conversion involves the use of the given molarity as a conversion factor.

$$50.0 \text{ g } C_8H_{10}N_4O_2 \times \frac{1 \text{ mole } C_8H_{10}N_4O_2}{194 \text{ g } C_8H_{10}N_4O_2} \times \frac{1 \text{ L } C_8H_{10}N_4O_2 \text{ solution}}{0.100 \text{ mole } C_8H_{10}N_4O_2}$$

Canceling units and doing the arithmetic gives

$$\frac{50.0 \times 1 \times 1}{194 \times 0.100} \text{ L } C_8H_{10}N_4O_2 \text{ solution} = 2.5773195 \text{ L } C_8H_{10}N_4O_2 \text{ solution}$$
(calculator answer)
$$= 2.58 \text{ L } C_8H_{10}N_4O_2 \text{ solution}$$
(correct answer)

PRACTICE EXERCISE 13.10

How many liters of 0.100 M NaOH solution can be prepared from 10.0 g of NaOH?

Ans. 2.50 L of solution.

Similar exercises: Problems 13.63 and 13.64

Molarity and mass percent are probably the two most commonly used concentration units. The need to convert from one to the other often arises. Such a conversion can easily be done provided the density of the solution is known.

EXAMPLE 13.11

A 12.0% (m/m) solution of HNO_3 has a density of 1.086 g/mL. What is the molarity of this solution?

Solution

We will need to calculate the moles of solute and liters of solution present in a sample of this solution. Since solution concentration is independent of sample size, we can use any size sample for our calculation. To simplify the math somewhat, we will choose a 100.0 g sample of solution.

STEP 1 *Moles of solute.* The given quantity is 100.0 g of solution, and the desired quantity is moles of HNO_3.

$$100.0 \text{ g solution} = ? \text{ moles } HNO_3$$

The pathway to be used in solving this problem is

$$\text{g solution} \longrightarrow \text{g solute} \longrightarrow \text{moles solute}$$

The known mass percent concentration will be the basis for the conversion factor that takes us from grams of solution to grams of solute.

$$100.0 \text{ g solution} \times \frac{12.0 \text{ g } HNO_3}{100 \text{ g solution}} \times \frac{1 \text{ mole } HNO_3}{63.0 \text{ g } HNO_3}$$

$$= 0.19047619 \text{ mole } HNO_3 \quad \text{(calculator answer)}$$
$$= 0.190 \text{ mole } HNO_3 \quad \text{(correct answer)}$$

STEP 2 *Liters of solution.* The density of the solution is used as a conversion factor in obtaining the volume of solution. The pathway for the calculation is

$$\text{g solution} \longrightarrow \text{mL solution} \longrightarrow \text{L solution}$$

The setup is

$$100.0 \text{ g solution} \times \frac{1 \text{ mL solution}}{1.086 \text{ g solution}} \times \frac{10^{-3} \text{ L solution}}{1 \text{ mL solution}}$$

$$= 0.09208103 \text{ L solution} \quad \text{(calculator answer)}$$
$$= 0.0921 \text{ L solution} \quad \text{(correct answer)}$$

STEP 3 *Molarity.* With both moles of solute and liters of solution known, the molarity is obtained by substitution into the defining equation for molarity:

$$\text{Molarity} = \frac{\text{moles } HNO_3}{\text{liters solution}} = \frac{0.190 \text{ mole } HNO_3}{0.0921 \text{ L solution}}$$

$$= 2.062975 \frac{\text{moles } HNO_3}{\text{L solution}} \quad \text{(calculator answer)}$$

$$= 2.06 \frac{\text{moles } HNO_3}{\text{L solution}} \quad \text{(correct answer)}$$

PRACTICE EXERCISE 13.11

A 3.60% (m/m) HCl solution has a density of 1.18 g/mL. What is the molarity of the solution?

Ans. 1.16 M HCl.

Similar exercises: Problems 13.65 and 13.66

Section 13.8 · Concentration: Molarity

Molar concentrations do not give information about the amount of *solvent* present. All that is known is that enough solvent is present to give a specific volume of *solution*. The amount of solvent present in a solution of a known molarity can be calculated if the density of the solution is known. Without the density it cannot be calculated.

EXAMPLE 13.12

A 1.350 M NaCl solution has a density of 1.054 g/mL at 20 °C. How many grams of solvent are present in $65\overline{0}$ mL of this solution?

Solution

To find the grams of solvent present we must first find the grams of solute (NaCl) and the grams of solution. The grams of solvent present is then obtained by calculating the difference.

$$\text{g solvent} = \text{g solution} - \text{g solute}$$

STEP 1 *Grams of solution.* The volume of solution is given. Density, used as a conversion factor, will enable us to convert this volume to grams of solution.

$$65\overline{0} \text{ mL solution} \times \frac{1.054 \text{ g solution}}{1 \text{ mL solution}} = 685.1 \text{ g solution} \quad \text{(calculator answer)}$$

$$= 685 \text{ g solution} \quad \text{(correct answer)}$$

STEP 2 *Grams of solute.* We will use the molarity of the solution as a conversion factor in obtaining the grams of solute. The setup for this calculation is similar to that in Example 13.9.

$$65\overline{0} \text{ mL solution} \times \frac{10^{-3} \text{ L solution}}{1 \text{ mL solution}} \times \frac{1.350 \text{ moles NaCl}}{1 \text{ L solution}} \times \frac{58.5 \text{ g NaCl}}{1 \text{ mole NaCl}}$$

$$= 51.33375 \text{ g NaCl} \quad \text{(calculator answer)}$$

$$= 51.3 \text{ g NaCl} \quad \text{(correct answer)}$$

STEP 3 *Grams of solvent.* The grams of solvent will be the difference in mass between the grams of solution and the grams of solute.

$$685 \text{ g solution} - 51.3 \text{ g solute} = 633.7 \text{ g solvent} \quad \text{(calculator answer)}$$

$$= 634 \text{ g solvent} \quad \text{(correct answer)}$$

PRACTICE EXERCISE 13.12

A 0.750 M $HC_2H_3O_2$ solution has a density of 1.01 g/mL at 25 °C. How many grams of solvent are present in $125\overline{0}$ mL of this solution? *Ans.* $12\overline{0}0$ g solvent.

Similar exercises: Problems 13.113 and 13.114

13.9 Concentration: Molality

Molality is a concentration unit based on a fixed amount of *solvent* and is used in areas where this is a concern. Despite this unit having a name very similar to molarity, molality differs distinctly from molarity; molarity is a unit based on a fixed amount of *solution* rather than a fixed amount of *solvent*. The **molality** of a solution, abbreviated m, is a ratio giving the number of moles of solute per kilogram of solvent.

$$\text{Molality } (m) = \frac{\text{moles of solute}}{\text{kilograms of solvent}}$$

Molality also finds use, in preference to molarity, in experimental situations where changes in temperature are of concern. Molality is a temperature-independent concentration unit; molarity is not. To be temperature independent, a concentration unit cannot involve a volume measurement. Volumes of solutions change (expand or contract) with changes in temperature. A change in temperature thus means a change in concentration, even though the amount of solute remains constant, if a concentration unit has a volume dependency. Volume changes caused by temperature change are usually very, very small; consequently, temperature independence or dependence is a factor in only the most precise experimental measurements.

Careful note should be taken of the fact that the same letter of the alphabet is used as an abbreviation for both molality and molarity—a lowercase m for molality (m) and a capitalized M for molarity (M).

In dilute aqueous solutions molarity and molality are practically identical in numerical value. This results from water having a density of 1.0 g/L. Molarity and molality have significantly different values when the solvent has a density that is not equal to unity or when the solution is concentrated.

EXAMPLE 13.13

Calculate the molality of a solution made by dissolving 10.00 g of $AgNO_3$ in 275.0 g of H_2O.

Solution
To calculate molality the number of moles of solute and the solvent mass in kilograms must be known.

In this problem the solvent mass is given, but in grams rather than kilograms. We, thus, need to change the grams unit to kilograms.

$$275.0 \text{ g } H_2O \times \frac{1 \text{ kg } H_2O}{10^3 \text{ g } H_2O} = 0.275 \text{ kg } H_2O \quad \text{(calculator answer)}$$

$$= 0.2750 \text{ kg } H_2O \quad \text{(correct answer)}$$

Information about the solute is given in terms of grams. We can calculate moles of solute from the given information by using molar mass (which is not given and must be calculated) as a conversion factor.

$$10.00 \text{ g AgNO}_3 \times \frac{1 \text{ mole AgNO}_3}{169.9 \text{ g AgNO}_3}$$

$$= 0.05885815 \text{ mole AgNO}_3 \quad \text{(calculator answer)}$$
$$= 0.05886 \text{ mole AgNO}_3 \quad \text{(correct answer)}$$

Substituting moles of solute and kilograms of solvent into the defining equation for molality gives

$$m = \frac{\text{moles solute}}{\text{kg solvent}}$$

$$= \frac{0.05886 \text{ mole AgNO}_3}{0.2750 \text{ kg H}_2\text{O}} = 0.21403636 \frac{\text{mole AgNO}_3}{\text{kg H}_2\text{O}} \quad \text{(calculator answer)}$$

$$= 0.2140 \frac{\text{mole AgNO}_3}{\text{kg H}_2\text{O}} \quad \text{(correct answer)}$$

Note that the units of molality are always moles per kilogram.

PRACTICE EXERCISE 13.13

Calculate the molality of a solution made by dissolving 25.0 g of K_2CO_3 in 725 g of H_2O.

Ans. 0.250 *m*.

Similar exercises: Problems 13.73 and 13.74

EXAMPLE 13.14

Calculate the number of grams of potassium hydroxide, KOH, that must be added to 25.0 g of water to prepare a 0.0100 *m* solution.

Solution

The given quantity is 25.0 g of H_2O and the desired quantity is grams of KOH.

$$25.0 \text{ g H}_2\text{O} = ? \text{ g KOH}$$

The pathway to be used in solving this problem is

$$\text{g solvent} \longrightarrow \text{kg solvent} \longrightarrow \text{moles solute} \longrightarrow \text{g solute}$$

The molality of the solution, which is given, will serve as a conversion factor to effect the change from kilograms of solvent to moles of solute.

The dimensional analysis setup for the problem is

$$25.0 \text{ g H}_2\text{O} \times \frac{1 \text{ kg H}_2\text{O}}{10^3 \text{ g H}_2\text{O}} \times \frac{0.0100 \text{ mole KOH}}{1 \text{ kg H}_2\text{O}} \times \frac{56.1 \text{ g KOH}}{1 \text{ mole KOH}}$$

The second conversion factor involves the numerical value of the molality and the third conversion factor is based on the molar mass of KOH.

Cancelling units and then doing the arithmetic gives

$$\frac{25.0 \times 1 \times 0.0100 \times 56.1}{10^3 \times 1 \times 1} \text{ g KOH} = 0.014025 \text{ g KOH} \quad \text{(calculator answer)}$$

$$= 0.0140 \text{ g KOH} \quad \text{(correct answer)}$$

PRACTICE EXERCISE 13.14

Calculate the number of grams of isopropyl alcohol (rubbing alcohol), C_3H_8O, which must be added to 275 g of water to prepare a 2.00 m solution of isopropyl alcohol.

Ans. 33.1 g C_3H_8O.

Similar exercises: Problems 13.75 and 13.76

Molal concentrations do not give information about the volume of solution present. All that is known is that a specific amount of solute has been dissolved in a definite mass of solvent. The volume of solution can be calculated if the density of the solution is known. Without the density it cannot be calculated.

EXAMPLE 13.15

Calculate the total mass, in grams, and the total volume, in milliliters, of a 1.20 m aqueous solution of $Pb(NO_3)_2$ containing 20.0 g of $Pb(NO_3)_2$. The density of the solution is 1.33 g/mL.

Solution

Solution mass: The total mass of the solution is the sum of the solute and solvent masses.

The solute mass is given as 20.0 g of $Pb(NO_3)_2$.

The solvent mass can be calculated from the solute mass by using molality as a conversion factor to convert from moles of $Pb(NO_3)_2$ to kilograms of solvent. The pathway is

$$\text{g Pb(NO}_3)_2 \longrightarrow \text{moles Pb(NO}_3)_2 \longrightarrow \text{kg H}_2\text{O} \longrightarrow \text{g H}_2\text{O}$$

The dimensional analysis setup for the calculation of solvent mass is

$$20.0 \text{ g Pb(NO}_3)_2 \times \frac{1 \text{ mole Pb(NO}_3)_2}{331 \text{ g Pb(NO}_3)_2} \times \frac{1 \text{ kg H}_2\text{O}}{1.20 \text{ moles Pb(NO}_3)_2} \times \frac{10^3 \text{ g H}_2\text{O}}{1 \text{ kg H}_2\text{O}}$$

Cancelling the units and doing the arithmetic gives

$$\frac{20.0 \times 1 \times 1 \times 10^3}{331 \times 1.20 \times 1} \text{ g H}_2\text{O} = 50.352467 \text{ g H}_2\text{O} \quad \text{(calculator answer)}$$

$$= 50.4 \text{ g H}_2\text{O} \quad \text{(correct answer)}$$

The total solution mass is the sum of the solute and solvent masses.

$$\underset{\text{solute}}{20.0 \text{ g}} + \underset{\text{solvent}}{50.4 \text{ g}} = \underset{\text{solution}}{70.4 \text{ g}} \quad \text{(calculator and correct answer)}$$

Solution volume: The solution volume can be obtained from the solution mass by using density as a conversion factor.

$$70.4 \text{ g solution} \times \frac{1 \text{ mL solution}}{1.33 \text{ g solution}} = 52.93233 \text{ mL solution} \quad \text{(calculator answer)}$$

$$= 52.9 \text{ mL solution} \quad \text{(correct answer)}$$

PRACTICE EXERCISE 13.15

Calculate the total mass, in grams, and the total volume, in milliliters, of a 2.90 m aqueous solution of potassium chromate, K_2CrO_4, containing 35.0 g of K_2CrO_4. The density of the solution is 1.40 g/mL.

Ans. 97.2 g, 69.4 mL.

Similar exercises: Problems 13.79 and 13.80

13.10 Dilution

A common problem encountered when working with solutions in the laboratory is that of diluting a solution of known concentration (usually called a stock solution) to a lower concentration. **Dilution** is the process in which more solvent is added to a solution in order to lower its concentration. Dilution always lowers the concentration of a solution. The same amount of solute is present, but it is now distributed in a larger amount of solvent (the original solvent plus the added solvent).

Since laboratory solutions are almost always liquids, dilution is normally a volumetric procedure. Most often, a solution of a specific molarity must be prepared by adding a predetermined volume of solvent to a specific volume of stock solution.

With molar concentration units, a very simple mathematical relationship exists between the volumes and molarities of the diluted and stock solutions. This relationship is derived from the fact that the same amount of solute is present in both solutions; only solvent is added in a dilution procedure.

$$\text{Moles solute}_{\text{stock solution}} = \text{moles solute}_{\text{diluted solution}}$$

The number of moles of solute in either solution is given by the expression

$$\text{Moles solute} = \text{molarity } (M) \times \text{liters of solution } (V)$$

(This equation is just a rearrangement of the defining equation for molarity to isolate moles of solute on the left side.) Substitution of this second expression into the first one gives the equation

$$M_s \times V_s = M_d \times V_d$$

In this equation M_s and V_s are the molarity and volume of the stock solution (the solution to be diluted) and M_d and V_d the molarity and volume of the solution resulting from the dilution. Because volume appears on both sides of the equation, any volume unit, not just liters, may be used as long as it is the same on both sides of the equation. Again, the validity of this equation is based on there being no change in the amount of solute present.

EXAMPLE 13.16

What is the molarity of the solution prepared by diluting 12.0 mL of 0.405 M NaCl solution to a final volume of 80.0 mL?

Solution
Three of the four variables in the equation

$$M_s \times V_s = M_d \times V_d$$

are known.

$$M_s = 0.405 \text{ M} \qquad M_d = ?\text{M}$$
$$V_s = 12.0 \text{ mL} \qquad V_d = 80.0 \text{ mL}$$

Rearranging the equation to isolate M_d on the left side and substituting the known variables into it gives

$$M_d = M_s \times \frac{V_s}{V_d}$$

$$= 0.405 \text{ M} \times \frac{12.0 \text{ mL}}{80.0 \text{ mL}} = 0.06075 \text{ M} \quad \text{(calculator answer)}$$

$$= 0.0608 \text{ M} \quad \text{(correct answer)}$$

Thus, the diluted solution's concentration is 0.0608 M.

PRACTICE EXERCISE 13.16

What is the molarity of the solution prepared by diluting 65 mL of 0.95 M Na_2SO_4 solution to a final volume of 135 mL?

Ans. 0.46 M Na_2SO_4.

Similar exercises: Problems 13.83 and 13.84

EXAMPLE 13.17

How much solvent must be added to 100.0 mL of 1.50 M $AgNO_3$ solution to decrease its concentration to 0.350 M?

Solution
The volume of solvent added is equal to the difference between the final and initial volumes. The initial volume is known. The final volume can be calculated using the equation

$$M_s \times V_s = M_d \times V_d$$

Once the final volume is known, the difference between the two volumes can be obtained.

Substituting the known quantities into the dilution equation, rearranged to isolate V_d on the left side, gives

$$V_d = V_s \times \frac{M_s}{M_d}$$

$$V_d = 100.0 \text{ mL} \times \frac{1.50 \cancel{M}}{0.350 \cancel{M}} = 428.57143 \text{ mL} \quad \text{(calculator answer)}$$

$$= 429 \text{ mL} \quad \text{(correct answer)}$$

The solvent added is

$$V_d - V_s = (429 - 100.0 \text{ mL}) = 329 \text{ mL} \quad \text{(calculator answer and correct answer)}$$

PRACTICE EXERCISE 13.17

How much solvent must be added to 50.0 mL of 2.20 M KCl solution to decrease its concentration to 0.0113 M?

Similar exercises:
Ans. 9680 mL. Problems 13.87 and 13.88

When two "like" solutions—that is, solutions that contain the same solute and the same solvent—of differing known molarities and volumes are mixed together, the molarity of the newly formed solution can be calculated by the same principles that apply in a simple dilution problem.

Again, the key concept involves the amount of solute present; it is constant. The sum of the amounts of solute present in the individual solutions prior to mixing is the same as the amount of solute present in the solution obtained from the mixing. No solute is lost or gained in the mixing process. Thus, we can write

$$\text{Moles solute}_{\text{first solution}} + \text{moles solute}_{\text{second solution}} = \text{moles solute}_{\text{combined solution}}$$

Substituting the expression ($M \times V$) for moles solute in this equation gives

$$(M_1 \times V_1) + (M_2 \times V_2) = M_3 \times V_3$$

where the subscripts 1 and 2 denote the solutions to be mixed and the subscript 3 is the solution resulting from the mixing. Again, this expression is valid only when the solutions that are mixed are "like" solutions.

EXAMPLE 13.18

What is the molarity of the solution obtained by mixing 50.0 mL of 3.75 M NaCl solution with 160.0 mL of 1.75 M NaCl solution?

Solution

Five of the six variables in the equation

$$(M_1 \times V_1) + (M_2 \times V_2) = M_3 \times V_3$$

are known.

$$M_1 = 3.75 \text{ M} \qquad V_1 = 50.0 \text{ mL}$$
$$M_2 = 1.75 \text{ M} \qquad V_2 = 160.0 \text{ mL}$$
$$M_3 = ? \text{ M} \qquad V_3 = 210.0 \text{ mL}$$

Note that in the mixing process we consider the volumes of the solution to be additive; that is,

$$V_3 = V_1 + V_2$$

This is a valid assumption for "like" solutions.

Solving our equation for M_3 and then substituting the known quantities into it gives

$$M_3 = \frac{(M_1 \times V_1) + (M_2 \times V_2)}{V_3}$$

$$= \frac{(3.75 \text{ M} \times 50.0 \text{ mL}) + (1.75 \text{ M} \times 160.0 \text{ mL})}{(210.0 \text{ mL})}$$

$$= 2.2261905 \text{ M} \quad \text{(calculator answer)}$$

$$= 2.23 \text{ M} \quad \text{(correct answer)}$$

PRACTICE EXERCISE 13.18

What is the molarity of the solution obtained by mixing 50.0 mL of 1.25 M NH_4Cl solution with 175 mL of 0.125 M NH_4Cl solution?

Ans. 0.375 M NH_4Cl.

Similar exercises: Problems 13.93 and 13.94

In the solution of Example 13.18 the given liquid volumes were considered additive. In Section 13.6, when discussing volume percent, it was stressed that volumes were not additive. Why the difference? Volumes of different liquids (Sec. 13.6) are not additive; volumes of the same liquid in Example 13.18 (water is the solvent in both samples) are additive.

13.11 Molarity and Chemical Equations

In Section 10.8 we were introduced to a general problem-solving procedure for setting up problems that involve chemical equations. With this procedure, if information is given about one reactant or product in a chemical reaction (number of grams, moles, or particles), similar information can easily be obtained for any other reactant or product.

In Section 12.13 this procedure was refined to allow us to do mass-to-volume or volume-to-mass calculations for reactions when at least one reactant or product is a gas.

In this section we further refine our problem-solving procedure in order to deal efficiently with reactions that occur in aqueous solution. Of primary importance to us in this new area of problem solving will be *solution volume*. In most situations, solution volume is more conveniently determined than solution mass.

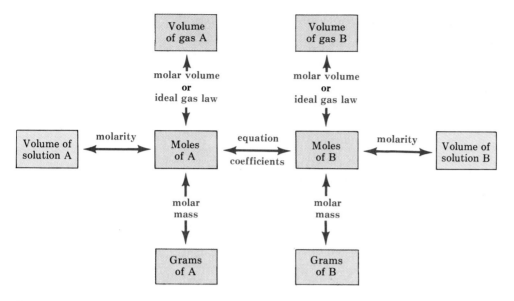

FIGURE 13.4 Conversion factor relationships needed to solve problems involving chemical reactions that occur in aqueous solution.

When solution concentrations are expressed in terms of molarity, a direct relationship exists between solution volume (in liters) and moles of solute present. The definition of molarity itself gives the relationship; molarity is the ratio of moles of solute to volume (in liters) of solution. Thus, molarity is the connection that links volume of solution to the other common problem-solving parameters, such as moles and grams. Figure 13.4 shows diagrammatically the place that volume of solution occupies, relative to other parameters, in the overall scheme of chemical-equation-based problem solving. This diagram is a simple modification of Figure 12.9; "volume of solution" boxes have replaced "particles" boxes. It is used in the same way as Figure 12.9 was.

EXAMPLE 13.19

What volume, in liters, of 2.50 M NaOH solution is needed to react completely with 0.300 L of 3.75 M H_3PO_4 solution according to the following equation?

$$3\,NaOH(aq) + H_3PO_4(aq) \longrightarrow Na_3PO_4(aq) + 3\,H_2O(l)$$

Solution

STEP 1 The given quantity is 0.300 L of H_3PO_4 solution, and the desired quantity is liters of NaOH solution.

$$0.300\,L\,H_3PO_4 = ?\,L\,NaOH$$

STEP 2 This problem is a "solution volume A" to "solution volume B" problem. The pathway used in solving it, in terms of Figure 13.4, is

```
Solution    molarity    Moles    equation       Moles    molarity    Solution
volume A   ─────────▶   of A    ──────────▶    of B    ─────────▶   volume B
                                coefficients
```

STEP 3   The dimensional analysis setup for the calculation is

$$0.300 \text{ L H}_3\text{PO}_4 \times \frac{3.75 \text{ moles H}_3\text{PO}_4}{1 \text{ L H}_3\text{PO}_4} \times \frac{3 \text{ moles NaOH}}{1 \text{ mole H}_3\text{PO}_4} \times \frac{1 \text{ L NaOH}}{2.50 \text{ moles NaOH}}$$

STEP 4   Combining all of the numerical factors gives

$$\frac{0.300 \times 3.75 \times 3 \times 1}{1 \times 1 \times 2.50} \text{ L NaOH} = 1.35 \text{ L NaOH}$$
(calculator and correct answer)

PRACTICE EXERCISE 13.19

What volume, in liters, of a 3.40 M KOH solution is needed to react completely with 0.100 L of a 6.72 M H_2SO_4 solution according to the following equation?

$$2 \text{ KOH}(aq) + \text{H}_2\text{SO}_4(aq) \longrightarrow \text{K}_2\text{SO}_4(aq) + 2 \text{ H}_2\text{O}(l)$$

Ans. 0.395 L KOH.   Similar exercises: Problems 13.95 and 13.96

EXAMPLE 13.20

How many grams of $BaCrO_4$ can be produced from the reaction of 1.05 L of 0.470 M $BaCl_2$ solution with an excess of 2.00 M K_2CrO_4 solution according to the following equation?

$$\text{BaCl}_2(aq) + \text{K}_2\text{CrO}_4(aq) \longrightarrow \text{BaCrO}_4(s) + 2 \text{ KCl}(aq)$$

Solution

STEP 1   The given quantity is 1.05 L of $BaCl_2$ solution, and the desired quantity is grams of $BaCrO_4$.

$$1.05 \text{ L BaCl}_2 = ? \text{ g BaCrO}_4$$

STEP 2   This is a "solution volume A" to "grams of B" problem. The pathway, in terms of Figure 13.4, is

```
Solution    molarity    Moles    equation       Moles    molar       Grams
volume A   ─────────▶   of A    ──────────▶    of B    ─────────▶   of B
                                coefficients            mass
```

STEP 3   The dimensional analysis setup for the calculation is

$$1.05 \text{ L BaCl}_2 \times \frac{0.470 \text{ mole BaCl}_2}{1 \text{ L BaCl}_2} \times \frac{1 \text{ mole BaCrO}_4}{1 \text{ mole BaCl}_2} \times \frac{253 \text{ g BaCrO}_4}{1 \text{ mole BaCrO}_4}$$

STEP 4   The answer, obtained from combining all of the numerical factors, is

$$\frac{1.05 \times 0.470 \times 1 \times 253}{1 \times 1 \times 1} \text{ g BaCrO}_4 = 124.8555 \text{ g BaCrO}_4$$
(calculator answer)

$$= 125 \text{ g BaCrO}_4 \text{ (correct answer)}$$

Note that the concentration of K_2CrO_4 solution, given as 2.00 M in the problem statement, did not enter into the calculation. This is because the K_2CrO_4 solution is present in excess; we know that we have enough of it. If a specific volume of K_2CrO_4 solution had been given in the problem statement, we would have had to determine the limiting reactant (K_2CrO_4 and $BaCl_2$) as the first step in working the problem. The concept of a limiting reactant was discussed in Section 10.9.

PRACTICE EXERCISE 13.20

How many grams of AgCl can be produced from the reaction of 1.00 L of 0.375 M $AgNO_3$ solution with an excess of 1.00 M NaCl solution according to the following equation?

$$AgNO_3(aq) + NaCl(aq) \longrightarrow AgCl(s) + NaNO_3(aq)$$

Ans. 53.6 g AgCl.

Similar exercises: Problems 13.97 and 13.98

EXAMPLE 13.21

What volume, in liters, of H_2 gas measured at STP can be produced from 0.575 L of 1.22 M HBr solution and an excess of Zn according to the following equation?

$$2 \text{ HBr}(aq) + \text{Zn}(s) \longrightarrow \text{ZnBr}_2(aq) + H_2(g)$$

Solution

STEP 1   The given quantity is 0.575 L of HBr solution, and the desired quantity is liters of H_2 gas at STP.

$$0.575 \text{ L HBr} = ? \text{ L } H_2 \text{ (at STP)}$$

STEP 2   This is a "solution volume A" to "gaseous volume B" problem. From Figure 13.4, the pathway for solving the problem is

Solution volume A →(molarity)→ Moles of A →(equation coefficients)→ Moles of B →(molar volume)→ Gaseous volume B

STEP 3   The dimensional analysis setup for the calculation is

$$0.575 \text{ L HBr} \times \frac{1.22 \text{ moles HBr}}{1 \text{ L HBr}} \times \frac{1 \text{ mole } H_2}{2 \text{ moles HBr}} \times \frac{22.4 \text{ L } H_2}{1 \text{ mole } H_2}$$

The last conversion factor is derived from the fact that 1 mole of any gas occupies 22.4 L at STP conditions (Sec. 12.10).

STEP 4 The result, obtained by combining all of the numerical factors, is

$$\frac{0.575 \times 1.22 \times 1 \times 22.4}{1 \times 2 \times 1} \text{ L H}_2 = 7.8568 \text{ L H}_2 \quad \text{(calculator answer)}$$

$$= 7.86 \text{ L H}_2 \quad \text{(correct answer)}$$

PRACTICE EXERCISE 13.21

What volume, in liters, of NO gas measured at STP can be produced from 1.75 L of 0.550 M HNO_3 solution and an excess of a 0.650 M H_2S solution according to the following reaction?

$$2 \text{ HNO}_3(aq) + 3 \text{ H}_2\text{S}(aq) \longrightarrow 2 \text{ NO}(g) + 3 \text{ S}(s) + 4 \text{ H}_2\text{O}(l)$$

Ans. 21.6 L NO.   Similar exercises: Problems 13.103 and 13.104

Key Terms

The new terms or concepts defined in this chapter are

aqueous solution (Sec. 13.2)   A solution in which water is the solvent.

concentrated solution (Sec. 13.2)   A solution that contains a large amount of solute relative to the amount that could dissolve.

concentration (Sec. 13.5)   The amount of solute present in a specified amount of solution.

dilute solution (Sec. 13.2)   A solution that contains a small amount of solute relative to the amount that could dissolve.

dilution (Sec. 13.10)   Process in which more solvent is added to a solution in order to lower its concentration.

immiscible substances (Sec. 13.2)   Liquid substances that do not dissolve in each other.

mass–volume percent (Sec. 13.6)   Mass of solute (in grams) divided by the total volume of solution (in milliliters) multiplied by 100.

miscible substances (Sec. 13.2)   Liquid substances that dissolve in any amount in each other.

molality (Sec. 13.9)   Concentration of a solution in terms of moles of solute per kilogram of solvent.

molarity (Sec. 13.8)   Concentration of a solution in terms of moles of solute per liter of solution.

partially miscible substances (Sec. 13.2)   Liquid substances that have limited solubility in each other.

parts per billion (Sec. 13.7)   Number of parts of solute per billion parts of solution.

parts per million (Sec. 13.7)   Number of parts of solute per million parts of solution.

percent by mass (Sec. 13.6)   Mass of solute divided by the total mass of solution multiplied by 100.

percent by volume (Sec. 13.6)   Volume of solute divided by the total volume of solution multiplied by 100.

saturated solution (Sec. 13.2)   A solution that contains the maximum amount of solute that can be dissolved under the conditions at which the solution exists.

solubility (Sec. 13.2)   The maximum amount of solute that will dissolve in a given amount of solvent under a fixed set of conditions.

solute (Sec. 13.1) Component of a solution present in a small amount relative to that of the solvent.
solution (Sec. 13.1) Homogeneous mixture of two or more substances.

solvent (Sec. 13.1) Component of a solution present in the greatest amount.
unsaturated solution (Sec. 13.2) A solution where less solute than the maximum amount possible is dissolved in the solution.

Practice Problems

Solution Terminology (Sec. 13.2)

13.1 Indicate which substance is the solvent in each of the following solutions.
  (a) a solution containing 10.0 g of NaCl and 100.0 mL of water
  (b) a solution containing 20.0 mL of ethyl alcohol and 15.0 mL of water

13.2 Indicate which substance is the solvent in each of the following solutions.
  (a) a solution containing 1.00 g of $AgNO_3$ and 10.0 mL of water
  (b) a solution containing 45.0 mL of acetone and 15.0 mL of water

13.3 Use the terms *soluble, insoluble, miscible,* or *immiscible* to describe the behavior of the following pairs of substances. More than one term may apply to a given case.
  (a) 75 mL of water and 1 g of table sugar are shaken together. The resulting solution is clear and colorless.
  (b) 25 mL of water and 10 mL of mineral oil are shaken together. The mixture is cloudy and gradually separates into two layers.

13.4 Use the terms *soluble, insoluble, miscible,* or *immiscible* to describe the behavior of the following pairs of substances. More than one term may apply to a given case.
  (a) 10 mL of water and 0.1 g of fine white sand are shaken together. The resulting mixture is very cloudy, and the sand rapidly settles to the bottom of the container.
  (b) 100 mL of water and 100 mL of ethyl alcohol are shaken together. A clear colorless solution results that with time shows no separation into layers.

13.5 Use Table 13.2 to determine whether each of the following solutions is saturated or unsaturated.
  (a) 1.94 g of $PbBr_2$ in 100 g of H_2O at 50 °C
  (b) 34.0 g of NaCl in 100 g of H_2O at 0 °C
  (c) 75.4 g of $CuSO_4$ in 200 g of H_2O at 100 °C
  (d) 0.540 g of Ag_2SO_4 in 50 g of H_2O at 50 °C

13.6 Use Table 13.2 to determine whether each of the following solutions is saturated or unsaturated.

  (a) 175 g of CsCl in 100 g of H_2O at 100 °C
  (b) 455 g of $AgNO_3$ in 100 g of H_2O at 50 °C
  (c) 2.16 g of Ag_2SO_4 in 200 g of H_2O at 50 °C
  (d) 0.97 g of $PbBr_2$ in 50 g of H_2O at 50 °C

13.7 Using the solubilities given in Table 13.2, characterize each of the following solids as insoluble, slightly soluble, soluble, or very soluble in water at the indicated temperature.
  (a) lead(II) bromide at 50 °C
  (b) cesium chloride at 0 °C
  (c) silver nitrate at 100 °C
  (d) silver sulfate at 0 °C

13.8 Using the solubilities given in Table 13.2, characterize each of the following solids as insoluble, slightly soluble, soluble, or very soluble in water at the indicated temperature.
  (a) copper(II) sulfate at 100 °C
  (b) lead(II) bromide at 0 °C
  (c) silver nitrate at 50 °C
  (d) sodium chloride at 0 °C

13.9 Based on the solubilities in Table 13.2, characterize each of the following solutions as dilute or concentrated.
  (a) 0.20 g of $CuSO_4$ dissolved in 100 g of H_2O at 100 °C
  (b) 1.50 g of $PbBr_2$ dissolved in 100 g of H_2O at 50 °C
  (c) 60 g of $AgNO_3$ dissolved in 100 g of H_2O at 50 °C
  (d) 0.50 g of Ag_2SO_4 dissolved in 100 g of H_2O at 0 °C

13.10 Based on the solubilities in Table 13.2, characterize each of the following solutions as dilute or concentrated.
  (a) 5.00 g of NaCl dissolved in 100 g of H_2O at 100 °C
  (b) 1.40 g of Ag_2SO_4 dissolved in 100 g of H_2O at 100 °C
  (c) 50.0 g of CsCl dissolved in 100 g of H_2O at 50 °C
  (d) 100.0 g of $AgNO_3$ dissolved in 100 g of H_2O at 100 °C

Solubility Rules (Sec. 13.4)

13.11 Predict whether the following solutes are very soluble or slightly soluble in water.

(a) O_2 (a nonpolar gas)
(b) CH_3OH (a polar liquid)
(c) CBr_4 (a nonpolar liquid)
(d) $AgCl$ (an ionic solid)

13.12 Predict whether the following solutes are very soluble or slightly soluble in water.
(a) NH_3 (a polar gas)
(b) N_2 (a nonpolar gas)
(c) C_6H_6 (a nonpolar liquid)
(d) $NaC_2H_3O_2$ (an ionic solid)

13.13 In each of the following sets of ionic compounds, identify the members of the set that are soluble in water.
(a) $NaCl$, Na_2SO_4, $NaNO_3$, Na_2CO_3
(b) K_2S, KOH, KNO_3, KBr
(c) Ag_2SO_4, $AgNO_3$, $AgCl$, $AgBr$
(d) CaS, $Ca(NO_3)_2$, $CaSO_4$, $Ca(C_2H_3O_2)_2$

13.14 In each of the following sets of ionic compounds, identify the members of the set that are soluble in water.
(a) NH_4Cl, NH_4NO_3, $NH_4C_2H_3O_2$, $(NH_4)_2SO_4$
(b) Li_2S, Li_2SO_4, $LiOH$, $LiC_2H_3O_2$
(c) PbS, $PbCl_2$, $Pb(NO_3)_2$, $PbSO_4$
(d) $Ba(OH)_2$, $BaCl_2$, $BaSO_4$, $Ba(NO_3)_2$

13.15 In each of the following sets of ionic compounds, identify the members of the set that are soluble in water.
(a) $Zn(NO_3)_2$, $Al(NO_3)_3$, $Be(NO_3)_2$, $LiNO_3$
(b) $AgC_2H_3O_2$, $NaC_2H_3O_2$, $Mg(C_2H_3O_2)_2$, $Pb(C_2H_3O_2)_2$
(c) K_2CO_3, $MgCO_3$, $NiCO_3$, $Al_2(CO_3)_3$
(d) MgS, Rb_2S, Al_2S_3, CaS

13.16 In each of the following sets of ionic compounds, identify the members of the set that are soluble in water.
(a) KCl, $CaCl_2$, Hg_2Cl_2, $MgCl_2$
(b) K_2SO_4, $FeSO_4$, $Fe_2(SO_4)_3$, $SrSO_4$
(c) $Be_3(PO_4)_2$, $AlPO_4$, $FePO_4$, $(NH_4)_3PO_4$
(d) $Cu(OH)_2$, $Be(OH)_2$, $Ca(OH)_2$, $Zn(OH)_2$

Mass Percent (Sec. 13.6)

13.17 Calculate the mass percent of solute in each of the following solutions.
(a) 7.37 g of NaCl dissolved in 95.0 g of H_2O
(b) 3.73 g of KBr dissolved in 131 g of H_2O
(c) 10.3 g of KBr dissolved in 53.0 g of solution
(d) 0.0513 mole of Na_2SO_4 dissolved in 100.0 g of H_2O

13.18 Calculate the mass percent of solute in each of the following solutions.
(a) 1.13 g of $AgNO_3$ dissolved in 20.0 g of H_2O
(b) 218 g of CsCl dissolved in 102 g of H_2O
(c) 10.3 g of KNO_3 dissolved in 95.2 g of solution
(d) 0.0010 mole of BaI_2 dissolved in 25.0 g of H_2O

13.19 Calculate the mass, in grams, of solute needed to prepare each of the following amounts of solution.
(a) a 1.20% (m/m) HCl solution containing 875 g of H_2O
(b) a 8.0% (m/m) NH_4Cl solution containing 125 g of H_2O
(c) 12.2 g of a 12.0% (m/m) NaCl solution
(d) 250.0 g of a 5.000% (m/m) $NaNO_3$ solution

13.20 Calculate the mass, in grams, of solute needed to prepare each of the following amounts of solution.
(a) a 45.0% (m/m) NH_4NO_3 solution containing 725 g of H_2O
(b) a 0.0100% (m/m) $CaCl_2$ solution containing 125 g of H_2O
(c) 32.00 g of a 2.000% (m/m) K_2SO_4 solution
(d) 225 g of a 30.0% (m/m) KI solution

13.21 Using Table 13.2, calculate the percent by mass of solute in each of the following saturated solutions.
(a) NaCl at 0 °C
(b) NaCl at 100 °C
(c) $AgNO_3$ at 0 °C
(d) $AgNO_3$ at 100 °C

13.22 Using Table 13.2, calculate the percent by mass of solute in each of the following saturated solutions.
(a) $CuSO_4$ at 0 °C
(b) $CuSO_4$ at 100 °C
(c) CsCl at 0 °C
(d) CsCl at 100 °C

13.23 How many grams of water must be added to 50.0 g of each of the following solutes in order to prepare a 5.00% (m/m) solution?
(a) NaCl
(b) KCl
(c) Na_2SO_4
(d) $LiNO_3$

13.24 How many grams of water must be added to 20.0 g of each of the following solutes in order to prepare a 2.00% (m/m) solution?
(a) NaOH
(b) LiBr
(c) Li_2SO_4
(d) $Ca(NO_3)_2$

Volume Percent (Sec. 13.6)

13.25 Calculate the volume percent of solute in each of the following solutions.
(a) 30.0 mL of methyl alcohol (CH_3OH) dissolved in enough water to give 500.0 mL of solution
(b) 3.75 mL of Br_2 dissolved in enough carbon tetrachloride to produce 96.0 mL of solution
(c) 60.0 mL of ethylene glycol [$C_2H_4(OH)_2$] dissolved in enough water to produce 354 mL of solution
(d) 455 mL of ethyl alcohol (C_2H_5OH) dissolved in enough water to produce 1.375 L of solution

13.26 Calculate the volume percent of solute in each of the following solutions.
(a) 60.0 mL of methyl alcohol (CH_3OH) dissolved in enough water to give 90.0 mL of solution

(b) 4.68 mL of Br_2 dissolved in enough carbon tetrachloride to produce 56.0 mL of solution
(c) 25.0 mL of ethylene glycol [$C_2H_4(OH)_2$] dissolved in enough water to produce 47.0 mL of solution
(d) 75.0 mL of ethyl alcohol (C_2H_5OH) dissolved in enough water to produce 0.688 L of solution

13.27 The final volume of a solution resulting from the addition of 50.0 mL of ethyl alcohol (C_2H_5OH) to 50.0 mL of water is 96.5 mL. What is the volume percent of ethyl alcohol in the solution?

13.28 The final volume of a solution resulting from the addition of 50.0 mL of methyl alcohol (CH_3OH) to 50.0 mL of water is 96.0 mL. What is the volume percent of water in the solution?

13.29 How much isopropyl alcohol (C_3H_8O), in milliliters, is needed to prepare 225 mL of a 2.25% (v/v) solution of isopropyl alcohol in water?

13.30 How much ethyl alcohol (C_2H_6O), in milliliters, is needed to prepare 125 mL of a 1.25% (v/v) solution of ethyl alcohol in water?

13.31 What volume of water, in quarts, is contained in 5.00 qt of a 25.0% (v/v) solution of water in acetone?

13.32 What volume of water, in gallons, is contained in 2.00 gal of a 15.0% (v/v) solution of water in methyl alcohol?

Mass–Volume Percent (Sec. 13.6)

13.33 Calculate the concentration, as mass–volume percent, for each of the following $MgCl_2$ solutions.
(a) 5.0 g is added to enough water to give $25\overline{0}$ mL of solution.
(b) 85.0 g is added to enough water to give 1.50 L of solution.
(c) 1.00 mole is added to enough water to give $20\overline{0}$ mL of solution.
(d) 0.025 mole is added to enough water to give 1.105 L of solution.

13.34 Calculate the concentration, as mass–volume percent, for each of the following $NaNO_3$ solutions.
(a) 1.00 g is added to enough water to give 75.0 mL of solution.
(b) 100.0 g is added to enough water to give 1.000 L of solution.
(c) 2.00 mole is added to enough water to give 10.00 L of solution.
(d) 0.00100 mole is added to enough water to give 10.0 mL of solution.

13.35 Calculate the concentration, as mass–volume percent NaCl, for a solution prepared by adding 30.0 g of NaCl to 70.0 g of water to give a solution with a density of 1.25 g/mL.

13.36 Calculate the concentration, as mass–volume percent $NaNO_3$, for a solution prepared by adding 10.0 g of $NaNO_3$ to 40.0 g of water to give a solution with a density of 1.11 g/mL.

13.37 How many grams of Na_2CO_3 are needed to prepare 7.000 L of a 3.000% (m/v) Na_2CO_3 solution?

13.38 How many grams of $Na_2S_2O_3$ are needed to prepare 4.000 L of a 2.000% (m/v) $Na_2S_2O_3$ solution?

13.39 How many grams of $(NH_4)_2SO_4$ are present in 50.0 L of a 7.50% (m/v) $(NH_4)_2SO_4$ solution?

13.40 How many grams of $Al(NO_3)_3$ are present in 25.0 L of a 12.5% (m/v) $Al(NO_3)_3$ solution?

13.41 How many milliliters of a 4.5% (m/v) KBr solution would contain 975 g of KBr?

13.42 How many milliliters of a 2.2% (m/v) KNO_3 solution would contain 625 g of KNO_3?

Parts per Million and Parts per Billion (Sec. 13.7)

13.43 Normal blood serum contains about $1\overline{0}$ mg of Ca^{2+} ion per $10\overline{0}$ mL of serum. Express this Ca^{2+} ion concentration in
(a) ppm (m/v)   (b) ppb (m/v)

13.44 A $25\overline{0}$ mL sample of well water is found to contain 0.000011 g of dissolved lead (Pb^{2+} ion). Express the Pb^{2+} ion concentration in the well water in
(a) ppm (m/v)   (b) ppb (m/v)

13.45 The concentration of sulfate ion, SO_4^{2-}, in seawater averages 2.71 g/kg. Express this concentration in
(a) ppm (m/m)   (b) ppb (m/m)

13.46 The concentration of chloride ion, Cl^-, in seawater averages 19.35 g/kg. Express this concentration in
(a) ppm (m/m)   (b) ppb (m/m)

13.47 A gaseous pollutant is present in air at a concentration of 0.30 ppm (v/v). What volume of pollutant, in milliliters, is contained in 5.000 L of air?

13.48 A gaseous pollutant is present in air at a concentration of 0.10 ppm (v/v). How many liters of air would be needed to extract 1.00 mL of the pollutant?

13.49 A typical concentration of nitrogen oxides in urban atmospheres is 0.053 ppm (v/v). Express this concentration in ppb (v/v).

13.50 A typical concentration of sulfur dioxide in urban atmospheres is 0.087 ppm (v/v). Express this concentration in ppb (v/v).

13.51 Determine how much nitrous oxide (N_2O), in grams, must be added to a 875-mL sample of air to give a N_2O

concentration of
(a) 3.6 ppm (m/v)
(b) 4.2 ppb (m/v)
(c) 6.8% (m/v)

13.52 Determine how much sulfur dioxide (SO_2), in grams, must be added to a 1.00-L sample of air to give a SO_2 concentration of
(a) 15.0 ppm (m/v)
(b) 1.0 ppb (m/v)
(c) 2.0% (m/v)

Molarity (Sec. 13.8)

13.53 What is the difference in meaning, if any, between the phrases "1 mole KCl" and "1 molar KCl"?

13.54 What is the difference in meaning, if any, between the phrases "1 molar KCl" and "1 M KCl"?

13.55 Calculate the molarity of the following solutions.
(a) 2.0 moles NaCl in 0.50 L of solution
(b) 13.7 g KNO_3 in 90.0 mL of solution
(c) 53.0 g $C_{12}H_{22}O_{11}$ in 1255 mL of solution
(d) 0.0010 mole $NaHCO_3$ in 3.03 mL of solution

13.56 Calculate the molarity of the following solutions.
(a) 1.20 moles NH_4Cl in 2.50 L of solution
(b) 0.523 mole KCl in 875 mL of solution
(c) 100.0 g $NaC_2H_3O_2$ in 5000 mL of solution
(d) 23.4 g $AgNO_3$ in 87.6 L of solution

13.57 Calculate the number of grams of solute in each of the following solutions.
(a) 3.00 L of a 6.00 M HNO_3 solution
(b) 50.0 mL of a 0.350 M NaCl solution
(c) 375 mL of a 1.75 M $BaCl_2$ solution
(d) 150.0 g of a 6.00 M H_2SO_4 solution with a density of 1.3 g/mL

13.58 Calculate the number of grams of solute in each of the following solutions.
(a) 2.50 L of a 3.00 M HCl solution
(b) 10.0 mL of a 0.0100 M KCl solution
(c) 867 mL of a 1.83 M $Ca(NO_3)_2$ solution
(d) 200.0 g of a 3.00 M H_3PO_4 solution with a density of 1.2 g/mL

13.59 Calculate the volume, in milliliters, of solution required to provide each of the following.
(a) 2.50 g of KBr from a 0.200 M KBr solution
(b) 125 g of $KC_2H_3O_2$ from a 7.83 M $KC_2H_3O_2$ solution
(c) 4.50 moles of KNO_3 from a 0.300 M KNO_3 solution
(d) 0.0025 mole of K_2CO_3 from a 9.6 M K_2CO_3 solution

13.60 Calculate the volume, in milliliters, of solution required to provide each of the following.

(a) 4.30 g of LiCl from a 0.0753 M LiCl solution
(b) 425 g of $LiNO_3$ from a 10.2 M $LiNO_3$ solution
(c) 3.75 moles of Li_2SO_4 from a 0.200 M Li_2SO_4 solution
(d) 0.0010 mole of LiOH from a 4.53 M LiOH solution

13.61 Calculate the volume, in milliliters, of 0.125 M Na_2SO_4 solution needed to provide each of the following.
(a) 10.0 g of Na_2SO_4
(b) 2.5 g of Na^+ ion
(c) 0.567 mole of Na_2SO_4
(d) 0.112 mole of SO_4^{2-} ion

13.62 Calculate the volume, in milliliters, of 1.25 M $Mg(NO_3)_2$ solution needed to provide each of the following.
(a) 15.7 g of $Mg(NO_3)_2$
(b) 3.57 g of Mg^{2+} ion
(c) 1.2 moles of $Mg(NO_3)_2$
(d) 0.57 mole of NO_3^- ion

13.63 How many liters of 0.750 M solution can be prepared from 75.0 g of each of the following solutes?
(a) $H_2C_2O_4$
(b) $Na_2C_2O_4$
(c) CaC_2O_4
(d) $Al_2(C_2O_4)_3$

13.64 How many liters of 1.25 M solution can be prepared from 60.0 g of each of the following solutes?
(a) H_3BO_3
(b) Na_3BO_3
(c) $Mg_3(BO_3)_2$
(d) $AlBO_3$

13.65 The density of a 68.0% (m/m) HNO_3 solution is 1.41 g/mL. What is the molarity of this solution?

13.66 The density of a 28.0% (m/m) NH_3 solution is 0.898 g/mL. What is the molarity of this solution?

13.67 The density of a 0.75 M $HC_2H_3O_2$ solution is 1.01 g/mL. What is the concentration of this solution expressed as mass percent?

13.68 The density of a 3.00 M $NaNO_3$ solution is 1.16 g/mL. What is the concentration of this solution expressed as mass percent?

13.69 What is the molarity of a 0.92% (m/v) NaCl solution?

13.70 What is the molarity of a 2.53% (m/v) $AlCl_3$ solution?

Molality (Sec. 13.9)

13.71 What is the similarity that exists in the defining equations for molality and molarity?

13.72 What is the difference that exists in the defining equations for molality and molarity?

13.73 Calculate the molality of each of the following solutions.
(a) 165 g of ethylene glycol ($C_2H_6O_2$) in 1.35 kg of water
(b) 3.75 moles of sodium sulfate (Na_2SO_4) in 455 g of water

(c) 0.0350 g of ammonium chloride (NH_4Cl) in 12.0 g of water
(d) 50.0 g of calcium chloride ($CaCl_2$) in enough water to give 543 mL of solution with a density of 1.10 g/mL

13.74 Calculate the molality of each of the following solutions
(a) 223 g of propylene glycol ($C_3H_8O_2$) in 1.75 kg of water
(b) 5.67 moles of potassium nitrate (KNO_3) in 1890 g of water
(c) 0.275 g of ammonium fluoride (NH_4F) in 55.0 g of water
(d) 40.0 g of magnesium bromide ($MgBr_2$) in enough water to give 675 mL of solution with a density of 1.20 g/mL

13.75 Calculate the number of grams of each solute that must be added to 75.0 g of water to prepare a 0.353 m solution.
(a) NaCl   (b) KCN   (c) $BeCl_2$ (d) K_2SO_4

13.76 Calculate the number of grams of each solute that must be added to 85.0 g of water to prepare a 0.153 m solution.
(a) KI   (b) NaSCN
(c) LiOH   (d) $Mg(NO_3)_2$

13.77 How many grams of water must be added to 80.0 g of sucrose ($C_{12}H_{22}O_{11}$) to prepare a 0.350 m solution?

13.78 How many grams of water must be added to 180.0 g of glucose ($C_6H_{12}O_6$) to prepare a 0.750 m solution?

13.79 Calculate the total mass, in grams, and the total volume, in milliliters, of a 2.16 m H_3PO_4 solution containing 52.0 g of solute. The density of the solution is 1.12 g/mL.

13.80 Calculate the total mass, in grams, and the total volume, in milliliters, of a 0.710 m $H_3C_6H_5O_7$ (citric acid) solution containing 23.0 g of solute. The density of the solution is 1.05 g/mL.

13.81 Calculate the molality of a 14.0% by mass HCl solution.

13.82 Calculate the molality of a 23.0% by mass HNO_3 solution.

Dilution (Sec. 13.10)

13.83 What is the molarity of the solution prepared by diluting 15.0 mL of 0.330 M KOH to each of the following final volumes?
(a) 25.0 mL   (b) 75.0 mL
(c) 457 mL   (d) 3.75 L

13.84 What is the molarity of the solution prepared by diluting 50.0 mL of 4.50 M NaOH to each of the following final volumes?
(a) 60.0 mL   (b) 125 mL
(c) $80\overline{0}$ mL   (d) 5.60 L

13.85 What is the molarity of the solution prepared by concentrating, by evaporation of solvent, 2875 mL of 0.525 M $NaNO_3$ solution to each of the following final volumes?
(a) 2275 mL   (b) 932 mL
(c) 322 mL   (d) 1.23 L

13.86 What is the molarity of the solution prepared by concentrating, by evaporation of solvent, 1325 mL of 1.34 M KNO_3 solution to each of the following final volumes?
(a) 1125 mL   (b) 833 mL
(c) 409 mL   (d) 0.780 L

13.87 In each of the following NaCl solutions, how many milliliters of water should be added to obtain a solution that has a concentration of 0.100 M?
(a) 50.0 mL of a 3.00 M solution
(b) 2.00 mL of a 1.00 M solution
(c) 1.45 L of a 6.00 M solution
(d) 75.0 mL of a 0.120 M solution

13.88 In each of the following $AgNO_3$ solutions, how many milliliters of water should be added to obtain a solution that has a concentration of 1.00 M?
(a) 20.0 mL of a 3.00 M solution
(b) 20.0 mL of a 10.0 M solution
(c) 3.67 L of a 1.20 M solution
(d) 575 mL of a 5.60 M solution

13.89 Determine how many milliliters of H_2O must be boiled off from 1.50 L of a 0.100 M NaCl solution to increase its concentration to
(a) 0.130 M   (b) 0.500 M
(c) 3.00 M   (d) 7.50 M

13.90 Determine how many milliliters of H_2O must be boiled off from 2.00 L of a 0.500 M Na_2SO_4 solution to increase its concentration to
(a) 0.505 M   (b) 0.600 M
(c) 1.00 M   (d) 7.80 M

13.91 What will be the final concentration of each of the following solutions if the volume of the solution is increased by 20.0 mL by adding water?
(a) 25.0 mL of 6.0 M Na_2SO_4
(b) 100.0 mL of 3.0 M K_2SO_4
(c) 0.155 L of 10.0 M CsCl
(d) 2.00 mL of 0.100 M $MgCl_2$

13.92 What will be the final concentration of each of the following solutions if the volume of the solution is increased by 20.0 mL by adding water?
(a) 50.0 mL of 2.0 M KNO_3
(b) 50.0 mL of 3.0 M $AgNO_3$
(c) 1.0000 L of 1.2131 M $NaNO_3$
(d) 1.000 mL of 1.000 M $LiNO_3$

13.93 What would be the molarity of a solution made by mixing 235 mL of 0.350 M KBr with 351 mL of 6.53 M KBr?

13.94 What would be the molarity of a solution made by mixing 10.0 mL of 0.100 M $Ca(NO_3)_2$ with 25.0 mL of 0.200 M $Ca(NO_3)_2$?

Molarity and Chemical Equations (Sec. 13.11)

13.95 What volume, in liters, of 1.00 M H_2S is needed to react completely with 0.500 L of 4.00 M HNO_3 according to the following equation?

$$2\ HNO_3(aq) + 3\ H_2S(aq) \rightarrow 2\ NO(g) + 3\ S(s) + 4\ H_2O(l)$$

13.96 What volume, in milliliters, of 0.300 M $AgNO_3$ is needed to react completely with 40.0 mL of 0.200 M K_3PO_4 according to the following equation?

$$3\ AgNO_3(aq) + K_3PO_4(aq) \rightarrow Ag_3PO_4(s) + 3\ KNO_3(aq)$$

13.97 How many grams of $PbCl_2$ can be produced from the reaction of 30.0 mL of 12.0 M NaCl with an excess of 3.00 M $Pb(NO_3)_2$ solution according to the following equation?

$$Pb(NO_3)_2(aq) + 2\ NaCl(aq) \rightarrow PbCl_2(s) + 2\ NaNO_3(aq)$$

13.98 How many grams of $Ca_3(PO_4)_2$ can be produced from the reaction of 2.50 L of 0.200 M $CaCl_2$ with an excess of 0.750 M H_3PO_4 according to the following equation?

$$3\ CaCl_2(aq) + 2\ H_3PO_4(aq) \rightarrow Ca_3(PO_4)_2(s) + 6\ HCl(aq)$$

13.99 What volume, in milliliters, of 0.50 M HBr is required to react with 18.0 g of zinc according to the following equation?

$$2\ HBr(aq) + Zn(s) \rightarrow ZnBr_2(aq) + H_2(g)$$

13.100 What volume, in milliliters, of 1.50 M HNO_3 is required to react with 100.0 g of Cu_2O according to the following equation?

$$14\ HNO_3(aq) + 3\ Cu_2O(s) \rightarrow 6\ Cu(NO_3)_2(aq) + 2\ NO(g) + 7\ H_2O(l)$$

13.101 What is the molarity of a 37.5-mL sample of H_2SO_4 solution that will completely react with 23.7 mL of 0.100 M NaOH according to the following equation?

$$H_2SO_4(aq) + 2\ NaOH(aq) \rightarrow Na_2SO_4(aq) + 2\ H_2O(l)$$

13.102 What is the molarity of a 50.0-mL sample of HNO_3 solution that will completely react with 40.0 mL of 0.200 M $Mg(OH)_2$ according to the following equation?

$$2\ HNO_3(aq) + Mg(OH)_2(aq) \rightarrow Mg(NO_3)_2(aq) + 2\ H_2O(l)$$

13.103 What volume, in liters, of H_2 gas measured at STP can be produced from 50.0 mL of a 6.0 M H_2SO_4 solution and an excess of Ni metal according to the following reaction?

$$Ni(s) + H_2SO_4(aq) \rightarrow NiSO_4(aq) + H_2(g)$$

13.104 What volume, in liters, of NO gas measured at STP can be produced from 50.0 mL of a 3.0 M HNO_3 solution and an excess of Cu_2O according to the following reaction?

$$14\ HNO_3(aq) + 3\ Cu_2O(s) \rightarrow 6\ Cu(NO_3)_2(aq) + 2\ NO(g) + 7\ H_2O(l)$$

13.105 What is the molarity of a 1.75-L $Ca(OH)_2$ solution that would completely react with 2.00 L of CO_2 gas measured at STP according to the following reaction?

$$CO_2(g) + Ca(OH)_2(aq) \rightarrow CaCO_3(s) + H_2O(l)$$

13.106 What is the molarity of a 5.00-L NaOH solution that would completely react with 4.00 L of CO_2 gas measured at STP according to the following reaction?

$$CO_2(g) + 2\ NaOH(aq) \rightarrow Na_2CO_3(aq) + H_2O(l)$$

Additional Problems

13.107 In which of the following pairs of compounds do both members of the pair have like solubility (both soluble or both insoluble) in water?
(a) K_2CO_3 and NH_4SCN (b) BaS and $NiCO_3$
(c) AgCl and $Al(OH)_3$ (d) $Zn(NO_3)_2$ and $PbSO_4$

13.108 In which of the following pairs of compounds do both members of the pair have like solubility (both soluble or both insoluble) in water?

(a) $Be(C_2H_3O_2)_2$ and $AgNO_3$
(b) $Ba_3(PO_4)_2$ and CuS
(c) $ZnCl_2$ and $Mg(OH)_2$
(d) $CaSO_4$ and $Al_2(CO_3)_3$

13.109 After all of the water is evaporated from 254 mL of a $AgNO_3$ solution, 45.2 g of $AgNO_3$ remains. What was the original concentration of the $AgNO_3$ solution expressed as mass–volume percent?

13.110 After all of the water is evaporated from 10.0 mL of a CsCl solution, 3.75 g of CsCl remains. What was the original concentration of the CsCl solution expressed as mass–volume percent?

13.111 What mass, in grams, of Na_2SO_4 would be required to prepare 425 mL of a 1.55% (m/m) Na_2SO_4 solution whose density is 1.02 g/mL?

13.112 What mass, in grams, of NaCl would be required to prepare 275 mL of a 30.0% (m/m) NaCl solution whose density is 1.18 g/mL?

13.113 A 3.000 M $NaNO_3$ solution has a density of 1.161 g/mL at 20 °C. How many grams of solvent are present in 1.375 L of this solution?

13.114 A 0.157 M NaCl solution has a density of 1.09 g/mL at 20 °C. How many grams of solvent are present in 80.0 mL of this solution?

13.115 A solution is made by diluting 225 mL of a 0.245 M aluminum nitrate, $Al(NO_3)_3$, solution with water to a final volume of 0.750 L. Calculate
(a) the molarities of Al^{3+} ion and NO_3^- ion in the original solution
(b) the molarities of $Al(NO_3)_3$, Al^{3+} ion, and NO_3^- ion in the diluted solution

13.116 A solution is made by diluting 315 mL of a 0.115 M potassium phosphate, K_3PO_4, solution with water to a final volume of 0.650 L. Calculate
(a) the molarities of K^+ ion and PO_4^{3-} ion in the original solution
(b) the molarities of K_3PO_4, K^+ ion, and PO_4^{3-} ion in the diluted solution

13.117 A solution is made by mixing 175 mL of 0.100 M K_3PO_4 with 27 mL of 0.200 M KCl. Assuming that the volumes are additive, what are the molar concentrations of the following ions in the new solution?
(a) K^+ ion (b) Cl^- ion (c) PO_4^{3-} ion

13.118 A solution is made by mixing 50.0 mL of 0.300 M Na_2SO_4 with 30.0 mL of 0.900 M K_2SO_4. Assuming that the volumes are additive, what are the molar concentrations of the following ions in the new solution?
(a) Na^+ ion (b) K^+ ion (c) SO_4^{2-} ion

13.119 A solute concentration is 3.74 ppm (m/m). What would this concentration be in the units mg of solute per kg of solution?

13.120 A solute concentration is 5.14 ppm (m/m). What would this concentration be in the units μg of solute per mg of solution?

13.121 A solution is prepared by dissolving 1.00 g of NaCl in enough water to make 10.00 mL of solution. A 1.00-mL portion of this solution is then diluted to a final volume volume of 10.00 mL. What is the molarity of the final NaCl solution?

13.122 A solution is prepared by dissolving 30.0 g of Na_2SO_4 in enough water to make 750.0 mL of solution. A 10.00-mL portion of this solution is then diluted to a final volume of 100.0 mL. What is the molarity of the final Na_2SO_4 solution?

13.123 How many milliliters of 38.0% (m/m) HCl (density 1.19 g/mL) are needed to make, using a dilution procedure, 1.00 L of 0.100 M HCl?

13.124 How many milliliters of 20.0% (m/m) NaCl (density 1.15 g/mL) are needed to make, using a dilution procedure, 3.50 L of 0.150 M NaCl?

13.125 How many grams of water should you add to 1.50 kg of a 1.23 m NaCl solution to reduce the molality to 1.00 m?

13.126 How many grams of water should you add to 0.650 kg of a 0.0883 m NaCl solution to reduce the molality to 0.0100 m?

13.127 An aqueous solution of oxalic acid ($H_2C_2O_4$) is 0.568 M and has a density of 1.022 g/mL. What is the molality of the solution?

13.128 An aqueous solution of citric acid ($H_3C_6H_5O_7$) is 0.655 M and has a density of 1.049 g/mL. What is the molality of the solution?

13.129 An aqueous solution of acetic acid ($HC_2H_3O_2$) is 0.796 m and has a density of 1.004 g/mL. What is the molarity of the solution?

13.130 An aqueous solution of tartaric acid ($H_2C_4H_4O_6$) is 0.278 m and has a density of 1.006 g/mL. What is the molarity of the solution?

13.131 The concentration of a KCl solution is 0.273 in terms of molality and 0.271 in terms of molarity. What is the density of the solution, in grams per milliliter?

13.132 The concentration of a $Pb(NO_3)_2$ solution is 0.953 in terms of molality and 0.907 in terms of molarity. What is the density of the solution, in grams per milliliter?

13.133 An aqueous solution having a density of 0.980 g/mL is prepared by dissolving 11.3 mL of CH_3OH (density 0.793 g/mL) in enough water to produce 75.0 mL of solution. Express the percent CH_3OH in this solution as
(a) % (m/v) (b) % (m/m) (c) % (v/v)

13.134 An aqueous solution having a density of 0.993 g/mL is prepared by dissolving 20.0 mL of C_2H_5OH (density 0.789 g/mL) in enough water to produce 85.0 mL of solution. Express the percent C_2H_5OH in this solution as
(a) % (m/v) (b) % (m/m) (c) % (v/v)

Cumulative Problems

13.135 Identify the insoluble substance(s) formed when each of the following pairs of soluble substances react in aqueous solution through a double-replacement reaction.
(a) NaCl and $AgNO_3$
(b) $Ba(C_2H_3O_2)_2$ and K_3PO_4
(c) $Pb(NO_3)_2$ and Ag_2SO_4
(d) $CuSO_4$ and BaS

13.136 Identify the insoluble substance(s) formed when each of the following pairs of soluble substances react in aqueous solution through a double-replacement reaction.
(a) $MgCl_2$ and $Ba(OH)_2$ (b) NH_4Cl and $Pb(NO_3)_2$
(c) MgS and Na_2CO_3 (d) $SrCl_2$ and Ag_2SO_4

13.137 How many liters of $NH_3(g)$ at 25 °C and 1.46 atm pressure are required to prepare 2.00 L of a 3.50 M solution of NH_3?

13.138 How many liters of HCl(g) at 35 °C and 1.05 atm pressure are required to prepare 4.00 L of a 0.500 M solution of HCl?

13.139 Calculate the theoretical yield, in grams, of AgCl formed from the reaction of 6.41 g of $ZnCl_2$ with 40.0 mL of a 0.404 M $AgNO_3$ solution according to the reaction

$$ZnCl_2(s) + 2\,AgNO_3(aq) \rightarrow Zn(NO_3)_2(aq) + 2\,AgCl(s)$$

13.140 Calculate the theoretical yield, in grams, of AgCl formed from the reaction of 10.00 g of KCl with 100.0 mL of a 2.50 M $AgC_2H_3O_2$ solution according to the reaction

$$KCl(s) + AgC_2H_3O_2(aq) \rightarrow KC_2H_3O_2(aq) + AgCl(s)$$

13.141 What mass, in grams, of $BaCrO_4$ would be produced by mixing $35\overline{0}$ mL of a 3.25 M $BaCl_2$ solution with $45\overline{0}$ mL of a 4.50 M K_2CrO_4 solution? The two solutions react according to the equation

$$BaCl_2(aq) + K_2CrO_4(aq) \rightarrow BaCrO_4(s) + 2\,KCl(aq)$$

13.142 What mass, in grams, of $BaSO_4$ would be produced by mixing 1.53 L of a 4.50 M Na_2SO_4 solution with 3.20 L of a $2.5\overline{0}$ M $Ba(NO_3)_2$ solution? The two solutions react according to the equation

$$Na_2SO_4(aq) + Ba(NO_3)_2(aq) \rightarrow 2\,NaNO_3(aq) + BaSO_4(s)$$

13.143 A 1.25 g *impure* sample of Na_2CO_3 is found to react completely with 70.0 mL of 0.125 M HCl. The equation for the reaction is

$$Na_2CO_3(s) + 2\,HCl(aq) \rightarrow 2\,NaCl(aq) + CO_2(g) + H_2O(l)$$

What is the mass percent Na_2CO_3 in the impure sample?

13.144 A 5.00 g *impure* sample of $CaCO_3$ is found to react completely with 100.0 mL of 0.100 M H_2SO_4. The equation for the reaction is

$$CaCO_3(s) + H_2SO_4(aq) \rightarrow CaSO_4(s) + CO_2(g) + H_2O(l)$$

What is the mass percent $CaCO_3$ in the impure sample?

13.145 Magnesium, calcium, and zinc all react with hydrochloric acid as follows (where M represents any of these metals).

$$M(s) + 2\,HCl(aq) \rightarrow MCl_2(aq) + H_2(g)$$

A sample of one of these metals reacts completely with the acid in 27.9 mL of 2.48 M HCl, and the resulting solution is evaporated to dryness. The residue MCl_2 has a mass of 4.72 g. What is the identity of the metal used?

13.146 Iron, nickel, and tin all react with hydrochloric acid as follows (where M represents any of these metals).

$$M(s) + 2\,HCl(aq) \rightarrow MCl_2(aq) + H_2(g)$$

A sample of one of these metals reacts completely with the acid in 34.2 mL of 4.00 M HCl, and the resulting solution is evaporated to dryness. The residue MCl_2 has a mass of 8.87 g. What is the identity of the metal used?

13.147 A quantity of sodium peroxide (Na_2O_2) is added to water and the following reaction occurs.

$$2\,Na_2O_2(s) + 2\,H_2O(l) \rightarrow 4\,NaOH(aq) + O_2(g)$$

If 70.0 mL of O_2 gas (at STP) and $15\overline{0}$ mL of NaOH solution are produced, what is the molarity of the NaOH solution?

13.148 A quantity of lithium nitride (Li_3N) is added to water, and the following reaction occurs.

$$Li_3N(s) + 3\,H_2O(l) \rightarrow 3\,LiOH(aq) + NH_3(g)$$

If 100.0 mL of NH_3 gas (at STP) and 255 mL of LiOH solution are produced, what is the molarity of the LiOH solution?

Grid Problems

13.149 Select from the grid *all* correct responses for each situation.

1. nitrate	2. chloride	3. carbonate
4. hydroxide	5. acetate	6. iodide
7. phosphate	8. sulfide	9. sulfate

(a) forms a soluble compound when paired with all positive ions
(b) forms a soluble compound when paired with most positive ions
(c) forms an insoluble compound when paired with most positive ions
(d) forms an insoluble compound when paired with the calcium ion
(e) forms a soluble compound when paired with the barium ion
(f) pairs of negative ions with identical solubility guidelines

13.150 Select from the grid *all* correct responses for each situation.

1. molality	2. mass-volume percent	3. ppm (v/v)
4. ppb (m/v)	5. molarity	6. percent by volume
7. percent by mass	8. ppm (m/m)	9. ppb (m/v)

(a) the numerator of the defining equation is moles of solute
(b) the denominator of the defining equation involves solution rather than solvent
(c) a power of ten is present in the defining equation
(d) the denominator of the defining equation has mass units
(e) the numerator of the defining equation has volume units
(f) both the numerator and denominator of the defining equation have specified units

13.151 Select from the grid *all* correct responses for each situation. All responses are solutions with the same solute and solvent present.

1. 20 mL 1.0 M NaCl	2. 60 mL 1.0 M NaCl	3. 80 mL 1.0 M NaCl
4. 20 mL 2.0 M NaCl	5. 60 mL 2.0 M NaCl	6. 120 mL 2.0 M NaCl
7. 20 mL 3.0 M NaCl	8. 40 mL 3.0 M NaCl	9. 80 mL 3.0 M NaCl

(a) solutions that can be diluted to produce a 1.5 M solution
(b) solutions for which addition of 20 mL of solvent will halve the concentration
(c) solutions for which removal of 15 mL of solvent (by evaporation) will produce a solution with a molarity greater than 3.0
(d) solutions for which doubling the volume (by solvent addition) will halve the concentration
(e) pairs of solutions that contain the same number of moles of solute
(f) pairs of solutions that when added together produce a 1.5 M solution

13.152 Select from the grid *all* correct responses for each situation. All responses involve aqueous solutions of NaOH.

1. 6 L solution 3 moles NaOH	2. 10.0 M NaOH, 100 mL solution	3. 1 L solution, 40.0 g NaOH
4. 1 g NaOH in 50 mL solution mixed with 3 g NaOH in 150 mL solution	5. 10.0 g NaOH 250 mL solution	6. 0.1 mole NaOH, 500 mL solution
7. 0.5 L solution 6.0 M NaOH	8. 500 mL 10.0 M NaOH mixed with 500 mL 2.00 M NaOH	9. 50.0 mL 0.1 M NaOH mixed with 50.0 mL 0.3 M NaOH

(a) pairs of solutions that have the same volume
(b) pairs of solutions that have the same number of grams of solute present
(c) pairs of solutions that have the same molarity
(d) solutions in which more than 2 moles of solute are present
(e) pairs of solutions that have same concentration in mass–volume percent of solute
(f) solutions in which the number of moles of solute present and the molarity are numerically equal

CHAPTER FOURTEEN

Acids, Bases, and Salts

14.1 Arrhenius Acid–Base Theory

Acids and bases are among the most common and important compounds known. Aqueous solutions of acids and bases are key materials in both biological systems and chemical industrial processes.

Historically, as early as the seventeenth century, acids and bases were recognized as important groups of compounds. Such early recognition was based on what the substances did rather than on their chemical composition.

Early known facts about acids include

1. Acids, when dissolved in water, have a sour taste. (The name acid comes from the Latin word *acidus*, which means "sour.")
2. Acids cause the dye litmus to change from blue to red. (Litmus is a naturally occurring vegetable dye obtained from lichens.)
3. When certain metals, such as zinc and iron, are placed in acids, they dissolve, liberating hydrogen gas.

Early known characteristics of bases include

1. Water solutions of bases feel slippery or soapy to the touch and have a bitter taste.
2. Bases cause the dye litmus to change from red to blue.
3. When fatty substances are placed in a base solution, they dissolve.

It was not until 1884 that acids and bases were defined in terms of chemical composition. In that year, the Swedish chemist Svante August Arrhenius (1859–1927) proposed that acids and bases be defined in terms of the species they form upon dissolution in water. His definitions are the simplest and most commonly used today. An **Arrhenius acid** is a hydrogen-containing compound that, in water, produces hydrogen ions (H^+). The acidic species in Arrhenius theory is, thus, the hydrogen ion. An **Arrhenius base** is a hydroxide-containing compound that, in water, produces hydroxide ions (OH^-). The basic species in Arrhenius theory is, thus, the hydroxide ion.

Two common examples of acids, according to the Arrhenius definitions, are the substances HNO_3 and HCl.

$$HNO_3(l) \xrightarrow{H_2O} H^+(aq) + NO_3^-(aq)$$

$$HCl(g) \xrightarrow{H_2O} H^+(aq) + Cl^-(aq)$$

Arrhenius acids in the pure state (not in solution) are covalent compounds; that is, they do not contain H^+ ions. The ions are formed, through a chemical reaction, when the acid is mixed with water. This chemical reaction, between water and the acid molecules, results in removal of H^+ ions from acid molecules.

Two common examples of Arrhenius bases are $NaOH$ and KOH.

$$NaOH(s) \xrightarrow{H_2O} Na^+(aq) + OH^-(aq)$$

$$KOH(s) \xrightarrow{H_2O} K^+(aq) + OH^-(aq)$$

Arrhenius bases are usually ionic compounds in the pure state, in direct contrast to acids. When such compounds dissolve in water, the ions separate to yield the OH^- ions.

14.2 Brønsted–Lowry Acid–Base Theory

Although widely used, Arrhenius acid–base theory has some shortcomings. Two disadvantages are that it is restricted to aqueous solution and it cannot explain why compounds like ammonia (NH_3), which do not contain hydroxide ion, produce a basic water solution.

In 1923, Johannes Nicolaus Brønsted (1879–1947), a Danish chemist, and Thomas Martin Lowry (1874–1936), a British chemist, independently and almost simultaneously proposed broadened definitions for acids and bases that applied in both aqueous and nonaqueous solution and that also explained how some non-hydroxide-containing substances, when added to water, produce basic solutions.

A **Brønsted–Lowry acid** is any substance that can donate a proton (H^+) to some other substance. A **Brønsted–Lowry base** is any substance that can accept a proton (H^+) from some other substance. In simpler terms, a Brønsted–Lowry *acid* is a *proton donor* (or hydrogen ion donor) and a Brønsted–Lowry *base* is a *proton acceptor* (or hydrogen ion acceptor). Note that in these definitions the terms *proton* and *hydrogen ion* are used interchangeably. This is acceptable because a H^+ ion is a hydrogen atom (proton plus electron) that has lost its electron; hence, it is a proton.

Three important additional concepts associated with Brønsted–Lowry acid–base theory are

1. Any chemical reaction involving a Brønsted–Lowry acid must also involve a Brønsted–Lowry base. You cannot have one without the other. Proton donation (from an acid) cannot occur unless an acceptor (a base) is present.
2. All the acids and bases included in the Arrhenius theory (Sec. 14.1) are also acids and bases according to the Brønsted–Lowry theory. However, the converse is not true; some substances not considered Arrhenius bases are Brønsted–Lowry bases.
3. The identity of the acidic species in *aqueous solution* is not the H^+ ion but rather the H_3O^+ ion. Hydrogen ions in solution react with water. The attraction between a hydrogen ion and polar water molecule is sufficiently strong to bond the hydrogen ion to a water molecule to form a hydronium ion (H_3O^+). The bond between them is a coordinate covalent bond (Sec. 7.12) because both electrons are furnished by the oxygen atom.

$$H^+ + :\underset{H}{\overset{..}{O}}-H \longrightarrow \left[H:\underset{H}{\overset{..}{O}}-H \right]^+$$
hydronium ion

The Brønsted–Lowry acid–base definitions can best be illustrated by example. Let us consider the formation reaction for hydrochloric acid, which involves the dissolving of hydrogen chloride gas in water.

$$\underset{\text{acid}}{HCl(g)} + \underset{\text{base}}{H_2O(l)} \longrightarrow H_3O^+(aq) + Cl^-(aq)$$

The HCl behaves as a Brønsted–Lowry acid by donating a proton to a water molecule. Note that a hydronium ion is formed as a result. The base in this reaction is water since it has accepted a proton; no hydroxide ions are involved. The Brønsted–Lowry definition of a base includes any species capable of accepting a proton; hydroxide ions can do this, but so can many other substances.

It is not necessary that a water molecule be one of the reactants in a Brønsted–Lowry acid–base reaction or that the reaction take place in the liquid state. An important application of Brønsted–Lowry acid–base theory is to gas-phase reactions. The white solid haze that often covers glassware in a chemistry laboratory results from the gas-phase reaction between HCl and NH_3.

$$\text{HCl(g)} + \text{NH}_3\text{(g)} \longrightarrow \text{NH}_4^+\text{(g)} + \text{Cl}^-\text{(g)} \longrightarrow \text{NH}_4\text{Cl(s)}$$
$$\text{acid} \qquad \text{base}$$

This is a Brønsted–Lowry acid–base reaction, since the HCl molecules donate protons to the NH_3, forming NH_4^+ and Cl^- ions. These ions instantaneously combine to form the white solid NH_4Cl.

Another example of a Brønsted–Lowry acid–base reaction involves the dissolving of ammonia (a nonhydroxide base) in water. In the following equation, note how hydroxide ion is produced as the result of the transfer of a proton from water (written as HOH) to the ammonia

$$\text{NH}_3\text{(g)} + \text{HOH(l)} \longrightarrow \text{NH}_4^+\text{(aq)} + \text{OH}^-\text{(aq)}$$
$$\text{base} \qquad \text{acid}$$

As an ionic solid dissolves in water to produce an aqueous solution, the solid ionic lattice breaks up, producing individual ions that are free to move about in the solution (Sec. 13.3). In such a situation these ions can function as Brønsted–Lowry acids or bases, as is illustrated in the following two equations.

$$\text{HCO}_3^-\text{(aq)} + \text{H}_2\text{O(l)} \longrightarrow \text{CO}_3^{2-}\text{(aq)} + \text{H}_3\text{O}^+\text{(aq)}$$
$$\text{acid} \qquad \text{base}$$

$$\text{H}_2\text{PO}_4^-\text{(aq)} + \text{OH}^-\text{(aq)} \longrightarrow \text{HPO}_4^{2-}\text{(aq)} + \text{H}_2\text{O(l)}$$
$$\text{acid} \qquad \text{base}$$

14.3 Conjugate Acids and Bases

For most Brønsted–Lowry acid–base reactions, a 100% proton transfer does not occur. Instead, a state of equilibrium (Sec. 11.12) is reached in which a forward and a reverse reaction are occurring at an equal rate.

The equilibrium mixture of a Brønsted–Lowry acid–base reaction always has *two* acids and *two* bases present. To illustrate this, consider the acid–base reaction involving hydrogen fluoride and water.

$$\text{HF(aq)} + \text{H}_2\text{O(l)} \rightleftharpoons \text{H}_3\text{O}^+\text{(aq)} + \text{F}^-\text{(aq)}$$

(The double arrows in this equation indicate a state of equilibrium—both a forward and a reverse reaction are occurring.) For the forward reaction, the HF molecules donate protons to water molecules. Thus, the HF is functioning as an acid and the H_2O is functioning as a base.

$$\text{HF(aq)} + \text{H}_2\text{O(l)} \longrightarrow \text{H}_3\text{O}^+\text{(aq)} + \text{F}^-\text{(aq)}$$
$$\text{acid} \qquad \text{base}$$

For the reverse reaction, the one going from right to left, a different picture emerges. Here, H_3O^+ is functioning as an acid (by donating a proton) and F^- behaves as a base (by accepting the proton).

$$H_3O^+(aq) + F^-(aq) \longrightarrow HF(aq) + H_2O(l)$$
$$\text{acid} \qquad \text{base}$$

The two acids and two bases involved in a Brønsted–Lowry equilibrium situation can be grouped into two conjugate acid–base pairs. A **conjugate acid–base pair** is two species that differ from each other by one proton. The two conjugate acid–base pairs in our example are

$$\underbrace{HF(aq) + H_2O(l) \rightleftharpoons H_3O^+(aq)}_{\text{conjugate pair}} + F^-(aq)$$

acid   base   acid   base

(conjugate pair: HF / H_3O^+ and H_2O / F^- — HF and F^- pair; H_2O and H_3O^+ pair)

The **conjugate base** of an acid is the species that remains when an acid loses a proton. The conjugate base of HF is F^-. The **conjugate acid** of a base is the species formed when a base accepts a proton. The H_3O^+ is the conjugate acid of H_2O. Every acid has a conjugate base, and every base has a conjugate acid. In general terms, these relationships can be diagramed as follows.

$$HA + B \rightleftharpoons HB^+ + A^-$$

acid   base   conjugate acid   conjugate base

EXAMPLE 14.1

Identify the conjugate acid–base pairs in the following reaction.

$$HCN(aq) + H_2O(l) \rightleftharpoons H_3O^+(aq) + CN^-(aq)$$

Solution

To determine the conjugate acid–base pairs, we look for formulas that differ only by one H^+ ion. For this reaction, one pair must be HCN and CN^-, and the other pair must be H_2O and H_3O^+. In each pair, the acid is the substance with one more hydrogen, so the two **acids** are HCN and H_3O^+, and the two **bases** are H_2O and CN^-.

$$HCN(aq) + H_2O(l) \rightleftharpoons H_3O^+(aq) + CN^-(aq)$$

conjugate pair: HCN / CN^-; H_2O / H_3O^+

PRACTICE EXERCISE 14.1

Identify the conjugate acid–base pairs in the following reaction.

$$HBr(aq) + H_2O(l) \rightleftharpoons H_3O^+(aq) + Br^-(aq)$$

Ans. HBr, Br$^-$; H$_3$O$^+$, H$_2$O.

Similar exercises: Problems 14.19 and 14.20

EXAMPLE 14.2

Write formulas for the following.
(a) The conjugate base of HSO$_4^-$
(b) The conjugate acid of HPO$_4^{2-}$

Solution

(a) A conjugate base can always be found by removing one H$^+$ ion from a given acid. Removing one H$^+$ (both the atom and the charge) from HSO$_4^-$ leaves SO$_4^{2-}$. Thus, SO$_4^{2-}$ is the conjugate base of HSO$_4^-$.
(b) A conjugate acid can always be found by adding one H$^+$ ion to a given base. Adding one H$^+$ (both the atom and the charge) to HPO$_4^{2-}$ produces H$_2$PO$_4^-$. Thus, H$_2$PO$_4^-$ is the conjugate acid of HPO$_4^{2-}$.

PRACTICE EXERCISE 14.2

Write formulas for the following.

(a) The conjugate base of HS$^-$.
(b) The conjugate acid of PO$_4^{3-}$.

Ans. (a) S^{2-}; (b) HPO$_4^{2-}$.

Similar exercises: Problems 14.9–14.12

Some molecules or ions are able to function as either an acid or a base, depending on the kind of substance with which they react. Such molecules are said to be amphoteric. An **amphoteric substance** can either lose or accept a proton and thus can function as either an acid or a base. (The term *amphoteric* comes from the Greek *amphoteres*, meaning "partly one and partly the other.") Just as an *amphibian* is an animal that lives partly on land and partly in the water, an amphoteric substance is sometimes an acid and sometimes a base.

Water is the most common example of an amphoteric substance. In the first of the following two reactions, water functions as a base and in the second as an acid.

$$HNO_3(l) + H_2O(l) \rightleftharpoons H_3O^+(aq) + NO_3^-(aq)$$
$$\text{acid} \qquad \text{base}$$

$$NH_3(g) + H_2O(l) \rightleftharpoons NH_4^+(aq) + OH^-(aq)$$
$$\text{base} \qquad \text{acid}$$

Another example of an amphoteric substance is the hydrogen carbonate ion.

$$HCO_3^-(aq) + OH^-(aq) \rightleftharpoons CO_3^{2-}(aq) + H_2O(l)$$
$$\text{acid} \qquad \text{base}$$

$$HCO_3^-(aq) + H_3O^+(aq) \rightleftharpoons H_2CO_3(aq) + H_2O(l)$$
$$\text{base} \qquad \text{acid}$$

14.4 Polyprotic Acids

Acids can be classified according to the number of hydrogen ions (protons) they can transfer per molecule during an acid–base reaction. A **monoprotic acid** is an acid that can transfer only one H^+ ion (proton) per molecule during an acid–base reaction. Hydrochloric acid (HCl) and nitric acid (HNO_3) are both monoprotic acids.

A **diprotic acid** is an acid that can transfer two H^+ ions (two protons) per molecule during an acid–base reaction. Sulfuric acid (H_2SO_4) and carbonic acid (H_2CO_3) are examples of diprotic acids. The transfer of protons for a diprotic acid always occurs in steps. For H_2SO_4, the two steps are as follows.

$$H_2SO_4(aq) + H_2O(l) \longrightarrow H_3O^+(aq) + HSO_4^-(aq)$$
$$HSO_4^-(aq) + H_2O(l) \longrightarrow H_3O^+(aq) + SO_4^{2-}(aq)$$

A few triprotic acids exist. A **triprotic acid** is an acid that can transfer three H^+ ions (three protons) per molecule during an acid–base reaction. Phosphoric acid, H_3PO_4, is the most common triprotic acid. The three proton-transfer steps for this acid are as follows.

$$H_3PO_4(aq) + H_2O(l) \longrightarrow H_3O^+(aq) + H_2PO_4^-(aq)$$
$$H_2PO_4^-(aq) + H_2O(l) \longrightarrow H_3O^+(aq) + HPO_4^{2-}(aq)$$
$$HPO_4^{2-}(aq) + H_2O(l) \longrightarrow H_3O^+(aq) + PO_4^{3-}(aq)$$

The general term **polyprotic acid** describes acids that can transfer two or more H^+ ions (protons) per molecule during an acid–base reaction.

The number of hydrogen atoms present in one molecule of an acid cannot always be used to classify the acid as mono-, di-, or triprotic. For example, a molecule of acetic acid contains four hydrogen atoms and yet it is a monoprotic acid. Only one of the hydrogen atoms in acetic acid is acidic. An **acidic hydrogen atom** is a hydrogen atom in an acid molecule that can be transferred to a base during an acid–base reaction.

Whether or not a hydrogen atom is acidic is related to its location in a molecule, that is, to which other atom it is bonded. Let us consider our previously mentioned acetic acid example in more detail by looking at the structure of this acid. A *structural equation* for the acidic behavior of acetic acid is

TABLE 14.1 Selected Common Mono-, Di-, and Triprotic Acids

Name	Formula	Classification	Number of Nonacidic Hydrogen Atoms	Common Occurrence
Acetic acid	$HC_2H_3O_2$	monoprotic	three	vinegar
Lactic acid	$HC_3H_5O_3$	monoprotic	five	sour milk, cheese; produced during muscle contraction
Salicylic acid	$HC_7H_5O_3$	monoprotic	five	present in chemically combined form in aspirin
Hydrochloric acid	HCl	monoprotic	zero	constituent of gastric juice; industrial cleaning agent
Nitric acid	HNO_3	monoprotic	zero	used in urinalysis test for protein; used in manufacture of dyes and explosives
Tartaric acid	$H_2C_4H_4O_6$	diprotic	four	grapes
Carbonic acid	H_2CO_3	diprotic	zero	carbonated beverages; produced in the body from carbon dioxide
Sulfuric acid	H_2SO_4	diprotic	zero	storage batteries; manufacture of fertilizer
Citric acid	$H_3C_6H_5O_7$	triprotic	five	citrus fruits
Boric acid	H_3BO_3	triprotic	zero	antiseptic eyewash
Phosphoric acid	H_3PO_4	triprotic	zero	found in dissociated form (HPO_4^{2-}, $H_2PO_4^-$) in intracellular fluid; component of DNA

Note the structure of the acetic acid molecule (reactant side of the equation); one hydrogen atom is bonded to an oxygen atom and the other three hydrogen atoms are each bonded to a carbon atom. It is only the hydrogen atom bonded to the oxygen atom that is acidic. The hydrogen atoms bonded to the carbon atom are too tightly held to be removed by reaction with water molecules. Water has very little effect on a carbon–hydrogen bond because it is essentially nonpolar (Sec. 7.10). On the other hand, the hydrogen bonded to oxygen is involved in a very polar bond because of oxygen's large electronegativity (Sec. 7.10). Water, which is a polar molecule, readily attacks polar bonds but has very little effect on nonpolar bonds.

We now see why the formula for acetic acid is usually written as $HC_2H_3O_2$ rather than as $C_2H_4O_2$. In the situation where some hydrogens are easily removed (acidic) and others are not (nonacidic), it is accepted procedure to write the acidic hydrogen first, separated from the other hydrogens in the formula. Citric acid, the principal acid in citrus fruits, is another example of an acid that contains both acidic and nonacidic hydrogens. Its formula, $H_3C_6H_5O_7$, indicates that three of the eight hydrogen atoms present in a molecule are acidic. Table 14.1 gives the formulas, classifications, and common occurrences of selected mono-, di-, and triprotic acids, many of which contain nonacidic hydrogen atoms.

We have focused our attention on acids in the preceding discussion. It should be noted that similar concepts can be applied to bases. From an Arrhenius standpoint, bases can release more than one hydroxide ion; for example, $Ca(OH)_2$ is a base that produces two OH^- ions per molecule. From a Brønsted–Lowry viewpoint, bases exist that can accept more than one proton, in a stepwise manner; for example, the PO_4^{3-} ion is a Brønsted–Lowry base that can ultimately accept three protons through reaction with three H_3O^+ ions.

$$PO_4^{3-} \xrightarrow{+H_3O^+} HPO_4^{2-} \xrightarrow{+H_3O^+} H_2PO_4^- \xrightarrow{+H_3O^+} H_3PO_4$$

14.5 Strengths of Acids and Bases

Brønsted–Lowry acids vary in their ability to transfer protons and produce hydronium ions in solution. Such acids are classified as strong or weak on the basis of the extent that proton transfer occurs in aqueous solution. A **strong acid** is a substance that transfers 100%, or very nearly 100%, of its acidic hydrogen atoms to water. Thus, if an acid is strong, almost all of the acid molecules present give up protons to water. This extensive transfer of protons produces many hydronium ions (the acidic species) within the solution. A **weak acid** is a substance that transfers only a small percentage of its acidic hydrogen atoms to water. The extent of proton transfer for weak acids is usually less than 5%. The actual percentage of molecules involved in proton transfer to water depends on the molecular structure of the acid; molecular polarity and the strength and polarity of individual bonds are particularly important factors in determining whether an acid is strong or weak.

A graphical representation of the differences between strong and weak acids, in terms of species present in solution, is given in Figure 14.1. The formula HA represents the acid and H_3O^+ and A^- are the products from the proton transfer to H_2O.

A 0.1 M solution of nitric acid (HNO_3) or sulfuric acid (H_2SO_4) when spilled on your clothes and not immediately washed off will "eat" holes in your clothing. If 0.1 M solutions of either acetic acid ($HC_2H_3O_2$) or carbonic acid (H_2CO_3) were spilled on your clothes, the previously noted "corrosive" effects would not be observed. Why? All four acid solutions are of equal concentration; all are 0.1 M solutions. The difference in behavior relates to the *strength* of the acids; nitric and sulfuric acids are *strong* acids, whereas acetic and carbonic acids are *weak* acids. The number

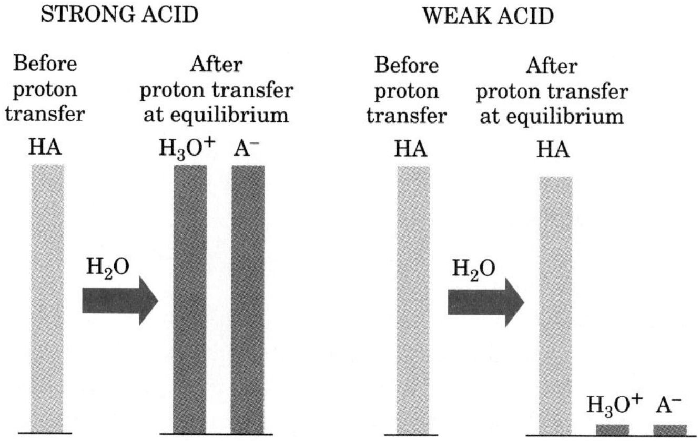

FIGURE 14.1 Many H_3O^+ ions are present in a solution of a strong acid, but only a few such ions are present in the solution of a weak acid.

TABLE 14.2 Commonly Encountered Strong Acids

Name[a]	Molecular Formula	Molecular Structure
Nitric acid	HNO_3	H—O—N—O with =O
Sulfuric acid	H_2SO_4	H—O—S(=O)(=O)—O—H
Perchloric acid	$HClO_4$	H—O—Cl(=O)(=O)—O
Hydrochloric acid	HCl	H—Cl
Hydrobromic acid	HBr	H—Br
Hydroiodic acid	HI	H—I

[a] Nomenclature for acids was discussed in Section 8.6.

of H_3O^+ ions (the active species) present in the strong acid solutions is many times greater than for the weak acid solutions even though all the solutions had the same number of acid molecules present (before reaction with water).

There are very few strong acids; the formulas and structures of the six most commonly encountered strong acids are given in Table 14.2. You should know the identity of these six strong acids; you will need such knowledge to write net ionic equations, the topic of Section 14.7.

The vast majority of acids that exist are weak acids. Familiar weak acids include acetic acid ($HC_2H_3O_2$), the acidic component of vinegar; boric acid (H_3BO_3), a common ingredient for eyewashes; and carbonic acid (H_2CO_3), found in carbonated beverages. Weak acids are not all equally weak; proton transfer occurs to a greater extent for some weak acids than for others. Table 14.3 gives percent proton-transfer values for selected weak acids. The calculational techniques needed to determine percent proton-transfer values, such as those in Table 14.3, will not be considered in this text.

For polyprotic acids the stepwise proton-transfer sequence that occurs (Sec. 14.4) can be used to determine relative acid strengths for the related acidic species. Consider the two-step proton-transfer process for carbonic acid.

$$H_2CO_3(aq) + H_2O(l) \longrightarrow H_3O^+(aq) + HCO_3^-(aq)$$
$$HCO_3^-(aq) + H_2O(l) \longrightarrow H_3O^+(aq) + CO_3^{2-}(aq)$$

The second proton is not as easily removed as the first because it must be pulled away from a negatively charged particle, HCO_3^-. Accordingly, HCO_3^- is a weaker acid than H_2CO_3. In general, each successive step in a stepwise proton-transfer process

TABLE 14.3 Percent Proton-Transfer Values for 1.0 M Solutions (at 25 °C) of Selected Weak Acids

Name of Acid	Formula	Percent Proton Transfer
Phosphoric acid	H_3PO_4	8.3
Nitrous acid	HNO_2	2.7
Hydrofluoric acid	HF	2.5
Acetic acid	$HC_2H_3O_2$	0.42
Carbonic acid	H_2CO_3	0.065
Dihydrogen phosphate ion	$H_2PO_4^-$	0.025
Hydrocyanic acid	HCN	0.0020
Hydrogen carbonate ion	HCO_3^-	0.00075
Hydrogen phosphate ion	HPO_4^{2-}	0.000047

occurs to a lesser extent than the previous step. Thus, for triprotic H_3PO_4, the parent H_3PO_4 species (first-step reactant) is a stronger acid than $H_2PO_4^-$ (second-step reactant), which in turn is a stronger acid than HPO_4^{2-} (third-step reactant). This ordering of the phosphoric acid-derived species is reflected in the values given in Table 14.3.

Just as there are strong acids and weak acids, there are also strong bases and weak bases. As with acids, there are only a few strong bases. Strong bases are limited to the hydroxides of groups IA and IIA of the periodic table listed in Table 14.4. Of the strong bases, only NaOH and KOH are commonly used in the chemical laboratory. The low solubility of the group IIA hydroxides in water limits their use. However, despite this low solubility, these hydroxides are still considered to be strong bases because whatever dissolves dissociates into ions 100%.

Only one weak base is routinely encountered—aqueous ammonia. It furnishes small amounts of OH^- ions through reaction with water molecules.

$$NH_3(g) + H_2O(l) \longrightarrow NH_4^+(aq) + OH^-(aq)$$

A solution of ammonia in water is most properly called *aqueous ammonia*, although it is also, somewhat erroneously, called ammonium hydroxide. Aqueous ammonia is the preferred designation, since most of the NH_3 present is in molecular form. Only a very few NH_3 molecules have reacted with the water to give ammonium (NH_4^+) and hydroxide (OH^-) ions.

TABLE 14.4 Common Strong Bases

Group IA Hydroxides	Group IIA Hydroxides
LiOH	
NaOH	
KOH	$Ca(OH)_2$
RbOH	$Sr(OH)_2$
CsOH	$Ba(OH)_2$

It is important to remember that the terms *strong* and *weak* apply to the extent of dissociation and not to the concentrations of acid or base. For example, stomach acid (gastric juice) is a *dilute* (not weak) solution of a strong acid; it is a 5% (m/m) solution of hydrochloric acid. On the other hand, a 35% (m/m) solution of hydrochloric acid would be considered to be a *concentrated* (not strong) solution of a strong acid.

14.6 Salts

The title of this chapter is "Acids, Bases, and Salts." In preceding sections, we have discussed acids and bases, but not salts. What is a salt? To a nonscientist the word *salt* connotes a white granular substance used as a seasoning for food. To the chemist it has a much broader meaning; sodium chloride, table salt, is only one of thousands of salts known to a chemist. "Pass the salt" is a very ambiguous request to a chemist.

From a chemical viewpoint, a **salt** is an ionic compound containing a metal or polyatomic ion as the positive ion and a nonmetal or polyatomic ion (except hydroxide) as the negative ion. (Ionic compounds containing hydroxide ion are bases rather than salts.)

Many salts occur in nature, and numerous others have been prepared in the laboratory. The wide variety of uses found for salts can be seen from Table 14.5, a listing of selected salts and their uses.

Much information concerning salts has been presented in previous chapters, although the term *salt* was not explicitly used in these discussions. Formula writing and nomenclature for binary ionic compounds (salts) was covered in Sections 7.5 and 7.6. Many salts, as shown in Table 14.5, contain polyatomic ions such as nitrate and sulfate. Such ions were discussed in Sections 7.8 and 8.4. The solubility of ionic compounds (salts) in water was the topic of Section 13.4.

TABLE 14.5 Some Common Salts and Their Uses

Name	Formula	Uses
Ammonium nitrate	NH_4NO_3	fertilizer; explosives manufacture
Barium sulfate	$BaSO_4$	X-rays of gastrointestinal tract
Calcium carbonate	$CaCO_3$	chalk; limestone
Calcium chloride	$CaCl_2$	drying agent for removal of small amounts of water
Iron(II) sulfate	$FeSO_4$	treatment for anemia
Potassium chloride	KCl	"salt" substitute for low sodium diets
Sodium chloride	NaCl	table salt; used as a deicer (to melt ice)
Sodium bicarbonate	$NaHCO_3$	ingredient in baking powder; stomach antacid
Sodium hypochlorite	NaClO	bleaching agent
Silver bromide	AgBr	light-sensitive material in photographic film
Tin(II) fluoride	SnF_2	toothpaste additive

In solution all common salts are dissociated into ions (Sec. 13.3). Even if a salt is only slightly soluble, the small amount that does dissolve completely dissociates. Thus, the terms weak and strong, used to denote qualitatively the percent dissociation of acids and bases, are not applicable to common salts. We do not use the terms *strong salt* and *weak salt*.

Acids, bases, and salts are related in that a salt is one of the products resulting from the reaction of an acid with a hydroxide base. This particular type of reaction, called neutralization, will be discussed in Section 14.8.

14.7 Ionic and Net Ionic Equations

Soluble strong acids, soluble strong bases, and soluble salts all dissociate 100% (or nearly 100%) in aqueous solution to produce ions (Secs. 14.5 and 14.6). It is extremely useful to discuss the reactions of such acids, bases, and salts in aqueous solution in terms of the ions present. This is most easily done using a new type of equation—a net ionic equation.

Up to this point in the text, most equations we have used have been *molecular equations*, equations where the complete formulas of all reactants and products are shown. From molecular equations, ionic equations may be written. An **ionic equation** is an equation in which the formulas of the predominant form of each compound in aqueous solution are used; dissociated compounds are written as ions, and undissociated compounds are written in molecular form. Net ionic equations are derived from ionic equations. A **net ionic equation** is an ionic equation from which nonparticipating (spectator) species have been eliminated.

The differences between molecular, ionic, and net ionic equations can best be illustrated by examples. Let us consider the chemical reaction that results when a solution of potassium chloride (KCl) is added to a solution of silver nitrate ($AgNO_3$). An insoluble salt, silver chloride (AgCl), is produced as a result of the mixing. The *molecular equation* for this reaction is

$$AgNO_3(aq) + KCl(aq) \longrightarrow KNO_3(aq) + AgCl(s)$$

Three of the four substances involved in this reaction—$AgNO_3$, KCl, and KNO_3—are soluble salts and thus exist in solution in ionic form. This is shown by writing the *ionic equation* for the reaction.

$$\underbrace{Ag^+(aq) + NO_3^-(aq)}_{AgNO_3 \text{ in ionic form}} + \underbrace{K^+(aq) + Cl^-(aq)}_{KCl \text{ in ionic form}} \longrightarrow \underbrace{K^+(aq) + NO_3^-(aq)}_{KNO_3 \text{ in ionic form}} + AgCl(s)$$

In this equation each of the three soluble salts is shown in dissociated (ionic) form rather than in undissociated form. A close look at this ionic equation shows that the potassium ions (K^+) and nitrate ions (NO_3^-) appear on both sides of the equation, indicating that they did not undergo any chemical change. In other words, they are *spectator ions*; they did not participate in the reaction. The *net ionic equation* for this reaction is written by dropping (canceling) all spectator ions from the ionic equation.

In our case, the net ionic equation becomes

$$Ag^+(aq) + Cl^-(aq) \longrightarrow AgCl(s)$$

This net ionic equation indicates that the product AgCl was formed by the reaction of silver ions (Ag^+) with chloride ions (Cl^-). It totally ignores the presence of those ions that are not taking part in the reaction. Thus, a net ionic equation focuses on only those species in a solution actually involved in a chemical reaction. It does not give all species present in the solution.

If you can write equations in molecular form, you will find it a straightforward process to convert these equations to net ionic form. Follow these three steps.

1. Check the given molecular equation to make sure that it is balanced.
2. Expand the molecular equation into an ionic equation.
3. Convert the ionic equation into a net ionic equation by eliminating spectator ions.

In expanding a molecular equation into an ionic equation (step 2) you must decide whether to write each reactant and product in dissociated (ionic) form or undissociated (molecular) form. The following rules serve as guidelines in making such decisions.

1. Soluble compounds that completely dissociate in aqueous solution are written in ionic form. They include
   (a) all soluble salts (see Sec. 13.4 for solubility rules).
   (b) all strong acids (see Table 14.3).
   (c) all strong bases (see Table 14.4).
2. Soluble weak acids and weak bases are written in molecular form, since they are incompletely ionized in solution and thus exist predominantly in the undissociated form. The following are considered to be weak acids or weak bases.
   (a) All acids not listed in Table 14.3 as strong acids. Common examples are HNO_2, HF, H_2S, $HC_2H_3O_2$, H_2CO_3, and H_3PO_4.
   (b) All bases not listed in Table 14.4 as strong bases. Aqueous ammonia (NH_3) is the most common weak base.
3. All insoluble substances (solids, liquids, and gases), whether ionic or covalent, exist as molecules or neutral ionic units and are written as such.
4. All soluble covalent substances, for example, carbon dioxide (CO_2) or sucrose ($C_{12}H_{22}O_{11}$), are written in molecular form.
5. If water, the solvent, appears in the equation, it is written in molecular form.

Now, we apply these guidelines by writing some net ionic equations.

EXAMPLE 14.3

Write the net ionic equation for the following aqueous solution reaction.

$$K_2CO_3 + CaCl_2 \longrightarrow CaCO_3 + KCl \quad \text{(unbalanced equation)}$$

Solution

STEP 1 To balance the given molecular equation, the coefficient 2 must be placed in front of KCl.

$$K_2CO_3 + CaCl_2 \longrightarrow CaCO_3 + 2\,KCl$$

STEP 2 A decision must be made to write each reactant and product in ionic or molecular form. Let us consider them one by one.

K_2CO_3: This compound is a salt. The solubility rules indicate that it is soluble. Thus, K_2CO_3 will be written in **ionic form**—$2\,K^+ + CO_3^{2-}$. Note that three ions (two K^+ ions and one CO_3^{2-} ion) are produced from the dissociation of one K_2CO_3 unit.

$CaCl_2$: This compound is also a soluble salt; it will be written in **ionic form**. Again, three ions are produced upon dissociation of one $CaCl_2$ unit—Ca^{2+} and $2\,Cl^-$.

$CaCO_3$: The solubility rules indicate that this compound is an insoluble salt. Thus, it will be written in **molecular form** in the ionic equation.

KCl: All potassium salts are soluble. Thus, KCl will be written in **ionic form**—$K^+ + Cl^-$.

The ionic equation will have the form

$$2\,K^+ + CO_3^{2-} + Ca^{2+} + 2\,Cl^- \longrightarrow CaCO_3 + 2\,K^+ + 2\,Cl^-$$

Note how the coefficient 2 in front of KCl in the molecular equation affects the ionic equation. The dissociation of two KCl units produces two K^+ ions and two Cl^- ions.

STEP 3 Inspection of the ionic equation shows that K^+ ions (two of them) and Cl^- ions (two of them) are spectator ions. Cancellation of these ions from the equation will give the net ionic equation.

$$\cancel{2\,K^+} + CO_3^{2-} + Ca^{2+} + \cancel{2\,Cl^-} \longrightarrow CaCO_3 + \cancel{2\,K^+} + \cancel{2\,Cl^-}$$

$$Ca^{2+} + CO_3^{2-} \longrightarrow CaCO_3$$

PRACTICE EXERCISE 14.3

Write the net ionic equation for the following aqueous solution reaction.

$$KCl + AgC_2H_3O_2 \longrightarrow KC_2H_3O_2 + AgCl$$

Ans. $Ag^+ + Cl^- \rightarrow AgCl$

Similar exercises:
Problems 14.49–14.54

EXAMPLE 14.4

Write the net ionic equation for the following aqueous solution reaction.

$$H_2S + AlCl_3 \longrightarrow Al_2S_3 + HCl \quad \text{(unbalanced equation)}$$

Solution

STEP 1 Balancing the molecular equation, we get

$$3\, H_2S + 2\, AlCl_3 \longrightarrow Al_2S_3 + 6\, HCl$$

STEP 2 The expansion of the molecular equation into an ionic equation is accomplished by the following analysis.

H_2S: This is a weak acid. All weak acids are written in **molecular form** in ionic equations.

$AlCl_3$: This is a soluble salt. All chloride salts are soluble, with three exceptions; this is not one of the exceptions. Soluble salts are written in **ionic form** in ionic equations.

Al_2S_3: This is an insoluble salt. All sulfides are insoluble except for groups IA and IIA and NH_4^+. Thus, Al_2S_3 will remain in **molecular form** in the ionic equation.

HCl: This compound is an acid. It is one of the six strong acids listed in Table 14.3. Strong acids are written in **ionic form**.

The ionic equation for the reaction is

$$3\, H_2S + 2\, Al^{3+} + 6\, Cl^- \longrightarrow Al_2S_3 + 6\, H^+ + 6\, Cl^-$$

Note again that the coefficients present in the balanced molecular equation must be taken into consideration when determining the total number of ions produced from dissociation. On dissociation an $AlCl_3$ unit produces four ions—one Al^{3+} ion and three Cl^- ions. This number must be doubled for the ionic equation since $AlCl_3$ carries the coefficient 2 in the balanced molecular equation. Similar considerations apply to HCl in this equation.

STEP 3 Inspection of the ionic equation shows that only Cl^- ions (six of them) are spectator ions. Cancellation of these ions from the equation will give the net ionic equation.

$$3\, H_2S + 2\, Al^{3+} + \cancel{6\, Cl^-} \longrightarrow Al_2S_3 + 6\, H^+ + \cancel{6\, Cl^-}$$
$$3\, H_2S + 2\, Al^{3+} \longrightarrow Al_2S_3 + 6\, H^+$$

PRACTICE EXERCISE 14.4

Write the net ionic equation for the following aqueous solution reaction.

$$H_2CO_3 + MgCl_2 \longrightarrow MgCO_3 + HCl \quad \text{(unbalanced equation)}$$

Ans. $H_2CO_3 + Mg^{2+} \rightarrow MgCO_3 + 2\, H^+$

Similar exercises: Problems 14.49–14.54

EXAMPLE 14.5

Write the net ionic equation for the following aqueous solution reaction.

$$H_2SO_4 + NaOH \longrightarrow Na_2SO_4 + H_2O \quad \text{(unbalanced equation)}$$

Solution

STEP 1   Balancing the given molecular equation, we get

$$H_2SO_4 + 2\ NaOH \longrightarrow Na_2SO_4 + 2\ H_2O$$

STEP 2   The expansion of the molecular equation into an ionic equation is based on the following analysis.

H_2SO_4: This compound is an acid. It is one of the six strong acids listed in Table 14.3. Strong acids are written in **ionic form** in ionic equations.

NaOH: This compound is a base. It is one of the strong bases listed in Table 14.4. Strong bases are written in **ionic form** in ionic equations.

Na_2SO_4: This compound is a soluble salt. All sodium salts are soluble. Thus, Na_2SO_4 is written in **ionic form**.

H_2O: This compound is a covalent compound; two nonmetals are present. Covalent compounds are always written in **molecular form**.

The ionic equation for the reaction, from the above information, is

$$2\ H^+ + SO_4^{2-} + 2\ Na^+ + 2\ OH^- \longrightarrow 2\ Na^+ + SO_4^{2-} + 2\ H_2O$$

STEP 3   Inspection of the ionic equation shows that SO_4^{2-} ions and Na^+ ions are spectator ions. Cancellation of these ions from the equation gives the net ionic equation.

$$2\ H^+ + \cancel{SO_4^{2-}} + \cancel{2\ Na^+} + 2\ OH^- \longrightarrow \cancel{2\ Na^+} + \cancel{SO_4^{2-}} + 2\ H_2O$$
$$2\ H^+ + 2\ OH^- \longrightarrow 2\ H_2O$$

The coefficients in the net ionic equation should be the smallest set of numbers that correctly balance the equation. In this case, all the coefficients are divisible by 2. Dividing by 2, we get

$$H^+ + OH^- \longrightarrow H_2O$$

The final net equation should always be checked to see that the coefficients are in the lowest possible integral ratio.

PRACTICE EXERCISE 14.5

Write the net ionic equation for the following aqueous solution reaction.

$$HNO_3 + Ba(OH)_2 \longrightarrow Ba(NO_3)_2 + H_2O \quad \text{(unbalanced equation)}$$

Ans. $H^+ + OH^- \rightarrow H_2O$ Similar exercises: Problems 14.49–14.54

14.8 Reactions of Acids

All acids have some unique properties that adapt them for use in specific situations. In addition, all acids have certain chemical properties in common, properties related to the presence of H_3O^+ ions in aqueous solution. In this section we consider three types of chemical reactions that acids characteristically undergo.

1. Acids react with active metals to produce hydrogen gas and a salt.
2. Acids react with bases to produce a salt and water.
3. Acids react with carbonates and bicarbonates to produce carbon dioxide, a salt, and water.

Reaction with Metals

Acids react with many, but not all, metals. When they do react, the metal dissolves and hydrogen gas (H_2) is liberated. In the reaction the metal atoms lose electrons and become metal ions. The lost electrons are taken up by the hydrogen ions (protons) of the acid; the hydrogen ions become electrically neutral, combine into molecules, and emerge from the reaction mixture as hydrogen gas. Illustrative of the reaction of an acid and a metal is the reaction between zinc and sulfuric acid.

$$\text{Molecular equation:} \quad Zn + H_2SO_4 \longrightarrow ZnSO_4 + H_2$$
$$\text{Net ionic equation:} \quad Zn + 2\,H^+ \longrightarrow Zn^{2+} + H_2$$

In terms of the reaction types discussed in Section 10.6 the reaction of an acid with a metal to produce hydrogen gas is a *single-replacement reaction*; the metal replaces the hydrogen from the acid. Recall, from Section 10.6, that a single-replacement reaction has the general form

$$X + YZ \longrightarrow Y + XZ$$

Metals can be arranged in a reactivity order based on their ability to react with acids. Such an ordering for the more common metals is given in Table 14.6. Any metal above hydrogen in the activity series will dissolve in an acid solution and form H_2. The closer a metal is to the top of the series, the more rapid the reaction. Those metals below hydrogen in the series do not dissolve in an acid to form H_2.

As noted in Table 14.6, the most active metals (those nearest the top in the activity series) also react with water. Again, hydrogen gas is produced. In the cases of potassium and sodium, the reaction is sometimes violent enough to cause explosions as the result of H_2 ignition. The equation for the reaction of potassium with water, which is also a single-replacement reaction, is

$$\text{Molecular equation:} \quad 2\,K + 2\,H_2O \longrightarrow 2\,KOH + H_2$$
$$\text{Net ionic equation:} \quad 2\,K + 2\,H_2O \longrightarrow 2\,K^+ + 2\,OH^- + H_2$$

Note that the resulting solution is basic when a metal reacts with water; hydroxide ions are present.

TABLE 14.6 Activity Series for Common Metals

	Metal	Symbol	Remarks
Increasing tendency to react ↑			
React with H^+ ions to liberate hydrogen gas	Potassium	K	react violently with cold water
	Sodium	Na	
	Calcium	Ca	reacts slowly with cold water
	Magnesium	Mg	
	Aluminum	Al	react slowly with hot water (steam)
	Zinc	Zn	
	Chromium	Cr	
	Iron	Fe	
	Nickel	Ni	
	Tin	Sn	
	Lead	Pb	
	Hydrogen	H	
Do not react with H^+ ions	Copper	Cu	
	Mercury	Hg	
	Silver	Ag	
	Platinum	Pt	
	Gold	Au	

Reaction with Bases

When acids and bases are mixed, they react with each other; their acidic and basic properties disappear, and we say that they have *neutralized* each other. **Neutralization** is the reaction between an acid and a base to form a salt and water. The hydrogen ions from the acid combine with the hydroxide ions from the base to form water. The salt formed contains the negative ion from the acid and the positive ion from the base. Neutralization is a *double-replacement* reaction (Sec. 10.6).

$$AX + BY \longrightarrow AY + BX$$
$$HCl + KOH \longrightarrow HOH + KCl$$
$$\text{acid} \quad \text{base} \quad \quad \text{water} \quad \text{salt}$$

Any time an acid is completely reacted with a base, neutralization occurs. It does not matter whether the acid and base are strong or weak. Sodium hydroxide (a strong base) and nitric acid (a strong acid) react as follows.

Molecular equation: $HNO_3 + NaOH \longrightarrow NaNO_3 + H_2O$

Net ionic equation: $H^+ + OH^- \longrightarrow H_2O$

The equations for the reaction of potassium hydroxide (a strong base) with hydrocyanic acid (a weak acid) are

Molecular equation: $HCN + KOH \longrightarrow KCN + H_2O$

Net ionic equation: $HCN + OH^- \longrightarrow CN^- + H_2O$

Note that in each case the products are a salt ($NaNO_3$ in the first reaction, KCN in the second) and water. Note also that the net ionic equations for the two neutralization reactions are different. In the second set of equations the acid must remain written in molecular form, since it is a weak acid.

Reaction with Carbonates and Bicarbonates

Carbon dioxide gas (CO_2), water, and a salt are always the products of the reaction of acids with carbonates or bicarbonates, as illustrated by the following equations.

Molecular equation: $\quad 2\ HCl + Na_2CO_3 \longrightarrow 2\ NaCl + CO_2 + H_2O$

Molecular equation: $\quad HCl + NaHCO_3 \longrightarrow NaCl + CO_2 + H_2O$

Baking powder is a mixture of a bicarbonate and an acid-forming solid. The addition of water to this mixture generates the acid that then reacts with the bicarbonate to release carbon dioxide into the batter. It is the generated carbon dioxide that causes the batter to rise. Baking soda is pure $NaHCO_3$. To cause it to release carbon dioxide, an acid-containing substance, such as buttermilk, must be added to it.

14.9 Reactions of Bases

The most important characteristic reaction of bases is their reaction with acids (neutralization), discussed in the preceding section. Another characteristic reaction, that of bases with certain salts, is discussed in Section 14.10.

Bases react with fats and oils and convert them into smaller, soluble molecules. For this reason most household cleaning products contain basic substances. Lye (impure NaOH) is an active ingredient in numerous drain cleaners. Also, many advertisements for liquid household cleaners emphasize the fact that aqueous ammonia (a weak base) is present in the product.

In Section 14.1 we noted that one of the general properties of bases is a "slippery or soapy" feeling to the touch. The bases themselves are not slippery; the slipperiness results as the bases react with fats and oils in the skin to form "slippery or soapy" compounds.

14.10 Reactions of Salts

Dissolved salts will react with metals, acids, bases, and other salts under specific conditions.

1. Salts react with some metals to convert the metallic ion of the salt to free metal and the free metal to its salt.
2. Salts react with some acid solutions to form other acids and salts.
3. Salts react with some base solutions to form other bases and salts.
4. Salts react with some solutions of other salts to form new salts.

The tendency for salts to react with metals is related to the relative positions of the two involved metals in the activity series (Table 14.6). In order for salts to react with acids, bases, or other salts, one of the reaction products must be (1) an insoluble salt, (2) a gas that is evolved from the solution, or (3) an undissociated soluble species, such as a weak acid or a weak base. The formation of any of these products serves as the driving force to cause the reaction to occur.

Reaction with Metals

If an iron nail is placed in a solution of copper sulfate ($CuSO_4$), metallic copper will be deposited on the nail and some of the iron will dissolve.

$$\text{Molecular equation:} \quad Fe(s) + CuSO_4(aq) \longrightarrow Cu(s) + FeSO_4(aq)$$
$$\text{Net ionic equation:} \quad Fe + Cu^{2+} \longrightarrow Cu + Fe^{2+}$$

One metal has replaced the other; a single-replacement reaction (Sec. 10.6) has occurred. This type of reaction will occur only if the metal going into solution is above the replaced metal in the activity series. Iron is above copper in the activity series and can replace it. If a strip of copper were placed in a solution of $FeSO_4$—just the opposite situation to what we have been discussing—no reaction would occur since copper is below iron in the activity series.

Reaction with Acids

In order for a salt to react with an acid, a new weaker acid, a new insoluble salt, or a gaseous compound must be one of the products.

An example of a reaction in which the formation of an *insoluble salt* is the driving force for the reaction to occur is

$$AgNO_3(aq) + HCl(aq) \longrightarrow AgCl(s) + HNO_3(aq)$$

This is a double-replacement reaction (Sec. 10.6); the silver and hydrogen have traded partners.

The conclusion that this reaction will occur comes from a consideration of the possible recombinations of the reacting species. In a solution made by mixing silver nitrate ($AgNO_3$) and hydrochloric acid (HCl), four kinds of ions are present initially (before any reaction occurs): Ag^+ and NO_3^- (since $AgNO_3$ is a soluble salt) and H^+ and Cl^- (since HCl is a strong acid). The question is whether these ions can get together in new appropriate combinations. The possible new combinations of oppositely charged ions are $H^+NO_3^-$ and Ag^+Cl^-. The first of these combinations would result in the formation of the strong acid HNO_3. Strong acids in solution exist in dissociated form; therefore, these ions will not combine. The second combination does occur because AgCl is an insoluble salt. Thus, the overall reaction takes place as a result of the formation of this insoluble salt. The net result of the reaction is that the original ions exchange partners.

The double-replacement reaction of sodium fluoride (a soluble salt) with hydrochloric acid (a strong acid) illustrates the case where formation of a new weaker acid is the driving force for the reaction.

$$\text{NaF(aq)} + \text{HCl(aq)} \longrightarrow \text{NaCl(aq)} + \text{HF(aq)}$$

Using an analysis pattern similar to that in the previous example, we find that four types of ions are present initially: Na^+ and F^- (from the soluble salt) and H^+ and Cl^- (from the strong acid). Possible new combinations are Na^+Cl^- and H^+F^-. Sodium chloride, the result of the first combination, will not form because this salt is soluble. The combination of H^+ ion with F^- ion does occur because it yields the weak acid HF. In solution weak acids exist predominantly in molecular form. In all reactions of this general type, the acid formed in the reaction must be weaker than the reactant acid. If the reactant acid is strong, as in this example, such a determination is obvious. If both the reactant and product acids are weak, information such as that given in Table 14.3 would be needed to predict which of the two acids is the weaker.

The most common type of reaction in which the driving force is the *evolution of a gas* involves a carbonate or bicarbonate. This type of reaction was discussed in Section 14.8.

Note that a reaction does not always occur when acid and salt solutions are mixed. Consider the possible reaction of NaCl and HNO_3 solutions. Initially four types of ions are present: Na^+ and Cl^- (from the soluble salt) and H^+ and NO_3^- (from the strong acid). The new combinations, if a reaction did occur, would be Na^+NO_3^- (a soluble salt) and H^+Cl^- (a strong acid). Since both of the products would exist in dissociated form in solution, no recombination of ions occurs; hence, no reaction occurs.

Reaction with Bases

The criteria for the reaction of bases with salts are similar to those for acid–salt reactions, except that weaker base formation replaces weaker acid formation as one of the three driving forces. An example of a base–salt reaction involving the formation of an insoluble salt is

$$\text{Ba(OH)}_2\text{(aq)} + \text{Na}_2\text{SO}_4\text{(aq)} \longrightarrow \text{BaSO}_4\text{(s)} + 2\,\text{NaOH(aq)}$$

The most common situation in which gas evolution is the driving force for base–salt reactions is where ammonium salts are involved. In such cases, ammonia gas is given off, as illustrated by the reaction of NH_4Cl and KOH.

$$\text{NH}_4\text{Cl(aq)} + \text{KOH(aq)} \longrightarrow \text{KCl(aq)} + \text{NH}_3\text{(g)} + \text{H}_2\text{O(l)}$$

Reaction of Salts with Each Other

Two different salt solutions will react when mixed, in a double-replacement reaction, only if an insoluble salt can be formed. Consider the following possible reactions.

$$\text{AgNO}_3\text{(aq)} + \text{NaCl(aq)} \longrightarrow \text{AgCl(s)} + \text{NaNO}_3\text{(aq)}$$
$$\text{KNO}_3\text{(aq)} + \text{NaCl(aq)} \longrightarrow \text{KCl(aq)} + \text{NaNO}_3\text{(aq)}$$

The first reaction occurs because AgCl is an insoluble salt. The second reaction does not occur since both of the possible products are soluble salts, which means there is no driving force for the reaction.

EXAMPLE 14.6

Write molecular, ionic, and net ionic equations for the reaction that occurs, if any, when 0.1 M solutions of the following substances are mixed.

(a) $Ni(NO_3)_2$ and Na_2S (b) $BaCl_2$ and H_2SO_4 (c) HNO_3 and $KC_2H_3O_2$

Solution

(a) Both of the reactants are soluble salts. Two different salt solutions will react when mixed only if an insoluble salt can be formed.

In a solution made by mixing $Ni(NO_3)_2$ and Na_2S, four kinds of ions are present initially (before any reaction occurs): Ni^{2+} and NO_3^- [from the $Ni(NO_3)_2$] and Na^+ and S^{2-} (from the Na_2S). The possible new combinations of oppositely charged ions are Ni^{2+} with S^{2-} and Na^+ with NO_3^-.

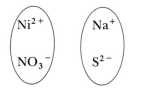
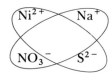

Original Ion Combinations   Possible New Combinations

The first of these new combinations, the formation of NiS, is the one that will be the driving force for the reaction to occur. NiS is an insoluble salt. The second new combination, the formation of $NaNO_3$, does not occur because $NaNO_3$ is a soluble salt and soluble salts exist in dissociated form in solution. The equations for the reaction are

Molecular: $Ni(NO_3)_2 + Na_2S \longrightarrow NiS + 2\ NaNO_3$
Ionic: $Ni^{2+} + 2\ \cancel{NO_3^-} + 2\ \cancel{Na^+} + S^{2-} \longrightarrow NiS + 2\ \cancel{Na^+} + 2\ \cancel{NO_3^-}$
Net ionic: $Ni^{2+} + S^{2-} \longrightarrow NiS$

(b) One of the reactants, $BaCl_2$, is a soluble salt and the other reactant, H_2SO_4, is a strong acid. Both reactants exist in solution in dissociated form; thus, four types of ions are present in the mixed solution (before any reaction occurs): Ba^{2+}, Cl^-, H^+, and SO_4^{2-}. The conclusion that a reaction will occur comes from a consideration of the possible new combinations of the reacting species.

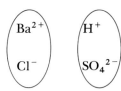
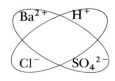

Original Ion Combinations   Possible New Combinations

The first of these new combinations, the formation of $BaSO_4$, is the driving force for the reaction to occur; $BaSO_4$ is an insoluble salt. The second recombination, the formation of HCl, does not occur because HCl is a strong acid and will exist in solution in dissociated form. The equations for the reaction are

Molecular: $\quad\quad\quad\quad BaCl_2 + H_2SO_4 \longrightarrow BaSO_4 + 2\ HCl$

Ionic: $\quad\quad Ba^{2+} + \cancel{2\ Cl^-} + \cancel{2\ H^+} + SO_4^{2-} \longrightarrow BaSO_4 + \cancel{2\ H^+} + \cancel{2\ Cl^-}$

Net ionic: $\quad\quad\quad\quad Ba^{2+} + SO_4^{2-} \longrightarrow BaSO_4$

(c) The reactants are a strong acid (HNO_3) and a soluble salt ($KC_2H_3O_2$). Both are dissociated in solution; hence, H^+, NO_3^-, K^+, and $C_2H_3O_2^-$ ions are present in the reaction mixture (before any reaction occurs). Possible new combinations of the reacting species are

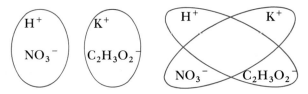

Weak acid formation is the driving force for the reaction; acetic acid ($HC_2H_3O_2$) forms from the combination of H^+ and $C_2H_3O_2^-$ ions. The K^+ and NO_3^- will not combine because a soluble salt would be the product. The equations for the reaction are

Molecular: $\quad\quad\quad\quad HNO_3 + KC_2H_3O_2 \longrightarrow HC_2H_3O_2 + KNO_3$

Ionic: $\quad\quad H^+ + \cancel{NO_3^-} + \cancel{K^+} + C_2H_3O_2^- \longrightarrow HC_2H_3O_2 + \cancel{K^+} + \cancel{NO_3^-}$

Net ionic: $\quad\quad\quad\quad H^+ + C_2H_3O_2^- \longrightarrow HC_2H_3O_2$

PRACTICE EXERCISE 14.6

Write molecular, ionic, and net ionic equations for the reaction that occurs, if any, when 0.1 M solutions of the following substances are mixed.

(a) $Fe(NO_3)_3$ and Na_3PO_4 (b) $CaCl_2$ and HNO_3 (c) HCl and Na_2S

Ans. (a) Molecular: $Fe(NO_3)_3 + Na_3PO_4 \rightarrow FePO_4 + 3\ NaNO_3$
Ionic: $Fe^{3+} + 3\ NO_3^- + 3\ Na^+ + PO_4^{3-} \rightarrow FePO_4 + 3\ Na^+ + 3\ NO_3^-$
Net ionic: $Fe^{3+} + PO_4^{3-} \rightarrow FePO_4$
(b) Molecular: $CaCl_2 + 2\ HNO_3 \rightarrow Ca(NO_3)_2 + 2\ HCl$
Ionic: $Ca^{2+} + 2\ Cl^- + 2\ H^+ + 2\ NO_3^- \rightarrow Ca^{2+} + 2\ NO_3^- + 2\ H^+ + 2\ Cl^-$
Net ionic: all species cancel (no reaction occurs)
(c) Molecular: $2\ HCl + Na_2S \rightarrow H_2S + 2\ NaCl$
Ionic: $2\ H^+ + 2\ Cl^- + 2\ Na^+ + S^{2-} \rightarrow H_2S + 2\ Na^+ + 2\ Cl^-$
Net ionic: $2\ H^+ + S^{2-} \rightarrow H_2S$

Similar exercises: Problems 14.67 and 14.68

14.11 Dissociation of Water

Normally, we think of water as a covalent, nondissociating substance. Experiments show, however, that in a sample of pure water a very small percentage of the water molecules have undergone dissociation to produce ions. The dissociation reaction may be thought of as involving the transfer of a proton from one water molecule to another (Brønsted–Lowry theory; Sec. 14.2)

$$H_2O + H_2O \longrightarrow H_3O^+ + OH^-$$

or simply as the dissociation of a single water molecule (Arrhenius theory; Sec. 14.1).

$$H_2O(HOH) \longrightarrow H^+ + OH^-$$

From either viewpoint, the net result is the formation of *equal amounts* of hydronium (hydrogen) ion and hydroxide ion.

The dissociation of water molecules is part of an equilibrium situation. Individual water molecules are continually dissociating. This process is balanced by hydroxide and hydronium ions recombining to form water at the same rate. At equilibrium, at 25 °C, the H_3O^+ and OH^- ion concentrations are each 1.00×10^{-7} M (0.000000100 M). This very, very small concentration is equivalent to there being one H_3O^+ and one OH^- ion present for every 550,000,000 undissociated water molecules. Even though the H_3O^+ and OH^- ion concentrations are very minute, they are important, as we shall shortly see.

Experimentally it is found that, at any given temperature, the product of the concentrations of H_3O^+ ion and OH^- ion in water is a constant. We can calculate the value of this constant at 25 °C, since we know that the concentration of each ion is 1.00×10^{-7} M at this temperature.

$$[H_3O^+] \times [OH^-] = \text{constant}$$
$$(1.00 \times 10^{-7}) \times (1.00 \times 10^{-7}) = 1.00 \times 10^{-14}$$

Ion product for water is the name given to the numerical value (1.00×10^{-14}) associated with the product of the H_3O^+ ion and OH^- ion molar concentrations in water. Note that the ion concentrations must be expressed in moles per liter (M) in order to obtain the value 1.00×10^{-14} for the ion product for water. In the general expression for the ion product for water,

$$[H_3O^+] \times [OH^-] = \text{ion product for water}$$

the brackets [ ] specifically denote ion concentrations in moles per liter.

The ion product expression for water is valid not only in pure water but also when solutes are present in the water. At all times, the product of the hydronium and hydroxide ion molarities in an aqueous solution, at 25 °C, must equal 1.00×10^{-14}. Thus if the $[H_3O^+]$ is increased by the addition of an acidic solute, the $[OH^-]$ must decrease until the expression

$$[H_3O^+] \times [OH^-] = 1.00 \times 10^{-14}$$

is satisfied. Similarly, if OH^- ions are added to the water, the $[H_3O^+]$ must correspondingly decrease. The extent of the decrease in $[H_3O^+]$ or $[OH^-]$, as the result of the addition of a quantity of the other ion, is easily calculated by the ion product expression.

EXAMPLE 14.7

Sufficient acidic solute is added to a quantity of water to produce $[H_3O^+] = 2.50 \times 10^{-2}$. What is the $[OH^-]$ in this solution?

Solution
The $[OH^-]$ can be calculated by the ion product expression for water. Solving this expression for $[OH^-]$ gives

$$[OH^-] = \frac{1.00 \times 10^{-14}}{[H_3O^+]}$$

Substituting into this expression the known $[H_3O^+]$ and doing the arithmetic gives

$$[OH^-] = \frac{1.00 \times 10^{-14}}{2.50 \times 10^{-2}} = 4 \times 10^{-13} \quad \text{(calculator answer)}$$

$$= 4.00 \times 10^{-13} \quad \text{(correct answer)}$$

PRACTICE EXERCISE 14.7

Sufficient acidic solute is added to a quantity of water to produce $[H_3O^+] = 4.50 \times 10^{-5}$. What is the $[OH^-]$ in this solution?   *Ans.* 2.22×10^{-10} M.

Similar exercises: Problems 14.73 and 14.74

In Section 14.2 we learned that the acidic species in aqueous solution is the H_3O^+ ion. Now that we have noted that a small amount of H_3O^+ ions is present in all aqueous solutions, even basic ones, a refining of our concepts of acids and bases is in order. An acid is a substance that increases the H_3O^+ ion concentration in water. All **acidic solutions** have a higher $[H_3O^+]$ than $[OH^-]$. In a similar manner, a base is a substance that increases the OH^- ion concentration in water. All **basic solutions** have a higher $[OH^-]$ than $[H_3O^+]$. In a **neutral solution**, the concentrations of both H_3O^+ ions and OH^- ions are equal. Table 14.7 summarizes the relationships between $[H_3O^+]$ and $[OH^-]$ that we have just considered.

TABLE 14.7 Relationships Between $[H_3O^+]$ and $[OH^-]$ in Aqueous Solutions at 25 °C

Neutral solution:	$[H_3O^+] = [OH^-] = 1.00 \times 10^{-7}$
Acidic solution:	$[H_3O^+]$ is greater than 1.00×10^{-7}
	$[OH^-]$ is less than 1.00×10^{-7}
Basic solution:	$[H_3O^+]$ is less than 1.00×10^{-7}
	$[OH^-]$ is greater than 1.00×10^{-7}

14.12 The pH Scale

Hydronium ion concentrations in aqueous solution range from relatively high values (10 M) to extremely small ones (10^{-14} M). It is somewhat inconvenient to work with numbers that extend over such a wide range; a hydronium ion concentration of 10 M is 1000 trillion times larger than a hydronium ion concentration of 10^{-14} M. The pH scale, proposed by the Danish chemist Sören Peter Lauritz Sörensen (1868–1939) in 1909, is a more practical way to handle such a wide range of numbers.

The **pH** of a solution is defined as the negative logarithm of the molar hydronium ion concentration. Expressed mathematically, the pH definition is

$$\text{pH} = -\log\,[\text{H}_3\text{O}^+]$$

Logarithms are simply exponents. The *common logarithm*, abbreviated log, which is the type of logarithm used in the definition of pH, is based on powers of 10. *For a number expressed in scientific notation that has a coefficient of 1, the log of that number is the value of the exponent.* For instance, the log of 1×10^{-8} is -8.0, and the log of 1×10^6 is 6.0. Table 14.8 gives more examples of the relationship between powers of 10 and logarithmic values for numbers in scientific notation whose coefficients are 1. Note from this table that log values may be either positive or negative depending on the sign of the exponent. (How we determine significant figures in logarithmic calculations is discussed later in this section.)

Integral pH Values

It is easy to calculate the pH value for a solution when the molar hydronium ion concentration is an exact power of 10, for example, 1.0×10^{-4}. In this situation the pH is given directly by the negative of the exponent value on the power of 10.

$$[\text{H}_3\text{O}^+] = 1.0 \times 10^{-x}$$
$$\text{pH} = x$$

Thus, if the hydronium ion concentration is 1.0×10^{-9}, the pH will be 9.00.

TABLE 14.8 Logarithm Values for Selected Numbers

Number	Number Expressed as a Power of 10	Common Logarithm
10,000	1×10^4	4.0
1,000	1×10^3	3.0
100	1×10^2	2.0
10	1×10^1	1.0
1	1×10^0	0.0
0.1	1×10^{-1}	-1.0
0.01	1×10^{-2}	-2.0
0.001	1×10^{-3}	-3.0
0.0001	1×10^{-4}	-4.0

This simple relationship between pH and power of 10 is obtained from the formal definition of pH as follows.

$$\begin{aligned}\text{pH} &= -\log\,[H_3O^+] \\ &= -\log\,[1.0 \times 10^{-x}] \\ &= -(-x) \\ &= x\end{aligned}$$

Again, it should be noted that this simple solution is valid only when the coefficient in the exponential expression for the hydronium ion concentration is 1.0. How the pH is calculated when the coefficient is not 1.0 will be covered later in this section.

EXAMPLE 14.8

Calculate the pH for each of the following solutions.

(a) $[H_3O^+] = 1 \times 10^{-5}$ **(b)** $[H_3O^+] = 1 \times 10^{-10}$
(c) $[OH^-] = 1 \times 10^{-6}$

Solution

(a) Let us use the formal definition of pH in obtaining this first pH value.

$$\text{pH} = -\log\,[H_3O^+]$$

This expression indicates that to obtain a pH we must first take the logarithm of the molar hydronium ion concentration and then change the sign of that logarithm.

The logarithm of 1×10^{-5} is a -5.0. Thus, we have

$$\begin{aligned}\text{pH} &= -\log(1 \times 10^{-5}) \\ &= -(-5.0) \\ &= 5.0\end{aligned}$$

(b) Let us use the shorter, more direct way for obtaining pH this time, the method based on the relationship

$$[H_3O^+] = 1 \times 10^{-x}$$
$$\text{pH} = x$$

Since the power of 10 is -10 in this case, the **pH will be 10.0**.

(c) The given quantity involves hydroxide ion rather than hydronium ion. Thus, we must first calculate the hydronium ion concentration, and then the pH.

$$[H_3O^+] = \frac{1.00 \times 10^{-14}}{[OH^-]} = \frac{1.00 \times 10^{-14}}{1 \times 10^{-6}}$$
$$= 1 \times 10^{-8} \quad \text{(calculator and correct answer)}$$

A solution with a hydronium ion concentration of 1×10^{-8} M will have a **pH of 8.0**.

PRACTICE EXERCISE 14.8

Calculate the pH for each of the following solutions.

(a) $[H_3O^+] = 1 \times 10^{-6}$ (b) $[H_3O^+] = 1 \times 10^{-12}$
(c) $[OH^-] = 1 \times 10^{-3}$

Ans. (a) 6.0; (b) 12.0; (c) 11.0.

Similar exercises: Problems 14.79–14.82

Since pH is simply another way of expressing hydronium ion concentration, acidic, neutral, and basic solutions can be identified by their pH values.

A neutral solution ($[H_3O^+] = 1.0 \times 10^{-7}$) has a pH of 7.00. Values of pH less than 7 correspond to acidic solutions. The lower the pH value, the greater the acidity. Values of pH greater than 7 represent basic solutions. The higher the pH value, the greater the basicity. The relationships between $[H_3O^+]$, $[OH^-]$, and pH are summarized in Table 14.9. Note that a change of *one* unit in pH corresponds to a *tenfold* increase or decrease in $[H_3O^+]$. Also note that *lowering* the pH always corresponds to *increasing* the H_3O^+ ion concentration.

Table 14.10 lists the pH values of a number of common substances. Except for gastric juices, most human body fluids have pH values within a couple of units of neutrality. Almost all foods are acidic. Tart taste is associated with food of low pH.

Nonintegral pH Values

Note that some of the pH values in Table 14.10 are nonintegral, that is, not whole numbers. Nonintegral pH values result from molar hydronium ion concentrations where the coefficient in the exponential expression for concentration has a value other than 1. For example, consider the following matchups between hydronium ion concentration and pH.

TABLE 14.9 The pH Scale

pH	$[H_3O^+]$	$[OH^-]$	
0.0	1	10^{-14}	
1.0	10^{-1}	10^{-13}	
2.0	10^{-2}	10^{-12}	
3.0	10^{-3}	10^{-11}	Acidic
4.0	10^{-4}	10^{-10}	
5.0	10^{-5}	10^{-9}	
6.0	10^{-6}	10^{-8}	
7.0	10^{-7}	10^{-7}	NEUTRAL
8.0	10^{-8}	10^{-6}	
9.0	10^{-9}	10^{-5}	
10.0	10^{-10}	10^{-4}	
11.0	10^{-11}	10^{-3}	Basic
12.0	10^{-12}	10^{-2}	
13.0	10^{-13}	10^{-1}	
14.0	10^{-14}	1	

TABLE 14.10 Approximate pH Values of Some Common Substances

	pH	
↑	0	1 M HCl (0.0)
	1	0.1 M HCl (1.0), gastric juice (1.6–1.8), lime juice (1.8–2.0)
	2	soft drinks (2.0–4.0), vinegar (2.4–3.4)
Acidic	3	grapefruit (3.0–3.3), peaches (3.4–3.6)
	4	tomatoes (4.0–4.4), human urine (4.8–8.4)
	5	carrots (4.9–5.3), peas (5.8–6.4)
	6	human saliva (6.2–7.4), cow's milk (6.3–6.6), drinking water (6.5–8.0)
NEUTRAL	7	pure water (7.0), human blood (7.35–7.45), fresh eggs (7.6–8.0)
	8	seawater (8.3), soaps, shampoos (8.0–9.0)
	9	detergents (9.0–10.0)
	10	milk of magnesia (9.9–10.1)
Basic	11	household ammonia (11.5–12.0)
	12	liquid bleach (12.0)
	13	0.1 M NaOH (13.0)
↓	14	1 M NaOH (14.0)

$$[H_3O^+] = 6.3 \times 10^{-5} \quad pH = 4.20$$
$$[H_3O^+] = 4.0 \times 10^{-5} \quad pH = 4.40$$
$$[H_3O^+] = 2.0 \times 10^{-5} \quad pH = 4.70$$

Obtaining nonintegral pH values like these from hydronium ion concentrations requires (1) an electronic calculator that allows for the input of exponential numbers and has a base 10 logarithm key (LOG), or (2) a table of logarithms.

In using an electronic calculator, you can obtain logarithm values simply by pressing the LOG key after having entered the number whose log value is desired. For pH, you must remember that after obtaining the log value you must change its sign because of the negative sign in the defining equation for pH.

Significant figure considerations for log values involve a concept not previously encountered. It can best be illustrated by considering some actual log values. Consider the following five related numbers and their log values.

Number	Logarithm
2.43×10^0	0.38560627
2.43×10^2	2.38560627
2.43×10^4	4.38560627
2.43×10^7	7.38560627
2.43×10^{11}	11.38560627

This tabulation shows that

1. The number to the left of the decimal point in each logarithm (called the *characteristic*) is related only to the exponent of the number whose logarithm was taken.

2. The number to the right of the decimal point in each logarithm (called the *mantissa*) is related only to the coefficient in the exponential notation form of the number. Since the coefficient is 2.43 in each case, the mantissas are all the same (0.38560627).

Combining generalizations (1) and (2) gives us the significant figure rule for logarithms. *In a logarithm the digits to the left of the decimal point are not counted as significant figures.* These digits relate to the placement of the decimal point in the number. From our previous significant figure work (Sec. 2.3) they are somewhat analogous to the leading zeros (which are not significant) in a number such as 0.0000243.

Thus, to be correct in significant figures with a logarithm the *coefficient* of the number whose logarithm has been taken and the *mantissa* of the logarithm must have the same number of digits. Rewriting the previous tabulation of logarithms to the correct number of significant figures gives

$$2.43 \times 10^0 \quad\quad 0.386$$
$$2.43 \times 10^2 \quad\quad 2.386$$
$$2.43 \times 10^4 \quad\quad 4.386$$
$$2.43 \times 10^7 \quad\quad 7.386$$
$$2.43 \times 10^{11} \quad\quad 11.386$$

$$\underbrace{}_{\text{3 digits}} \quad\quad \underbrace{}_{\text{3 digits}}$$

Now we can understand why the following number–logarithm relationships were given, without explanation at the time, earlier in this section.

$$\log 1 \times 10^{-4} = -4.0$$
$$\log 1.0 \times 10^{-9} = -9.00$$

The first exponential number has one significant figure, and the second one has two significant figures.

If a table of logarithm values is available, nonintegral pH values can be found without an electronic calculator. Table 14.11 is a compilation of such values to two significant figures. To obtain a logarithm from such a table you need to use the

TABLE 14.11 Two-Place Logarithm Table

N	0.0	0.1	0.2	0.3	0.4	0.5	0.6	0.7	0.8	0.9
1	0.00	0.04	0.08	0.11	0.15	0.18	0.20	0.23	0.26	0.28
2	0.30	0.32	0.34	0.36	0.38	0.40	0.42	0.43	0.45	0.46
3	0.48	0.49	0.51	0.52	0.53	0.54	0.56	0.57	0.58	0.59
4	0.60	0.61	0.62	0.63	0.64	0.65	0.66	0.67	0.68	0.69
5	0.70	0.71	0.72	0.72	0.73	0.74	0.75	0.76	0.76	0.77
6	0.78	0.79	0.79	0.80	0.81	0.81	0.82	0.83	0.83	0.84
7	0.85	0.85	0.86	0.86	0.87	0.88	0.88	0.89	0.89	0.90
8	0.90	0.91	0.91	0.92	0.92	0.93	0.93	0.94	0.94	0.95
9	0.95	0.96	0.96	0.97	0.97	0.98	0.98	0.99	0.99	1.00

mathematical fact that the logarithm of a product is the sum of the logarithms of each factor in the product, that is,

$$\log(n \times p) = \log n + \log p$$

Example 14.9 illustrates the calculation of a nonintegral pH value using an electronic calculator and also a logarithm table and the mathematical identity just given.

EXAMPLE 14.9

Calculate the pH of a solution with $[H_3O^+] = 6.2 \times 10^{-4}$.

Solution
With an electronic calculator, we first enter number 6.2×10^{-4} into the calculator. We then use the LOG key to obtain the logarithm value, -3.2076083.

$$\begin{aligned}
\text{pH} &= -\log(6.2 \times 10^{-4}) \\
&= -(-3.2076083) \\
&= 3.2076083 \quad \text{(calculator answer)} \\
&= 3.21 \quad \text{(correct answer)}
\end{aligned}$$

The given hydronium ion concentration has two significant figures. Therefore, the logarithm should have two significant figures, as 3.21 does. In a logarithm only the digits to the right of the decimal place are considered significant.

The same pH value, 3.21, can be obtained by using a logarithm table (Table 14.11) and the identity

$$\log(n \times p) = \log n + \log p$$

Substituting the known hydronium ion concentration into the formal definition for pH gives

$$\begin{aligned}
\text{pH} &= -\log(6.2 \times 10^{-4}) \\
&= -(\log 6.2 + \log 10^{-4}) \\
&= -\log 6.2 - \log 10^{-4}
\end{aligned}$$

Table 14.11 is used to obtain the value for the term log 6.2. First find, in the table, the number 6 in the column headed N. Then move horizontally to the right until you reach the column headed 0.2. The numerical value found at the intersection of row 6 and column 0.2 is 0.79, which is the log of 6.2.

Finding the log of 10^{-4}, the other log term in the pH expression, is easy. For a power of ten, the logarithm value is simply the value of the exponent. It is -4.

Therefore, we have

$$\begin{aligned}
\text{pH} &= -\log 6.2 - \log 10^{-4} \\
&= -0.79 - (-4) \\
&= -0.79 + 4 \\
&= 3.21 \quad \text{(calculator and correct answer)}
\end{aligned}$$

PRACTICE EXERCISE 14.9

Calculate the pH of a solution with $[H_3O^+] = 7.9 \times 10^{-11}$. Ans. 10.10.
Similar exercises: Problems 14.87 and 14.88

Now that we know the procedures involved in obtaining nonintegral pH values, it is easy to see why hydrogen ion concentrations such as 1.0×10^{-3} and 1.0×10^{-6} give integral pH values. Referring to Table 14.11, we see that the log of 1.0 is 0.00. Thus for a hydrogen ion concentration of 1.0×10^{-4} we have

$$\begin{aligned} pH &= -\log(1.0 \times 10^{-4}) \\ &= -(\log 1.0 + \log 10^{-4}) \\ &= -\log 1.0 - \log 10^{-4} \\ &= 0.00 - (-4) \\ &= 4.00 \end{aligned}$$

It is frequently necessary to calculate the hydronium ion concentration for a solution from its pH value. This type of calculation, which is the reverse of that just illustrated, is shown in Example 14.10.

EXAMPLE 14.10

The pH of a solution is 6.80. What is the molar hydronium ion concentration for this solution?

Solution

Because the pH is between 6 and 7, we know immediately that $[H_3O^+]$ will be between 10^{-6} and 10^{-7} M. From the defining equation for pH, we have

$$\begin{aligned} pH &= -\log [H_3O^+] = 6.80 \\ \log [H_3O^+] &= -6.80 \end{aligned}$$

To find $[H_3O^+]$ we need to determine the *antilog* of -6.80.

How an antilog is obtained using a calculator depends on the type of calculator you have. Many calculators have an antilog function (sometimes labeled INV log) that performs this operation. If this key is present, then

1. Enter the number -6.80. Note that it is the *negative* of the pH that is entered into the calculator.
2. Press the INV log key (or an inverse key and then a log key). The result is the desired hydronium ion concentration.

$$\begin{aligned} \log [H_3O^+] &= -6.80 \\ \text{antilog } [H_3O^+] &= 1.5848931 \times 10^{-7} \quad \text{(calculator answer)} \\ [H_3O^+] &= 1.6 \times 10^{-7} \quad \text{(correct answer)} \end{aligned}$$

Remember, that the original pH value was a two-significant-figure pH.

Some calculators use a 10^x key to perform the antilog operation. Use of this key is based on the mathematical identity

$$\text{antilog } X = 10^x$$

For our case this means

$$\text{antilog } -6.80 = 10^{-6.80}$$

If the 10^x key is present, then

1. Enter the number -6.80 (the negative of the pH).
2. Press the function key 10^x. The result is the desired hydronium ion concentration.

$$[H_3O^+] = 10^{-6.80} = 1.5848931 \times 10^{-7} \quad \text{(calculator answer)}$$
$$= 1.6 \times 10^{-7} \quad \text{(correct answer)}$$

To solve this problem using a table of logarithms, we begin with the rearranged form of the defining equation for pH.

$$\log [H^+] = -\text{pH}$$

Since the pH is given as 6.80, we have

$$\log [H^+] = -6.80$$

The pH value is next separated into a decimal part and a whole number part.

$$\log [H^+] = -0.80 - 6$$

We now add 1 to the decimal part and subtract 1 from the whole number part.

$$\log [H^+] = -0.80 + 1 - 6 - 1$$
$$= 0.20 - 7$$

The purpose of this "+1 and −1" technique is to convert the decimal part of the logarithm from a negative to a positive number. The number corresponding to the positive decimal part of the logarithm can be found by using Table 14.11, and the negative whole number part of the logarithm can be converted directly into a power of 10.

From Table 14.11, we find that a log value of 0.20 corresponds to the number 1.6, since the value 0.20 is in the first row and the 0.6 column. Thus, we have

$$\log [H^+] = 0.2 - 7.0$$
$$[H^+] = 1.6 \times 10^{-7} \quad \text{(calculator and correct answer)}$$

PRACTICE EXERCISE 14.10

The pH of a solution is 8.40. What is the molar hydronium ion concentration for this solution?   *Ans.* 4.0×10^{-9} M.

Similar exercises: Problems 14.91–14.94

14.13 Hydrolysis of Salts

Salts contain neither H_3O^+ ions nor OH^- ions (Sec. 14.6). Consequently, we might expect that aqueous solutions of salts would be neutral (pH = 7). Surprisingly, many salt solutions are not neutral; they have pH values greater than 7 (basic) or less than 7 (acidic). How can this be? In the dissolution of certain salts, a phenomenon called salt hydrolysis causes pH changes. **Salt hydrolysis** is a reaction in which a salt interacts with water to produce an acidic or basic solution. Salt hydrolysis reactions are important in numerous settings, including many involving the human body.

In order for a solution to be acidic or basic, an excess of H_3O^+ ion or OH^- ion must be present (Sec. 14.10). An explanation of why the addition of a salt to a solution can cause excess H_3O^+ or OH^- ions involves a consideration of the following points.

1. Any aqueous salt solution contains four, rather than two, ions. Positive and negative ions are produced by the salt as it dissolves (Sec. 14.6). Also present are small numbers (1×10^{-7} M) of H_3O^+ and OH^- ions, which are produced by the dissociation of water (Sec. 14.11). The numbers of H_3O^+ and OH^- ions are equal because proton transfer between two water molecules always produces one ion of each type.

$$H_2O + H_2O \longrightarrow H_3O^+ + OH^-$$

2. The positive ion from the salt reacts with the OH^- ions from the water under certain conditions. The stimulus (or driving force) for such a reaction is the production of a weak base, a substance that exists predominantly in molecular form. Weak base formation reduces the amount of free OH^- ion present; the OH^- ions are tied up in molecules. The H_3O^+ ions present in the solution are unaffected by this weak base formation and outnumber the OH^- ions; this situation is characteristic of an acidic solution.

3. The negative ion from the salt reacts with the H_3O^+ ions from the water under certain conditions. The driving force for such a reaction is the production of a weak acid. Like weak bases, weak acids are substances that exist predominantly in molecular form. Thus formation of a weak acid reduces the amount of free H_3O^+ ion present; the H_3O^+ ions are tied up in molecules. The OH^- ions present in the solution are unaffected by this weak acid formation and outnumber the H_3O^+ ions; this situation is characteristic of a basic solution.

Using the preceding considerations, let us now consider the hydrolysis of a specific salt, NaCN. Will the hydrolysis of this salt produce an acidic or basic solution? We first write the formulas of the four ions present in the solution.

$$Na^+ \quad CN^-$$
$$OH^- \quad H_3O^+$$

Will Na^+ ions tie up OH^- ions? The answer is "no" because NaOH is a strong base, and strong bases exist in ionic form in solution (Sec. 14.5). Will CN^- ions tie up H_3O^+ ions? The answer is "yes" because HCN is a weak acid. (Recall the strong and weak acids and bases that were discussed in Sec. 14.5.) After hydrolysis, the predominant species present in solution will be

$$Na^+ \qquad HCN$$
$$OH^-$$

After hydrolysis, the solution will be basic. There are free OH^- ions present, while most of the H_3O^+ ions have been converted to HCN molecules and water.

$$H_3O^+ + CN^- \longrightarrow HCN + H_2O$$

EXAMPLE 14.11

Predict whether solutions of each of the following soluble salts will be acidic, basic, or neutral

(a) sodium acetate, $NaC_2H_3O_2$
(b) ammonium chloride, NH_4Cl
(c) potassium chloride, KCl
(d) ammonium fluoride, NH_4F

Solution

A familarity with the strengths of common acids and bases (weak or strong) is the key to solving hydrolysis problems (Sec. 14.5).

(a) The four ions present in solution are

$$Na^+ \qquad C_2H_3O_2^-$$
$$OH^- \qquad H_3O^+$$

The Na^+ ion will not hydrolyze because the base that would be formed, NaOH, is a strong base. The $C_2H_3O_2^-$ ion does hydrolyze, forming the weak acid $HC_2H_3O_2$.

$$Na^+ \qquad HC_2H_3O_2$$
$$OH^-$$

The solution will be **basic** because of the presence of the free OH^- ions.

(b) This time, the four ions present in solution are

$$NH_4^+ \qquad Cl^-$$
$$OH^- \qquad H_3O^+$$

The NH_4^+ ion will hydrolyze because this hydrolysis produces the weak base aqueous ammonia. The Cl^- ion does not hydrolyze because HCl is a strong acid. The solution will be **acidic**. The OH^- ions have been tied up as the result of aqueous ammonia formation, but the H_3O^+ ions are still "free."

(c) Potassium chloride is a salt that does not hydrolyze. The four ions initially present are

$$K^+ \qquad Cl^-$$
$$OH^- \qquad H_3O^+$$

The possible products from the hydrolysis are KOH and HCl; KOH is a strong base and HCl is a strong acid. There is no driving force for hydrolysis, and the solution remains **neutral**.

(d) It is possible for both the positive and the negative ions from a salt to undergo hydrolysis at the same time. This is the situation for NH_4F. The four ions initially present are

$$NH_4^+ \qquad F^-$$
$$OH^- \qquad H_3O^+$$

The hydrolysis products are aqueous ammonia (a weak base) and hydrofluoric acid (a weak acid). In a double hydrolysis situation such as this, it is not easy to predict whether the resulting solution will be acidic or basic. It depends on the relative strengths of the weak acid and weak base produced. This type of strength comparison is beyond the scope of this text, and you will not be expected to predict acidity or basicity in double hydrolysis situations. However, you should be able to identify double hydrolysis situations.

PRACTICE EXERCISE 14.11

Predict whether solutions of each of the following soluble salts will be acidic, basic, or neutral.

(a) ammonium bromide, NH_4Br (b) potassium carbonate, K_2CO_3
(c) sodium chloride, NaCl

Ans. (a) acidic; (b) basic; (c) neutral

Similar exercises: Problems 14.99 and 14.100

A set of simple rules for determining whether or not a salt will hydrolyze is easily formulated if salts are visualized as the products of the neutralization reaction between an acid and a base (Sec. 14.8).

$$\text{Acid} + \text{base} \longrightarrow \text{salt} + \text{water}$$

From this viewpoint, four categories of salts exist. They are formed from the reaction of

1. A strong acid and a strong base.
2. A strong acid and a weak base.
3. A weak acid and a strong base.
4. A weak acid and a weak base.

Salts that will hydrolyze are obtained from a neutralization that involves either a weak acid or a weak base, or both. Table 14.12 gives further information about the hydrolytic behavior of the various types of salts. Note from the table that in all cases the ions that hydrolyze are those that were originally part of a weak acid or weak base. Category 2 salts always produce acidic solutions; formation of a weak base is the driving force for the hydrolysis. Category 3 salts always produce basic solutions; formation of a weak acid is the driving force for the hydrolysis. Category 4 salts always give a double hydrolysis situation; both a weak acid and a weak base are formed.

TABLE 14.12 Hydrolytic Behavior of Various Types of Salts

Type of Salt	Hydrolysis Behavior	pH of Solution
1. Negative ion from strong acid Positive ion from strong base	neither ion hydrolyzes	neutral
2. Negative ion from strong acid Positive ion from weak base	positive ion hydrolyzes	acidic
3. Negative ion from weak acid Positive ion from strong base	negative ion hydrolyzes	basic
4. Negative ion from weak acid Positive ion from weak base	both positive and negative ions hydrolyze	acidic, basic, or neutral depending on the relative strengths of the weak acid and base

Generally, hydrolysis reactions take place only to a small extent, about 2–5%. However, in some cases, this is sufficient to cause very significant changes in pH, as can be seen from the pH values in Table 14.13.

14.14 Buffer Solutions

As we discussed in the last section, certain salts that hydrolyze can change the pH of water. In this section, we will consider a second phenomenon involving salts and their effect on solution pH. Certain *combinations* of compounds can protect the pH of a solution from change. These combinations, which always involve at least one salt, are called *buffers*, and the solutions containing them are called *buffer solutions*. A **buffer solution** is a solution that resists a change in pH when small amounts of acid or base are added to it. A **buffer** is the solute (or solutes) present in a buffer solution that causes it to be resistant to a change in pH.

Buffers contain two species: (1) a substance to react with and remove added base, and (2) a substance to react with and remove added acid. The "chemical com-

TABLE 14.13 Approximate pH of Selected 0.1 M Aqueous Salt Solutions

Salt	Formula	pH of Solution	Category of Salt
Ammonium nitrate	NH_4NO_3	5.1	strong acid–weak base
Ammonium nitrite	NH_4NO_2	6.3	weak acid–weak base
Ammonium acetate	$NH_4C_2H_3O_2$	7.0	weak acid–weak base
Sodium chloride	NaCl	7.0	strong acid–strong base
Sodium fluoride	NaF	8.1	weak acid–strong base
Sodium acetate	$NaC_2H_3O_2$	8.9	weak acid–strong base
Ammonium cyanide	NH_4CN	9.3	weak acid–weak base
Sodium cyanide	NaCN	11.1	weak acid–strong base

bination" that meets these requirements is a weak acid and a salt of its conjugate base or a weak base and salt of its conjugate acid. Thus, buffers involve conjugate acid–base pairs (Sec. 14.3). Such pairs commonly employed as buffers include $HC_2H_3O_2/C_2H_3O_2^-$, NH_4^+/NH_3, $H_2PO_4^-/HPO_4^{2-}$, and H_2CO_3/HCO_3^-.

Consider a buffer solution containing approximately equal concentrations of acetic acid (a weak acid) and sodium acetate (a salt of this weak acid). This solution resists pH change by the following mechanisms.

1. When a small amount of a strong acid such as HCl is added to the solution, the added H_3O^+ ions react with the acetate ions from the sodium acetate to give acetic acid.

$$H_3O^+ + C_2H_3O_2^- \longrightarrow HC_2H_3O_2 + H_2O$$

Most of the added H_3O^+ ions are tied up in acetic acid molecules, and the pH changes very little.

2. When a small amount of a strong base such as NaOH is added to the solution, the added OH^- ions react with the acetic acid (neutralization) to give acetate ions and water.

$$OH^- + HC_2H_3O_2 \longrightarrow C_2H_3O_2^- + H_2O$$

Most of the added OH^- ions are converted to water, and the pH changes only slightly.

The reactions that are responsible for the buffering action in the acetic acid–acetate ion system can be summarized as follows.

$$C_2H_3O_2^- \underset{OH^-}{\overset{H_3O^+}{\rightleftharpoons}} HC_2H_3O_2$$

Note that one member of the buffer pair removes excess H_3O^+ ion and the other removes excess OH^- ion. The buffering action always results in the active species being converted to its partner species.

Buffer systems have their limits. If large amounts of H_3O^+ or OH^- are added to a buffer, the buffering capacity can be exceeded; then the buffer system is overwhelmed, and the pH changes. For example, if large amounts of H_3O^+ were added to the acetate–acetic acid buffer just discussed, the H_3O^+ ion would react with acetate ion until the acetate was depleted. Then the pH would begin to drop as free H_3O^+ ions accumulate in the solution.

14.15 Buffers in the Human Body

Buffer solutions play an important role in the functioning of the human body. All body fluids have definite pH values that must be maintained within very narrow ranges because living cells are extremely sensitive to even slight changes in pH. The protection against pH change is provided by buffers, which can be referred to as "chemical shock absorbers" or "chemical sponges" because of their key protective role.

Blood is a vital buffer solution. Even small departures from the normal pH range of blood (7.35–7.45) can cause serious illness, and death can result from variations that exceed a few tenths of a pH unit. This situation results because many of the key reactions that take place in blood are enzyme catalyzed and reach optimum conditions only within the narrow normal pH range. Altering the pH slows down or stops the action of the enzymes.

The major buffer system in blood is composed of carbonic acid (H_2CO_3) and bicarbonate salts such as sodium bicarbonate ($NaHCO_3$). Certain proteins and, to a small extent, hydrogen phosphate ions also help buffer blood. The carbonic acid–bicarbonate buffering system in blood operates in the following manner. Any H_3O^+ formed in the blood reacts with bicarbonate ion to give carbonic acid.

$$H_3O^+ + HCO_3^- \longrightarrow H_2CO_3 + H_2O$$

Carbonic acid is an unstable acid that readily decomposes to give carbon dioxide and water.

$$H_2CO_3 \longrightarrow H_2O + CO_2$$

Excess carbon dioxide in the blood, which is formed from this decomposition, is removed from the blood by the lungs and is exhaled.

The presence of OH^- in the blood is not a common occurrence. If it does occur, the carbonic acid–bicarbonate buffer adjusts for its presence through the reaction

$$H_2CO_3 + OH^- \longrightarrow HCO_3^- + H_2O$$

Excess HCO_3^- ions can be eliminated from the body through the kidneys. Figure 14.2 summarizes the workings of the carbonic acid–bicarbonate ion buffer system in the blood. The capacity of the carbonic acid–bicarbonate buffer system to handle increases in OH^- ion is not as great as its capacity to handle increases in H_3O^+ ion. This is because the ratio of HCO_3^- ion to H_2CO_3 in blood is about 20 to 1.

A very important buffer system within cells is the dihydrogen phosphate ($H_2PO_4^-$)–hydrogen phosphate (HPO_4^{2-}) system. The control of pH within cellular fluids is maintained by the reaction of OH^- ion with $H_2PO_4^-$ and the reaction of H_3O^+ with HPO_4^{2-}.

$$H_2PO_4^- + OH^- \longrightarrow HPO_4^{2-} + H_2O$$
$$HPO_4^{2-} + H_3O^+ \longrightarrow H_2PO_4^- + H_2O$$

The overall dihydrogen phosphate–hydrogen phosphate buffering action can be summarized as

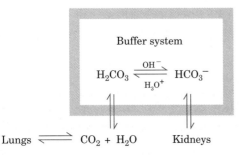

FIGURE 14.2 The carbonic acid–bicarbonate ion buffer system in human blood.

$$H_2PO_4^- \xrightleftharpoons[H_3O^+]{OH^-} HPO_4^{2-}$$

The normal hydrogen phosphate–dihydrogen phosphate ion ratio in cellular fluids is about 4 to 1. Thus, the phosphate buffer system is better equipped to handle influxes of acid than influxes of base. Significant amounts of acids (up to 10 moles/day) are produced in a human body as the result of normal metabolic reactions. For example, lactic acid ($HC_3H_5O_3$) is produced in muscle tissue during exercise.

Under normal conditions, the body's carbonate and phosphate buffer systems are adequate to maintain pH in the normal range. However, under certain stress conditions, the buffer systems can be temporarily overwhelmed. When this happens, compensatory mechanisms involving the lungs and kidneys help to return the pH to normal. Both the lungs and kidneys play a role in pH control at all times, but this role is more important during periods of stress.

Acidosis is a body condition in which the pH of blood drops from its normal value of 7.4 to 7.1–7.2. Various factors can cause acidosis, including emphysema-caused hypoventilation (breathing too little), congestive heart failure, diabetes mellitus, excess loss of bicarbonate ion in severe diarrhea, or decreased excretion of hydrogen ions in kidney failure. A temporary condition of acidosis can result from prolonged, intensive exercise. The body reacts to alleviate this acidosis condition in two ways. Excess carbon dioxide (formed from the decomposition of carbonic acid) is expelled by increasing the rate of respiration. Also, kidney system changes occur that increase excretion of H_3O^+ and retention of HCO_3^-; this results in acidic urine.

Alkalosis is a body condition in which the pH of blood increases from its normal value of 7.4 to a value of 7.5. Alkalosis can result from hyperventilation (excess breathing) caused by anxiety or hysteria or from extreme fever, severe vomiting, or exposure to high altitude (altitude sickness). The body's responses to alkalosis include a decrease in respiration rate (less expulsion of carbon dioxide by the lungs) and an increase in HCO_3^- excretion by the kidneys, resulting in alkaline urine. Alkalosis is not as common as acidosis.

14.16 Acid–Base Titrations

Determining the concentration of acid or base in a solution is a regular activity in many laboratories. The concentration of an acid or base in a solution and the solution's pH are two different entities. The pH of a solution gives information about the concentration of hydrogen (hydronium) ions in solution. Only dissociated molecules influence the pH value. The concentration of an acid or base solution gives information about the *total number* of acid or base molecules present; both dissociated and undissociated molecules are counted.

The procedure most frequently used to determine the concentration of an acidic or basic solution is that of titration. **Titration** is the gradual adding of one solution to another until the solute in the first solution has reacted completely with the solute in the second solution.

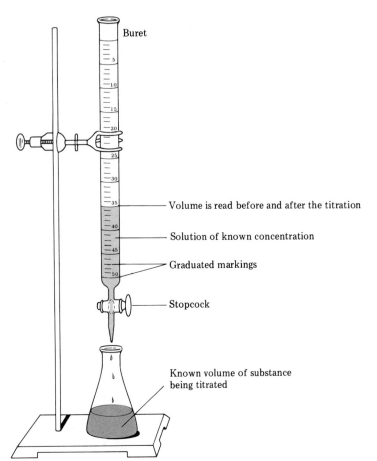

Figure 14.3 Use of a buret in a titration procedure.

Suppose we want to determine the concentration of an acid solution by titration. We would first measure out a *known volume* of the acid solution into a beaker or flask. We would then slowly add a solution of base of *known concentration* to the flask or beaker by means of a buret (see Fig. 14.3). Base addition continues until all the acid has completely reacted with added base. The *volume of base* needed to reach this point is obtained from the buret readings. Knowing the original volume of acid, the concentration of the base, and the volume of added base, we can calculate the concentration of the acid (Sec. 14.17).

In order to complete a titration sucessfully, we must be able to detect when the reaction between acid and base is complete. One way to do this is to add an indicator to the solution being titrated. An **indicator** is a compound that exhibits different colors depending on the pH of its surroundings. Typically, an indicator is one color in basic solutions and another color in acidic solutions. An indicator is selected that will change color at a pH corresponding as nearly as possible to the pH of the solution when the titration is complete. This pH can be calculated ahead of time based on the identities of the acid and base involved in the titration.

14.17 Acid–Base Titration Calculations Using Normality

The concentration unit normality is the most convenient unit to work with when doing acid–base calculations. This is a concentration unit that is very closely related to molarity (Sec. 13.8). The difference between the two units involves the expression of the amount of solute present. Moles of solute are specified for molarity and equivalents of solute are specified for normality. The **normality** of a solution, which is designated by N, is a ratio giving the number of equivalents of solute per liter of solution.

$$\text{Normality} = \frac{\text{equivalents of solute}}{\text{liters of solution}}$$

Normality takes into account the fact that all acids and bases do not yield the same number of hydronium or hydroxide ions per molecule on dissociation. Even though their molar concentrations are equal, 0.10 M H_2SO_4 (which can supply two H_3O^+ ions per molecule) will neutralize twice as much base as will the same volume of 0.10 M HNO_3 (which can supply only one H_3O^+ ion per molecule). However, equal volumes of 0.10 N H_2SO_4 and 0.10 N HNO_3 will neutralize exactly the same amount of base.

The key to comprehending normality concentration units is an understanding of the concept of equivalents. Definitions of equivalent for use in acid–base work are as follows. An **equivalent of acid** is the quantity of acid that will supply 1 mole of H_3O^+ ion during an acid–base reaction. Similarly, an **equivalent of base** is the quantity of base that will react with 1 mole of H_3O^+ ion during an acid–base reaction. Examples 14.12 and 14.13 illustrate the use of these two definitions for an equivalent in problem-solving contexts.

EXAMPLE 14.12

Determine the number of equivalents of acid or base in each of the following samples.

(a) 1 mole of H_3PO_4 (b) 3 moles of $Ba(OH)_2$ (c) 2 moles of HNO_3

Solution

(a) The equation for the complete dissociation of H_3PO_4 is

$$H_3PO_4 + 3 H_2O \longrightarrow 3 H_3O^+ + PO_4^{3-}$$

Thus, 1 mole of H_3PO_4 gives up 3 moles of H_3O^+ ion upon complete reaction. Three moles of H_3O^+ ion is equal to 3 equiv of H_3O^+ ion by definition. Therefore,

$$1 \text{ mole } H_3PO_4 \times \frac{3 \text{ equiv } H_3PO_4}{1 \text{ mole } H_3PO_4} = 3 \text{ equiv } H_3PO_4$$

(b) The equation for the complete dissociation of $Ba(OH)_2$ is

$$Ba(OH)_2 \longrightarrow Ba^{2+} + 2\ OH^-$$

Thus, 1 mole of $Ba(OH)_2$ yields 2 moles of OH^- ion (which can react with 2 moles of H_3O^+). Therefore

$$3\ \text{moles } Ba(OH)_2 \times \frac{2\ \text{equiv } Ba(OH)_2}{1\ \text{mole } Ba(OH)_2} = 6\ \text{equiv } Ba(OH)_2$$

(c) The equation for the complete dissociation of HNO_3 is

$$HNO_3 + H_2O \longrightarrow H_3O^+ + NO_3^-$$

One mole of acid yields only 1 mole (or equiv) of H_3O^+ ion. Therefore,

$$2\ \text{moles } HNO_3 \times \frac{1\ \text{equiv } HNO_3}{1\ \text{mole } HNO_3} = 2\ \text{equiv } HNO_3$$

PRACTICE EXERCISE 14.12

Determine the number of equivalents of acid or base in each of the following samples.

(a) 1 mole of H_2CO_3 (b) 1 mole of KOH   (c) 2 moles of $HC_2H_3O_2$

Ans. (a) 2 equiv H_2CO_3; (b) 1 equiv KOH; (c) 2 equiv $HC_2H_3O_2$.

Similar exercises: Problems 14.113 and 14.114

EXAMPLE 14.13

Calculate the normality of a solution made by dissolving 25.0 g of NaOH in enough water to give 500.0 mL of solution.

Solution

To calculate normality, we need to know the number of equivalents of solute and solution volume (in liters).
  For NaOH,

$$1\ \text{equiv} = 1\ \text{mole}$$

since 1 mole of NaOH yields 1 mole of OH^- ion,

$$NaOH \longrightarrow Na^+ + OH^-$$

and 1 mole of OH^- reacts with 1 mole of H_3O^+ ion. (Hydronium and hydroxide ions always react in a one-to-one ratio.)
  The number of equivalents of NaOH in 25.0 g of NaOH is obtained by converting the grams of NaOH to moles of NaOH and then using the above mole–equivalent relationship as a conversion factor to obtain equivalents.

$$25.0\ \text{g NaOH} \times \frac{1\ \text{mole NaOH}}{40.0\ \text{g NaOH}} \times \frac{1\ \text{equiv NaOH}}{1\ \text{mole NaOH}} = 0.625\ \text{equiv NaOH}$$

(calculator and correct answer)

The solution volume of 500.0 mL, changed to liter units, is 0.5000 L.

Section 14.17 · Acid–Base Titration Calculations Using Normality

Both quantities called for in the defining equation for normality, equivalents of solute and liters of solution, are now known. Therefore, substituting into the defining equation, we get

$$N = \frac{0.625 \text{ equiv NaOH}}{0.5000 \text{ L solution}} = 1.25 \frac{\text{equiv NaOH}}{\text{L solution}}$$

$$= 1.25 \text{ N} \quad \text{(calculator and correct answer)}$$

PRACTICE EXERCISE 14.13

Calculate the normality of a solution made by dissolving 5.00 g of $Sr(OH)_2$ in enough water to give 750.0 mL of solution

Ans. 0.109 N $Sr(OH)_2$.

Similar exercises: Problems 14.115 and 14.116

In the laboratory, solution concentrations are specified in terms of molarity rather than normality. To convert acid or base molarities to normalities (or vice versa) is a very simple operation. Only the conversion factor relating moles to equivalents is needed.

EXAMPLE 14.14

Express the following molar concentrations as normalities.

(a) 0.030 M H_3PO_4 (b) 0.10 M $Ba(OH)_2$ (c) 6.0 M HNO_3

Solution

(a) A 0.030 M H_3PO_4 solution contains 0.030 mole of H_3PO_4 per liter of solution. Using this fact and the fact (from Example 14.12) that

$$1 \text{ mole } H_3PO_4 = 3 \text{ equiv } H_3PO_4$$

we calculate the normality as follows.

$$\frac{0.030 \text{ mole } H_3PO_4}{1 \text{ L solution}} \times \frac{3 \text{ equiv } H_3PO_4}{1 \text{ mole } H_3PO_4} = \frac{0.09 \text{ equiv } H_3PO_4}{\text{L solution}} \quad \text{(calculator answer)}$$

$$= 0.090 \text{ N} \quad \text{(correct answer)}$$

Therefore, the solution is 0.090 N.

(b) A 0.10 M $Ba(OH)_2$ solution contains 0.10 mole of $Ba(OH)_2$ per liter of solution. This information, coupled with the fact (from Example 14.12) that

$$1 \text{ mole } Ba(OH)_2 = 2 \text{ equiv } Ba(OH)_2$$

enables us to calculate the normality as follows.

$$\frac{0.10 \text{ mole } Ba(OH)_2}{1 \text{ L solution}} \times \frac{2 \text{ equiv } Ba(OH)_2}{1 \text{ mole } Ba(OH)_2} = \frac{0.2 \text{ equiv } Ba(OH)_2}{\text{L solution}}$$

(calculator answer)

$$= 0.20 \text{ N } Ba(OH)_2$$

(correct answer)

(c) For 6.0 M HNO$_3$, we calculate the normality, using reasoning similar to that in parts (a) and (b), as follows.

$$\frac{6.0 \text{ moles HNO}_3}{1 \text{ L solution}} \times \frac{1 \text{ equiv HNO}_3}{1 \text{ mole HNO}_3} = \frac{6 \text{ equiv HNO}_3}{\text{L solution}} \quad \text{(calculator answer)}$$

$$= 6.0 \text{ N HNO}_3 \quad \text{(correct answer)}$$

PRACTICE EXERCISE 14.14

Express the following molar concentrations as normalities.

(a) 0.050 M H$_2$CO$_3$ (b) 1.50 M KOH   (c) 0.500 M HC$_2$H$_3$O$_2$ Similar exercises:
Ans. (a) 0.10 N H$_2$CO$_3$; (b) 1.50 N KOH; (c) 0.500 N HC$_2$H$_3$O$_2$.   Problems 14.117 and 14.118

Molarities can also be converted to normalities, without use of dimensional analysis, by the following equations.

$$N_{acid} = M_{acid} \times \text{number of acidic hydrogens in acid}$$
$$N_{base} = M_{base} \times \text{number of hydroxide ions in base}$$

Both of these equations are based on the Arrhenius system for defining acids and bases. It is not recommended that you use these "shortcut" equations until you thoroughly understand the dimensional analysis approach to obtaining normalities (Example 14.14).

In an acid–base titration, the chemical reaction that occurs is that of neutralization (Sec. 14.8). The H$_3$O$^+$ ions from the acid react with the OH$^-$ ions from the base to produce water.

$$H_3O^+ + OH^- \longrightarrow 2\, H_2O$$

These two ions always react in a one-to-one ratio; thus, 1 mole of H$_3$O$^+$ ions always reacts with 1 mole of OH$^-$ ions. Since 1 mole of each of these ions is an equivalent, it follows that at the endpoint of an acid–base titration equal numbers of equivalents of acid and base have been consumed; that is,

Equiv of reacted acid = equiv of reacted base

A simple equation, whose derivation is based on the fact that equal numbers of equivalents of acid and base are consumed during an acid–base titration, is employed in titration calculations. It is

$$N_{acid} \times V_{acid} = N_{base} \times V_{base}$$

The derivation of this equation is as follows. The defining equations for the normality of the acid and base are

$$N_{acid} = \frac{E_{acid}}{V_{acid}} \quad \text{and} \quad N_{base} = \frac{E_{base}}{V_{base}}$$

where E is the equivalents of solute and V is the volume of solution (in liters). Each of these equations can be rearranged as follows.

Section 14.17 · Acid–Base Titration Calculations Using Normality

$$E_{acid} = N_{acid} \times V_{acid}$$
$$E_{base} = N_{base} \times V_{base}$$

At the endpoint of the titration, where $E_{acid} = E_{base}$, we can therefore write

$$N_{acid} \times V_{acid} = N_{base} \times V_{base}$$

The volumes V_{acid} and V_{base} can be expressed in liters or milliliters, provided the same unit is used for both. Since burets are calibrated in milliliters, the milliliter is the unit most often used in titration calculations.

The data obtained from an acid–base titration (Sec. 14.12) are (1) the volume of base used, (2) the volume of acid used, and (3) the concentration of the acid or base used to titrate the solution of unknown concentration. These quantities are three of the four variables in the equation $N_{acid} \times V_{acid} = N_{base} \times V_{base}$. Hence, the fourth variable, the concentration of the acid or base that was titrated, can easily be calculated.

EXAMPLE 14.15

What volume of 0.500 M H_3PO_4 would be needed to completely neutralize 275 mL of 0.125 M $Ba(OH)_2$ solution?

Solution

The concentrations of the two solutions must be converted from molarities to normalities. Since H_3PO_4 yields three H_3O^+ ions per acid molecule, 1 mole of H_3PO_4 contains 3 equiv of acid; hence, the normality of the H_3PO_4 solution is three times the molarity.

$$\frac{0.500 \text{ mole } H_3PO_4}{1 \text{ L solution}} \times \frac{3 \text{ equiv } H_3PO_4}{1 \text{ mole } H_3PO_4} = \frac{1.5 \text{ equiv } H_3PO_4}{\text{L solution}} \quad \text{(calculator answer)}$$

$$= 1.50 \text{ N } H_3PO_4 \quad \text{(correct answer)}$$

For $Ba(OH)_2$, the normality will be double the molarity, since $Ba(OH)_2$ yields two OH^- ions per molecule on dissociation.

$$\frac{0.125 \text{ mole } Ba(OH)_2}{1 \text{ L solution}} \times \frac{2 \text{ equiv } Ba(OH)_2}{1 \text{ mole } Ba(OH)_2} = \frac{0.25 \text{ equiv } Ba(OH)_2}{\text{L solution}}$$
$$\text{(calculator answer)}$$

$$= 0.250 \text{ N } Ba(OH)_2$$
$$\text{(correct answer)}$$

Substituting the known values in the equation

$$V_{acid} = \frac{N_{base} \times V_{base}}{N_{acid}}$$

gives

$$V_{acid} = \frac{(0.250 \text{ N})(275 \text{ mL})}{(1.50 \text{ N})} = 45.833333 \text{ mL} \quad \text{(calculator answer)}$$

$$= 45.8 \text{ mL} \quad \text{(correct answer)}$$

PRACTICE EXERCISE 14.15

What volume of 0.500 M HNO_3 would be needed to completely neutralize 375 mL of 0.515 M $Ba(OH)_2$ solution?   Ans.   772 mL HNO_3.

Similar exercises: Problems 14.123 and 14.124

EXAMPLE 14.16

In an acid–base titration, 32.7 mL of 0.105 M NaOH is required to neutralize 50.0 mL of HNO_3. What are the normality and the molarity of the HNO_3 solution?

Solution

We will use the equation

$$N_{acid} \times V_{acid} = N_{base} \times V_{base}$$

to solve this problem. Three of the four quantities in this equation are known. The volumes of acid and base are, respectively, 50.0 and 32.7 mL. The normality of the base is not directly given, but can easily be calculated from the molarity of the base. Since 1 mole of NaOH contains 1 equiv of OH^- ions, the normality of the base will be the same as the molarity; that is, the base is not only 0.105 M but also 0.105 N.

Substitution of these known values into the equation, rearranged so that N_{acid} is isolated on the left side, gives

$$N_{acid} = \frac{N_{base} \times V_{base}}{V_{acid}}$$

$$= 0.105 \text{ N} \times \frac{32.7 \text{ mL}}{50.0 \text{ mL}} = 0.06867 \text{ N} \quad \text{(calculator answer)}$$

$$= 0.0687 \text{ N} \quad \text{(correct answer)}$$

The molarity of HNO_3, which was also asked for in this problem, will have the same numerical value as the normality, since HNO_3 is a monoprotic acid, a situation for which equivalents and moles are identical.

PRACTICE EXERCISE 14.16

In an acid–base titration, 50.2 mL of 0.252 M NaOH is required to neutralize 32.7 mL of H_2SO_4. What are the normality and the molarity of the H_2SO_4 solution?   Ans.   0.387 N and 0.194 M H_2SO_4.

Similar exercises: Problems 14.127 and 14.128

14.18 Acid and Base Stock Solutions

Acids and bases are used so often in most laboratories that stock solutions (Sec. 13.10) of the most common ones are made readily available at each work space. The concentrations of such solutions are traditionally the same from laboratory to laboratory and

TABLE 14.14 Concentrations of Common Laboratory Stock Solutions of Acids and Bases

Label Designation	Chemical Formula	Concentration	
		Molarity	Normality
Acids			
Dilute hydrochloric acid	HCl	6	6
Concentrated hydrochloric acid	HCl	12	12
Dilute nitric acid	HNO_3	6	6
Concentrated nitric acid	HNO_3	16	16
Dilute sulfuric acid	H_2SO_4	3	6
Concentrated sulfuric acid	H_2SO_4	18	36
Dilute acetic acid	$HC_2H_3O_2$	6	6
Concentrated acetic acid[a]	$HC_2H_3O_2$	18	18
Bases			
Dilute aqueous ammonia[b]	$NH_3(aq)$	6	6
Concentrated aqueous ammonia[b]	$NH_3(aq)$	15	15
Dilute sodium hydroxide[c]	NaOH	6	6

[a] Often labeled "glacial acetic acid."
[b] Often labeled "ammonium hydroxide."
[c] Often labeled simply "sodium hydroxide."

are given in Table 14.14. Ordinarily the concentrations of the stock solutions are not given on their containers. Only the name of the acid or base and the term dilute (dil) or concentrated (conc) are found. It is assumed that students or researchers know what is implied by the designations dil and conc in each specific case; that is, that they know the information found in Table 14.14. Note from Table 14.14 that the designations dil and conc do not have a constant meaning in terms of molarity. For example, concentrated solutions of sulfuric acid, nitric acid, and hydrochloric acid have molarities of 18, 16, and 12, respectively. There is not as much variation in the meaning of the term dilute; except for sulfuric acid, all of the listed dilute solutions are 6 M. Note that all the listed dilute solutions, including sulfuric acid, are 6 N.

Key Terms

The new terms or concepts defined in this chapter are

acidic hydrogen atom (Sec. 14.4)   A hydrogen atom in an acid molecule that can be transferred to a base during an acid–base reaction.

acidic solution (Sec. 14.11)   A solution that has a higher H_3O^+ concentration than OH^- concentration.

acidosis (Sec. 14.15)   A body condition in which the pH of blood drops from its normal value of 7.4 to 7.1–7.2.

alkalosis (Sec. 14.15)   A body condition in which the pH of blood increases from its normal value of 7.4 to a value of 7.5.

amphoteric substance (Sec. 14.3)   A substance that can either lose or accept protons and thus can function as either an acid or a base.

Arrhenius acid (Sec. 14.1)   A hydrogen-containing compound that, in water, produces hydrogen ions (H^+).

Arrhenius base (Sec. 14.1)   A hydroxide-containing compound that, in water, produces hydroxide ions (OH^-).

basic solution (Sec. 14.11) A solution that has a higher OH^- concentration than H_3O^+ concentration.

Brønsted–Lowry acid (Sec. 14.2) Any substance that can donate a proton (H^+) to some other substance.

Brønsted–Lowry base (Sec. 14.2) Any substance that can accept a proton (H^+) from some other substance.

buffer (Sec. 14.14) The solute (or solutes) present in a buffer solution that cause it to be resistant to a change in pH.

buffer solution (Sec. 14.14) A solution that resists a change in pH when small amounts of acid or base are added to it.

conjugate acid (Sec. 14.3) The species formed when a Brønsted–Lowry base accepts a proton.

conjugate acid–base pair (Sec. 14.3) Two species that differ from each other by one proton (H^+).

conjugate base (Sec. 14.3) The species that remains when a Brønsted–Lowry acid loses a proton.

diprotic acid (Sec. 14.4) An acid that can transfer two H^+ ions (two protons) per molecule during an acid–base reaction.

equivalent of acid (Sec. 14.17) Quantity of an acid that will supply 1 mole of H_3O^+ ion during an acid–base reaction.

equivalent of base (Sec. 14.17) Quantity of a base that will react with 1 mole of H_3O^+ ion during an acid–base reaction.

indicator (Sec. 14.16) A compound that exhibits different colors depending on the pH of its surroundings.

ionic equation (Sec. 14.7) An equation in which the formulas of dissociated compounds are written in terms of ions and the formulas of undissociated compounds are written in terms of molecules.

ion product for water (Sec. 14.11) Name given to the numerical value 1.00×10^{-14} which is the product of the H_3O^+ ion and OH^- ion molar concentrations in water.

monoprotic acid (Sec. 14.4) An acid that can transfer only one H^+ ion (proton) per molecule during an acid–base reaction.

net ionic equation (Sec. 14.7) An ionic equation from which nonparticipating (spectator) species have been eliminated.

neutralization (Sec. 14.8) Reaction between an acid and a base to form a salt and water.

neutral solution (Sec. 14.11) A solution in which the concentrations of H_3O^+ ion and OH^- ion are equal.

normality (Sec. 14.17) Concentration of a solution in terms of equivalents of solute per liter of solution.

pH (Sec. 14.12) The negative logarithm of the molar hydronium ion concentration.

polyprotic acid (Sec. 14.4) An acid that can transfer two or more H^+ ions (protons) per molecule during an acid–base reaction.

salt (Sec. 14.6) An ionic compound containing a metal or polyatomic ion as the positive ion and a nonmetal or polyatomic ion (except hydroxide) as the negative ion.

salt hydrolysis (Sec. 14.13) A reaction in which a salt interacts with water to produce an acidic or basic solution.

strong acid (Sec. 14.5) An acid that transfers 100%, or very nearly 100%, of its acidic hydrogen atoms to water.

titration (Sec. 14.16) A procedure in which one solution is gradually added to another until the solute in the first solution has reacted completely with the solute in the second solution.

triprotic acid (Sec. 14.4) An acid that can transfer three H^+ ions (three protons) per molecule during an acid–base reaction.

weak acid (Sec. 14.5) An acid that transfers only a small percentage of its acidic hydrogen atoms to water.

Practice Problems

Acid–Base Definitions (Secs. 14.1 and 14.2)

14.1 What are the Arrhenius definitions for an acid and a base?

14.2 What are the Brønsted–Lowry definitions for an acid and a base, and what advantages does their use offer over the Arrhenius definitions?

14.3 Write equations for the dissociation of the following Arrhenius acids and bases in water.
(a) HBr (hydrobromic acid)
(b) $HClO_2$ (chlorous acid)
(c) LiOH (lithium hydroxide)
(d) $Ba(OH)_2$ (barium hydroxide)

14.4 Write equations for the dissociation of the following Arrhenius acids and bases in water.
(a) $HClO_3$ (chloric acid)
(b) HI (hydroiodic acid)
(c) H_2SO_4 (sulfuric acid)
(d) CsOH (cesium hydroxide)

14.5 Identify the Brønsted–Lowry acid and base in the following reactions.
(a) $HBr + H_2O \rightarrow H_3O^+ + Br^-$
(b) $H_2O + N_3^- \rightarrow HN_3 + OH^-$
(c) $H_2O + H_2S \rightarrow H_3O^+ + HS^-$
(d) $HS^- + H_2O \rightarrow H_3O^+ + S^{2-}$

14.6 Identify the Brønsted–Lowry acid and base in the following reactions.
(a) $NH_3 + H_3PO_4 \rightarrow NH_4^+ + H_2PO_4^-$
(b) $H_3O^+ + OH^- \rightarrow H_2O + H_2O$
(c) $HSO_4^- + H_2O \rightarrow SO_4^{2-} + H_3O^+$
(d) $H_2O + S^{2-} \rightarrow HS^- + OH^-$

14.7 Write equations to illustrate the acid–base reactions that can take place between the following Brønsted–Lowry acids and bases.
(a) acid, HOCl; base, NH_3
(b) acid, $HClO_4$; base, H_2O
(c) acid, H_2O; base, NH_2^-
(d) acid, $HC_2O_4^-$; base, H_2O

14.8 Write equations to illustrate the acid–base reactions that can take place between the following Brønsted–Lowry acids and bases.
(a) acid, H_3O^+; base, NH_3
(b) acid, H_2CO_3; base, H_2O
(c) acid, H_2O; base, F^-
(d) acid, $H_2PO_4^-$; base, H_2O

Conjugate Acids and Bases (Sec. 14.3)

14.9 Give the formula of the conjugate base of each of the following Brønsted–Lowry acids.
(a) HNO_2 (b) H_3PO_4 (c) HPO_4^{2-} (d) NH_4^+

14.10 Give the formula of the conjugate base of each of the following Brønsted–Lowry acids.
(a) H_2S (b) $HClO_2$ (c) $H_2PO_4^-$ (d) PH_4^+

14.11 Give the formula of the conjugate acid of each of the following Brønsted–Lowry bases.
(a) ClO_2^- (b) HSO_4^- (c) SO_3^{2-} (d) PH_3

14.12 Give the formula of the conjugate acid of each of the following Brønsted–Lowry bases.
(a) ClO^- (b) HS^- (c) NH_2^- (d) NH_3

14.13 Classify each of the following species as a Brønsted–Lowry acid or base or both.
(a) NO_2^- (b) CN^- (c) $H_2PO_4^-$ (d) H_2CO_3

14.14 Classify each of the following species as a Brønsted–Lowry acid or base or both.
(a) HS^- (b) CO_3^{2-} (c) H_2SO_3 (d) $HClO$

14.15 In each of the following reactions, decide whether the first species listed is a Brønsted–Lowry acid or base.
(a) $H_2O + S^{2-} \rightarrow HS^- + OH^-$
(b) $H_2O + HNO_2 \rightarrow H_3O^+ + NO_2^-$
(c) $HCl + CN^- \rightarrow Cl^- + HCN$
(d) $HCO_3^- + OH^- \rightarrow CO_3^{2-} + H_2O$

14.16 In each of the following reactions, decide whether the first species listed is a Brønsted–Lowry acid or base.
(a) $H_2O + H_2S \rightarrow H_3O^+ + HS^-$
(b) $PO_4^{3-} + H_2O \rightarrow HPO_4^{2-} + OH^-$
(c) $HNO_2 + F^- \rightarrow NO_2^- + HF$
(d) $H_2CO_3 + H_2PO_4^- \rightarrow HCO_3^- + H_3PO_4$

14.17 Write equations to illustrate the acid–base reactions that can take place between the following Brønsted–Lowry acids and bases.
(a) acid, HClO; base, H_2O
(b) acid, $HClO_4$; base, NH_3
(c) acid, H_3O^+; base, OH^-
(d) acid, H_3O^+; base, NH_2^-

14.18 Write equations to illustrate the acid–base reactions that can take place between the following Brønsted–Lowry acids and bases.
(a) acid, $HClO_2$; base, H_2O
(b) acid, HNO_3; base, NH_2^-
(c) acid, NH_4^+; base, HS^-
(d) acid, H_3O^+; base, PO_4^{3-}

14.19 Identify the two conjugate acid–base pairs involved in each of the following reactions.
(a) $H_2C_2O_4 + ClO^- \rightleftharpoons HC_2O_4^- + HClO$
(b) $HSO_4^- + H_2O \rightleftharpoons H_3O^+ + SO_4^{2-}$
(c) $HPO_4^{2-} + NH_4^+ \rightleftharpoons NH_3 + H_2PO_4^-$
(d) $HCO_3^- + H_2O \rightleftharpoons OH^- + H_2CO_3$

14.20 Identify the two conjugate acid–base pairs involved in each of the following reactions.
(a) $SO_4^{2-} + H_2O \rightleftharpoons HSO_4^- + OH^-$
(b) $CN^- + H_2O \rightleftharpoons HCN + OH^-$
(c) $HSO_4^- + HCO_3^- \rightleftharpoons SO_4^{2-} + H_2CO_3$
(d) $H_3PO_4 + PO_4^{3-} \rightleftharpoons H_2PO_4^- + HPO_4^{2-}$

Polyprotic Acids (Sec. 14.4)

14.21 Classify each of the following acids as monoprotic, diprotic, or triprotic.
(a) $HClO_4$ (perchloric acid)
(b) $H_2C_2O_4$ (oxalic acid)
(c) $HC_2H_3O_2$ (acetic acid)
(d) $HC_4H_7O_2$ (butyric acid)

14.22 Classify each of the following acids as monoprotic, diprotic, or triprotic.
(a) H_2SO_4 (sulfuric acid)
(b) H_3PO_4 (phosphoric acid)
(c) $H_3C_6H_5O_7$ (citric acid)
(d) $HC_7H_5O_3$ (salicylic acid)

14.23 Write equations for the stepwise proton-transfer process that occurs in aqueous solution for each of the following acids.
(a) $H_2C_2O_4$ (oxalic acid)
(b) $H_3C_6H_5O_7$ (citric acid)

14.24 Write equations for the stepwise proton-transfer process that occurs in aqueous solution for each of the following acids.
(a) H_2CO_3 (carbonic acid)
(b) $H_2C_4H_4O_6$ (tartaric acid)

14.25 The formula for lactic acid is preferably written as $HC_3H_5O_3$ rather than as $C_3H_6O_3$. Explain why this is so.

14.26 The formula for malonic acid is preferably written as $H_2C_3H_2O_4$ rather than as $C_3H_4O_4$. Explain why this is so.

14.27 Pyruvic acid, which is produced in metabolic reactions within the human body, has the following structure.

$$\begin{array}{c} H\ \ O\ \ O \\ |\ \ \ ||\ \ \ || \\ H-C-C-C-O-H \\ | \\ H \end{array}$$

Would you predict this acid to be mono-, di, tri, or tetraprotic? Give your reasoning for your answer.

14.28 Succinic acid, a biologically important substance, has the following structure.

$$\begin{array}{c} O\ \ H\ \ H\ \ O \\ ||\ \ |\ \ |\ \ || \\ H-O-C-C-C-C-O-H \\ |\ \ | \\ H\ \ H \end{array}$$

How many acidic hydrogen atoms are present in the structure? Give your reasoning for your answer.

Strengths of Acids and Bases (Sec. 14.5)

14.29 What is the distinction between a strong acid and a weak acid?

14.30 What is the relationship between the extent of proton transfer for an acid in aqueous solution and the strength of the acid?

14.31 What is the distinction between the terms *weak acid* and *dilute acid*?

14.32 What is the distinction between the terms *strong acid* and *concentrated acid*?

14.33 Classify each of the following as a weak acid or a strong acid.
(a) HI   (b) H_3BO_3 (c) H_2SO_4 (d) HCN

14.34 Classify each of the following as a weak acid or a strong acid.
(a) HCl   (b) HNO_3 (c) H_2CO_3 (d) H_3PO_4

14.35 List the chemical formulas of six common strong acids.

14.36 List the chemical formulas of eight common strong bases.

14.37 With the help of Table 14.3 when necessary, indicate which acid in each of the following pairs of acids is the stronger.
(a) HNO_3 and HNO_2 (b) $HClO_4$ and $HC_2H_3O_2$
(c) H_3PO_4 and HCN   (d) H_2CO_3 and HF

14.38 With the help of Table 14.3 when necessary, indicate which acid in each of the following pairs of acids is the stronger.
(a) H_2SO_4 and H_2SO_3 (b) HCl and HF
(c) HCN and HNO_2 (d) $HC_2H_3O_2$ and H_3PO_4

Salts (Sec. 14.6)

14.39 Identify each of the following substances as an acid, a base, or a salt.
(a) NH_4Cl (b) HCl   (c) KCl   (d) $NaNO_3$

14.40 Identify each of the following substances as an acid, a base, or a salt.
(a) NaOH   (b) Na_2SO_4 (c) K_3PO_4 (d) NH_4CN

14.41 Give the formula and name of the positive and negative ions present in each of the following salts.
(a) $CaSO_4$ (b) Li_2CO_3 (c) NaBr   (d) Al_2S_3

14.42 Give the formula and name of the positive and negative ions present in each of the following salts.
(a) NH_4Cl (b) $Be(C_2H_3O_2)_2$
(c) AgI   (d) KNO_3

14.43 Indicate whether each of the salts in problem 14.41 is soluble or insoluble in water.

14.44 Indicate whether each of the salts in problem 14.42 is soluble or insoluble in water.

14.45 Write a balanced equation for the dissociation in water of each of the following soluble salts into ions.
(a) $NaNO_3$ (b) $CuSO_4$
(c) $BaCl_2$ (d) K_3PO_4

14.46 Write a balanced equation for the dissociation in water of each of the following soluble salts into ions.
(a) $Ba(NO_3)_2$
(b) $Al_2(SO_4)_3$
(c) NH_4CN
(d) KF

Ionic and Net Ionic Equations (Sec. 14.6)

14.47 What are the differences between molecular and net ionic equations?

14.48 What are the differences between ionic and net ionic equations?

14.49 Classify the following equations for reactions occurring in aqueous solution as molecular, ionic, or net ionic.
(a) $MgCO_3 + 2\ HBr \rightarrow MgBr_2 + H_2O + CO_2$
(b) $2\ OH^- + H_2CO_3 \rightarrow CO_3^{2-} + 2\ H_2O$
(c) $K^+ + OH^- + H^+ + I^- \rightarrow K^+ + I^- + H_2O$
(d) $Ag^+ + Cl^- \rightarrow AgCl$

14.50 Classify the following equations for reactions occurring in aqueous solution as molecular, ionic, or net ionic.
(a) $CaCO_3 + 2\ H^+ + 2\ NO_3^- \rightarrow Ca^{2+} + 2\ NO_3^- + H_2O + CO_2$
(b) $Ni + Cu^{2+} \rightarrow Cu + Ni^{2+}$
(c) $NaCl + AgNO_3 \rightarrow NaNO_3 + AgCl$
(d) $Ca^{2+} + 2\ OH^- \rightarrow Ca(OH)_2$

14.51 Write a balanced net ionic equation for each of the following reactions, each of which occurs in aqueous solution.
(a) $FeS + 2\ HCl \rightarrow FeCl_2 + H_2S$
(b) $HC_2H_3O_2 + NaOH \rightarrow NaC_2H_3O_2 + H_2O$
(c) $Zn + H_2SO_4 \rightarrow ZnSO_4 + H_2$
(d) $HCl + KOH \rightarrow KCl + H_2O$

14.52 Write a balanced net ionic equation for each of the following reactions, each of which occurs in aqueous solution.
(a) $Pb(NO_3)_2 + (NH_4)_2S \rightarrow 2\ NH_4NO_3 + PbS$
(b) $2\ Al + 6\ HBr \rightarrow 2\ AlBr_3 + 3\ H_2$
(c) $2\ K + 2\ H_2O \rightarrow 2\ KOH + H_2$
(d) $Cl_2 + 2\ NaBr \rightarrow 2\ NaCl + Br_2$

14.53 Write a balanced net ionic equation for each of the following reactions, each of which occurs in aqueous solution.
(a) $Ba(NO_3)_2 + K_2SO_4 \rightarrow 2\ KNO_3 + BaSO_4$
(b) $Ba(OH)_2 + H_2SO_4 \rightarrow BaSO_4 + 2\ H_2O$
(c) $HNO_3 + NaOH \rightarrow NaNO_3 + H_2O$
(d) $AgNO_3 + NaBr \rightarrow AgBr + NaNO_3$

14.54 Write a balanced net ionic equation for each of the following reactions, each of which occurs in aqueous solution.
(a) $Na_2CO_3 + Ca(OH)_2 \rightarrow CaCO_3 + 2\ NaOH$
(b) $2\ NaNO_3 + H_2SO_4 \rightarrow 2\ HNO_3 + Na_2SO_4$
(c) $HBr + NH_3 \rightarrow NH_4Br$
(d) $3\ NaClO \rightarrow NaClO_3 + 2\ NaCl$

Reactions of Acids and Bases (Secs. 14.8 and 14.9)

14.55 On the basis of the activity series for metals, predict whether each of the following metals will react with the indicated substance to form H_2. If the answer is yes, then write a net ionic equation for the reaction. (Note that Cu, Sn, Zn, and Fe all form +2 ions in solution.)
(a) Cu and H_2SO_4
(b) Sn and HNO_3
(c) Zn and HCl
(d) Fe and hot H_2O

14.56 On the basis of the activity series for metals, predict whether each of the following metals will react with the indicated substance to form H_2. If the answer is yes, then write a net ionic equation for the reaction. (Note that Na and Ag form +1 ions and Al and Au form +3 ions in solution.)
(a) Na and H_2O
(b) Au and HCl
(c) Al and HNO_3
(d) Ag and H_2SO_4

14.57 Write balanced net ionic equations showing the total neutralization between the following acids and bases.
(a) HBr and NaOH
(b) $HC_2H_3O_2$ and KOH
(c) H_2S and $Ba(OH)_2$
(d) H_3PO_4 and LiOH

14.58 Write balanced net ionic equations showing the total neutralization between the following acids and bases.
(a) HCl and $Sr(OH)_2$
(b) HNO_3 and LiOH
(c) H_2SO_4 and $Mg(OH)_2$
(d) $HClO_4$ and KOH

14.59 Give the formulas of the acid and the base needed to prepare the following salts by neutralization reactions.
(a) NaF
(b) Li_3PO_4
(c) $MgCl_2$
(d) $Al_2(SO_4)_3$

14.60 Give the formulas of the acid and the base needed to prepare the following salts by neutralization reactions.
(a) $BaCO_3$
(b) KNO_3
(c) NaCN
(d) $Ca_3(PO_4)_2$

14.61 Write a molecular equation for the action of HCl on each of the following. (Zn metal forms a +2 ion in solution.)
(a) Zn
(b) NaOH
(c) Na_2CO_3
(d) $NaHCO_3$

14.62 Write a molecular equation for the action of HNO_3 on each of the following. (Ni metal forms a +2 ion in solution.)
(a) Ni
(b) KOH
(c) Li_2CO_3
(d) $LiHCO_3$

Reactions of Salts (Sec. 14.10)

14.63 On the basis of the activity series for metals, predict whether zinc metal will dissolve in the following aqueous

salt solutions. If the answer is yes, then write a net ionic equation for the reaction. (Zn forms a +2 ion in solution.)
(a) $Pb(NO_3)_2$ (b) $NaCl$
(c) $AgNO_3$ (d) $FeCl_3$

14.64 On the basis of the activity series for metals, predict whether nickel metal will dissolve in the following aqueous salt solutions. If the answer is yes, then write a net ionic equation for the reaction. (Ni forms a +2 ion in solution.)
(a) $Zn(NO_3)_2$ (b) $MgCl_2$
(c) $AgC_2H_3O_2$ (d) $SnSO_4$

14.65 Indicate the driving force (condition) that causes each of the following acid–salt reactions to occur.
(a) $Ba(NO_3)_2 + H_2SO_4 \rightarrow BaSO_4 + 2\,HNO_3$
(b) $3\,CaCl_2 + 2\,H_3PO_4 \rightarrow Ca_3(PO_4)_2 + 6\,HCl$
(c) $AgC_2H_3O_2 + HCl \rightarrow AgCl + HC_2H_3O_2$
(d) $K_2CO_3 + 2\,HNO_3 \rightarrow 2\,KNO_3 + CO_2 + H_2O$

14.66 Indicate the driving force (condition) that causes each of the following acid–salt reactions to occur.
(a) $NaCN + HCl \rightarrow NaCl + HCN$
(b) $3\,MgSO_4 + 2\,H_3PO_4 \rightarrow Mg_3(PO_4)_2 + 3\,H_2SO_4$
(c) $Li_2CO_3 + 2\,HBr \rightarrow 2\,LiBr + CO_2 + H_2O$
(d) $Na_3PO_4 + 3\,HCl \rightarrow 3\,NaCl + H_3PO_4$

14.67 Write molecular, ionic, and net ionic equations for the reaction, if any, that occurs when 0.1 M solutions of the following substances are mixed.
(a) $Co(NO_3)_2$ and $(NH_4)_2S$
(b) HBr and $NaC_2H_3O_2$
(c) $NaCl$ and H_2SO_4
(d) $BeSO_4$ and H_3PO_4

14.68 Write molecular, ionic, and net ionic equations for the reaction, if any, that occurs when 0.1 M solutions of the following substances are mixed.
(a) H_2CO_3 and $Ca(OH)_2$ (b) H_2SO_4 and Na_3PO_4
(c) K_2S and $AgC_2H_3O_2$ (d) $Cu(NO_3)_2$ and Li_2SO_4

Hydronium Ion and Hydroxide Ion Concentrations (Sec. 14.11).

14.69 What is the concentration of hydronium and hydroxide ion in pure water at 25 °C?

14.70 What is the numerical value of the ion product for water?

14.71 Calculate the H_3O^+ ion concentration of a solution if the OH^- ion concentration is
(a) 4.0×10^{-3} M (b) 6.5×10^{-7} M
(c) 8.0×10^{-10} M (d) 2.0×10^{-6} M

14.72 Calculate the H_3O^+ ion concentration of a solution if the OH^- ion concentration is
(a) 2.0×10^{-2} M (b) 4.0×10^{-7} M
(c) 8.0×10^{-8} M (d) 1.0×10^{-11} M

14.73 Calculate the hydroxide ion concentration of a solution if the hydronium ion concentration is
(a) 2.3×10^{-3} M (b) 4.6×10^{-5} M
(c) 7.5×10^{-7} M (d) 9.9×10^{-9} M

14.74 Calculate the hydroxide ion concentration of a solution if the hydronium ion concentration is
(a) 1.7×10^{-4} M (b) 7.7×10^{-6} M
(c) 3.0×10^{-8} M (d) 1.5×10^{-11} M

14.75 How is an acidic aqueous solution defined in terms of H_3O^+ ion and OH^- ion concentrations?

14.76 How is a basic aqueous solution defined in terms of H_3O^+ ion and OH^- ion concentrations?

14.77 Indicate whether solutions with the following characteristics are acidic, basic, or neutral.
(a) $[H_3O^+] = 4.7 \times 10^{-8}$ (b) $[H_3O^+] = 4.3 \times 10^{-5}$
(c) $[OH^-] = 3.2 \times 10^{-4}$ (d) $[OH^-] = 1.0 \times 10^{-7}$

14.78 Indicate whether solutions with the following characteristics are acidic, basic, or neutral.
(a) $[H_3O^+] = 7.5 \times 10^{-8}$ (b) $[H_3O^+] = 1.0 \times 10^{-7}$
(c) $[OH^-] = 2.0 \times 10^{-3}$ (d) $[OH^-] = 3.5 \times 10^{-9}$

The pH Scale (Sec. 14.12)

14.79 Calculate the pH of solutions with the following hydronium ion concentrations.
(a) 1×10^{-5} M (b) 1×10^{-8} M
(c) 0.000001 M (d) 0.00000000001 M

14.80 Calculate the pH of solutions with the following hydronium ion concentrations.
(a) 1×10^{-2} M (b) 1×10^{-6} M
(c) 0.00001 M (d) 0.000000001 M

14.81 Calculate the pH of solutions with the following hydroxide ion concentrations.
(a) 1×10^{-5} M (b) 1×10^{-10} M
(c) 0.001 M (d) 0.0000001 M

14.82 Calculate the pH of solutions with the following hydroxide ion concentrations.
(a) 1×10^{-12} M (b) 1×10^{-2} M
(c) 0.0001 M (d) 0.000000001 M

14.83 How is an acidic aqueous solution defined in terms of pH?

14.84 How is a basic aqueous solution defined in terms of pH?

14.85 Indicate whether the following solutions are very acidic, moderately acidic, slightly acidic, neutral, slightly basic, moderately basic, or very basic.
(a) saliva, pH = 6.9
(b) seawater, pH = 8.5
(c) lime juice, pH = 1.8
(d) desert soil, pH = 9.9

14.86 Indicate whether the following solutions are very acidic, moderately acidic, slightly acidic, neutral, slightly basic, moderately basic, or very basic.
(a) forest soil, pH = 5.3
(b) blood plasma, pH = 7.5
(c) household ammonia, pH = 11.7
(d) gastric juice, pH = 1.0

14.87 Calculate the pH of solutions with the following hydronium ion concentrations.
(a) 5×10^{-3} M (b) 7×10^{-4} M
(c) 7×10^{-7} M (d) 2×10^{-10} M

14.88 Calculate the pH of solutions with the following hydronium ion concentrations.
(a) 4×10^{-4} M (b) 4×10^{-8} M
(c) 3×10^{-7} M (d) 9×10^{-11} M

14.89 Calculate the pH of solutions with the following hydronium ion concentrations.
(a) 2.2×10^{-5} M (b) 2.20×10^{-5} M
(c) 2.200×10^{-5} M (d) 2.200×10^{-6} M

14.90 Calculate the pH of solutions with the following hydronium ion concentrations.
(a) 3.7×10^{-6} M (b) 3.70×10^{-6} M
(c) 3.700×10^{-6} M (d) 3.700×10^{-8} M

14.91 What is the molar hydronium ion concentration in solutions with each of the following pH values?
(a) 5.2 (b) 5.0 (c) 7.1 (d) 3.3

14.92 What is the molar hydronium ion concentration in solutions with each of the following pH values?
(a) 8.2 (b) 8.0 (c) 4.7 (d) 6.2

14.93 What is the molar hydronium ion concentration in solutions with each of the following pH values?
(a) 2.44 (b) 4.44 (c) 7.44 (d) 7.460

14.94 What is the molar hydronium ion concentration in solutions with each of the following pH values?
(a) 5.37 (b) 2.37 (c) 10.37 (d) 10.350

14.95 A teacher once mistakenly told a class that the $[H_3O^+]$ increased as the pH increased. Explain what is wrong with this statement.

14.96 Solution A has a pH of 4 and solution B has a pH of 6. Which solution is more acidic? How many times more acidic is the one solution than the other?

14.97 Solution A has $[OH^-] = 4.3 \times 10^{-4}$. Solution B has $[H_3O^+] = 7.3 \times 10^{-10}$.
(a) Which solution is more basic?
(b) Which solution has the lower pH?

14.98 Solution A has $[H_3O^+] = 2.7 \times 10^{-6}$. Solution B has $[OH^-] = 4.5 \times 10^{-8}$.
(a) Which solution is more acidic?
(b) Which solution has the higher pH?

Hydrolysis of Salts (Sec. 14.13)

14.99 Predict whether each of the following aqueous salt solutions will be acidic, basic, or neutral.
(a) $NaNO_3$ (b) NaCN (c) NH_4Cl (d) LiCl

14.100 Predict whether each of the following aqueous salt solutions will be acidic, basic, or neutral.
(a) $KC_2H_3O_2$ (b) NH_4NO_3 (c) Na_2CO_3 (d) NaF

14.101 Identify the ion (or ions) present in each of the following salts that will undergo hydrolysis in aqueous solution.
(a) NaF (b) NH_4Cl
(c) KCN (d) Na_3BO_3

14.102 Identify the ion (or ions) present in each of the following salts that will undergo hydrolysis in aqueous solution.
(a) $LiC_2H_3O_2$ (b) $(NH_4)_3PO_4$
(c) Na_2CO_3 (d) KNO_3

14.103 It is useful in hydrolysis discussions to classify salts into four categories. Describe each of the four categories.

14.104 It is useful in hydrolysis discussions to classify salts into four categories. Give an example of a salt that belongs to each category.

14.105 Explain why category 1 salts do not hydrolyze.

14.106 Explain why category 4 salts give acidic, basic, or neutral solutions depending upon the individual salt involved.

Buffer Solutions (Secs. 14.14 and 14.15)

14.107 Predict whether each of the following pairs of substances could function as a buffer system in aqueous solution.
(a) HCl and NaCl (b) HCl and HCN
(c) HCN and KCN (d) NaCN and KCN

14.108 Predict whether each of the following pairs of substances could function as a buffer system in aqueous solution.
(a) NaOH and NaCl
(b) $HC_2H_3O_2$ and $KC_2H_3O_2$
(c) NaCl and NaCN
(d) HCN and $HC_2H_3O_2$

14.109 Write an equation for each of the following buffering actions.
(a) the response of $H_2PO_4^-/HPO_4^{2-}$ buffer to the addition of H_3O^+ ions
(b) the response of H_2CO_3/HCO_3^- buffer to the addition of OH^- ions
(c) the response of H_2CO_3/HCO_3^- buffer to the addition of H_3O^+ ions

14.110 Write an equation for each of the following buffering actions.
(a) the response of HCN/CN$^-$ buffer to the addition of OH$^-$ ions
(b) the response of H$_2$PO$_4^-$/HPO$_4^{2-}$ buffer to the addition of OH$^-$ ions
(c) the response of HCN/CN$^-$ buffer to the addition of H$_3$O$^+$ ions

14.111 Characterize each of the following human body fluid pH values as acidosis conditions, alkalosis conditions, or neither.
(a) 7.35   (b) 7.53   (c) 7.41   (d) 7.11

14.112 Characterize each of the following human body fluid pH values as acidosis conditions, alkalosis conditions, or neither.
(a) 7.39   (b) 7.07   (c) 7.52   (d) 7.21

Normality of Acids and Bases (Sec. 14.17)

14.113 How many equivalents of acid or base are present in each of the following samples of acid or base?
(a) 1.00 mole NaOH   (b) 1.50 moles H$_3$PO$_4$
(c) 3.00 moles HC$_2$H$_3$O$_2$ (d) 34.5 g HNO$_3$

14.114 How many equivalents of acid or base are present in each of the following samples of acid or base?
(a) 2.50 moles Ba(OH)$_2$ (b) 1.25 moles H$_2$C$_2$O$_4$
(c) 1.50 moles HCl   (d) 85.0 g H$_2$SO$_4$

14.115 Calculate the normality of each of the following solutions.
(a) 250.0 mL of solution containing 5.00 g of NaOH
(b) 750.0 mL of solution containing 10.0 g of H$_3$PO$_4$
(c) 100.0 mL of solution containing 0.250 mole of H$_2$SO$_4$
(d) 2.00 L of solution containing 0.400 equivalent of H$_2$C$_2$O$_4$

14.116 Calculate the normality of each of the following solutions.
(a) 100.0 mL of solution containing 1.000 g of Ba(OH)$_2$
(b) 250.0 mL of solution containing 5.00 g of HNO$_3$
(c) 57.0 mL of solution containing 0.100 mole of HCl
(d) 3.75 L of solution containing 7.00 equivalents of H$_2$SO$_4$

14.117 Express the following molarities as normalities.
a) 0.0025 M H$_2$SO$_4$ (b) 1.70 M HNO$_3$
(c) 0.645 M NaOH   (d) 6.0 M HC$_2$H$_3$O$_2$

14.118 Express the following molarities as normalities.
(a) 0.0010 M Ba(OH)$_2$ (b) 3.0 M H$_3$PO$_4$
(c) 1.712 M HCl   (d) 1.000 M H$_2$CO$_3$

14.119 Express the following normalities as molarities.
(a) 0.0030 N H$_2$SO$_4$ (b) 2.80 N HNO$_3$
(c) 0.800 N NaOH   (d) 4.0 N HC$_2$H$_3$O$_2$

14.120 Express the following normalities as molarities.
(a) 0.100 N Ba(OH)$_2$ (b) 1.2 N H$_3$PO$_4$
(c) 2.000 N HCl   (d) 0.500 N H$_2$CO$_3$

14.121 How many grams of oxalic acid (H$_2$C$_2$O$_4$) are needed to produce 400.0 mL of 0.0500 N H$_2$C$_2$O$_4$ solution?

14.122 How many grams of lactic acid (HC$_3$H$_5$O$_3$) are needed to produce 1.20 L of 1.000 N HC$_3$H$_5$O$_3$ solution?

Acid–Base Titration Calculations (Sec. 14.17)

14.123 What volume, in milliliters, of a 0.100 N NaOH solution would be needed to neutralize each of the following acid samples?
(a) 10.00 mL of 0.350 N H$_2$SO$_4$
(b) 50.00 mL of 1.500 N H$_3$PO$_4$
(c) 5.00 mL of 0.500 N HNO$_3$
(d) 75.00 mL of 0.00030 N HCl

14.124 What volume, in milliliters, of a 2.00 N KOH solution would be needed to neutralize each of the following acid samples?
(a) 50.00 mL of 6.50 N HCl
(b) 5.00 mL of 0.0500 N H$_2$SO$_4$
(c) 250.0 mL of 1.00 N H$_3$PO$_4$
(d) 75.00 mL of 2.50 N HNO$_3$

14.125 What volume, in milliliters, of 2.50 M aqueous ammonia [NH$_3$(aq)] would be needed to neutralize each of the following acid samples?
(a) 10.00 mL of 2.00 M HNO$_3$
(b) 20.00 mL of 6.00 M H$_2$SO$_4$
(c) 30.00 mL of 7.50 M H$_3$PO$_4$
(d) 100.0 mL of 0.100 M HCl

14.126 What volume, in milliliters, of a 1.30 M LiOH solution would be needed to neutralize each of the following acid samples?
(a) 7.50 mL of 0.100 M H$_3$PO$_4$
(b) 75.00 mL of 0.100 M H$_2$SO$_4$
(c) 150.0 mL of 7.50 M HNO$_3$
(d) 100.0 mL of 12.0 M HCl

14.127 34.5 mL of 0.102 N NaOH is required to neutralize each of the following acid solution samples. What are the normality and molarity of each of the acid samples?
(a) 25.0 mL of H$_2$SO$_4$
(b) 20.0 mL of HClO
(c) 20.0 mL of H$_3$PO$_4$
(d) 10.0 mL of HNO$_3$

14.128 21.4 mL of 0.198 N NaOH is required to neutralize each of the following acid solution samples. What are the normality and molarity of each of the acid samples?
(a) 30.0 mL of H$_2$CO$_3$
(b) 25.0 mL of H$_2$C$_2$O$_4$

(c) 25.0 mL of $HC_2H_3O_2$
(d) 20.0 mL of HCl

Acid and Base Stock Solutions (Sec. 14.18)

14.129 Do the designations "dilute acid" and "concentrated acid," when used in describing laboratory stock solutions, have a constant meaning? Explain.

14.130 Dilute stock solutions of the acids hydrochloric, nitric, and sulfuric have the same normality but not the same molarity. Explain why.

14.131 What is the molarity of each of the following stock solutions?
(a) dilute H_2SO_4 (b) dilute HCl
(c) concentrated HNO_3 (d) dilute NaOH

14.132 What is the molarity of each of the following stock solutions?
(a) dilute HNO_3 (b) concentrated NH_3
(c) concentrated H_2SO_4 (d) dilute $HC_2H_3O_2$

Additional Problems

14.133 Identify each of the following species as the conjugate base of a strong acid or the conjugate base of a weak acid.
(a) Br^- (b) CN^- (c) $H_2PO_4^-$ (d) NO_3^-

14.134 Identify each of the following species as the conjugate base of a strong acid or the conjugate base of a weak acid.
(a) $C_2H_3O_2^-$ (b) Cl^- (c) F^- (d) HPO_4^{2-}

14.135 For each of the following amphoteric substances write the two equations needed to describe its behavior in aqueous solution.
(a) HPO_4^{2-} (b) OH^-

14.136 For each of the following amphoteric substances write the two equations needed to describe its behavior in aqueous solution.
(a) NH_3 (b) $H_2PO_4^-$

14.137 A solution has a pH of 5.3. What must be the pH of another solution so that the $[H_3O^+]$ is six times as large?

14.138 A solution has a pH of 5.7. What must be the pH of another solution so that the $[H_3O^+]$ is one-sixth as large?

14.139 A solution has a pH of 2.2. Another solution has a pH of 4.5. How many times greater is the $[H_3O^+]$ in the first solution than the second one?

14.140 A solution has a pH of 3.4. Another solution has a pH of 6.7. How many times greater is the $[H_3O^+]$ in the first solution than the second one?

14.141 What would be the pH of a solution that contains 0.1 mole each of the solutes NaCl, HNO_3, HCl, and NaOH in enough water to give 3.00 L of solution?

14.142 What would be the pH of a solution that contains 0.1 mole each of the solutes NaBr, HBr, KOH, and NaOH in enough water to give 5.00 L of solution?

14.143 Arrange the following 0.1 M aqueous solutions in order of decreasing pH: NH_4Br, $Ba(OH)_2$, $HClO_4$, K_2SO_4, and LiCN.

14.144 Arrange the following 0.1 M aqueous solutions in order of increasing pH: HBr, $HC_2H_3O_2$, $NH_4C_2H_3O_2$, KOH, and $Ca(NO_3)_2$.

14.145 Identify the buffer systems (conjugate acid–base pairs) present in a solution containing equal molar amounts of $HC_2H_3O_2$, $NaHCO_3$, $KC_2H_3O_2$, and Na_2CO_3.

14.146 Identify the buffer systems (conjugate acid–base pairs) present in a solution containing equal molar amounts of HCN, Na_2HPO_4, KH_2PO_4, and KCN.

14.147 A 20.00-mL sample of a solution of citric acid ($H_3C_6H_5O_7$) was titrated with a 0.250 N NaOH solution. A total of 27.86 mL of NaOH was required.
(a) What was the normality of the $H_3C_6H_5O_7$ solution?
(b) How many equivalents of $H_3C_6H_5O_7$ were present in the 20.0-mL sample?
(c) How many moles of $H_3C_6H_5O_7$ were present in the 20.0-mL sample?
(d) What was the molarity of the $H_3C_6H_5O_7$ solution?

14.148 A 20.00-mL sample of a solution of lactic acid ($HC_3H_5O_3$) was titrated with a 0.100 N $Ba(OH)_2$ solution. A total of 17.03 mL of $Ba(OH)_2$ was required.
(a) What was the normality of the $HC_3H_5O_3$ solution?
(b) How many equivalents of $HC_3H_5O_3$ were present in the 20.0-mL sample?
(c) How many moles of $HC_3H_5O_3$ were present in the 20.0-mL sample?
(d) What was the molarity of the $HC_3H_5O_3$ solution?

Cumulative Problems

14.149 Name each of the following species.
(a) conjugate base of hydroiodic acid
(b) conjugate base of the dihydrogen phosphate ion
(c) conjugate acid of the oxide ion
(d) conjugate acid of water

14.150 Name each of the following species.
(a) conjugate base of perchloric acid
(b) conjugate base of the bicarbonate ion
(c) conjugate acid of the hydroxide ion
(d) conjugate acid of ammonia

14.151 Write electron-dot structures for hydrocyanic acid and its conjugate base.

14.152 Write electron-dot structures for hypochlorous acid and its conjugate base.

14.153 What is the pH of a solution obtained by mixing 245 mL of 0.30 N HNO_3 with 135 mL of 0.50 M HCl. Assume the volumes are additive.

14.154 What is the pH of a solution obtained by mixing 475 mL of 0.50 M HNO_3 with 225 mL of 0.30 N HCl. Assume the volumes are additive.

14.155 A solution of a monoprotic strong acid with a pH of 4.5 is diluted to three times its original volume. What is the normality of the resulting solution?

14.156 A solution of a monoprotic strong acid with a pH of 5.6 is diluted to four times its original volume. What is the normality of the resulting solution?

14.157 The pH of a hydrochloric acid solution with a density of 1.10 g/mL is 3.4. Calculate the concentration of the acid in
(a) molarity units   (b) mass percent units

14.158 The pH of a nitric acid solution with a density of 1.20 g/mL is 2.7. Calculate the concentration of the acid in
(a) molarity units   (b) mass percent units

14.159 How many ions are present in a 236-mL sample of a HNO_3 solution with a pH of 2.37 to which 0.100 mole of Na_2SO_4 has been added and dissolved?

14.160 How many ions are present in a 435-mL sample of a HCl solution with a pH of 1.54 to which 0.050 mole of $Mg(NO_3)_2$ has been added and dissolved?

14.161 An impure 1.00-g sample of the monoprotic potassium hydrogen phthalate ($KHC_8H_4O_4$) is dissolved in water and titrated with 32.3 mL of 0.1000 N NaOH solution. Calculate the mass percent $KHC_8H_4O_4$ in the impure sample.

14.162 An impure 1.00-g sample of the diprotic oxalic acid ($H_2C_2O_4$) is dissolved in water and titrated with 17.6 mL of 0.200 N NaOH solution. Calculate the mass percent $H_2C_2O_4$ in the impure sample.

Grid Problems

14.163 Select from the grid *all* correct responses for each situation.

1. H_2S	2. CO_3^{2-}	3. OH^-
4. HCO_3^-	5. HS^-	6. HPO_4^{2-}
7. $H_2PO_4^-$	8. H_3PO_4	9. S^{2-}

(a) species whose conjugate acid is also present in the grid
(b) species whose conjugate base is also present in the grid
(c) species that cannot function as Brønsted–Lowry acids
(d) species that would be expected to exhibit amphoteric behavior
(e) pairs of species that constitute a conjugate acid–base pair
(f) pairs of species whose reaction would produce S^{2-} as one of the products

14.164 Select from the grid *all* correct responses for each situation.

1. pH = 5.0	2. [OH⁻] = 10^{-7}	3. pH = 3.0
4. $[H_3O^+] = 10^{-10}$	5. $[H_3O^+] = 10^{-5}$	6. $[OH^-] = 10^{-14}$
7. $[H_3O^+] = [OH^-]$	8. $[H_3O^+] = 2[OH^-]$	9. $[H_3O^+] = 10[OH^-]$

(a) solutions that are neither acidic nor basic
(b) solutions with a pH of less than 4
(c) solutions that are acidic
(d) pairs of solutions with the same pH
(e) pairs of solutions with the same hydroxide ion concentration
(f) pairs of solutions whose acidity differs by a factor of 100

14.165 Select from the grid *all* correct responses for each situation.

1. 1.5 M H_3PO_4	2. 4.0 N HNO_3	3. 1.0 M KOH
4. 2.0 N NaOH	5. 4.0 M H_2SO_4	6. 2.0 M $Ca(OH)_2$
7. 3.0 N HCl	8. 1.0 N $Ba(OH)_2$	9. 1.5 M H_2CO_3

(a) solutions for which normality and molarity are the same
(b) solutions for which the normality is double the molarity
(c) pairs of solutions with the same molarity
(d) pairs of solutions with the same normality
(e) pairs of solutions where equal volumes will completely neutralize each other
(f) pairs of solutions where equal volumes contain the same number of equivalents of solute

14.166 Select from the grid *all* correct responses for each situation.

1. 0.1 M $HC_2H_3O_2$	2. 0.1 M H_2S	3. 0.1 M $Mg(C_2H_3O_2)_2$
4. 0.1 M $Ca(NO_3)_2$	5. 0.1 M HCl	6. 0.1 M K_2SO_4
7. 0.1 M $AgNO_3$	8. 0.1 M NH_4Cl	9. 0.1 M NaOH

(a) solutions that would be expected to be neutral
(b) solutions that would have a pH greater than 7.0
(c) compounds that would be written in ionic form in a net ionic equation
(d) pairs of solutions that would react to give an insoluble salt
(e) pairs of solutions where equal volumes would completely neutralize each other
(f) pairs of solutions that, when combined, could func- function as a buffer

CHAPTER FIFTEEN

Oxidation and Reduction

15.1 Oxidation–Reduction Terminology

Oxidation–reduction reactions are a very important class of chemical reactions. They occur all around us and even within us. The bulk of the energy needed for the functioning of all living organisms, including humans, is obtained from food via oxidation–reduction processes. Such diverse phenomena as the electricity obtained from a battery to start a car, the use of natural gas to heat a home, iron rusting, illumination from a flashlight, and the functioning of antiseptic agents to kill or prevent the growth of bacteria all involve oxidation–reduction reactions. In short, knowledge of this type of reaction is fundamental to understanding many biological and technological processes.

The terms oxidation and reduction, like the terms acid and base (Sec. 14.1), have several definitions. Historically, the word *oxidation* was first used to describe the reaction of a substance with oxygen. According to this original definition, each of the following reactions involves oxidation.

Section 15.1 · Oxidation–Reduction Terminology

$$4\ Fe + 3\ O_2 \longrightarrow 2\ Fe_2O_3$$
$$S + O_2 \longrightarrow SO_2$$
$$CH_4 + 2\ O_2 \longrightarrow CO_2 + 2\ H_2O$$

The substance on the far left in each of these equations is said to have been *oxidized.*

Originally, the term *reduction* referred to processes where oxygen was removed from a compound. A particularly common type of reduction reaction, according to this original definition, is the removal of oxygen from a metal oxide to produce the free metal.

$$CuO + H_2 \longrightarrow Cu + H_2O$$
$$2\ Fe_2O_3 + 3\ C \longrightarrow 4\ Fe + 3\ CO_2$$

The term reduction comes from the reduction in mass of the metal-containing species; the metal has a mass less than that of the metal oxide.

Today the words oxidation and reduction are used in a much broader sense. Current definitions include the previous examples but also much more. It is now recognized that the changes brought about in a substance from reaction with oxygen can also be caused by reaction with numerous non-oxygen-containing substances. For example, consider the following reactions.

$$2\ Mg + O_2 \longrightarrow 2\ MgO$$
$$Mg + S \longrightarrow MgS$$
$$Mg + F_2 \longrightarrow MgF_2$$
$$3\ Mg + N_2 \longrightarrow Mg_3N_2$$

In each of these reactions magnesium metal is converted to a magnesium compound that contains Mg^{2+} ions. The process is the same—the changing of magnesium atoms, through the loss of two electrons, to magnesium ions; the only difference is the identity of the substance that causes magnesium to undergo the change. All of these reactions are considered to involve oxidation by the modern definition. **Oxidation** is the process whereby a substance in a chemical reaction loses one or more electrons. The modern definition for reduction involves the use of similar terminology. **Reduction** is the process whereby a substance in a chemical reaction gains one or more electrons.

Oxidation and reduction are complementary processes rather than isolated phenomena. They *always* occur together; you cannot have one without the other. If electrons are lost by one species, they cannot just disappear; they must be gained by another species. Electron transfer, then, is the basis for oxidation and reduction. The collective term **oxidation–reduction reaction** is used to describe any reaction involving the transfer of electrons between reactants. This designation is often shortened to simply **redox reaction**.

There are two different ways of looking at the reactants in a redox reaction. First, the reactants can be viewed as being acted upon. From this viewpoint one reactant is *oxidized* (the one that loses electrons) and one is *reduced* (the one that gains electrons). Second, the reactants can be looked at as bringing about the reaction. In this approach the terms *oxidizing agent* and *reducing agent* are used. An **oxidizing agent**

TABLE 15.1 Oxidation–Reduction Terminology in Terms of Loss and Gain of Electrons

Terms Associated with the Loss of Electrons	*Terms Associated with the Gain of Electrons*
process of oxidation	process of reduction
substance oxidized	substance reduced
reducing agent	oxidizing agent

causes oxidation by accepting electrons from the other reactant. Such acceptance, the gain of electrons, means that the oxidizing agent itself is reduced. Similarly, the **reducing agent** causes reduction by providing electrons for the other reactant to accept. As a result of providing electrons, the reducing agent itself becomes oxidized. Note, then, that the reducing agent and substance oxidized are one and the same, as are the oxidizing agent and substance reduced.

The terms oxidizing agent and reducing agent sometimes cause confusion because the oxidizing agent is not oxidized (it is reduced) and the reducing agent is not reduced (it is oxidized). By simple analogy, a travel agent is not the one who takes a trip—he or she is the one who causes the trip to be taken.

Table 15.1 summarizes the terms presented in this section.

15.2 Oxidation Numbers

Oxidation numbers are used to help determine whether oxidation or reduction has occurred in a reaction, and if such is the case, the identity of the oxidizing and reducing agents. Formally defined, an **oxidation number** is the charge that an atom appears to have when the electrons in each bond it is participating in are assigned to the more electronegative of the two atoms involved in the bond.*

Consider an HCl molecule, a molecule in which there is one bond involving two shared electrons.

$$H \overset{..}{\underset{..}{\overset{\times}{C}l}}:$$

According to the definition for oxidation number, the electrons in this bond are assigned to the chlorine atom (the more electronegative atom; Sec. 7.10). This results in the chlorine atom having one more electron than a neutral Cl atom; hence, the oxidation number of chlorine is -1 (one extra electron). At the same time, the H atom in the HCl molecule has one less electron than a neutral H atom; its electron was given to the chlorine. This electron deficiency of one results in an oxidation number of $+1$ for hydrogen.

As a second example, consider the molecule CF_4.

---

*In some textbooks the term *oxidation state* is used in place of oxidation number. In other textbooks the two terms are used interchangeably. We will use oxidation number.

Fluorine is more electronegative than carbon. Hence, the two shared electrons in each of the four carbon–fluorine bonds are assigned to the fluorine atom. Each F atom thus gains an extra electron, resulting in F having a -1 oxidation number. The carbon atom loses a total of four electrons, one to each F atom, as a result of the electron "assignments." Hence, its oxidation number is $+4$, indicating the loss of the four electrons.

As a third example, consider the N_2 molecule, a molecule where like atoms are involved in a triple bond.

$$:N:::N:$$

Since the identical atoms are of equal electronegativity, the shared electrons are "divided" equally between the two atoms; each N receives three of the bonding electrons to count as its own. This results in each N atom having five valence electrons (three from the triple bond and two nonbonding electrons), the same number of valence electrons as in a neutral N atom. Hence, the oxidation number of N in N_2 is zero.

Before going any further in our discussion of oxidation numbers, it should be noted that *calculated* oxidation numbers are *not* actual charges on atoms. This is why the phrase "appears to have" is found in the definition of oxidation number given at the start of this section. In assigning oxidation numbers, we assume when we give the bonding electrons to the more electronegative element that each bond is ionic (complete transfer of electrons). We know that this is not always the case. Sometimes it is a good approximation, sometimes it is not. Why, then, do we do this when we know that it does not always correspond to reality? Oxidation numbers, as we shall see shortly, serve as a very convenient device for "keeping track" of electron transfer in redox reactions. Even though they do not always correspond to physical reality, they are very, very useful entities.

In principle, the procedures used to determine oxidation numbers for the atoms in the molecules HCl, CF_4, and N_2 can be used to determine oxidation numbers in all molecules. However, the procedures become very laborious in many cases, especially when complicated electron-dot structures are involved. In practice, an alternative, much simpler procedure that does not require electron-dot structures is used to obtain oxidation numbers. This alternative procedure is based on a set of operational rules that are consistent with and derivable from the general definition for oxidation numbers. The operational rules are

RULE 1 The oxidation number of an atom in its elemental state is zero.

For example, the oxidation number of Cu is zero, and the oxidation number of Cl in Cl_2 is zero.

RULE 2 The oxidation number of any monoatomic ion is equal to the charge on the ion.

For example, the Na$^+$ ion has an oxidation number of $+1$, and the S^{2-} ion has an oxidation number of -2.

RULE 3 The oxidation numbers of groups IA and IIA elements are always $+1$ and $+2$, respectively.

RULE 4 The oxidation number of fluorine is always -1 and that of the other group VIIA elements (Cl, Br, and I) is usually -1.

The exception for these latter elements is when they are bonded to more electronegative elements. In this case they are assigned positive oxidation numbers.

RULE 5 The usual oxidation number for oxygen is -2.

The exceptions occur when oxygen is bonded to the more electronegative fluorine (O then is assigned a positive oxidation number) or found in compounds containing oxygen–oxygen bonds (peroxides). In peroxides the oxidation number -1 is assigned to oxygen. Peroxides form only between oxygen and hydrogen (H_2O_2), group IA elements (Na_2O_2, etc.), and group IIA elements (BaO_2, etc.).

RULE 6 The usual oxidation number for hydrogen is $+1$.

The exception occurs in hydrides, compounds where hydrogen is bonded to a metal of lower electronegativity. In such compounds hydrogen is assigned an oxidation number of -1. Examples of hydrides are NaH, CaH_2, and LiH.

RULE 7 The algebraic sum of the oxidation numbers of all atoms in a neutral molecule must be zero.

RULE 8 The algebraic sum of the oxidation numbers of all atoms in a polyatomic ion is equal to the charge on the ion.

The use of these rules is illustrated in Example 15.1.

EXAMPLE 15.1

Assign oxidation numbers to each element in the following chemical species.

(a) NO_2 (b) P_2F_4 (c) $K_2Cr_2O_7$ (d) ClO_2^-

Solution

(a) **Oxygen** has an oxidation number of -2 (rule 5). The oxidation number of N can be calculated by rule 7. Letting x equal the oxidation number of N, we have

$$\begin{aligned}
&\text{N:} \quad 1 \text{ atom} \times (x) \quad = \quad x \\
&\text{O:} \quad 2 \text{ atoms} \times (-2) = -4 \\
&\hphantom{\text{O: 2 atoms} \times (-2)} \text{sum} = \hphantom{-}0 \quad \text{(rule 7)}
\end{aligned}$$

Solving for x algebraically, we get

$$x + (-4) = 0$$
$$x = +4$$

Consequently, the oxidation number of **nitrogen** is +4 in the compound NO_2.

(b) **Fluorine** has an oxidation number of -1 (rule 4). Rule 7 will allow us to calculate the oxidation number of P; the sum of the oxidation numbers must be zero. Letting x equal the oxidation number of P, we have

$$\begin{aligned} \text{F:} \quad & 4 \text{ atoms} \times (-1) = -4 \\ \text{P:} \quad & 2 \text{ atoms} \times (x) = 2x \\ & \phantom{2 \text{ atoms} \times (x)}\text{sum} = \overline{0} \quad \text{(rule 7)} \end{aligned}$$

Solving for x algebraically, we get

$$2x + (-4) = 0$$
$$x = +2$$

Thus, the oxidation number of **phosphorus** in P_2F_4 is +2. Note that the oxidation number of P is not +4 (the calculated charge associated with two P atoms). Oxidation number is always specified on a *per atom* basis.

(c) **Potassium** has an oxidation number of +1 (rule 3), and **oxygen** has an oxidation number of -2 (rule 5). Letting x equal the oxidation number of chromium and using rule 7, we get

$$\begin{aligned} \text{K:} \quad & 2 \text{ atoms} \times (+1) = +2 \\ \text{Cr:} \quad & 2 \text{ atoms} \times (x) = 2x \\ \text{O:} \quad & 7 \text{ atoms} \times (-2) = -14 \\ & \phantom{2 \text{ atoms} \times (x)}\text{sum} = 0 \quad \text{(rule 7)} \end{aligned}$$

Solving for x algebraically, we get

$$(+2) + 2x + (-14) = 0$$
$$x = +6$$

Thus, the oxidation number of **chromium** in $K_2Cr_2O_7$ is +6.

(d) According to rule 8, the sum of the oxidation numbers must equal -1, the charge on this polyatomic ion. The oxidation number of **oxygen** is -2 (rule 5). Chlorine will have a positive oxidation number, since it is bonded to a more electronegative element (rule 4). Letting x equal the oxidation number of chlorine, we have

$$\begin{aligned} \text{Cl:} \quad & 1 \text{ atom} \times (x) = x \\ \text{O:} \quad & 2 \text{ atoms} \times (-2) = -4 \\ & \phantom{2 \text{ atoms} \times (x)}\text{sum} = -1 \quad \text{(rule 8)} \end{aligned}$$

Solving for x algebraically, we get

$$x + (-4) = -1$$
$$x = +3$$

Thus, **chlorine** has an oxidation number of +3 in this ion.

PRACTICE EXERCISE 15.1

Assign oxidation numbers to each element in the following chemical species.

(a) SO_2 (b) NH_4^+ (c) $NaNO_3$ (d) N_2F_4

Ans. (a) +4 for S, −2 for O; (b) −3 for N, +1 for H; (c) +1 for Na, +5 for N, −2 for O; (d) +2 for N, −1 for F.

Similar exercises: Problems 15.15 and 15.16

Many elements display a range of oxidation numbers in their various compounds. For example, nitrogen exhibits oxidation numbers ranging from −3 to +5 in various compounds. Selected examples are

NH_3	N_2H_4	N_2O	NO	N_2O_3	NO_2	HNO_3
−3	−2	+1	+2	+3	+4	+5

As shown in this listing of nitrogen-containing compounds, the oxidation number of an atom is written *underneath* the atom in the formula. This convention is used to avoid confusion with the charge on an ion.

Although not common, nonintegral oxidation numbers are possible. For example, the oxidation number of iron in the compound Fe_3O_4 is +2.67. The oxidation number of the oxygens in the compound add up to −8. Therefore, the iron atoms must have an oxidation number sum of +8. Dividing +8 by 3 (the number of iron atoms) gives +2.67.

Oxidizing and reducing agents can be defined in terms of changes in oxidation numbers. The **oxidizing agent** in a redox reaction is the substance that *contains* the atom that shows a decrease in oxidation number. Since the oxidizing agent is the substance reduced in a reaction, reduction involves a decrease in oxidation number; the oxidation number is reduced (decreased) in a reduction. The **reducing agent** in a redox reaction is the substance that *contains* the atom that shows an increase in oxidation number. Since the reducing agent is the substance oxidized in a reaction, oxidation involves an increase in oxidation number.

Table 15.2 summarizes the relationships between oxidation–reduction terms and oxidation number changes. A comparison of Table 15.2 with Table 15.1 shows that the loss of electrons and oxidation number increases are synonymous as are the gain of electrons and oxidation number decreases. The fact that the oxidation number becomes more positive (increases) as electrons are lost is consistent with our understanding of the proton–electron charge relationships in an atom.

TABLE 15.2 Oxidation–Reduction Terminology in Terms of Oxidation Number Change

Terms Associated with an Increase in Oxidation Number	*Terms Associated with a Decrease in Oxidation Number*
process of oxidation	process of reduction
substance oxidized	substance reduced
reducing agent	oxidizing agent

EXAMPLE 15.2

Determine oxidation numbers for each atom in the following reactions, and identify the oxidizing and reducing agents.

(a) $2 SO_2 + O_2 \longrightarrow 2 SO_3$
(b) $2 Fe_2O_3 + 3 C \longrightarrow 4 Fe + 3 CO_2$
(c) $S_2O_8{}^{2-} + 2 I^- \longrightarrow I_2 + 2 SO_4{}^{2-}$

Solution

The oxidation numbers are calculated by the methods illustrated in Example 15.1.

(a) $2 SO_2 + O_2 \longrightarrow 2 SO_3$
    $+4-20+6-2$
    rules 5, 7 rule 1   rules 5, 7

The oxidation number of S has increased from $+4$ to $+6$. Therefore, the substance that contains S, **SO_2**, has been oxidized and is the **reducing agent**.

The oxidation number of the O in O_2 has decreased from 0 to -2. Therefore, the **O_2** has been reduced and is the **oxidizing agent**.

(b) $2 Fe_2O_3 + 3 C \longrightarrow 4 Fe + 3 CO_2$
    $+3\ -200+4-2$
    rules 5, 7 rule 1   rule 1 rules 5, 7

The oxidation number of C has increased from 0 to $+4$. An increase in oxidation number is associated with oxidation. Therefore, the element **C**, since it has been oxidized, is the **reducing agent**.

The oxidation number of Fe has decreased from $+3$ to 0. Since a decrease in oxidation number is associated with reduction, the **Fe_2O_3**, the iron-containing compound, is the **oxidizing agent**.

(c) $S_2O_8{}^{2-} + 2 I^- \longrightarrow I_2 + 2 SO_4{}^{2-}$
    $+7-2-10+6-2$
    rules 5, 8   rule 2   rule 1 rules 5, 7

The oxidation number of I has increased from -1 to 0. Thus, **I^-**, the iodine-containing reactant, has been oxidized and is the **reducing agent**.

The oxidation number of S has decreased from $+7$ to $+6$. Thus, **$S_2O_8{}^{2-}$**, the sulfur-containing reactant, has been reduced and is the **oxidizing agent**.

PRACTICE EXERCISE 15.2

Determine oxidation numbers for each atom in the following equations, and identify the oxidizing and reducing agents.

(a) $2 NO + O_2 \longrightarrow 2 NO_2$
(b) $Zn + 2 HCl \longrightarrow ZnCl_2 + H_2$

(c) $Pb + Cu^{2+} + SO_4^{2-} \longrightarrow Cu + PbSO_4$

Ans. (a) $2\,NO + O_2 \rightarrow 2\,NO_2$
$+2-2\ \ \ \ 0\ \ \ \ \ +4-2$
reducing agent: NO; oxidizing agent: O_2

(b) $Zn + 2\,HCl \rightarrow ZnCl_2 + H_2$
$0\ \ \ +1-1\ \ +2-1\ \ \ 0$
reducing agent: Zn; oxidizing agent: HCl

(c) $Pb + Cu^{2+} + SO_4^{2-} \rightarrow Cu + PbSO_4$
$0\ \ \ \ +2\ \ \ \ +6-2\ \ \ \ \ 0\ \ +2+6-2$
reducing agent: Pb; oxidizing agent: Cu^{2+}

Similar exercises:
Problems 15.23 and 15.24

15.3 Types of Chemical Reactions

Two classification systems for chemical reactions are in common use. We have now encountered both of them.

The first system, presented initially in Section 10.6, recognizes four types of reactions.

1. Synthesis ($X + Y \longrightarrow XY$)
2. Decomposition ($XY \longrightarrow X + Y$)
3. Single-replacement ($X + YZ \longrightarrow Y + XZ$)
4. Double-replacement ($AX + BY \longrightarrow AY + BX$)

The second system involves two reaction types.

1. Oxidation–reduction (or redox)
2. Non-oxidation–reduction (or non-redox)

As we have just learned (Sec. 15.2), reactions in which oxidation numbers change are called oxidation–reduction reactions. When there are no changes in oxidation numbers we have a non-redox reaction.

These two classification systems are not mutually exclusive and are commonly used together. For example, a particular reaction may be characterized as a single-replacement redox reaction.

Synthesis reactions with only elements as reactants are always oxidation–reduction reactions. Oxidation number changes must occur because all elements (the reactants) have an oxidation number of zero and all of the constituent elements of a compound *cannot* have oxidation numbers of zero. Synthesis reactions in which compounds are the reactants may or may not be redox reactions.

$$S + O_2 \longrightarrow SO_2 \quad \text{(redox synthesis)}$$
$$K_2O + H_2O \longrightarrow 2\,KOH \quad \text{(non-redox synthesis)}$$
$$2\,NO + O_2 \longrightarrow 2\,NO_2 \quad \text{(redox synthesis)}$$

Both redox and non-redox decomposition reactions are common. At sufficiently high temperatures all compounds can be broken down (decomposed) into their constituent elements. Such reactions, where only elements are the products, are always redox reactions. Decomposition reactions where compounds are the products are most often non-redox reactions.

$$2\ CuO \longrightarrow 2\ Cu + O_2 \quad \text{(redox decomposition)}$$
$$2\ KClO_3 \longrightarrow 2\ KCl + O_2 \quad \text{(redox decomposition)}$$
$$CaCO_3 \longrightarrow CaO + CO_2 \quad \text{(non-redox decomposition)}$$

Single-replacement reactions are always redox reactions. By definition, an element and a compound are reactants and an element and a compound are products. The elements always undergo oxidation number change. Two of the reaction types studied in Chapter 14 are redox single-replacement reactions—the reaction between an acid and an active metal (Sec. 14.8) and the reaction between a metal and an aqueous salt solution (Sec. 14.10).

Double-replacement reactions generally involve acids, bases, and salts in aqueous solution. In such reactions ions, which maintain their identity, are generally trading places. Such reactions will always be non-redox reactions. All acid–base neutralization reactions (Sec. 14.8) are non-redox double-replacement reactions.

Combustion reactions (Sec. 10.5) are always redox reactions. However, as mentioned in Section 10.5, they do not fit any of the four general reaction patterns of synthesis, decomposition, single-replacement, and double-replacement.

EXAMPLE 15.3

Classify the following reactions as redox or non-redox. Further classify them, when possible, as synthesis, decomposition, single replacement, or double replacement.

(a) $2\ KNO_3 \longrightarrow 2\ KNO_2 + O_2$
(b) $Zn + 2\ AgNO_3 \longrightarrow Zn(NO_3)_2 + 2\ Ag$
(c) $CH_4 + 2\ O_2 \longrightarrow CO_2 + 2\ H_2O$
(d) $Ni(NO_3)_2 + 2\ NaOH \longrightarrow Ni(OH)_2 + 2\ NaNO_3$

Solution

The oxidation numbers are calculated by the methods illustrated in Example 15.1.

(a) $2\ KNO_3 \longrightarrow 2\ KNO_2 + O_2$
 +1 +5 −2 $\qquad$ +1 +3 −2 $\quad$ 0
 rules 3, 5, 7 $\quad$ rules 3, 5, 7 $\ $ rule 1

The oxidation number of N decreases from +5 to +3; the oxidation number for some O atoms increases from −2 to 0. The reaction is a **redox** reaction. Since two substances are produced from a single substance, it is also a decomposition reaction. We thus have a **redox decomposition** reaction.

(b) $Zn + 2\ AgNO_3 \longrightarrow Zn(NO_3)_2 + 2\ Ag$
 0 $\quad$ +1 +5 −2 $\qquad$ +2 +5 −2 $\quad$ 0
 rule 1 $\ $ rules 5, 7, 8 $\quad$ rules 5, 7, 8 $\ $ rule 1

This is a **redox** reaction; zinc is oxidized, silver is reduced. Having an element and a compound as reactants and an element and compound as products is a characteristic of a single-replacement reaction. That is the type of reaction we have here: zinc and silver are exchanging places. We thus have a **redox single-replacement** reaction.

(c) $\text{CH}_4 + 2\,\text{O}_2 \longrightarrow \text{CO}_2 + 2\,\text{H}_2\text{O}$

$\quad\;$ +4 −1 $\quad$ 0 $\quad\quad$ +4 −2 $\quad$ +1 −2
$\quad$ rules 6, 7 $\;$ rule 1 $\;\;$ rules 5, 7 $\;$ rules 5, 6

This is a **redox** reaction; the oxidation numbers of both carbon and oxygen change. This reaction does not fit any of the four reaction patterns of synthesis, decomposition, single replacement, and double replacement. (It is a **combustion** reaction.)

(d) $\text{Ni(NO}_3)_2 + 2\,\text{NaOH} \longrightarrow \text{Ni(OH)}_2 + 2\,\text{NaNO}_3$

$\quad$ +2 +5 −2 $\quad$ +1 −2 +1 $\quad\;$ +2 −2 +1 $\quad$ +1 +5 −2
$\quad$ rules 5, 7, 8 $\;$ rules 3, 5, 7 $\quad$ rules 5, 7, 8 $\;$ rules 3, 5, 7

This is a **non-redox** reaction; there are no oxidation number changes. The reaction is also a double-replacement reaction; nickel and sodium are changing places, that is, "swapping partners." Thus we have a **non-redox double-replacement** reaction

PRACTICE EXERCISE 15.3

Classify the following reactions as redox or non-redox. Further classify them, when possible, as synthesis, decomposition, single-replacement, or double-replacement.

(a) $\text{Ni} + \text{S} \longrightarrow \text{NiS}$
(b) $\text{Fe}_2\text{O}_3 + 3\,\text{C} \longrightarrow 2\,\text{Fe} + 3\,\text{CO}$
(c) $\text{C}_3\text{H}_8 + 5\,\text{O}_2 \longrightarrow 3\,\text{CO}_2 + 4\,\text{H}_2\text{O}$
(d) $\text{H}_2\text{SO}_4 + 2\,\text{NaOH} \longrightarrow \text{Na}_2\text{SO}_4 + 2\,\text{H}_2\text{O}$

Ans. (a) redox synthesis; (b) redox single replacement; (c) redox; (d) non-redox double replacement.

Similar exercises: Problems 15.27 and 15.28

15.4 Balancing Oxidation–Reduction Equations

Balancing an equation is not a new topic to us. In Section 10.3 we learned how to balance equations by the *inspection method*. With this method, we start with the most complicated compound within the equation and balance one of the elements in it. Then we balance the atoms of a second element, then third, and so on until all elements are balanced. This inspection procedure is a useful method for balancing simple equations with small coefficients. However, it breaks down when applied to complicated equations.

Equations for redox reactions are often quite complicated and contain numerous reactants and products and large coefficients. Trying to balance redox equations such as

$$\text{PH}_3 + \text{CrO}_4^{2-} + \text{H}_2\text{O} \longrightarrow \text{P}_4 + \text{Cr(OH)}_4^- + \text{OH}^-$$

or
$$As_4O_6 + MnO_4^- + H_2O \longrightarrow AsO_4^{3-} + H^+ + Mn^{2+}$$

by inspection is a tedious, time-consuming, frustrating experience. Balancing such equations is, however, easily accomplished by systematic equation-balancing procedures that use oxidation numbers and focus on the fact that the numbers of electron lost and gained in a redox reaction must be equal.

Two distinctly different approaches for systematically balancing redox equations are in common use; the oxidation-number method and the ion–electron method. Each method has advantages and disadvantages. We will consider both methods.

15.5 Using the Oxidation-Number Method to Balance Redox Equations

A useful feature of oxidation numbers is that they provide a rather easy method for balancing redox equations. The steps involved in their use in this balancing process are as follows.

STEP 1   Assign oxidation numbers to all atoms in the equation and determine which atoms are undergoing a change in oxidation number.

STEP 2   Determine the magnitude of the change in oxidation number *per atom* for the elements undergoing a change in oxidation number.

Draw a large bracket from the element in the reactant to the element in the product, and write the increase or decrease in oxidation number at the middle of the bracket. (See the examples that follow.)

STEP 3   **When more than one atom of an element that changes oxidation number is present in a formula unit (of either reactant or product), determine the change in oxidation number per *formula unit*.**

Indicate this change per formula unit by multiplying the oxidation number change per atom, already written on the brackets, by an appropriate factor.

STEP 4   Determine multiplying factors that make the total increase in oxidation number equal to the total decrease in oxidation number.

Place them on the bracket also.

STEP 5   Place in front of the oxidizing and reducing agents and their products in the equation coefficients that are consistent with the total number of atoms of the elements undergoing oxidation-number change.

STEP 6   Balance all other atoms in the equation except those of hydrogen and oxygen.

In doing this, do not alter the coefficients determined in the previous step.

STEP 7   Balance the charge (the sum of all the ionic charges) so that it is the same on both sides of the equation by adding H^+ or OH^- ions.

This step is necessary only when dealing with net ionic equations describing aqueous solution reactions. If the reaction takes place in acidic solution, add H^+ ion to the side deficient in positive charge. If the reaction takes place in basic solution, add OH^- ions to the side deficient in negative charge.

STEP 8 Balance the oxygen atoms.

For net ionic equations, H_2O must usually be added to the appropriate side of the equation to achieve oxygen balance. Water is, of course, present in all aqueous solutions and can be either a reactant or a product.

STEP 9 Balance the hydrogen atoms.

If you have correctly carried out all of the previous steps, the hydrogens should automatically have been balanced. If hydrogens do not balance, you have made a mistake in a previous step. Check your work.

Now let us consider some examples where these rules are applied. The first two examples will involve molecular equations. The third example involves a net ionic equation. In balancing net ionic equations, any H_2O, H^+, or OH^- present is usually left out of the unbalanced equation that we start with and then added as needed during the balancing process.

EXAMPLE 15.4

Balance the following molecular redox equation by the oxidation-number method of balancing.

$$Fe + O_2 + HCl \longrightarrow FeCl_3 + H_2O$$

Solution

STEP 1 We identify the elements being oxidized and reduced by assigning oxidation numbers.

$$\begin{array}{cccccc} Fe & + & O_2 & + & HCl & \longrightarrow & FeCl_3 & + & H_2O \\ 0 & & 0 & & +1\;-1 & & +3\;-1 & & +1\;-2 \end{array}$$

Iron (Fe) and oxygen (O) are the elements that undergo oxidation-number change.

STEP 2 The change in oxidation number *per atom* is shown by drawing brackets connecting the oxidizing and reducing agents to their products and indicating the change at the middle of the bracket.

$$\begin{array}{cccccc} 0 & & (+3) & & +3 & \\ \overline{Fe} & + & O_2 & + & HCl \longrightarrow & \overline{FeCl_3} & + & H_2O \\ 0 & & \underline{(-2)} & & -2 & \end{array}$$ change in oxidation number per atom

STEP 3 For Fe the change in oxidation number per formula unit is the same as the change per atom, since both Fe and $FeCl_3$, the two iron-containing species, contain only one Fe atom. For O the change in oxidation number per formula unit will be double the change per atom since O_2 contains two atoms. The change per formula unit is indicated by multiplying the per atom change by an appropriate numerical factor, which is 2 in this case.

$$\overset{(+3)}{\overline{Fe} + \underset{2(-2)}{\underline{O_2 + HCl}} \longrightarrow \overline{FeCl_3} + \underline{H_2O}} \quad \text{change in oxidation number per formula unit}$$

STEP 4   For Fe, the total increase in oxidation number per formula unit is $+3$. For oxygen, the total decrease in oxidation number per formula unit is -4. To make the increase equal to the decrease, we must multiply the oxidation-number change for the element oxidized (Fe) by 4 and the oxidation-number change for the element reduced (O) by 3. This will make the increase and decrease both numerically equal to 12.

$$\overset{4(+3)}{\overline{Fe} + \underset{3[2(-2)]}{\underline{O_2 + HCl}} \longrightarrow \overline{FeCl_3} + \underline{H_2O}} \quad \text{oxidation-number increase equals oxidation-number decrease}$$

STEP 5   We are now ready to place coefficients in the equation in front of the oxidizing and reducing agents and their products. The bracket notation indicates that four Fe atoms undergo an oxidation-number change. Place the coefficient 4 in front of both Fe and $FeCl_3$. The bracket notation also indicates that six O atoms (3×2) undergo an oxidation-number decrease of two units. Thus, we need six oxygen atoms on each side. Place the coefficient 3 in front of O_2 (6 atoms of O), and the coefficient 6 in front of H_2O (6 atoms of O).

$$4\,Fe + 3\,O_2 + HCl \longrightarrow 4\,FeCl_3 + 6\,H_2O$$

The equation is only partially balanced at this point; only Fe and O atoms are balanced.

STEP 6   We next balance the element Cl (by inspection). There are twelve Cl atoms on the right side. Thus, to obtain twelve Cl atoms on the left side we place the coefficient 12 in front of HCl.

$$4\,Fe + 3\,O_2 + 12\,HCl \longrightarrow 4\,FeCl_3 + 6\,H_2O$$

STEP 7   This step is not needed when the equation is a molecular equation.

STEP 8   In this particular equation the O atoms are already balanced since the O underwent an oxidation-number change. There are six oxygen atoms on each side of the equation.

STEP 9   If all of the previous procedures (steps) have been carried out correctly, the H atoms should automatically balance. They do. There are twelve H atoms on each side of the equation. The balanced equation is thus

$$4\,Fe + 3\,O_2 + 12\,HCl \longrightarrow 4\,FeCl_3 + 6\,H_2O$$

PRACTICE EXERCISE 15.4

Balance the following molecular redox equation by the oxidation-number method of balancing.

$$IF_5 + Fe \longrightarrow IF_3 + FeF_3$$

Ans. $3\ IF_5 + 2\ Fe \rightarrow 3\ IF_3 + 2\ FeF_3$

Similar exercises: Problems 15.31 and 15.32

EXAMPLE 15.5

Balance the following molecular redox equation by the oxidation-number method of balancing.

$$CuO + NH_3 \longrightarrow Cu + N_2 + H_2O$$

Solution

STEP 1

$$CuO + NH_3 \longrightarrow Cu + N_2 + H_2O$$
$$+2\ -2 \quad -3\ +1 \quad\quad 0 \quad\ 0 \quad\ +1\ -2$$

The two elements undergoing oxidation-number change are Cu and N.

STEP 2

$$\overset{+2\quad\quad\quad\quad (-2)\quad\ 0}{CuO + NH_3 \longrightarrow Cu + N_2 + H_2O}$$
$$\underset{-3\quad\quad (+3)\quad 0}{}$$

change in oxidation number per atom

STEP 3

$$\overset{(-2)}{CuO + NH_3 \longrightarrow Cu + N_2 + H_2O}$$
$$\underset{2(+3)}{}$$

change in oxidation number per formula unit

The nitrogen oxidation-number change per atom had to be multiplied by 2 since there are two nitrogen atoms per molecule in N_2. Thus, a minimum of two N atoms must undergo an oxidation-number increase. This illustrates that *both* reactant and product formulas must be considered when determining the change in oxidation number per formula unit.

STEP 4

$$\overset{3(-2)}{CuO + NH_3 \longrightarrow Cu + N_2 + H_2O}$$
$$\underset{2(+3)}{}$$

oxidation-number increase equals oxidation-number decrease

By multiplying the Cu per formula unit oxidation-number decrease of -2 by 3 we make the oxidation-number increase and decrease equal; both are at six units.

STEP 5 The coefficients for CuO, NH_3, Cu, and N_2 in the equation are determined from the bracket information, which indicates that three Cu atoms undergo an oxidation number change for every two N atoms that change.

$$3 \text{ CuO} + 2 \text{ NH}_3 \longrightarrow 3 \text{ Cu} + 1 \text{ N}_2 + \text{H}_2\text{O}$$

STEP 6 The only atoms left to balance are H and O

STEP 7 This step is not needed for a molecular equation.

STEP 8 We balance the oxygen by placing the coefficient 3 in front of the H_2O on the right side.

$$3 \text{ CuO} + 2 \text{ NH}_3 \longrightarrow 3 \text{ Cu} + 1 \text{ N}_2 + 3 \text{ H}_2\text{O}$$

STEP 9 If all of the procedures in previous steps have been carried out correctly, the H atoms should automatically balance. They do. There are six H atoms on each side of the equation.

$$3 \text{ CuO} + 2 \text{ NH}_3 \longrightarrow 3 \text{ Cu} + \text{N}_2 + 3 \text{ H}_2\text{O}$$

PRACTICE EXERCISE 15.5

Balance the following molecular redox equation by the oxidation-number method of balancing.

$$\text{NF}_3 + \text{AlCl}_3 \longrightarrow \text{N}_2 + \text{Cl}_2 + \text{AlF}_3$$

Ans. $2 \text{ NF}_3 + 2 \text{ AlCl}_3 \rightarrow \text{N}_2 + 3 \text{ Cl}_2 + 2 \text{ AlF}_3$

Similar exercises: Problems 15.33–15.36

EXAMPLE 15.6

Balance the following net ionic redox equation by the oxidation-number method of balancing.

$$\text{Au} + \text{Cl}^- + \text{NO}_3^- \longrightarrow \text{AuCl}_4^- + \text{NO}_2$$

This reaction occurs in acidic solution.

Solution

STEP 1
$$\begin{array}{ccccc} \text{Au} + & \text{Cl}^- + & \text{NO}_3^- \longrightarrow & \text{AuCl}_4^- + & \text{NO}_2 \\ 0 & -1 & +5\ -2 & +3\ -1 & +4\ -2 \end{array}$$

STEP 2 The two elements undergoing oxidation-number change are Au and N.

$$\begin{array}{c} 0(+3)+3 \\ \text{Au} + \text{Cl}^- + \text{NO}_3^- \longrightarrow \text{AuCl}_4^- + \text{NO}_2 \\ +5(-1)+4 \end{array}$$ change in oxidation number per atom

STEP 3 For both Au and N the oxidation-number change per formula unit is the same as per atom. Both Au and $AuCl_4^-$ contain only one Au atom; similarly, both NO_3^- and NO_2 contain only one N atom.

STEP 4 By multiplying the N oxidation-number decrease by 3 we make the oxidation-number increase and decrease per formula unit the same— three units.

$$\underset{3(-1)}{\underbrace{Au + Cl^- + \overset{+3}{\overbrace{NO_3^- \longrightarrow AuCl_4^-}} + NO_2}} \quad \text{oxidation-number increase equals oxidation-number decrease}$$

STEP 5   The bracket notation indicates that three N atoms and one Au atom undergo an oxidation-number change. Translating this information into coefficients, we get

$$1\, Au + Cl^- + 3\, NO_3^- \longrightarrow 1\, AuCl_4^- + 3\, NO_2$$

STEP 6   We next balance Cl (by inspection). The coefficient 4 is needed on the left side in front of Cl^-. This gives four Cl atoms on each side of the equation.

$$1\, Au + 4\, Cl^- + 3\, NO_3^- \longrightarrow 1\, AuCl_4^- + 3\, NO_2$$

STEP 7   Since this is a net ionic equation, the charges must balance; that is, the sum of the ionic charges of all species on each side of the equation must be equal. (They do not have to add up to zero; they just have to be equal.) In acidic solution, which is the case in this example, charge balance is accomplished by adding H^+ ion.

As the equation now stands, we have a charge of -7 on the left side (three nitrate ions each with a -1 charge and four chloride ions each with a -1 charge) and a charge of -1 on the right side (one $AuCl_4^-$ ion). By adding six H^+ to the left side we balance the charge at -1.

$$-7 + (+6) = -1$$

The equation, at this point, becomes

$$1\, Au + 4\, Cl^- + 3\, NO_3^- + 6\, H^+ \longrightarrow 1\, AuCl_4^- + 3\, NO_2$$

STEP 8   The oxygen atoms are balanced through the addition of H_2O molecules. There are nine O atoms on the left side (three NO_3^- ions) and only six on the right side (three NO_2 molecules). Addition of three H_2O molecules to the right side will balance the O atoms at nine per side.

$$1\, Au + 4\, Cl^- + 3\, NO_3^- + 6\, H^+ \longrightarrow 1\, AuCl_4^- + 3\, NO_2 + 3\, H_2O$$

STEP 9   The H atoms automatically balance at six atoms on each side. This is our double check that previous steps have been correctly carried out. The balanced net ionic equation is thus

$$Au + 4\, Cl^- + 3\, NO_3^- + 6\, H^+ \longrightarrow AuCl_4^- + 3\, NO_2 + 3\, H_2O$$

PRACTICE EXERCISE 15.6

Balance the following net ionic redox equation by the oxidation-number method of balancing.

$$UO_2^+ + Cr_2O_7^{2-} \longrightarrow UO_2^{2+} + Cr^{3+}$$

This reaction occurs in acidic solution.

Ans. $6\,UO_2^+ + Cr_2O_7^{2-} + 14\,H^+ \rightarrow 6\,UO_2^{2+} + 2\,Cr^{3+} + 7\,H_2O$

Similar exercises: Problems 15.17 and 15.18

15.6 Using the Ion–Electron Method to Balance Redox Equations

A second method for balancing redox equations is the *ion–electron method*. In this method, two separate *partial equations*, called half-reactions, are constructed. One half-reaction describes the oxidation process, the other half-reaction the reduction process. The balanced half-reactions are added together to get the desired balanced redox equation.

The division of the original unbalanced equation into two parts (half-reactions) is done to simplify the process of balancing. It is artificial. The half-reactions do not really take place alone; we cannot have oxidation without reduction. Because of its dependence on half-reactions, the ion–electron method is also often referred to as the *half-reaction method*.

The name "ion–electron" for the method draws attention to the facts that (1) the method is used predominantly for balancing equations of reactions occurring in aqueous solution where ions are reactants and products and (2) electrons are explicitly shown as reactants or products in the balanced half-reactions. Concerning this latter point, we note that the electrons in the balanced half-reactions cancel out when the half-reactions are combined to give the total reaction. Thus, the equations obtained by the ion–electron method of balancing are identical in appearance to those obtained by the oxidation-number method of balancing.

As we did with the oxidation-number method, we will break the ion–electron balancing process into a series of steps.

STEP 1 Write the equation in net ionic form if it is not already in that form, and then determine which substances are oxidized and reduced by assigning an oxidation number to each atom or ion.

STEP 2 Write two skeletal partial equations: an *oxidation half-reaction* that includes the formula of the substance containing the element oxidized along with the formulas of the oxidation products, and a *reduction half-reaction* that includes the formula of the substance containing the element reduced along with the formulas of the reduction products. Except for hydrogen and oxygen, the same elements must appear on both sides of the given half-reaction.

STEP 3 Balance each half-reaction with respect to the element oxidized or reduced with appropriate coefficients.

STEP 4 Balance each half-reaction with respect to all other elements present except for hydrogen and oxygen.

STEP 5 Balance each half-reaction with respect to oxygen.

How this is accomplished depends on whether the solution in which the reaction is occurring is acidic, basic, or neutral.

In acidic and neutral solutions, oxygen balance is achieved by adding H_2O molecules, as a source of oxygen, to the side of the equation deficient in oxygen.

In basic solution, oxygen balance is achieved by adding, for each deficient oxygen, two OH^- ions to the side deficient in oxygen and one H_2O molecule to the other side of the equation. (The reason for this more complicated procedure is that the two species used to achieve hydrogen–oxygen balance in basic solution, OH^- and H_2O, both contain hydrogen and oxygen.)

STEP 6 Balance each half-reaction with respect to hydrogen.

Again, how this is done depends on whether the solution is acidic or basic.

In acidic solution, hydrogen balance is achieved by adding H^+ ion to the side deficient in hydrogen.

In basic solution, hydrogen balance is achieved by adding, for each deficient hydrogen, one H_2O molecule to the side deficient in H and then one OH^- ion to the other side. (At this point H_2O and OH^- ions may be present on both sides of the half-reaction, since they were also added in step 5; if so, cancel out what you can. For example, if four H_2O are present on the left and two H_2O on the right, then cancel two H_2O from each side to leave two H_2O remaining on the left.)

STEP 7 Balance each half-reaction with respect to charge by determining the total charge on each side of the equation and then adding electrons to the side with the more positive total charge such that the charges become equal.

This step completes the balancing process for the half-reactions.

STEP 8 Multiply each balanced half-reaction by the smallest whole number that will make the total number of electrons lost equal to the total numbers of electrons gained.

The oxidation half-reaction will contain the electrons lost, and the reduction half-reaction the electrons gained. All coefficients in the balanced half-reactions are affected by the multipliers.

STEP 9 Add the two half-reactions together to obtain the final balanced equation.

The electrons will *always* cancel out at this point since in the previous step electron loss was made equal to electron gain.

STEP 10 Cancel out any other species besides electrons common to both sides of the equation.

In some, but not all, cases some cancellation of H_2O molecules, H^+ ions, or OH^- ions is possible.

STEP 11 Check to see that the equation coefficients are in the lowest possible ratio. Reduce the ratio if needed.

Examples 15.7–15.9 illustrate how the above instructions for balancing redox equations are applied.

EXAMPLE 15.7

Balance the following net ionic equation by the ion–electron method of balancing.

$$Cu + NO_3^- \longrightarrow Cu^{2+} + NO_2 \quad \text{(acidic solution)}$$

Solution

STEP 1 Assigning oxidation numbers, we get

$$Cu + NO_3^- \longrightarrow Cu^{2+} + NO_2$$
$$0 +5\ -2 +2 +4\ -2$$

Copper is oxidized, increasing in oxidation number from 0 to +2. Nitrogen is reduced, decreasing in oxidation number from +5 to +4.

STEP 2 The skeleton half-reactions for oxidation and reduction are

$$\text{Oxidation:} \quad Cu \longrightarrow Cu^{2+}$$
$$\text{Reduction:} \quad NO_3^- \longrightarrow NO_2$$

STEP 3 In both half-reactions, the element being oxidized or reduced is already balanced—one atom of Cu on both sides in the first half-reaction and one atom of N on both sides in the second.

STEP 4 There are no other elements present except oxygen.

STEP 5 The problem statement indicates that the reaction is occurring in acidic solution. Thus, H_2O molecules will be used to achieve any needed O balance.

The oxidation half-reaction does not need to be balanced with respect to oxygen; no oxygen is present.

$$\text{Oxidation:} \quad Cu \longrightarrow Cu^{2+}$$

To achieve oxygen balance in the reduction half-reaction we add an H_2O molecule to the right side.

$$\text{Reduction:} \quad NO_3^- \longrightarrow NO_2 + H_2O$$

STEP 6 No H balance is required for the oxidation half-reaction.

$$\text{Oxidation:} \quad Cu \longrightarrow Cu^{2+}$$

Since we are in acidic solution, H balance in the reduction half-reaction is achieved through use of H^+ ions. Two H^+ ions are added to the left side.

$$\text{Reduction:} \quad NO_3^- + 2\,H^+ \longrightarrow NO_2 + H_2O$$

STEP 7 The half-reactions are balanced with respect to charge by adding electrons.

For the oxidation half-reaction, there is no charge on the left side and a charge of +2 (from the copper ion) on the right side. Addition of two electrons to the right side will balance the charge at zero on each side of the half-reaction.

Oxidation: $Cu \longrightarrow Cu^{2+} + 2\,e^-$

For the reduction half-reaction, the two positive charges (the H^+ ions) and one negative charge (the NO_3^- ion) give a net charge of +1 on the left side. There is no charge associated with the species on the right side. Hence, addition of one electron to the left side will balance the charges at zero.

Reduction: $NO_3^- + 2\,H^+ + e^- \longrightarrow NO_2 + H_2O$

STEP 8   The oxidation half-reaction involves two electrons, and the reduction half-reaction contains one electron. To make the number of electrons in the two half-reactions equal, the reduction half-reaction is multiplied by the factor 2.

Oxidation: $Cu \longrightarrow Cu^{2+} + 2\,e^-$
Reduction: $2(NO_3^- + 2\,H^+ + e^- \longrightarrow NO_2 + H_2O)$

Note that when we multiply the reduction half-reaction by 2, all of the coefficients in the equation will be doubled.

STEP 9   Adding the two half-reactions together, we get

Oxidation: $Cu \longrightarrow Cu^{2+} + \cancel{2\,e^-}$
Reduction: $2\,NO_3^- + 4\,H^+ + \cancel{2\,e^-} \longrightarrow 2\,NO_2 + 2\,H_2O$
$\overline{Cu + 4\,H^+ + 2\,NO_3^- \longrightarrow Cu^{2+} + 2\,NO_2 + 2\,H_2O}$

The electrons must always cancel at this stage in the balancing process. If they did not cancel that would mean that electron loss was not equal to electron gain. Having these two quantities equal is an absolute requirement in a redox process.

STEP 10   There are no additional species to cancel.

STEP 11   The coefficients are already in the smallest possible whole number ratio. Thus, the equation in step 9 is the final balanced equation.

PRACTICE EXERCISE 15.7

Balance the following net ionic equation by the ion–electron method of balancing.

$$Fe^{2+} + Cr_2O_7^{2-} \longrightarrow Fe^{3+} + Cr^{3+} \quad \text{(acidic solution)}$$

Ans. $6\,Fe^{2+} + Cr_2O_7^{2-} + 14\,H^+ \rightarrow 6\,Fe^{3+} + 2\,Cr^{3+} + 7\,H_2O$

Similar exercises: Problems 15.45–15.48

EXAMPLE 15.8

Balance the following net ionic equation by the ion–electron method of balancing.

$$SeO_3^{2-} + Cl_2 \longrightarrow SeO_4^{2-} + Cl^- \quad \text{(basic solution)}$$

Solution

STEP 1

$$SeO_3^{2-} + Cl_2 \longrightarrow SeO_4^{2-} + Cl^-$$
$$+4-2 \qquad 0 \qquad +6-2 \qquad -1$$

Selenium is oxidized, increasing in oxidation number from $+4$ to $+6$. Chlorine is reduced, decreasing in oxidation number from 0 to -1.

STEP 2 The skeleton half-reactions for oxidation and reduction are

$$\text{Oxidation:} \quad SeO_3^{2-} \longrightarrow SeO_4^{2-}$$
$$\text{Reduction:} \quad Cl_2 \longrightarrow Cl^-$$

STEP 3 In the oxidation half-reaction the Se is already balanced.

$$\text{Oxidation:} \quad SeO_3^{2-} \longrightarrow SeO_4^{2-}$$

To balance the Cl in the reduction half-reaction the coefficient 2 must be added on the right side:

$$\text{Reduction:} \quad Cl_2 \longrightarrow 2\,Cl^-$$

STEP 4 There are no other elements present besides oxygen.

STEP 5 The problem statement indicates that the reaction is occurring in basic solution. In basic solution, oxygen balance is achieved by using both OH^- ions and H_2O molecules. We add two OH^- ions to the side deficient in O for every needed O, and then we add one H_2O molecule to the other side of the equation for every two OH^- ions that were added.

For the oxidation half-reaction there is an O deficiency of one on the left side. Thus, we add two OH^- ions to the left side and one H_2O molecule to the right side.

$$\text{Oxidation:} \quad SeO_3^{2-} + 2\,OH^- \longrightarrow SeO_4^{2-} + H_2O$$

The reduction half-reaction is oxygen-independent; hence, no balance is needed.

$$\text{Reduction:} \quad Cl_2 \longrightarrow 2\,Cl^-$$

STEP 6 The procedure for balancing H in basic solution is not needed.

For the oxidation half-reaction, the H is already balanced. It balanced at the same time the O was balanced. Such "automatic" H balancing when the O is balanced is usually not the case.

$$\text{Oxidation:} \quad SeO_3^{2-} + 2\,OH^- \longrightarrow SeO_4^{2-} + H_2O$$

The reduction half-reaction does not contain any hydrogen.

$$\text{Reduction:} \quad Cl_2 \longrightarrow 2\,Cl^-$$

STEP 7 Adding electrons to achieve a charge balance gives us

$$\text{Oxidation:} \quad SeO_3^{2-} + 2\,OH^- \longrightarrow SeO_4^{2-} + H_2O + 2\,e^-$$
$$\text{Reduction:} \quad Cl_2 + 2\,e^- \longrightarrow 2\,Cl^-$$

Note that the charge need not balance at zero. In this example the oxidation half-reaction charge balances at -4 and the reduction half-reaction charge balances at -2.

STEP 8 Both the oxidation and reduction half-reactions involve two electrons. Thus, we already have electron loss equal to electron gain and do not need to multiply the equations by any small whole numbers to achieve electron balance.

$$\text{Oxidation:} \quad SeO_3^{2-} + 2\,OH^- \longrightarrow SeO_4^{2-} + H_2O + 2\,e^-$$
$$\text{Reduction:} \quad Cl_2 + 2\,e^- \longrightarrow 2\,Cl^-$$

STEP 9 Adding the two half-reactions together, we get

$$\text{Oxidation:} \quad SeO_3^{2-} + 2\,OH^- \longrightarrow SeO_4^{2-} + H_2O + \cancel{2\,e^-}$$
$$\text{Reduction:} \quad Cl_2 + \cancel{2\,e^-} \longrightarrow 2\,Cl^-$$
$$\overline{SeO_3^{2-} + Cl_2 + 2\,OH^- \longrightarrow SeO_4^{2-} + 2\,Cl^- + H_2O}$$

The electrons canceled as they should have.

STEP 10 There are no additional species to cancel.

STEP 11 The coefficients are already in the smallest possible whole number ratio. Thus, the equation in step 9 is the final balanced equation.

PRACTICE EXERCISE 15.8

Balance the following net ionic equation by the ion–electron method of balancing.

$$Zn + MnO_4^- \longrightarrow Zn(OH)_2 + MnO_2 \quad \text{(basic solution)}$$

Ans. $3\,Zn + 2\,MnO_4^- + 4\,H_2O \rightarrow 3\,Zn(OH)_2 + 2\,MnO_2 + 2\,OH^-$

Similar exercises: Problems 15.49–15.52

EXAMPLE 15.9

Balance the following net ionic equation using the ion–electron method of balancing.

$$As_4O_6 + MnO_4^- \longrightarrow H_3AsO_4 + Mn^{2+} \quad \text{(acidic solution)}$$

Solution

STEP 1

$$As_4O_6 + MnO_4^- \longrightarrow H_3AsO_4 + Mn^{2+}$$
$$\;\;+3\;-2 \quad\;\; +7\;-2 \quad\;\; +1\;+5\;-2 \quad\;\; +2$$

Arsenic is oxidized, increasing in oxidation number from $+3$ to $+5$. Manganese is reduced, decreasing in oxidation number from $+7$ to $+2$.

STEP 2 The skeleton half-reactions are

$$\text{Oxidation:} \quad As_4O_6 \longrightarrow H_3AsO_4$$
$$\text{Reduction:} \quad MnO_4^- \longrightarrow Mn^{2+}$$

STEP 3 To balance As in the oxidation half-reaction, add the coefficient 4 on the right side.

$$\text{Oxidation:} \quad As_4O_6 \longrightarrow 4\,H_3AsO_4$$

The Mn in the reduction half-reaction is already balanced.

$$\text{Reduction:} \quad MnO_4^- \longrightarrow Mn^{2+}$$

STEP 4 There are no other elements present besides H and O.

STEP 5 The problem statement indicates that the reaction occurs in acidic solution. Therefore, use H_2O molecules to achieve oxygen balance. Both half-reactions need water added to them.

$$\text{Oxidation:} \quad As_4O_6 + 10\,H_2O \longrightarrow 4\,H_3AsO_4$$
$$\text{Reduction:} \quad MnO_4^- \longrightarrow Mn^{2+} + 4\,H_2O$$

STEP 6 In acidic solution, H balance is achieved by using H^+ ions. Both half-reactions need H^+ ions added to them.

$$\text{Oxidation:} \quad As_4O_6 + 10\,H_2O \longrightarrow 4\,H_3AsO_4 + 8\,H^+$$
$$\text{Reduction:} \quad MnO_4^- + 8\,H^+ \longrightarrow Mn^{2+} + 4\,H_2O$$

STEP 7 Charge balance is achieved using electrons as a source of negative charge. Both half-reactions require the addition of electrons.

$$\text{Oxidation:} \quad As_4O_6 + 10\,H_2O \longrightarrow 4\,H_3AsO_4 + 8\,H^+ + 8\,e^-$$
$$\text{Reduction:} \quad MnO_4^- + 8\,H^+ + 5\,e^- \longrightarrow Mn^{2+} + 4\,H_2O$$

STEP 8 Eight electrons are lost in the oxidation half-reaction, and five electrons are gained in the reduction half-reaction. To make the total number of electrons lost equal to the total number gained, we must multiply the oxidation half-reaction by 5 and the reduction half-reaction by 8. (The lowest common multiple of an 8 and a 5 is 40.)

$$\text{Oxidation:} \quad 5(As_4O_6 + 10\,H_2O \longrightarrow 4\,H_3AsO_4 + 8\,H^+ + 8\,e^-)$$
$$\text{Reduction:} \quad 8(MnO_4^- + 8\,H^+ + 5\,e^- \longrightarrow Mn^{2+} + 4\,H_2O)$$

Remember that the multiplying coefficient affects all of the substances in the half-reaction and not just the electrons.

STEP 9 Adding the two half-reactions together, we get

Oxidation: $\quad 5\,As_4O_6 + 50\,H_2O \longrightarrow 20\,H_3AsO_4 + 40\,H^+ + \cancel{40\,e^-}$
Reduction: $\;\; 8\,MnO_4^- + 64\,H^+ + \cancel{40\,e^-} \longrightarrow 8\,Mn^{2+} + 32\,H_2O$
$\overline{}$
$\quad 5\,As_4O_6 + 50\,H_2O + 8\,MnO_4^- + 64\,H^+ \longrightarrow$
$\qquad\qquad\qquad 20\,H_3AsO_4 + 40\,H^+ + 8\,Mn^{2+} + 32\,H_2O$

The 40 electrons cancel out.

STEP 10   Both H_2O and H^+ appear on both sides of the equation. We can cancel 32 H_2O molecules from each side of the equation, leaving 18 H_2O on the left side and no H_2O on the right side. We can also cancel 40 H^+ ions from each side of the equation, leaving 24 H^+ ions on the left side and none on the right side.

$$5\,As_4O_6 + 8\,MnO_4^- + 18\,H_2O + 24\,H^+ \longrightarrow 20\,H_3AsO_4 + 8\,Mn^{2+}$$

STEP 11   The coefficients are in the lowest possible ratio. The equation in step 10 is therefore the final balanced equation.

PRACTICE EXERCISE 15.9

Balance the following net ionic equation using the ion–electron method of balancing.

$$HNO_2 + Cr_2O_7^{2-} \longrightarrow Cr^{3+} + NO_3^- \quad \text{(acidic solution)}$$

Ans. $3\,HNO_2 + Cr_2O_7^{2-} + 5\,H^+ \to 3\,NO_3^- + 2\,Cr^{3+} + 4\,H_2O$

Similar exercises: Problems 15.45–15.48

In each of the three examples we have just considered, the oxidation and reduction half-reactions were simultaneously balanced. This approach was used in the examples to enable us to make comparisons. In practice, particularly when you are thoroughly familiar with the balancing procedure, one half-reaction is usually completely balanced before work begins on balancing the other half-reaction. Usually, it is better to work on just one reaction at a time.

A comparison of the two methods for balancing redox equations is in order. Basic to each method is being able to recognize the elements involved in the actual oxidation–reduction process. The oxidation–number method works on the principle that the increase in oxidation number must equal the decrease in oxidation number. The ion–electron method involves equalizing the number of electrons lost by the substance oxidized with the number of electrons gained by the substance reduced.

The oxidation–number method is usually faster, particularly for simple equations. This potential speed is considered the major advantage of the oxidation–number method.

A major advantage of the ion–electron method is that all of the coefficients in the equation are generated by systematic procedures. In the oxidation–number method the coefficients of the reducing and oxidizing agents and their products are the only ones generated by systematic procedures; the others are determined by inspection. Another advantage of the ion–electron method is its focus on electron transfer. This feature becomes particularly important in electrochemistry. In this field it is most useful to discuss chemical reactions in terms of half-reactions occurring at different locations (electrodes) in an electrochemical cell (Sec. 15.8).

15.7 Disproportionation Reactions

A disproportionation reaction is a special type of oxidation–reduction reaction. A **disproportionation reaction** is a reaction in which some atoms of a single element in a reactant are oxidized and others are reduced. For such reactant behavior to be possible, the reactant must contain an element that is capable of having at least three oxidation numbers—its original number plus one higher and one lower oxidation number. Note that any given atom is not both oxidized and reduced. Some atoms are oxidized, and other atoms of the same element are reduced.

An example of a disproportionation reaction is

$$3\, Br_2 + 3\, H_2O \longrightarrow HBrO_3 + 5\, HBr$$

Note that two bromine-containing products have been produced from one bromine-containing reactant. The reactant bromine atoms have an oxidation number of zero. Bromine in $HBrO_3$ has a $+5$ oxidation number (it has been oxidized), and bromine in HBr has a -1 oxidation number (it has been reduced).

$$3\, Br_2 + 3\, H_2O \longrightarrow HBrO_3 + 5\, HBr$$
$$ 0 +5 -1$$

Thus, some of the reactant bromine atoms have been oxidized, while others have been reduced. A disproportionation reaction has taken place.

Example 15.10 shows how the procedures for balancing redox equations (Sec. 15.5 and 15.6) are slightly modified to balance disproportionation reaction equations.

EXAMPLE 15.10

Balance the following disproportionation redox reaction using (a) the oxidation-number method and (b) the ion–electron method.

$$MnO_4^{2-} \longrightarrow MnO_2 + MnO_4^- \quad \text{(acidic solution)}$$

Solution

(a) *Oxidation-Number Method*

STEP 1 In assigning oxidation numbers we immediately become aware that this is a disproportionation reaction. Manganese is the only element for which an oxidation-number change occurs.

$$MnO_4^{2-} \longrightarrow MnO_2 + MnO_4^-$$
$$+6\ -2 +4\ -2 +7\ -2$$

STEP 2 Since the species MnO_4^{2-} is undergoing both oxidation and reduction, for balancing purposes we will write it twice on the reactant side of the equation. With the MnO_4^{2-} in two places, brackets can then be drawn in the "normal" manner to connect the substances involved in oxidation and reduction.

$$\overset{+6}{\underset{+6}{\underbrace{MnO_4^{2-}}}} + \overset{(-2)}{\underset{(+1)}{\underbrace{MnO_4^{2-} \longrightarrow MnO_2}}} + \overset{+4}{MnO_4^-} \quad \overset{+7}{} \quad \text{change in oxidation number per atom}$$

(Later in the balancing process we will recombine the MnO_4^{2-} ions into one location.)

STEP 3 The change in oxidation number per formula unit in both cases is the same as per atom.

STEP 4 By multiplying the oxidation-number increase by 2 we equalize the oxidation-number increase and decrease per formula unit.

$$\overset{(-2)}{\underset{2(+1)}{\underbrace{MnO_4^{2-} + MnO_4^{2-} \longrightarrow MnO_2}}} + MnO_4^- \quad \text{oxidation-number increase equals oxidation-number decrease}$$

STEP 5 The bracket notation indicates that two Mn atoms undergo an increase in oxidation number for every one that undergoes a decrease in oxidation number. Translating this information into equation coefficients gives

$$MnO_4^{2-} + 2\ MnO_4^{2-} \longrightarrow MnO_2 + 2\ MnO_4^-$$

Now that the equation coefficients for the substance involved in oxidation and reduction, MnO_4^{2-}, have been determined, we can combine the MnO_4^{2-} ions into one location, reversing the process carried out in step 2.

$$3\ MnO_4^{2-} \longrightarrow MnO_2 + 2\ MnO_4^-$$

STEP 6 The only atoms left to balance are oxygen atoms.

STEP 7 Since this is a net ionic equation, charge must be balanced. In acidic solution, which is the case here, we balance the charge by adding H^+ ion. As the equation now stands, we have a charge of -6 on the left side (three MnO_4^{2-} ions) and a charge of -2 on the right side (two MnO_4^- ions). By adding four H^+ ions to the left side of the equation we balance the charge at -2.

$$3\ MnO_4^{2-} + 4\ H^+ \longrightarrow MnO_2 + 2\ MnO_4^-$$

STEP 8 Oxygen atom balance is achieved through the addition of H_2O molecules. There are 12 oxygen atoms on the left side (three MnO_4^{2-} ions) and 10 oxygen atoms on the right side. Addition of two H_2O molecules to the right side will balance the O atoms at 12 on each side.

$$3\ MnO_4^{2-} + 4\ H^+ \longrightarrow MnO_2 + 2\ MnO_4^- + 2\ H_2O$$

STEP 9 The hydrogen atoms should automatically balance. They do, at 4 atoms on each side.

$$3\ MnO_4^{2-} + 4\ H^+ \longrightarrow MnO_2 + 2\ MnO_4^- + 2\ H_2O$$

Section 15.7 · Disproportionation Reactions

(b) *Ion–Electron Method*

STEP 1 Assignment of oxidation numbers is the same as in part (a).

$$MnO_4^{2-} \longrightarrow MnO_2 + MnO_4^-$$
$$+6 \; -2 \qquad +4 \; -2 \quad +7 \; -2$$

Manganese is undergoing both oxidation and reduction.

STEP 2 The skeleton half-reactions for oxidation and reduction are

Oxidation: $MnO_4^{2-} \longrightarrow MnO_4^-$
Reduction: $MnO_4^{2-} \longrightarrow MnO_2$

Note how disproportionation is handled at this point. The substance undergoing disproportionation appears as a reactant in both the oxidation and reduction half-reactions.

STEP 3 In both half-reactions the element manganese is already balanced.

STEP 4 Oxygen is the only other element present.

STEP 5 Since this reaction occurs in acidic solution, we use H_2O molecules to achieve oxygen balance. They are needed only in the reduction half-reaction.

Oxidation: $MnO_4^{2-} \longrightarrow MnO_4^-$
Reduction: $MnO_4^{2-} \longrightarrow MnO_2 + 2\,H_2O$

STEP 6 In acidic solution, H balance is achieved using H^+ ions. Again, only the reduction half-reaction will need such ions.

Oxidation: $MnO_4^{2-} \longrightarrow MnO_4^-$
Reduction: $MnO_4^{2-} + 4\,H^+ \longrightarrow MnO_2 + 2\,H_2O$

STEP 7 Charge balance is achieved using electrons as a source of negative charge. Both half-reactions require the addition of electrons; this will always be the case.

Oxidation: $MnO_4^{2-} \longrightarrow MnO_4^- + e^-$
Reduction: $MnO_4^{2-} + 4\,H^+ + 2\,e^- \longrightarrow MnO_2 + 2\,H_2O$

STEP 8 The oxidation half-reaction contains one electron, and the reduction half-reaction two electrons. The numbers of electrons lost and gained are made equal by multiplying the oxidation half-reaction by the factor 2.

Oxidation: $2(MnO_4^{2-} \longrightarrow MnO_4^- + e^-)$
Reduction: $MnO_4^{2-} + 4\,H^+ + 2\,e^- \longrightarrow MnO_2 + 2\,H_2O$

STEP 9 Adding the two half-reactions together, we get

Oxidation: $2\,MnO_4^{2-} \longrightarrow 2\,MnO_4^- + \cancel{2\,e^-}$
Reduction: $MnO_4^{2-} + 4\,H^+ + \cancel{2\,e^-} \longrightarrow MnO_2 + 2\,H_2O$
$\overline{ 3\,MnO_4^{2-} + 4\,H^+ \longrightarrow 2\,MnO_4^- + MnO_2 + 2\,H_2O}$

STEP 10 There are no additional species to cancel.
STEP 11 The coefficients are already in the smallest possible whole number ratio. Thus, the equation in step 9 is the final balanced equation.

PRACTICE EXERCISE 15.10

Balance the following disproportionation redox reaction using (a) the oxidation number method and (b) the ion-electron method.

$$NO_2 \longrightarrow NO_3^- + NO \quad \text{(acidic solution)}$$

Ans. $3\,NO_2 + H_2O \rightarrow 2\,NO_3^- + NO + 2\,H^+$

Similar exercises: Problems 15.53–15.56

15.8 Some Important Oxidation–Reduction Reactions

In this section we consider two important applications of redox reactions.

1. A *spontaneous* oxidation–reduction reaction can be used to convert chemical energy into electrical energy. For this to occur, a redox reaction must be carried out in a specially designed apparatus called a *galvanic cell.*
2. A *nonspontaneous* oxidation–reduction reaction can be caused to occur by using electrical energy to produce chemical energy. Such a process is called *electrolysis,* and the apparatus in which the reaction is carried out is called an *electrolytic cell.*

Galvanic Cells

When a strip of zinc metal is placed in a solution of copper(II) sulfate, a source of Cu^{2+} ion, a coating of copper metal forms on the zinc strip (see Fig. 15.1). At the same time this occurs, some of the zinc dissolves to give Zn^{2+} ions in solution. The reaction occurring is

$$Zn(s) + Cu^{2+}(aq) \longrightarrow Zn^{2+}(aq) + Cu(s)$$

The sulfate ions (SO_4^{2-}) present in the copper(II) sulfate solution remain unaffected by this change.

This reaction is an example of a spontaneous oxidation–reduction reaction. When zinc metal and a solution of Cu^{2+} ions come into contact with each other, the Zn metal is spontaneously oxidized and the Cu^{2+} ions are reduced; a direct transfer of electrons from the zinc atoms to the Cu^{2+} ions occurs. The products of this reaction are copper atoms (Cu^{2+} ions that have gained electrons) and Zn^{2+} ions (zinc atoms that have lost electrons). Heat is liberated as the reaction proceeds, as evidenced by a slight warming of the solution.

By rearranging the reactants for this spontaneous redox reaction, one can obtain energy in the form of electricity rather than heat. The desired arrangement, a simple

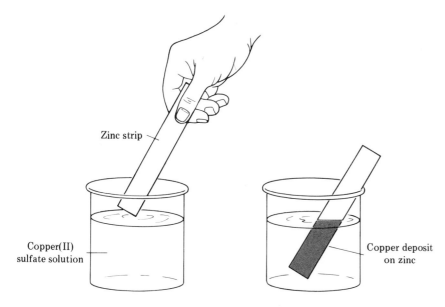

FIGURE 15.1 A spontaneous redox reaction occurs when zinc metal is placed in a solution of Cu^{2+} ions. A coating of copper metal quickly deposits on the zinc.

galvanic cell, is shown in Figure 15.2. A **galvanic cell** is an apparatus in which a spontaneous redox reaction is used to convert chemical energy to electrical energy.

The galvanic cell of Figure 15.2 has two compartments separated by a porous partition. One compartment contains a strip of zinc metal immersed in a solution of zinc sulfate ($ZnSO_4$), and the other contains a strip of copper metal immersed in a solution of copper(II) sulfate ($CuSO_4$). The porous partition prevents the solutions from mixing completely, but it does allow for passage of ions from one compartment to the other, a necessity for proper operation of the cell. The two strips of metal, called *electrodes*, are connected by a wire. This wire allows electrons to be transferred from one electrode to the other. As the spontaneous reaction between Zn and Cu^{2+} ions occurs, the flow of electrons through the wire can be demonstrated by placing a light bulb in the external circuit (see Fig. 15.2). The light bulb glows.

What is actually happening in the cell to cause the light bulb to glow?

1. Electrons are produced at the zinc electrode through the process of oxidation.

$$Zn(s) \longrightarrow Zn^{2+}(aq) + 2\ e^-$$

2. The electrons pass from the zinc electrode to the copper electrode through the external circuit (the wire).
3. Electrons enter the copper electrode and are accepted by Cu^{2+} ions in solution adjacent to the electrode. This is a reduction reaction.

$$Cu^{2+}(aq) + 2\ e^- \longrightarrow Cu(s)$$

4. To complete the circuit, ions (both positive and negative) move through the solution, passing through the porous membrane as needed.

Special names are given to the two electrodes in a galvanic cell; one is called the *cathode* and the other is the *anode*. The **cathode** is the electrode at which reduction

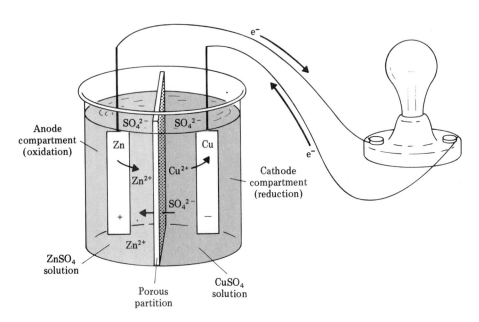

FIGURE 15.2 A zinc–copper galvanic cell. The electrical energy produced by this cell is generated by the spontaneous redox reaction

$$Zn(s) + Cu^{2+}(aq) \rightarrow Zn^{2+}(aq) + Cu(s)$$

takes place. Here electrons enter the galvanic cell from the external circuit. The **anode** is the electrode at which oxidation takes place. Here electrons leave the cell for the external circuit. In the cell now under discussion, the copper electrode is the cathode and the zinc electrode is the anode.

Many students have a hard time remembering the relationship between cathode–anode and oxidation–reduction. A mnemonic device can be helpful here. The two words that begin with vowels (anode and oxidation) go together, and the two words beginning with consonants (cathode and reduction) go together.

In principle, any spontaneous oxidation–reduction reaction can be used to build a galvanic cell, and many such cells have been studied in the laboratory. A selected few of such galvanic cells are now major commercial entities. We will discuss two that are commercially important: (1) the dry cell and (2) the lead storage battery.

The Dry Cell

The *dry cell* is widely used in flashlights, portable radios and tape recorders, and battery-powered toys; it is often referred to as a flashlight battery. Two versions of the dry cell are marketed: an acidic version and an alkaline version.

The *acidic version* of the dry cell contains a zinc outer surface (covered with cardboard or paint for protection) that acts as the anode and a carbon (graphite) rod in contact with a moist paste that serves as the cathode (see Fig. 15.3). The paste (the cell is not truly dry) is a mixture of solid MnO_2, solid NH_4Cl, and graphite powder (C) moistened with water. In the operation of the cell, Zn is oxidized to

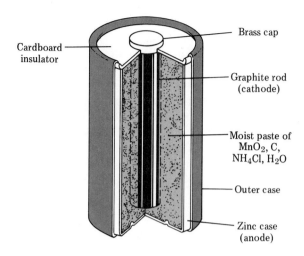

FIGURE 15.3 The acidic Zn–MnO$_2$ dry cell is commonly known as a flashlight battery.

Zn^{2+} ion, and MnO$_2$ (a paste component) is reduced to Mn$_2$O$_3$. The reduction takes place at the interface between the carbon cathode and the paste; the inert carbon electrode conducts the electrons to the external circuit. The electrode reactions are

Anode (oxidation):

$$\text{Zn(s)} \longrightarrow \text{Zn}^{2+}\text{(aq)} + 2\,e^-$$

Cathode (reduction):

$$2\,\text{MnO}_2\text{(s)} + 2\,\text{NH}_4^+\text{(aq)} + 2\,e^- \longrightarrow \text{Mn}_2\text{O}_3\text{(s)} + 2\,\text{NH}_3\text{(aq)} + \text{H}_2\text{O}$$

The useful life of acidic dry cells can be shortened if the slightly acidic paste corrodes the zinc can. A protective paper is inserted between the paste and zinc to help minimize this problem.

In the *alkaline version* of the dry cell, the solid NH$_4$Cl is replaced with KOH, and a steel rod rather than a graphite one is the cathode. The anode reaction still involves oxidation of zinc, but the zinc is present as a powder in a gel formulation. The cathode reaction also still involves the reduction of MnO$_2$. Equations for the electrode reactions are

Anode (oxidation):

$$\text{Zn(s)} + 2\,\text{OH}^-\text{(aq)} \longrightarrow \text{ZnO(s)} + \text{H}_2\text{O(l)} + 2\,e^-$$

Cathode (reduction):

$$2\,\text{MnO}_2\text{(s)} + \text{H}_2\text{O(l)} + 2\,e^- \longrightarrow \text{Mn}_2\text{O}_3\text{(s)} + 2\,\text{OH}^-\text{(aq)}$$

The alkaline dry cell costs roughly three times as much to produce as the acidic version. A major cost factor is the more elaborate internal construction needed to prevent leakage of the KOH solution. These cells provide up to 50% more total energy than the less expensive acidic model because they maintain usable voltage over a larger fraction of the lifetime of the cathode and anode materials. Miniature alkaline cells find extensive use in calculators, watches, and camera exposure controls.

Lead Storage Battery

The lead storage battery provides the starting power for automobiles. A 12-volt lead storage battery, the standard size, consists of six galvanic cells connected together (see Fig. 15.4). Each cell generates 2 volts.

Both electrodes in a lead storage battery involve the metal lead. Lead serves as the anode, and lead coated with lead dioxide serves as the cathode. Glass fiber spacers separate electrodes to prevent them from touching each other. The electrodes are immersed in a 38% (m/m) sulfuric acid (H_2SO_4) solution. Details of a single cell of a lead storage battery are shown in Figure 15.5.

The electrode reactions when a lead storage battery is used to supply power (discharging) are

Anode (oxidation):

$$Pb(s) + SO_4^{2-}(aq) \longrightarrow PbSO_4(s) + 2\,e^-$$

Cathode (reduction):

$$PbO_2(s) + 4\,H^+(aq) + SO_4^{2-}(aq) + 2\,e^- \longrightarrow PbSO_4(s) + 2\,H_2O(l)$$

The chemical results of discharge are a buildup of $PbSO_4$ on the electrodes and a decrease in the density of the sulfuric acid. [The state of charge of a lead storage battery can thus be checked by a service station attendant by measuring the density (Sec. 3.8) of the sulfuric acid.]

Unlike dry cells, a lead storage battery can be recharged. This reverse process, which is nonspontaneous, must be started by an external source of energy; in the

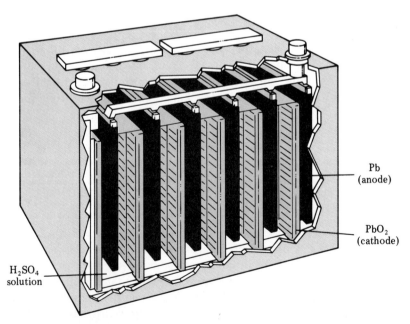

FIGURE 15.4 The six galvanic cells in a 12-volt automobile battery.

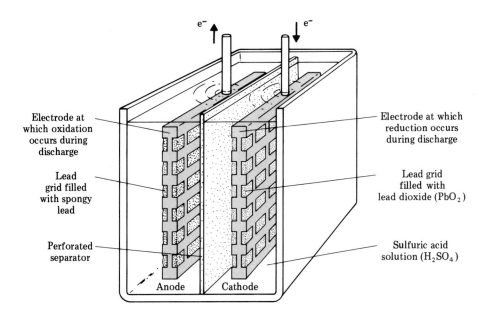

FIGURE 15.5 A single lead storage battery cell.

automobile the external energy source is an alternator driven by the automobile engine.

In recharging, the $PbSO_4$ on the electrodes (formed during discharge) is converted back to Pb at one electrode and to PbO_2 at the other. The electrode reactions are the reverse of what occurs during discharge.

Theoretically a lead storage battery should be rechargeable indefinitely. In practice, such batteries have a lifetime of 3–5 yr because small amounts of lead sulfate continually fall from the electrodes (to the bottom of the cell) as a result of "road shock" and chemical side reactions. Eventually the electrodes lose so much lead sulfate that the recharging process is no longer effective.

In "standard" lead storage batteries, water must be added to the individual cells on a regular basis. Recharging the battery, besides converting $PbSO_4$ back to Pb and PbO_2, also decomposes small amounts of water to give H_2 and O_2; hence, the H_2O must be replenished. Because of the possible presence of H_2 gas in a lead storage battery, a person should wear glasses (for eye protection) when releasing the cap of a battery since escaping gas can force sulfuric acid out. In addition, a person should not smoke while doing this, since hydrogen gas is flammable and forms explosive mixtures with oxygen. Newer automobile batteries have electrodes made of an alloy of calcium and lead. The presence of the calcium minimizes the decomposition of water during recharging. Thus, batteries with these alloy electrodes can be sealed; there is no need to add water.

Electrolytic Cells

The application of electrical energy from an external power source can be used to cause a nonspontaneous redox reaction to occur. The charging of the lead storage battery previously discussed is an example of this. The general term for such a process is electrolysis. **Electrolysis** is the process in which electrical energy is used to produce a nonspontaneous redox reaction. During recharging the lead storage battery is functioning as an electrolytic cell. An **electrolytic cell** is an apparatus in which chemical change is caused to occur through the application of electrical energy.

Electrolysis has a number of important commercial applications, including (1) production of important industrial chemicals, (2) electrorefining and purification of metals, and (3) electroplating. Let us consider examples in all three of these areas.

Chlorine gas, hydrogen gas, and sodium hydroxide—three important industrial chemicals—can be produced simultaneously from the electrolysis of a concentrated aqueous sodium chloride solution (saltwater or brine solution). The type of electrolytic cell needed is shown in Figure 15.6.

In this cell the two electrodes are inert; that is, they do not participate in the redox reactions themselves but serve as surfaces on which the redox reactions occur. As with galvanic cells, reduction occurs at the cathode and oxidation at the anode.

The negative ions in the solution, the Cl^- ions, react at the anode, where they give up electrons and are reduced to produce Cl_2 gas.

$$\text{Anode (oxidation):} \quad 2\,Cl^-(aq) \longrightarrow Cl_2(g) + 2\,e^-$$

The positive ions in the solution, the Na^+ ions, you would expect to react at the cathode, where they could pick up electrons to produce Na metal atoms. However, this process does not occur. Instead, water, the solvent in the solution, reacts at the cathode; it is more easily reduced than Na^+ ion. The cathode reaction is

$$\text{Cathode (reduction):} \quad 2\,H_2O(l) + 2\,e^- \longrightarrow H_2(g) + 2\,OH^-(aq)$$

This reduction yields H_2 gas as well as OH^- ions. The OH^- ions remain in solution, producing a solution that now contains Na^+ ions and OH^- ions; our solution of sodium chloride has been changed to one of sodium hydroxide. Thus, three sub-

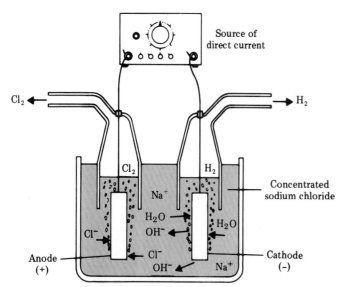

FIGURE 15.6 The electrolysis of aqueous sodium chloride solution (saltwater brine) produces hydrogen gas, chlorine gas, and sodium hydroxide solution.

stances—Cl_2 gas, H_2 gas, and NaOH solution—result from the electrolysis of a concentrated NaCl solution.

Metal purification frequently depends on electrolysis. All copper used in electrical wire is electrolytically purified; impurities cut the electrical conductance of the wire. In an electrolytic cell used to purify copper, large slabs of impure copper (obtained from the reduction of copper ores) serve as anodes, and thin sheets of very pure copper serve as cathodes. The solution in which the electrodes are immersed is an acidic copper(II) sulfate solution. As the cell is operated, the anodes (the impure copper) decrease in size and the cathodes (the pure copper) increase in size (see Fig. 15.7). What is happening? At the anodes, oxidation of Cu causes it to dissolve. The reaction is

$$\text{Anode (oxidation):} \quad Cu(s) \longrightarrow Cu^{2+}(aq) + 2\,e^-$$

As the anodic copper dissolves, the impurities present also go into solution (are also oxidized) or fall to the bottom of the cell. At the cathode, reduction causes copper to come out of solution.

$$\text{Cathode (reduction):} \quad Cu^{2+}(aq) + 2\,e^- \longrightarrow 2\,Cu(s)$$

The net effect of this oxidation–reduction process is that copper is transferred via solution from one electrode (the impure one) to the other (the pure one). The electricity supplied to the cell is set at a value that allows copper ions to be reduced but is not sufficient to reduce any dissolved impurities. Thus, only copper, and not impurities, deposits on the cathode.

Electroplating is the deposition of a thin layer of a metal on an object through the process of electrolysis. Electroplated objects are common in our society. Jewelry is plated with silver and gold. Tableware is often plated with silver. Gold-plated electrical contacts are used extensively. "Tin cans" are actually steel cans with a thin coating of tin. Chromium-plated steel automobile bumpers have been used for many years. The thin metallic layer deposited during electroplating is generally only 0.001–0.002 in. thick.

Figure 15.8 shows a typical apparatus used for electroplating items with silver. The object to be plated is made the cathode in a solution containing Ag^+ ions. The

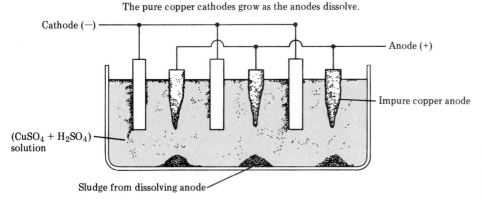

FIGURE 15.7 Cross section of an electrolytic cell for purifying copper.

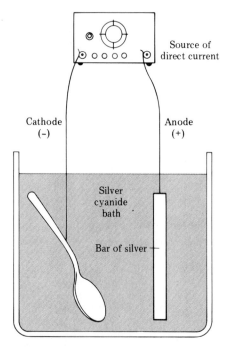

FIGURE 15.8 An apparatus for electroplating silver.

anode is a bar of the plating metal, silver in this case. At the cathode, Ag^+ ions from solution are deposited as metallic silver.

$$\text{Cathode (reduction):} \quad Ag^+(aq) + e^- \longrightarrow Ag(s)$$

At the anode, silver from that electrode is oxidized to give Ag^+ ions in solution; this replenishes the supply of Ag^+ ions in solution needed for the plating process.

$$\text{Anode (oxidation):} \quad Ag(s) \longrightarrow Ag^+(aq) + e^-$$

The electroplating bath, the solution around the electrodes, usually contains other chemicals besides the plating metal. For example, silver plating is usually done from a solution containing both AgCN and KCN.

Key Terms

The new terms or concepts defined in this chapter are

anode (Sec. 15.8)   The electrode in an electrochemical cell at which oxidation takes place.

cathode (Sec. 15.8)   The electrode in an electrochemical cell at which reduction takes place.

disproportionation reaction (Sec. 15.7)   A redox reaction in which some atoms of a single element in a reactant are oxidized and others are reduced.

electrolysis (Sec. 15.8)   The process in which electrical energy is used to produce a nonspontaneous redox reaction.

electrolytic cell (Sec. 15.8)   An apparatus in which chemical change is caused to occur through the application of electrical energy.

electroplating (Sec. 15.8)   The deposition of a thin layer of metal on an object through the process of electrolysis.

galvanic cell (Sec. 15.8) An apparatus in which a spontaneous redox reaction is used to convert chemical energy to electrical energy.

ion–electron method for balancing redox equations (Sec. 15.6) A method for balancing redox equations that employs two partial equations, an oxidation half-reaction and a reduction half-reaction.

oxidation (Sec. 15.1) Process whereby a substance in a chemical reaction loses one or more electrons.

oxidation number (Sec. 15.2) Charge that an atom appears to have when the electrons in each bond it is participating in are assigned to the more electronegative of the two atoms involved in the bond.

oxidation-number method for balancing redox equations (Sec. 15.5) A method for balancing redox equations in which oxidation numbers are assigned to each element present and oxidation-number increase is made to equal oxidation-number decrease.

oxidation–reduction reaction (Sec. 15.1) A reaction in which there is a transfer of electrons between reactants.

oxidizing agent (Secs. 15.1 and 15.2) The substance that accepts electrons in a redox reaction or the substance that contains the atom that shows a decrease in oxidation number in a redox reaction.

redox reaction (Sec. 15.1) A shortened designation for an oxidation–reduction reaction.

reducing agent (Secs. 15.1 and 15.2) The substance that provides electrons in a redox reaction or the substance that contains the atom that shows an increase in oxidation number in a redox reaction.

reduction (Sec. 15.1) The process whereby a substance in a chemical reaction gains one or more electrons.

Practice Problems

Oxidation–Reduction Terminology (Secs. 15.1 and 15.2)

15.1 Oxidation–reduction reactions are often called electron-transfer reactions. Explain why.

15.2 Oxidation and reduction occur simultaneously in a chemical reaction. Explain why.

15.3 Give definitions of *oxidation* in terms of
(a) loss or gain of electrons
(b) increase or decrease in oxidation number

15.4 Give definitions of *reduction* in terms of
(a) loss or gain of electrons
(b) increase or decrease in oxidation number

15.5 Give definitions of *oxidizing agent* in terms of
(a) loss or gain of electrons
(b) increase or decrease in oxidation number
(c) substance oxidized or substance reduced

15.6 Give definitions of *reducing agent* in terms of
(a) loss or gain of electrons
(b) increase or decrease in oxidation number
(c) substance oxidized or substance reduced

15.7 In each of the following statements, choose the word in parentheses that best completes the statement.
(a) An element that has lost electrons in a redox reaction is said to have been (oxidized, reduced).
(b) Reduction always results in an (increase, decrease) in the oxidation number.
(c) The substance oxidized in a redox reaction is the (oxidizing, reducing) agent.
(d) The reducing agent (gains, loses) electrons during a redox reaction.

15.8 In each of the following statements, choose the word in parentheses that best completes the statement.
(a) The reducing agent causes an (increase, decrease) in the oxidation number of the oxidizing agent in a redox reaction.
(b) The oxidizing agent (gains, loses) electrons during a redox reaction.
(c) Oxidation always results in an (increase, decrease) in the oxidation number.
(d) An element that has gained electrons in a redox reaction is said to have been (oxidized, reduced).

Assignment of Oxidation Numbers (Sec. 15.2)

15.9 What is the oxidation number of the indicated element in each of the following substances?
(a) P in PF_3 (b) S in H_2S
(c) N in HNO_3 (d) Na in NaOH

15.10 What is the oxidation number of the indicated element in each of the following substances?
(a) N in N_2O (b) P in P_4O_6
(c) S in H_2SO_4 (d) K in $KClO_3$

15.11 What is the oxidation number of the indicated element in each of the following ions?
(a) N in N^{3-} (b) Ca in Ca^{2+}
(c) Al in $AlCl_4^-$ (d) P in PH_4^+

15.12 What is the oxidation number of the indicated element in each of the following ions?
(a) S in S^{2-} (b) K in K^+
(c) P in PO_4^{3-} (d) Cl in ClF_2^+

15.13 What is the oxidation number of Mn in each of the following manganese-containing substances?
(a) MnO (b) MnO_2
(c) K_2MnO_4 (d) Mn_2O_7

15.14 What is the oxidation number of Cr in each of the following chromium-containing substances?
(a) Cr_2O_3 (b) $KCrO_2$
(c) CrF_5 (d) $BeCr_2O_7$

15.15 What is the oxidation number of each element in each of the following substances?
(a) P_4O_{10} (b) S_2O
(c) $H_4P_2O_7$ (d) $(NH_4)_3PO_4$

15.16 What is the oxidation number of each element in each of the following substances?
(a) CO_2 (b) Cl_2O_7
(c) HPO_3 (d) $(NH_4)_2CO_3$

15.17 Indicate whether oxygen has a -2, -1, or positive oxidation number in each of the following oxygen-containing species.
(a) Na_2O (b) OF_2 (c) Na_2O_2 (d) BaO

15.18 Indicate whether oxygen has a -2, -1, or positive oxidation number in each of the following oxygen-containing species.
(a) O_2F_2 (b) SO_3 (c) BaO_2 (d) BeO

15.19 Indicate whether hydrogen has a $+1$ or -1 oxidation number in each of the following hydrogen-containing species.
(a) NaH (b) CH_4 (c) HCl (d) CaH_2

15.20 Indicate whether hydrogen has a $+1$ or -1 oxidation number in each of the following hydrogen-containing species.
(a) H_2Se (b) N_2H_2 (c) KH (d) MgH_2

Oxidizing and Reducing Agents (Sec. 15.2)

15.21 For each of the following unbalanced equations, indicate whether the sulfur-containing species has been oxidized, reduced, or neither oxidized nor reduced.
(a) $S + O_2 \rightarrow SO_2$
(b) $H_2SO_3 + I_3^- \rightarrow HSO_4^- + I^-$
(c) $Ba^{2+} + SO_4^{2-} \rightarrow BaSO_4$
(d) $MnO_4^- + H_2S \rightarrow Mn^{2+} + S$

15.22 For each of the following unbalanced equations, indicate whether the nitrogen-containing species has been oxidized, reduced, or neither oxidized nor reduced.
(a) $N_2 + H_2 \rightarrow NH_3$
(b) $NO + O_2 \rightarrow NO_2$
(c) $Al + N_2 \rightarrow AlN$
(d) $N_2O_4 \rightarrow 2 NO_2$

15.23 For each of the following unbalanced equations, identify the oxidizing agent and the reducing agent.
(a) $H_2 + Cl_2 \rightarrow HCl$
(b) $SO_2 + O_2 \rightarrow SO_3$
(c) $HBr + Mg \rightarrow MgBr_2 + H_2$
(d) $H_2 + FeCl_3 \rightarrow HCl + FeCl_2$

15.24 For each of the following unbalanced equations, identify the oxidizing agent and the reducing agent.
(a) $H_2S + HNO_3 \rightarrow S + NO + H_2O$
(b) $Zn + Cu^{2+} \rightarrow Zn^{2+} + Cu$
(c) $K_2S + I_2 + KOH \rightarrow K_2SO_4 + KI + H_2O$
(d) $MnO_4^- + H^+ + Hg \rightarrow Mn^{2+} + H_2O + Hg_2^{2+}$

15.25 For each of the following reactions, identify the substance oxidized, the substance reduced, the oxidizing agent, and the reducing agent.
(a) $2 Al + 3 Cl_2 \rightarrow 2 AlCl_3$
(b) $2 ReCl_5 + SbCl_3 \rightarrow 2 ReCl_4 + SbCl_5$
(c) $Fe_2O_3 + 2 Al \rightarrow Al_2O_3 + 2 Fe$
(d) $Zn + CuCl_2 \rightarrow ZnCl_2 + Cu$

15.26 For each of the following reactions, identify the substance oxidized, the substanced reduced, the oxidizing agent, and the reducing agent.
(a) $2 NiS + 3 O_2 \rightarrow 2 NiO + 2 SO_2$
(b) $WO_3 + 3 H_2 \rightarrow W + 3 H_2O$
(c) $4 NH_3 + 3 O_2 \rightarrow 2 N_2 + 6 H_2O$
(d) $K_2Cr_2O_7 + 14 HI \rightarrow 2 CrI_3 + 2 KI + 3 I_2 + 7 H_2O$

Types of Chemical Reactions (Sec. 15.3)

15.27 Characterize each of the following reactions using one selection from the choices *redox* and *non-redox* combined with one selection from the choices *synthesis, decomposition, single replacement,* and *double replacement*.
(a) $SO_3 + H_2O \rightarrow H_2SO_4$
(b) $2 H_2 + O_2 \rightarrow 2 H_2O$
(c) $2 AgNO_3 + K_2SO_4 \rightarrow Ag_2SO_4 + 2 KNO_3$
(d) $2 KClO_3 \rightarrow 2 KCl + 3 O_2$

15.28 Characterize each of the following reactions using one selection from the choices *redox* and *non-redox* combined with one selection from the choices *synthesis, decomposition, single replacement,* and *double replacement*.
(a) $2 HgO \rightarrow 2 Hg + O_2$
(b) $Al(OH)_3 + 3 HCl \rightarrow AlCl_3 + 3 H_2O$

(c) $Mg + 2\,HCl \rightarrow MgCl_2 + H_2$
(d) $2\,SO_2 + O_2 \rightarrow 2\,SO_3$

15.29 Characterize each of the following reactions as (1) a redox reaction, (2) a non-redox reaction, or (3) "can't classify" because of insufficient information.
(a) a synthesis reaction in which both reactants are elements
(b) a combustion reaction
(c) a decomposition reaction in which the products are all compounds
(d) a decomposition reaction in which an element and a compound are products

15.30 Characterize each of the following reactions as (1) a redox reaction, (2) a non-redox reaction, or (3) "can't classify" because of insufficient information.
(a) a synthesis reaction in which one reactant is an element and the other is a compound
(b) an acid–base neutralization reaction
(c) a decomposition reaction in which the products are all elements
(d) a single–replacement reaction involving an active metal and an acid

Balancing Redox Equations: Oxidation-Number Method
(Sec. 15.5)

15.31 Balance the following equations by the oxidation-number method.
(a) $F_2 + H_2O \rightarrow HF + O_2$
(b) $BaSO_4 + C \rightarrow BaS + CO$
(c) $I_2O_5 + CO \rightarrow I_2 + CO_2$
(d) $Br_2 + H_2O + SO_2 \rightarrow HBr + H_2SO_4$

15.32 Balance the following equations by the oxidation-number method.
(a) $PH_3 + NO_2 \rightarrow H_3PO_4 + N_2$
(b) $Mg + AgNO_3 \rightarrow Ag + Mg(NO_3)_2$
(c) $H_2 + KClO \rightarrow KCl + H_2O$
(d) $Bi(OH)_3 + K_2SnO_2 \rightarrow Bi + K_2SnO_3 + H_2O$

15.33 Balance the following equations by the oxidation-number method.
(a) $PbO_2 + Sb + NaOH \rightarrow PbO + NaSbO_2 + H_2O$
(b) $S + H_2O + Pb(NO_3)_2 \rightarrow Pb + H_2SO_3 + HNO_3$
(c) $SnSO_4 + FeSO_4 \rightarrow Sn + Fe_2(SO_4)_3$
(d) $MnO_4^- + SO_2 + H_2O \rightarrow Mn^{2+} + SO_4^{2-} + H^+$

15.34 Balance the following equations by the oxidation-number method.
(a) $As_4O_6 + Cl_2 + H_2O \rightarrow H_3AsO_4 + HCl$
(b) $Na_2TeO_3 + NaI + HCl \rightarrow NaCl + Te + H_2O + I_2$
(c) $KMnO_4 + AsH_3 + H_2SO_4 \rightarrow H_3AsO_4 + MnSO_4 + H_2O + K_2SO_4$
(d) $Br_2 + SO_2 + H_2O \rightarrow H^+ + Br^- + SO_4^{2-}$

15.35 Balance the following equations by the oxidation-number method.
(a) $NaNO_3 + Pb \rightarrow NaNO_2 + PbO$
(b) $NF_3 + AlCl_3 \rightarrow N_2 + Cl_2 + AlF_3$
(c) $H_2S + HNO_3 \rightarrow S + NO + H_2O$
(d) $S^{2-} + Cl_2 + OH^- \rightarrow SO_4^{2-} + Cl^- + H_2O$

15.36 Balance the following equations by the oxidation-number method.
(a) $H_2O_2 + HI \rightarrow I_2 + H_2O$
(b) $PbO_2 + H_2SO_4 + Mn(NO_3)_2 \rightarrow PbSO_4 + HNO_3 + HMnO_4 + H_2O$
(c) $HNO_3 + I_2 \rightarrow NO_2 + H_2O + HIO_3$
(d) $SnO_2^{2-} + Bi^{3+} + OH^- \rightarrow SnO_3^{2-} + Bi + H_2O$

15.37 Balance the following equations by the oxidation-number method. All reactions occur in acidic solution.
(a) $H_2S + NO_3^- \rightarrow S + NO$
(b) $Cu + NO_3^- \rightarrow Cu^{2+} + NO_2$
(c) $I^- + HNO_2 \rightarrow I_2 + NO$
(d) $Mn^{2+} + BiO_3^- \rightarrow MnO_4^- + Bi^{3+}$

15.38 Balance the following equations by the oxidation-number method. All reactions occur in acidic solution.
(a) $Cu + NO_3^- \rightarrow Cu^{2+} + NO$
(b) $ClO_3^- + I^- \rightarrow Cl^- + I_2$
(c) $AsH_3 + Ag^+ \rightarrow As_4O_6 + Ag$
(d) $IO_3^- + N_2H_4 \rightarrow I^- + N_2$

15.39 Balance the following equations by the oxidation-number method. All reactions occur in basic solution.
(a) $S^{2-} + Cl_2 \rightarrow SO_4^{2-} + Cl^-$
(b) $SO_3^{2-} + CrO_4^{2-} \rightarrow Cr(OH)_4^- + SO_4^{2-}$
(c) $MnO_4^- + IO_3^- \rightarrow MnO_2 + IO_4^-$
(d) $I_2 + Cl_2 \rightarrow H_3IO_6^{2-} + Cl^-$

15.40 Balance the following equations by the oxidation-number method. All reactions occur in basic solution.
(a) $Zn + MnO_4^- \rightarrow Zn(OH)_2 + MnO_2$
(b) $NO_2^- + Al \rightarrow NH_3 + AlO_2^-$
(c) $NO_2^- + MnO_4^- \rightarrow NO_3^- + MnO_2$
(d) $Al + NO_3^- \rightarrow Al(OH)_4^- + NH_3$

Balancing Redox Equations: Ion–Electron Method
(Sec. 15.6)

15.41 Balance the following half-reactions occurring in an acidic solution.
(a) $MnO_4^- \rightarrow Mn^{2+}$
(b) $H_2S \rightarrow S$
(c) $S_2O_3^{2-} \rightarrow S_2O_6^{2-}$
(d) $Pt + Cl^- \rightarrow PtCl_6^{2-}$

15.42 Balance the following half-reactions occurring in an acidic solution
(a) $ClO_3^- \rightarrow Cl^-$
(b) $SO_4^{2-} \rightarrow SO_2$

(c) $Cr_2O_7^{2-} \to Cr^{3+}$
(d) $SO_2Cl_2 \to SO_3^{2-} + Cl^-$

15.43 Balance the following half-reactions occurring in a basic solution.
(a) $MnO_4^- \to MnO_2$
(b) $F_2 \to F^-$
(c) $PH_3 \to P_4$
(d) $CN^- \to CNO^-$

15.44 Balance the following half-reactions occurring in a basic solution.
(a) $BrO_4^- \to Br^-$
(b) $Zn \to Zn(OH)_4^{2-}$
(c) $As \to AsO_3^{3-}$
(d) $CNS^- \to CN^- + SO_4^{2-}$

15.45 Balance each of the following redox reactions by the ion–electron method. Each reaction occurs in acidic solution.
(a) $Zn + Ag^+ \to Zn^{2+} + Ag$
(b) $SO_4^{2-} + Zn \to Zn^{2+} + SO_2$
(c) $NO_2 + HOCl \to NO_3^- + Cl^-$
(d) $MnO_4^- + H_2C_2O_4 \to Mn^{2+} + CO_2$

15.46 Balance each of the following redox reactions by the ion–electron method. Each reaction occurs in acidic solution.
(a) $As + NO_3^- \to H_3AsO_3 + NO$
(b) $PH_3 + I_2 \to H_3PO_2 + I^-$
(c) $XeO_3 + I^- \to Xe + I_3^-$
(d) $ReO_2 + Cl_2 \to HReO_4 + Cl^-$

15.47 Balance each of the following redox reactions by the ion–electron method. Each reaction occurs in acidic solution.
(a) $Fe^{2+} + MnO_4^- \to Fe^{3+} + Mn^{2+}$
(b) $S_2O_3^{2-} + ClO^- \to Cl^- + S_4O_6^{2-}$
(c) $H_3PO_2 + Cr_2O_7^{2-} \to H_3PO_4 + Cr^{3+}$
(d) $P_4 + NO_3^- \to H_3PO_4 + NO$

15.48 Balance each of the following redox reactions by the ion–electron method. Each reaction occurs in acidic solution.
(a) $PbO_2 + Cl^- \to PbCl_2 + Cl_2$
(b) $Ag + NO_3^- \to NO_2 + Ag^+$
(c) $Te + NO_3^- \to TeO_2 + NO$
(d) $ClO_3^- + Fe^{2+} \to Fe^{3+} + Cl^-$

15.49 Balance each of the following redox reactions by the ion–electron method. Each reaction occurs in basic solution.
(a) $MnO_4^- + ClO_2^- \to MnO_2 + ClO_4^-$
(b) $Al + NO_3^- \to NH_3 + AlO_2^-$
(c) $BrO_3^- + I^- \to Br^- + IO^-$
(d) $ClO_3^- + N_2H_4 \to NO + Cl^-$

15.50 Balance each of the following redox reactions by the ion–electron method. Each reaction occurs in basic solution.
(a) $NiO_2 + Mn(OH)_2 \to Mn_2O_3 + Ni(OH)_2$
(b) $PH_3 + CrO_4^{2-} \to Cr(OH)_4^- + P_4$
(c) $Mo(OH)_3 + BrO^- \to MoO_4^{2-} + Br^-$
(d) $S^{2-} + I_2 \to SO_4^{2-} + I^-$

15.51 Balance each of the following redox reactions by the ion–electron method. Each reaction occurs in basic solution.
(a) $O_2 + N_2H_4 \to N_2 + H_2O_2$
(b) $I_2 + Cl_2 \to H_3IO_6^{2-} + Cl^-$
(c) $Bi(OH)_3 + SnO_2^{2-} \to SnO_3^{2-} + Bi$
(d) $Cr(OH)_3 + BrO^- \to CrO_4^{2-} + Br^-$

15.52 Balance each of the following redox reactions by the ion–electron method. Each reaction occurs in basic solution.
(a) $NiO_2 + Fe \to Ni(OH)_2 + Fe(OH)_3$
(b) $Al + PO_3^- \to PH_3 + AlO_2^-$
(c) $MnO_4^- + C_2O_4^{2-} \to MnO_2 + CO_2$
(d) $Bi(OH)_3 + Sn(OH)_3^- \to Sn(OH)_6^{2-} + Bi$

Balancing Redox Reactions: Disproportionation Reactions (Sec. 15.7)

15.53 Balance each of the following redox reactions by the oxidation-number method.
(a) $HNO_2 \to NO + NO_3^-$ (acidic solution)
(b) $Cl_2 \to Cl^- + ClO^-$ (acidic solution)
(c) $S \to S^{2-} + SO_3^{2-}$ (basic solution)
(d) $Br_2 \to BrO_3^- + Br^-$ (basic solution)

15.54 Balance each of the following redox reactions by the oxidation-number method.
(a) $NO + NO_3^- \to N_2O_4$ (acidic solution)
(b) $H_5IO_6 + I^- \to I_2$ (acidic solution)
(c) $P_4 \to HPO_3^{2-} + PH_3$ (basic solution)
(d) $HClO_2 \to ClO_2 + Cl^-$ (basic solution)

15.55 Balance each of the redox reactions in Problem 15.53 using the ion–electron method.

15.56 Balance each of the redox reactions in Problem 15.54 using the ion–electron method.

Important Oxidation–Reduction Processes (Sec. 15.8)

15.57 What processes occur at the anode and cathode in a galvanic cell?

15.58 What processes occur at the anode and cathode in an electrolytic cell?

15.59 When a strip of copper is dipped into a solution of silver nitrate (Ag^+ ions and NO_3^- ions), the copper is silver-plated. Write a chemical equation to describe this process.

15.60 When a strip of zinc is dipped into a solution of nickel(II) nitrate (Ni^{2+} ions and NO_3^- ions), the zinc is nickel-plated. Write a chemical equation to describe this process.

15.61 What are the anode and cathode reactions during the operation of an acidic dry cell?

15.62 What are the anode and cathode reactions during the operation of an alkaline dry cell?

15.63 What are the anode and cathode reactions during the discharging of a lead storage battery?

15.64 What are the anode and cathode reactions during the charging of a lead storage battery?

15.65 Why does the density of the H_2SO_4 in a lead storage battery decrease as the cell discharges?

15.66 Why is it not possible to recharge a lead storage battery an infinite number of times?

15.67 Write an equation for what happens at the anode during the electrolysis of a concentrated aqueous NaCl solution.

15.68 Write an equation for what happens at the cathode during the electrolysis of a concentrated aqueous NaCl solution.

Additional Problems

15.69 The noble gas xenon forms compounds with the elements oxygen and/or fluorine. In such compounds xenon is always the central atom and all other atoms are bonded to it. Assign an oxidation number to xenon in the following such compounds.
 (a) XeO_4 (b) $XeOF_4$ (c) XeO_2F_2 (d) XeO_3F_2

15.70 Phosphorus forms a number of oxides. Assign an oxidation number to phosphorus in each of the following phosphorus oxides.
 (a) P_4O_6 (b) P_4O_7 (c) P_4O_9 (d) P_4O_{10}

15.71 Determine the oxidation numbers of all elements in each of the following *ionic* compounds.
 (a) $Mn(CO_3)_2$ (b) $Pb(S_2O_3)_2$
 (c) $HgCr_2O_7$ (d) Cu_2SO_3

15.72 Determine the oxidation numbers of all elements in each of the following *ionic* compounds.
 (a) $Ni_3(PO_4)_2$ (b) $Hg_2(BrO_3)_2$
 (c) $Fe_2(CrO_4)_3$ (d) $Cu(N_3)_2$

15.73 Identify the following species for the redox reaction

$$2\ MnO_4^- + 3\ SO_3^{2-} + H_2O \rightarrow 2\ MnO_2 + 3\ SO_4^{2-} + 2\ OH^-$$

 (a) oxidizing agent
 (b) substance reduced
 (c) substance that is the electron donor
 (d) substance that contains the element that increases in oxidation number

15.74 Identify the following species for the redox reaction

$$2\ MnO_2 + 4\ KOH + O_2 \rightarrow 2\ K_2MnO_4 + 2\ H_2O$$

 (a) reducing agent
 (b) substance oxidized
 (c) substance that is the electron acceptor
 (d) substance that contains the element that decreases in oxidation number

15.75 Write balanced equations for all possible redox reactions obtainable by combining the following balanced half-reactions in sets of two. The half-reactions must be used as written; they cannot be reversed in direction.

$$2\ H_2O + PH_3 \rightarrow H_3PO_2 + 4\ H^+ + 4\ e^-$$
$$3\ H_2O + As \rightarrow H_3AsO_3 + 3\ H^+ + 3\ e^-$$
$$MnO_4^- + 8\ H^+ + 5\ e^- \rightarrow Mn^{2+} + 4\ H_2O$$
$$SO_4^{2-} + 4\ H^+ + 2\ e^- \rightarrow SO_2 + 2\ H_2O$$

15.76 Write balanced equations for all possible redox reactions obtainable by combining the following balanced half-reactions in sets of two. The half-reactions must be used as written; they cannot be reversed in direction.

$$4\ OH^- + ClO_2^- \rightarrow ClO_4^- + 2\ H_2O + 4\ e^-$$
$$2\ H_2O + MnO_4^- + 3\ e^- \rightarrow MnO_2 + 4\ OH^-$$
$$6\ H_2O + NO_3^- + 8\ e^- \rightarrow NH_3 + 9\ OH^-$$
$$4\ OH^- + Al \rightarrow AlO_2^- + 2\ H_2O + 3\ e^-$$

15.77 Write the two balanced half-reactions associated with the following redox reaction.

$$4\ Zn + 10\ HNO_3 \rightarrow 4\ Zn(NO_3)_2 + NH_4NO_3 + 3\ H_2O$$

15.78 Write the two balanced half-reactions associated with the following redox reaction.

$$14\ HNO_3 + 3\ Cu_2O \rightarrow 6\ Cu(NO_3)_2 + 2\ NO + 7\ H_2O$$

Cumulative Problems

15.79 Classify each of the following as an acid–base reaction or as an oxidation–reduction reaction. If it is an acid–base reaction, identify the acid; if it is an oxidation–reduction reaction, identify the oxidizing agent.
(a) $2\,HNO_3 + 3\,H_2S \rightarrow 3\,S + 2\,NO + 4\,H_2O$
(b) $HCl + NaCN \rightarrow NaCl + HCN$
(c) $2\,HI + H_2O_2 \rightarrow I_2 + 2\,H_2O$
(d) $H_2SO_4 + 2\,NaOH \rightarrow Na_2SO_4 + 2\,H_2O$

15.80 Classify each of the following as an acid–base reaction or as an oxidation–reduction reaction. If it is an acid–base reaction, identify the acid; if it is an oxidation–reduction reaction, identify the oxidizing agent.
(a) $7\,HI + H_5IO_6 \rightarrow 4\,I_2 + 6\,H_2O$
(b) $H_2CO_3 + 2\,KOH \rightarrow K_2CO_3 + 2\,H_2O$
(c) $5\,HClO_2 + NaOH \rightarrow 4\,ClO_2 + 3\,H_2O + NaCl$
(d) $H_2C_2O_4 + NaClO \rightarrow NaHC_2O_4 + HClO$

15.81 Convert each of the following balanced molecular redox equations to balanced net ionic redox equations.
(a) $SnSO_4(aq) + 2\,FeSO_4(aq) \rightarrow Sn(s) + Fe_2(SO_4)_3(aq)$
(b) $PH_3(g) + 2\,NO_2(g) \rightarrow H_3PO_4(aq) + N_2(g)$
(c) $S(s) + 3\,H_2O(l) + 2\,Pb(NO_3)_2 \rightarrow$
  $2\,Pb(s) + H_2SO_3(aq) + 4\,HNO_3(aq)$
(d) $4\,Zn(s) + 10\,HNO_3(aq) \rightarrow$
  $4\,Zn(NO_3)_2(aq) + NH_4NO_3(aq) + 3\,H_2O(l)$

15.82 Convert each of the following balanced molecular redox equations to balanced net ionic redox equations.
(a) $NaNO_3(aq) + Pb(s) \rightarrow NaNO_2(aq) + PbO(s)$
(b) $3\,H_2S(aq) + 2\,HNO_3(aq) \rightarrow$
  $3\,S(s) + 2\,NO(g) + 4\,H_2O(l)$
(c) $2\,HNO_3(aq) + SO_2(g) \rightarrow H_2SO_4(aq) + 2\,NO_2(g)$
(d) $10\,FeSO_4(aq) + 2\,KMnO_4(aq) + 8\,H_2SO_4(aq) \rightarrow$
  $5\,Fe_2(SO_4)_3(aq) + 2\,MnSO_4(aq) + K_2SO_4(aq) + 8\,H_2O(l)$

15.83 Balance each of the following ionic redox reactions.
(a) hydrosulfuric acid plus dichromate ion produces chromium(III) ion plus sulfur (acidic solution)
(b) chlorate ion plus iodine produces iodate ion plus chloride ion (acidic solution)
(c) sulfide ion plus bromine produces sulfate ion plus bromide ion (basic solution)
(d) nitrogen dioxide disproportionates to produce nitrate ion plus nitrite ion (basic solution)

15.84 Balance each of the following ionic redox reactions.
(a) iron(II) ion plus permanganate ion produces iron(III) ion plus manganese(II) ion (acidic solution)
(b) iodine plus sulfur dioxide produces iodide ion plus sulfate ion (acidic solution)
(c) manganese(II) hydroxide plus nickel(IV) oxide produces manganese(III) oxide plus nickel(II) hydroxide (basic solution)
(d) chlorine disproportionates to produce chlorate ion and chloride ion (basic solution)

15.85 The amount of I_3^-(aq) in a solution can be determined by reacting it with a solution containing $S_2O_3^{2-}$(aq).

$$I_3^-(aq) + 2\,S_2O_3^{2-}(aq) \rightarrow 3\,I^-(aq) + S_4O_6^{2-}(aq)$$

Calculate the molarity of I_3^- in a solution given that 43.2 mL of 0.300 M $S_2O_3^{2-}$ solution reacts with a 20.0-mL sample of the I_3^- solution.

15.86 Oxalic acid, $H_2C_2O_4$ reacts with chromate ion, CrO_4^{2-}, in acidic solution as follows.

$$3\,H_2C_2O_4(aq) + 2\,CrO_4^{2-}(aq) + 10\,H^+(aq) \rightarrow$$
$$6\,CO_2(g) + 2\,Cr^{3+}(aq) + 8\,H_2O(l)$$

If 35.0 mL of an oxalic acid solution reacts completely with 25.0 mL of a 0.100 M CrO_4^{2-} solution, what is the molarity of the oxalic acid solution?

15.87 The amount of ozone, O_3, in polluted air can be determined in a two-step process. First the ozone is reacted with a solution containing iodide ion that is acidic.

$$O_3(g) + 2\,I^-(aq) + 2\,H^+(aq) \rightarrow O_2(g) + I_2(s) + H_2O(l)$$

The iodine so produced is then reacted with a thiosulfate, $S_2O_3^{2-}$, solution.

$$2\,S_2O_3^{2-}(aq) + I_2(s) \rightarrow S_4O_6^{2-}(aq) + 2\,I^-(aq)$$

If 18.03 mL of a 0.00200 M $S_2O_3^{2-}$ solution completely reacts with the I_2 produced by a 28.09-g sample of polluted air, calculate the O_3 concentration, in ppm (m/m), in the sample of air.

15.88 The active ingredient in household bleach is the hypochlorite ion, ClO^-. The concentration of this ion in bleach can be determined using a two-step process. First, the bleach is reacted with an I^-(aq) solution.

$$ClO^-(aq) + 2\,I^-(aq) + 2\,H^+(aq) \rightarrow$$
$$I_2(s) + Cl^-(aq) + H_2O(l)$$

The iodine so produced is then reacted with a thiosulfate, $S_2O_3^{2-}$, solution.

$$I_2(s) + 2\,S_2O_3^{2-}(aq) \rightarrow 2\,I^-(aq) + S_4O_6^{2-}(aq)$$

A 50.00 g sample of a certain household bleach is found to react completely with 42.5 mL of a 0.0150 M $S_2O_3^{2-}$ solution. Calculate the concentration in mass percent of ClO^- ion in the bleach.

Grid Problems

15.89 Select from the grid *all* correct responses for each situation.

1. H_3PO_2	2. $H_5P_3O_{10}$	3. P_4O_6
4. Ca_3P_2	5. PO_4^{3-}	6. P_4O_7
7. PH_4^+	8. P_2F_4	9. P_4O_8

(a) substances in which oxygen has an oxidation number of −2
(b) substances for which the algebraic sum of all oxidation numbers is zero
(c) substances for which the algebraic sum of the *negative* oxidation numbers is −8
(d) substances in which phosphorus has a negative oxidation number
(e) substances in which an element has a fractional oxidation number
(f) pairs of substances in which P has the same oxidation number

15.90 Select from the grid *all* correct responses for each situation.

1. O_3	2. H_3O^+	3. O_2^+
4. H_2O_2	5. BaO_2	6. CaH_2
7. Na_2HPO_4	8. NH_4NO_3	9. $Al(NO_2)_3$

(a) substances in which oxygen does not have its most common oxidation number
(b) substances in which hydrogen does not have its most common oxidation number
(c) substances in which a zero oxidation number is present
(d) substances for which the algebraic sum of oxidation numbers does not add up to zero
(e) species where two different negative oxidation numbers are present
(f) species where two elements have the same oxidation number

15.91 Select from the grid *all* correct responses for each situation.

1. $S \rightarrow SO_2$	2. $SO_2 \rightarrow SO_4^{2-}$	3. $S^{2-} \rightarrow S$
4. $S^{2-} \rightarrow SO_4^{2-}$	5. $SO_4^{2-} \rightarrow S_2O_3^{2-}$	6. $SO_2 \rightarrow SO_3$
7. $SO_3 \rightarrow SO_3^{2-}$	8. $SO_2 \rightarrow S^{2-}$	9. $SO_3 \rightarrow SO_4^{2-}$

(a) a change in which S decreases in oxidation number
(b) a change in which S is oxidized
(c) a change in which S is the reducing agent
(d) a change in which S is the oxidizing agent
(e) a change in which S is reduced
(f) a change in which S is neither oxidized or reduced

15.92 Select from the grid *all* correct responses for each situation. All reactions and half-reactions occur only in the direction indicated.

1. partial equation	2. non-redox reaction	3. disproportionation reaction
4. oxidizing agent present	5. loss of electrons occurs	6. oxidation-number increase occurs
7. reducing agent present	8. gain of electrons occurs	9. oxidation occurs

(a) pertains to the half-reaction $2 Cl^- \rightarrow Cl_2 + 2 e^-$
(b) pertains to the reaction $2 NO + O_2 \rightarrow 2 NO_2$
(c) pertains to the half-reaction $Cu^{2+} + 2 e^- \rightarrow Cu$
(d) pertains to the reaction $2 Fe^{3+} + Sn^{2+} \rightarrow 2 Fe^{2+} + Sn^{4+}$
(e) pertains to the reaction $CaO + CO_2 \rightarrow CaCO_3$
(f) pertains to the reaction
$3 NO_2 + H_2O \rightarrow 2 NO_3^- + NO + 2 H^+$

APPENDIX

Mathematical Review

A.1 Basic Mathematical Operations

Early in the history of mathematics it was noticed that it was more efficient to "count by threes or fours or fives, etc." than to "count by ones." This led to the concept of multiplication, which is repeated addition. **Multiplication** is the addition of a number or quantity to itself a certain number of times. Multiplying 7 times 3 means 7 *added* 3 *times*, which gives 21.

Various ways exist for representing the multiplication process. Each of the following five notations mean *3 times b*.

$$3 \times b \quad 3 \cdot b \quad 3(b) \quad (3)(b) \quad 3b$$

The sign × is the most used symbol for multiplication. However, a dot and parentheses are also in common use. Notice that in the last notation, $3b$, no symbol is present to indicate multiplication. The term *product* is used to denote the answer obtained by multiplying two or more numbers together.

Multiplication of a series of terms by a given quantity is denoted by parentheses, as in the expression

$$3(a + b + 2c)$$

Each term within the parentheses is to be multiplied by 3, giving as an answer the expression $3a + 3b + 6c$.

Division can be considered to be the reverse process of multiplication. **Division** is the process of finding out how many times one number or quantiy is contained in another.

Three common notations are in use for indicating division.

$$a \div b \qquad \frac{a}{b} \qquad a/b$$

All mean a divided by b. The result of dividing one number into another is called the *quotient*.

In problems stated in words, the term *per* is an indication of division. For example, density, which is defined as mass *per* unit volume, is written as

$$\text{Density} = \frac{\text{mass}}{\text{volume}}$$

In working problems it is sometimes necessary to *invert* or *take the reciprocal of* a number. Both of these expressions mean the same thing, namely, to divide that number into 1. Thus, the number 3 when inverted becomes $\frac{1}{3}$, and the reciprocal of the number 8 is $\frac{1}{8}$.

A.2 Fractions

A **fraction** is an arithmetic expression showing that one number is to be divided by another number. Some examples of fractions are $\frac{2}{5}, \frac{3}{7}, \frac{8}{3}$, and $\frac{1}{10}$.

In a fraction the number above the division line is called the *numerator*, and the number below the line the *denominator*.

$$\text{numerator} \longrightarrow \frac{a}{b} \longleftarrow \text{denominator}$$

Fractions are classified into two categories: proper and improper. A *proper fraction* is one in which the numerator is smaller than the denominator. An *improper fraction* has a larger numerator than its denominator. The fraction $\frac{5}{9}$ is a proper fraction, whereas the fraction $\frac{11}{9}$ is an improper fraction. Improper fractions are often encountered in scientific calculations. There is nothing "wrong" with an improper fraction, and most often it is used "as is" in chemical calculations.

All fractions can be converted to decimal numbers simply by carrying out the implied division. For example,

$$\frac{5}{7} = 5 \div 7 = 0.714$$

EXAMPLE A.1

Classify each of the following fractions as proper or improper, and express it as a decimal number to three significant figures.

(a) $\dfrac{5}{11}$ (b) $\dfrac{11}{5}$ (c) $\dfrac{17}{12}$ (d) $\dfrac{1}{7}$

Solution

(a) Since the numerator is smaller than the denominator, this is a **proper** fraction. Its decimal value is

$$\frac{5}{11} = 5 \div 11 = 0.455$$

(b) Since the numerator is larger than the denominator, this is an **improper** fraction. Its decimal value is

$$\frac{11}{5} = 11 \div 5 = 2.20$$

(c) This is an **improper** fraction whose decimal value is

$$\frac{17}{12} = 17 \div 12 = 1.42$$

(d) This is a **proper** fraction whose decimal value is

$$\frac{1}{7} = 1 \div 7 = 0.143$$

Note that proper fractions always give decimal numbers with values less than 1, and improper fractions always have decimal values greater than 1.

A *ratio* conveys the same information as a fraction. If the ratio of nickels to dimes in a pile of coins is 3 to 2, we are mathematically saying that

$$\frac{\text{Nickels}}{\text{Dimes}} = \frac{3}{2}$$

The product obtained from multiplying two fractions is another fraction whose numerator is the product of the two given numerators and whose denominator is the product of the two given denominators. In general terms we have

$$\frac{a}{b} \times \frac{c}{d} = \frac{a \times c}{b \times d} = \frac{ac}{bd}$$

Division of fractions is most easily handled by converting the division operation into multiplication. This is accomplished by inverting (turning upside down) the fraction to the right of the division sign and then following the rules for multiplication. In general terms, division proceeds as follows.

A.4 Appendix · Mathematical Review

$$\frac{a/b}{c/d} = \frac{a}{b} \div \frac{c}{d} = \frac{a}{b} \times \frac{d}{c} = \frac{a \times d}{b \times c} = \frac{ad}{bc}$$

Division sign has been changed to multiplication ⟶   ⟵ This fraction has been inverted.

EXAMPLE A.2

Carry out the following multiplications of fractions.

(a) $\dfrac{3}{4} \times \dfrac{5}{8}$ (b) $\dfrac{2}{3} \times \dfrac{6}{7}$ (c) $\dfrac{1}{3} \times \dfrac{2}{3} \times \dfrac{4}{5}$

Solution

(a) Fractions are multiplied together by multiplying the numerators together and the denominators together.

$$\frac{3}{4} \times \frac{5}{8} = \frac{3 \times 5}{4 \times 8} = \frac{15}{32}$$

(b) Multiplying the numerators together and then multiplying the denominators together gives

$$\frac{2}{3} \times \frac{6}{7} = \frac{2 \times 6}{3 \times 7} = \frac{12}{21}$$

(c) The product of three fractions is obtained in the same manner as the product of two fractions. All numerators are multiplied together, and all denominators are multiplied together.

$$\frac{1}{3} \times \frac{2}{3} \times \frac{4}{5} = \frac{1 \times 2 \times 4}{3 \times 3 \times 5} = \frac{8}{45}$$

EXAMPLE A.3

Carry out the following division operations.

(a) $\dfrac{3}{7} \div \dfrac{2}{5}$ (b) $\dfrac{2/3}{3/5}$

Solution

(a) We convert the division operation into a multiplication by inverting the fraction $\tfrac{2}{5}$ and replacing the division sign with a multiplication sign.

$$\frac{3}{7} \div \frac{2}{5} = \frac{3}{7} \times \frac{5}{2} = \frac{3 \times 5}{7 \times 2} = \frac{15}{14}$$

multiplication sign ⟶   ⟵ inverted fraction

(b) Inverting the fraction to the right of the division sign, $\frac{3}{5}$ in this case, gives

$$\frac{2/3}{3/5} = \frac{2}{3} \div \frac{3}{5} = \frac{2}{3} \times \frac{5}{3} = \frac{2 \times 5}{3 \times 3} = \frac{10}{9}$$

Adding and subtracting fractions is usually more complicated than multiplying and dividing them. This is because fractions must have the same denominator, called a *common denominator*, before they can be added or subtracted. By analogy, to add values for coins of different denominations, we first must express them as equivalent amounts with a common denomination and then add; for example, 2 dimes and 1 quarter would be expressed as 20 cents and 25 cents, respectively, and then added, obtaining the sum, 45 cents. The actual numerical examples of addition and subtraction of fractions that follow should clarify this "common denominator process."

EXAMPLE A.4

Carry out the following additions and subtractions of fractions.

(a) $\dfrac{2}{3} + \dfrac{3}{4}$ (b) $\dfrac{7}{8} - \dfrac{1}{4}$ (c) $\dfrac{7}{6} + \dfrac{2}{3} - \dfrac{1}{2}$

Solution

(a) Examining the denominators (3 and 4) of the two fractions, we note that the smallest number that each can be divided into evenly is the number 12; that is, each can be converted to 12 by multiplying it by some whole number.

$$\underbrace{3}_{\text{denominator of first fraction}} \times 4 = 12 \qquad \underbrace{4}_{\text{denominator of second fraction}} \times 3 = 12$$

Thus, 12 is the common denominator for 3 and 4.

The numerator and denominator of each fraction are both multiplied by the number that will convert the denominator of that particular fraction to the common denominator, which is 12 in this case.

$$\frac{2}{3} \times \frac{4}{4} = \frac{8}{12} \qquad \text{and} \qquad \frac{3}{4} \times \frac{3}{3} = \frac{9}{12}$$

Now that both fractions have the same denominator, we can proceed to add them together by adding numerators and retaining the denominator.

$$\frac{8}{12} + \frac{9}{12} = \frac{17}{12}$$

(b) The common denominator for 8 and 4 is 8.

$$8 \times 1 = 8 \qquad \text{and} \qquad 4 \times 2 = 8$$

Expressing both fractions in eighths, we obtain

$$\frac{7}{8} \times \frac{1}{1} = \frac{7}{8} \quad \text{and} \quad \frac{1}{4} \times \frac{2}{2} = \frac{2}{8}$$

Now that both fractions have the same denominator, we can perform the subtraction by subtracting numerators and maintaining the denominator.

$$\frac{7}{8} - \frac{2}{8} = \frac{5}{8}$$

(c) This problem requires that we find a common denominator for three numbers (6, 3, and 2). This common denominator is 6, since both 3 and 2 will divide evenly into 6. Thus, changing all fractions to sixths we have

$$\frac{7}{6} \times \frac{1}{1} = \frac{7}{6}$$

$$\frac{2}{3} \times \frac{2}{2} = \frac{4}{6}$$

$$\frac{1}{2} \times \frac{3}{3} = \frac{3}{6}$$

Carrying out the indicated arithmetic gives

$$\frac{7}{6} + \frac{4}{6} - \frac{3}{6} = \frac{7+4-3}{6} = \frac{8}{6}$$

A.3 Positive and Negative Numbers

All numbers have both a magnitude and a sign. The sign can be either positive or negative. If a sign is not shown, it is assumed that it is positive.

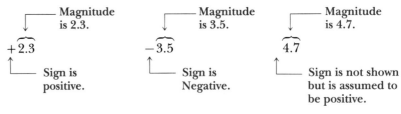

Most students feel comfortable with the use of positive numbers, but are much less certain when negative numbers are encountered. Hence, a review of the use of negative numbers is in order. The effects that negative numbers have on the processes of multiplication, division, addition, and subtraction will be considered.

The sign associated with the product of the multiplication of two signed numbers is positive if both numbers have the same sign and negative if they have opposite signs. Thus,

Positive × positive = positive
Positive × negative = negative
Negative × positive = negative
Negative × negative = positive

In division the same "sign rules" as those for multiplication are used. That is, the quotient of two signed numbers is positive if both numbers have the same sign and negative if they have opposite signs.

EXAMPLE A.5

Carry out the following multiplications and divisions, each of which involves signed numbers.

(a) $(+3) \times (-4)$ (b) $(-2) \times (-3)$ (c) $(-4)/(2)$ (d) $(+9)/(+3)$

Solution

(a) The sign of the product will be negative because the two numbers have opposite signs.
$$(+3) \times (-4) = -12$$

(b) The product of two negative numbers is a positive number; if both numbers have the same sign the product is positive.
$$(-2) \times (-3) = +6$$

(c) The "sign rules" for division are the same as those for multiplication. A "negative" divided by a "positive" is a "negative."
$$(-4)/(2) = -2$$

(d) Since both numbers have the same sign, the quotient will be positive.
$$(+9)/(+3) = +3$$

When adding numbers, if the signs of the two numbers are the same, add the numbers and keep that same sign. When the signs of the two numbers to be added are not the same, subtract the number of smaller magnitude from the one of larger magnitude and keep the sign of the larger number.

Subtraction of signed numbers is most easily handled by converting the subtraction operation into an addition. This is accomplished by changing the sign of the *number being subtracted* and then following the rules for addition.

Sign indicating subtraction is changed to addition.
$$(-3) - (+7) = (-3) + (-7) = -10$$
Sign of number being subtracted is changed.

EXAMPLE A.6

Carry out the following additions, each of which involves signed numbers.

(a) $(+3) + (+3)$ (b) $(-7) + (-2)$
(c) $(-6) + (+9)$ (d) $(-4) + (+3)$

Solution

(a) Since both numbers are positive, the sum will be positive.
$$(+3) + (+3) = +6$$

(b) Since both numbers are negative, the sum will be negative.
$$(-7) + (-2) = -9$$

(c) The signs of the two numbers are not the same. Thus, the number of smaller magnitude, 6, is subtracted from the number of larger magnitude, 9, and the sign of the larger number, plus, is attached to the answer.
$$(-6) + (+9) = +3$$

(d) Again, the signs of the two numbers are not the same. The smaller number, 3, is subtracted from 4, and the sign of the larger number, minus, is used.
$$(-4) + (+3) = -1$$

EXAMPLE A.7

Carry out the following subtractions, each of which involves signed numbers.

(a) $(-2) - (-5)$ (b) $(+5) - (-4)$ (c) $(+2) - (+5)$

Solution

(a) The -5 is changed to a $+5$, and the two numbers are added.
$$(-2) - (-5) = (-2) + (+5) = +3$$

(b) The -4 is changed to a $+4$, and the two numbers are added.
$$(+5) - (-4) = (+5) + (+4) = +9$$

(c) The $+5$ is changed to a -5, and the two numbers are added.
$$(+2) - (+5) = (+2) + (-5) = -3$$

Remember that in adding the two numbers in each of the above examples, the rules for addition shown in Example A.6 were used.

A.4 Solving Algebraic Equations

A common type of mathematical problem encountered in chemistry involves finding the value of an unknown. Such problems almost always involve manipulating an equation by simple algebraic techniques.

Section A.4 · Solving Algebraic Equations

To solve for an unknown in an algebraic equation we must change the equation in such a way that the desired unknown is "isolated" on one side of the equation. This can be accomplished in a number of ways; the method used is determined by the nature of the equation. Equation manipulation procedures include

1. Adding the same quantity to both sides of an equation.
2. Subtracting the same quantity from both sides of an equation.
3. Multiplying both sides of an equation by the same quantity.
4. Dividing both sides of an equation by the same quantity.

All four of these procedures encompass the same fundamental mathematical concept: *The validity of an equation is maintained as long as whatever is done to one side of the equation is also done to the other side of the equation.* Illustrative of this mathematical principle is the procedure of multiplying both sides of the equation.

$$a = b$$

by the number 3 to give the equation

$$3a = 3b$$

This latter equation is "just as valid" as the original equation.

Example A.8 contains five examples of algebraic manipulation of equations by the previously listed procedures either singly or in combination.

EXAMPLE A.8

Solve each of the following equations for the quantity b.

(a) $ab = 3c$ (b) $\dfrac{a}{c} = \dfrac{b}{d}$ (c) $\dfrac{3b}{c} = a$

(d) $b + c = 4a$ (e) $\dfrac{3(b-c)}{2} = a$

Solution

(a) Solving the equation

$$ab = 3c$$

for b requires only one step. Divide both sides of the equation by the quantity a.

$$\frac{\cancel{a}b}{\cancel{a}} = \frac{3c}{a}$$

The a's on the left cancel, leaving only b on the left side of the equation.

$$b = \frac{3c}{a}$$

(b) To solve the equation

$$\frac{a}{c} = \frac{b}{d}$$

for b we must eliminate d from the right side of the equation. To do that we multiply both sides of the equation by the quantity d.

$$d\left(\frac{a}{c}\right) = d\left(\frac{b}{d}\right)$$

The d's on the right cancel, giving the equation

$$d\left(\frac{a}{c}\right) = b$$

Note the procedure for removing a quantity from the denominator of one side of an equation. You multiply both sides of the equation by that quantity.

(c) To solve the equation

$$\frac{3b}{c} = a$$

for the quantity b we first multiply both sides of the equation by c.

$$c\left(\frac{3b}{c}\right) = ac$$

The c's on the left cancel, giving

$$3b = ac$$

Next, we divide both sides of the equation by 3:

$$\frac{3b}{3} = \frac{ac}{3}$$

Canceling the 3's on the left gives

$$b = \frac{ac}{3}$$

(d) The equation

$$b + c = 4a$$

is different from the previous three examples in that it contains a "plus" sign. To remove c from the left side of the equation, thus isolating b, requires a new procedure. We subtract the quantity c from both sides of the equation.

$$b + c - c = 4a - c$$

The $+c$ and $-c$ on the left add to give zero, leaving the equation

$$b = 4a - c$$

(e) To isolate the quantity b from the equation

$$\frac{3(b-c)}{2} = a$$

requires three steps. First, we multiply both sides of the equation by 2.

$$2\left[\frac{3(b-c)}{2}\right] = 2a$$

The 2's on the left cancel, giving

$$3(b-c) = 2a$$

Second, we divide both sides of the equation by 3.

$$\frac{3(b-c)}{3} = \frac{2a}{3}$$

Canceling the 3's on the left yields the equation

$$b - c = \frac{2a}{3}$$

Finally, we add the quantity c to both sides of the equation.

$$b - c + c = \frac{2a}{3} + c$$

The c's on the left add to zero, giving the equation

$$b = \frac{2a}{3} + c$$

Answers to Odd-Numbered Problems

Chapter 1

1. c, b, e, a, and d. **3.** (a) correct; (b) incorrect; (c) incorrect; (d) correct. **5.** (a) false; (b) false; (c) true; (d) true. **7.** (a) law; (b) hypothesis; (c) fact; (d) hypothesis. **9.** (a) 4 is eliminated; (b) 1 is eliminated; (c) 1 and 4 are eliminated; (d) 1 and 2 are eliminated. **11.** It is published so that other scientists have the opportunity to critique and check both the data and experimental design. **13.** Scientific laws are discovered by research. Researchers have no control over what the laws turn out to be. Societal laws are arbitrary conventions that can be and are changed when necessary. **15.** All theories in science are considered provisional and subject to change in light of new experimental observations. **17.** A hypothesis is a tentative model or picture that offers an explanation for a law. A theory is a hypothesis that has been tested and validated over a long period of time.

Chapter 2

1. Student A: low precision, low accuracy; student B: high precision, high accuracy; student C: high precision, low accuracy. **3.** Results obtained using a high-precision, poorly calibrated instrument would give precision but not accuracy. **5.** (a) estimated; (b) exact; (c) estimated; (d) exact. **7.** The uncertainty in the first measurement involves tenths and that in the second measurement involves hundredths. **9.** (a) 2563 ± 1; (b) $134,001 \pm 1$; (c) 0.00357 ± 0.00001; (d) 12.00 ± 0.01. **11.** Leading zeros are never significant. Confined zeros are always significant. Trailing zeros are significant only if (1) there is a decimal point present in the number or (2) they carry overbars. **13.** (a) 4; (b) 5; (c) 4; (d) 6. **15.** (a) 2; (b) 6; (c) 8; (d) 7. **17.** (a) 2; (b) 4; (c) 5; (d) 6. **19.** (a) 4; (b) 4; (c) 3; (d) 3. **21.** 6. **23.** (a) 3; (b) 4; (c) 2; (d) 5. **25.** (a) ± 0.01; (b) ± 100; (c) ± 10; (d) ± 1. **27.** (a) 0.351; (b) 653,900; (c) 22.555; (d) $1\overline{0}00$. **29.** (a) 0.035; (b) 2.50; (c) 1,500,000; (d) $1\overline{0}0$. **31.** (a) 0.12; (b) 120,000; (c) 12; (d) 0.00012. **33.** (a) 1; (b) 3; (c) 2; (d) 3. **35.** (a) 0.0143; (b) 1350; (c) 3500; (d) 27. **37.** (a) 761; (b) 480; (c) 0.982; (d) 0.25. **39.** (a) 3.9; (b) 4.84; (c) 63; (d) 1. **41.** (a) 162; (b) 9.3; (c) 1261; (d) 20.0. **43.** (a) 957.0; (b) 343; (c) 1200; (d) 132. **45.** If the answer is less than 1000 it contains 3 significant figures and if it is 1000 or greater it contains 4 significant figures. **47.** 9.38 in. **49.** (a) 267.3; (b) 260,000; (c) 201.3; (d) 3.8. **51.** (a) $3 \times 3 \times 3 \times 3$; (b) $2 \times 2 \times 2 \times 2 \times 2 \times 2$; (c) $10 \times 10 \times 10 \times 10 \times 10$; (d) $1/(10 \times 10)$. **53.** (a) 10^7; (b) 10^{-1}; (c) 10^{-3}; (d) 10^{-7}. **55.** (a) 2.345×10^1; (b) 1.207×10^2; (c) 3.4×10^{-3}; (d) 1.005×10^{-1}. **57.** (a) 2.3100×10^2; (b) 2.31×10^8; (c) 1.000×10^{-4}; (d) 1.0010×10^{-1}. **59.** (a) 0.00000170; (b) 573; (c) 0.555; (d) 60,012. **61.** (a) 23,000; (b) $23,\overline{0}00$; (c) $23,\overline{00}0$; (d) 23,000.00. **63.** (a) 10^8; (b) 10^{-8}; (c) 10^2; (d) 10^{-2}. **65.** (a) 6.0×10^{12}; (b) 4.14×10^{-5}; (c) 4.24×10^1; (d) 2.99×10^4. **67.** (a) 1.44×10^3; (b) 3.71×10^{-3}; (c) 2.70×10^{-1}; (d) 6.0×10^8. **69.** (a) 10^2; (b) 10^8; (c) 10^{-8}; (d) 10^{-2}. **71.** (a) 2.0×10^2; (b) 1.369×10^9; (c) 3.29×10^4; (d) 2×10^{-11}. **73.** (a) 2.00×10^5; (b) 9.32×10^0; (c) 9.637×10^2;

(d) 1.05×10^{-5}. **75.** (a) 10^1; (b) 10^{-1}; (c) 10^{20}; (d) 10^2.
77. (a) 1.5×10^0; (b) 6.7×10^{-1}; (c) 8.51×10^{-19}; (d) 8×10^{15}. **79.** (a) 6.43×10^3; (b) 3.20×10^3; (c) 3.7113×10^{-3};
(d) 6.57×10^6. **81.** (a) 1.93×10^{-3}; (b) 2.48×10^4;
(c) 3.09×10^4; (d) 5.11×10^1. **83.** (a) 1.4047×10^6;
(b) 3.575×10^5; (c) 3.82×10^5; (d) 8.249×10^6.
85. (a) 3.630502; (b) 3.6305; (c) 3.630; (d) 3.63. **87.** (a) 3;
(b) 4; (c) 4; (d) 4. **89.** (a) no; (b) yes; (c) yes; (d) yes.
91. (a) 6.326×10^5; (b) 3.13×10^{-1}; (c) 6.300×10^7;
(d) 5.000×10^{-1}. **93.** (a) 27; (b) 0.9; (c) 180; (d) 0.07.
95. An exact number is a whole number; it can not possess decimal digits. **97.** (a) 370; (b) -370; (c) 0.41; (d) 4.2.
99. (a) 2.07×10^2, 243, 1.03×10^3; (b) 2.11×10^{-3}, 0.0023, 3.04×10^{-2}; (c) 23,000, 9.67×10^4, 2.30×10^5;
(d) 0.000014, 0.00013, 1.5×10^{-4}. **101.** (a) 1.5×10^{-29};
(b) 5.8×10^{-11}; (c) 2.794×10^{21}; (d) 2.191×10^{-7}.
103. (a) 1, 2, and 9; (b) 1, 2, 3, 4, and 5; (c) 5, 6, 8, and 9;
(d) 4, 5, 6, and 8; (e) 3, 4, 5, and 6; (f) 7 and 9. **105.** (a) 1 and 9, 2 and 8, 3 and 7, 4 and 6; (b) 1 and 6, 2 and 5, 3 and 4; (c) 3 and 6; (d) 1 and 3, 2 and 4, 3 and 5, 4 and 6, 5 and 7, 6 and 8, 7 and 9; (e) all choices; (f) 1, 3, 5, and 7.

Chapter 3

1. (a) kilo; (b) micro; (c) milli; (d) centi. **3.** (a) mm;
(b) μL; (c) kg; (d) Mg. **5.** (a) n (nano); (b) μ (micro);
(c) k (kilo); (d) G (giga). **7.** (a) centiliter; (b) micrometer;
(c) gigameter; (d) milligram. **9.** (a) volume; (b) length;
(c) volume; (d) area. **11.** (a) milligram, centigram, decigram; (b) micrometer, millimeter; kilometer; (c) ng, cg, kg; (d) pL, GL, TL. **13.** (a) 1 inch; (b) 1 meter;
(c) 1 pound; (d) 1 gallon. **15.** (a) $2\overline{0}$ mm^2; (b) 53.2 cm^3;
(c) 1.7 km^2; (d) 3.6. **17.** (a) 20.4 cm^2; (b) 32 m^2;
(c) 65.88 mm^2; (d) 8.3 mm^2. **19.** (a) 9.5 cm^3; (b) 130 mm^3;
(c) 2.789×10^6 mm^3; (d) 370 cm^3. **21.** Since they are derived from exact definitions, they contain an unlimited number of significant figures. **23.** (a) 1 mg/10^{-3} g and 10^{-3} g/1 mg; (b) 1 kL/10^3 L and 10^3 L/1 kL;
(c) 1 pt/0.4732 L and 0.4732 L/1 pt; (d) 1 in./2.54 cm and 2.54 cm/1 in. **25.** (a) 4 significant figures; (b) exact;
(c) 4 significant figures; (d) exact. **27.** (a) 4.832×10^6 μg;
(b) 4.832×10^{-9} Gg; (c) 4.832×10^{-3} kg; (d) 4.832×10^2 cg.
29. (a) 3.12 km; (b) 3.3×10^{-11} km; (c) 1.3×10^{-1} km;
(d) 3×10^2 km. **31.** (a) 300 mL; (b) 3 km; (c) 3×10^6 μg;
(d) 3×10^{-3} kg. **33.** 2.1×10^{26} oz. **35.** 1100 in./sec.
37. 2.30×10^{-9} g/mL. **39.** 1.6×10^{-3} in. **41.** 193 lb.
43. 1.436 gal. **45.** (a) 5.2 cm^2; (b) 0.81 in.^{2}. **47.** 51 ft^3.
49. (a) 1.23 gal; (b) 1.234 gal; (c) 1.2337 gal; (d) 1.23371 gal.
51. 13.55 g/cm^3. **53.** 1.039 g/mL. **55.** 2.163 g/mL.
57. 47.5 g. **59.** 27.1 mL. **61.** 61 lb. **63.** 266 g.
65. 22.3 g. **67.** 3.0. **69.** 66.9 g. **71.** 41 °C. **73.** -361.1 °F.

75. (a) 504 K; (b) 504.9 K; (c) 504.89 K; (d) 276.1 K.
77. 7 °F. **79.** -10 °C. **81.** -60 °F. **83.** Energy is the capacity to do work. **85.** (a) light and heat; (b) light and heat; (c) chemical; (d) light and heat. **87.** Kinetic energy is energy of motion, and potential energy is stored energy resulting from an object's position, condition, and/or composition. **89.** (a) potential; (b) kinetic; (c) potential;
(d) potential. **91.** 1 cal = 4.184 J. **93.** (a) 2.29×10^6 J;
(b) 547 kcal; (c) 5.47×10^5 cal; (d) 547 Cal. **95.** gold.
97. (a) 1310 J; (b) 91,000 J; (c) 1600 J; (d) 630 J.
99. 3.99 °C. **101.** 1.58 g. **103.** (a) 5.2 J/°C; (b) 15.3 J/°C;
(c) 37 J/°C; (d) 167 J/°C. **105.** (a) 0.444 J/g · °C;
(b) 0.239 J/g · °C. **107.** 0.39 J/g · °C. **109.** 47 g.
111. 1.425 g. **113.** no. **115.** 9.3×10^7 mi. **117.** 994 g.
119. 4900 lb. **121.** 17 in. **123.** 6.3 g. **125.** 2.1 hr.
127. (a) yes; (b) no. **129.** 0.76 g/cm^3. **131.** 0.786 J/g · °C.
133. 25 kJ. **135.** (a) 5.00×10^{-1} g/mL; (b) 5.0×10^{-1} g/mL; (c) 5.0000×10^{-1} g/mL; (d) 5.000×10^{-1} g/mL.
137. (a) 8000 cm^3; (b) $8\overline{0}00$ cm^3; (c) $8\overline{00}0$ cm^3; (d) $8\overline{000}$ cm^3.
139. (a) 5.1×10^{-2} lb; (b) 5.071×10^{-2} lb; (c) 5.07064×10^{-2} lb; (d) 5.0706320×10^{-2} lb. **141.** 3.2500 m.
143. (a) 2, 4, and 6; (b) 2 and 4; (c) 5, 8, and 9; (d) 4 and 7, 5 and 8, 7 and 9; (e) 4 and 7, 5 and 8, 7 and 9; (f) 2 and 8, 3 and 6, 4 and 7. **145.** (a) 2 and 4; (b) 1 and 5;
(c) 3; (d) 8; (e) 6; (f) 9.

Chapter 4

1. (a) definite; (b) indefinite; (c) indefinite. **3.** (a) solid;
(b) solid; (c) liquid; (d) gas. **5.** Some solids, upon heating, decompose rather than melt. **7.** Physical properties are observable without changing a substance into another substance; chemical properties are observed as a substance undergoes change(s) in chemical composition.
9. (a) physical; (b) physical; (c) physical; (d) chemical.
11. (a) physical; (b) chemical; (c) physical; (d) chemical.
13. (a) physical; (b) physical; (c) physical; (d) chemical.
15. It means that the composition does not change.
17. (a) physical; (b) physical; (c) chemical; (d) physical.
19. (a) physical; (b) physical; (c) chemical; (d) chemical.
21. (a) freezing; (b) condensation; (c) sublimation;
(d) evaporation. **23.** (a) heterogeneous mixture, homogeneous mixture; (b) homogeneous mixture.
25. (a) homogeneous mixture; (b) homogeneous mixture;
(c) pure substance; (d) heterogeneous mixture.
27. (a) heterogeneous mixture; (b) homogeneous mixture;
(c) homogeneous mixture; (d) heterogeneous mixture.
29. (a) homogeneous mixture, 1 phase; (b) homogeneous mixture, 1 phase; (c) heterogeneous mixture, 2 phases;
(d) heterogeneous mixture, 2 phases. **31.** (1) physical combination of two or more substances, (2) variable

composition, (3) properties vary as composition varies, and (4) components can be separated using physical means. **33.** A compound, but not an element, can be broken down into two or more simpler substances by chemical means. **35.** (a) compound; (b) compound; (c) no classification possible; (d) no classification possible. **37.** (a) true; (b) false; (c) false; (d) false. **39.** (a) A (no classification possible), B (no classification possible), C (compound); (b) D (compound), E (no classification possible), F (no classification possible), G (no classification possible). **41.** first box: mixture; second box: compound. **43.** yes—a compound is the chemical combination of two or more elements, and a mixture can be the physical combination of two or more elements. **45.** (a) 13; (b) 21; (c) 48; (d) 27. **47.** (a) hydrogen and helium; (b) oxygen and iron; (c) oxygen and silicon; (d) hydrogen and oxygen. **49.** (a) silver; (b) gold; (c) calcium; (d) sodium; (e) phosphorus, (f) silver. **51.** (a) Sn; (b) Cu; (c) Al; (d) B; (e) Ba; (f) Ar. **53.** boron, carbon, fluorine, hydrogen, iodine, potassium, nitrogen, oxygen, phosphorus, sulfur, uranium, vanadium, tungsten, yttrium. **55.** (a) fluorine; (b) zinc; (c) potassium; (d) sulfur. **57.** (a) unnilseptium (Uns); (b) ununtrilium (Uut); (c) ununpentium (Uup); (d) unbiunium (Ubu). **59.** (a) compound; (b) mixture; (c) compound; (d) mixture. **61.** (a) homogeneous mixture; (b) element; (c) heterogeneous mixture; (d) homogeneous mixture. **63.** (a) no; (b) yes; (c) yes; (d) no. **65.** (a) solid; (b) state determination not possible; (c) gas; (d) state determination not possible. **67.** (a) same unknown; (b) no. **69.** compound. **71.** likely a mixture of A and B. **73.** (a) 6, 7, 8, and 9; (b) 2; (c) 1 and 5; (d) 2 and 4; (e) 6, 7, 8, and 9; (f) 1 and 5. **75.** (a) 1 and 3; (b) 2, 5, 6, 7, and 9; (c) 8; (d) 2, 5, 6, and 7; (e) 2, 5, 6, 7, and 9; (f) 4 and 5.

Chapter 5

1. (c) **3.** An atom is the smallest particle of an element that can exist and still have the properties of the element. A molecule is a group of two or more atoms tightly bound together. **5.** (a) false—molecules must contain two or more atoms; (b) true; (c) false—some compounds have molecules as their basic unit, while others have ions as their basic unit; (d) true; (e) true. **7.** approximately 10^{-8} centimeter. **9.** Both the first and second letters of the formula are capitalized. **11.** (a) compound; (b) compound; (c) element; (d) element. **13.** (a) $C_{20}H_{30}O$; (b) H_2SO_4; (c) P_4; (d) HCN. **15.** (a) 2; (b) 1; (c) 3; (d) 3. **17.** (a) NH_4ClO_4; (b) $Al_2(SO_4)_3$; (c) $Ba(ClO_2)_2$; (d) K_3PO_4. **19.** (a) H_3PO_4; (b) $SiCl_4$; (c) NO_2; (d) H_2O_2. **21.** On a relative scale where the mass of an electron is 1 unit, the proton is 1837 and the neutron 1839. **23.** the protons. **25.** The numbers of protons and electrons are equal. **27.** It is negligible when compared to that of the proton or neutron. **29.** (a) electron; (b) proton; (c) proton, neutron; (d) neutron. **31.** (a) false; (b) false; (c) false. **33.** The positive and negative charges present are equal. **35.** When an obstacle is placed between the cathode and the opposite glass wall, a sharp shadow of the obstacle is cast on the wall. **37.** a discharge tube in which the cathode is a metal plate with a large number of holes drilled in it. **39.** proton and electron. **41.** A thin piece of gold foil is bombarded with alpha particles. A fluorescent circular screen encompassing the gold foil detects the alpha particles. **43.** An atom is a sphere of positive electricity with small negative electrons attached to its surface. **45.** (a) number of protons or number of electrons; (b) number of neutrons. **47.** (a) $Z = 2$ and $A = 4$; (b) $Z = 4$ and $A = 9$; (c) $Z = 5$ and $A = 9$; (d) $Z = 28$ and $A = 58$. **49.** (a) p = 8, n = 8, and e = 8; (b) p = 8, n = 10, and e = 8; (c) p = 20, n = 24, and e = 20; (d) p = 100, n = 157, and e = 100. **51.** Since every element has a unique atomic number, $^{15}_7N$ and ^{15}N are interchangeable. Since elements do not have unique mass numbers, $^{15}_7N$ and $_7N$ are not interchangeable. **53.** $^{54}_{26}Fe$, $^{56}_{26}Fe$, $^{57}_{26}Fe$, and $^{58}_{26}Fe$. **55.** (a) p = 24, n = 29, and e = 24; (b) p = 44, n = 59, and e = 44; (c) p = 101, n = 155, and e = 101; (d) p = 16, n = 18, and e = 16. **57.** (a) $^{34}_{15}P$; (b) $^{18}_8O$; (c) $^{54}_{24}Cr$; (d) $^{197}_{79}Au$. **59.** (a) isotopes; (b) isobars; (c) neither; (d) isotopes. **61.** (a) Li; (b) B; (c) F; (d) Al. **63.** (a) 5; (b) 1, 2. **65.** (a) same number of protons, same number of electrons, and different numbers of neutrons; (b) numbers of protons differ by one; numbers of electrons differ by one; and numbers of neutrons differ by one. **67.** A = 8.00 bebs, B = 4.00 bebs, and C = 2.00 bebs. **69.** 224.8 lb. **71.** 35.46 amu. **73.** 47.88 amu. **75.** 107 amu. **77.** 6.1968. **79.** The number 12.011 applies to naturally occurring carbon, a mixture of ^{12}C, ^{13}C, and ^{14}C. **81.** (a) chlorine ($Z = 17$) and fluorine ($Z = 9$); (b) calcium ($Z = 20$) and nitrogen ($Z = 7$); (c) nickel ($Z = 28$), iodine ($Z = 53$), and oxygen ($Z = 8$); (d) sodium ($Z = 11$) and sulfur ($Z = 16$). **83.** Since the atomic mass and mass number of ^{39}K are close in value, this isotope is the most abundant. **85.** (a) ^8_5B; (b) $^{12}_5B$; (c) $^{12}_5B$; (d) $^{16}_5B$. **87.** (a) Ar, K, Ca, Sc, Ti; (b) Sc, Ca, Ti, K, Ar; (c) same as (a); (d) Sc, Ti, Ca, K, Ar. **89.** (a) different; (b) same; (c) same. **91.** no—the number of protons can never be greater than the number of protons plus neutrons (mass number). **93.** (a) 4, 9, 11, 13, and 15; (b) 8, 10, 12, 14, 18, and 19. **95.** (a) 5 and 8; (b) 36 and 40. **97.** (a) 36.756 "new" amu; (b) 67.117 "new" amu. **99.** 370 atoms $^{17}_8O$, 2000 atoms $^{18}_8O$. **101.** 1638 protons. **103.** 92 subatomic particles. **105.** 1.90×10^{23} Au atoms. **107.** 2.84×10^9 mi. **109.** 0.022%. **111.** 2.4 mi. **113.** (a) 2 and 3; (b) 5 and 7; (c) 1, 4, 5, 7, 8, and 9; (d) 8 and 9; (e) 6 and 7; (f) 4 and 8, 5 and 7. **115.** (a) 2, 5, 6, 8, and

9; (b) all choices; (c) 1, 3, 4, and 7; (d) 7 and 8; (e) 6 and 7, 8 and 9; (f) 4 and 5, 6 and 8, 7 and 9.

Chapter 6

1. When elements are arranged in order of increasing atomic number, elements with similar chemical properties occur at periodic intervals. **3.** A group is a vertical column of elements within the periodic table. **5.** (a) period 2, group IA; (b) period 3, group VIA; (c) period 5, group IB; (d) period 7, group VIB. **7.** (a) Ga; (b) Zr; (c) Li; (d) Cl. **9.** (a) group; (b) periodic law; (c) periodic law; (d) period. **11.** elemental symbol, atomic number, and atomic mass. **13.** a property that can have only certain values. **15.** (a) orbital; (b) orbital; (c) shell; (d) shell. **17.** (a) five; (b) $5s$, $5p$, $5d$, and $5f$; (c) 1, 3, 5, and 7; (d) 50; (e) 2, 6, 10, and 14. **19.** (a) true; (b) true; (c) true; (d) true; (e) false. **21.** (a) spherical; (b) spherical; (c) dumb-bell; (d) cloverleaf. **23.** (a) 10; (b) 2; (c) 8; (d) 32. **25.** (a) possible; (b) impossible; (c) impossible; (d) possible. **27.** The Aufbau principle states that an electron normally occupies the lowest energy subshell available. An Aufbau diagram is a listing of subshells arranged in a manner that shows their relative energies. **29.** (a) $1s$ (lowest), $3s$ (highest); (b) $3s$ (lowest), $3d$ (lowest); (c) $4d$ (lowest), $4f$ (highest); (d) $3d$ (lowest), $7s$ (highest). **31.** The number 4 indicates the shell, the letter p the subshell, and the superscript 6 the occupancy of the subshell. **33.** (a) $1s^2 2s^2 2p^4$; (b) $1s^2 2s^2 2p^6 3s^2$; (c) $1s^2 2s^2 2p^6 3s^2 3p^3$; (d) $1s^2 2s^2 2p^6 3s^2 3p^6 4s^2$. **35.** (a) $1s^2 2s^2 2p^6 3s^2 3p^6 4s^2 3d^3$; (b) $1s^2 2s^2 2p^6 3s^2 3p^6 4s^2 3d^{10} 4p^4$; (c) $1s^2 2s^2 2p^6 3s^2 3p^6 4s^2 3d^{10} 4p^6 5s^2 4d^{10} 5p^6 6s^1$; (d) $1s^2 2s^2 2p^6 3s^2 3p^6 4s^2 3d^{10} 4p^6 5s^2 4d^{10} 5p^6 6s^2 4f^{14} 5d^{10} 6p^6 7s^2$. **37.** (a) O; (b) K; (c) Fe; (d) Zn. **39.** (a) 2, 8, 3; (b) 2, 8, 8; (c) 2, 8, 4; (d) 2, 2. **41.** (a–d) orbital diagrams. **43.** (a–d) orbital diagrams.

45. (a) 2; (b) 2; (c) 0; (d) 3. **47.** (a) paramagnetic; (b) paramagnetic; (c) diamagnetic; (d) diamagnetic. **49.** (a) no; (b) yes; (c) no; (d) yes. **51.** (a) p area; (b) d area; (c) s area; (d) p area. **53.** (a) $3p$; (b) $5p$; (c) $5d$; (d) $5s$. **55.** (a) d^5; (b) s^1; (c) p^6; (d) p^1. **57.** (a) Al; (b) Li; (c) La; (d) Sc. **59.** (a) Kr; (b) Li; (c) K; (d) Lu. **61.** (a) Fe; (b) Rb; (c) Se; (d) Unq. **63.** (a) $1s^2 2s^2 2p^6 3s^2 3p^6 4s^2$; (b) $1s^2 2s^2 2p^6 3s^2 3p^6 4s^2 3d^{10} 4p^3$; (c) $1s^2 2s^2 2p^6 3s^2 3p^2$; (d) $1s^2 2s^2 2p^6 3s^2 3p^6 4s^2 3d^{10} 4p^6 5s^2 4d^3$. **65.** (a) 4; (b) 9; (c) 2; (d) 0. **67.** (a) noble gas; (b) transition element; (c) representative element; (d) inner-transition element. **69.** (a) nonmetal; (b) metal; (c) metal; (d) nonmetal. **71.** (a) Zn and Zr; (b) none; (c) Ge and Sr; (d) As. **73.** (a) Mg; (b) Au; (c) S; (d) Rb. **75.** (a) F; (b) P; (c) Zn; (d) Cl. **77.** (a) N; (b) Ga; (c) K; (d) Rb. **79.** (a) dependent on; (b) independent of; (c) independent of. **81.** (a) 5; (b) 7; (c) 4; (d) 36. **83.** (a) 2; (b) 4; (c) 1; (d) 2. **85.** (a) $2s$ subshell has a maximum occupancy of 2 electrons; (b) $2p$ subshell can accommodate 6 electrons; (c) $2p$ subshell fills after the $2s$; (d) $3p$ and $4s$ subshells fill after the $3s$. **87.** (a) Si: $1s^2 2s^2 2p^6 3s^2 3p^2$; (b) C: $1s^2 2s^2 2p^2$; (c) N: $1s^2 2s^2 2p^3$; (d) N: $1s^2 2s^2 2p^3$. **89.** (a) Be; (b) Be; (c) Ne; (d) Ar. **91.** (a) period 3, group IA; (b) period 3, group IIIA; (c) period 4, group IIIB; (d) period 4, group VIIA. **93.** carbon—the two $2p$ electrons go into separate equivalent orbitals rather than into the same orbital. **95.** (a) groups IVA and VIA; (b) group VB and middle column of group VIIIB; (c) group IVB and last column of group VIIIB; (d) group IA. **97.** (a) Po; (b) Cr; (c) elements 88–109; (d) elements 12–18. **99.** (a) F; (b) Ag. **101.** $1s^2 2s^2 2p^6 3s^2 3p^6 4s^2$. **103.** the same. **105.** 54. **107.** 38. **109.** (a) 5; (b) 5 and 9; (c) 4 and 8; (d) 2; (e) 1, 3, 7, and 9; (f) 2, 3, 5, 6, and 7. **111.** (a) 2, 3, 4, 6, and 7; (b) 1, 4, and 7; (c) 5, 6, and 7; (d) 4 and 7; (e) 3 and 4, 6 and 7; (f) 1 and 4, 1 and 7, 2 and 6, 3 and 9, 5 and 8.

Chapter 7

1. (a) 2; (b) 7; (c) 3; (d) 5. **3.** (a) 1; (b) 8; (c) 2; (d) 7. **5.** The group number and number of valence electrons are the same. **7.** (a) $1s^2 2s^2 2p^2$; (b) $1s^2 2s^2 2p^5$; (c) $1s^2 2s^2 2p^6 3s^2$; (d) $1s^2 2s^2 2p^6 3s^2 3p^3$. **9.** (a) Mg·; (b) ·P·; (c) K·; (d) ·Br: **11.** (a) B; (b) C; (c) O; (d) Li. **13.** The outermost s and p subshells are completely filled with electrons. **15.** (a) O^{2-}; (b) Mg^{2+}; (c) F^-; (d) Al^{3+}. **17.** (a) Ca^{2+}; (b) O^{2-}; (c) Na^+; (d) Al^{3+}. **19.** (a) 15 p and 18 e; (b) 7 p and 10 e; (c) 12 p and 10 e; (d) 3 p and 2 e. **21.** Li denotes a lithium atom, and Li^+ denotes a lithium ion. **23.** (a) +2; (b) −1; (c) −3; (d) +1. **25.** (a) two lost; (b) two gained; (c) one lost; (d) two lost. **27.** (a) loss; (b) loss; (c) loss; (d) gain. **29.** (a) Cl; (b) Na; (c) Al; (d) S. **31.** (a) Ne; (b) Ar; (c) Ar; (d) Ne. **33.** (a) Al^{3+} and F^-; (b) I^-, Ba^{2+}, and Cs^+; (c) S^{2-}

and P^{3-}, Rb^+ and Sr^{2+}; (d) Ga^{3+}, Cl^-, P^{3-}, and K^+. **35.** If a Mg atom lost three electrons, it would not have a noble gas electron configuration.

37. (a) Na· ⤴ :Ö: → 2 Na$^+$ + O^{2-} → Na$_2$O; Na·

(b) Li· ⤴ :N̈: → 3 Li$^+$ + N^{3-} → Li$_3$N; Li·

(c) Cȧ ⤴ :S̈: → Ca^{2+} + S^{2-} → CaS;

(d) Mg· ⤴ :C̈l: → Mg^{2+} + 2 Cl$^-$ → MgCl$_2$.
    ⤵ :C̈l:

39. (a) Li· ⤴ :C̈l: → Li$^+$ + Cl$^-$ → LiCl;

(b) Mg· ⤴ :S̈: → Mg^{2+} + S^{2-} → MgS;

(c) Ga· ⤴ :N̈: → Ga^{3+} + N^{3-} → GaN;

(d) K· ⤴ :Ö: → 2 K$^+$ + O^{2-} → K$_2$O.
    K·

41. (a) CaCl$_2$; (b) CaO; (c) Ca$_3$P$_2$; (d) CaS. **43.** (a) K$_2$S; (b) Li$_2$S; (c) Al$_2$S$_3$; (d) BeS. **45.** (a) X$_2$Y; (b) XY$_3$; (c) X$_3$Y; (d) YX$_2$. **47.** (a) Al^{3+} and O^{2-}; (b) Ca^{2+} and S^{2-}; (c) Li$^+$ and Br$^-$; (d) Na$^+$ and N^{3-}. **49.** It represents the simplest ratio in which ions combine. **51.** (a) NaOH; (b) NH$_4$NO$_3$; (c) CaSO$_4$; (d) Mg(CN)$_2$. **53.** (a) AlPO$_4$; (b) Al$_2$(CO$_3$)$_3$; (c) Al(ClO$_3$)$_3$; (d) Al(C$_2$H$_3$O$_2$)$_3$.

55. (a) :B̈r:B̈r: (b) H:P̈:H (c) :C̈l:N̈:C̈l: (d) :F̈:C̈:F̈:
                    H   :C̈l: :F̈:

57. (a) :F̈:S̈i:F̈: (b) H:B̈r: (c) H:T̈e: (d) :C̈l:F̈:
         :F̈:   H   H

59. (a) H:C̈:C̈:H (b) H:N̈:N̈:H
           H H   H H

(c) :F̈:C̈:H   (d) H:C̈:C̈:C̈l:
       :F̈:   H:C̈l:

61. (a) O; (b) O; (c) S; (d) Mg. **63.** (a) Na, Mg, Al, and P; (b) I, Br, Cl, and F; (c) As, P, S, and O; (d) Ca, Ge, C, and O. **65.** (a) O; (b) N; (c) H; (d) Br. **67.** (a) H—Br, H—Cl, and H—O; (b) O—F, O—P, and O—Al; (c) Br—Br, H—Cl, and B—N; (d) C—N, Cl—F, and Al—Cl. **69.** (a) polar covalent; (b) ionic; (c) polar covalent; (d) nonpolar covalent. **71.** (a) 22.0%; (b) 6.0%; (c) 70.0%; (d) 30.0%.

73. (a) H:C::C::C:H (b) :F̈:N::N:F̈:
              H H

(c) H:C̈::Ö: (d) H:C::C:B̈r:
       H   H :B̈r:

75. (a) H:C̈:C:::N: (b) H:C:::C:C̈:H
              H   H

(c) :N:::C:C:::N: (d) H:N̈:C:::N:
                            H

77. a bond in which both electrons of a shared pair come from one of the two atoms involved in the bond.

79. (a) ×C×::Ö: (b) H:C̈l:Ö×
                           ××

(c) H×Ö×P×Ö×H (d) H:Ö:B̈r×Ö:
       ×Ö×   :Ö:
        H

81. two or more electron-dot structures for a molecule or ion that have the same arrangement of atoms, contain the same number of electrons, and differ in the location of the electrons.

83. (a) :Ö:O: ↔ :Ö::Ö   (b) :F̈:N::Ö: ↔ :F̈:N:Ö:
            :Ö: :Ö: :Ö: :Ö:

85. (a) O=C(O)O   (b) O=P(O)(O)O

87. (a) 24; (b) 26; (c) 30; (d) 32.

89. (a) :C̈l:S̈i:C̈l: (b) :F̈:P̈:F̈:
            :C̈l: :F̈:
            ..C̈l..

(c) H:Ö:Ö:H (d) H:C̈:Ö:H
                     H
                     H

91. (a) [:F̈:N:F̈:]$^+$ (b) [H:Be:H]$^{2-}$
            :F̈:   H
                                 H

(c) [:Ö:C̈l:Ö:]$^-$ (d) [:Ö:I:Ö:]$^-$
         :Ö: :Ö:
                         :Ö:

93. (a) H:C̈:N::Ö: ⟷ H:C̈:N:Ö:
           H:Ö: H Ö:

(b) :N:::N:Ö: ⟷ :N̈::N::Ö: ⟷

:N̈:N:::O:

(c) [:Ö:C::Ö: :Ö:]²⁻ ⟷ [:Ö::C:Ö: :Ö:]²⁻ ⟷

[:Ö:C:Ö: :Ö:]²⁻

(d) [:S̈:C:::N:]⁻ ⟷ [:S̈::C::N̈:]⁻ ⟷

[:S:::C:N̈:]⁻

95. (a) 1 bonding pair, 3 nonbonding pairs; (b) 3 bonding pairs, 2 nonbonding pairs; (c) 1 bonding pair, 3 nonbonding pairs; (d) 3 bonding pairs; 2 nonbonding pairs. **97.** 1. **99.** yes—polar bonds that "cancel" as the result of molecular symmetry may be present. **101.** (a) polar; (b) polar; (c) polar; (d) nonpolar. **103.** (a) polar bonds, nonpolar molecule; (b) polar and nonpolar bond, polar molecule; (c) polar bonds, polar molecule; (d) nonpolar bonds, nonpolar molecule. **105.** (a) angular; (b) linear; (c) linear; (d) angular. **107.** (a) trigonal pyramidal; (b) linear; (c) trigonal planar; (d) tetrahedral. **109.** (a) tetrahedral; (b) trigonal planar; (c) tetrahedral; (d) tetrahedral. **111.** (a) angular; (b) trigonal pyramidal; (c) trigonal pyramidal; (d) angular. **113.** (a) angular; (b) linear; (c) angular; (d) linear (diatomic molecule). **115.** (a) $1s^22s^22p^6$; (b) $1s^22s^22p^6$; (c) $1s^22s^22p^63s^23p^6$; (d) $1s^22s^22p^63s^23p^6$. **117.** (a) $2s^22p^6$; (b) $2s^22p^6$; (c) $3s^23p^6$; (d) $3s^23p^6$. **119.** (a) not stable—ionic charge does not add up to zero; (b) not stable—ionic charge does not add up to zero; (c) stable; (d) not stable—ionic charge does not add up to zero. **121.** (a) LiBr; (b) Be_3P_2; (c) MgO; (d) CaS. **123.** (a) N; (b) O; (c) F; (d) F. **125.** P_2Br_4 is a molecular (covalent) compound, and Ca_2Br_4 is an ionic compound. Ionic compound formulas give the smallest whole-number ratio of ions present ($CaBr_2$). **127.** (a) Cl; (b) Cl. **129.** (a) not enough electrons; (b) too many electrons. **131.** (a) covalent; (b) ionic; (c) covalent; (d) covalent. **133.** The bonding is covalent within the polyatomic ions and ionic between ions.

(a) 2[Cs⁺] [:Ö: :Ö:S:Ö: :Ö:]²⁻ (b) [H H:N̈:H H]⁺ [:Ö:N::Ö: :Ö:]⁻

135. AB, AC, BD, and AD. **137.** (a) same; (b) different; (c) different; (d) different. **139.** (a) tetrahedral electron pairs, tetrahedral geometry; (b) tetrahedral electron pairs, trigonal pyramidal geometry; (c) linear electron pairs, linear geometry; (d) tetrahedral electron pairs; tetrahedral geometry. **141.** (a) 109°; (b) 120°. **143.** (a) polar; (b) nonpolar; (c) polar; (d) polar. **145.** A is Mg, D is S, formula is MgS. **147.** A is Al, D is N, formula is AlN.
149. :F̈:Ö:F̈:
151. A is H, D is O, x is 2, y is 1, formula is H_2O. **153.** A is Be, D is F, electron-dot structure is

[:F̈: :F̈:Be:F̈: :F̈:]²⁻

155. (a) 1, 2, 3, 4, and 5; (b) 1, 5, 7, and 8; (c) 1, 2, 3, 4, 5, and 7; (d) 6, 8, and 9; (e) 2, 3, 4, 5, 8, and 9; (f) 1, 5, and 7. **157.** (a) 1 and 9, 3 and 7, 3 and 9, 4 and 9, 7 and 8, 8 and 9; (b) 3 and 8; (c) 1 and 6, 2 and 3, 2 and 6, 2 and 8, 2 and 9, 3 and 5, 4 and 6, 5 and 6, 5 and 8, 5 and 9, 6 and 7; (d) 3 and 6, 3 and 8, 3 and 9, 6 and 8, 6 and 9, 8 and 9; (e) 3 and 9, 8 and 9; (f) 2 and 9, 5 and 9, 6 and 7.

Chapter 8

1. (a) binary; (b) ternary; (c) ternary; (d) binary. **3.** (a) molecular; (b) ionic; (c) ionic; (d) ionic. **5.** (a) ionic; (b) molecular; (c) molecular; (d) molecular. **7.** (a) fixed; (b) variable; (c) variable, (d) fixed. **9.** (a) fixed; (b) variable; (c) variable; (d) fixed. **11.** (a) +1; (b) +1; (c) +3; (d) +3. **13.** (a) nitride; (b) chloride; (c) sulfide; (d) bromide. **15.** (a) magnesium; (b) potassium; (c) cobalt(II); (d) cobalt(III). **17.** (a) lithium oxide; (b) magnesium sulfide; (c) zinc chloride; (d) silver bromide. **19.** The name of $FeBr_2$ must contain a Roman numeral. **21.** (a) +2; (b) +4; (c) +2; (d) +4. **23.** (a) gold(III) sulfide; (b) copper(I) sulfide; (c) lead(IV) sulfide; (d) nickel(II) sulfide; **25.** (a) yes; (b) yes; (c) no; (d) no. **27.** (a) silver chloride; (b) cobalt(III) chloride; (c) aluminum chloride; (d) lead(IV) chloride. **29.** (a) copper(II) oxide and copper(I) oxide; (b) manganese(II) oxide and manganese(III) oxide; (c) tin(II) sulfide and tin(IV) sulfide; (d) lead(II) bromide and lead(IV) bromide. **31.** (a) iron(II) oxide and iron(III) oxide; (b) manganese(II) sulfide and manganese(III) sulfide; (c) cobalt(II) bromide and cobalt(III) bromide; (d) copper(I) iodide and copper(II) iodide. **33.** (a) plumbic oxide; (b) auric chloride; (c) iron(III) iodide; (d) tin(II) bromide. **35.** (a) FeS; (b) SnS_2; (c) Li_2S; (d) ZnS. **37.** (a) KI; (b) $CoBr_2$; (c) FeI_2; (d) Ba_3P_2. **39.** (a) nitrate; (b) sulfate; (c) hydroxide; (d) cyanide. **41.** (a) permanganate; (b) borate; (c) thiosulfate; (d) azide.

43. PO_4^{3-}, HPO_4^{2-}, and $H_2PO_4^-$. **45.** (a) differ in oxygen content; (b) no difference (two names for the same ion); (c) one contains O and the other S; (d) differ in chromium and oxygen content. **47.** (a) none; (b) phosphate; (c) sulfate; (d) nitrate. **49.** (a) magnesium carbonate; (b) zinc sulfate; (c) beryllium nitrate; (d) silver phosphate. **51.** (a) +2; (b) +2; (c) +1; (d) +2. **53.** (a) iron(II) sulfate; (b) manganese(II) nitrate; (c) nickel(II) cyanide; (d) copper(II) hydroxide. **55.** (a) iron(III) carbonate and iron(II) carbonate; (b) gold(I) sulfate and gold(III) sulfate; (c) tin(II) hydroxide and tin(IV) hydroxide; (d) chromium(III) acetate and chromium(II) acetate. **57.** (a) gold(III) nitrate; (b) silver sulfate; (c) aluminum carbonate; (d) lead(II) hydroxide. **59.** (a) ammonium chloride; (b) ammonium sulfate; (c) copper(II) phosphate; (d) sodium phosphate. **61.** (a) Ag_2CO_3; (b) $AuNO_3$; (c) $Cr_2(SO_4)_3$; (d) $NH_4C_2H_3O_2$. **63.** (a) 7; (b) 5; (c) 3; (d) 10. **65.** (a) tetraphosphorus decoxide; (b) sulfur tetrafluoride; (c) carbon tetrabromide; (d) chlorine dioxide. **67.** (a) hydrogen sulfide; (b) hydrogen fluoride; (c) ammonia; (d) methane. **69.** (a) ICl; (b) NCl_3; (c) SF_6; (d) OF_2. **71.** (a) PH_3; (b) HBr; (c) C_2H_6; (d) H_2Te. **73.** (a) no; (b) yes; (c) yes; (d) no. **75.** (a) oxy; (b) nonoxy; (c) oxy; (d) nonoxy. **77.** (a) hydrochloric acid; (b) hydrocyanic acid; (c) chlorous acid; (d) sulfuric acid. **79.** (a) H_2CO_3; (b) HNO_3; (c) $HBrO_3$; (d) $HClO_3$. **81.** (a) $HBrO_4$; (b) HIO_2; (c) HNO_3; (d) HBrO. **83.** (a) HCN; (b) HCl; (c) $HClO_3$; (d) $HBrO_2$. **85.** (a) hypozoooous acid; (b) perzooic acid; (c) zooous acid. **87.** (a) binary ionic; (b) binary molecular; (c) ternary ionic; (d) oxyacid. **89.** (a) Ca_3N_2; (b) $Ca(NO_3)_2$; (c) $Ca(NO_2)_2$; (d) $Ca(CN)_2$. **91.** (a) K_3P; (b) K_3PO_4; (c) K_2HPO_4; (d) KH_2PO_4. **93.** (a) N_2O, CO_2; (b) NO_2, SO_2; (c) SF_2, SCl_2; (d) N_2O_3. **95.** (a) $CaCO_3$, HNO_2; (b) $NaClO_4$, $NaClO_3$; (c) $HClO_2$, HClO; (d) Li_2CO_3, Li_3PO_4. **97.** (a) sodium nitrate; (b) aluminum sulfide, magnesium nitride, beryllium phosphide; (c) iron(III) oxide; (d) gold(I) chlorate. **99.** (a) $Ni_2(SO_4)_3$; (b) Ni_2O_3; (c) $Ni_2(C_2O_4)_3$; (d) $Ni(NO_3)_3$. **161.** NO_2O_2. **103.** (a) $MgCl_2$; (b) OF_2. **105.** (a) x is 4; silicon tetrachloride; (b) x is 2; magnesium chloride; (c) x is 3, potassium nitride; (d) x is 3, nitrogen trichloride. **107.** beryllium bromate. **109.** aluminum nitride. **111.** carbon dioxide. **113.** beryllium cyanide. **115.** (a) 4, 5, 6, and 9; (b) all except 1; (c) 6; (d) 3 and 8; (e) 4 and 9; (f) 4, 6, 8, and 9. **117.** (a) 1, 2, 5, 6, and 9; (b) 8; (c) 4, 6, and 7; (d) 3; (e) 8; (f) 6.

Chapter 9

1. For each sample the mass percent A is 57.81% and the mass percent B is 42.2%. **3.** experiments 1 and 3. **5.** 100.0 g NH_3 produced, 7.2 g H left unreacted. **7.** (a) 156.71 amu; (b) 151.92 amu; (c) 342.34 amu; (d) 354.48 amu. **9.** (a) 164.10 amu; (b) 323.5 amu; (c) 324.17 amu; (d) 227.15 amu. **11.** $y = 4$. **13.** 62.1 amu. **15.** (a) 39.34% Na, 60.66% Cl; (b) 4.90% H, 17.48% B, 77.62% O; (c) 74.02% C, 8.71% H, 17.27% N; (d) 74.25% C, 7.81% H, 12.99% N, 4.946% O. **17.** $CaSO_4$, CaC_2O_4, $Ca_3(PO_4)_2$, $CaCO_3$. **19.** 69.95% Fe, 30.05% O. **21.** 25.9% N, 74.08% O. **23.** 32.37% Na, 22.57% S, 45.06% O. **25.** 363.1 g N. **27.** (a) 1.50×10^{24} atoms Zn; (b) 1.50×10^{24} atoms Au; (c) 1.50×10^{24} atoms Ag; (d) 1.50×10^{24} atoms Mn. **29.** (a) 1.20×10^{24} atoms B; (b) 8.73×10^{23} atoms Na; (c) 6.02×10^{22} atoms C; (d) 5.27×10^{23} atoms U. **31.** (a) 9.03×10^{23} molecules CO_2; (b) 3.01×10^{23} molecules NH_3; (c) 1.40×10^{24} molecules PF_3; (d) 6.715×10^{23} molecules N_2H_4. **33.** (a) 1.20×10^{24} units CaO; (b) 1.20×10^{24} units MgO; (c) 3.13×10^{24} units Na_2SO_4; (d) 2.07×10^{23} units KOH. **35.** (a) 63.5 g Cu; (b) $15\overline{0}$ g As; (c) 27.0 g Be; (d) 128 g S. **37.** (a) 78.01 g $Al(OH)_3$; (b) 100.9 g Mg_3N_2; (c) 187.6 g $Ca(NO_3)_2$; (d) 325.8 g La_2O_3. **39.** (a) 172 g Li_2CO_3; (b) 31.1 g LiF; (c) 113 g Li_3N; (d) 17.4 g LiOH. **41.** (a) 607.9 g UF_6; (b) 120.2 g B_2O_3; (c) 79.58 g C_2H_6O; (d) 558.7 g $Pb(C_2H_5)_4$. **43.** 79.89 g/mole. **45.** 3.16×10^{-23} g. **47.** phosphorus. **49.** (a) same; (b) different; (c) same; (d) same. **51.** (a) more atoms; (b) fewer atoms; (c) the same; (d) more atoms. **53.** (a) 31.07 g S; (b) 8.730 g Be; (c) 230.5 g U; (d) 27.21 g Si. **55.** 3 moles H/1 mole H_3PO_4, 1 mole H_3PO_4/3 moles H, 1 mole P/1 mole H_3PO_4, 1 mole H_3PO_4/1 mole P, 4 moles O/1 mole H_3PO_4, 1 mole H_3PO_4/4 moles O. **57.** (a) 4.00 moles N, 6.00 moles O; (b) 0.100 mole S, 0.100 mole O, 0.200 mole Cl; (c) 10.5 moles Ca, 7.00 moles P, 28.0 moles O; (d) 1 mole C, 2 moles O. **59.** (a) 2.50 moles S; (b) 7.50 moles S; (c) 5.00 moles S; (d) 10.0 moles S. **61.** (a) 8.00 moles atoms; (b) 16.0 moles atoms; (c) 18.0 moles atoms; (d) 40.0 moles atoms. **63.** (a) N_2O_3; (b) N_2O_3; (c) N_2O_4; (d) N_2O. **65.** (a) 2.25×10^{22} atoms Au; (b) 1.01×10^{23} atoms Ba; (c) 1.86×10^{20} atoms H; (d) 1.485×10^{24} atoms Be. **67.** (a) 1.37×10^{23} molecules CO_2; (b) 3.54×10^{23} molecules H_2O_2; (c) 1.13×10^{24} molecules NH_3; (d) 3.951×10^{23} molecules Cl_2O_7. **69.** (a) 65.38 g Zn; (b) 21.30 g S; (c) 2.830 g NH_3; (d) 2.614×10^{-19} g SO_2. **71.** (a) 5.325×10^{-23} g S; (b) 7.951×10^{-23} g Ti; (c) 4.995×10^{-23} g C_2H_6; (d) 1.664×10^{-22} g C_7H_{16}. **73.** (a) 60.7 g Cl; (b) 921.93 g Cl; (c) 9.73 g Cl; (d) 17.1 g Cl. **75.** (a) 1.71×10^{23} atoms P; (b) 3.38×10^{23} atoms P; (c) 9.84×10^{22} atoms P; (d) 1.23×10^{23} atoms P. **77.** (a) 0.0718 g C; (b) 29.4 g C; (c) 2.668 g C; (d) 567.90 g C. **79.** (a) 5.32×10^{-13} g THC; (b) 2.53×10^{-14} g THC; (c) $1.48 \times$

10^{-14} g THC; (d) 9.01×10^{-15} g THC. **81.** (a) 1.746 moles atoms; (b) 3.505×10^{23} atoms C; (c) 2.866 g O; (d) 2.696×10^{22} molecules Treflan. **83.** (a) 0.03118 mole H_2S; (b) 2.347×10^{21} molecules S_8; (c) 3.040 g $S_3N_3O_3Cl_3$; (d) 1.878×10^{22} atoms S. **85.** (a) C_5H_{11}; (b) C_7H_{16}; (c) ClO_3; (d) $PNCl_2$. **87.** (a) Na_2S; (b) $KMnO_4$; (c) H_2SO_4; (d) $C_2H_3O_5N$. **89.** (a) 3 to 5; (b) 6 to 3; (c) 6 to 10 to 13; (d) 4 to 8 to 3. **91.** (a) P_2O_5; (b) Mg_3N_2; (c) $Na_2S_2O_3$; (d) $Mg_2P_2O_7$. **93.** CoO, Co_2O_3. **95.** C_2H_6S. **97.** BrF_5. **99.** C_5H_6O. **101.** (a) B_2H_6; (b) SO_2F_2; (c) C_8H_{18}; (d) $S_4N_4F_4$. **103.** C_3H_8O. **105.** CH_2O, $C_3H_6O_3$. **107.** 2.56×10^{24} atoms O. **109.** (a) 8.3%; (b) 64.285714%; (c) 64.285714%. **111.** 3.71. **113.** 20.1 g Be. **115.** (a) CrO_3; (b) CrO_3. **117.** 216 g. **119.** 107.9 g/mole. **121.** 10.81 g/mole, B **123.** (a) Na_3AlF_6; (b) SO_2; (c) $BaCO_3$; (d) HOCl. **125.** (a) CH_4S; (b) 7.221 g. **127.** $x = 3$. **129.** 40.08 g/mole. **131.** $C_3H_8O_2$. **133.** 11.3 g/cm^3. **135.** 22.9 cm^3. **137.** (a) 148.3 g/mole; (b) 65.02 g/mole; (c) 96.69 g/mole; (d) 117.5 g/mole. **139.** 650 J. **141.** (a) 1.40×10^{24} units $Al_2(SO_4)_3$; (b) 2.81×10^{24} Al^{3+} ions; (c) 4.21×10^{24} SO_4^{2-} ions; (d) 7.01×10^{24} ions. **143.** 4.51×10^{24} Cl^- ions. **145.** 6.82×10^{26} atoms Ni. **147.** 81.0 mL. **149.** 1.39×10^{24} atoms. **151.** 6.67×10^{22} atoms Cu. **153.** 2.73×10^{26}. **155.** (a) 2 and 3, 5 and 6; (b) 2 and 3, 5 and 6; (c) 3 and 9, 5 and 6, 7 and 8; (d) 3 and 9, 5 and 6, 7 and 8; (e) all pairs; (f) 2 and 3, 5 and 6. **157.** (a) 1, 5, 7, and 8; (b) 1, 2, 4, 8, and 9; (c) 2 and 4, 6 and 7; (d) 2 and 4, 6 and 7; (e) 1 and 9; 2 and 4, 3 and 5, 6 and 7; (f) 1 and 9, 2 and 8, 3 and 5, 6 and 7.

Chapter 10

1. 3.0 g D. **3.** 2.2 g CO_2. **5.** H_2, O_2, N_2, F_2, Cl_2, Br_2, I_2. **7.** (a) (s) means solid, and (aq) means aqueous solution; (b) (g) means gas, (l) means liquid, and (aq) means aqueous solution. **9.** an arrow. **11.** law of conservation of mass. **13.** (a) balanced; (b) unbalanced; (c) balanced; (d) unbalanced. **15.** (a) $N_2 + O_2 \rightarrow 2$ NO; (b) $2 SO_2 + O_2 \rightarrow 2 SO_3$; (c) $U + 3 F_2 \rightarrow UF_6$; (d) $4 Al + 3 O_2 \rightarrow 2 Al_2O_3$. **17.** (a) $4 Mg + Fe_3O_4 \rightarrow 4 MgO + 3 Fe$; (b) $Na_2NH + 2 H_2O \rightarrow NH_3 + 2 NaOH$; (c) $4 NH_3 + 5 O_2 \rightarrow 4 NO + 6 H_2O$; (d) $Cl_2 + H_2O \rightarrow HCl + HClO$. **19.** (a) $3 PbO + 2 NH_3 \rightarrow 3 Pb + N_2 + 3 H_2O$; (b) $2 NaHCO_3 + H_2SO_4 \rightarrow Na_2SO_4 + 2 H_2O + 2 CO_2$; (c) $TiO_2 + C + 2 Cl_2 \rightarrow TiCl_4 + CO_2$; (d) $2 NBr_3 + 3 NaOH \rightarrow N_2 + 3 NaBr + 3 HOBr$. **21.** (a) $Ca(OH)_2 + 2 HNO_3 \rightarrow Ca(NO_3)_2 + 2 H_2O$; (b) $BaCl_2 + (NH_4)_2SO_4 \rightarrow BaSO_4 + 2 NH_4Cl$; (c) $2 Fe(OH)_3 + 3 H_2SO_4 \rightarrow Fe_2(SO_4)_3 + 6 H_2O$; (d) $Na_3PO_4 + 3 AgNO_3 \rightarrow 3 NaNO_3 + Ag_3PO_4$. **23.** (a) $H_2O + C \rightarrow CO + H_2$; (b) $C_2H_6O + 3 O_2 \rightarrow 2 CO_2 + 3 H_2O$; (c) $2 Al + 3 S \rightarrow Al_2S_3$; (d) $H_2SO_4 + Ca(OH)_2 \rightarrow CaSO_4 + 2 H_2O$. **25.** (a) $C_2H_4 + 3 O_2 \rightarrow 2 CO_2 + 2 H_2O$; (b) $C_3H_8 + 5 O_2 \rightarrow 3 CO_2 + 4 H_2O$; (c) $2 C_3H_6 + 9 O_2 \rightarrow 6 CO_2 + 6 H_2O$; (d) $2 C_4H_6 + 11 O_2 \rightarrow 8 CO_2 + 6 H_2O$. **27.** (a) $C_4H_8O_2 + 5 O_2 \rightarrow 4 CO_2 + 4 H_2O$; (b) $C_3H_6O + 4 O_2 \rightarrow 3 CO_2 + 3 H_2O$; (c) $2 C_5H_{10}O_2 + 13 O_2 \rightarrow 10 CO_2 + 10 H_2O$; (d) $4 C_6H_{13}O + 37 O_2 \rightarrow 24 CO_2 + 26 H_2O$. **29.** (a) $PbCO_3 \rightarrow PbO + CO_2$; (b) $Co_2(CO_3)_3 \rightarrow Co_2O_3 + 3 CO_2$. **31.** (a) $4 C_2H_7N + 19 O_2 \rightarrow 8 CO_2 + 14 H_2O + 4 NO_2$; (b) $CH_4S + 3 O_2 \rightarrow CO_2 + 2 H_2O + SO_2$. **33.** (a) synthesis; (b) synthesis; (c) double replacement; (d) single replacement. **35.** (a) double replacement; (b) synthesis; (c) double replacement; (d) decomposition. **37.** (a) 4 molecules NH_3 react with 3 molecules O_2 to produce 2 molecules N_2 and 6 molecules H_2O; (b) 4 moles NH_3 react with 3 moles O_2 to produce 2 moles N_2 and 6 moles H_2O. **39.** 3 moles HNO_2/2 moles NO, 3 moles HNO_2/1 mole HNO_3, 3 moles HNO_2/1 mole H_2O, 2 moles NO/1 mole HNO_3, 2 moles NO/1 mole H_2O, 1 mole HNO_3/1 mole H_2O, and the reciprocal of each of these factors. **41.** (a) 1.50 moles H_2O; (b) 0.375 mole N_2; (c) 0.750 mole Mg; (d) 0.375 mole H_2. **43.** (a) 2.00 moles NaN_3; (b) 9.00 moles CO; (c) 6.00 moles NH_2Cl; (d) 2.00 moles $C_3H_5O_9N_3$. **45.** (a) 1.68 moles O_2; (b) 1.18 moles H_2O; (c) 2.58 moles NH_3; (d) 0.509 mole NO. **47.** (a) 6.12 moles products; (b) 4.38 moles products; (c) 2.62 moles products; (d) 4.81 moles products. **49.** (a) 0.5764 g C_7H_{16}; (b) 1.475 g HCl; (c) 11.98 g Na_2SO_4; (d) 1.064 g Na_2CO_3. **51.** (a) 7.25 g N_2H_4; (b) 38.4 g N_2; (c) 22.8 g H_2O; (d) 333 g $\underline{H_2O_2}$. **53.** 6.866 g K_3AsO_4. **55.** 52.83 g Cr. **57.** (a) 300 g $CaCO_3$; (b) 0.0498 g $CaCO_3$; (c) 1.12 g $CaCO_3$; (d) 2.08×10^{-20} g $CaCO_3$. **59.** (a) 0.2977 moles Na_2SiO_3; (b) 79.95 g HF; (c) 9.864×10^{21} molecules H_2SiF_6; (d) 65.55 g HF. **61.** 13.3 g K. **63.** 7.23 g MgO. **65.** nuts. **67.** 213 kits. **69.** (a) N_2; (b) Mg; (c) Mg; (d) N_2. **71.** (a) 3.6 g NH_3; (b) 36.4 g NH_3; (c) 44.9 g NH_3; (d) 68 g NH_3. **73.** 4.16 g Fe_2O_3. **75.** 22.1 g $NiCl_2$. **77.** 2.57 g SF_4, 3.21 g S_2Cl_2, 5.57 g NaCl. **79.** no H_2S, 27 g O_2. **81.** In reactions where the molar ratio of reactants is the same as the balanced equation coefficient ratio for reactants all reactants are depleted at the same time. **83.** 30.8%. **85.** (a) 208 g Al_2S_3; (b) 60.1%. **87.** 78.80%. **89.** 50.8 g ZnS. **91.** 43.5 g PCl_5. **93.** 54.7 g SO_2. **95.** (a) 24.2 g HCl; (b) 79.2 g $MgCl_2$. **97.** 39.1 g SO_2. **99.** 343 g $KClO_3$. **101.** 6.69 g $AgNO_3$. **103.** 8.33 g N_2, 21.4 g H_2O, 45.2 g Cr_2O_3. **105.** 80.1 g NO. **107.** 52.48 g C_8H_{10}. **109.** 44.5% $CuSO_4$. **111.** 92.8% Ag_2O. **113.** (a) 75.4% IF_5; (b) 75.4% Br_2. **115.** 3.678 tons Cu ore. **117.** 38.3% $CaCO_3$. **119.** (a) $CaO + SO_3 \rightarrow CaSO_4$; (b) $NH_4NO_3 \rightarrow N_2O + 2 H_2O$; (c) $Al + 3 AgNO_3 \rightarrow Al(NO_3)_3 + 3 Ag$; (d) $HBr + KOH \rightarrow KBr + H_2O$. **121.** C_3H_6. **123.** $2 Cu_2S + 3 O_2 \rightarrow 2 Cu_2O + 2 SO_2$. **125.** $2 C_6H_6 + 15 O_2 \rightarrow 12 CO_2 + 6 H_2O$. **127.** 0.16 ton

CaSO$_3$. **129.** 27.04 moles electrons. **131.** 2.039 moles Cl-containing products. **133.** 8.138 × 10^{24} positive ions. **135.** 87.3 mL NH$_3$. **137.** 0.0789 g HNO$_3$. **139.** (a) 2; (b) 1, 5, 6, 7, 8, and 9; (c) 9; (d) 1; (e) 6, 7, 8, and 9; (f) 3, 4, 5, and 6. **141.** (a) 4 and 5; (b) 4; (c) 3 and 6; (d) 2, 3, 5, 6, 7, and 8; (e) 8 and 9; (f) all choices.

Chapter 11

1. (a) gaseous; (b) liquid; (c) gaseous; (d) gaseous.
3. definite shape, slightly higher density, slightly lower compressibility and thermal expansion. **5.** through collisions with each other. **7.** potential energy.
9. Disruptive force magnitude increases with temperature and cohesive force magnitude is independent of temperature. **11.** gaseous state. **13.** (a) The particles of a gas are widely separated because cohesive forces between particles have been overcome by kinetic energy. There are relatively few particles in a given volume (a low density). (b) A temperature increase causes increased vibrational motion for the particles present in a liquid. This causes a slight expansion of the liquid. (c) The strong cohesive forces present hold the particles in essentially fixed positions.
15. (a) endothermic; (b) endothermic; (c) exothermic.
17. (a) 7180 J; (b) 590 J; (c) 79,100 J; (d) 1950 J.
19. (a) 16,700 J; (b) 3250 J. **21.** 1.92. **23.** 4.06 moles.
25. A by 6 J/g. **27.** 73$\overline{0}$ J. **29.** (a) 7730 J; (b) 236,000 J; (c) 2250 J; (d) 83,$\overline{0}$00 J. **31.** 21,400 J. **33.** (a) 2500 J; (b) 118,000 J; (c) 153,000 J; (d) 155,000 J. **35.** Because of higher kinetic energy at the higher temperature, it is easier for molecules to attain the minimum kinetic energy needed to overcome attractive forces. **37.** liquid temperature and liquid surface area. **39.** In an uninsulated container heat transfer from the surroundings into the container causes the temperature to remain constant. **41.** temperature and strength of intermolecular forces (nature of the liquid).
43. A larger volume will enable more molecules to be present in the vapor at equilibrium. However, the larger number of molecules spread over a larger volume produces the same pressure as a small number of molecules in a small volume. **45.** The substance with the highest vapor pressure, CS$_2$, is the most volatile. **47.** Boiling point varies with external pressure; this results in many different boiling points. The normal boiling point is the boiling point at one specific pressure. **49.** At the higher elevation the boiling point of water will be lower; the maximum water temperature (its boiling point) will be lower, and the egg will cook at a slower rate. **51.** The boiling point does not change since it depends only on external (atmospheric) pressure. **53.** molecules of the liquid in the vapor state.
55. Intermolecular forces are forces between molecules; intramolecular forces are forces within molecules (chemical bonds). **57.** the presence of polar molecules. **59.** the stronger the intermolecular forces, the higher the boiling point. **61.** a covalent bond—the strongest hydrogen bond is about one-tenth as strong as a covalent bond. **63.** Cl$_2$— London force strength depends on molecular size, and Cl$_2$ is the larger molecule. **65.** CH$_4$ is a nonpolar molecule, and CH$_3$F is a polar molecule; attractive forces between polar molecules are stronger than between nonpolar molecules.
67. 4 (see Figure 11.15). **69.** It is more difficult for molecules to escape from the condensed state when hydrogen bonding is present. The water will have to be heated to a higher temperature (because of the hydrogen bonds) to bring its vapor pressure up to the point where boiling occurs. **71.** When water molecules are hydrogen-bonded, they are farther apart than when they are not hydrogen-bonded. The temperature of 4 °C is the trade-off point between random motion from kinetic energy and orientation from hydrogen bonding. Below that temperature density increases rather than decreases as hydrogen bonding causes the molecules to move farther apart. **73.** Hydrogen bonding between water molecules causes ice to have an "open" structure that is less dense than liquid water.
75. Hydrogen bonding increases significantly the ability of water molecules to absorb heat energy from the surroundings when evaporating and release energy when condensing. This large absorption and release of energy produce a temperature-moderating effect on climate around large bodies of water. **77.** Molecules that occupy liquid surface positions experience an imbalance in intermolecular attractions that pulls them toward the interior of the liquid. Surface tension is a measure of this force imbalance.
79. (a) nonpolar molecules; (b) positive and negative ions.
81. (a) electronstatic attractions between ions; (b) dipole–dipole attractions and London forces. **83.** (a) extremely high melting points, nonvolatile; (b) variable melting points, low volatility. **85.** Crystalline solids have a regular three-dimensional arrangement of the particles present. Amorphous solids have a random, nonrepetitive three-dimensional arrangement of particles present. **87.** (a) NaCl, since it is an ionic compound; (b) SiO$_2$, since it is a macromolecular compound; (c) Cu, since it is a molecular solid; (d) MgF$_2$, since interionic attractions are stronger because of the presence of +2 ions. **89.** Vaporizing phase takes longer. **91.** 9.52 g ice. **93.** 38.3 °C.
95. 0.421 J/g °C. **97.** 1400 J/g. **99.** 0.775 g/mL.
101. 30.1 amu. **103.** 9.45 kJ. **105.** 2370 J. **107.** 2.9 × 10^7 gal/day. **109.** (a) 1, 3, 4, 7, and 8; (b) 3, 4, and 7; (c) 2, 5, 6, and 9; (d) 1, 2, 5, 6, 8, and 9; (e) 3, 4, and 7; (f) 3, 4, and 7. **111.** (a) 1, 2, 3, 5, 6, 7, and 9; (b) 1 and 3, 5 and 9; (c) 4, 6, and 8; (d) 1, 3, 5, and 9; (e) 2, 3, 6, and 8; (f) 1, 2, 4, 5, 6, 7, and 8.

Chapter 12

1. (a) 347 mm Hg; (b) 521 mm Hg; (c) 39,100 mm Hg.
3. (a) 769.9 mm Hg; (b) 14.87 psi; (c) 1.013 atm.   **5.** (a) the level will drop; (b) no change in level.   **7.** 999 mm Hg.
9. $P_1V_1 = P_2V_2$.   **11.** (a) 723 mm Hg; (b) 138 mm Hg; (c) 1040 mm Hg; (d) 2840 mm Hg.   **13.** (a) 1.42 L; (b) 49.2 L; (c) 1.12 atm; (d) 4.41 atm.   **15.** 6.6 atm.   **17.** 7%.
19. If the temperature increases the volume will also increase and if the temperature decreases the volume will also decrease.   **21.** (a) 3.64 L; (b) 6.79 L; (c) 2.24 L; (d) 2.95 L.   **23.** (a) 294 °C; (b) −225.2 °C; (c) 332 °C; (d) 2670 °C.   **25.** 26.7 mL.   **27.** 73 °C.   **29.** $P_1/T_1 = P_2/T_2$.
31. (a) 3.12 atm; (b) 1.23 atm; (c) 7.62 atm; (d) 4.75 atm.
33. (a) 711 °C; (b) 1480 °C; (c) 3490 °C; (d) 533 °C.
35. 3.1 atm.   **37.** $V_1 = V_2P_2T_1/P_1T_2$.   **39.** (a) 5.90 L; (b) 3700 mL; (c) 2.11 atm; (d) −171 °C.   **41.** 471 °C.
43. (a) 472 mm Hg; (b) 141 °C.   **45.** (a) 4.00 L; (b) 4.00 L; (c) 1.00 L; (d) 24.0 L.   **47.** (a) 0.873 L; (b) 0.320 L; (c) 0.361 L; (d) 4.20 L.   **49.** 10.3 L; (b) 105 mL; (c) 34.8 L; (d) 375 mL.   **51.** (a) 37.5 L O_2; (b) 5.0 L CO_2.
53. (a) 0.433 L C_3H_8; (b) 0.325 L C_3H_8.   **55.** 0.817 L.
57. 1.37 L (the same volume).   **59.** 1.5 g He.   **61.** N_2O.
63. (a) 52.2 L H_2; (b) 52.2 L CH_4; (c) 52.2 L SO_2; (d) 52.2 L F_2.   **65.** (a) 19.6 L N_2; (b) 32.3 L NH_3; (c) 12.5 L CO_2; (d) 7.73 L Cl_2.   **67.** (a) 3.88 g H_2; (b) 1.78 g NO; (c) 45.2 g CO_2; (d) 6.97 g CH_4.
69. (a) 0.179 g/L; (b) 1.25 g/L; (c) 1.21 g/L; (d) 2.14 g/L.
71. 26.0 g/mole.   **73.** 2.51 L Xe.   **75.** (a) 2.24 L; (b) 2.52 L; (c) 0.323 L; (d) 2.20 L.   **77.** 397 °C.   **79.** (a) 57.5 L; (b) 0.193 mole.   **81.** 0.0824 atm · L/mole · K   **83.** 27.1 L.
85. 38.6 g N_2.   **87.** (a) 24.0 g; (b) 0.0495 g; (c) 5.71 g; (d) 0.0669 g.   **89.** 13.2 g/mole.   **91.** (a) 2.08 g/L; (b) 0.297 g/L; (c) 4.05 g/L; (d) 3.27 g/L.   **93.** (a) 2.31 g/L; (b) 23.1 g/L; (c) 1.74 g/L; (d) 7.43 g/L.   **95.** 56.0 g/mole.
97. 0.469 atm.   **99.** 80.0 L NO.   **101.** 13.7 L O_2.
103. 37.6 g NO_2.   **105.** 12.2 L H_2.   **107.** 78.1 L NO_2.
109. 1005 mm Hg.   **111.** 98 mm Hg.   **113.** (a) 4.8 atm; (b) 4.8 atm; (c) 4.8 atm; (d) 0.15 atm.   **115.** (a) 0.60 atm; (b) 0.40 atm; (c) 0.400 atm; (d) 0.400 atm.   **117.** 5.2 atm.
119. 0.71.   **121.** (a) 726 mm Hg; (b) 617 mm Hg; (c) 722 mm Hg; (d) 690 mm Hg.   **123.** 13.5 L.
125. 3.48 L NH_3.   **127.** 0.15 g O_2.   **129.** 25%.   **131.** 50.0%.
133. 1200 mm Hg.   **135.** 2.5 atm.   **137.** 1240 mm Hg.
139. 36.6 mL.   **141.** (a) 1.14; (b) 1.14.   **143.** 2.69×10^{22} molecules SO_2.   **145.** 3.3×10^{13} molecules O_2.
147. 75 amu.   **149.** $x = 3$.   **151.** 2.71 kg/m³.   **153.** C_4H_{10}.
155. 99.1 g/mole, $C_4H_4Cl_2$.   **157.** 1.06 g H_2O.
159. 0.0870 m³ HCl.   **161.** 2.0 atm.   **163.** 19,200 breaths.
165. (a) 2 and 4, 3 and 9; (b) 1, 2, 3, 4, 6, 7, and 9; (c) 1, 2, 3, 4, 8, and 9; (d) all choices; (e) 1, 5, 7 and 8; (f) 1, 5, and 8.   **167.** (a) 1 and 6; (b) 3 and 5; (c) 1, 4, 6, 7, 8, and 9; (d) 1 and 6, 3 and 5, 7 and 8; (e) 1 and 6, 3 and 5, 7 and 8; (f) 1, 3, 4, 5, 6, 7, 8, and 9.

Chapter 13

1. (a) water; (b) ethyl alcohol.   **3.** (a) soluble; (b) immiscible.
5. (a) saturated; (b) unsaturated; (c) unsaturated; (d) saturated.   **7.** (a) soluble; (b) very soluble; (c) very soluble; (d) slightly soluble.   **9.** (a) dilute; (b) concentrated; (c) dilute; (d) concentrated.   **11.** (a) slightly soluble; (b) very soluble; (c) slightly soluble; (d) slightly soluble.   **13.** (a) all choices; (b) all choices; (c) Ag_2SO_4 and $AgNO_3$; (d) CaS, $Ca(NO_3)_2$, and $Ca(C_2H_3O_2)_2$.   **15.** (a) all choices; (b) all choices; (c) K_2CO_3; (d) MgS, Rb_2S, and CaS.   **17.** (a) 7.20% (m/m); (b) 2.77% (m/m); (c) 19.4% (m/m); (d) 6.79% (m/m).   **19.** (a) 10.6 g HCl; (b) 11 g NH_4Cl; (c) 1.46 g NaCl; (d) 12.50 g $NaNO_3$.   **21.** (a) 26.3% (m/m); (b) 28.5% (m/m); (c) 55.0% (m/m); (d) 90.5% (m/m).   **23.** (a) 950 g; (b) 950 g; (c) 950 g; (d) 950 g.   **25.** (a) 6.00% (v/v); (b) 3.91% (v/v); (c) 16.9% (v/v); (d) 33.1% (v/v).
27. 51.8% (v/v).   **29.** 5.06 mL C_3H_8O.   **31.** 1.25 qt H_2O.
33. (a) 2.0% (m/v); (b) 5.67% (m/v); (c) 47.6% (m/v); (d) 0.22% (m/v).   **35.** 37.5% (m/v).   **37.** 210.0 g Na_2CO_3.
39. 3750 g $(NH_4)_2SO_4$.   **41.** 22,000 mL.   **43.** (a) 100 ppm (m/v); (b) 100,000 ppb (m/v).   **45.** (a) 2710 ppm (m/m); (b) 2,710,000 ppb (m/m).   **47.** 0.0015 mL pollutant.
49. 53 ppb (v/v).   **51.** (a) 0.0032 g N_2O; (b) 3.7×10^{-6} g N_2O; (c) 60 g N_2O.   **53.** "1 mole KCl" denotes 74.6 g KCl; "1 molar KCl" denotes a solution containing 74.6 g of KCl per liter.   **55.** (a) 4.0 M; (b) 1.51 M; (c) 0.124 M; (d) 0.33 M.   **57.** (a) 1130 g HNO_3; (b) 1.02 g NaCl; (c) 137 g $BaCl_2$; (d) 68 g H_2SO_4.
59. (a) 105 mL; (b) 163 mL; (c) 15,000 mL; (d) 0.26 mL.
61. (a) 563 mL; (b) 435 mL; (c) 4540 mL; (d) 896 mL.
63. (a) 1.11 L; (b) 0.746 L; (c) 0.781 L; (d) 0.314 L.
65. 15.2 M.   **67.** 4.5% (m/m).   **69.** 0.157 M.   **71.** Both have moles of solute in the numerator of the defining equation.   **73.** (a) 1.97 m; (b) 8.24 m; (c) 0.0545 m; (d) 0.823 m.   **75.** (a) 1.55 g NaCl; (b) 1.72 g KCN; (c) 2.12 g $BeCl_2$; (d) 4.61 g K_2SO_4.   **77.** 668 g H_2O.
79. 298 g, 266 mL.   **81.** 4.46 m.   **83.** (a) 0.198 M; (b) 0.0660 M; (c) 0.0108 M; (d) 0.00132 M.
85. (a) 0.663 M; (b) 1.62 M; (c) 4.69 M; (d) 1.23 M.
87. (a) 1450 mL; (b) 18.0 mL; (c) 85,600 mL; (d) 15.0 mL.
89. (a) 350 mL; (b) 1200 mL; (c) 1450 mL; (d) 1480 mL.
91. (a) 3.3 M; (b) 2.5 M; (c) 8.86 M; (d) 0.00909 M.
93. 4.04 M.   **95.** 3.00 L H_2S.   **97.** 50.0 g $PbCl_2$.
99. 1100 mL HBr.   **101.** 0.0315 M.   **103.** 6.7 L H_2.
105. 0.0510 M.   **107.** (a) both soluble; (b) unlike; (c) both insoluble; (d) unlike.   **109.** 17.8% (m/v).
111. 6.72 g Na_2SO_4.   **113.** 1245 g.   **115.** (a) 0.245 M Al^{3+},

0.735 M NO_3^-; (b) 0.0735 M $Al(NO_3)_3$, 0.0735 M Al^{3+}, 0.2205 M NO_3^-. **117.** (a) 0.29 M K^+; (b) 0.027 M Cl^-; (c) 0.0866 M PO_4^{3-}. **119.** 3.74 mg solute/kg solution. **121.** 0.171 M. **123.** 8.06 mL. **125.** 340 g H_2O. **127.** 0.585 m. **129.** 0.763 M. **131.** 1.01 g/mL. **133.** (a) 11.9% (m/v); (b) 12.2% (m/m); (c) 15.1% (v/v). **135.** (a) AgCl; (b) $Ba_3(PO_4)_2$; (c) $PbSO_4$; (d) $BaSO_4$ and CuS. **137.** 117 L NH_3. **139.** 2.32 g AgCl. **141.** 288 g $BaCrO_4$. **143.** 37.1% (m/m). **145.** zinc. **147.** 0.0833 M NaOH. **149.** (a) 1 and 5; (b) 2, 6, and 9; (c) 3, 4, 7, and 8; (d) 3, 7 and 9; (e) 1, 2, 4, 5, 6, and 8; (f) 1 and 5, 2 and 6, 3 and 7. **151.** (a) 4, 5, 6, 7, 8, and 9; (b) 1, 4 and 7; (c) 1, 4, 7, 8, and 9; (d) all choices; (e) 2 and 7, 5 and 8, 6 and 9; (f) 1 and 4, 2 and 5, 2 and 9.

Chapter 14

1. An Arrhenius acid is a substance that releases H^+ ions in aqueous solution. An Arrhenius base is a substance that releases OH^- ions in aqueous solution. **3.** (a) $HBr \rightarrow H^+ + Br^-$; (b) $HClO_2 \rightarrow H^+ + ClO_2^-$; (c) $LiOH \rightarrow Li^+ + OH^-$; (d) $Ba(OH)_2 \rightarrow Ba^{2+} + 2\,OH^-$. **5.** (a) HBr (acid), H_2O (base); (b) H_2O (acid), N_3^- (base); (c) H_2S (acid), H_2O (base); (d) HS^- (acid), H_2O (base). **7.** (a) $HOCl + NH_3 \rightarrow NH_4^+ + OCl^-$; (b) $HClO_4 + H_2O \rightarrow H_3O^+ + ClO_4^-$; (c) $H_2O + NH_2^- \rightarrow NH_3 + OH^-$; (d) $HC_2O_4^- + H_2O \rightarrow H_3O^+ + C_2O_4^{2-}$. **9.** (a) NO_2^-; (b) $H_2PO_4^-$; (c) PO_4^{3-}; (d) NH_3. **11.** (a) $HClO_2$; (b) H_2SO_4; (c) HSO_3^-; (d) PH_4^+. **13.** (a) base; (b) base; (c) both; (d) acid. **15.** (a) acid; (b) base; (c) acid; (d) acid. **17.** (a) $HClO + H_2O \rightarrow H_3O^+ + ClO^-$; (b) $HClO_4 + NH_3 \rightarrow NH_4^+ + ClO_4^-$; (c) $H_3O^+ + OH^- \rightarrow H_2O + H_2O$; (d) $H_3O^+ + NH_2^- \rightarrow H_2O + NH_3$. **19.** (a) $H_2C_2O_4/HC_2O_4^-$, $HClO/ClO^-$; (b) HSO_4^-/SO_4^{2-}, H_3O^+/H_2O; (c) $H_2PO_4^-/HPO_4^{2-}$, NH_4^+/NH_3; (d) H_2CO_3/HCO_3^-, H_2O/OH^-. **21.** (a) monoprotic; (b) diprotic; (c) monoprotic; (d) monoprotic. **23.** (a) $H_2C_2O_4 + H_2O \rightarrow H_3O^+ + HC_2O_4^-$, $HC_2O_4^- + H_2O \rightarrow H_3O^+ + C_2O_4^{2-}$; (b) $H_3C_6H_5O_7 + H_2O \rightarrow H_3O^+ + H_2C_6H_5O_7^-$, $H_2C_6H_5O_7^- + H_2O \rightarrow H_3O^+ + HC_6H_5O_7^{2-}$, $HC_6H_5O_7^{2-} + H_2O \rightarrow H_3O^+ + C_6H_5O_7^{3-}$. **25.** The formula is written as $HC_3H_5O_3$ to indicate that only 1 of the 6 hydrogen atoms present is acidic. **27.** monoprotic—hydrogens bonded to carbon atoms are not acidic. **29.** A strong acid dissociates 100% or very nearly 100% in solution; a weak acid dissociates only slightly in solution. **31.** The term *weak* applies to the extent of dissociation; the term *dilute* refers to the concentration. **33.** (a) strong; (b) weak; (c) strong; (d) weak. **35.** HNO_3, H_2SO_4, $HClO_4$, HCl, HBr, and HI. **37.** (a) HNO_3; (b) $HClO_4$; (c) H_3PO_4; (d) HF. **39.** (a) salt; (b) acid; (c) salt; (d) salt. **41.** (a) Ca^{2+} (calcium ion) and SO_4^{2-} (sulfate ion); (b) Li^+ (lithium ion) and CO_3^{2-} (carbonate ion); (c) Na^+ (sodium ion) and Br^- (bromide ion); (d) Al^{3+} (aluminum ion) and S^{2-} (sulfide ion). **43.** (a) insoluble; (b) soluble; (c) soluble; (d) insoluble. **45.** (a) $NaNO_3 \rightarrow Na^+ + NO_3^-$; (b) $CuSO_4 \rightarrow Cu^{2+} + SO_4^{2-}$; (c) $BaCl_2 \rightarrow Ba^{2+} + 2\,Cl^-$; (d) $K_3PO_4 \rightarrow 3\,K^+ + PO_4^{3-}$. **47.** A molecular equation contains complete formulas for all reactants and products; in an ionic equation the formulas of compounds that dissociate into ions in aqueous solution are written in ionic form. **49.** (a) molecular; (b) net ionic; (c) ionic; (d) net ionic. **51.** (a) $FeS + 2\,H^+ \rightarrow Fe^{2+} + H_2S$; (b) $HC_2H_3O_2 + OH^- \rightarrow C_2H_3O_2^- + H_2O$; (c) $Zn + 2\,H^+ \rightarrow Zn^{2+} + H_2$; (d) $H^+ + OH^- \rightarrow H_2O$. **53.** (a) $Ba^{2+} + SO_4^{2-} \rightarrow BaSO_4$; (b) $Ba^{2+} + 2\,OH^- + 2\,H^+ + SO_4^{2-} \rightarrow BaSO_4 + 2\,H_2O$; (c) $H^+ + OH^- \rightarrow H_2O$; (d) $Ag^+ + Br^- \rightarrow AgBr$. **55.** (a) no reaction; (b) $Sn + 2\,H^+ \rightarrow Sn^{2+} + H_2$; (c) $Zn + 2\,H^+ \rightarrow Zn^{2+} + H_2$; (d) no reaction. **57.** (a) $H^+ + OH^- \rightarrow H_2O$; (b) $HC_2H_3O_2 + OH^- \rightarrow C_2H_3O_2^- + H_2O$; (c) $H_2S + OH^- \rightarrow S^{2-} + 2\,H_2O$; (d) $H_3PO_4 + 3\,OH^- \rightarrow PO_4^{3-} + 3\,H_2O$. **59.** (a) HF and NaOH; (b) H_3PO_4 and LiOH; (c) HCl and $Mg(OH)_2$; (d) H_2SO_4 and $Al(OH)_3$. **61.** (a) $Zn + 2\,HCl \rightarrow ZnCl_2 + H_2$; (b) $HCl + NaOH \rightarrow NaCl + H_2O$; (c) $2\,HCl + Na_2CO_3 \rightarrow 2\,NaCl + CO_2 + H_2O$; (d) $HCl + NaHCO_3 \rightarrow NaCl + CO_2 + H_2O$. **63.** (a) $Zn + Pb^{2+} \rightarrow Pb + Zn^{2+}$; (b) no reaction; (c) $Zn + 2\,Ag^+ \rightarrow Zn^{2+} + 2\,Ag$; (d) $3\,Zn + 2\,Fe^{3+} \rightarrow 3\,Zn^{2+} + 2\,Fe$. **65.** (a) insoluble salt formation; (b) insoluble salt formation; (c) insoluble salt formation; (d) evolution of a gas. **67.** (a) $Co(NO_3)_2 + (NH_4)_2S \rightarrow CoS + 2\,NH_4NO_3$, $Co^{2+} + 2\,NO_3^- + 2\,NH_4^+ + S^{2-} \rightarrow CoS + 2\,NH_4^+ + 2\,NO_3^-$, $Co^{2+} + S^{2-} \rightarrow CoS$; (b) $HBr + NaC_2H_3O_2 \rightarrow HC_2H_3O_2 + NaBr$, $H^+ + Br^- + Na^+ + C_2H_3O_2^- \rightarrow HC_2H_3O_2 + Na^+ + Br^-$, $H^+ + C_2H_3O_2^- \rightarrow HC_2H_3O_2$; (c) no reaction; (d) $3\,BeSO_4 + 2\,H_3PO_4 \rightarrow Be_3(PO_4)_2 + 3\,H_2SO_4$, $3\,Be^{2+} + 3\,SO_4^{2-} + 2\,H_3PO_4 \rightarrow Be_3(PO_4)_2 + 6\,H^+ + 3\,SO_4^{2-}$, $3\,Be^{2+} + H_3PO_4 \rightarrow Be_3(PO_4)_2 + 6\,H^+$. **69.** 1.00×10^{-14} M. **71.** (a) 2.5×10^{-12} M; (b) 1.5×10^{-8} M; (c) 1.2×10^{-5} M; (d) 5.0×10^{-9} M. **73.** (a) 4.3×10^{-12} M; (b) 2.2×10^{-10} M; (c) 1.3×10^{-8} M; (d) 1.0×10^{-6} M. **75.** The H_3O^+ concentration is greater than the OH^- concentration. **77.** (a) basic; (b) acidic; (c) basic; (d) neutral. **79.** (a) 5.0; (b) 8.0; (c) 6.0; (d) 11.0. **81.** (a) 9.0; (b) 4.0; (c) 11.0; (d) 7.0. **83.** pH is less than 7. **85.** (a) slightly acidic; (b) moderately basic; (c) very acidic; (d) moderately basic. **87.** (a) 2.3; (b) 3.2; (c) 6.2; (d) 9.7. **89.** (a) 4.66; (b) 4.658; (c) 4.6576; (d) 5.6576. **91.** (a) 6×10^{-6} M; (b) 1×10^{-5} M; (c) 8×10^{-8} M; (d) 5×10^{-4} M. **93.** (a) 3.6×10^{-3} M; (b) 3.6×10^{-5} M; (c) 3.6×10^{-8} M; (d) 3.47×10^{-8} M. **95.** As H_3O^+ concentration increases, the pH decreases. **97.** (a) A; (b) B. **99.** (a) neutral; (b) basic; (c) acidic; (d) neutral. **101.** (a) F^-; (b) NH_4^+; (c) CN^-; (d) BO_3^{3-}. **103.** salts formed from the reactions of (1) a strong acid and a strong base, (2) a strong acid and a weak base, (3) a weak acid and a strong base, and (4) a weak acid and a weak base. **105.** Neither a weak acid nor a weak base can be formed. **107.** (a) no; (b) no; (c) yes; (d) no. **109.** (a) $HPO_4^{2-} + H_3O^+ \rightarrow H_2PO_4^- + H_2O$; (b) $H_2CO_3 + OH^- \rightarrow HCO_3^- + H_2O$; (c) $HCO_3^- + H_3O^+ \rightarrow H_2CO_3 + H_2O$. **111.** (a) neither; (b) alkalosis; (c) neither; (d) acidosis. **113.** (a) 1.00 equiv base; (b) 4.50 equiv acid; (c) 3.00 equiv acid; (d) 0.548 equiv acid. **115.** (a) 0.500 N; (b) 0.408 N; (c) 5.00 N; (d) 0.200 N. **117.** (a) 0.0050 N; (b) 1.70 N; (c) 0.645 N; (d) 6.0 N. **119.** (a) 0.0015 M; (b) 2.80 M; (c) 0.800 M; (d) 4.0 M. **121.** 0.900 g $H_2C_2O_4$. **123.** (a) 35.0 mL; (b) $75\overline{0}$ mL; (c) 25.0 mL; (d) 0.22 mL. **125.** (a) 8.00 mL; (b) 96.0 mL; (c) $27\overline{0}$ mL; (d) 4.00 mL. **127.** (a) 0.141 N, 0.0705 M; (b) 0.176 N, 0.176 M; (c) 0.176 N, 0.0587 M; (d) 0.352 N, 0.352 M. **129.** no—the concentrations for "concentrated acids" vary from acid to acid. The same is true for "dilute acids," although most dilute acids are 6 M. **131.** (a) 3 M; (b) 6 M; (c) 16 M; (d) 6 M. **133.** (a) strong; (b) weak; (c) weak; (d) strong. **135.** (a) $HPO_4^{2-} + H_2O \rightarrow H_3O^+ + PO_4^{3-}$, $HPO_4^{2-} + H_2O \rightarrow H_2PO_4^- + OH^-$; (b) $OH^- + H_2O \rightarrow H_3O^+ + O^{2-}$, $OH^- + H_2O \rightarrow H_2O + OH^-$. **137.** 4.5. **139.** 200. **141.** 1.5. **143.** $Ba(OH)_2$, LiCN, K_2SO_4, NH_4Br, $HClO_4$. **145.** $HC_2H_3O_2/C_2H_3O_2^-$ and HCO_3^-/CO_3^{2-}. **147.** (a) 0.348 N; (b) 0.00696 equiv $H_3C_6H_5O_7$; (c) 0.00232 mole $H_3C_6H_5O_7$; (d) 0.116 M. **149.** (a) iodide ion; (b) hydrogen phosphate ion; (c) hydroxide ion; (d) hydronium ion. **151.** $H:C:::N:/[:C:::N:]^-$. **153.** 0.43. **155.** 0.00001 N. **157.** (a) 0.0004 M; (b) 0.0013% (m/m). **159.** 1.82×10^{23} ions. **161.** 65.9% (m/m). **163.** (a) 2, 5, 6, 7, and 9; (b) 1, 4, 5, 7, and 8; (c) 2 and 9; (d) 3, 4, 5, 6, and 7; (e) 1 and 5, 2 and 4, 5 and 9, 6 and 7, 7 and 8; (f) 2 and 5, 3 and 4, 4 and 5, 5 and 6, 5 and 7, 5 and 8. **165.** (a) 2, 3, 4, and 7; (b) 5, 6, 8, and 9; (c) 1 and 9, 2 and 5, 4 and 6; (d) 2 and 6, 3 and 8, 7 and 9; (e) 2 and 6; (f) 2 and 6, 3 and 8, 7 and 9.

Chapter 15

1. Oxidation is the loss of one or more electrons by a substance, and reduction is the gain of one or more electrons by a substance. Thus, electron transfer is the basis for oxidation and reduction. **3.** (a) Oxidation occurs when an atom loses electrons. (b) Oxidation occurs when the oxidation number of an atom increases. **5.** (a) An oxidizing agent gains electrons from another substance. (b) An oxidizing agent contains the atom that shows an oxidation number decrease. (c) An oxidizing agent is itself reduced. **7.** (a) oxidized; (b) decrease; (c) reducing agent; (d) loses. **9.** (a) +3; (b) −2; (c) +5; (d) +1. **11.** (a) −3; (b) +2; (c) +3; (d) −3. **13.** (a) +2; (b) +4; (c) +6; (d) +7. **15.** (a) +5 for P and −2 for O; (b) +1 for S and −2 for O; (c) +1 for H, +5 for P, and −2 for O; (d) −3 for N, +1 for H, +5 for P, and −2 for O. **17.** (a) −2; (b) +2; (c) −1; (d) −2. **19.** (a) −1; (b) +1; (c) +1; (d) −1. **21.** (a) oxidized; (b) oxidized; (c) neither; (d) oxidized. **23.** (a) Cl_2 is the oxidizing agent, and H_2 is the reducing agent; (b) O_2 is the oxidizing agent, and SO_2 is the reducing agent; (c) HBr is the oxidizing agent, and Mg is the reducing agent; (d) $FeCl_3$ is the oxidizing agent, and H_2 is the reducing agent. **25.** (a) Cl_2 (oxidizing agent, reduced) and Al (reducing agent, oxidized); (b) $ReCl_5$ (oxidizing agent, reduced) and $SbCl_3$ (reducing agent, oxidized); (c) Fe_2O_3 (oxidizing agent, reduced) and Al (reducing agent, oxidized); (d) $CuCl_2$ (oxidizing agent, reduced) and Zn (reducing agent, oxidized). **27.** (a) non-redox synthesis; (b) redox synthesis; (c) non-redox double replacement; (d) redox decomposition. **29.** (a) redox; (b) redox; (c) can't classify; (d) redox. **31.** (a) $3 F_2 + 3 H_2O \rightarrow 6 HF + O_3$; (b) $BaSO_4 + 4 C \rightarrow BaS + 4 CO$; (c) $I_2O_5 + 5 CO \rightarrow I_2 + 5 CO_2$; (d) $Br_2 + 2 H_2O + SO_2 \rightarrow 2 HBr + H_2SO_4$. **33.** (a) $3 PbO_2 + 2 Sb + 2 NaOH \rightarrow 3 PbO + 2 NaSbO_2 + H_2O$; (b) $S + 3 H_2O + 2 Pb(NO_3)_2 \rightarrow 2 Pb + H_2SO_3 + 4 HNO_3$; (c) $SnSO_4 + 2 FeSO_4 \rightarrow Sn + Fe_2(SO_4)_3$; (d) $2 MnO_4^- + 5 SO_2 + 2 H_2O \rightarrow 2 Mn^{2+} + 5 SO_4^{2-} + 4 H^+$. **35.** (a) $NaNO_3 + Pb \rightarrow NaNO_2 + PbO$; (b) $2 NF_3 + 2 AlCl_3 \rightarrow N_2 + 3 Cl_2 + 2 AlF_3$; (c) $3 H_2S + 2 HNO_3 \rightarrow 3 S + 2 NO + 4 H_2O$; (d) $S^{2-} + 4 Cl_2 + 8 OH^- \rightarrow SO_4^{2-} + 8 Cl^- + 4 H_2O$. **37.** (a) $2 H^+ + 3 H_2S + 2 NO_3^- \rightarrow 3 S + 2 NO + 4 H_2O$; (b) $4 H^+ + Cu + 2 NO_3^- \rightarrow Cu^{2+} + 2 NO_2 + 2 H_2O$; (c) $2 H^+ + 2 I^- + 2 HNO_2 \rightarrow I_2 + 2 NO + 2 H_2O$; (d) $14 H^+ + 2 Mn^{2+} + 5 BiO_3^- \rightarrow 2 MnO_4^- + 5 Bi^{3+} + 7 H_2O$. **39.** (a) $8 OH^- + S^{2-} + 4 Cl_2 \rightarrow SO_4^{2-} + 8 Cl^- + 4 H_2O$; (b) $5 H_2O + 3 SO_3^{2-} + 2 CrO_4^{2-} \rightarrow 2 Cr(OH)_4^- + 3 SO_4^{2-} + 2 OH^-$; (c) $H_2O + 2 MnO_4^- + 3 IO_3^- \rightarrow 2 MnO_2 + 3 IO_4^- + 2 OH^-$; (d) $18 OH^- + I_2 + 7 Cl_2 \rightarrow 2 H_3IO_6^{2-} + 14 Cl^- + 6 H_2O$. **41.** (a) $8 H^+ + MnO_4^- + 5 e^- \rightarrow Mn^{2+} + 4 H_2O$; (b) $H_2S \rightarrow S + 2 H^+ + 2 e^-$; (c) $S_2O_3^{2-} + 3 H_2O \rightarrow S_2O_6^{2-} + 6 H^+ + 6 e^-$; (d) $Pt + 6 Cl^- \rightarrow PtCl_6^{2-} + 4 e^-$. **43.** (a) $MnO_4^- + 2 H_2O + 3 e^- \rightarrow MnO_2 + 4 OH^-$; (b) $F_2 + 2 e^- \rightarrow 2 F^-$; (c) $4 PH_3 + 12 OH^- \rightarrow P_4 + 12 H_2O + 12 e^-$; (d) $CN^- + 2 OH^- \rightarrow CNO^- + H_2O + 2 e^-$. **45.** (a) $Zn + 2 Ag^+ \rightarrow Zn^{2+} + 2 Ag$; (b) $Zn + SO_4^{2-} + 4 H^+ \rightarrow Zn^{2+} + SO_2 + 2 H_2O$; (c) $2 NO_2 + H_2O + HOCl \rightarrow 2 NO_3^- + 3 H^+ + Cl^-$; (d) $5 H_2C_2O_4 + 2 MnO_4^- + 6 H^+ \rightarrow 10 CO_2 + 2 Mn^{2+} + 8 H_2O$. **47.** (a) $5 Fe^{2+} + MnO_4^- + 8 H^+ \rightarrow 5 Fe^{3+} + Mn^{2+} + 4 H_2O$; (b) $2 S_2O_3^{2-} + ClO^- + 2 H^+ \rightarrow S_4O_6^{2-} + Cl^- + H_2O$; (c) $3 H_3PO_2 + 2 Cr_2O_7^{2-} + 16 H^+ \rightarrow 3 H_3PO_4 + 4 Cr^{3+} + 8 H_2O$; (d) $3 P_4 + 20 NO_3^- + 8 H_2O + 20 H^+ \rightarrow 12 H_3PO_4 + 20 NO$. **49.** $3 ClO_2^- + 2 H_2O + 4 MnO_4^- \rightarrow 3 ClO_4^- + 4 MnO_2 + 4 OH^-$; (b) $5 OH^- + 8 Al + 2 H_2O + 3 NO_3^- \rightarrow 8 AlO_2^- + 3 NH_3$; (c) $3 I^- + BrO_3^- \rightarrow 3 IO^- + Br^-$; (d) $3 N_2H_4 + 4 ClO_3^- \rightarrow 6 NO + 4 Cl^- + 6 H_2O$. **51.** (a) $N_2H_4 + 2 O_2 \rightarrow N_2 + 2 H_2O_2$; (b) $I_2 + 7 Cl_2 + 18 OH^- \rightarrow 2 H_3IO_6^{2-} + 14 Cl^- + 6 H_2O$; (c) $3 SnO_2^{2-} +$

$2\text{ Bi(OH)}_3 \to 3\text{ SnO}_3^{2-} + 2\text{ Bi} + 3\text{ H}_2\text{O}$; (d) $2\text{ Cr(OH)}_3 + 3\text{ BrO}^- + 4\text{ OH}^- \to 2\text{ CrO}_4^{2-} + 3\text{ Br}^- + 5\text{ H}_2\text{O}$. **53.** (a) $3\text{ HNO}_2 \to 2\text{ NO} + \text{NO}_3^- + \text{H}_2\text{O} + \text{H}^+$; (b) $\text{H}_2\text{O} + \text{Cl}_2 \to \text{Cl}^- + \text{ClO}^- + 2\text{ H}^+$; (c) $6\text{ OH}^- + 3\text{ S} \to 2\text{ S}^{2-} + \text{SO}_3^{2-} + 3\text{ H}_2\text{O}$; (d) $6\text{ OH}^- + 3\text{ Br}_2 \to \text{BrO}_3^- + 5\text{ Br}^- + 3\text{ H}_2\text{O}$. **55.** (a) $3\text{ HNO}_2 \to \text{NO}_3^- + 2\text{ NO} + \text{H}^+ + \text{H}_2\text{O}$; (b) $\text{Cl}_2 + \text{H}_2\text{O} \to \text{ClO}^- + 2\text{ H}^+ + 2\text{ Cl}^-$; (c) $6\text{ OH}^- + 3\text{ S} \to \text{SO}_3^{2-} + 2\text{ S}^{2-} + 3\text{ H}_2\text{O}$; (d) $3\text{ Br}_2 + 6\text{ OH}^- \to \text{BrO}_3^- + 5\text{ Br}^- + 3\text{ H}_2\text{O}$. **57.** Reduction occurs at the cathode and oxidation at the anode. **59.** $\text{Cu(s)} + 2\text{ Ag}^+(\text{aq}) \to 2\text{ Ag(s)} + \text{Cu}^{2+}(\text{aq})$ **61.** anode: $\text{Zn(s)} \to \text{Zn}^{2+}(\text{aq}) + 2\text{ e}^-$; cathode: $2\text{ MnO}_2(\text{s}) + 2\text{ NH}_4^+(\text{aq}) + 2\text{ e}^- \to \text{Mn}_2\text{O}_3(\text{s}) + 2\text{ NH}_3(\text{aq}) + \text{H}_2\text{O(l)}$. **63.** anode: $\text{Pb(s)} + \text{SO}_4^{2-}(\text{aq}) \to \text{PbSO}_4(\text{s}) + 2\text{ e}^-$; cathode: $\text{PbO}_2(\text{s}) + 4\text{ H}^+(\text{aq}) + \text{SO}_4^{2-}(\text{aq}) + 2\text{ e}^- \to \text{PbSO}_4(\text{s}) + 2\text{ H}_2\text{O(l)}$. **65.** The H_2SO_4 is converted to PbSO_4, which is deposited on the electrodes. **67.** $2\text{ Cl}^-(\text{aq}) \to \text{Cl}_2(\text{g}) + 2\text{ e}^-$. **69.** (a) +8; (b) +6; (c) +6; (d) +8. **71.** (a) +4 for Mn, +4 for C, −2 for O; (b) +4 for Pb, +2 for S, −2 for O; (c) +2 for Hg, +6 for Cr, −2 for O; (d) +1 for Cu, +4 for S, −2 for O. **73.** (a) MnO_4^-; (b) MnO_4^-; (c) SO_3^{2-}; (d) SO_3^{2-}. **75.** (1) $5\text{ PH}_3 + 4\text{ MnO}_4^- + 12\text{ H}^+ \to 5\text{ H}_3\text{PO}_4 + 4\text{ Mn}^{2+} + 6\text{ H}_2\text{O}$, (2) $\text{PH}_3 + 2\text{ SO}_4^{2-} + 4\text{ H}^+ \to \text{H}_3\text{PO}_4 + 2\text{ SO}_2 + 2\text{ H}_2\text{O}$, (3) $5\text{ As} + 3\text{ MnO}_4^- + 3\text{ H}_2\text{O} + 9\text{ H}^+ \to 5\text{ H}_3\text{AsO}_3 + 3\text{ Mn}^{2+}$, (4) $2\text{ As} + 3\text{ SO}_4^{2-} + 6\text{ H}^+ \to 2\text{ H}_3\text{AsO}_3 + 3\text{ SO}_2$. **77.** $\text{Zn} \to \text{Zn}^{2+} + 2\text{ e}^-$, $\text{NO}_3^- + 10\text{ H}^+ + 8\text{ e}^- \to \text{NH}_4^+ + 3\text{ H}_2\text{O}$. **79.** (a) redox, HNO_3 is the oxidizing agent; (b) acid–base, HCl is the acid; (c) redox, H_2O_2 is the oxidizing agent; (d) acid–base, H_2SO_4 is the acid. **81.** (a) $\text{Sn}^{2+} + 2\text{ Fe}^{2+} \to \text{Sn} + 2\text{ Fe}^{3+}$; (b) $\text{PH}_3 + 2\text{ NO}_2 \to \text{H}_3\text{PO}_4 + \text{N}_2$; (c) $\text{S} + 3\text{ H}_2\text{O} + 2\text{ Pb}^{2+} \to \text{Pb} + \text{H}_2\text{SO}_3 + 4\text{ H}^+$; (d) $4\text{ Zn} + 10\text{ H}^+ + \text{NO}_3^- \to 4\text{ Zn}^{2+} + \text{NH}_4^+ + 3\text{ H}_2\text{O}$. **83.** (a) $8\text{ H}^+ + 3\text{ H}_2\text{S} + \text{Cr}_2\text{O}_7^{2-} \to 2\text{ Cr}^{3+} + 3\text{ S} + 7\text{ H}_2\text{O}$; (b) $3\text{ H}_2\text{O} + 5\text{ ClO}_3^- + 3\text{ I}_2 \to 6\text{ IO}_3^- + 5\text{ Cl}^- + 6\text{ H}^+$; (c) $8\text{ OH}^- + \text{S}^{2-} + 4\text{ Br}_2 \to \text{SO}_4^{2-} + 8\text{ Br}^- + 4\text{ H}_2\text{O}$; (d) $2\text{ OH}^- + 2\text{ NO}_2 \to \text{NO}_3^- + \text{NO}_2^- + \text{H}_2\text{O}$. **85.** 0.324 M I_3^-. **87.** 30.8 ppm (m/m). **89.** (a) 1, 2, 3, 5, 6, and 9; (b) 1, 2, 3, 4, 6, 8, and 9; (c) 5; (d) 4 and 7; (e) 6; (f) 2 and 5, 4 and 7. **91.** (a) 5, 7, and 8; (b) 1, 2, 3, 4, and 6; (c) 1, 2, 3, 4, and 6; (d) 5, 7, and 8; (e) 5, 7, and 8; (f) 9.

Index

A **boldfaced** term is defined on the indicated page.

Accuracy, 12
  relationship to precision, 13
Acid(s)
  acidic and nonacidic hydrogens in, 533–34
  Arrhenius definition for, 528
  Brønsted–Lowry definition for, 529
  characteristics of, early known, 527
  conjugate, 531–32
  diprotic, 533
  monoprotic, 533
  neutralization of, 545–46
  nonoxy, nomenclature for, 280
  oxy, nomenclature for, 280–84
  polyprotic, 533–35
  reactions of, 544–46
  reactions with bases, 545–46
  reactions with carbonates and bicarbonates, 546
  reactions with metals, 544–45
  reactions with salts, 547–48
  stock solutions of, 574–75
  strengths of, 535–38
  strong, formulas for, 536
  titration of, 567–68
  triprotic, 533
  uses and occurrence of common, 534
  weak, relative strengths of, 536–38, 540
  See also Acidic solution(s)
Acidic hydrogen atom, 533
Acidic solution(s), 552
  hydrogen ion concentrations of, 552
  pH of, 555
  See also Acid(s)
Acidosis, 567
Activity series, for metals, 545
Actual yield, 370
  use of, in problem solving, 371–72
Addition
  scientific notation and, 38–40
  significant figures and, 21–22, 24–26
Air, composition of clean, dry, 473
Alkalosis, 567
Amorphous solids, 419
Amphoteric substance(s), 532
  examples of, 532–33
Anode, 616
Aqueous solution, 489
Area, 54
  formulas for calculation of, 54–55
Arrhenius, Svante August, 528
Arrhenius acid, 528
  See also Acid(s)
Arrhenius base, 528
  See also Base(s)
Atom(s), 128
  charged (ions), 209–10
  diamagnetic, 183
  electrical neutrality of, 138
  electron-dot structures for, 206–207
  limit of chemical subdivision and, 132
  mass of, 129–30
  nucleus of, 137–38
  number of subatomic particles within, 143–44

Atom(s) *(cont.)*
  paramagnetic, 183
  radii of, 194–96
  relative masses of, 149–51
  size of, 129–30
  subatomic particles and, 136–38
Atomic mass(es), 149
  biannual revisions in, 155
  calculation of, 152–54
  of elements, 165, *inside front and back covers*
  relative scale for, 149–51
  sequence of, within periodic table, 167–68
  use of, in formula masses, 298–301
  values, precision of, 155
Atomic mass unit, 151
Atomic nucleus, *See* Nucleus
Atomic number(s), 143
  definition of an element in terms of, 145
  of elements, 165, *inside front and back covers*
  number of electrons and, 143
  number of protons and, 143
Atomic orbital, *See* Electron orbital(s)
Atomic radii, chemical periodicity and, 194–96
Atomic theory of matter, 129
  isotopes and, 145
  statements of, 129
Aufbau diagram, 176
  electron configurations and, 176–80
Aufbau principle, 176
  electron configurations and, 176–80
Avogadro, Amedeo, 305, 453
Avogadro's law, 453
  calculations involving, 454–55
  mathematical form of, 454
Avogadro's number, 305
  as a conversion factor, 305–306
  the mole and, 305

Balanced chemical equation, 345
  procedures used in obtaining a, 345–50
  See also Chemical equation(s)
Barometer, 431
Base(s)
  Arrhenius definition for, 528
  Brønsted–Lowry definition for, 529

  characteristics of, early known, 528
  conjugate, 531–32
  neutralization of, 545–46
  reactions of, 546
  reactions with acids, 545–46
  reactions with salts, 548
  stock solutions of, 574–75
  strengths of, 537
  strong, formulas for, 537
  titration of, 567
  weak, 537, 540
Basic solution(s), 552
  hydroxide ion concentrations of, 552
  pH of, 555
  See also Base(s)
Berzelius, Jöns Jakob, 118
Binary compound, 266
  See also Compound(s)
Binary ionic compound, 266
Binary molecular compound, 276
Boiling, 408
  process explained using kinetic molecular theory, 408–409
Boiling point, 409
  factors affecting magnitude of, 409–10
  intermolecular force strength and, 412–13
  normal, 409
  water, at various pressures, 409–10
Bond length, 235
Bond polarity, 227
  See also Polarity
Bond strength, 235
Boyle, Robert, 435
Boyle's law, 435
  applications involving, 440
  calculations involving, 437–39
  explained using kinetic molecular theory, 439–40
  mathematical form of, 435, 437
Brønsted, Johannes Nicholas, 528
Brønsted–Lowry acid(s), 529
  conjugate acid–base pairs and, 530–33
  conjugate base of, 531–32
  important concepts concerning, 529–30
  See also Acid(s)
Brønsted–Lowry base(s), 529
  conjugate acid of, 531–32
  conjugate acid–base pairs and, 530–33
  See also Base(s)

Buffer, 564
Buffer solution(s), 564
  human body and, 565–67
  pH change and, 564–65

Calorie, 87
Canal rays, 141
  See also Discharge tube experiments
Cathode, 140, 615–16
Cathode rays, 139–40
  See also Discharge tube experiments
Celsius, Anders, 81
Chadwick, James, 136
Changes of state
  condensation, 107, 395, 398
  deposition, 107, 395
  endothermic, 395
  energy and, 395–404
  evaporation, 107, 395, 398, 404–407
  exothermic, 395
  freezing, 107, 395, 404–407
  heating curves and, 395–96
  melting, 107, 395
  relationships between, 107, 395
  sublimation, 107, 395
Charles, Jacques Alexander Cesar, 440
Charles's law, 440
  applications involving, 444
  calculations involving, 442
  explained using kinetic molecular theory, 443–44
  importance of Kelvin temperature in, 443
  mathematical form of, 441–42
Chemical bond(s), 204
  polarity of, 227–31
  types of, 204–205
  See also Covalent bond(s); Ionic bond(s)
Chemical change, 106
  characteristics of, 106–108
Chemical energy, 86
Chemical equation(s), 344
  balancing of, 345–50, 597–614
  calculations based upon, 356–75, 512–16
  coefficients in, 345–46
  conventions in writing, 344–45
  ionic, 539–40
  limiting reactant in, 365–70
  macroscopic level interpretation of, 356–57

  mass–mass conversions using, 360–62
  microscopic level interpretation of, 356
  net ionic, 539–43
  redox, balancing of, 597–614
  special symbols used in, 350–51
  use of, in gas law calculations, 466–69
  use of, in stoichiometric calculations, 360–70
Chemical formula(s), 133
  calculations based upon, 298–302
  determination of, 325–30
  empirical, 324–30
  macroscopic level interpretation of, 316–17
  meaning of, 133–36
  microscopic level interpretation of, 316
  molecular, 324–25, 330–31
  parentheses in, 134–35
  subscripts in, 133–35
  writing of, from names, 286–87
Chemical nomenclature, 265
  See also Nomenclature
Chemical periodicity, 193
  atomic radii and, 194–96
  electronegativity and, 226–27
  metallic character and, 193–94
  nonmetallic character and, 193–94
Chemical properties, 105
  characteristics of, 105–106
Chemical reaction(s), 108
  acids, 544–46
  bases, 546
  classes of, 353–56
  combustion, 351–52
  consecutive, calculations involving, 373–75
  conservation of mass in, 342–44
  decomposition, 354, 594
  disproportionation, 611–14
  double-replacement, 354–55, 594–95
  non-oxidation–reduction, 594–96
  oxidation–reduction, 586–88
  reactants and products in, 342
  redox, 587
  salt hydrolysis, 561–64
  salts, 546–50
  simultaneous, calculations involving, 372–73
  single-replacement, 354, 544, 594–95
  synthesis, 353–54, 594
  theoretical yield of product(s), 370–72

Chemical stoichiometry, 359–60
Chemical symbol(s), 118
    listing of, 118, *inside front and back covers*
    non-English names and, 119
    provisional, 120–21
Chemistry, 102
    scope of, 3
Coefficient(s), 345
    determination of, in chemical equations, 345–50
    mole–mole conversions using, 357–59
Combined gas law, 445
    calculations involving, 447–49
    mathematical form of, 445–46
Combustion reaction(s), 351
    examples of, 352–53
Common name, 278
    See also Nomenclature
Compound(s), 113
    binary, types of, 266–68
    binary ionic, formulas for, 211–18
    binary molecular, 276
    characteristics of, 113–14
    covalent, 220–24, 231–35
    distinguishing between ionic and covalent, 266
    formulas for, determination of, 325–30
    formulas for, meaning of, 133–36
    heteroatomic molecules and, 131
    ionic, polyatomic-ion containing, 218–20
    molecular and ionic, 131–32
    natural and synthetic, 133
    nomenclature for, 265–87
    percent composition for, 301–303
    properties of, contrasted with mixtures, 115
    solubility rules, 491–93
    ternary, 266
    See also Covalent compound(s); Ionic compound(s)
Compressibility, 389
Concentrated solution, 489
Concentration, 493
    defining equations for, 493
    molality, 506–509
    molarity, 500–505
    normality, 569–74
    parts per billion, 498–500
    parts per million, 498–500
    percent solute, 494–97
    units, for solutions, 493–512
Condensation, 107, 395
Conjugate acid, 531
    See also Acid(s)
Conjugate acid–base pair(s), 531
    determination of, 530–33
Conjugate base, 531
    See also Base(s)
Conservation of energy, law of, 86
Conservation of mass, law of, 342–44
Conversion factor(s), 59
    Avogadro's number as a, 305–306
    calculations based upon, 356–59
    construction of, 59–60
    density as a, 76–78
    English–English, 61
    equation coefficients as, 357–59
    "equivalence" versus "equality" in, 78
    formula subscripts as, 317–18
    metric–English, 61–62
    metric–metric, 60
    molar mass as a, 310–12
    molar volume as a, 455–56
    molarity as a, 502–505, 512–16
    percent as a, 78–81
    significant figures within, 60–61
    use of, in dimensional analysis, 62–74
Coordinate covalent bond(s), 234
    electron-dot structures and, 234–35
    See also Chemical bond(s)
Counting by weighing, 314
    particles in a sample and, 313–16
Covalent bond(s), 221
    coordinate, 234–35
    double, 231–33
    electron sharing and, 221–24
    electronegativity differences and, 227–31
    ionic and covalent character of, 229–31
    length of, 235
    nonpolar, 227–31
    octet rule and, 222–24
    polar, 227–31
    polarity of, 227–31
    resonance and, 235–36
    single, 220–24, 231
    strength of, 235
    triple, 231–33

Covalent compound(s)
  binary, nomenclature for, 276–79
  electron-dot structures and, 221–24
  See also Compound(s)
Crystal lattice, 419
Crystal lattice sites, 419–21
Crystalline solids, 419

Dalton, John, 128, 469
Dalton's law of partial pressures, 469
  applications of, 417–75
  calculations involving, 470–71, 474–75
  explained using kinetic molecular theory, 469
  mathematical form of, 470
Davy, Humphry, 139
Decomposition reaction(s), 354
  examples of, 354–55, 594–95
Definite proportions, law of, 296–98
Density, 74
  calculation of, using ideal gas law, 464–65
  calculation of, using mass and volume, 75–76
  calculation of, using molar volume and molar mass, 458–59
  "lighter" and "heavier" and, 75
  units for, 74
  use of, as a conversion factor, 76–78
  values, tables of, 75
  water, at various temperatures, 415–16
Deposition, 107, 395
Diamagnetic atom, 183
Dilute solution, 489
Dilution, 509
  calculations involving, 510–12
  stock solutions and, 509
Dimensional analysis, 62
  conversion factors and, 62–74
  use of, in solving problems, 63–74
Dipole–dipole interactions, 410
  See also Intermolecular force(s)
Diprotic acid, 533
  See also Acid(s)
Discharge tube experiments, 139–41
  canal rays and, 140–41
  cathode rays, and 139–40
Disproportionation reaction, 611
Distinguishing electron, 184
  periodic table relationships and, 184–91

  See also Electron configuration(s)
Division
  scientific notation and, 35–38
  significant figures and, 21–24
Double covalent bond(s), 231
  electron dot structures and, 232–33
  See also Covalent bonds
Double-replacement reaction(s), 354
  examples of, 354–55, 545, 595
Dry cells, 616–17

Electrolysis, 620
Electrolytic cell(s), 620
  commercial applications of, 620–22
  electroplating and, 619–20
Electron(s), 136
  atomic number and, 143
  bonding pairs, and molecular geometry, 246–48
  characteristics of, 136
  discharge-tube experiments and, 139–41
  location within the atom, 137–38
  nonbonding pairs, and molecular geometry, 248–49
  paired and unpaired, 181–83
  quantized energy of, 169–70
  valence, 206–207
Electron configuration(s), 176
  abbreviated form of, 180
  Aufbau diagram and, 176–80
  distinguishing electron and, 184–91
  notation used in, 176
  periodic law and, 183–84
  periodic table and, 184–91
  valence electrons and, 205–206
  writing, 176–80
Electron-dot structure(s), 206
  covalent bonding and, 221–24, 232–35
  ionic bonding and, 213–15
  oxidation numbers and, 588–89
  periodic table and, 206–207
  resonance and, 235–36
  systematic procedures for writing, 236–42
  valence electrons and, 206–07
Electron orbital(s), 173
  Hund's rule and, 180–83
  notation for, 173
  number of electrons within, 173
  order of filling, 176–80
  overlapping of, 221–24

Electron orbital(s) *(cont.)*
  shapes of, 174–75
Electron shell(s), 170
  notation for, 171
  number of electrons in, 171
  number of subshells within, 172
Electron subshell(s), 172
  notation for, 172–73
  number of electrons in, 172–73
  number of orbitals within, 173–74
  order of filling, 176–80
  relative energies of, 176–77
Electronegativity, 226
  bond polarity and, 227–31
  values, periodic trends in, 226–27
  values, table of, 226
Electroplating, 619
Electrostatic interactions, 391
  potential energy and, 391
Element(s), 113, 145
  abundances of, 116–17
  atomic masses of, 165, *inside front and back covers*
  atomic numbers of, 165, *inside front and back covers*
  characteristics of, 113–14
  classification systems for, 191–93
  discovery of, 116
  homoatomic molecules and, 131
  inner-transition, 191–92
  metallic, 193–94
  metalloidal, 194–95
  naming of, 117–21
  natural versus synthetic, 116
  noble gas, 191–92
  nonmetallic, 193–94
  number of, 113, 116
  provisional names for, 119–21
  representative, 191–92
  symbols for, 118–21
  synthetic, 116
  transition, 191–92
Empirical formula(s), 324
  calculation of, from combustion analysis data, 328–30
  calculation of, from mass data, 326–28
  calculation of, from percent composition data, 325–26
  relationship to molecular formulas, 324–25
  See also Chemical formula(s)

Endothermic change of state, 395
Energy, 85
  changes of state and, 395–404
  chemical, 86
  conservation of, law of, 86
  forms of, 85–86
  heat, calculations involving, 89–93, 397–404
  heat, units for, 87–88
  kinetic, 86
  potential, 86
  types of, 86–87
Equation(s), *See* Chemical equation(s)
Equilibrium
  conjugate acid–base pairs and, 530–32
  liquid–vapor, 406–408
  solute–solvent, 489
  state of, 407
Equivalent(s)
  acid–base calculations and, 569–70
  definition of, for acids, 569
  definition of, for bases, 569
  normality and, 569–71
Equivalent of acid, 569
  See also Equivalent(s)
Equivalent of base, 569
  See also Equivalent(s)
Errors
  measurement, types of, 13
  random, 13
  systematic, 13
Evaporation, 107, 395, 404
  cooling effect of, 405
  explained using kinetic molecular theory, 404–407
  in a closed container, 406–407
  rate of, 404
Exothermic change of state, 395
Experiment(s), 5
  scientific method and, 4, 6–7
Exponent(s), 28
  meaning of negative, 28–29

Fact(s), 5
  publishing of, 5
  scientific method and, 5
Fahrenheit, Gabriel, 82
Fixed-charge metal(s), 266–67
  compounds containing, nomenclature for, 268–69
  identity of, 266–67

Formula mass(es), 298
   calculation of, from atomic masses, 298–301
   significant figures in, 299–301
Formula(s), *See* Chemical, Empirical, and Molecular formula(s)
Freezing, 107, 395
Fusion, heat of, 397–98

Galvanic cell(s), 615
   commercial applications of, 616–19
   dry cells, types of, 616–17
   examples of, 614–19
   lead storage battery, 618–19
Gas(es)
   collection of, by water displacement, 473–75
   mixtures of, 469–75
   molar volume of, 455
   molecular form of elemental, 345
   physical properties of selected common, 429–30
   See also Noble gases
Gas law(s), 430
   variables in, 430–35
   See also Avogadro's law; Boyle's law; Charles's law; Combined gas law; Dalton's law of partial pressures; Gay-Lussac's laws; Ideal gas law
Gaseous state, 103
   compared to other physical states, 394–95
   explained using kinetic molecular theory, 393–94
   macroscopic properties of, 389–90
Gay-Lussac, Joseph Louis, 446, 451
Gay-Lussac's law, 444
   applications involving, 444
   calculations involving, 444–45
   explained using kinetic molecular theory, 444
   mathematical form of, 446
Gay-Lussac's law of combining volumes, 451
   calculations involving, 452–53
   equation coefficients and, 451–52
Goldstein, Eugene, 136, 140
Gram, 52
   conversion factors involving, 62
Group(s), 167
   within periodic table, 167

Heat capacity, 91
   relationship to specific heat, 91
   use of in calculations, 91–93
Heat of condensation, 398
Heat energy, 87–93, 397–404
Heat of fusion, 397
   use of in calculations, 398
   values for, table of, 397
Heat of solidification, 397
Heat of vaporization, 398
   use of in calculations, 398–99
   values for, table of, 399
Heating curve, 395–96
Heteroatomic molecule(s), 131
   compounds and, 131
   See also Molecule(s)
Heterogeneous mixture(s), 111
   characteristics of, 111
   See also Mixture(s)
Homoatomic molecule(s), 131
   elements and, 131
   See also Molecule(s)
Homogeneous mixture(s), 111
   characteristics of, 113
   See also Mixture(s)
Hund, Frederick, 181
Hund's rule, 181
   orbital diagrams and, 180–83
Hydrogen bond(s), 411
   effects of, on general physical properties, 411, 413–19
   effects of, on properties of water, 413–19
   strengths of, 411
   See also Intermolecular force(s)
Hypothesis, 6
   "proving" of, 6–7
   scientific method and, 6–7

Ideal gas, 463
Ideal gas constant, 460–61
Ideal gas law, 460
   calculations involving, 461–63
   calculations involving modified forms of, 464–65
   constant, values for, 460–61
   equations derived from, 463–64
   mass–volume conversions using, 466–69
   mathematical form of, 460
   relationship to individual gas laws, 460

Immiscible substances, 489
Indicator, 568
Inner-transition elements, 191
   location within the periodic table, 191–92
Intermolecular force(s), 410
   dipole–dipole interactions, 410–11
   hydrogen bonds, 411
   liquid state and, 411–13
   London forces, 411–13
   solution formation and, 490–91
   strength of, boiling points and, 412–13
   types of, 410–13
Ion(s), 208
   combining ratios for, 215–17
   isoelectronic, 212–13
   monoatomic, charges on, 212
   nomenclature for monoatomic, 268–73
   nomenclature for polyatomic, 273–76
   notation for, 209–10
   spectator, 539–40
   subatomic particles and, 209–10
   symbols for 209–10
Ion product for water, 551
   defining equation for, 551–52
Ionic bond(s), 211
   electron transfer and, 211
   octet rule and, 213–15
Ionic compound(s)
   binary, formulas for, 215–17
   binary, nomenclature for, 268–73
   charge neutrality for, 215–17
   electron-dot structures for, 213–15
   electron transfer process and, 211–15
   polyatomic-ion containing, 218–20
   polyatomic-ion containing, nomenclature for, 273–76
   structure of, 217–18
   See also Compound(s)
Ionic equation(s), 539
   See also Net ionic equation(s)
Ionic solid(s), 217, 420
   characteristics of, 217–18
   crystal lattice for, 420–21
Isobar(s), 147
   occurrence of, for selected elements, 147–48
Isoelectronic species, 212
Isotope(s), 145
   mass spectrometer and, 156
   notation for, 145
   number of, for elements, 1–20, 147–48
   percent abundances of, 145–46
   postulates of atomic theory and, 145
IUPAC rules, 265–66
   See also Nomenclature

Joule, 87
   relationship to other energy units, 87–88
Joule, James Prescott, 87

Kelvin, William, 81
Kinetic energy, 86
   "disruptive forces" and, 391–392
   formula for calculation of, 391
   role in determining physical states, 391–94
   temperature dependency of, 392
Kinetic molecular theory, 390
   boiling process and, 408–409
   Boyle's law and, 439–40
   Charles's law and, 443–44
   Dalton's law of partial pressures and, 469
   evaporation and, 404–407
   gaseous state properties and, 393–94
   Guy-Lussac's law and, 444
   liquid state properties and, 392–93
   solid state properties and, 392
   statements of, 390–92
   vapor pressure and, 406–407

Law(s), 5
   scientific method and, 5–6
   "scientific" versus "societal," 6
   See also Avogadro's law; Boyle's law; Charles's law; Combined gas law; Dalton's law of partial pressures; Gay-Lussac's law; Gay-Lussac's law of combining volumes; Ideal gas law; *following entries;* Periodic law.
Law of conservation of energy, 86
Law of conservation of mass, 343
   illustrations of, 343
Law of definite proportions, 296
   illustrations of, 296–98
Lead storage battery, 618–19
Length
   metric and English units compared, 51

metric–English conversion factors, 62
metric units for, 50–51
Lewis, Gilbert Newton, 207
Limiting reactant, 365
concept of, 365–67
determination of, 366–70
Liquid state, 103
compared to other physical states, 394–95
explained using kinetic molecular theory, 392–93
intermolecular forces and, 411–13
macroscopic properties of, 389–90
Liter, 55
conversion factors involving, 62
London, Fritz, 411
London forces, 411
effects of, on physical properties, 411–13
factors determining strength of, 412–13
See also Intermolecular force(s)
Lowry, Thomas Martin, 528

Macromolecular solid(s), 420
crystal lattice for, 420–21
Macroscopic level interpretation of a chemical formula, 316–17
See also Chemical formula(s)
Manometer, 432
Mass, 52
conservation of, in a chemical reaction, 342–44
mass–mass conversions, 322–23, 360–63
mass–mole conversions, 307–13
mass–particle conversion, 316–18
mass–volume conversions using ideal gas law, 466–69
measurement of, 53–54
metric and English units compared, 52
metric–English conversion factors, 62
metric units for, 52
relative, 149–51
weight and, differences between, 52–53
Mass number(s), 143
nonuniqueness of, for elements, 144
number of neutrons and, 143
values for elements, 1–20, 147–48
Mass percent, *See* Percent by mass

Mass spectrometer, isotopic abundances and, 156
Mass–volume percent, 497
calculations involving, 497
defining equation for, 497
Matter, 102
changes in, classification of, 106–108
classifications of, 109–16
physical states of, 103–104, 387–90
properties of, 104–06
Measurement(s)
accuracy of, 12–13
errors in, 13
importance of, 11–12
precision of, 12–13
significant figures and, 14–19
uncertainty in, 13–16
Melting, 107, 395
Mendeleev, Dmitri Ivanovich, 164
Metal(s), 191
activity series for, 545
chemical periodicity of properties, 193–94
fixed-charge, 266–69
periodic table location of, 192–93
properties of, 191–92
reactions with acids, 544–45
reactions with salts, 547
variable-charge 267–68, 270–73
Metal-foil experiments, 141–43
nucleus and, 142–43
Metallic solid(s), 420
crystal lattice for, 420–21
Metalloid(s), 194
periodic table location of, 194–95
Meter, 50
conversion factors involving, 62
Metric system
advantages of, 49
length units in, 50–51
mass units in, 52–54
prefixes used in, 50
volume units in, 54–57
Meyer, Julius Lothar, 164
Microscopic level interpretation of a chemical equation, 356
See also Chemical equations
Microscopic level interpretation of a chemical formula, 316
See also Chemical formula(s)
Miscible substances, 489

Mixture(s), 105
  characteristics of, 109–113
  heterogeneous, 111
  homogeneous, 111–13
  properties of, contrasted with compounds, 115
  types of, 111–13
Molality, 506
  calculations involving, 506–509
  defining equation for, 506
Molar interpretation of a chemical equation, 356
  See also Chemical equation(s)
Molar mass
  calculation of, from atomic masses, 307–10
  calculation of, using ideal gas law, 464–65
  contrasted with molar volume, 455
Molar mass of an element, 308
Molar mass of a compound, 309
Molar volume, 455
  calculations involving, 456–60
  contrasted with molar mass, 455
  use of, in calculating density, 458–59
  use of, as a conversion factor, 455–56
Molarity, 500
  calculations involving, using chemical equations, 512–16
  calculations involving, using chemical formulas, 500–505
  conversion to normality, 571–72
  defining equation for, 500
  use as a mole–volume conversion factor, 502–505, 512–16
Mole(s), 304, 313
  Avogadro's number and, 305
  as a counting unit, 303–307
  mass of a, 307–13
  mole–mass conversions, 310–13
  mole–mole conversions using formula subscripts, 317–18
  mole–particle conversion, 305–307
  mole–volume conversions using molarity, 502–505, 512–16
  "size" of, 307
  use of, in chemical calculations, 305–307, 310–13, 318–24
Mole fraction, 472
Molecular compound(s), *See* Covalent compound(s)

Molecular formula(s), 324
  calculation of, 330–31
  relationship to empirical formulas, 324–25
  See also Chemical formula(s)
Molecular geometry, 242
  prediction of, using VSEPR theory, 246–54
Molecular polarity
  factors determining, 242–46
  role of geometry in, 242–46
Molecule(s), 131
  atoms within, number of, 131
  classification of, 131
  geometry of, from VSEPR theory, 246–54
  heteroatomic, 131
  homoatomic, 131
  limit of physical subdivision, 132
  polarity of, 242–46
Monoatomic ion(s), 218
  charges on, 212
  nomenclature, 268–73
Monoprotic acid, 533
  See also Acid(s)
Multiple covalent bond(s), 231
  See also Covalent bond(s)
Multiplication
  scientific notation and, 33–35
  significant figures and, 21–24

Net ionic equation(s), 539
  examples of, 540–43
  rules for writing, 540
  solubility rules and, 540
Neutral solution(s), 552
  hydrogen ion concentrations of, 552
  pH of, 555
  See also Solution(s)
Neutralization, 545
Neutron(s), 136
  characteristics of, 136
  location within the atom, 137–38
Noble gases, 191
  electron configurations of, 207–208
  location within periodic table, 191–92
  octet rule and, 207–208
Nomenclature
  common names and, 278–79
  covalent compounds, binary, 276–79
  elements, 117–21

ionic compounds, binary, 268–73
ionic compounds, polyatomic-ion containing, 273–76
IUPAC rules and, 265–66
molecular compounds, binary, 276–79
nonoxyacids, 280
oxyacids, 280–84
polyatomic ions, common, 273–76
rules, summary of, 284–87

Nonmetal(s), 192
chemical periodicity of properties, 193–94
periodic table location of, 192–93
properties of, 192

Nonoxyacid(s), 280
See also Acid(s)

Nonpolar covalent bond(s), 227
See also Covalent bonds

Nonpolar molecular solid(s), 420
crystal lattice for, 420–21

Normal boiling point, 409

Normality, 569
calculations involving, 570–74
conversion to molarity, 571–72
defining equation for, 569

Nucleons, 137

Nucleus, 137
characteristics of, 137–38
discovery of, 141–43
location within an atom, 137–38
metal-foil experiments and, 141–43
size of, 138
subatomic particles within, 137

Numbers
counted, 13
defined, 13
exact, significant figures and, 26–27
"rounding off" of, 20–21
scientific notation for, 27–32

Octet rule, 208
covalent bonds and, 222–24
ionic bonds and, 211–15

Orbital, See Electron orbital

Orbital diagram(s), 181
mechanics of writing, 180–83
unpaired electrons and, 181–83

Order of magnitude, 29

Oxidation, 587
defined, in terms of electron loss, 587–88
defined, in terms of oxidation number increase, 592

Oxidation number(s), 588
calculation of, 590–92
electron–dot structures and, 588–89
rules for determining, 589–90
use in balancing redox equations, 597–603

Oxidation–reduction equation(s)
disproportionation reactions, 611–14
ion-electron method of balancing, 603–10
oxidation number method of balancing, 597–603

Oxidation–reduction reaction(s), 587
classified according to type, 594–96
nonspontaneous, 614
spontaneous, 614

Oxidizing agent, 587–88, 592
defined in terms of electron gain, 587–88
defined in terms of oxidation number decrease, 592
identification of, 593–94

Oxyacid(s), 280
See also Acid(s)

Paramagnetic atom, 183

Partial pressure, 469
Dalton's law of, 469–75
use of, in calculations, 470–75

Partially miscible substances, 489

Particle(s), counting of, by weighing, 313–16

Parts per billion, 498
calculations involving, 499–500
defining equation for, 498

Parts per million, 498
calculations involving, 499–500
defining equations for, 498

Pauling, Linus, 226

Percent, 78
use of, as a conversion factor, 78–81

Percent by mass, 494
calculations involving, 494–95, 503–504
defining equation for, 494

Percent by volume, 496
calculations involving, 496–97
defining equation for, 496

Percent composition, 301
    calculation of, from chemical formula, 301–302
    calculation of, from mass data, 302–303
Percent yield, 370
    calculation of, 371–72
Period(s), 167
    within periodic table, 167
Periodic law, 164
    discovery of, 165–65
    electron configuration and, 183–84
Periodic table, 165
    areas of, 184–85
    atomic mass sequence within, 167–68
    commonly used form of, 165, *inside front cover*
    electron configurations and, 184–91
    electron-dot structures and, 206–207
    groups of, 167
    information found on, 166
    location of inner-transition elements in, 191–92
    location of metals in, 192–93
    location of nonmetals in, 192–93
    location of representative elements in, 191–92
    "long" and "short" forms of, 167–68
    metalloids and, 194–95
    periods of, 167
    valence electrons and, 206–207
pH, 552
    acidic, basic, and neutral solutions, 555
    calculation of, integral values, 553–55
    calculation of, nonintegral values, 555–60
    change in, salt hydrolysis and, 561–64
    defining equation for, 553
    hydrogen and hydroxide ion concentrations and, 554–60
    regulation of, using buffers, 564–67
    of solutions, 553–60
    values, for common substances, 556
    values, significant figures and, 556–57
Physical change, 106
    characteristics of, 106–107
Physical properties, 105
    characteristics of, 105–106
Physical state(s)
    change in, terminology for, 107
    characteristics of, 103–104
    indication of, in chemical equations, 351
    role of kinetic and potential energy in determining, 391–94
    temperature ranges of, for selected substances, 388
Polar covalent bond(s), 227
    electronegativity differences and, 227–29
    ionic and covalent character of, 229–31
    See also Chemical bond(s)
Polar molecular solid(s), 420
    crystal lattice for, 420–21
Polarity
    bond, 227–231
    electronegativities and, 227–31
    molecular, 242–46
Polyatomic ion(s), 218
    common, names and formulas of, 274
    formula writing conventions, 218–20
    nomenclature for, 273–76
Polyprotic acid, 533
    See also Acid(s)
Potential energy, 86
    "cohesive" forces and, 391–92
    factors influencing magnitude of, 391
    role in determining physical states, 391–94
    temperature dependence of, 392
Precision, 12
    accuracy and, relationship between, 13
    significant figures and, 15–16
Pressure, 431
    atmospheric, measurement of, 431–32
    effect on boiling point, 409–10
    as a gas law variable, 431–35
    measurement of, with barometer, 431–32
    measurement of, with manometer, 432–33
    partial, definition of, 469
    partial, use of, in calculations, 470–75
    standard, for gases, 449
    units for, 431–35
Product(s), 342
Properties, 104
    chemical, 105–106
    physical, 105–106
    quantized, 170
    types of, 105–106
    uses for, 105

Proton(s), 136
  atomic number and, 143
  characteristics of, 136
  discharge-tube experiments and, 140–41
  location within the atom, 137–38
Proust, Joseph Louis, 296
Pure substance(s), 110
  characteristics of, 110
  types of, 113–15

Quantized properties, 170

Random errors, 13
Reactant(s), 342
Redox reaction, 587
  See also Oxidation–reduction reaction(s)
Reducing agent, 588, 592
  defined in terms of electron loss, 588
  defined in terms of oxidation number increase, 592
  identification of, 593–94
Reduction, 587
  defined, in terms of electron gain, 587–88
  defined, in terms of oxidation number decrease, 592
Relative mass(es), for atoms, 149–51
Representative elements, 191
  location within periodic table, 191–92
Resonance structure(s), 236
  electron-dot structures for, 235–36
  reasons for, 235–36
Rounding off, 20
  rules for, 20
Rutherford, Ernest, 141

Salt(s), 538
  hydrolysis of, 561–64
  production of, from neutralization, 545–46
  reactions of, 546–50
  reactions with acids, 547–48
  reactions with bases, 548
  reactions with metals, 547
  reactions with other salts, 548–50
  uses of, common, 538
Salt hydrolysis, 561
  pH changes associated with, 561–64
Saturated solution(s), 489
  equilibrium and, 489
  See also Solution(s)
Schrödinger, Erwin, 169
Science, 1
  scientific disciplines within, 1–2
Scientific disciplines, 1
  boundaries between, 1–2
  relationship between chemistry and other, 1–2
  relationship to technology, 2–3
Scientific method, 4
  procedural steps in, 4
  terminology associated with, 4–8
Scientific notation, 27
  addition and subtraction in, 38–40
  advantages of, 27
  converting to decimal notation from, 31–32
  division in, 35–38
  multiplication in, 33–35
  significant figures in, 29–31
  writing numbers in, 29–31
Semiconductor, 194
Significant figure(s), 14
  addition and subtraction and, 21–22, 24–26
  bar notation for zeros, 17–19
  conversion factors and, 60–61
  electronic calculators and, 19–20
  exact numbers and, 26–27
  multiplication and division and, 21–24
  pH values and, 556–57
  plus-minus notation and, 15
  precision and, 15–16
  "rounding off" and, 20–21
  rules for determining, 17–19
  scientific notation numbers and, 29–31
  zeros as, 17–19
Single covalent bond(s), 231
  electron-dot structures and, 220–24
  See also Covalent bonds
Single-replacement reaction(s), 354
  examples of, 354–55, 544, 594–95
Solid state, 103
  amorphous solids, 419
  compared to other physical states, 394–95
  crystal lattice in, 419–21
  crystalline solids, 419–21
  explained using kinetic molecular theory, 392
  lattice sites in, 419–21

Solid state (*cont.*)
  macroscopic properties of, 389–90
  types of crystalline solids, 419–21
Solubility, 488
  factors influencing, 492
  for selected compounds, 488
  numerical values for, various
    compounds, 488
  qualitative terms for, 488
Solubility rules
  ionic compounds, 492–93
  molecular compounds, 491–93
  net ionic equations and, 540
  table of, ionic compounds, 492
Solute, 486
Solution(s), 486
  aqueous, 489
  buffer, 564–67
  components of, 486–87
  concentrated, 489
  concentration units for, 493–512
  dilute, 489
  dilution of, 457–60
  formation of, interparticle interactions
    and, 489–91
  miscibility of components, 489
  neutral, 552, 555
  pH of, 553–60
  saturated, 489
  stock, diluting of, 509–12
  terminology used in describing, 486,
    488–89
  types of two-component, 486–87
  unsaturated, 489
Solvent, 486
Sörensen, Sören Peter Lauritz, 553
Specific heat, 89
  units for, 89
  use of, in calculations, 89–93
Standard pressure, 449
Standard temperature, 449
State of equilibrium, 407
  See also Equilibrium
Strong acid(s), 535
  net ionic equations and, 540
  See also Acid(s)
Strong base(s)
  net ionic equations and, 540
  See also Base(s)
Subatomic particles, 136
  arrangement within atoms, 137–38

  discovery of, 139–43
  ions and, 209–10
  properties of, 136–37
  types of, 136–37
Sublimation, 107, 395
Subscript(s)
  formula, interpretation of, 133–35
  formula, mole–mole conversions using,
    316–18
  use of, in chemical formulas, 133–35
Subshell, *See* Electron subshell
Subtraction
  scientific notation and, 38–40
  significant figures and, 21–22, 24–26
Surface tension, 418
Synthesis reaction(s), 353
  examples of 354, 594
Synthetic element(s), 116
Systematic errors, 13
Systematic name, 278
  See also Nomenclature

Technology, 2
  relationship to scientific disciplines, 2–3
Temperature, 81
  absolute zero value of, 82
  Celsius scale for, 81
  change in as a substance is heated,
    395–96
  constancy of, during changes of state,
    396
  conversions between different scales,
    82–85
  energy and, 392
  Fahrenheit scale for, 82
  as a gas law variable, 431
  Kelvin scale for, 81–82
  scales for, 81–82
  solubility and, 488
  standard, for gases, 449
  vapor pressure magnitude and, 407–408
Ternary compound, 266
  See also Compound(s)
Theoretical yield, 370
  calculation of, 371–72
Theory, 7
  "imperfect" nature of, 8
  misuse of the term, 8

scientific method and, 7–8
See also Atomic theory of matter;
Kinetic molecular theory
Thermal expansion, 389
Thomson, Joseph John, 136, 140
Thomson, William, 443
Titration, 567
acid–base concentrations and, 567–74
indicator use in, 568
Transition elements, 191
location within periodic table, 191–92
Triple covalent bond(s), 231
electron-dot structures and, 231–33
See also Covalent bond(s)
Triprotic acid, 533
See also Acid(s)

Units
mathematical operations and, 57–58
metric system, 48–56
Unsaturated solution, 489

Valence electron(s), 205
electron configurations and, 205–206
electron-dot structures and, 206–207
Valence-shell electron-pair repulsion
theory, *See* VSEPR theory
Vapor, 406
Vapor pressure, 407
explained using kinetic molecular
theory, 406–407
factors affecting magnitude of, 407–408
of water, at selected temperatures, 407, 474
Variable-charge metal(s), 267
compounds containing, nomenclature
for, 270–73
identity of, 267–68
Volatile substance, 407
Volume, 54
formulas for calculation of, 54–55
as a gas law varible, 431
metric and English units compared, 56
metric–English conversion factors, 62
metric units for, 55–57
Volume percent, *See* Percent by volume

VSEPR theory, 246
applications of, 251–54
bonding electron pairs and, 246–48
guidelines used in, 246–49
nonbonding electron pairs and, 248–49

Water
boiling point, effects of hydrogen
bonding on, 414
boiling point, at various pressures, 409–10
crystalline structure of ice, 416–17
density, effects of hydrogen bonding
on, 415–18
density, at various temperatures, 415–16
dissociation of, 551–52
ion product for, 551–52
partial pressure of, in gaseous
mixtures, 473–75
properties of, effects of hydrogen
bonding on, 413–19
solubility of salts in, 492–93
surface tension, effects of hydrogen
bonding on, 418–19
thermal properties, effects of hydrogen
bonding on, 415
vapor pressure, at selected
temperatures, 407, 474
vapor pressure, effects of hydrogen
bonding on, 413–14
Weak acid(s), 535
net ionic equations and, 540
See also Acid(s)
Weak base(s)
net ionic equations and, 540
See also Base(s)
Weight, 52
mass and, differences between, 52–53
measurement of, 53

Yield
actual, 370–72
percent, 370–72
theoretical, 370–72

Atomic Numbers and Atomic Masses of the Elements

Atomic masses are based on $^{12}_{6}C$. Numbers in parentheses are the mass numbers of the most stable isotopes of radioactive elements.

Element	Symbol	Atomic Number	Atomic Mass	Element	Symbol	Atomic Number	Atomic Mass
Actinium	Ac	89	227.0278	Europium	Eu	63	151.96
Aluminum	Al	13	26.98154	Fermium	Fm	100	(257)
Americium	Am	95	(243)	Fluorine	F	9	18.998403
Antimony	Sb	51	121.757	Francium	Fr	87	(223)
Argon	Ar	18	39.948	Gadolinium	Gd	64	157.25
Arsenic	As	33	74.9216	Gallium	Ga	31	69.723
Astatine	At	85	(210)	Germanium	Ge	32	72.59
Barium	Ba	56	137.33	Gold	Au	79	196.9665
Berkelium	Bk	97	(247)	Hafnium	Hf	72	178.49
Beryllium	Be	4	9.01218	Helium	He	2	4.002602
Bismuth	Bi	83	208.9804	Holmium	Ho	67	164.9304
Boron	B	5	10.811	Hydrogen	H	1	1.00794
Bromine	Br	35	79.904	Indium	In	49	114.818
Cadmium	Cd	48	112.41	Iodine	I	53	126.9045
Calcium	Ca	20	40.078	Iridium	Ir	77	192.22
Californium	Cf	98	(251)	Iron	Fe	26	55.847
Carbon	C	6	12.011	Krypton	Kr	36	83.80
Cerium	Ce	58	140.12	Lanthanum	La	57	138.9055
Cesium	Cs	55	132.9054	Lawrencium	Lr	103	(260)
Chlorine	Cl	17	35.453	Lead	Pb	82	207.2
Chromium	Cr	24	51.996	Lithium	Li	3	6.941
Cobalt	Co	27	58.9332	Lutetium	Lu	71	174.967
Copper	Cu	29	63.546	Magnesium	Mg	12	24.305
Curium	Cm	96	(247)	Manganese	Mn	25	54.9380
Dysprosium	Dy	66	162.50	Mendelevium	Md	101	(258)
Einsteinium	Es	99	(252)	Mercury	Hg	80	200.59
Erbium	Er	68	167.26	Molybdenum	Mo	42	95.94